14th International Workshop on Advanced Computing and Analysis Techniques in Physics Research 2011

(ACAT 2011)

Journal of Physics: Conference Series Volume 368

Uxbridge, London, United Kingdom
5-7 September 2011

ISBN: 978-1-62276-785-4
ISSN: 1742-6588

Printed from e-media with permission by:

Curran Associates, Inc.
57 Morehouse Lane
Red Hook, NY 12571

CURRAN ASSOCIATES INC. proceedings.com

Some format issues inherent in the e-media version may also appear in this print version.

Printed by Curran Associates, Inc. (2013)

Institute of Physics
Dirac House, Temple Back
Bristol BS1 6BE UK

Phone: 44 1 17 929 7481
Fax: 44 1 17 920 0979

techtracking@iop.org

Additional copies of this publication are available from:

Curran Associates, Inc.
57 Morehouse Lane
Red Hook, NY 12571 USA
Phone: 845-758-0400
Fax: 845-758-2634
Email: curran@proceedings.com
Web: www.proceedings.com

TABLE OF CONTENTS

COMPUTING TECHNOLOGIES FOR PHYSICS RESEARCH

DATA ANALYSIS – ALGORITHMS AND TOOLS

COMPUTATIONS IN THEORETICAL PHYSICS – TECHNIQUES AND METHODS

Preface

This volume of *Journal of Physics: Conference Series* is dedicated to scientific contributions presented at the 14th International Workshop on Advanced Computing and Analysis Techniques in Physics Research (ACAT 2011) which took place on 5–7 September 2011 at Brunel University, UK.

The workshop series, which began in 1990 in Lyon, France, brings together computer science researchers and practitioners, and researchers from particle physics and related fields in order to explore and confront the boundaries of computing and of automatic data analysis and theoretical calculation techniques. It is a forum for the exchange of ideas among the fields, exploring and promoting cutting-edge computing, data analysis and theoretical calculation techniques in fundamental physics research.

This year's edition of the workshop brought together over 100 participants from all over the world. 14 invited speakers presented key topics on computing ecosystems, cloud computing, multivariate data analysis, symbolic and automatic theoretical calculations as well as computing and data analysis challenges in astrophysics, bioinformatics and musicology. Over 80 other talks and posters presented state-of-the art developments in the areas of the workshop's three tracks: Computing Technologies, Data Analysis Algorithms and Tools, and Computational Techniques in Theoretical Physics. Panel and round table discussions on data management and multivariate data analysis uncovered new ideas and collaboration opportunities in the respective areas.

This edition of ACAT was generously sponsored by the Science and Technology Facility Council (STFC), the Institute for Particle Physics Phenomenology (IPPP) at Durham University, Brookhaven National Laboratory in the USA and Dell.

We would like to thank all the participants of the workshop for the high level of their scientific contributions and for the enthusiastic participation in all its activities which were, ultimately, the key factors in the success of the workshop.

Further information on ACAT 2011 can be found at http://acat2011.cern.ch

Dr Liliana Teodorescu
Brunel University

Committees and Sponsors

Organising Committee

Liliana Teodorescu (Brunel University)
Ivan D Reid (Brunel University)
Alina Grigoras (CERN)

Track Convenors

Computing Technology for Physics Research
- David Britton
- Jerome Lauret
- Axel Naumann

Data Analysis – Algorithms and Tools
- Thomas Speer
- Pedro Teixeira-Dias

Computations in Theoretical Physics – Techniques and Methods
- Nigel Glover
- Gudrun Heinrich

International Advisory Committee

Pushpalatha Bhat (Fermilab)
Fawzi Boudjema (LAPTH France)
David Britton (Glasgow University)
Federico Carminati (CERN)
Denis Oliveira Damazio (BNL)
Bruce Denby (Université Pierre et Marie Curie)
Prabhakar S Dhekne (Bhabha Atomic Research Centre)
Nigel Glover (IPPP-Durham)
Gudrun Heinrich (MPI-Munich)
Viacheslav Ilyin (MSU)
Sverre Jarp (CERN)
Toshiaki Kaneko (KEK)
Matthias Kasemann (DESY)
Andrei Kataev (INR)
Christian Kiesling (MPI-Munich)
Yoshimasa Kurihara (KEK)
Jerome Lauret (BNL)
Denis Perret-Gallix (IN2P3/CNRS) (Chair)
Sudhir Raniwala (Rajasthan University)
Tord Riemann (DESY)
Jose Seixas (Rio de Janeiro Federal University)
Yoshimitsu Shimizu (Sokendai)
Thomas Speer (Brown University)
Liliana Teodorescu (Brunel University)
Peter Uwer (Humboldt University)
Jos Vermaseren (NIKHEF)
Monique Werlen (LAPTH)

Sponsors

Science & Technology Facilities Council

www.ippp.dur.ac.uk

Science and Technology Facilities Council, UK

Institute for Particle Physics Phenomenology, Durham University, UK

BROOKHAVEN NATIONAL LABORATORY

DELL

Brookhaven National Laboratory, USA

Dell

Peer review statement

All papers published in this volume of *Journal of Physics: Conference Series* have been peer reviewed through processes administered by the proceedings Editors. Reviews were conducted by expert referees to the professional and scientific standards expected of a proceedings journal published by IOP Publishing.

The ADAM project: a generic web interface for retrieval and display of ATLAS TDAQ information.

A Harwood[1], G Lehmann Miotto[2], L Magnoni[2], W Vandelli[2] and D Savu[2]

[1] University of the West of England, Bristol, United Kingdom
[2] CERN, European Laboratory for Particle Physics, Geneva, Switzerland

E-mail: adam2.harwood@uwe.ac.uk, giovanna.lehmann@cern.ch, luca.magnoni@cern.ch, wainer.vandelli@cern.ch, dan.savu@cern.ch

Abstract. This paper describes a new approach to the visualization of information about the operation of the ATLAS Trigger and Data Acquisition system. ATLAS is one of the two general purpose detectors positioned along the Large Hadron Collider at CERN. Its data acquisition system consists of several thousand computers interconnected via multiple gigabit Ethernet networks, that are constantly monitored via different tools. Operational parameters ranging from the temperature of the computers to the network utilization are stored in several databases for later analysis. Although the ability to view these data-sets individually is already in place, currently there is no way to view this data together, in a uniform format, from one location. The ADAM project has been launched in order to overcome this limitation. It defines a uniform web interface to collect data from multiple providers that have different structures. It is capable of aggregating and correlating the data according to user defined criteria. Finally, it visualizes the collected data using a flexible and interactive front-end web system. Structurally, the project comprises of 3 main levels of the data collection cycle: The Level 0 represents the information sources within ATLAS. These providers do not store information in a uniform fashion. The first step of the project was to define a common interface with which to expose stored data. The interface designed for the project originates from the Google Data Protocol API. The idea is to allow read-only access to data providers, through HTTP requests similar in format to the SQL query structure. This provides a standardized way to access this different information sources within ATLAS. The Level 1 can be considered the engine of the system. The primary task of the Level 1 is to gather data from multiple data sources via the common interface, to correlate this data together, or over a defined time series, and expose the combined data as a whole to the Level 2 web interface. The Level 2 is designed to present the data in a similar style and aesthetic, despite the different data sources. Pages can be constructed, edited and personalized by users to suit the specific data being shown. Pages can show a collection of graphs displaying data potentially coming from multiple sources. The project as a whole has a great amount of scope thanks to the uniform approach chosen for exposing data, and the flexibility of the Level 2 in presenting results. The paper will describe in detail the design and implementation of this new tool. In particular we will go through the project architecture, the implementation choices and the examples of usage of the system in place within the ATLAS TDAQ infrastructure.

1. Introduction

This paper presents the ADAM project, a new approach for aggregating and visualizing ATLAS operational data. In the first section we describe the project architecture, then we present the details of the ADAM interface, a lightweight HTTP interface for data collection from multiple providers, and finally we describe the ADAM visualization dashboard, a web portal to gather and visualize data from different data sources.

1.1. ATLAS TDAQ infrastructure
ATLAS [1] (A Toroidal LHC Apparatus) is a particle physics experiment at the Large Hadron Collider at CERN. The Trigger and Data Acquisition (TDAQ) system of the ATLAS experiment is responsible for selecting and transferring ATLAS experimental data from the detector to the mass storage system. To cope with the very high data rate produced by the detector a complex and distributed computing infrastructure has been built. More than 200 switches and routers interconnect around 2000 hosts to build a real-time filtering system for particle collision events [2]. On top of the hardware infrastructure, over 20.000 applications are responsible for analyzing, filtering and moving event data to permanent storage [3].

1.2. Motivation
The maximization of the ATLAS data-taking efficiency requires operators and experts to precisely know the TDAQ system status in order to be able to understand and solve any unexpected behavior. The system status is defined by a large set of operational parameters, collected by different tools and provided via different means, such as web pages, graphical user interfaces or ad-hoc APIs. Moreover, the retrieved data have to be correlated, on the basis of desired criteria, and finally visualized.

The main objectives of the ADAM project are:

- the definition of a standard way of accessing data gathered by various TDAQ information storage systems;
- the correlation of pieces of information provided by different sources;
- the creation of a single point for data aggregation and visualization.

2. The ADAM project
The goal of the ADAM project (ATLAS Data API Mechanism) is to provide a uniform interface for information retrieval from several applications (data providers) and to expose a unique access point for data querying and collection. Data providers are heterogeneous data sources that offer several kind of information, potentially very different in type and structure. The complexity of each data provider implementation is however hidden, since their capabilities are exposed through the ATLAS Data API (ADAM interface), a SQL-like web interface. Leveraging on the common interface, a unified web portal can gather and visualize data from different providers. By using the web portal, users and experts have access to summary graphs that can be dynamically customized.

2.1. Architecture
The ADAM architecture is structured as a 3 layer application, as shown in figure 1.

- **Level 0** includes all the ADAM data providers.
- **Level 1** is an intermediate component able to gather and correlate data from different data sources.
- **Level 2** is a visualization layer implemented as a web dashboard where users can query and visualize data coming from multiple data providers.

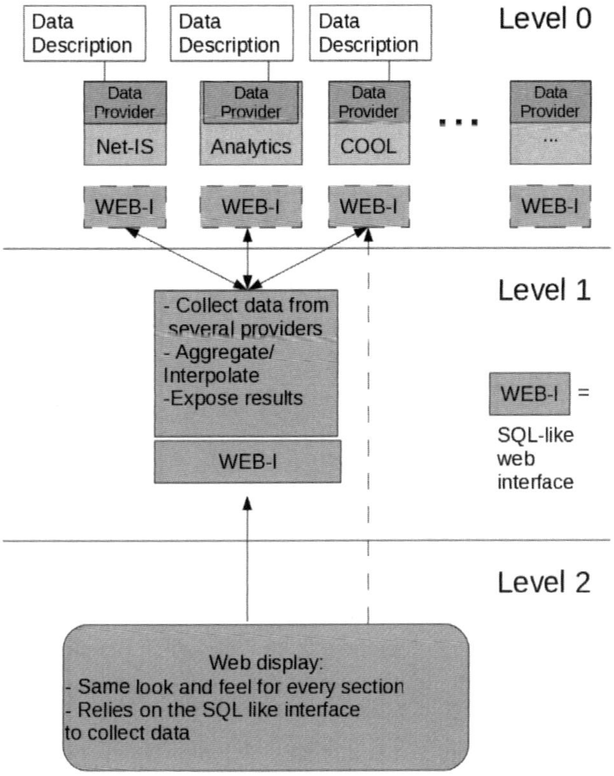

Figure 1. The ADAM project architecture.

2.2. Data providers

An application displaying ad-hoc or historical statistics about a system has either to monitor all the parameters by itself or to rely on the information gathered by other applications (data providers). Usually these data providers use custom methods for data collection and a variety of technologies for data storage and data retrieval (RRD files [4], relational databases, custom user interfaces, APIs etc). This is an expected situation given the complexity of the system and the long development time-line. For efficient log-back analysis, system wide statistic generation or problem investigation operators and experts need a unified visualization system. The ADAM project defines a common web-based SQL-like interface that every data provider has to implement. As a result, the common interface hides the complexity of the provider implementations giving users a simple but very powerful language to collect the desired information. The language, described in section 3, allows data providers to expose their data in various formats. The default format is XML (Extensible Markup Language), where results are represented as structured data suitable for machine-based processing. Optionally, data provider can return CSV (Comma Separated Value) documents or JSON (JavaScript Object Notation) data, useful for web technologies integration. Through the web-based approach a data collection exchange is composed of the following steps, as shown in figure 2:

(i) The data provider exposes a URL, called the data source URL;

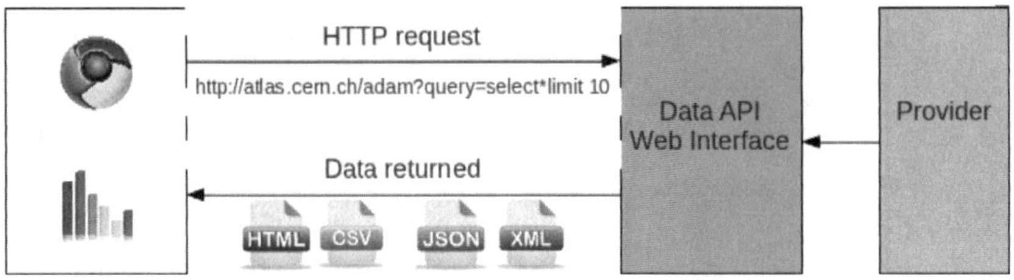

Figure 2. Client request via the ADAM interface with query criteria.

(ii) The client makes a HTTP request with optional parameters that describe the query string and the required result format;

(iii) The data provider receives and parses the request, as described in section 3.1;

(iv) The data provider prepares the data in the requested format (XML, CSV, JSON);

(v) The data provider creates a HTTP response that includes the serialized data and other response parameters, and sends it back to the client.

Currently, the following data providers expose a ADAM-compliant interface.

COOL COOL [5] is the condition database of the ATLAS experiment. Operational data are stored indexed by interval of validity. Information are retrieved via a programmatic API available in C/C++ and Python or via an ad-hoc web interface.

Net-IS The Net-IS (Networking Integrated System) [6] is the main tool used for collecting statistics of the network infrastructure. Net-IS provides system metrics (e.g. CPU load, system temperature), environmental conditions (e.g. rack power, ambient temperature) and a reduced set of data taking parameters allowing the correlation of network related problems with external factors. The tool is designed to efficiently handle time-series data and uses RRD files for raw data and a relational database for metadata storage.

MRS Message analytics The MRS analytics [7] data source provides an aggregate and effective view on the flow of messages sent by applications during data-taking runs. Messages are sent via the Message Reporting facility (MRS) [8].

In order to facilitate and stimulate the development of ADAM-compliant providers, a Java-based library targeting relation databases has been developed. The library allows to expose any table (relying on the JDBC abstraction) through a web interface compatible with the ADAM interface.

2.3. Level 1

The subsequent step in the system status analysis requires not only to collect information but also to correlate and unify data access across different providers. Because of the ADAM unified interface an intermediate aggregation and analysis engine, so called Level 1, has been developed. The Level 1 is placed between clients and data providers and is able to gather data from multiple sources and to perform basic correlation functions. For example, the Level 1 is able to correlate data sets over the desired criteria, such as time, or to extrapolate values in order to unify time resolution. The Level 1 capabilities are exposed via the same ADAM interface, and users can query the Level 1 asking a desired set of data independently from the data provider offering the information. Handling a data request in the Level 1 requires:

(i) Parse the HTTP request identifying the data sources involved;
(ii) Forward the request and retrieve data from each provider;
(iii) Perform aggregation, correlation and combination functions on the data;
(iv) Return the combined data in the required format.

3. ADAM interfacc

The ADAM interface is the core of the ADAM project. It is a HTTP-based interface that unifies the way information is exposed from different data providers. It has a simple request format based on HTTP GET requests and it offers a powerful SQL-like protocol to express query on data with desired criteria, inspired by the Google Charts project [9].

3.1. Request format

A client can send a HTTP GET request with several parameters, including custom elements and an optional query string. Representing requests as HTTP URIs, rather than as HTTP headers or as part of the payload, has several benefits such as direct linking and persistence.

Table 1. ADAM interface request parameters.

Parameter name	Key:Value	Description
op	-	A set of colon-delimited key/value pairs for standard or custom properties. Pairs are separated by semicolons.
-	**reqId**:*value*	A numeric identifier for the request, to be sent back in the response. It allows multiple requests to be sent in parallel.
-	**hash**:*value*	A hash of the query results sent to the client in any previous requests to this data source. This is an optimization to avoid sending identical data to a client twice.
-	**out:**[**xml**|**json**|**csv**]	A string describing the format of the returned data.
query	-	A query written in ADAM Query Language, described in section 3.2, specifying how to filter, sort, or otherwise manipulate the returned data.

The request parameters shown in table 1 are composed to build the desired data query. Some examples of possible requests are given below. A simple request with no parameters:

- `http://dataprovider.cern.ch/resource`

This is a request with the `op` parameter and two properties:

- `http://dataprovider.cern.ch/resource?op=reqId:0;hash:4641982796834063168`

Request with a minimal query string:

- `http://dataprovider.cern.ch/resource?query=select time, CPU limit 10`

3.2. The Query language
The query language used when communicating with data providers can be defined as a dilute dialect of the SQL query language. As presented in table 2, the simple query style of the SQL language is preserved as well as the capability of expressing basic data manipulation functions (e.g. MIN, MAX, AVG, COUNT). On the other hand, the SQL statements are meant to address one database. In the ADAM project the concept is therefore extended to allow one SQL-like statement to query multiple data providers. This allows the Level 1 to parse a multi provider query, retrieve data from every provider and aggregate the result.

Table 2. ADAM query language clauses.

Clause	Description
select	Select which information to return and in which order.
from	Select which data provider to query.
where	Returns only the information that matches the condition.
group by	Aggregate values across rows.
order by	Sort the result in the specific order.
limit	Limit the number of returned rows.
correlate by	Only for the Level 1. Request to correlate result from multiple data provider on the specified criteria, such as time.

The following is an example of a multiple data sources query:

- `SELECT datasource1:table2:field1, datasource2:table2:field2 FROM datasource1:table2:field1, datasource2:table2:field2 WHERE globalTime < 20-10-2011 16:20 AND globalTime > 20-10-2011 15:20`

In this way multiple tables from differing sources can be selected, using a common time format.

3.3. Describe capabilities
A dedicated feature is built into the ADAM data provider interface to allow users to introspect the data source architectures, internal structures and supported data formats. Very similar to the `SQL DESCRIBE` command, the feature has two basic variants:

- `describe table`
- `describe provider`

The behaviour of the `describe table` command matches the equivalent SQL statement (for MySQL databases). It returns the list of field properties within the specified table.

The `describe provider` returns information on the data source capabilities in terms of supported ADAM query functionalities. For example, a data source can expose the capability of performing `MAX` and `MIN` calculations.

4. The ADAM visualization dashboard
A web-based visualization dashboard, that gathers and aggregates information from ADAM-capable data sources, has been implemented with the aim of providing a unique entry point to query and visualize operational data. The dashboard, which can be easily customized by users both in the layout as well as in the information displayed, has been developed on top of

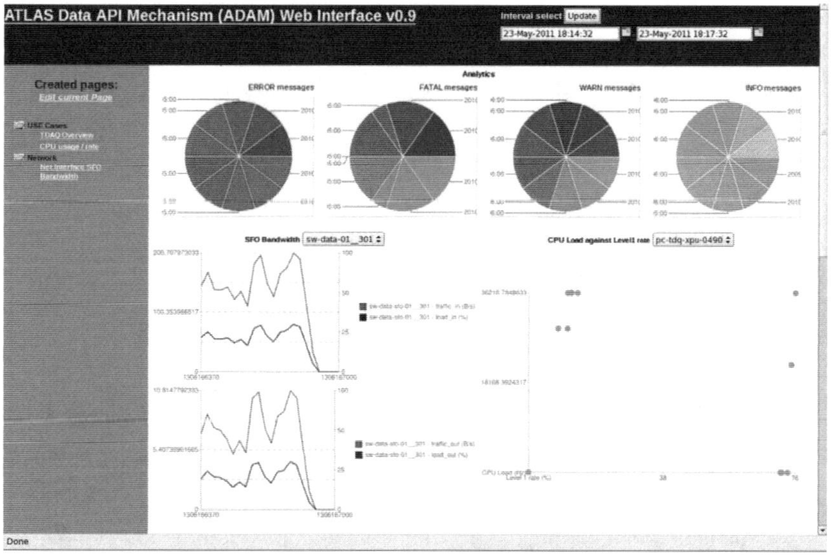

Figure 3. The ADAM visualization dashboard.

the Django framework [10] and is written in the Python scripting language. The web-site is based on a series of customizable HTML components and tables representing the set of basic visualization elements and it is completely dynamic with respect to the layout. Every HTML component can be mapped with an ADAM query and a visualization option. The query can be directed to individual providers or to Level 1, for more advanced aggregation, normalization or analysis features.

Graph to query mapping The dashboard, as shown in figure 3, presents information aggregated in tables and charts of different types. Users can customize visualization elements both in type and content. The supported charts are the ones provided by the Google Charts Image tool [11], such as pie charts, column/bar charts or more advanced options such as heat maps or composite graphs. Thanks to the common ADAM interface the content of visualization elements can be associated to data via HTTP queries. The latter are executed only at rendering time on user browsers. The dashboard allows customization of the query, supporting all the ADAM language features.

User customizable The dashboard is structured in pages and tabs whose layouts can be dynamically composed using the basic visualization elements. The dashboard layout is therefore user dependent: dedicated views can be created with the aim of tackling specific area of the TDAQ system.

4.1. Command line shell

Similarly to the command line interfaces provided by several database vendors, an ADAM command line shell has been developed. The shell allows to interact with ADAM-capable data providers. It is based on the ADAM query language, offering auto completion and syntax checking functionalities. It is a fast and effective means to build and validate queries and to access ADAM capabilities in a programmatic way.

5. Conclusions

The main goal of the ADAM project is to propose a new and effective approach to gather and correlate information from many different data providers. The ADAM interface is a lightweight HTTP-based interface with a very powerful SQL-like query language. The uniform interface allows for loose coupling of clients and data providers. In fact, different data-provider back-end technologies are abstracted using the same SQL-like paradigm. The ADAM visualization dashboard is a unique entry point to collect and visualize data of different nature. The main areas of research for the future are the finalization of the web-based visualization, offering dedicated layout for specific use cases, and the expansion of the gathering and processing capabilities, including more ADAM-compatible data providers.

References

[1] ATLAS Collaboration 2008 *The ATLAS Experiment at the CERN Large Hadron Collider.* J. Inst. vol 3, S08003
[2] Ciobotaru M, Leahu L, Martin B, Meirosu C and Stancu S 2006 Proc. of the Int. Conf. on Computing in High Energy and Nuclear Physics (Mumbai, India)
[3] ATLAS Collaboration 2003 *ATLAS High-Level Trigger, Data Acquisition and Controls Technical Design Report* http://cdsweb.cern.ch/record/616089/
[4] Oetiker T The RRD Tool http://oss.oetiker.ch/rrdtool
[5] Valassi A and et al 2010 *LCG Persistency Framework (CORAL, COOL, POOL): Status and Outlook* Proc. Int. Conf. on Computing in High Energy and Nuclear Physics (Taipei, Taiwan)
[6] Savu D, Al-Shabibi A, Martin B, Sjoen R, Batraneanu S and Stancu S *Efficient Network Monitoring for Large Data Acquisition Systems* Proc. of the Int. Conf. ICAPELCS
[7] Magnoni L, Miotto G L and Sloper J 2010 *The TDAQ Analytics Dashboard: a real-time web application for the ATLAS TDAQ control infrastructure* Proc. of the Int. Conf. on Computing in High Energy and Nuclear Physics (Taipei, Taiwan)
[8] Fedorko I *The Message Reporting System of the ATLAS DAQ System* Proc. of the Int. Conf. on Astroparticle, Particle, Space Physics Detectors and Medical Physics Applications (Como, Italy)
[9] Google 2010 Google Charts http://code.google.com/apis/chart/
[10] The Django software foundation Django framework https://www.djangoproject.com
[11] Google 2010 Google Charts Image http://code.google.com/apis/chart/image

Journal of Physics: Conference Series **368** (2012) 012002 doi:10.1088/1742-6596/368/1/012002

A persistent back-end for the ATLAS TDAQ online information service (P-BEAST)

Alexandru D Sicoe[1], Giovanna Lehmann Miotto[1], Luca Magnoni[1], Serguei Kolos[2], Igor Soloviev[2]

[1] CERN, CH-1211 Geneve 23, CH

[2] University of California Irvine, Irvine, CA 92697, USA

E-mail: asicoe@cern.ch

Abstract. This paper describes P-BEAST, a highly scalable, highly available and durable system for archiving monitoring information of the trigger and data acquisition (TDAQ) system of the ATLAS experiment at CERN. Currently this consists of 20,000 applications running on 2,400 interconnected computers but it is foreseen to grow further in the near future. P-BEAST stores considerable amounts of monitoring information which would otherwise be lost. Making this data accessible, facilitates long term analysis and faster debugging. The novelty of this research consists of using a modern key-value storage technology (Cassandra) to satisfy the massive time series data rates, flexibility and scalability requirements entailed by the project. The loose schema allows the stored data to evolve seamlessly with the information flowing within the Information Service. An architectural overview of P-BEAST is presented alongside a discussion about the technologies considered as candidates for storing the data. The arguments which ultimately lead to choosing Cassandra are explained. Measurements taken during operation in production environment illustrate the data volume absorbed by the system and techniques for reducing the required Cassandra storage space overhead.

1. Introduction

ATLAS is one of several experiments built along the Large Hadron Collider at CERN, Geneva. Its aim is to measure particle production when protons collide at a very high center of mass energy, thus mimicking the behavior of matter a few instants after the Big Bang. The detecting techniques used for this purpose are very sophisticated and the amount of digitized data created by the sensing elements requires a very large data acquisition system [1]. This consists of a few tens of thousands of applications that operate on the physics data since the moment it leaves the detector until it is archived on disk. All these applications publish their monitoring information in a central repository called the ATLAS on-line Information Service [2].

The experiment is successfully taking data since the end of 2008 and the trigger and data acquisition is now in production. Among the current development efforts is the addition of easy to use and intuitive tools to aid experts monitor different components or subsystems. P-BEAST is an example of such a tool. It archives monitoring data at unprecedented levels of granularity giving experts access to detailed information about what happened in the system at a certain point in the past.

1.1. Dataset and workload considerations
The Information Service (IS) represents the data source. It is important to analyse the behaviour of this system in order to understand the information volumes it is capable of generating, the way this data can be acquired and the dynamics of the data. IS consists of a configurable set of server processes where most TDAQ applications publish histograms and state variables. Examples of such variables are the total system memory used by a particular application or the number of physics events processed by an event filter application.

IS servers have two main types of clients: information providers and information receivers. As already mentioned, the providers are applications running in the TDAQ infrastructure. They can create new information objects in the IS repository or update existing ones. Receivers can subscribe to get updates of the information objects. The communication between the IS repository and information receivers/providers is implemented on top of CORBA [3].

Most TDAQ applications have several variables to publish. These are grouped in structures called information objects. An information object is the main unit of information passed between IS clients and servers. It has a unique name (based on the name of the source application), a fixed set of fields (one for each variable of the source application), a type and a microsecond precision time stamp. When such a TDAQ application is started, it first registers its information objects with a certain IS server. As the objects are updated, they are sent to the IS server. The rate at which this occurs varies widely: some are regular and others are highly dependant on runtime conditions thus exhibiting transient behavior with peaks of up to several tens of thousands of information object updates per second. Another key aspect to point out is that all the fields of an information object are sent in every update regardless of whether they changed their value or not. This leads to data duplication that has to be accounted for.

Currently the load of monitoring data from all 20,000 TDAQ applications is spread across a total of around 230 IS servers, depending on configuration. About half of them contain histograms, which are already being archived via different means. The remaining ones contain the operational monitoring data, most of which is volatile. At present the focus lies on the 87 servers which are used by the central DAQ and Trigger systems, though an extension of the archiving for sub-detector specific information could be envisaged in the future. The total number of information objects published within these 87 servers is approximately 200,000. On average there are 10 fields in each object giving a total of 2,000,000 variables that need to be persisted for long term access. These variables are of primitive types directly mapped to C++ basic types (8,16,32,64 bit signed and unsigned integers; boolean; float; double; string) and arrays of those.

2. Requirements
The goal of the P-BEAST project is to design and implement a generic mechanism to store a subset of operational monitoring data of the ATLAS TDAQ system into a suitable database and provide an efficient means for retrieving that data. The reason why this data should be persisted is to allow the data flow experts to analyze the quality of data taking a posteriori, accessing by means of specialized dashboards, information that is stored in a permanent location.

2.1. Users
P-BEAST will be used by any system expert who wants to have a more holistic view of certain interesting elements of the TDAQ data flow that took place in the past (days, weeks, months, years). Such functionality is useful for off line analysis: to understand the behavior of different parts of the system, to make comparisons between data taking sessions of the ATLAS detector, to investigate problems, to correlate data etc. This data can serve as evidence of system operation in reports, presentations or papers.

2.2. Functional requirements

2.2.1. Insertion path P-BEAST has to take into account the structural evolution of the data published within the Information Service. Since the information objects are completely configurable they can change over time. A recent example is the implementation of a logical subdivision of racks which lead to lots of changes in the information object names. The addition/deletion of fields of an information object, from one data taking session of ATLAS to another, is also possible.

The operational data must be permanently persisted. Data up to several months has to be readily available, while data older than that has to be moved to CERN's long term storage system: CASTOR [4]. The stored information should not loose its granularity over time.

P-BEAST should be able to replicate/transfer data outside of the ATLAS enclosed network where the data providers (IS servers) reside, in order to allow for accessibility.

2.2.2. Retrieval path The system needs to be able to handle random time based queries. Given the names of the information objects and fields together with a time interval, the requested data (plus associated meta data) has to be returned in an efficient manner. With this time series data at hand, the client, which could for instance be a web dashboard [5], can correlate and aggregate different data streams prior to display.

3. System architecture

Figure 1 shows the architectural view of P-BEAST: insertion and retrieval parts are clearly separated.

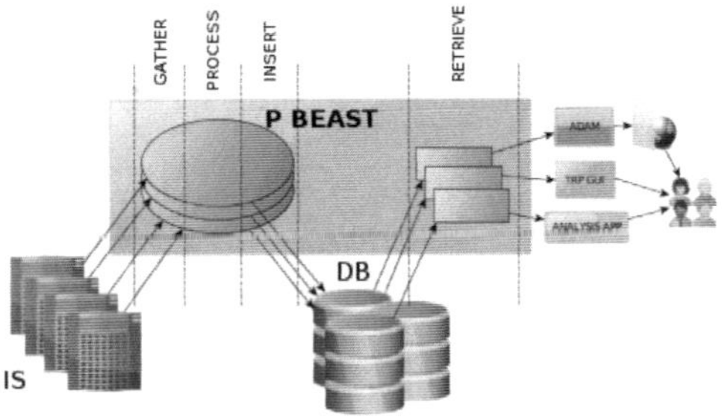

Figure 1: P-BEAST architecture.

3.1. The insertion part

Consists of multiple information gathering agents. These are completely independent and configurable applications that perform identical tasks: gathering, processing (decomposing, unmarshalling, filtering) and database insertion. At start up, each of them makes an initial subscription to a subset of IS servers, defined in their configuration file, and then waits to receive information objects. Each individual field is converted to the appropriate type and passed to a collection of filters. These share a common interface which allows them to be plugged in (the filter class names are specified in the configuration file) such that new filter implementations can be added without the need of modifying application code. Filters can be configured to target individual fields within a unique IS object or can have a broader reach.

Currently there are two filter implementations employed. The default is discarding duplicates generated by successive updates of information objects containing the same field value. This is a characteristic of the Information Service that simply resends the entire information object even if none of its constituents changed value since the last update. The second filter implementation applies smoothing to numerical values only. A certain field value is rejected if it is within a predefined range from the last value inserted. This range is defined in terms of percentages.

Field values that are accepted by all configured filters are placed in a batch. When the batch reaches a certain size it is passed to a database access object that performs a batch insertion into the database.

3.2. The retrieval part

Is based on a similar concept as the insertion part with multiple independent clients. Each client will access the database through a common retrieval API. In the simplest scenario, the functionality provided by the API can be used by a custom analysis application. More advanced clients would receive HTTP requests (like the ones issued by the ADAM dashboard [5]) which they would map to P-BEAST API calls. The data returned would be arranged in a client specified format (JSON, XML, CSV etc.).

4. Database platform considerations

Initially relational database technologies were considered as the storage platform for P-BEAST. A simple exercise in mapping IS information to the relational model lead to the conclusion that the fixed tables cannot provide the flexibility necessary to easily and efficiently track information object structure over long periods of time. In order for this to be attained, de-normalization has to be employed. This technique places considerable burdens on maintenance. The complexity of retrieval queries grows and performance decreases as the data set grows larger. Moreover, the high rates of insertion are generally problematic for relational databases. The solution is scaling up by using more powerful machines in terms of RAM and CPU. For extremely large datasets this strategy becomes costly.

A whole range of alternatives were evaluated. Part of them belong to a technological trend called NoSQL [6] (e.g. Cassandra [7][8], HBase [9], MongoDB [10]). This term brings together a plethora of distributed structured storage systems that are very different in implementation and data models but which share a few common aspects like: flexible schema, horizontal scalability, non-ACID [11] etc, which can bring an advantage to P-BEAST in terms of ease of development, scalability and performance. Due to the abundance of solutions it was found that the fastest way to choosing a final candidate for the storage component of P-BEAST was to filter them against a set of criteria: open source, good fit to the use case of P-BEAST, read/write performance, community activity, ease of deployment and maintenance.

RRDTool [12], Graphite [13] and OpenTSDB [14] are specifically built for handling monitoring time series data but were ruled out for their lack of generality (only support numerical values) and loss of granularity. KDB [15] is a highly performing tool used for storing and processing financial time series data. Unfortunately it is a commercial product and just the 32 bit version is available openly with not much active development. HDF5 [16] is a file system with a library that provides functionality to access the files. It has a rich data set that supports multidimensional arrays but does not have any data management functionality. MongoDB is a document-oriented storage system that provides very high schema flexibility. It can be used as a fast storage system for analytics data but at the moment when it was evaluated it was not durable, meaning that data could be potentially lost after a machine failure because it is initially cached in RAM. Also, MongoDB performs really well only if the indexes fit entirely in memory. As the data size of IS is large this was most likely going to cause difficulties and would entail scaling up.

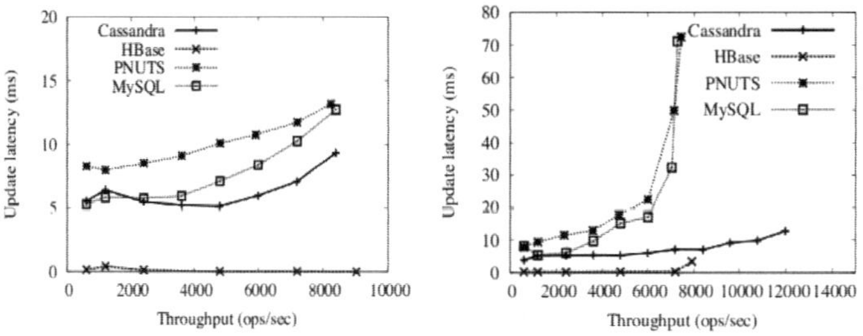

(a) (b)

Figure 2: Update (write) latencies - from [17] p.9: a) Work load: 95% reads / 5% updates, b) Work load: 50% reads / 50% updates.

In evaluating Cassandra and HBase it was found that the primary use case of Cassandra, high availability and write performance with optimized queries for retrieving slices of values, is a better match for P-BEAST than that of HBase which is more focused on real time queries and performing computation across large data sets. The relative performance of the two is shown by a study published by Yahoo! [17] containing benchmarks under different load conditions. These were obtained using their cloud serving benchmarking platform [17] which is available for free and can be easily adapted to stress test any storage system.

Figure 2 shows the update (write) latency as a function of the load placed on each platform, for two different work loads: read intensive (95% of the operations are reads) and balanced (reads and writes occur in the same proportion). The second workload is closer to the intended usage of the platforms because the use case of P-BEAST entails writing for the majority of the time. In the read intensive workload (figure 2a) clearly HBase performs better but in the workload with more writes (figure 2b) even though HBase outperforms Cassandra's to start with, it proves incapable of handling loads higher than 8,000 operations/sec. Cassandra's write latency remains stable across higher loads. This is a desired feature to have in order to satisfy the potentially growing loads generated by the ATLAS Information Service. For the read latency under the same two workloads (figure 3), Cassandra presents superior performance showing it is a better fit for content retrieval.

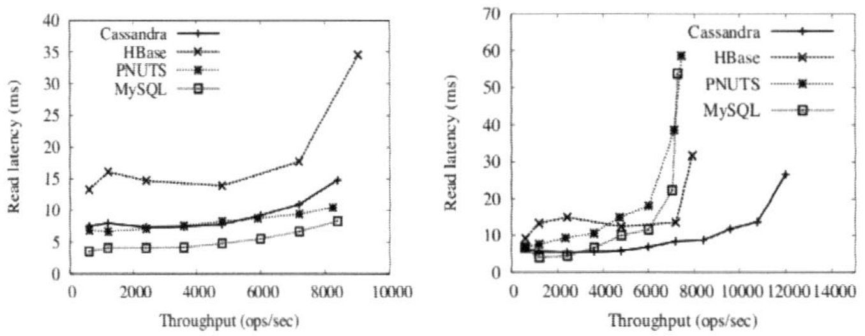

(a) (b)

Figure 3: Read latencies - from [17] p.8: a) Work load: 95% reads / 5% updates, b) Work load: 50% reads / 50% updates.

The main reasons why Cassandra is in a better position than the others to become the platform of choice are:

- Cassandra offers the best fit to the use cases of P-BEAST

  - Lots of writes with significalntly less read operations
  - Can absorb peaks in the write rates [18]
  - Designed to handle a very high write load (thousands of operations per second) [19] from multiple clients

- Easy to increase performance by just adding more machines (scales linearly with the number of machines assuming that keys are randomly distributed which is the case in real data sets)
- Ease of deployment (facilitates the fast prototyping approach used for P-BEAST)
- Low maintenance requirements (important for the long term intended use)

Cassandra is a key value storage system that operates as a distributed hash map. Due to its simple data model, it has a strong use case for storing large amounts of time series data (used by Cloudkick [19] for monitoring cloud infrastructure). Each value inserted has to have an associated key and a name. The key identifies the row where the name value pair is placed. Each row stores name-value pairs sorted by the name. This essentially provides two indexes for retrieving values: an index on row keys and one on names inside a row. Data is distributed within a cluster of nodes by applying MD5 function to the key. Each cluster node is assigned a particular range of MD5 hash values. All the name value pairs of a particular row will be stored on the node responsible for the range where that row key hashed to.

Cassandra was inspired from Amazon's Dynamo [20] where it got the distributed ring arrangement and its high availability. Furthermore it borrowed the storage data model and basic insertion and retrieval mechanisms from Google's Big Table [21]. It uses the SEDA [22] architecture for multi threading which allows more control on the computation process. To communicate with any Cassandra node there is an API on top of Apache's Thrift [23] protocol which allows remote procedural calls and facilitates communication between applications written in different languages.

5. Results

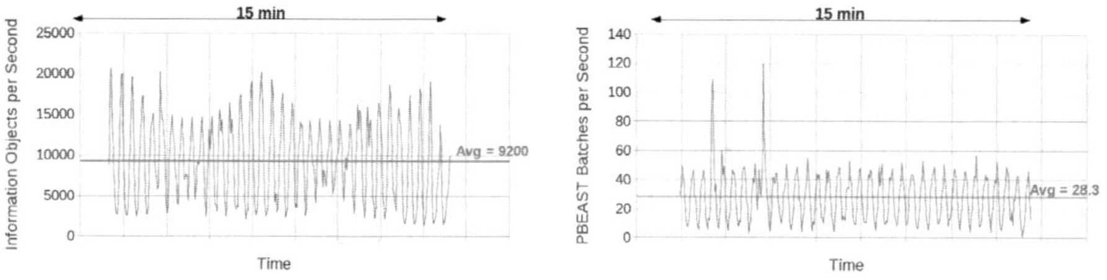

(a) (b)

Figure 4: Insertion rates measured while runnning during ATLAS recording physics data: a) Update transactions per second, b) Batch insertions per sec.

Deploying P-BEAST (with six gathering instances) in the TDAQ infrastructure while ATLAS is recording physics events reveals the rates shown in figure 4. The gathering agents are configured with only the default duplicates filter and send data to a three node Cassandra cluster (each node has a 4 core processor, 4 GB of RAM and separate rotational disks for the commit log and the data directories). Figure 4a represents a measurement of the total update transactions per second received by all six P-BEAST gathering instances from the 87 IS servers of interest. An update transaction corresponds to an information object being transferred over the network to P-BEAST. The plot shows the irregular traffic that the Information Service generates with an average of 9,200 updates per second. This input load is transformed by filtering and batching into a load presented to Cassandra depicted in figure 4b. On average 28.3 batches/sec are inserted. With a measured average of 226KB of data per batch, the Cassandra cluster thus handles an average of 6.4 MB/sec of input data.

The Cassandra meta data can have a significant impact on the required storage space. The storage space needed for the ATLAS partition during a test was estimated based on standard formulas [24] for calculating Cassandra overhead for rows and name value pairs (essentially for the native indexes offered, bloom filters etc). The formulas described in table 1 were fed with several statistics gathered during 15 minute tests: the total number of rows created during the test (TNR), the total number of name value pairs inserted (TNC), the average number of name value pairs per row (NC), the average sizes of row keys and of names (RKS).

Table 1: Cassandra storage space: 100 sec vs 1000 sec time bins.

Measurement	100 sec bins	1000 sec bins
Raw data inserted[Sum (size of name + size of value)]	1.5 GB	1.3 GB
Name value pairs overhead [TNC x 15]	287.9 MB	398.49 MB
Row header overhead [TNR x 23]	103.5 MB	51.93 MB
Row bloom filter[TNR*(2 + (8 + ceiling((NC * 4 + 20) / 8)))]	67.94 MB	42.2 MB
SSTable index[TNR * (10 + RKS)]	783.5 MB	386.5 MB
SSTable bloom filter[(TNR * 15 + 20)/8]	8.4 MB	4.2 MB
Base Storage (BS)	2.7 GB	2.2 GB
Raw data inserted (% BS)	54%	60%
Column Overhead (% BS)	10%	18%
Row overhead (%BS)	4%	2%
SSTableOverhead (%BS)	29%	18%

These calculations were performed in order to test the impact of various Cassandra schema parameters on the storage efficiency. The schema employed uses Cassandra rows as time bins. The time bins are implemented in the following manner: for each new field of an incoming information object, a Cassandra row key is computed by concatenating the 8th (for 100 sec bins) or 7th (for 1000 sec bins) most significant digits of the object time stamp, to the object name and the field name. The effect of computing the row keys in this way is that the value of a certain field is inserted in the same row if two successive updates are received within the same time bin, essentially stacking name value pairs in the same row.

By varying the length of the bins from 100 second to 1000 second, the number of rows is reduced while the average number of name value pairs per row is increased. The results of storage space estimation is depicted in table 1. The same test configuration described earlier was used to perform these tests. The figures in the table clearly show that by enabling longer time bins the row and SSTable[3] overhead is reduced by almost half. This change also has a positive effect on retrieval query performance because it improves data locality (the name-value pairs in one row are always stored on the same machine in the cluster).

Despite metadata optimizations, the amount of raw data being stored is still high. In order to further reduce the intake of information smoothing filters are applied in addition to the default one (duplicates filter). Figure 5 shows the estimated base storage requirements using the same techniques described before. The data rates are lower because the tests were performed on a more limited infrastructure. The procedure adopted was to run the system, in the same configuration (6 P-BEAST gathering agents and 3 node Cassandra cluster), in 4 separate 12 minute tests.

[3] An SSTable is the on disk data representation in Cassandra.

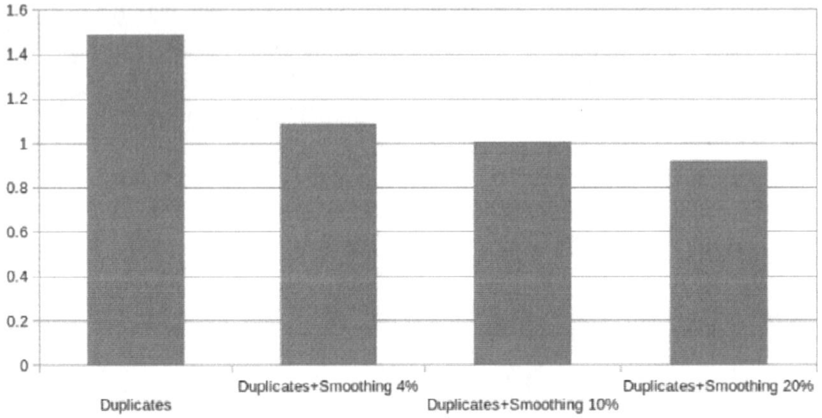

Figure 5: Estimated Cassandra storage space for different degrees of smoothing.

Each test had increasingly restrictive smoothing applied, ranging from no smoothing to 20% smothing. The effect of smoothing is to decrease the storage space up to 38%.

6. Conclusions

P-BEAST is a system that maintains an archive of operational monitoring data which would otherwise be lost. Access to this history helps data flow experts debug problems occurring within the infrastructure and make analysis of past behavior.

The writing part of P-BEAST has been tested in the production environment. On-going tests within the TDAQ infrastructure indicate that P-BEAST is capable of absorbing the load produced by the 87 servers of interest of the ATLAS TDAQ Information Service, thanks to the intermediate buffering and to Cassandra's extremely high write performance. The results presented here show the data rates handled and the possible optimizations that can be performed with respect to storage space occupancy. The raw input data rates can be further controlled by applying smoothing. The smoothing factors are completely configurable which allow probing the limits of relevancy of applying this kind of filtering for different sources of data in the future.

Further work will involve finishing insertion tests/optimizations and the long term storage space estimates. These will feed into the proposal for acquisition of new machines to deploy the Cassandra cluster on. Retrieval performance measurements will follow as well as development of the reading clients.

References

[1] Jenni P, Nessi M, Nordberg M and Smith K 2003 *ATLAS high-level trigger, data-acquisition and controls:Technical Design Report* (Geneva: CERN) pp 6–9, 23–31 http://cdsweb.cern.ch/record/616089/
[2] Jenni P, Nessi M, Nordberg M and Smith K 2003 *ATLAS high-level trigger, data-acquisition and controls:Technical Design Report* (Geneva: CERN) pp 161–162 http://cdsweb.cern.ch/record/616089/
[3] Group O M 2011 Corba http://www.corba.org/
[4] CERN 2011 Cern advanced storage manager http://castor.web.cern.ch/castor/
[5] Harwood A and Magnoni L 2011 *The ADAM project: a generic web interface for retrieval and display of ATLAS TDAQ information* Proc. Int. Ws. on Advanced Computing And Analysis Techniques In Physics Research (Uxbridge, UK)
[6] Prof Edlich S 2011 Nosql http://nosql-database.org/
[7] Foundation T A S 2011 Cassandra http://cassandra.apache.org/
[8] Lakshman A and Malik P 2010 *SIGOPS Oper. Syst. Rev.* **44** 35–40 ISSN 0163-5980
[9] Foundation T A S 2011 Hbase http://hbase.apache.org/
[10] 10gen 2011 Mongodb http://www.mongodb.org/

[11] Gray J 1981 *Proc. Int. Conf. on Very Large Data Bases - Volume 7* VLDB '81 (VLDB Endowment) pp 144–154
[12] Oetiker T 2011 Rrd tool http://oss.oetiker.ch/rrdtool/
[13] Graphite 2011 Graphite http://graphite.wikidot.com/
[14] OpenTSDB 2011 Opentsdb http://opentsdb.net/
[15] Systems K 2011 Kdb http://kx.com/
[16] NCSA 2011 Hdf5 http://www.hdfgroup.org/HDF5/
[17] Cooper B F, Silberstein A, Tam E, Ramakrishnan R and Sears R 2010 *Proc. of the 1st ACM Sym. on Cloud computing* SoCC '10 (New York, NY, USA: ACM) pp 143–154 ISBN 978-1-4503-0036-0
[18] Hewitt E *Cassandra: The Definitive Guide* (O'Reilly Media, Inc.)
[19] Cloudkick 2011 Cassandra blog http://www.cloudkick.com/blog/
[20] DeCandia G, Hastorun D, Jampani M, Kakulapati G, Lakshman A, Pilchin A, Sivasubramanian S, Vosshall P and Vogels W 2007 *SIGOPS Oper. Syst. Rev.* **41** 205–220 ISSN 0163-5980
[21] Chang F, Dean J, Ghemawat S, Hsieh W C, Wallach D A, Burrows M, Chandra T, Fikes A and Gruber R E 2008 *ACM Trans. Comput. Syst.* **26** 4:1–4:26 ISSN 0734-2071
[22] Welsh M, Culler D and Brewer E 2001 *SIGOPS Oper. Syst. Rev.* **35** 230–243 ISSN 0163-5980
[23] Foundation T A S 2011 Thrift http://thrift.apache.org/
[24] Todd B 2011 Cassandra storage sizing http://btoddb-cass-storage.blogspot.com/

Journal of Physics: Conference Series **368** (2012) 012003 doi:10.1088/1742-6596/368/1/012003

Online measurement of LHC beam parameters with the ATLAS High Level Trigger

E Strauss

SLAC National Accelerator Laboratory, 2575 Sand Hill Road, Menlo Park, CA 94025-7015

E-mail: estrauss@slac.stanford.edu

Abstract. We present an online measurement of the LHC beamspot parameters in ATLAS using the High Level Trigger (HLT). When a significant change is detected in the measured beamspot, it is distributed to the HLT. There, trigger algorithms like b-tagging which calculate impact parameters or decay lengths benefit from a precise, up-to-date set of beamspot parameters. Additionally, online feedback is sent to the LHC operators in real time. The measurement is performed by an algorithm running on the Level 2 trigger farm, leveraging the high rate of usable events. Dedicated algorithms perform a full scan of the silicon detector to reconstruct event vertices from registered tracks. The distribution of these vertices is aggregated across the farm and their shape is extracted through fits every 60 seconds to determine the beamspot position, size, and tilt. The reconstructed beamspot values are corrected for detector resolution effects, measured *in situ* using the separation of vertices whose tracks have been split into two collections. Furthermore, measurements for individual bunch crossings have allowed for studies of single-bunch distributions as well as the behavior of bunch trains. This talk will cover the constraints imposed by the online environment and describe how these measurements are accomplished with the given resources. The algorithm tasks must be completed within the time constraints of the Level 2 trigger, with limited CPU and bandwidth allocations. This places an emphasis on efficient algorithm design and the minimization of data requests.

1. Introduction

In 2011, ATLAS collected over 5 fb^{-1} of data from collisions produced by the Large Hadron Collider (LHC) at center-of-mass energy $\sqrt{s}$ = 7 TeV. The collisions within the luminous region, or beamspot, follow a Gaussian distribution along the x-, y-, and z-axis[1] determined by the interaction region of the two beams. Interaction Point (IP) parameters typical of the results reported in this paper are listed in table 1. As the LHC lacks instrumentation close to the IP, the ATLAS detector provides the best measurement of the beamspot parameters. Not only is this valuable monitoring information for the LHC operators, it is crucial for running several ATLAS triggers with high efficiency and purity.

[1] The ATLAS reference system is a Cartesian right-handed coordinate system, with nominal collision point at the origin. The anti-clockwise beam direction defines the positive z-axis, with the x-axis pointing to the center of the LHC ring. The pseudo-rapidity is defined as η = -ln(tan($\theta/2$), where the polar angle θ is taken with respect to the positive z direction.

Table 1: The typical beam parameters during 2011 LHC operation.

$\sqrt{s}$ (TeV)	7.0
Typical Number of Protons per Bunch	$1.2 - 1.5 \times 10^{11}$
Half-Crossing Angle of the Beams	120 μrad
Typical IP β-function (m)	1.0 and 1.5
Typical Number of Bunches	1000 - 1300
Peak Instantaneous Luminosity ($\mathrm{cm}^{-2}\mathrm{s}^{-1}$)	3.65×10^{33}
Typical Number of Collisions per Bunch Crossing	10 to 20

Using tracks measured by the ATLAS Inner Detector (ID), we reconstruct the position of primary vertices in three dimensions. The mean of the vertex positions in x, y, and z mirrors the center of the luminous region as it drifts over the course of a fill. The widths of the Gaussian reflect the size of the luminous region, and can track the evolution of the transverse and longitudinal emittances. The measurements we provide must therefore be made both precisely, and quickly enough to track the changes in these parameters.

As we will discuss in later sections, the second level (L2) of the ATLAS High Level Trigger (HLT) provides an excellent environment for making these measurements. Its technical constraints require the use of highly efficient algorithms in order to maximally exploit each event. Despite these constraints, the high rate of events and access to tracking information makes the HLT the best place to perform such a complex measurement with the high frequency and low latency we require. As we will discuss in later chapters, the L2 environment allows for very rapid feedback of the beamspot parameters at the ATLAS interaction point, with precisions at the μm level. With longer time intervals it is also possible to measure the beamspot position and width for each of the bunch crossing with approximately five percent precision.

2. The ATLAS Inner Detector

The primary vertices in the event are reconstructed using the ATLAS ID tracking [1]. The ID has three sub-detector components. The algorithm uses information from the two inner-most, the silicon pixel detector, and the silicon strip detector (SCT), for their excellent spatial resolution. The pixel detector has three layers which sit just outside of the LHC beam-pipe, and 3 disks on each end-cap. Its hit resolution is about 10 μm in $r\phi$, and about 115 μm in z. Just outside of this, the SCT is comprised of 4 cylindrical layers in the barrel and 9 disks on each end-cap. The SCT hit resolution is about 17 μm in $r\phi$ and about 850 μm in z. The ID is immersed in a 2 T solenoidal field, which bends the tracks for accurate p_T measurements. All told, the pixel and SCT covers a radius of approximately 600 cm and measure tracks up to pseudo-rapidity $|\eta| < 2.5$.

3. The ATLAS High Level Trigger

The ATLAS trigger system [2] is comprised of three levels; one hardware (L1), and two software (L2 and EF). The rate of events written to tape is maintained at approximately 300 Hz, five orders of magnitude down from the approximately 20 MHz produced by the LHC during the 2011 data-taking period. L2, where we perform our measurement, is capable of accepting a rate of approximately 75 kHz; of this about 1.5 kHz is allocated to our trigger algorithm.

When the Central Trigger Processor determines that a relevant L1 trigger has fired, an instance of the trigger algorithm gets run by one of the CPUs on a farm of over 6,000 processors. The result is histogramed locally, and aggregated every 60 seconds across the farm. They are then stored, and made available for processing by tools running within the ATLAS Data Acquisition

(DAQ) infrastructure [3].

4. Beamspot Algorithm

The beamspot parameters are determined by reconstructing the distribution of primary vertices produced by the hard-scatter of the beams. Events with these primary vertices are collected by the trigger chain at an input rate of approximately 1.5 kHz. The beamspot measurement is made in several steps:

Tracking in which the hits from the pixel and SCT detectors are used to measure the position and momentum of tracks.

Vertexing in which the tracks are clustered into common primary vertices.

Histograming in which the distributions of the vertices in x and y directions are stored by the trigger farm nodes before being aggregated.

Measuring in which the aggregated distributions are used to extract the beamspot position, width, and tilt versus time, correcting for resolution effects. This is done in multiple passes, with various time windows, to achieve the statistics necessary for monitoring, as well as global and per-bunch-crossing measurements.

Distribution in which the parameters are published for monitoring, distributed to the HLT nodes as input for other triggers, and broadcast to the LHC operators for feedback.

4.1. Tracking and Vertex Reconstruction

Trigger algorithms at L2 typically operate on regions of interest, identified at L1, to keep the data-requests from the detector read out system low. The beamspot algorithm requests information about all tracks in the event and performs a "full-scan" of all the hits in the ID.

Fast custom algorithms are used to create 3D space-points from these hits before fitting the tracks with a Kalman filter. These algorithms have been tuned to perform this operation very quickly to stay within the stringent L2 timing requirements. The L2 tracking manages to maintain efficiencies over 90%s compared to the offline track selection.

Although the clustering and vertexing portion of the algorithm is very fast, typically taking approximately 250 μs per event, the reconstruction of tracks from the ID takes 300 ms on average and requires relatively large transfers of detector data. For this reason, the trigger steering has been optimized to run the beamspot algorithm first, such that the ID data can be cached for other algorithms.

The trigger algorithm first applies some basic quality cuts to the tracks from the previous step. Those which satisfy the cuts are clustered by impact parameters. The true distribution of the impact parameter along the z axis is much larger than the ID z resolution. As the distribution of tracks is much broader than the transverse distributions, this clustering is a quick and efficient way of grouping the tracks, simplifying the vertexing and reducing the mis-assignment of tracks.

After being clustered, the vertices are measured by a rapid vertex fitter [4]. The fitter uses the track covariance directly, avoiding costly matrix inversions and can reconstruct a vertex in under a millisecond.

The luminosity attained by the LHC in 2011 has been at the cost of "pile-up". As the beam intensities have increased, so have the number of interactions per bunch crossing, averaging around 20 in the last few weeks of the 2011 run. Though this provides the opportunity to make multiple measurements per event, the higher occupancy increases both the event size and the time it takes to reconstruct all tracks. Worse still, the close proximity of multiple vertices introduces possible systematic biases when tracks are assigned to the wrong vertex. The z window in which tracks are clustered has been tightened to cope with the higher density of vertices. Over the course of 2011 the L1 seed was switched from a Minimum Bias trigger, to a three jet trigger. The three jet trigger has a lower rate, but more tracks per vertex on average.

Raising the p_T threshold on the tracks fed to the vertex fitter has helped to remove background, and raising the required number of tracks per vertex has improved the average uncertainty on the vertex position. These tighter requirements help to maintain a high quality of data produced by our algorithm, despite harder operational conditions.

4.2. Histograming the Luminous Region

The spatial distributions of the vertices are stored locally in histograms by each of the over 6000 core processes on the trigger farm which run the beamspot algorithm. As described in Section 3, these are then collected across the farm. Here, they are available to a tool in the DAQ infrastructure that uses them to determine the beamspot parameters. By summing consecutively produced histograms, the tool can make measurements for arbitrary time periods. The position can be precisely measured as the mean of a Gaussian function, fit independently to the x-, y-, and z-projections of the vertex distribution, as shown in figure 1. The width of the distribution of vertex positions is the beamspot width, convolved with the detector resolution. We describe in the next section how a resolution-corrected measurement is made. The tilt with respect to the ATLAS coordinate system is measured from the slope of the xy and yz profiles.

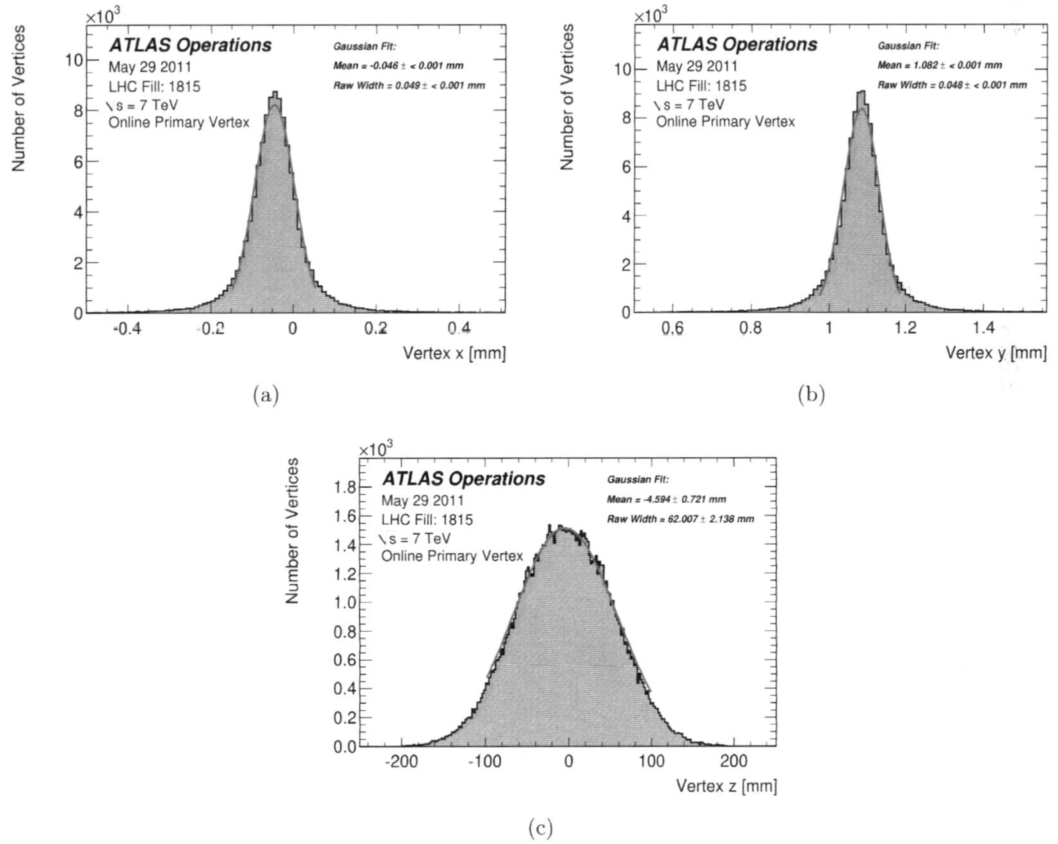

Figure 1: The x (a), y (b), and z (c) distribution of primary vertices reconstructed online with ≥ 6 tracks in the High Level Trigger in 1 minute of data-taking.

4.3. In Situ Resolution Correction

The true beamspot width in 2011 was roughly 15 to 45 μm in x and y, and 25 to 75 mm in z. The z vertex resolution is on the order of 100 μm, so the resolution is not a significant consideration in the measurement. This is not the case in x and y where the resolution is on the order of 10 μm.

A split-vertex method is used to remove the effect of detector resolution on the beamspot width measurement in all three directions. The tracks which have been used to fit the primary vertices are sorted in ϕ, and then assigned alternately into two sub-clusters for a new round of vertex fitting. These tracks are expected to come from the same true vertex, the differences in their reconstructed vertex is assumed to be entirely due to resolution effects. The true beamspot width, σ_{beamspot}, can therefore be measured as $\sigma_{\text{beamspot}} = \sqrt{\sigma^2_{\text{raw}} - \sigma^2_{\text{res}}}$ where σ_{raw} is the width of the primary vertex distribution and σ^2_{res} is $1/\sqrt{2}$ times the width of the displacement between the two split-vertices. Not all vertices are created equal though, the more tracks in the vertex, the smaller the uncertainty. The primary vertex distributions and the split-vertex displacements are stored in 2D histograms against the number of tracks per-vertex so that the correction can be applied as a function of track multiplicity. Figure 2a shows the narrowing distribution at higher track multiplicities as well as the rapid drop-off in statistics.

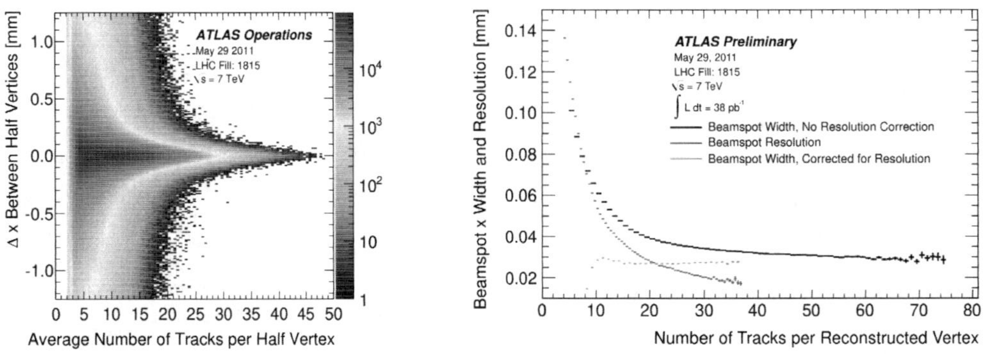

(a) (b)

Figure 2: The displacements in x between each pair of separately reconstructed split-vertices as a function of the number of tracks, measured online in the High Level Trigger (a). The observed width of the primary vertex distribution, the measured resolution, and the resolution-corrected width, all as a function of the number of tracks per vertex in x (b).

The split-vertex method has a few limitations. The measured resolution is biased towards larger values when the true beamspot width is less than half the size of the resolution. This can be compensated for by restricting ourselves to vertices with large track-multiplicities. At the beginning of the year we required six tracks per vertex, as the beam widths narrowed, we raised the threshold to ten tracks per vertex.

Unfortunately, the split-vertices probe the resolution for vertices with half as many tracks as the primary vertex. For a sample of reconstructed vertices with up to n_{Max} tracks per vertex, in the best case all those above $n_{\text{Max}}/2$ will effectively be thrown out as we do not have an estimate of their resolution. Figure 2b demonstrates this. It shows the width of the primary vertex width, the resolution measurement, and the resolution-corrected beamspot width as a function of track multiplicity for the x-axis. The primary vertex width distribution extends up to 75 tracks per reconstructed vertex, whereas the resolution and corrected width only extend up to 36.

Finally, the systematic uncertainties and background effects at low track multiplicities can cause large deviations in the resolution-corrected width measurement. As the size of these effects diminish, the corrected width distribution flattens, typically as it approaches the size of the resolution measurement. At higher track multiplicities, the statistical variations in the resolution measurement become more significant.

Despite these limitations, the values produced by the online algorithm have been found to be in good agreement with those expected based on LHC instrumentation, as well as those measured with an alternate method offline [5]. Figure 3 gives a sense of the small statistical uncertainty achieved even on short time intervals.

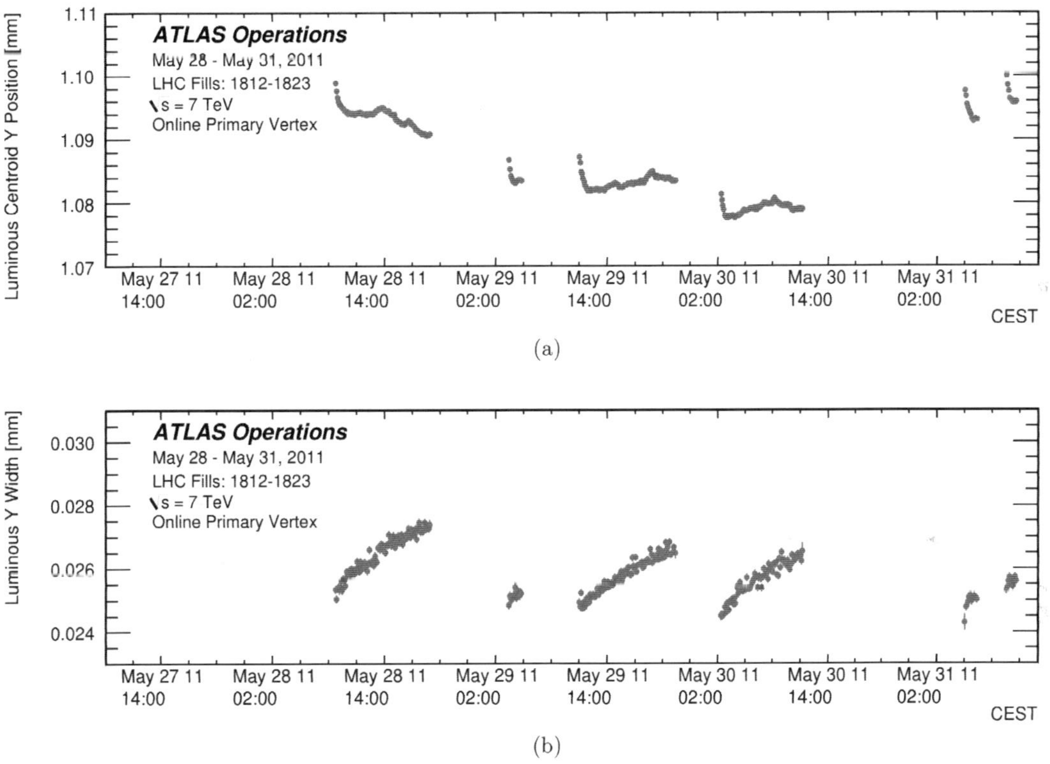

Figure 3: Time-variation of the luminous centroid position (a) and resolution-corrected width (b) in y measured in the High Level Trigger every five minutes, for six separate LHC fills recorded over the span of four days.

4.4. Measurement and Re-distribution

As the online track reconstruction is performed in a window around the stored beamspot position, incorrect values can significantly decrease the track reconstruction efficiency. Jet triggers which calculate b-jet probability [6] are another major client of the beamspot parameters. These typically use the transverse impact parameter significance of tracks within a jet to identify those which were produced by the decay of a b-quark.

A mechanism was developed to issue automatic updates to the HLT conditions database. The values measured using 10 minutes of integrated data are compared against those currently in the database. If the newly computed values and the database values differ significantly, the new values are published to the database for the the HLT nodes to retrieve. Such an update is issued if any the following conditions are met:

- the position along any of the three axes changes by more than 10%,
- the width along any of the three axes changes by more than 10%,
- or there is a reduction of 50% in the statistical uncertainty of any of the values.

Typically, this means frequent updates at the start of the run, as the resolution is boot-strapped, followed by a period of infrequent updates, dominated by the beam drift and emittance blow-up.

4.5. Measurements Per Bunch Crossing

A third set of measurements is made every 20 minutes for the LHC operators. The longer integration time is needed to accumulate enough statistics to measure the beamspot parameters for each of the approximately 1300 individual bunch crossings (BC) to sub-micron precision on the position. This is a feature which only the online beamspot tools are able to provide due to our high rate of usable events. The large statistics required place it just at the limit of our capabilities, requiring approximately 1 Hz per BC. Even so, there are not enough events to measure the per-BC parameters as a function of track multiplicity. A coarser method is used, in which the vertex parameters are histograms in 1D, summing all track multiplicities. An example of such a measurement can be seen in figure 4 where distinct structures are visible. In particular the vertical position of the interaction-point shows variations of up to 5 μm and repeating patterns across the injected bunch trains [7, 8].

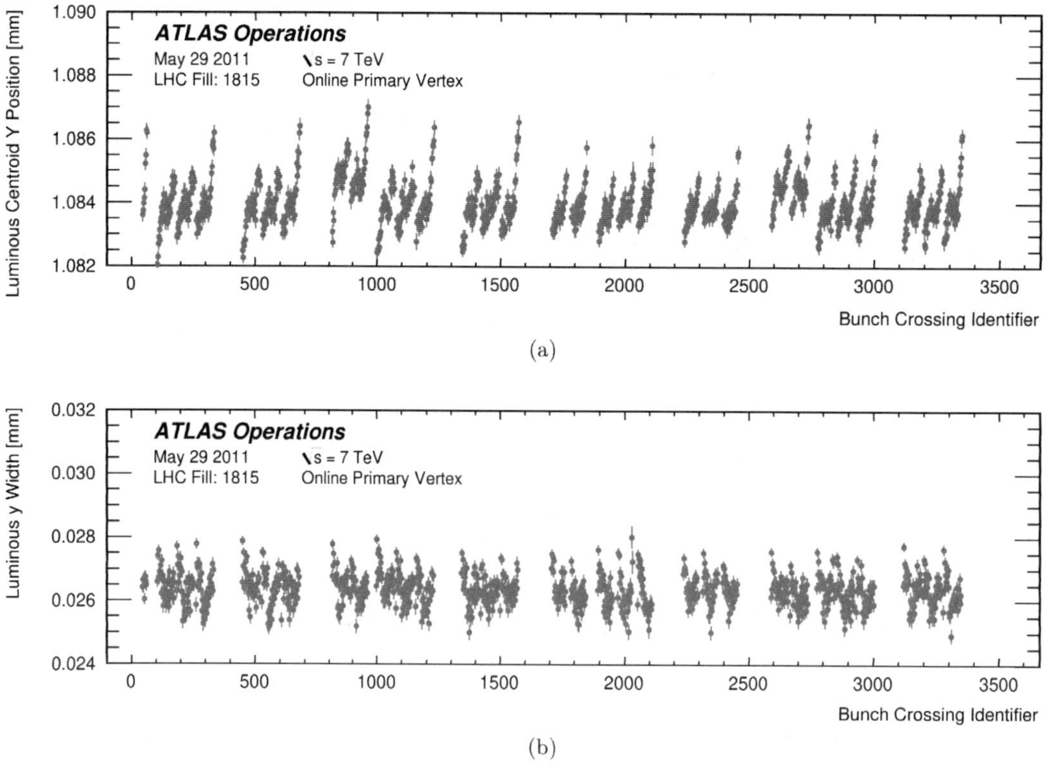

Figure 4: The luminous centroid position (a) and width (b) in y measured in the High Level Trigger farm for each of the 1024 colliding bunch crossings separately.

Using a single Gaussian fit to describe the sample introduces some systematic biases, as the primary vertex distribution is actually a sum of Gaussian distributions where events with fewer tracks dominate but have broader widths. The resolution correction is applied by producing a 1D histogram of split-vertex differences, where each event is given a weight such that the track multiplicity distribution of the split-vertices mirrors that of the primary vertices. This ensures an equivalent event mixture is used in both measurements, reducing the size of the systematic bias.

5. Pre-fit Optimizations

The beamspot values are computed every minute for monitoring purposes, every 10 minutes for the HLT, and every 20 minutes for the LHC operators. Between these three sets of processes, the beamspot tool must perform on the order of 200 fits per minute. At this rate, even simple Gaussian fits become quite costly. We apply various cost-cutting approaches to keep the processing time reasonable.

(i) Fits are only performed on histograms containing at least 100 entries. This requirement ensures a reasonable statistical uncertainty on the fit parameters, and saves us from wasting time on histograms whose entries are too sparse.

(ii) Fits are only performed within ± 1.8 RMS of the histogram mean. The distribution outside of this interval contains a larger fraction of background events which are not Gaussianly distributed.

(iii) Histograms are re-binned before the fit, such that the average error per bin in the core is less than 20%. In sparsely populated histograms, too-high granularity causes the fit to take longer to converge. The re-binning also smooths out the effects on the RMS of a few bins far out in the tails. A more stringent requirement on the statistical uncertainty would cause further re-binning which might wipe out details about the shape of the distribution.

6. Conclusions

We have used the ATLAS ID and HLT to measure the beamspot parameters in quasi-real-time. This feat required several components to work: optimized data-requests from the ID, fast reconstruction algorithms in the HLT, aggregation of histograms from the trigger farm at fixed time intervals, and rapid measurements by a tool running in the DAQ system. By measuring the parameters every 60s we provide precise monitoring of the beamspot to ATLAS. On a longer time-scale, the parameters are distributed to the HLT to maintain high tracking and b-jet tagging efficiency by continuously adjusting the vertex position. The LHC operators benefit from our tool as well, as the the global beamspot and the beamspot per-bunch-crossing are published for them every twenty minutes. The results from the online determination have been cross-checked against expectations from the parameters of the two beams, as well as the offline beamspot reconstruction results. Data taking in 2011 provided unique challenges and development is ongoing on the algorithm and the data selection to prepare for those which 2012 will bring.

References

[1] Aad G *et al.* 2008 *Journal of Instrumentation* **3** S08003
[2] Aad G *et al.* 2011 *submitted to EPJC*
[3] Cogan J 2011 *13th International Conference on Accelerator and Large Experimental Physics Control Systems*
[4] Emeliyanov D 2007 *11th International Workshop on Advanced Computing and Analysis Techniques in Physics Research* 058
[5] Bartoldus R *et al.* 2010 *MOPEC008, 1st International Particle Accelerator Conference: IPAC10*
[6] Coccaro A, on behalf of the ATLAS Collaboration 2012 *these proceedings*
[7] Schaumann M and Fernandez R A 2011 *WEPC081, 2nd International Particle Accelerator Conference: IPAC11*
[8] Herr W, Alemany R, Giachino R, Papotti G, Pieloni T, Calaga R, Laface E and Schaumann M 2011 *WEODA01, 2nd International Particle Accelerator Conference: IPAC11*

ACAT 2011 IOP Publishing
Journal of Physics: Conference Series **368** (2012) 012004 doi:10.1088/1742-6596/368/1/012004

The AAL project: automated monitoring and intelligent analysis for the ATLAS data taking infrastructure

A Kazarov[1], G Lehmann Miotto[2] and L Magnoni[2]

[1] CERN, on leave from PNPI, St.Petersburg, Russian Federation
[2] CERN, European Laboratory for Particle Physics (CERN), Geneva, Switzerland

E-mail: andrei.kazarov@cern.ch, giovanna.lehmann@cern.ch, luca.magnoni@cern.ch

Abstract. The Trigger and Data Acquisition (TDAQ) system of the ATLAS experiment at CERN is the infrastructure responsible for collecting and transferring ATLAS experimental data from detectors to the mass storage system. It relies on a large, distributed computing environment, including thousands of computing nodes with thousands of application running concurrently. In such a complex environment, information analysis is fundamental for controlling applications behavior, error reporting and operational monitoring. During data taking runs, streams of messages sent by applications via the message reporting system together with data published from applications via information services are the main sources of knowledge about correctness of running operations. The flow of data produced (with an average rate of O(1-10KHz)) is constantly monitored by experts to detect problem or misbehavior. This requires strong competence and experience in understanding and discovering problems and root causes, and often the meaningful information is not in the single message or update, but in the aggregated behavior in a certain time-line. The AAL project is meant at reducing the man power needs and at assuring a constant high quality of problem detection by automating most of the monitoring tasks and providing real-time correlation of data-taking and system metrics. This project combines technologies coming from different disciplines, in particular it leverages on an Event Driven Architecture to unify the flow of data from the ATLAS infrastructure, on a Complex Event Processing (CEP) engine for correlation of events and on a message oriented architecture for components integration. The project is composed of 2 main components: a core processing engine, responsible for correlation of events through expert-defined queries and a web based front-end to present real-time information and interact with the system. All components works in a loose-coupled event based architecture, with a message broker to centralize all communication between modules. The result is an intelligent system able to extract and compute relevant information from the flow of operational data to provide real-time feedback to human experts who can promptly react when needed. The paper presents the design and implementation of the AAL project, together with the results of its usage as automated monitoring assistant for the ATLAS data taking infrastructure.

1. Introduction

This paper presents the AAL project, a new tool meant at providing automated monitoring and intelligent analysis for the ATLAS data taking infrastructure. First we introduce the data taking facilities of the ATLAS experiment at CERN and the motivation which drove this new development. In section 2 we describe the AAL project, the main functionalities and the different technologies we are using. In section 3 we present the AAL engine architecture, in section 4 the visualization strategies we implemented and finally in section 5 we describe how the tool has been adopted and the results and conclusions.

1.1. ATLAS TDAQ system

ATLAS [1] (A Thoroidal LHC Apparatus) is a particle physics experiment at the Large Hadron Collider at CERN. The Trigger and Data Acquisition (TDAQ) system of ATLAS is responsible for selecting and transferring ATLAS experimental data from the detector to the mass storage system. To cope with the very high data rate produced by the detector a complex and distributed computing infrastructure has been built. More than 200 switches and routers interconnect around 2000 hosts to build a real-time filtering system for particle collision events. On top of the hardware infrastructure there are over 20.000 applications responsible of analyzing, filtering and moving event data to permanent storage [2].

1.2. Motivation: improving monitoring with automation and event correlation

The TDAQ system is operated by a non-expert shift crew, assisted by a set of experts providing knowledge for specific components. The daily work of operators is made of procedures to run the system, periodic checks and controls on system status, well defined reaction in case of known problems and interaction with experts in case of non standard issues. The evaluation of correctness of running operations requires strong competence and experience in understanding problems and root causes, and often the meaningful information is not in the single message or update, but in the aggregated behavior in a certain time-line. The AAL project automatizes checks and controls, detects complex event patterns over time and helps operators with more pertinent and meaningful information for system analysis.

2. The AAL project

The AAL project is meant at assuring a constant high quality of problem detection by automating most of the monitoring tasks and providing real-time correlation of operational data and system metrics. AAL guides operators and experts in their daily work, for this reason is also known as *"The shifter assistant"*. It performs a real-time analysis of the whole TDAQ system, detecting problematic situations and misbehaviour and producing **alerts** suggesting expected reactions. This project combines technologies coming from different disciplines, in particular it leverages on an Event Driven Architecture to unify the flow of data to be monitored, on a Complex Event Processing (CEP) engine for real time correlation of events and pattern recognition and on a Message Queuing system for components integration and communication. The picture in figure 1 presents an overview of the system together with the three operational stages: **information gathering**, **information processing** and **result distribution**.

2.1. Information gathering

Evaluation of TDAQ working conditions requires data collection from multiple data sources, each one with different technology, publication mechanism and format. AAL is able to gather data from these different information providers in order to detect problems and failures. Farm metrics, such as machines status and problems, are collected from the Nagios tool [3]. TDAQ operational data is gathered via the streams of messages produced by applications via the Message Reporting System (MRS) [4] and from the information published via the Information Services (IS) by running applications. As illustrated in figure 1, system metrics and operational data can be abstracted as a flow of information composed on many single events over time. Moreover, AAL is also able to retrieve more static information at run time in a pull fashion via reader plug-ins, such as the case for system topology from the configuration database.

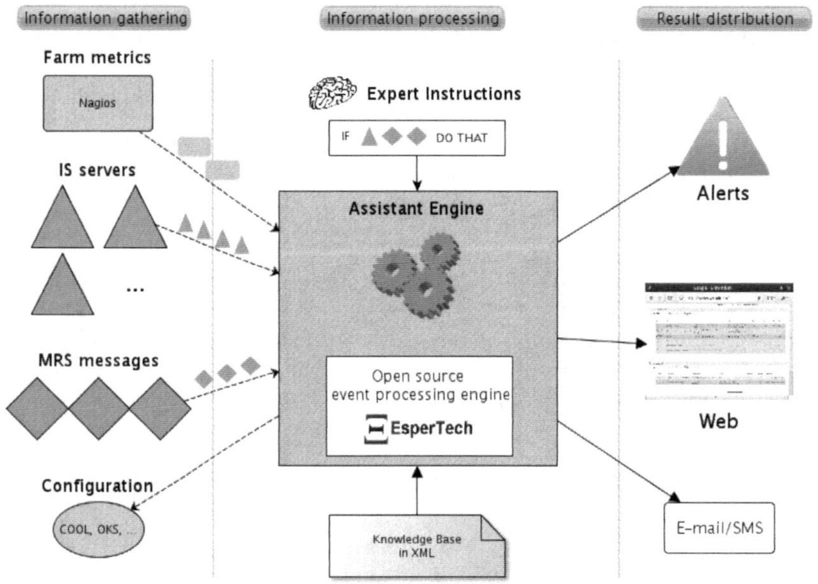

Figure 1. Overview of the AAL project.

2.2. Information processing

Information processing is the key functionality of the tool. To detect problems and to perform root cause analysis of failures AAL needs to correlate at real-time possibly different streams of events, detecting specific time-based patterns previously defined by experts. The main requirements for the processing system are:

- Real-time processing for fast error detection and reporting;
- Fine-grained information processing for specific event(s) detection;
- Events aggregation for detection of patterns over time or other properties;
- Effective representation of patterns and correlations, similar to the SQL query language for database systems.

We found a very effective solution coming from the Complex Event Processing field of research, mainly adopted in financial analysis and advanced system monitoring.

2.2.1. Complex Event Processing Complex Event Processing (CEP) [5] is one of the fundamental techniques in the so called Operational Intelligent procedures [6], where the goal is to to provide insight into business operations by running query analysis against live feeds and event data. In the last decade it has been adopted in very different contexts, from financial analysis to business process management and intelligent system monitoring. CEP systems are meant to analyze and correlate streams of event, but the great advantages they offer is to abstract events structure and to provide a standard language to express pattern and correlation as queries.

2.2.2. Esper In the CEP world there are several solutions, both commercial and open source, offering event processing functionalities based on SQL-like processing languages to express patterns and rules. Esper [7] is a Complex Event Processing engine written in Java that offers a very powerful Event Processing Language (EPL) with built-in capabilities for analysis over time and event correlation. Esper is able to handle very high flow of data, thanks to a non-persistent approach where events pass through memory and the engine continuously check for the matching criteria expressed by queries. The continuous queries approach perfectly matches with our real-time requirements and this, together with the powerful functionalities offered by the EPL language, convinced us to adopt Esper for building the information processing engine of AAL. EPL queries are used to express pattern describing problems and failures. The figure 2 illustrate graphical examples of types of patterns detect by AAL relying on Esper capabilities.

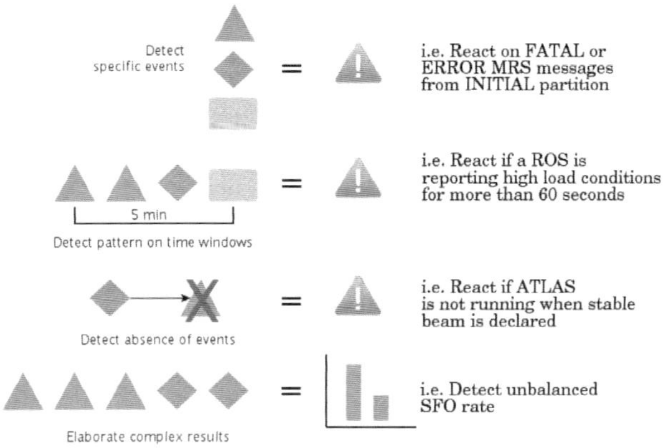

Figure 2. Simple and complex patterns detected by the AAL engine.

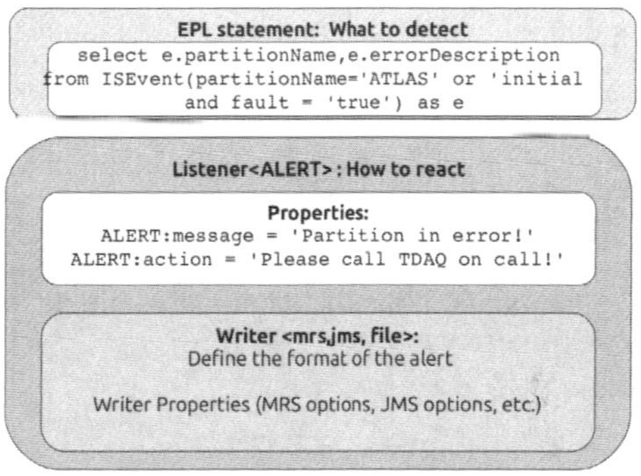

Figure 3. Directive schema.

2.3. Knowledge engineering: directives and alerts

TDAQ system experts have to feed AAL with instructions about what situations to detect and suggestions on how to react. This knowledge engineering process is fundamental to build and maintain an up to date data set of problems and errors. These instructions, named *directives*, are codified in XML documents and loaded by the engine at configuration time. Every directive generates one or more alerts. An alert is composed of:

- **Problem description**: brief description of the problematic situations.
- **Reaction**: expected reaction to be performed by the operator.
- **Domain(s)**: one or more category defining who should care about the problem. This information is mainly used for routing of the alerts at visualization.
- **Events details**: events that triggered the alerts.

A directive, as represented in figure 3, is composed by two main elements: the *query*, that is the Esper EPL statement that defines the event pattern to react on and the *listener*. The listener defines alert details and one or more writers responsible for alert distribution. The same alert can be produced by the engine in different formats by dedicated *writers*, as described in section 3.2.

2.4. Alerts visualization and distribution

For a fast and effective reaction on system failures alerts have to be distributed in several ways to operators and experts. The main goal is to build an effective real time visualization for operators in the ATLAS control room, that need to promptly receive information with problem details and expected reactions. Moreover, additional alert distribution strategies exist for remote monitoring purpose. This has been accomplished with a generic architecture of the AAL engine, as presented in section 3, and by a message oriented communication, presented in section 4.

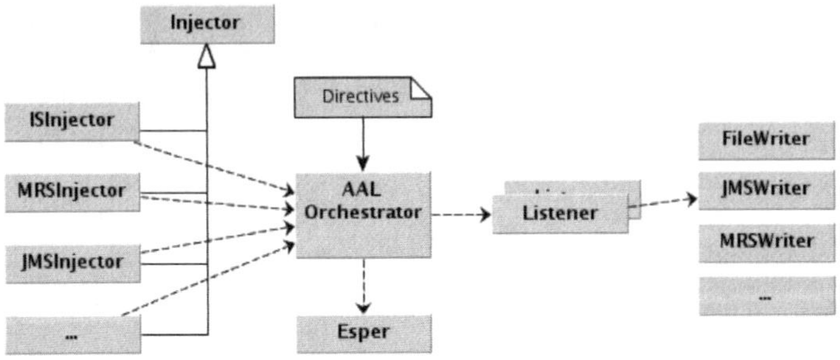

Figure 4. The AAL engine architecture.

2.5. AAL components

The AAL project is composed of the following components:

- **The AAL engine**: a standalone Java-based service responsible for data collection and information processing;
- **The AAL web frontend**:a web application to visualize and interact with alerts;
- **A message broker** for components communication.

3. The AAL engine

The main role of the AAL engine is to orchestrate the interactions between information providers and the complex event processing engine, as shown in figure 4. It parses the knowledge base at configuration time and it initializes all the corresponding queries and statements in Esper. It then handles all the information gatherer components and create the writers and listeners ready to produce alerts when patterns are detected. The flow of events coming from all data sources is then constantly processed by the Esper engine and every time an event matches a directive query or pattern the proper alert is created and distributed.

3.1. Injectors

The set of components responsible for data collection are called *injectors*, as shown in figure 4. Injectors implement a generic design that allows to extend the system with new information sources without affecting the collection logic. Injectors specialization exists for the different data source types but they expose functionalities via the same interface to the engine. For data sources with high rate streams of event, like the IS servers that can reach up to 50 KHz of information updates, injectors leverages on the CORBA-based IPC mechanism provided by TDAQ infrastructure [8]. In other cases, like for the Nagios metrics, we hide the complexity of the data source by using a message queuing system to collect and distribute information.

3.2. Writers

The *writer* components are responsible to codify and distribute alerts in specific formats. A directive specifies one or more *writers* elements to define how the generated alert have to be propagated. The default format is XML (Extensible Markup Language), suitable for machine-based processing and distributed via the message queuing as described in section 4.1. But alerts can also be written to file in more human-readable format or can be sent as standard TDAQ application messages.

3.2.1. JMS writer To build a scalable and modular system we adopted a message oriented architecture where the communication take place via exchanging messages. A message queuing system, or message broker, is used to distribute alerts offering a publish/subscribe interface. More information about the chosen tool is given in section 4.1. In order to be independent from the message broker implementation the JMS writer send alert as a XML message using the generic Java Message Service (JMS) API [9].

3.2.2. MRS writer To be compatible with the TDAQ architecture alerts can be distributed also as MRS messages relying on the existing message reporting facilities [4].This functionality is provided by the MRS writer. This integration allow to use the alert produced by AAL to trigger automatic reaction and recovery procedures via the existing error recovery facilities [10].

3.2.3. Mail and RSS When a critical situation requires a prompt reaction from users and experts alerts can be explicitly sent as e-mail to a set of addresses and can also be delivered as sms thanks to the mail to sms service provided by CERN. Moreover, AAL relies on the message broker functionalities to send alerts as a RSS feed stream, easy to integrate in any mail reader or feed reader application.

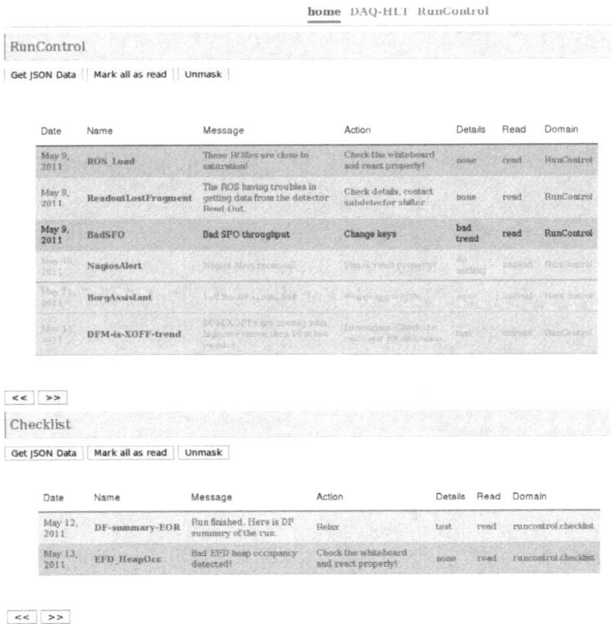

Figure 5. Web application for alerts visualization.

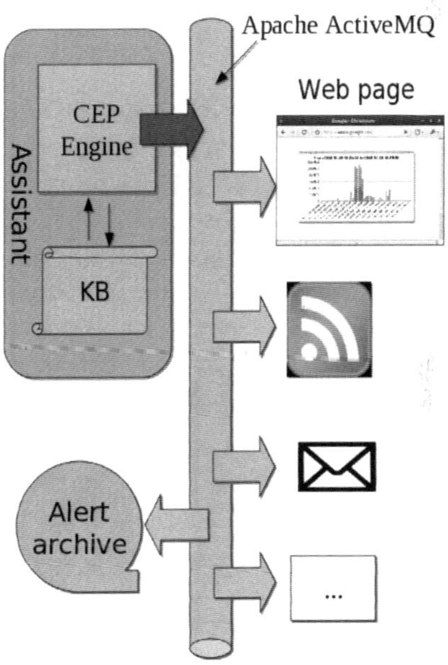

Figure 6. Message oriented communication for modules interaction.

4. Web based visualization

Building a web based visualization service was the natural choice to gain independence from user location, hardware and software, all critical features for a monitoring system. A Rich Internet web Application (RIA) has been built allowing users to perform actions on the page and interact with alerts visualized. The web application is developed using the Django framework [11]. Alerts are collected by an archiver plug-in connected to the message broker via the Streaming Text Orientated Messaging Protocol (STOMP) protocol. Once received, alerts are archived in a SQLite database and made available to the web application. Web pages are automatically updated when new messages arrive, providing a real-time feedback in case of problems.

Users can interact with the web page by making the following actions:

- Alerts can be marked as processed when the problem is addressed;
- Classes of alerts can be masked for a certain time interval;
- Select and sort alerts per properties (Date, Name, etc.).

Web pages have a basic structure composed by several sections, as presented in figure 5. Every section is a table visualizing alerts that match with one or more criteria. The structure is easily customizable, so experts can built their own page layout, defining which alerts they want to receive in every sections.

4.1. Message oriented architecture

The figure 6 presents the different software involved in building the web visualization: a web application, an alert archiver, the RSS server and possibly other generic alert consumers. As introduced in the previous sections AAL relies on a message oriented communication where a message queuing system, or message broker, distributes alerts offering a publish/subscribe interface. We chose ActiveMQ as open source message broker from the Apache software foundation [12]. It is compatible with the JMS [9] interface, the standard for message oriented solutions, it supports multiple wire protocols and multiple network protocols. Moreover, it provides several cross language clients for non-Java applications.

4.1.1. Network of message brokers The AAL engine runs on a dedicated host on the TDAQ computing farm of the ATLAS experiment. Access to this computing facility is highly restricted and direct communications are not possible from the general public CERN network. While shifters operate from a control room inside this restricted area, experts are in most cases outside this facility. To allow remote monitoring alerts produced by AAL have to be made available both inside and outside the confined network. This is accomplished by a master-slave configuration of 2 ActiveMQ instances. The master is running inside the TDAQ farm and it receives all the alerts from the engine. It then forwards all messages to the slave instance, running on the public network, via an outgoing TCP connection.

5. Use cases

The AAL tool is used in production to assist monitoring and operational procedures during ATLAS data taking runs. More than 100 directives has been coded gathering requirements from very different TDAQ domains. The following are some practical situations detected by the tool and the corresponding EPL statements:

- analysis of problems reported by Nagios (the farm monitoring tool):
  - `select * from NagiosAlert(hostname regexp '(?!pc-atlas-pub-[0-9]+.*') group by hostname output every 1 minute`
- detect bunch of error messages as symptom of an issue that impact most of TDAQ applications, such as the lost of connectivity for a read-out machine (ROS):
  - `select ROS from ROS.win:time(40 sec).std:groupwin(ROS, Message) having size > 100 output first every 120 seconds`
- detect critical amount of read-out errors on the detectors (not easy to spot by a human eye):
  - `select Crate, NofBadLinks from SCTLinkErrorsCounter having NofBadLinks > 400 and PARTITION_IS_RUNNING output every 2 minute`

6. Conclusion

Effective monitoring and promptly error detection are fundamental properties for maximizing the data taking efficiency of high energy physics experiments. The main goal of the AAL project is to assist ATLAS operators with automated analysis of system conditions and intelligent reasoning for root-cause problems analysis. AAL has been implemented leveraging on modern event correlation techniques combined with message driven architecture for components integration and web technology for alerts visualization. The tool has been quickly adopted in the ATLAS TDAQ community and it gathered the interest from other groups with similar monitoring requirements, such as networking and system administration.

The main area of research for the future is to investigate the possibility to build a learning module to detect anomalies in an unsupervised manner, as proven by the HOLMES project [13]. Furthermore, the web interface will be extended to include directives management and system administration together with improved user experience.

References

[1] ATLAS Collaboration 2008 *The ATLAS Experiment at the CERN Large Hadron Collider.* J. Inst., 3, S08003
[2] ATLAS Collaboration 2003 *ATLAS High-Level Trigger, Data Acquisition and Controls Technical Design Report* http://cdsweb.cern.ch/record/616089/
[3] Nagios Enterprises http://nagios.org/
[4] Fedorko I 2007 *The Message Reporting System of the ATLAS DAQ System* Proc. Conf. on Astroparticle, Particle, Space Physics Detectors and Medical Physics Applications (Como, Italy)
[5] Luckham D 2002 *The Power of Events: An Introduction to Complex Event Processing in Distributed Enterprise Systems* (New York: Addison-Wesley Professional)
[6] Revised Selected Papers 2008 *Business Intelligence for the Real-Time Enterprise* Sec. Int. Workshop BIRTE (Auckland, New Zealand)
[7] Bernhardt T The Esper project, http://esper.codehaus.org/
[8] Liko, D et al 2004 *Control in the ATLAS TDAQ System.* Computing in High Energy Physics and Nuclear Physics. Interlaken, Switzerland
[9] Oracle-Sun The Java Message Service API http://docs.oracle.com/cd/B1409919/jms.htm
[10] Corso-Radu A, Garcia R M, Kazarov A, Miotto G L, Magnoni L and Sloper J 2010 *Applications of expert system technology in the Atlas TDAQ controls framework* Int. Conf. on Software and Data Technologies. (Athens, Grece)
[11] The Django software foundation Django framework https://www.djangoproject.com
[12] The Apache software foundation Activemq http://activemq.apache.org/
[13] Dos Santos Teixeira P H, Clemente R G, Kaiser R A and Vieira D A *HOLMES: An event-driven solution to monitor data centers through continuous queries and machine learning.* Proc. Int. Conf. DEBS10 (Cambridge, UK)

Journal of Physics: Conference Series **368** (2012) 012005 doi:10.1088/1742-6596/368/1/012005

Advances in service and operations for ATLAS data management

Graeme A Stewart[1], Vincent Garonne[1], Mario Lassnig[1], Angelos Molfetas[1], Martin Barisits[1], Donal Zhang[2], Ivan Calvet[1], Thomas Beermann[1], Fernando Barreiro Megino[1], Andrii Tykhonov[3], Simone Campana[1], Cedric Serfon[4], Danila Oleynik[5], Artem Petrosyan[5] for The ATLAS Collaboration

[1]CERN, CH-1211, Genève 23 Switzerland
[2]Institute of High Energy Physics, Chinese Academy of Sciences, Beijing, China
[3]Jožef Stefan Institute, Ljubljana, Slovenia
[4]Ludwig-Maximilians-Universität München, Munich, Germany
[5]Joint Institute for Nuclear Research, Dubna, Russia

E-mail: `graeme.andrew.stewartcern.ch`

Abstract. ATLAS has recorded almost 5PB of RAW data since the LHC started running at the end of 2009. Many more derived data products and complimentary simulation data have also been produced by the collaboration and, in total, 70PB is currently stored in the Worldwide LHC Computing Grid by ATLAS. All of this data is managed by the ATLAS Distributed Data Management system, called Don Quixote 2 (DQ2). DQ2 has evolved rapidly to help ATLAS Computing operations manage these large quantities of data across the many grid sites at which ATLAS runs and to help ATLAS physicists get access to this data. In this paper we describe new and improved DQ2 services: popularity; space monitoring and accounting; exclusion service; cleaning agents; deletion agents. We describe the experience of data management operation in ATLAS computing, showing how these services enable management of petabyte scale computing operations. We illustrate the coupling of data management services to other parts of the ATLAS computing infrastructure, in particular showing how feedback from the distributed analysis system in ATLAS has enabled dynamic placement of the most popular data, helping users and groups to analyse the increasing data volumes on the grid.

1. Introduction

The ATLAS experiment[1] at the Large Hadron Collider (LHC) is a general purpose particle physics detector designed to investigate physics at the energy frontier. The ATLAS detector is capable of operating at the 20MHz collision rate of the LHC, however, online trigger systems reduce the rate of data taking to offline to 200-400Hz. Even with this drastic reduction ATLAS records a huge amount of data – more that 5PB of raw collision data have been taken since the LHC started running. This data is first processed by the ATLAS Tier-0[2], which runs first pass reconstruction of events and writes the data to tape at CERN. It then registers this data in the ATLAS Distributed Data Management system (DDM)[3]. DDM organises, transfers and manages ATLAS data across more than a hundred individual grid sites that are part of the Worldwide LHC Grid[4] in accordance with the policies established in the ATLAS Computing

Model[5]. This not only covers raw data, but the entire lifecycle of derived data products for the collaboration physics groups and individual physicists.

In this paper we give an overview of ATLAS Distributed Data Management, covering core concepts (section 2), performance with LHC data (section 3) and then, in some detail, new services and features (section 4).

2. ATLAS distrubuted data management

2.1. Data model

ATLAS data, like most high energy physics experiments, mostly consists of persistified C++ objects. These objects are stored together in files, usually corresponding to many detector events. DDM considers files to be its elemental unit (thus, DDM has no knowledge of sub-file structures). However, files themselves are usually defined by operational considerations and rarely correspond to a complete set of data of interest to the user of the system. For this reason DDM allows the aggregation of files into *datasets*. Datasets are the operational unit of replication for DDM – they may be transferred to grid sites, whereas single files may not. Datasets in DDM may contain files that are in other datasets, i.e., datasets may overlap. This is illustrated in figure 1.

Datasets may be *open* (able to have new files added to them) or *closed* (no new files may be added). Closed datasets may have new versions created, which might have a different file content; however, datasets which are *frozen* cannot have files added or new versions created. These datasets are immutable (excepting that files which have been permanently lost can be removed).

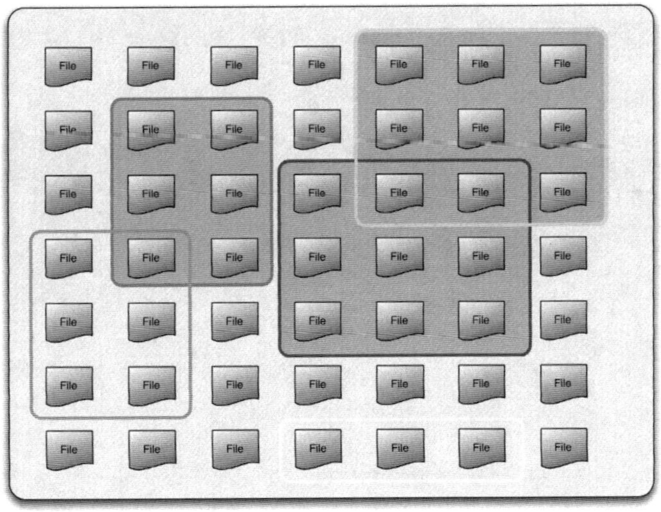

Figure 1. ATLAS Data Management core concepts. Individual files are collected into datasets (coloured outlines). Datasets may overlap and contain the same files (e.g., green and blue). Datasets themselves may be aggregated into containers, e.g., the red, blue and green datasets form the grey container.

There is a further level of aggregation provided by DDM. This is the concept of a *container*, which is a collection of datasets (see figure 1). Containers are not units of replication, but allow large blocks of data, which could not be replicated to single site, to be described in the system.

In practice, most dataset overlaps and aggregation into containers are hierarchical: e.g., monte-carlo production will use small datasets, referring to a few jobs processed at a site; these are then aggregated into the main dataset, which refers to a particular task in the ATLAS production system (all of the small datasets overlap with this main dataset). Then these main datasets may be added to a container, where the output of several similar simulation tasks is gathered.

2.2. Data management responsibilities
The ATLAS Distributed Data Management system is charged with managing and organising ATLAS data for the collaboration. In particular this means:

- Registering and cataloging ATLAS data.
  - Registration of datasets and containers.
  - Registration of files into datasets.
  - Registration of datasets into containers.
- Transferring data between sites.
  - Registering the data transfer request.
  - Allowing requests to be queried and cancelled.
- Delete replicas from sites.
- Ensure dataset consistency on sites.
  - In particular manage file loses on sites.
- Enforce ATLAS Computing Model policies.

The current implementation of DDM is called DQ2 (*Don Quixote 2*).

2.3. DQ2 architecture
The architecture of DQ2 is based on a service stack – clients are at the top and use a well defined API to inteact with the system. Internally, the system core utilises an Oracle database for all state. The access points for the central serices are stateless apache web servers, which then talk to the database, which allows horizontal scaling of this component. To interact with the grid, the *Site Services* components shield the central services of the system from dependencies on any particular grid implementation or technology. However, there are certain basic common libraries which are shared between DQ2 components. This is illustrated in figure 2.

2.4. Scope and users
As ATLAS is a collaboration of many thousands of physicists, managing data across hundreds of sites and running hundreds of thousands of jobs a day, DDM has a wide base of clients. The majority of calls to DDM (see section 3) are from other ATLAS systems, e.g., the ATLAS PanDA workload management system[6]. Thus, a few clients which produce the majority of the DDM load. However, DDM also supports a user base of thousands of individuals: about 500 unique users each day, 1000 each week and 1500 every month.

3. Scaling and performance
Since the start of LHC data taking the total amount of data managed by DQ2 has grown steadily, from about 10PB in 2008 to more than 70PB in summer 2011 (figure 3).

Likewise, in figure 4, the evolution of the total number of managed files is shown.

The figures show that the current implementation is scaling with data volume. However, this scaling is achieved at the cost of considerable tuning efforts, especially in the area of interactions with Oracle. Indeed, to achieve sufficient performance with Oracle, a de-normalised schema has

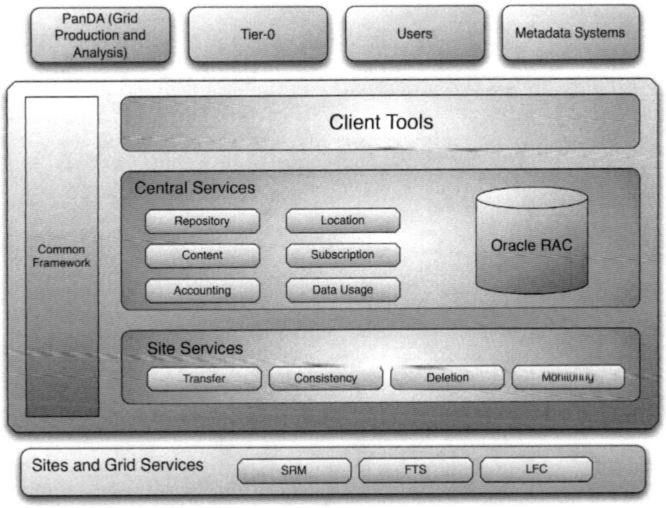

Figure 2. Schematic illustration of DQ2 architecture.

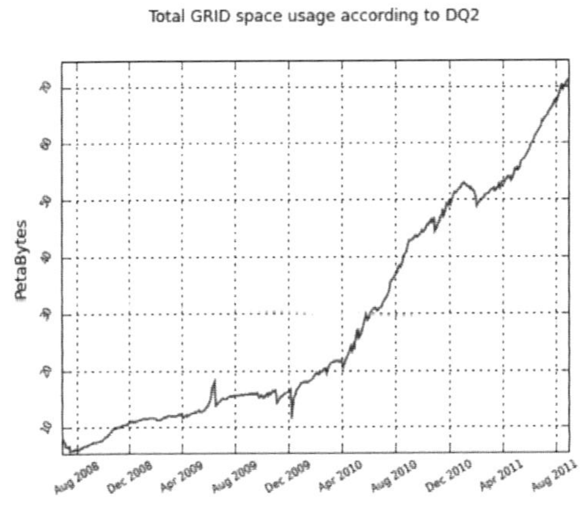

Figure 3. Total size of data managed by DDM from LHC statup. Note the sharp increase in data taking rate once the LHC centre of mass energy reached 7TeV in spring 2010.

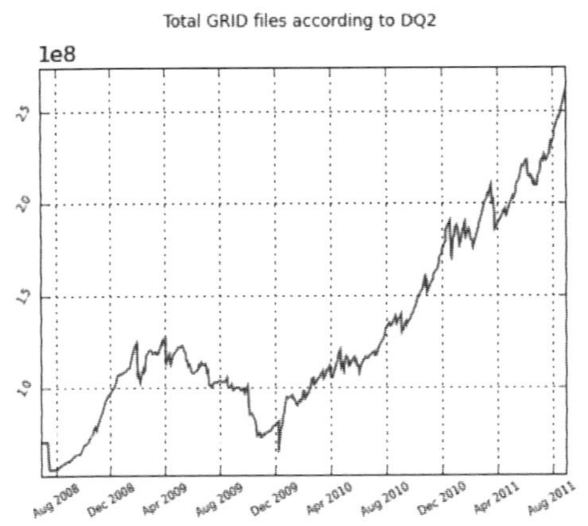

Figure 4. Total number of files managed by DDM from LHC statup. Note the sharp increase in data taking rate once the LHC centre of mass energy reached 7TeV in spring 2010.

had to be adopted, where columns are replicated using triggers in order to reduce the number of table joins needed for popular queries. The queries also require significant numbers of Oracle hints to ensure that the database executes them in an efficient manner.

Currently DQ2 has a load of about 14M reads and 0.6M writes per day.

As the amount of data managed by DQ2 has increased, so have the transfer rates of data around the grid. In figure 6 it can be seen that transfer rates average 2GB/s continuously for ATLAS. Peaks of up to 10GB/s have been observed after reprocessing campaigns, when large amounts of data need to be replicated across the grid.

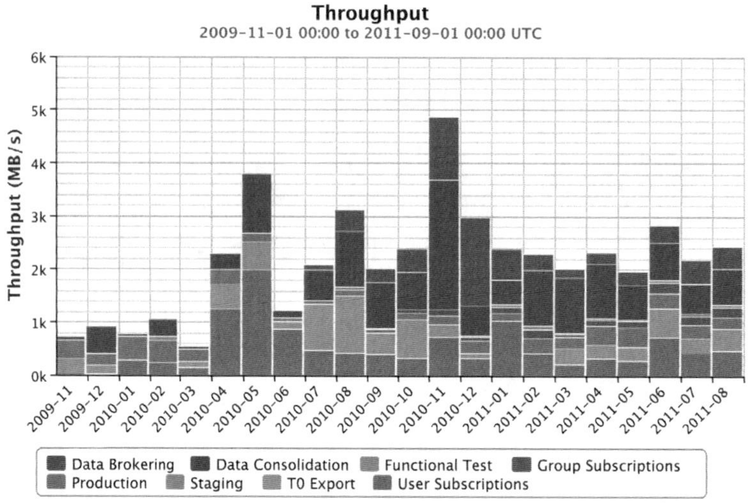

Figure 5. ATLAS grid data transfer rates for different activities since LHC startup. An increase in data rate is observed once 7TeV data taking began.

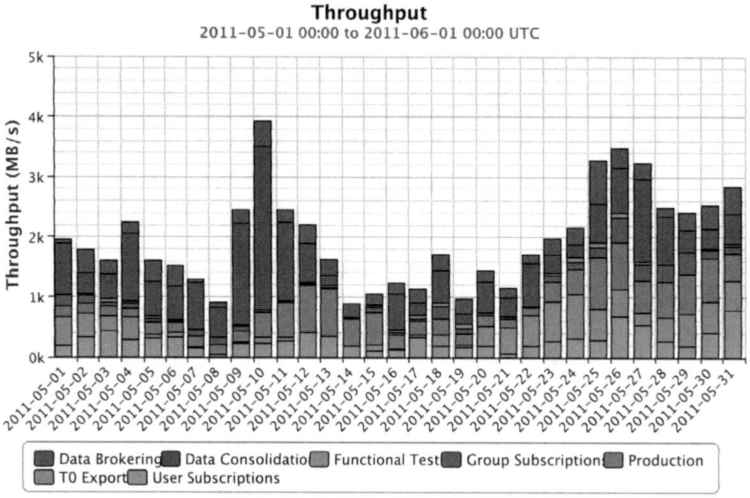

Figure 6. Data transfer rates per activity for May 2011. The system has a continuous substantial workload.

4. New services and features

As DQ2 has evolved to become a more complete system for managing ATLAS data throughout its life a number of new services have been introduced, which we now describe.

4.1. Tracer

One of the key problems facing a large collaboration like ATLAS in an inhomogeneous resource scarce environment is to determine which data should be widely replicated because it is, or it likely to be, popular and which data should replicated less.

In order to achieve an objective measure of this the DDM team introduced the tracer system[7, 8]. The tracer has an API library that is used by DQ2 clients to send a callback when data is accessed on the grid. Data pertaining to the dataset, file, site (both local and remote), user, activity and timestamps are included in the tracer message, which is sent via http. These messages are collected via the tracer service, which then inserts them into the DQ2 database.

The tracer service is optimised to accept a high rate of data insertion (more than 5M per day) and recent work has investigated the use of the Cassandra NoSQL database to buffer insertions before doing a bulk insert of data into Oracle. This helps the system to scale.

4.2. Popularity

The DQ2 tracer keeps a very fine granularity for data access – each single access results in multiple database row entries. This makes it unsuitable for querying directly, as data aggregation would be very slow. For this reason the DQ2 *popularity* service[7] was introduced. This service runs over the tracer data each day and provides summaries of data access per local or remote site, user, dataset, etc.

The popularity service can be queried by clients via a web interface or an API to discover popular and unpopular data on sites (e.g., see figure 7). When taken globally such information is valuable for knowing what data might be deleted across the grid in order to free up space on sites. This is used to form part of the ATLAS's dynamic data placement system. The ATLAS PanDA workload management system[6] measures the instantaneous popularity of data based on the jobs that users have submitted and, if the number of replicas held in the system is below a threshold, extra replicas are created via a DDM subscription request. Conversely, the popularity service is used to decide which replicas of unpopular data to delete at each site, allowing more popular data to be sent to replace it.

4.3. Deletion

While deletion might seem like a fairly straight-forward activity on the surface, in a complex distributed environment, such as that managed by DDM, it is far from trivial. Dataset deletion requests on a particular site need to be done with care to ensure that:

- The dataset replica entry is deleted from the DDM central catalog.
- Corresponding files are physically deleted from storage.
- All file replica locations are removed properly from the local file catalog.

Each of these steps might fail, so it is necessary for the deletion service to have an internal state engine that records the state of deletion for any dataset at a particular site. It is also necessary to throttle both catalog and physical deletion requests for files in order to prevent external services from being overwhelmed. In addition, there is also the additional complexity of overlapping datasets. If two datasets share files on a site and only one of them is deleted then the shared files should not be deleted, otherwise the remaining dataset would become incomplete. This requires some care in mapping dataset deletion requests onto file deletions.

Journal of Physics: Conference Series **368** (2012) 012005 doi:10.1088/1742-6596/368/1/012005

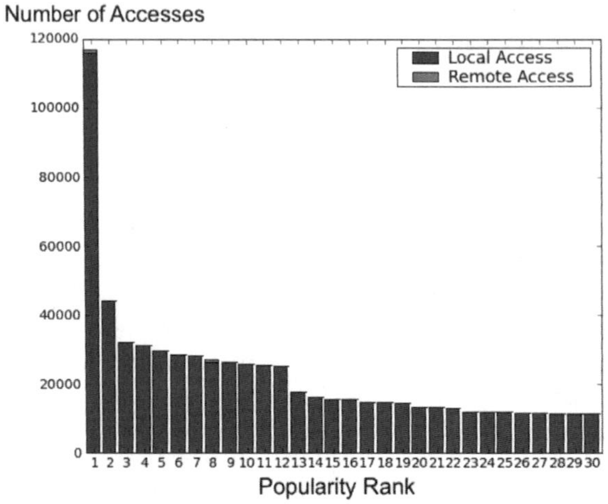

Dataset listing

No	Dataset	Accesses
1	mc11_7TeV.105200.T1_McAtNlo_Jimmy.merge.AOD.e835_s1272_s1274_r3043_r29	116916
2	mc11_7TeV.107650.AlpgenJimmyZeeNp0_pt20.merge.AOD.e835_s1299_s1300_r30	44277
3	mc09_7TeV.105830.JF17_herwig_jet_filter.merge.AOD.e507_s765_s767_r1215	32284
4	mc11_7TeV.107660.AlpgenJimmyZmumuNp0_pt20.merge.AOD.e835_s1299_s1300_r	31253
5	data11_7TeV.00191190.physics_Muons.merge.AOD.f413_m1019	29835
6	mc11_7TeV.107670.AlpgenJimmyZtautauNp0_pt20.merge.AOD.e835_s1299_s1300	28593
7	mc11_7TeV.106046.PythiaZee_no_filter.merge.AOD.e815_s1272_s1274_r3043_	28211
8	user.olezenin.mc11_7TeV.126185.Pythia8Zmumu.merge.AOD.e984_a131_s1353_	27312
9	mc11_7TeV.126184.Pythia8Zee.merge.AOD.e984_a131_s1353_a141_r2900_tid59	26518

Figure 7. Results of a typical query to the popularity service, giving the grid-wide popularity of AOD datasets over a 30 day period, ranked by number of accesses.

Deletion rates in ATLAS can reach significant levels, with millions of file deletions per day and terabytes of data being cleaned. Figure 8 shows a typical example.

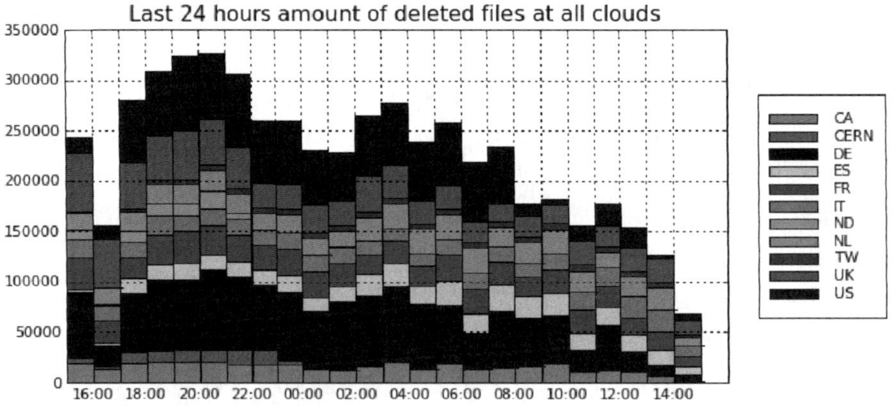

Figure 8. Deletion across the ATLAS clouds over a 24 hour period. Deletion rates of more than 300k per hour are frequent.

Analysis of this level of deletion shows that it primarily arises from the use of intermediate files in processing chains. e.g., simulation can only produce small detector hits files as jobs are limited by cpu/runtime considerations to only process about 50 events; these small hits files are then merged to larger files and afterwards the original hits are deleted. A considerable load also comes from the deletion of user outputs from transient disk areas.

4.4. Consistency

Grid operations across the large number of sites used by an experiment such as ATLAS are far from trivial (see, e.g., [9]). Data loss at sites occurs frequently and recovery from this state must be an automated process to be efficacious. In addition consistency problems can occur between the different catalog layers in the system, which need to be rectified.

The consistency service[10] has been developed in order to achieve this goal. This service operates by allowing storage URLs which have been lost to the site to be declared via a dedicated API. The consistency service will then set about re-establishing the consistency of the data on the site by:

- Declaring affected datasets on a site to be *incomplete*, so that they will not be used for data processing.
- Cleaning the storage and file catalog namespaces of lost files.
- Re-subscribing files which have other replicas elsewhere back to the site.
- Deleting files which have no other replicas from the definition of the affected datasets (thus permanently lost).
- Declaring datasets to be *complete* again as a site after consistency is re-established.

The consistency service also keeps records of which files have been lost and which recovery actions have been taken. This allows a historical view of data loss on the grid and for sites with particular problems to be identified.

4.5. Accounting

Knowledge of how storage space is used on the grid is vital for the effective management of computing operations. This problem is complicated by the number of different dimensions along which ownership can be measured. e.g., data type (RAW, ESD, AOD) is orthogonal to project (data11_7TeV, mc10), which are both orthogonal to accounting by the data owner.

For this reason the accounting system, which was previously based on pattern matching, has been replaced with a system based on key-value pairs. Thus a wildcard query which previously might have been `'data10.*.ESD.*' + 'CERN'` can be expressed as `{'project':'data10', 'type':'ESD', 'location':'CERN'`. The new system allows arbitrary combinations of key-value pairs to be specified, which offers considerably more flexibility than the old pattern based approach. Once a set of key-values has been established the system will query these periodically. In this way historical data will be built up and trends can be analysed.

An initial implementation is available based on an Oracle backend, but an assessment of NoSQL backends, which are inherently more suitable to such key-value stores, is underway. This approach would offer the additional benefit of reducing the overall load on Oracle.

5. Conclusions and future directions

The ATLAS Distributed Data Management project has provided ATLAS with a functioning system in which to organise and manage data during LHC running. New services have been introduced to help automate the data lifecycle and to improve the robustness of the system to real operating conditions. Introducing and managing these new features, while keeping the system stable have taken considerable and continual efforts of developers and database experts, but the multi-petabytes of data moved and stored shows that the system has scaled well to date.

However, it is also true that some conceptual and design limitations have arisen in the current system, which now hamper future development. For example, dataset versions (section 2.1), while supported, are largely unused and have not proved to be really necessary; however, the inclusion of versioning in the current schema introduces an indirection between a dataset and its constituent files that requires a table join with a consequent performance impact in the database layer. Therefore the ATLAS DDM group are undertaking a review of DQ2, looking forward to the long LHC shutdown of 2013-2014. A new version of DDM, *Rucio*, is planned for 2013 to take ATLAS forward into the next years of high luminosity LHC running.

References

[1] The ATLAS Collaboration 2008 *Journal of Instrumentation* **3** S08003
[2] Elsing M, Goossens L, Nairz A and Negri G 2010 *Journal of Physics: Conference Series* **219** 072011
[3] Branco M, Cameron D, Gaidioz B, Garonne V, Koblitz B, Lassnig M, Rocha R, Salgado P and Wenaus T 2008 *Journal of Physics: Conference Series* **119** 062017
[4] Knobloch J et al 2005 *LHC Computing Grid Technical Design Report* Tech. Rep. CERN-LHCC-2005-024 CERN
[5] Jones R and Barberis D 2008 *Journal of Physics: Conference Series* **119** 072020
[6] Maeno T, De K, Wenaus T, Nilsson P, Stewart G A, Walker R, Stradling A, Caballero J, Potekhin M, Smith D and Collaboration T A 2011 *Journal of Physics: Conference Series* **331** 072024
[7] Molfetas A, Megino F B, Tykhonov A, Lassnig M, Garonne V, Barisits M, Campana S, Dimitrov G, Jezequel S, Ueda I and Viegas F T A 2011 *Journal of Physics: Conference Series* **331** 062018
[8] Lassnig M, Garonne V, Branco M and Molfetas A 2010 *Journal of Physics: Conference Series* **219** 062054
[9] I Ueda and the ATLAS collaboration 2011 *Journal of Physics: Conference Series* **331** 072034
[10] Serfon C, Garonne V and The ATLAS Collaboration 2011 *Journal of Physics: Conference Series* **331** 072065

Development of noSQL data storage for the ATLAS PanDA Monitoring System

M Potekhin on behalf of the ATLAS Collaboration

Brookhaven National Laboratory, Upton, NY11973, USA

E-mail: potekhin@bnl.gov

Abstract. For several years the PanDA Workload Management System has been the basis for distributed production and analysis for the ATLAS experiment at the LHC. Since the start of data taking PanDA usage has ramped up steadily, typically exceeding 500k completed jobs/day by June 2011. The associated monitoring data volume has been rising as well, to levels that present a new set of challenges in the areas of database scalability and monitoring system performance and efficiency. These challenges are being met with a R&D effort aimed at implementing a scalable and efficient monitoring data storage based on a noSQL solution (Cassandra). We present our motivations for using this technology, as well as data design and the techniques used for efficient indexing of the data. We also discuss the hardware requirements as they were determined by testing with actual data and realistic loads.

1. Introduction

PanDA is a Workload Management System built around the concept of Pilot Frameworks [1]. In this approach, workload is assigned to successfully activated and validated Pilot Jobs, which are lightweight processes that probe the environment and act as a 'smart wrapper' for the payload. This 'late binding' of workload jobs to processing slots prevents latencies and failures in slot acquisition from impacting the jobs, and maximizes the flexibility of job allocation to globally distributed and heterogeneous resources [2]. The system was developed in response to computing demands of the ATLAS collaboration at the LHC, and for a number of years now has been the backbone of ATLAS data production and user analysis.

A central and crucial component of the PanDA architecture is its database (hosted on an Oracle RDBMS), which at any given time reflects the state of both pilot and payload jobs, as well as stores a variety of vital configuration information. It is driven by the *PanDA Server*, which is implemented as an Apache-based Python application and performs a variety of brokerage and workload management tasks, as well as advanced pre-emptive data placement at sites, as required by the workload.

By accessing the central database, another PanDA component – the *PanDA Monitoring System* (or simply *PanDA Monitor*) – offers to its users and operators a comprehensive and coherent view of the system and job execution, from high level summaries to detailed drill-down job diagnostics. Some of the entities that are monitored in the system are:

- Pilot Job Generators (schedulers)
- Queues (abstractions of physical sites)
- Pilot Jobs (or simply "pilots")
- Payload Jobs (the actual workload)

All of these are related to each other by reference, i.e. an instance of the Pilot Job Generator is submitting pilots to a particular queue, a queue in turn can contain any number of pilots in various stages of execution, and payload jobs are mapped to successfully activated, running pilots. Payload jobs are also associated with their input/output datasets (details of data handling in PanDA go beyond the scope of this paper). The purpose of the PanDA Monitoring System is to present these objects and their logical associations to the user with optimal utility.

2. Characteristics of the data and current design of the PanDA monitor

As already mentioned, the PanDA Monitor handles a variety of data, and perhaps the most important and widely accessed objects stored in its database are entries describing payload jobs. We used this type of data as the most relevant case study in considering our database strategies going forward. Each job entry is mapped onto a row in the Oracle table, with roughly a hundred attributes (columns).

The monitor is currently implemented as a Python application embedded in the Apache server. Database access is done via the *cx_Oracle* Python library. The Oracle server is handling loads of the order of 10^3 queries of varying complexity per second. There are more than 500,000 new entries generated daily, which results in multi-terabyte accumulated data load. Despite ongoing query optimization effort, there are capability and performance limits imposed by available resources of the current Oracle installation and nature of the queries themselves.

Since RDBMS features such as the ability to perform join operation across several tables can be costly in terms of performance, most of the widely accessed data stored in the PanDA Oracle database is de-normalized. For that reason, there is slight redundancy in the data and joins are not used in the application.

For performance and scalability reasons, the data stored in Oracle is partitioned into "live" and "archive" tables. The former contains the state of live objects (such as job being queued or executed in one of the clouds), while the latter contains a much larger set of data which is final (read-only, such as parameters of fully completed jobs) and has been migrated from the live table after a certain period of time after becoming static. In addition, we note that monitoring applications do not require guaranteed "hard consistency" of the data, which would be another crucial motivation to use a RDBMS. In other words, the time window between the instants when information is updated in the database by its client application, and when this is reflected in results of subsequent queries, is not required to be zero (which is achieved in RDBMS by making the "write" operation synchronous).

3. Motivation for noSQL and choice of platform

As explained above, we actually don't have compelling reasons to store the monitoring data in a RDBMS. In situations like this, a variety of so-called *noSQL* database solutions are often employed by major Web services that require extreme degrees of availability, resilience and scalability. Examples include Google, Amazon, Facebook, Twitter, Digg and many others, as well as "brick and mortar" businesses. In order to leverage this new technology, we initiated a R&D project aiming at applying this approach to elements of the PanDA monitoring system to make it future-proof and gain relevant experience going forward.

The "noSQL" term applies to a broad class of database management systems that differ from RDBMS in some significant aspects. These systems include such diverse categories as Document Store, Graph Databases, Tabular, Key-Value Store and others. Based on the patterns observed in queries in PanDA, a conclusion was made that either a Key-Value or a Tabular Store would be an adequate architecture in the context of PanDA monitoring system. A significant portion of all queries done in PanDA is performed using a unique identifier each entry has, therefore representing a classic case of Key-Value query. Other types of queries are done by indexing relevant columns, which is a general approach that will still work with noSQL solutions which support indexing.

We considered two of the most popular platforms: Apache Cassandra (Key-Value) and Apache Hbase (Tabular). The latter relies on Hadoop Distributed Filesystem. To cut on the learning curve, and for ease of operation, it was decided to choose Cassandra [3], which does not have such dependency.

In addition, Cassandra was already deployed and undergoing evaluation by ATLAS personnel, giving us leverage in quickly acquiring our own R&D platform.

4. Cassandra characteristics

A quick description of Cassandra would be as follows: it's a truly distributed, schema-less, eventually consistent, highly scalable key-value storage system.

The *key*, in this context, can be of almost any data type, as Cassandra handles it as an array of bytes. The *value* is a collection of elements called *columns* (again, stored as arrays of bytes). The column is the smallest unit of data in Cassandra and consists of a *name*, *value* and a timestamp (used by Cassandra internally). This gives the developer flexibility in modeling objects in the data. Aggregation of *columns* with the same key is often called a row, and it can be thought of as a dictionary describing an object. A collection of rows pertaining to the same application is termed a *column family* and serves as a very distant approximation of what is called a table in the traditional RDBMS – note however that Cassandra rows can be drastically different from each other inside a single column family, while rows in a RDBS table follow the same schema.

By design, the *write* operation is durable: once a write is completed (and this becomes known to the client requesting this procedure), data will survive many modes of hardware failure. While an instance of Cassandra can be successfully run on a single node, its advantages are best utilized when using a cluster of nodes, possibly distributed across more than one data center, which brings the following features:

- Incremental and linear scalability – capacity can be added with no downtime
- Automatic load balancing
- Fault tolerance, with data replication within the cluster as well as across data centers

Cassandra also features built-in caching capabilities which will not be discussed here for sake of brevity. The application is written in Java, which makes it highly portable. There are client libraries for Cassandra, for almost any language platform in existence. In our case, we chose *Pycassa* Python module to interface Cassandra.

5. Data Design and migration scheme

Design of PanDA data for storage in the Cassandra system underwent a few iterations which included storage of data as *csv* (comma-separated values) data chunks, "bucketing" of data in groups with sequential identifiers (to facilitate queries of serially submitted jobs) etc. After performance evaluation, a simple scheme was adopted where a single job entry maps onto a Cassandra row. The row key then is the same as used as primary index in the Oracle database, which is the unique ID automatically assigned to each job by the PanDA system.

At the time of this writing, parts of the archived data in PanDA are mirrored on Amazon S3 system in *csv* format for backup and R&D purposes, grouped in segments according to the date of the last update. This body of data was used as input for populating a Cassandra instance as it effectively allows us to avoid placing an extra load on the production RDBMS during testing, is globally and transparently available and does not require DB-specific tools to access data (all data can be downloaded by the client application from a specific URL using tools like *curl*, *wget* etc).

In this arrangement, a multithreaded client application (written in Python) pulls the data from Amazon, optionally caches it on the local file system, parses the *csv* format, and feeds it to the Cassandra cluster.

6. Characteristics of the cluster used in the project

In order to achieve optimal performance, Cassandra makes full use of the resources of the machine on which it is deployed. It is therefore not optimal to place a Cassandra system on a cluster of virtual machines, since there is nothing to be gained from sharing redundant resources. For that reason, after initial testing on a 4-VM cluster at CERN, a dedicated cluster was built at Brookhaven National

Laboratory, with the following characteristics: 3 nodes with 24 cores, 48GB RAM and 1TB of attached SSD storage each. Its first version relied on rotational media (6 spindles per node in RAID0 configuration), but testing showed that it didn't scale up to performance levels of the PanDA Oracle instance, therefore storage was upgraded to SSDs. Main results of testing will be given below. The Oracle cluster used in PanDA had a total spindle count of 110 (Raptor-class rotational media). Its performance would doubtless benefit from using SSDs as well, but clearly at this scale this wouldn't be economical.

It's worth noting that Cassandra use of resources is radically different between read and write operations – the former is I/O bound, while the latter is CPU intensive. Depending on application, it's important to provide sufficient resources in each domain.

Monitoring load on the nodes as well as client machines that load and index data on the cluster is an important part of regular operations, troubleshooting and maintenance. We used the Ganglia system to accomplish that.

7. Indexing

Efficient use of data makes it necessary to create an effective indexing scheme in the data. Analysis of actual queries done against PanDA database shows that in addition to a simple row key query they often contain selection based in values in a few columns. This knowledge can be utilized in building a Cassandra-based application, by creating composite indexes mapped to such queries. This can be done in one of two ways:

1. Relying on "native" indexing in Cassandra
2. Creating additional lookup tables by a separate client application

If one goes with the former, it effectively necessitates creation of composite columns in the column families. This is akin to duplication of parts of data and leads to inflation of disk space associated with the data. On the other hand, there is no extra code to be written or run against the database, as the cluster will handle indexing transparently and asynchronously.

Producing indexes "by hand" (the second option) gives the operator complete control over when and how the data is indexed, and saves disk space since no extra columns are attached to column families. In the end, driven by consideration of referential integrity when retiring parts of data, it was decided to use the former option in creating indexes, trading disk space for ease of operation and maintenance.

Indexes were designed as follows: we identified queries most frequently used in the PanDA monitor. As an example, there are queries that select data based on the following job attributes: *computingsite, prodsourcelabel, date*. A new column is created in each row, with a name that indicates its composite nature: "**computingsite+prodsourcelabel+date**". The value of such column may then look like "ANALY_CERN+user+20110601". This augmentation of data can be done either at loading stage, or by running a separate process at a later time, at discretion of the operator. When and if this column is declared as index for Cassandra (which is typically done via its CLI), the cluster will start building corresponding data structures, automatically and asynchronously. To utilize the index, client application must determine that the user's request contains requisite query elements, and then a query is run on the cluster by selecting values of the composite column according to the user's request. Seven indexes constructed in such a way cover a vast majority of all queries done in the Monitor. The remaining small fraction can be handled by using individual "single" column indexes in combination, since this is also allowed in Cassandra. It won't be as fast as a composite index since it will involve iterations over a collection of objects instead of a seek and fetch, however preliminary testing shows that the performance will still be acceptable, given relative scarcity of such queries.

8. Results of testing

A few client applications, written in Python, were created for purposes of data loading and indexing, as well as performance and scalability tests. The Cassandra cluster was populated with real PanDA data, and the data load was equal to one year worth of job records, covering the period of June 2010 to

May 2011, in order to be approximately the same as in testing done with previous configurations. Based on observed query patterns, two main categories of performance and scalability tests were performed, along with comparisons with analogous queries against the original Oracle database:

- Random seek test, whereby a long series of queries were extracting individual data for randomly selected jobs from the past year data sample
- Indexed queries, when a number of representative complex queries were performed using composite indexes as described in the section above

In order to judge the scalability of the system, queries of both classes were run concurrently in multiple client threads, simulating real operational conditions of the database. Our capability to fully stress the cluster was limited to approximately 10 threads, by the available CPU resources on the client machine, however the results extracted provide useful guidance nevertheless.

In the random seek test, Cassandra's time per extracted entry was 10 ms, with 10 concurrent clients. This translates into 1000 queries per second, which is about twice the rate currently experienced by our Oracle database. An analogous test done against Oracle (from a Python client) yielded roughly 100ms per query.

In the case of indexed queries, Cassandra's time per entry was 4ms. This result is counterintuitive at first, when compared to the random query metric, because indexed queries result in additional disk operations (the index must be read before the actual data extraction). However this result makes sense if one considers that multiple rows extracted in the query were packaged more efficiently in the network layer during data transmission from the cluster to the client, i.e. effective overhead per entry was lower. A comparable Oracle test returned a range of values from 0.5ms to 15 ms.

The queries that were run for comparison purposes against Oracle and Cassandra were effectively identical in scope and logic, i.e. full content of the row was pulled from the database in each case.

When increasing the load further, we found that the Cassandra cluster will adequately handle loads of at least 1500 queries per second, which is approximately 3 times more than current query load in the PanDA Monitor. Actual timing results were less relevant because of the client machine limitations mentioned above, i.e. we were not able to reach stress limits of the Cassandra cluster at this time.

9. Conclusions

We identified a "noSQL" database system Cassandra as a promising technology platform to ensure scalability and performance of the PanDA monitor database, operating under conditions of consistently high data and query load. A few versions of data design and indexing were evaluated. Hardware requirements were confirmed via testing of the Cassandra cluster under realistic conditions and benchmarking it against Oracle RDBMS. A simple but effective technique of composite index creation was employed to guarantee high performance in most popular queries generated in the PanDA monitor. Quantitative scalability and performance test results are favorable and pave the way for integration of this new noSQL data store with PanDA monitor.

References

[1] Maeno T 2008 PanDA: distributed production and distributed analysis system for ATLAS *J. Phys.: Conf. Ser.* **119** 062036

[2] Nilsson P 2008 Experience from a pilot based system for ATLAS *J. Phys.: Conf. Ser.* **119** 062038

[3] Hewitt E 2010 *Cassandra: The definitive guide* (O'Reilly Media)

Integrating Amazon EC2 with the CMS production framework

Andrew Melo[1,2], Paul Sheldon[1]

[1] Vanderbilt University, 6301 Stevenson Center, Nashville, TN, 37235, USA

[2] Corresponding author

andrew.m.melo@vanderbilt.edu

Abstract. As cloud middleware and cloud providers have become more robust, various experiments with experience in Grid submission have begun to investigate the possibility of taking previously Grid-Enabled applications and making them compatible with Cloud Computing. Successful implementation will allow for dynamic scaling of the available hardware resources, providing access to peak-load handling capabilities and possibly resulting in lower costs to the experiment. Here we discuss current work within the CMS collaboration at the LHC to both perform computation on EC2, both for production and analysis use-cases. We also discuss break-even points between dedicated and cloud resources using real-world costs derived from a CMS site.

1. Introduction

Large international scientific collaborations face significant computing challenges; multi-Petabyte data sets that require petaflops of computing as well as globally distributed users and computing resources. To meet those computing challenges, collaborations have invested heavily in dedicated computing resources. Such an approach requires the expense and careful management of floor space, power, and cooling, the use of grid technologies and high performance networks, and the implementation of highly developed strategies for data movement and placement.

The Compact Muon Solenoid (CMS) experiment at the Large Hadron Collider (LHC), for example, processes several petabytes of data yearly, using approximately 45,000 cores purchased and maintained by the experiment and spread globally across the grid. Core usage is highly managed to insure that the highest priority experimental goals are met. There is some fluctuation in demand: the experiment doesn't log data year-round and user demand has significant variation depending on the schedule of conferences. The experiment makes opportunistic use of shared resources such as those offered by various academic computing centers/collaborations. However, fixed operational costs and asset depreciation reduce the benefits and cost-effectiveness of increased hardware capacity when such hardware is not being used regularly.

Recently, a proliferation of "computing as a utility" or "cloud computing" services have made available to users the ability to purchase CPU time at an hourly rate, with little to no financial commitment. Service providers are often established internet businesses, such as Google and Amazon, that have built extremely large compute server facilities, initially for their own purposes. Amazon's Elastic Compute Cloud (EC2) is a commercial on-demand compute service where a user can rent virtual machines by the hour and merely pay the hourly rate plus bandwidth charges. The costs of

housing, powering, cooling, maintaining, and managing this hardware is borne by the provider. A growing number of varied users are sourcing their short-term computing needs to EC2 and similar services, when it would otherwise be difficult, expensive and time-consuming to roll out more capacity for what may be a transient need [1].
Cloud computing resources free the user from the burden of housing, servicing, and managing expensive and sometimes unreliable equipment. They also allow the user to purchase only those compute cycles and bits of storage needed to fulfill their specific goals or mission. In principle these can be purchased just-in-time as needed. Economies of scale may lead to a lower support cost per unit of hardware, and these lower costs may be passed on to the user.

There are also potential disadvantages to cloud computing services. The provider must balance availability with inactivity – if it over buys hardware then it has increased its cost without recovery, and these costs must be passed on to the consumer. If it under-buys hardware, then an important user advantage, being able to get access to resources when needed, is compromised. Necessary profit margins required by a provider also increase cost of service. Economies of scale may not dramatically lower support costs, since large scale infrastructure increases management and maintenance complexity, as does the need to support a user base with broad requirements and usage patterns. Supporting infrastructure that has only one "user" (like a dedicated CMS site) is likely to be significantly easier to manage and less expensive.

We have investigated using EC2 to supplement or replace the compute capacity of CMS by renting out virtual machines and integrating them with a virtual site consisting of solely cloud resources.

2. Methodology

We sought to provide tools and the framework for end users to integrate cloud resources to their workflows as seamlessly as possible. To do this, we created a virtual Compute Element (CE) and registered it with the Open Science Grid (OSG) and CMS. On the CE, we installed and configured a number of services required to accept jobs from the grid and submit them to the cloud. Standard GRAM, Condor, and Gridftp services were installed and client scripts were created to automate starting and stopping virtual machines (VMs) on EC2. Once the user creates their VMs, the VMs are programmed to bootstrap themselves and connect back to the Condor master located on the CE, allowing them to join the dynamic pool of VMs available to the CE. Then the user can submit jobs using the standard CMS analysis tools (such as CRAB) to the virtual site where the jobs would be scheduled to run within EC2.

To lower the amount of maintenance required, we use the ready-built virtual machine provided by CERN (CernVM). Maintained by CERN, it has a virtual, distributed filesystem that stores the CMS software (CMSSW) installs. This allows us to decouple the release schedule for our VM slave from the release schedule for CMSSW. An additional advantage of CernVM is that it comes with all the necessary CMS-specific prerequisites installed, some of which aren't provided in base Redhat linux install.

An important issue with this architecture is that by default, any user can submit jobs to the virtual site and have their jobs run another users' VMs. In a shared-use environment, such as an academic cluster, there is an expectation of a sharing of resources. Under this architecture, users (or groups of users) are paying for their CPU time directly from Amazon and have an expectation to be able to utilize the resources they've paid for.

Our solution to this issue is a combination of Condor logic and webpages that let the owner of the VM configure an Access Control List (ACL) of allowed users to run on their resources. Each job received from the grid by GRAM has its Distinguished Name (DN) attached to it. The DN is the identity of the user, verified cryptographically with a grid certificate signed by a certificate service (for example, DOEGrids or CERN Certification Authority). By using some scripts and custom Condor logic on the slaves, we are able to restrict jobs to only run on specific nodes, preventing use of others' resources.

Originally, we had intended to create a virtual Storage Element (SE) using the same ACL structure. However, in July 2011, Amazon changed their bandwidth pricing and made inbound transfers free of charge. Simultaneously, CMS rolled out xrootd to each of their larger sites. Xrootd is a remote file access protocol that lets applications access ROOT files over the Wide Area Network (WAN). Nearly 100% of CMS' datasets are available remotely through xrootd obviating the need to stage-in input datasets. A lot of work has been done in CMS to optimize this remote I/O, and it is reported that even transatlantic I/O (from Lincoln, NE, USA to CERN) only suffers a 10% performance hit. Given these advances, we decided to leave the SE as a possible future enhancement to this framework.

Once caveat within CMS is that the various frameworks (such as CRAB for analysis or WMAgent for production) perform checks to verify that data is actually at a site's SE before jobs can be sent to the CE. Since we've opted to not attach storage to our virtual SE, this site mapping would fail on the virtual CE. Fortunately, patches exist to relax this restriction, allowing jobs to run on CEs, even if the data isn't there. The xrootd service is intended as a fallback; since WAN network links are relatively expensive, it's not efficient for the experiment to provide the functionality by default. For our testing, we patched our analysis tool to ignore data locality, enabling our jobs to be scheduled even if the requested data isn't stored locally.

To test our architecture, we ran jobs which would take data produced from the experiment, scan each event for certain characteristics, and write out a subset of each event to a different file if it matched. Each job was run on both EC2 and Vanderbilt and transferred their outputs to Fermilab. To isolate the performance of the xrootd infrastructure, we ran the jobs at Vanderbilt twice: once reading data locally and once reading data over the WAN from a different site using xrootd. At Amazon, the nodes were the "large" instance with 7.5GB of memory and 2 virtual cores. At Vanderbilt, the nodes were typically 8 core Xeon processors clocked at 2.27GHz with 32GB of memory.

3. Results

Once everything was configured, a number of EC2 instances were started and dummy CMSSW jobs were executed to pre-warm the CernVMFS caches. Then, analysis jobs were executed at both EC2 and Vanderbilt. Conveniently, this approach left the client-side interface unchanged (except for the patches we applied), so we were able to use our standard performance measurements to compare the two sites.

A direct comparison between running at EC2 and running with dedicated resources is difficult to replicate due to load on external resources such as storage elements and network links. Additionally, even on idealized resources, different workloads will demonstrate different performance due to differing read access patterns and amount of CPU time required. With those caveats, we observed a processing rate (in seconds/1000 events) of 288 at EC2, 265 at Vanderbilt using local storage, and 310 at Vanderbilt using remote storage. This rate is calculated from the walltime of the jobs and doesn't include time to stage the outputs back to Fermilab. The stage-out times were roughly comparable in all cases, possibly due to the small size of the outputs and the large overhead in setting up connections using the SRM protocol.

4. Costs

An important consideration when evaluating cloud resources compared to dedicated resources is the costs involved. Since the pricing model is variable based on utilization, there is a natural break-even point when it would be more cost-effective to purchase dedicated hardware and vice-versa. Amazon has recently added more pricing structures ranging from a zero upfront cost/high hourly cost model to a high upfront cost/low hourly cost model. Clients can choose to purchase reserved instances for a period of one or three years which grants a lower hourly rate for those resources. The Vanderbilt site provides dedicated resources at a cost of $213.60 per core per year. This amortizes the purchase price of the hardware over three years, and includes all power, cooling, facilities, network bandwidth, and the costs of support, maintenance, and administration. The hardware includes the compute node and an appropriate share of the costs for internal cluster networking, racks, cluster system services, and shared disk space.

By comparing the efficiencies of the two sites and applying the current prices, we can make a rough comparison between the cost-effectiveness of the two models. At $213.60 per core per year, Vanderbilt can perform this analysis at an approximate cost of $1.79 per 1 million events. Similarly, using on-demand instances, one can perform the same analysis using EC2 at an approximate cost of $13.60 per million events. However, with dedicated resources a year commitment is required and if they are utilized less than 14.5% throughout that year, it would be more cost-effective to simply purchase resources with EC2. This is a break-even point between the two approaches and is roughly the point where a user no longer would want to purchase dedicated hardware to handle bursts of demand.

A user could save some money by purchasing reserved instances, but only at the price tier with the lowest up-front costs. The "medium" and "heavy" utilization reserved instances have up-front yearly and triennial costs which already exceed the cost of dedicated hardware even before additional hourly usage fees are incurred.

Perhaps counter-intuitively, reserved instances make the break-even point worse. The break-even point between dedicated hardware and on-demand EC2 resources is at 1,200 hours/year, but the same break-even point between dedicated hardware and Amazon's light reserved instances is less than 300/hours year. Currently, this means that for the same amount of money, one could either purchase dedicated hardware for a year, use 1,200 hours of on-demand EC2 time, or less than 300 hours of reserved instance time. With reserved instances, the break-even point for EC2 drops to 3.5% for the one-year option and 5.5% for the three-year option.

5. Conclusions

Scientific computing requires an enormous amount of compute resources. These resources have traditionally been provided by the scientific collaborations themselves, requiring a significant commitment in equipment, manpower and facilities. Additionally, many scientific collaborations have bursty resource requirements, often related to the operation of their experiment and the schedule of scientific conferences. This leads to either resources being under-utilized or incurred delays due to a lack of resources when demand increases.

Cloud computing, or computing as a service, has potential to solve these problems. By allowing users and organizations to purchase compute time on-demand with little to no commitment, it is possible to meet rapidly changing demand without the financial commitment of purchasing hardware that will remain largely idle and unused.

We used Amazon's EC2 service to perform physics analyses with data from the CMS experiment. By registering a grid site and configuring a compute element, we were able to almost seamlessly integrate EC2 with the CMS' analysis tools. We also created an ACL system to ensure that each virtual machine would only be used by authorized users. Through our tests, we show these resources are technically feasible but are not cost-effective unless the resources are going to be used less than 14.5% per year. Any greater than that and dedicated hardware begins to become the more cost-effective option.

Acknowledgements

The authors would like to think Burt Holzman for his helpful discussions about Condor and grid software. We would also like to thank Brian Bockleman for his assistance with xrootd.

References

[1] http://aws.amazon.com/solutions/case-studies

ACAT 2011 IOP Publishing
Journal of Physics: Conference Series **368** (2012) 012008 doi:10.1088/1742-6596/368/1/012008

Automated quality monitoring and validation of the CMS reconstruction software

Danilo Piparo
CERN, CH-1211 Genve 23, Switzerland
E-mail: danilo.piparo@cern.ch

Abstract. A crucial component of the CMS Software is the reconstruction, which translates the signals coming from the detector's readout electronics into concrete physics objects such as leptons, photons and jets. Given its relevance for all physics analyses, the behaviour and quality of the reconstruction code must be carefully monitored. In particular, the compatibility of its outputs between subsequent releases and the impact of the usage of new algorithms must be carefully assessed. The automated procedure adopted by CMS to accomplish this ambitious task and the innovative tools developed for that purpose are presented. The whole chain of steps is illustrated, starting from the application testing over large ensembles of datasets to emulate Tier-0, Tier-1 and Tier-2 environments, to the collection of the physical quantities in the form of several hundred thousand histograms, to the estimation of their compatibility between releases, to the final production and publication of reports characterised by an efficient representation of the information.

1. Introduction

The CMS Software (CMSSW) is a highly modular framework written in C++, which is characterised by a code base of more than two million lines and about 250 active developers constantly maintain and improve it. One of its crucial components is the reconstruction. It ensures that the electronic signals coming from all the sub-detectors are correctly transposed into energy depositions and then physics objects like photons, muons, jets, tracks and vertices. These objects are the ones then used for the data analysis. The feature sets of such a huge software project are grouped into release cycles and different cycles can coexist. The release schedule is tight, indeed:

- Pre-releases are provided once a week in order to consolidate the state of the code, test interdependencies among software components.
- Releases are cut approximately once per month. They close a development cycle, are used for central processing, Monte Carlo production and analysis.
- Amendment Releases are made available to solve specific issues and supersede the previous releases.

An automated and central validation of physics output of all releases is needed. The first step to achieve this ambitious goal is to provide for each release sufficiently large data and Monte Carlo samples for several processes, the so-called release validation datasets [1].

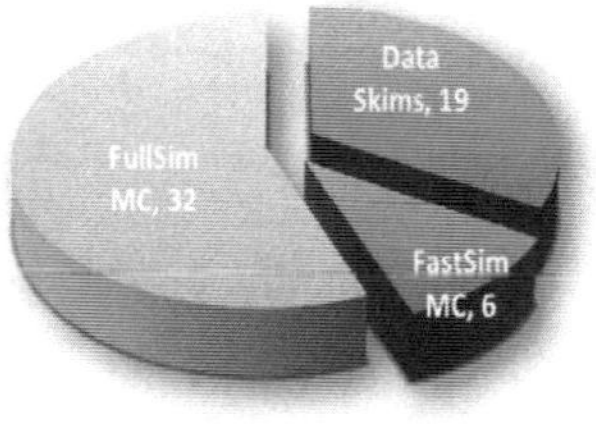

Figure 1. Composition of RelVal datasets. Several different types of events coming from real data, fully and fast simulated Monte Carlo samples are studied.

2. Production of release validation samples

As soon as any kind of CMSSW release is available, different types of samples are processed. The first group is represented by several categories of data subsets (also called "skims") obtained by selecting and grouping the events characterised by particular distinctive signatures. The second group consists in fully and fast simulated Monte Carlo events mimicking relevant processes like top quark and electroweak bosons production or important signatures like high p_T photons or leptons. This procedure guarantees in the first place that all components work together without failures during execution in a large scale environment. Moreover, this broad spectrum of signatures and event types gives the possibility not only to validate the physics objects and CMS sub-detectors but also to check the correctness of the alignment and calibration procedures.

Presently, to cover the quoted use cases, several processing workflows are submitted to a batch system (see figure 2):

- 19 different data skims.
- 32 full simulation Monte Carlo datasets.
- 6 fast simulation Monte Carlo datasets.

To deliver such an amount of data, dedicated resources are exploited at the FNAL farm (up to 1000 batch slots). This setup allows to provide all the necessary datasets within a 24 hours latency. For a given produced sample, the CMS data quality monitoring (DQM) infrastructure allows to collect the relevant quantities in a set of monitoring histograms (about 250.000) and save them in output ROOT [3] files [2]. This step represents an important data reduction with respect to the initial events n-tuples.

3. Data quality monitoring

Data quality monitoring is essential for the detector and operation efficiency. It allows to certify recorded data and produced Monte Carlo samples, allowing to spot possible problems or unwanted features. The CMS DQM provides tools for creation, transfer and archival of histograms and scalar monitor elements. In addition, visualisation of monitoring results, standardisation and integration into CMS software releases is guaranteed.

The histograms filled during the data processing are stored in ROOT files. At the end of the processing chains, these files are uploaded to the DQM servers and their content indexed in a database for performance and concurrent request handling. One interface to this database is a web GUI (see [4]) for the visualisation of the data quality monitoring. The state of the user session and application logic are held on the web server and users actions are mapped directly to http API calls.

The plots containing the monitoring histograms are created on demand with a special renderer which allows to perform quite sofisticated manipulations. For example, it can display a single

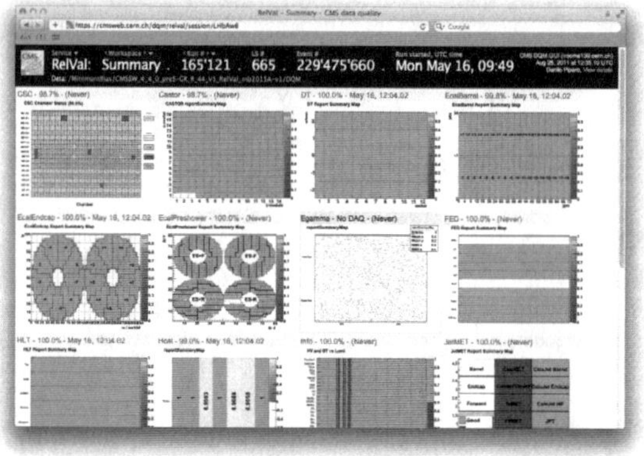

Figure 2. A snapshot of the CMS DQM web graphical user interface. The plots are created on the fly accessing the necessary information which is stored in an internal database, avoiding to rely on the ROOT I/O.

histogram or overlay multiple histograms, access basic ROOT draw options and manipulate plot axes settings.

CMS deploys four production DQM server instances. Three are used to monitor the data being recorded at CMS point 5, to check the reconstructed data coming from the Tier-0 and Tier-1s and to survey the data at the CMS CERN Analysis Facility. The fourth instance is exclusively dedicated to the release validation samples.

All CMS validation suites, for both physics objects and subdetectors, are based on the DQM technology

4. Relmon tool

RelMon is a tool to compare two sets of histograms, stored in ROOT files. Highly modular, it is written in Python and is interfaced to ROOT via the PyROOT bindings. RelMon is not dependent on any component of the CMS software.

The agreement between pairs of histograms is quantified with a statistical test. Predefined tests are available (Chi-square, Kolmogorov-Smirnov, Bin-by-Bin) and new user-defined compatibility criteria can be easily implemented. Pairs of corresponding histograms are selected by name. Moreover, the information about the outcome of the statistical tests is aggregated following the directory tree present in the ROOT files. Since the operations performed can be quite CPU intensive, the tool can exploit multiple cores to speed up the execution of the comparisons.

The outcome of the comparison of the two sets can be produced in the form of a minimalist ASCII output or an elegant browsable web report featuring effective representation of the information, intuitive diagrams such as gauges, pie charts and bar charts and images of histograms overlays to inspect the details of the comparisons (see figures 3 and 4).

RelMon provides also CMS specific features. For example an interface to fetch the histograms directly from the DQM server over an authenticated connection instead of ROOT files, a command-line tool to compare all the DQM output ROOT files coming from release validation productions or two given CMSSW releases (figure 4) and interface to DQM rendering to visualise large amounts of histograms comparisons.

RelMon allows to easily compare large sets of histograms according to a predefined criterion and produce convenient reports to spot all the differences. Moreover it is suited to be deployed in order to setup a central validation mechanism.

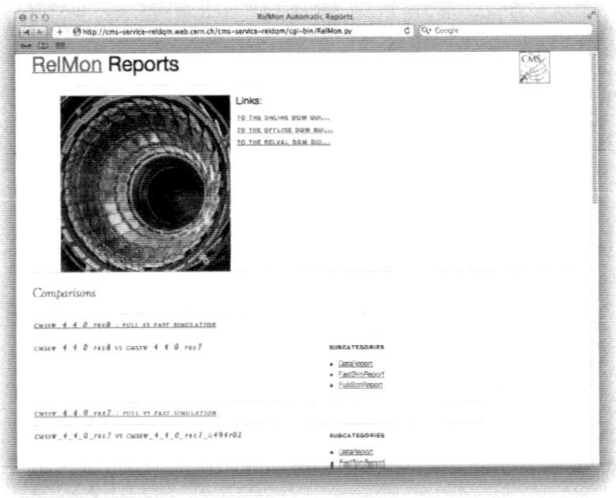

Figure 3. Central page where all the regression tests between CMSSW releases are stored.

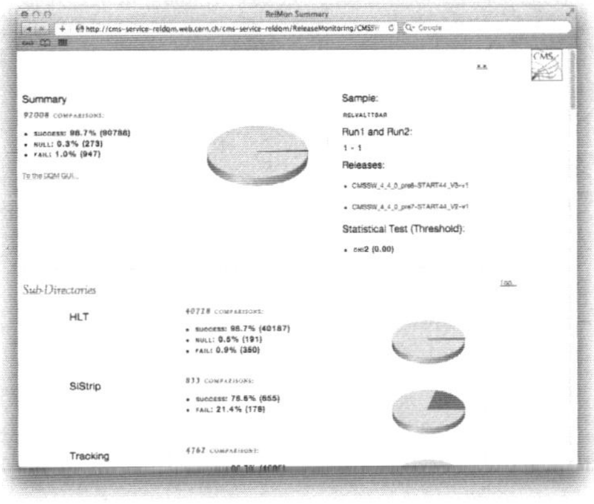

Figure 4. Piecharts and graphs are used in order to provide an overview of the global agreement between histogram sets.

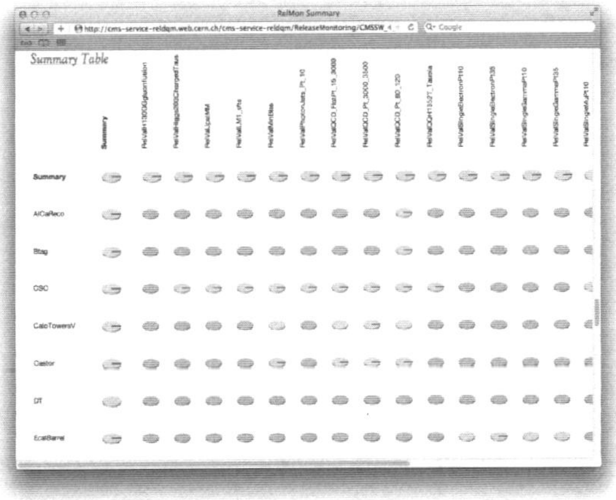

Figure 5. With the relmon reports, a complete overview about the Physics performance of a CMSSW release can be given. Large samples of histograms are compared, considering several datasets at the time.

5. Central validation

In CMS, the automatic generation of RelMon reports is triggered for every CMSSW release. Regression tests are performed across all data, full and fast simulation Monte Carlo DQM histogram samples coming from a given release and the preceding one, used as a reference. In addition, a comparison between full and fast simulation is performed. The latency between the arrival of the ROOT files containing the DQM histograms and the appearance of the RelMon report on the web is $\sim$ 30 minutes, even though more than one million histograms are compared.

This strategy exposes many potentially critical aspects of the physics output produced by the two releases: anomalies can be immediately pin-pointed. With this strategy, reconstruction coordinators gain broad overview of the overall physics performance of the release and the experts of the single physics objects and detector performance groups can immediately give their feedback about possible issues.

6. Conclusions

CMSSW is a huge software project, a fundamental part of which is represented by the reconstruction. The stability of the physics performance must be checked for each new release. In order to achieve this goal, CMS provides large groups of Data and Monte Carlo samples to perform release validation. The study of these samples, which aims to the quality monitoring and validation of the CMS reconstruction software, is performed exploiting the DQM infrastructure and RelMon. This strategy has proven to be very effective to spot anomalies and unwanted features within a very restricted timescale.

References

[1] The CMS Collaboration 2010 Validation of software releases for CMS *J. Phys. Conf. Ser.* **219** 042040
[2] The CMS Collaboration 2010 CMS data quality monitoring: systems and experiences *J. Phys. Conf. Ser.* **219** 082005
[3] R. Brun et al. 1997 ROOT - An Object Oriented Data Analysis Framework *Nucl. Inst. Meth. In Phys.* **A 389** pp 81-86
[4] The CMS Collaboration 2010 CMS data quality monitoring web service *J. Phys. Conf. Ser.* **219** 072055

ACAT 2011 IOP Publishing
Journal of Physics: Conference Series **368** (2012) 012009 doi:10.1088/1742-6596/368/1/012009

LHCb distributed computing operations

F Stagni[1], R Santinelli[2], M Cattaneo[3], S Roiser[4]

[1,3] PH Department, CH-1211 Geneva 23 Switzerland
[2,4] IT Department, CH-1211 Geneva 23 Switzerland

E-mail: [1]`federico.stagni@cern.ch`

E-mail: [2]`roberto.santinelli@cern.ch`

E-mail: [3]`marco.cattaneo@cern.ch`

E-mail: [4]`stefan.roiser@cern.ch`

Abstract. In the Grid world, there are many tools for monitoring both activities and infrastructure. The huge amount of information available needs to be well organized, especially considering the pressing need for prompt reaction in case of problems impacting the activities of a large Virtual Organization. Such activities include data taking, data reconstruction, data reprocessing and user analysis. The monitoring system for the LHCb Grid Computing relies on many heterogeneous and independent sources of information. These offers different views for a better understanding of problems, while an operations team follow defined procedures that have been put in place to handle them. This work summarizes the state-of-the-art of LHCb Grid operations, emphasizing the reasons that brought to various choices, and what are the tools currently in use to run our daily activities. We highlight the most common problems experienced across years of activities on the WLCG infrastructure, the services with their criticality, the procedures in place, the relevant metrics, the tools available and the ones still missing.

1. Introduction

This paper presents how LHCb organized its distributed computing operations, in light of the resources, the tools and the activities. The reader is expected to be familiar with the Tier levels of the MONARC model [1]. This paper is organized as follows: section 2 presents the LHCb distributed computing activities. Section 3 presents the organization given to the operations team. Section 4 introduces the tools used for monitoring, management, reporting and testing. Conclusions are given in section 5.

2. Distributed computing activities

The LHCb Distributed Computing Activities are based on the experiment Computing Model [2]. Even if the computing model has been recently subject to some relaxation, in its fundamentals it is still in use. The original computing model was to use Tier 2 for simulations activities, and Tier 0 and Tier 1 for real data manipulations: today, some Tier 2 are actively used for part of our real data reprocessing activity. Apart from few exceptions, computing resources are mainly provided by WLCG [3].

The distributed computing activities can be split between data manipulation, handled in the following section, and data management, explained in section 2.2.

2.1. Data manipulation

We split Data Manipulation activities into production and non-production activities. Production activities include:

- **Prompt Data Reconstruction**: data coming directly from the experiment is distributed, as *RAW* files, to the six Tier 1 centers participating to the LHCb Grid, and reconstructed. The model dictates that a copy of each RAW file is stored at CERN, which is the Tier 0, and a replica is at a Tier 1. The distribution of the RAW files and their subsequent reconstruction is driven by shares, agreed between the experiment and the resource providers.
- **Data Reprocessing**: this activity is done from time to time, usually at the end of a data talking period. From a workflow point of view, it is identical to the data reconstruction process. Its objective is to provide a second reconstruction of already reconstructed RAW files. Data Reprocessing is usually a highly-demanding activity, in terms of CPU and storage usage. For this reason, the used resources can be different from the ones used in normal data reconstruction. For the reprocessing of 2011 data, some Tier 2 have been used in addition to Tier 0 and Tier 1.
- **Data Stripping**: stripping is the name LHCb gave to the selection of interesting events. Data Stripping comes just after Data Reconstruction or Data Reprocessing, and it is usually a much lighter activity with respect to computing needs. It's anyway stressing for the storage resources, due to the high I/O involved. When reconstructed data is stripped, a number of *streams* are created, each containing a selection of physics events.
- **Calibration**: from the operational point of view, calibration is a lightweight activity that follows the same workflow of Data Stripping, thus starts from already reconstructed data. Sometimes, though, calibration activities take a very high priority and should be done as fast as possible.
- **Monte Carlo (MC) simulations**: the simulation of real collisions started way before real data recording, and it is still a dominant activity for what regards purely the number of jobs handled by the system. Even if simulated events are continuosly produced, large MC campaigns are commonly scheduled during longer shut-down periods, when LHCb is not recording collisions. Since Tier 0 and Tier 1 centers are for most of the time busy with real data processing, Tier 2 centers represent the main resource when coming to production of simulated events. This does not exclude using Tiers 0 and 1 resources.

Some of these production activities also include a merging phase: for example, streamed data that is produced by stripping jobs is usually stored into small files. Thus, it is convenient to merge them into larger files that are then stored in the Grid Storage Elements. While merging is computationally a light activity, it can very well represent a stress in terms of usage of storage resources, especially when the number of files to merge is large.

Non-production activities include:

- **User data analysis**: user analysis jobs represent a good fraction of the total amount of jobs handled by the system. User jobs, when compared to production jobs, are usually less demanding in terms of computing and storage. User jobs have a priority over production jobs, and when a large number of them are queued, they may represent an operational burden, slowing down if not completely blocking part of the production activities. For this reason, pressure has recently put on analysis working groups to integrate as much as possible their activities in the production framework. This would allow the operations team to better control the stress put on computing and storage resources, depending on the relative priorities of the various scheduled production activities. Working Groups will also gain advantages when coming to data provenance, recording and distribution.

- **Monitoring and testing**: continuous testing is done using test jobs. Some of these test jobs are sent by WLCG, and are VO-agnostic. LHCb complements the tests with VO-specific ones.

2.2. Data management

Data Management activities are usually a direct consequence of the production activities of Data Manipulation. Tier 0 and Tier 1 centers provide LHCb with storage resources, that are split in three *space tokens*: *user*, *tape*, and *disk*. Staging data from tape to disk is an important part of daily activities, which requires a careful monitoring. Data management activities include:

- **Replication**: data is replicated, on disks, as soon as it is produced. The number of replicas produced follows the computing model. There are usually more replicas of real data than replicas of simulated data.
- **Archival**: at least one copy of each data that is produced is archived, on a tape system. This custodial copy is not meant to be used by users or productions. Archival is a safe measure introduced not long ago in the computing model.
- **Removal**: data is from time to time selectively removed, to free disk space for newer data. Sometimes, the number of replicas of older datasets is reduced, based on a simple popularity algorithm.
- **Checks**: these checks include consistency, for example between the file catalogs in use, or with the real content of the storage elements, and user quota checks.

Even if a large part of these Data Managements operations are automatically executed, data managers are in charge of performing part of these activities, especially for what regards reducing the number of replicas, or removing datasets.

3. The operations team

The distributed computing operations team comprises a varying number of shifters, while few roles are covered continuosly by a small set of experts. Most of the members of the team meet each other at the daily LHCb offline operations meeting to discuss progress, problems encountered, and to take decisions on short term operational issues. Figure 1 shows the connections between the roles of the operations team.

Within this section, we explain each role's duties, its connection with the other members, a bit of history behind it, and who is covering each of them.

- **GEOC (Grid Expert On Call)**: this role is covered on a week-long shift base by a small set of experts (not more than 5). The GEOC is a central role in the daily operations schedule. A GEOC acts first of all as an information collector, is in charge of running the daily operations meeting, and to represent LHCb at the daily WLCG operations meeting. When issues have to be discussed with resource owners, GEOC deals with site managers and with LHCb contacts at the Tier 1. She receives input from the computing shifters, but it is also their main support when in need of an expert opinion. Other shifters, like distributed analysis support shifters, are referring to the GEOC when a grid-releated problem can't be simply solved. A GEOC is first of all an expert shifter, and since LHCb main resource provider is WLCG, she has to have a good knowledge of how LHCb is using WLCG resources. A GEOC is not an all-round expert, is in fact a connection role inside the operations team.
- **Production Managers**: the role of production manager is covered on a six months base. In today's team, there is a production manager in charge of running Monte Carlo simulations, and a second has been introduced since when real data taking started, and runs and controls

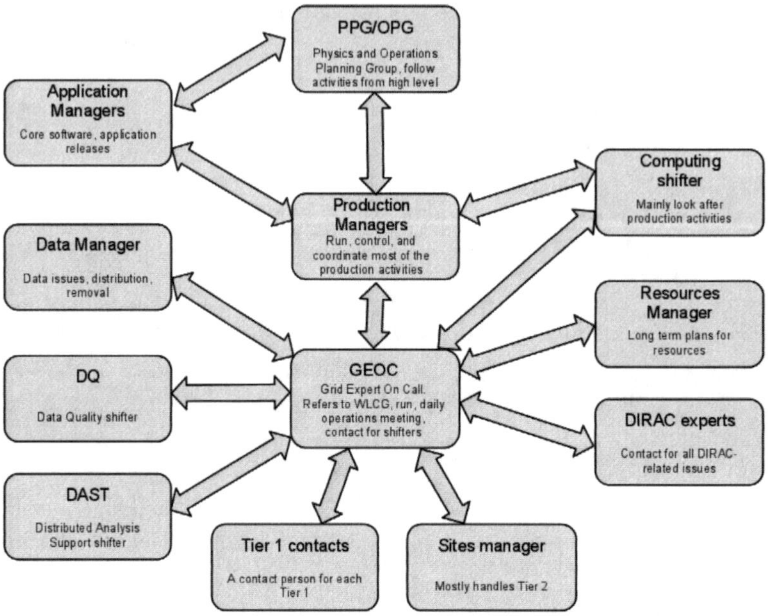

Figure 1. Organization of the Operations team.

all the other production activities. All real data activities are highly stressing in terms of computing and storage, and also very delicate: thus, the role of the real data production manager is fundamental for the collaboration. Production managers are in close contact not only with the computing shifters, but also with application managers, and translates in production terms the decisions taken by Physics and Operations Planning Groups.

- **Computing shifters**: acting as helpers of the GEOC and production managers, computing shifters are selected from a pool that range from 20 to 30 people. A shift lasts 4 to 7 days. Shifters' reports are produced daily, and are publicly available on the LHCb computing pages.
- **Data Quality (DQ) shifters**: as well as computing shifters, DQ shifters are selected from a pool that range from 20 to 30 users with the same 4 to 7 days turnover. DQ shifters constitute one of the link between Online and Offline activities. Their role is most of all the flagging of real data based on quality parameters, making the data available for its processing.
- **Distributed Analysis Support (DAST)**: users' issues are answered in first place by the DAST shifter. DAST shifters are selected from a pool of less than 10 experienced users, and rotate each 7 days.
- **Data Manager**: this role, as well as the ones that follow, is not subject to any turnover. The Data Manager became operationally an important role since when real data taking started. Tools for bulk dataset handling have been recently developed, as well as consistency checks and automated monitoring.
- **Sites manager**: mostly dedicated to the handling of Tier 2 resources, the sites manager adds and removes sites and resources from the production mask, depending on their availability.
- **Tier 1 contact persons**: at each Tier 1, LHCb has a contact person. Even if a contact

person doesn't usually dedicate 100% of her time to LHCb computing issues, they become useful when issues with the site arise.

- **Application Managers**: each LHCb data processing application has a manager, who is in charge of formally specifying the application versions to be used by Data Manipulation productions at any time.
- **Resources Mananager**: defines long term planning of resources.
- **LHCbDIRAC experts**: being LHCbDIRAC the tool LHCb developed for handling its distributed computing activities, some DIRAC developers are also part of the operations team.

4. Used tools

DIRAC (Distributed Infrastructure with Remote Agent Control) [4] is the software used by LHCb for its distributed computing activities. It has been initially developed as a LHCb-specific project, but many efforts have been made to re-engineering it into a generic framework, capable to serve the distributed computing needs of a number of Virtual Organizations. Now DIRAC, which is developed in python, is a community Grid solution, offering powerful job submission functionalities, and a developer-friendly way to create services and agents. After this complex reorganization, the LHCb-specific code resides in the LHCbDIRAC extension while a core, VO-agnostic, DIRAC project has been disentangled. Even if DIRAC and LHCbDIRAC are now following distinct development cycles, for simplicity within this paper we refer to DIRAC and LHCbDIRAC as synonyms. DIRAC includes many components, some of which are used daily in the operations.

DIRAC is not the only tool used, but has a central role in all LHCb activities. In what follows, some of its components are highlighted.

We split operations actions into 4 areas: *monitoring*, *management*, *reporting*, and *testing*.

4.1. Monitoring

Monitoring is the source of the majority of operation activities. The proliferation of monitoring tools brought the VO administrators to interact with a variety of dispersed information somehow difficult to handle successfully. This is one of the main reasons to promote initiatives for aggregating various infrastructure monitoring data, which is a commonly recognized requirement expressed by many Virtual Organizations. DIRAC philosophy has always been of creating a single system as a central point of control, and also the development of the monitoring infrastructure falls into this paradigm. For completeness, the tools used are:

- **DIRAC portal**: highly integrated, and redundant, the DIRAC web portal provides the majority of the information monitored daily by the team. Users and productions jobs, data replications and removals, as well as sites monitoring, historical accounting for jobs and data, are all available via the portal. As an of example, figure 2 shows the *productions monitoring page*, as well as some accounting plots.
- **DIRAC notifications**: DIRAC is also used for sending notifications, usually in the form of e-mails. Notifications come from monitoring agents. Examples of such notifications are, for example, sent to users who are over-using their storage quota.
- **SLS**: Service Level Status (SLS) is web-based tool that dynamically shows availability of IT services. LHCb uses SLS mostly for infrastructure monitoring.
- **GOCDB**: The Grid Operations Center Database (GOCDB) contains general information about the sites participating to the production Grid that are part of WLCG. A visualisation portal is provided for storing and presenting Grid topology. Downtime announcement are sent regularly as mails.

- **Dashboard**[5]: extensively used by different VOs, for different purposes, dashboard offers a toolkit for monitoring both community activities and status of resources aggregating them in useful views. LHCb uses it only for a very specific monitoring, namely the Nagios [6] testing infrastructure, which provides VO agnostic testing of grid services.

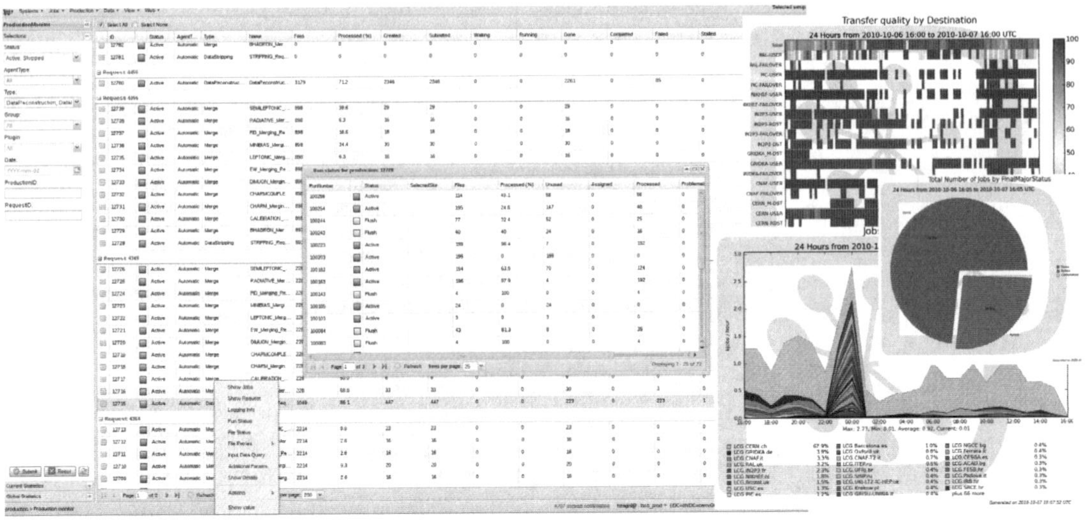

Figure 2. DIRAC Production monitoring page.

4.2. Management of operational parameters

There are few places where operational parameters are set. A change of some parameters can have a drastic effect on how the resources are used: this highlights once more how vital is the experience of the members of the operations team.

- **DIRAC CS**: The DIRAC Configuration Service is the source of a very large part of configuration parameters. Reliable, easily accessible, integrated and redundant, its parameters are used at the client and server side, while each and every job runs with a dump of its content. A selected number of experts are capable to modify the parameters.
- **VO ID Card**: expresses the minimal requirements each site has to fullfill to be part of the LHCb Grid, like memory or CPU requirements.
- **DIRAC Production Parameters**: every time a new production is created, a number of parameters are set, influencing how the jobs are created, the input and output files treated, and the CPU requested.

Once the Grid operations within a team are established with clear procedures in place, running the system becomes a routine job for an experienced Grid administrator. Taking a decision or applying a procedure should be as much as possible an automatic operation. We refer to such operations as policies. As of today, a tool to evaluate and enforce such policies is missing. The lack of automation in the management of such policies always requires a Grid operations administrator to be available in order to react manually when changes happen in the system. LHCb is integrating within LHCbDIRAC an autonomic system that will help the operation team automating certain operations.

4.3. Reporting
Reporting of progress, issues, or simply as knowledge spreading, is an often undervalued activity. It's instead an important part of the daily duties of an operations team. The tools used by LHCb comprise:

- **eLog**: an electronic Log application is used for maintaining an operational logbook.
- **GGUS**: the Global Grid User Support (GGUS), part of EGI, the European Grid Infrastructure, is the official tool for VO support and issues reporting.
- **Twiki**: a wiki implementation, mostly used at CERN.
- **Savannah**: bugs reporting, discussions with developers.

4.4. Testing
Testing of resources, and application stacks, are performed regurarly:

- **Nagios**: Nagios tests are sent continuously at each Grid site as part of the monitoring and testing activity.
- **Production Validation**: when a new LHCb software stack is available, a validation production is performed. A validation production is an integration test which proved to be very useful for both the distributed computing team and the application developers.

5. Conclusions
In this paper we presented the computing activities performed by the LHCb offline operations team. We presented the team organization, and the monitoring system for the LHCb Grid Computing, which relies on many heterogeneous and independent sources of information. This work summarized the state-of-the-art of LHCb Grid operations. We showed the reasons that brought to various choices and what are the tools currently in use to run our daily activities.

References
[1] Aderholz M e a 2000 Models of networked analysis at regional centres for lhc experiments (monarc), phase 2 report, 24th march 2000 Tech. Rep. CERN-LCB-2000-001. KEK-2000-8 CERN Geneva
[2] Brook N 2004 Lhcb computing model Tech. Rep. LHCb-2004-119. CERN-LHCb-2004-119 CERN Geneva
[3] Burke S, Campana S, Lanciotti E, Mendez Lorenzo P, Miccio V, Nater C, Santinelli R and Sciaba A 2009 *gLite 3.2 User Guide*
[4] Tsaregorodtsev A and al 2008 *Journal of Physics: Conference Series* **119** 062048
[5] Andreeva J, Gaidioz B, Herrala J, Maier G, Rocha R, Saiz P and Sidorova I 2009 *Grid Computing* ed Lin S and Yen E (Springer US) pp 131–139 ISBN 978-0-387-78417-5 10.1007/978-0-387-78417-5_12
[6] Barth W 2006 *Nagios. System and Network Monitoring* u.s. ed ed (No Starch Press) ISBN 1593270704

ACAT 2011 IOP Publishing
Journal of Physics: Conference Series **368** (2012) 012010 doi:10.1088/1742-6596/368/1/012010

The LHCb DIRAC-based production and data management operations systems

F Stagni[1], P Charpentier[2]
[1,2]PH Department, CH-1211 Geneva 23 Switzerland
E-mail: [1]`federico.stagni@cern.ch`
E-mail: [2]`philippe.charpentier@cern.ch`

On behalf of LHCb Collaboration

Abstract. The LHCb computing model was designed in order to support the LHCb physics program, taking into account LHCb specificities (event sizes, processing times etc...). Within this model several key activities are defined, the most important of which are real data processing (reconstruction, stripping and streaming, group and user analysis), Monte-Carlo simulation and data replication. In this contribution we detail how these activities are managed by the LHCbDIRAC Data Transformation System. The LHCbDIRAC Data Transformation System leverages the workload and data management capabilities provided by DIRAC, a generic community grid solution, to support data-driven workflows (or DAGs). The ability to combine workload and data tasks within a single DAG allows to create highly sophisticated workflows with the individual steps linked by the availability of data. This approach also provides the advantage of a single point at which all activities can be monitored and controlled. While several interfaces are currently supported (including python API and CLI), we will present the ability to create LHCb workflows through a secure web interface, control their state in addition to creating and submitting jobs. To highlight the versatility of the system we present in more detail experience with real data of the 2010 and 2011 LHC run.

1. Introduction

The Production system transforms physics requests into physics events, implementing the LHCb computing model. The heart of the system is the Transformation System. The LHCbDIRAC Transformation system is a key component for all the distributed computing activities of LHCb. It is a reliable, performant, and easy to extend system. Together with the Bookkeping system and the Requests Management system, it provides what is called the Production System. This paper explains how the Production System is developed and used, giving an architectural and an operational perspective. It will also show experience from 2010 and 2011 Grid processing.

This paper is organized as follows: section 2 introduces DIRAC and LHCbDIRAC. Section 3 explains the architecture of the production system, as a whole, introducing the concepts used, together with a short description of their technical implementation. Within section 4, we give a brief summary of how the operations team uses the system for its daily activities. Section 5 shows some results from past year's run, and final remarks are given in section 6.

2. LHCbDIRAC

DIRAC (Distributed Infrastructure with Remote Agent Control) [1] is a community Grid solution. Developed in python, it offers powerful job submission functionalities, and a developer-friendly way to create services and agents. A DIRAC service exposes an XML-RPC implementation; a DIRAC agent is a stateless light-weight component (comparable to a cron-job).

DIRAC has been initially developed as a LHCb-specific project, but many efforts have been made to re-engineering it into a generic framework, capable to serve the distributed computing needs of a number of Virtual Organizations. After this complex reorganization, the LHCb-specific code resides in the LHCbDIRAC extension while a core, VO-agnostic, DIRAC project has been disentangled. In this way, other VOs, like Belle II [2], have developed their custom extensions to this core DIRAC framework. DIRAC is a collection of sub-systems, each constituted of services. Sub-systems are, for example, the WMS (Workload Management System) or the DMS (Data Management System). Each system comprises a generic part, and a VO-specific part.

LHCbDIRAC extends DIRAC to handle all the distributed computing activities of LHCb. Such activities include real data processing (reconstruction, stripping and streaming), Monte-Carlo simulation and data replication. Other activities are groups and user analysis, data management, resources management and monitoring, data provenance, accounting for user and production jobs. While DIRAC and LHCbDIRAC follow independent release cycles, every LHCbDIRAC is built on top of an existing DIRAC release.

3. The production system

The LHCb distributed computing activities include production and non-production activities. Within this paper, we have no interest in non-production activities, which include *user analysis*, and *monitoring and testing*. The production activities can be split between data manipulation, and data management. Data manipulation comprises *Reconstruction*, *Reprocessing*, *Stripping* (which is the name LHCb gave to a selection of events), *Calibration* and *Montecarlo simulations*. Data management activities are usually a direct consequence of the production activities of Data Manipulation. They include: *Replication*, *Archival* and *Removal*. These data management activities share a large fraction of the DIRAC and LHCbDIRAC code used for data manipulations.

Within this section, we introduce the concepts used in the system, and their implementations: DIRAC workflows, the DIRAC and LHCbDIRAC transformation system, the LHCb bookkeeping and production requests system.

3.1. DIRAC workflows

A workflow is, by definition, a sequence of connected steps. DIRAC provides an implementation of the workflow concepts in one of its "Core" packages, with the declared scope of running "complex" jobs, i.e. jobs who run one application after another, whose input/outputs are usually directly connected to each other. The implementation comes in the interchangeable formats of an XML file, or an extended python dictionary. All production jobs are described using a DIRAC workflow.

As can be seen in figure 1, each worfklow is composed of steps, and each step include a number of modules. These workflow modules are connected to python modules, that are executed in the specified order by the jobs. Parameters can be specified at any level: workflow, step, and even modules.

Figure 1. A workflow is composed by steps, that include modules.

3.2. DIRAC transformation system

The DIRAC Transformation System is used for handling "repetitive" work. It has two main uses: the first is for creating job productions, and the second for data management operations. When a new "Production Jobs" transformation is requested, a number of transformation tasks are dynamically created, based on the input files (if present), and the "plugin" specified. A "plugin" specifies the way the tasks are created, and can decide where the jobs will run, or how many input files can go into the same task. Each of the tasks pertaining to the same transformation will run the DIRAC workflow specified when the transformation is created.

A task, when inside the transformation system, is not yet a Grid job nor a data management operation. For this last step, agents are defined, that upon inspections of the Transformation System tables, will submit the tasks either to the Workload Management System, or the Data Management System. The Transformation System, unless when being used just for simulation activities, can not live as a stand-alone system: whenever there are input data to be handled, external Metadata and Replica catalogs are needed. For LHCb, the Metadata and Replica Catalogs are the LHCb Bookkeeping, presented in the next section, and the LFC catalog [3]. Figure 2 shows some components of the system implementation.

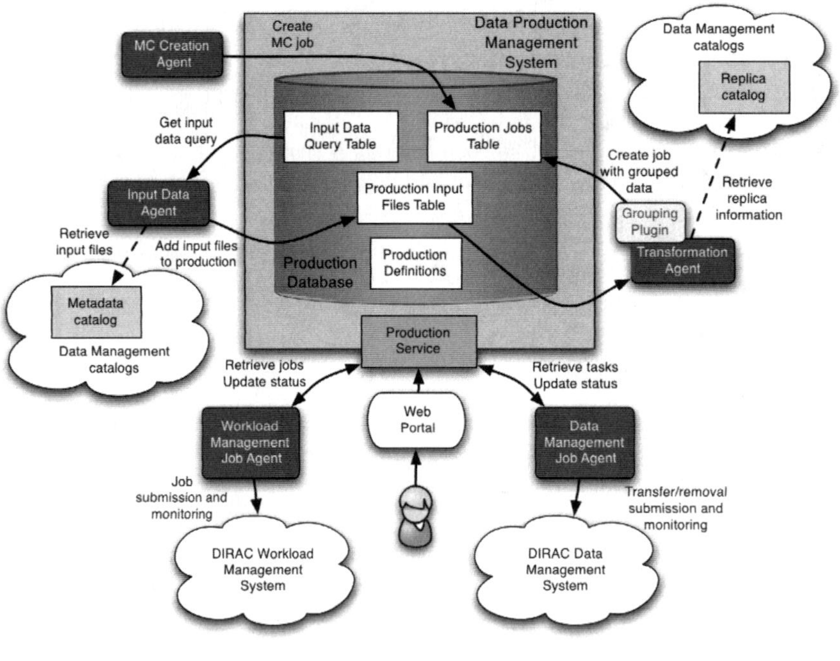

Figure 2. Schematic view of the DIRAC Transformation System.

3.3. LHCb bookkeping

The LHCb Bookkeping System is integrated within LHCbDIRAC, and it is the LHCb data recording and data provenance tool. The Bookkeping System is widely used inside the Production System, by a large fraction of the members of the Operations team, and by all the physicists doing data analysis within the collaboration. Being recognized as a key system by the collaboration, its design and implementation were subject to important modifications throughout the years.

The bookkeeping is not necessarily a tool for doing distributed computing. Among its functions, users and machines are retrieving datasets for analysis and productions, together with metdata like data taking and simulation conditions, event types and file types. The bookkeeping is a read-only database for general users.

Unlike the other DIRAC system, where MySQL is the prime choice as RDBMS, the CERN Oracle backend has been chosen instead. Figure 3 shows the main components within this system.

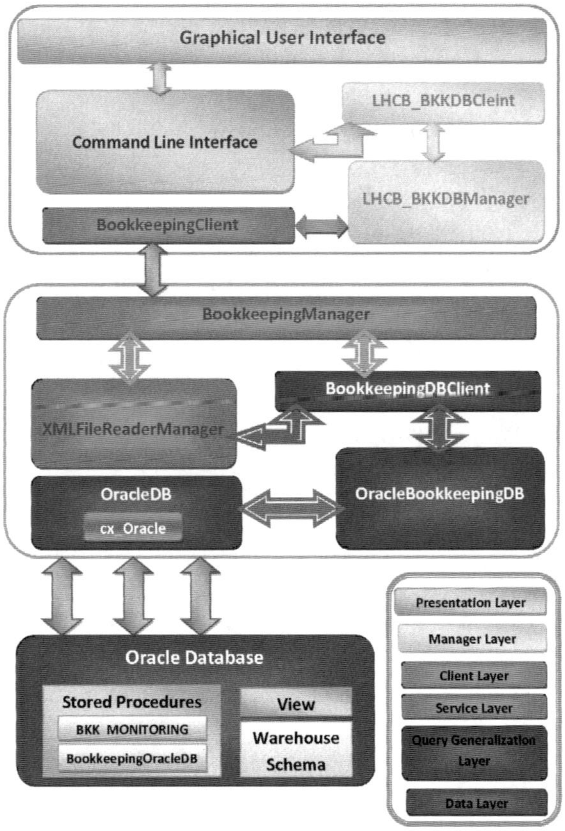

Figure 3. Schematic view of the LHCbDIRAC Bookkeping System. The complexity of the system deserves an explanation per se. A number of layers are present, as shown in the legend in the bottom-right corner of the figure. Within this paper, we just observe that the data recorded represent the first source of information for production activities within LHCb.

3.4. LHCbDIRAC production request system

The *Production Request System* is a way to expose the users to the production system. It is an LHCbDIRAC developement, and represents first of all a formalization of the processing activities. A production is a combination of *steps*, and a production request can be created by any users, provided that there are formal *steps* definition in the steps database. Figure 4 shows this simple concept.

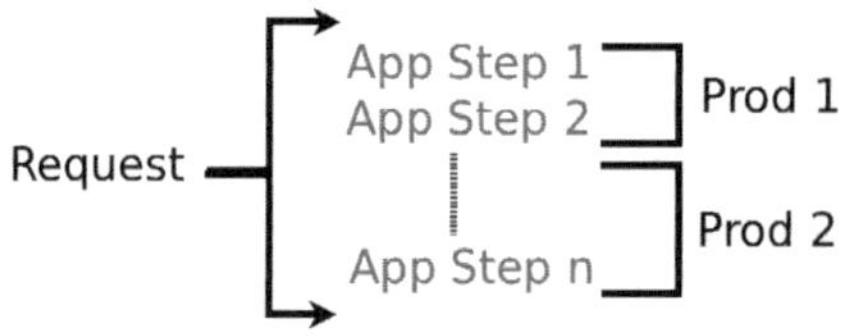

Figure 4. A request is composed by a number of steps.

Creating a step, a production request, and subsequently launching such production is done using a web interface, integrated in the LHCbDIRAC web portal, in a user-friendly way. The production request web page also offers a simple monitoring of the status of each request: for example, the percentage of processed events, as requested, is reported and publicly available. Such information is also used, for simulation productions, to trigger their automatic completion.

4. Operating productions

Putting all the pieces together, operationally means a workflow composed of the following steps:

(i) *Application managers* create usable steps, where all the options are set;

(ii) *Production Managers* combine the available steps to create production requests;

(iii) The same *Production Managers* create real productions starting from such requests, using production templates;

(iv) The Productions are followed by the computing shifters, the *GEOC (Grid Expert On Call)* and by the same Production Managers.

A Data Manager creates bulk replication, deletion or archival of datasets using the transformation system. There are tools provided for massive final checks, that for example allows to know if the correct number of replicas are registered in the catalogs and on the storage services. Many interfaces are provided: GUI/CLI used in all steps. In figure 5 all the concepts discussed are put together.

5. The system at work in 2010 and 2011

In figure 6, we summarize the results from 2010 processing, using four plots. A few notable points that we can extrapolate are:

- *Many user jobs*: analysis jobs from users represent a large fraction of the distributed activities of 2010.
- *MC productions*: the use of the production system was mostly used for simulation campaigns.
- *Activities*: even if peaks of 50k jobs/day were reached, the system sustained an average of 20k jobs/day.
- *Use of Tier 0*: CERN resources were enough for one fifth of the total amount of resources used.

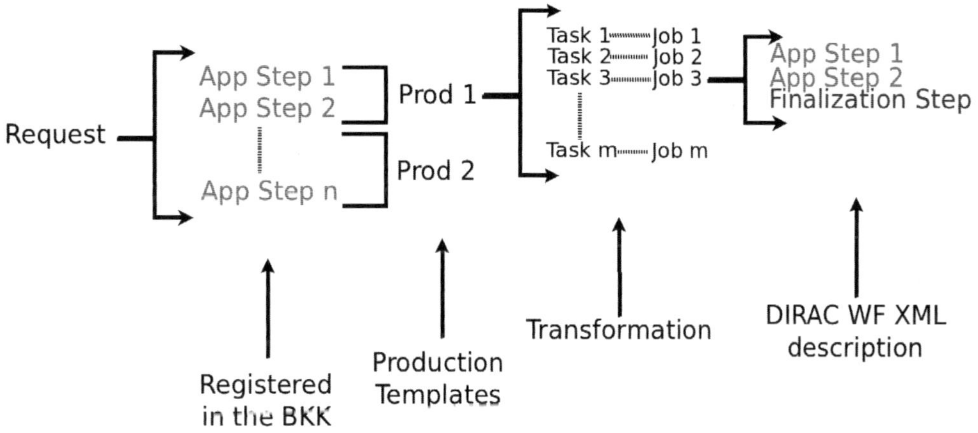

Figure 5. Putting all the concepts together: from a Request to a Production job.

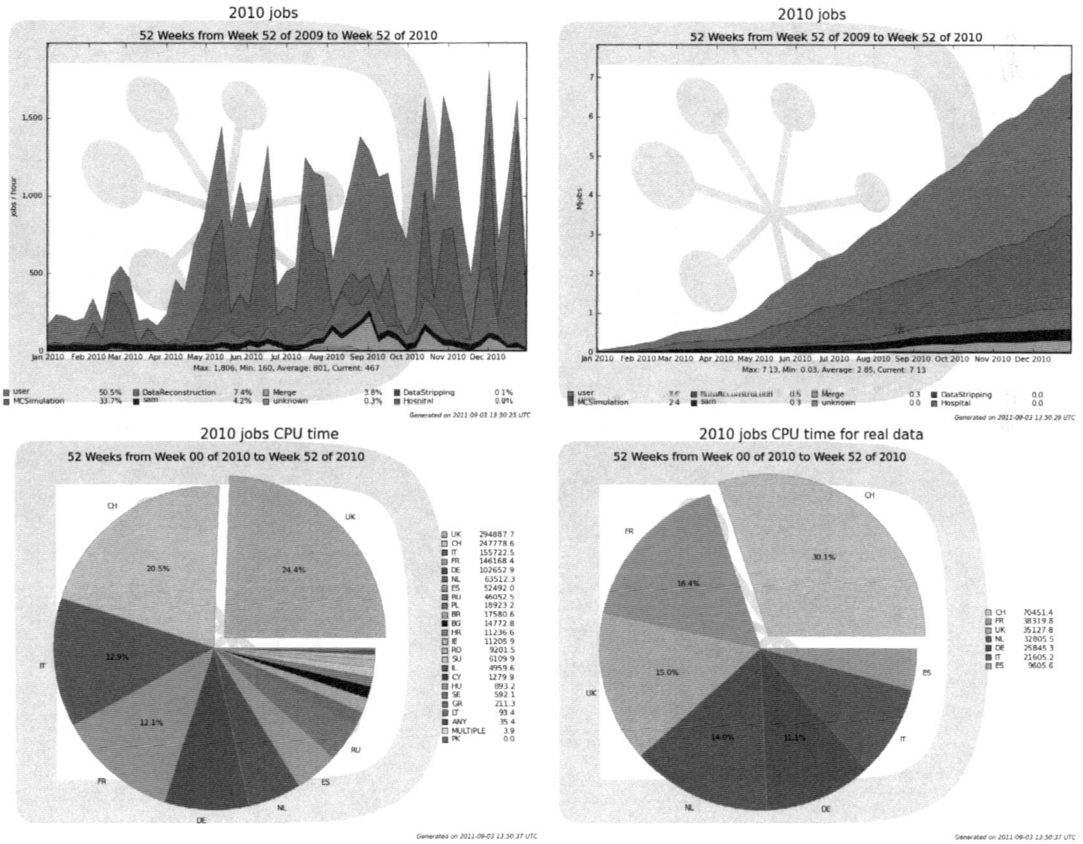

Figure 6. Some plots showing the results of data processing in 2010. Clockwise: the rate of jobs per hour in the system by job type, the cumulative number of jobs managed throughout the year, the percentage of CPU used for real data processing, and the percentage of CPU used by country by the totality of the jobs.

Figure 7 shows four plots related to part of the 2011 activities. What can be seen is that:

- *MC productions*: a huge simulation campaign has started at the beginning of the year, and continued steadily through the year.
- *Activities*: the system sustained an average of 40k jobs/day, which is twice what was done during the previous year.
- *Efficiency*: the wasted CPU time has been less than 5%. This is due to the high efficiency of the LHCb software, and LHCbDIRAC itself.
- *Use of resource*: mostly due to the simulation campaign, all Tier 2 have been intensively used.

It has to be noted that these plots for 2011 covers only the first 34 weeks of the year. During the last months of 2011, a *reprocessing* activity started. Such activity used resources from Tier 1 and Tier 2.

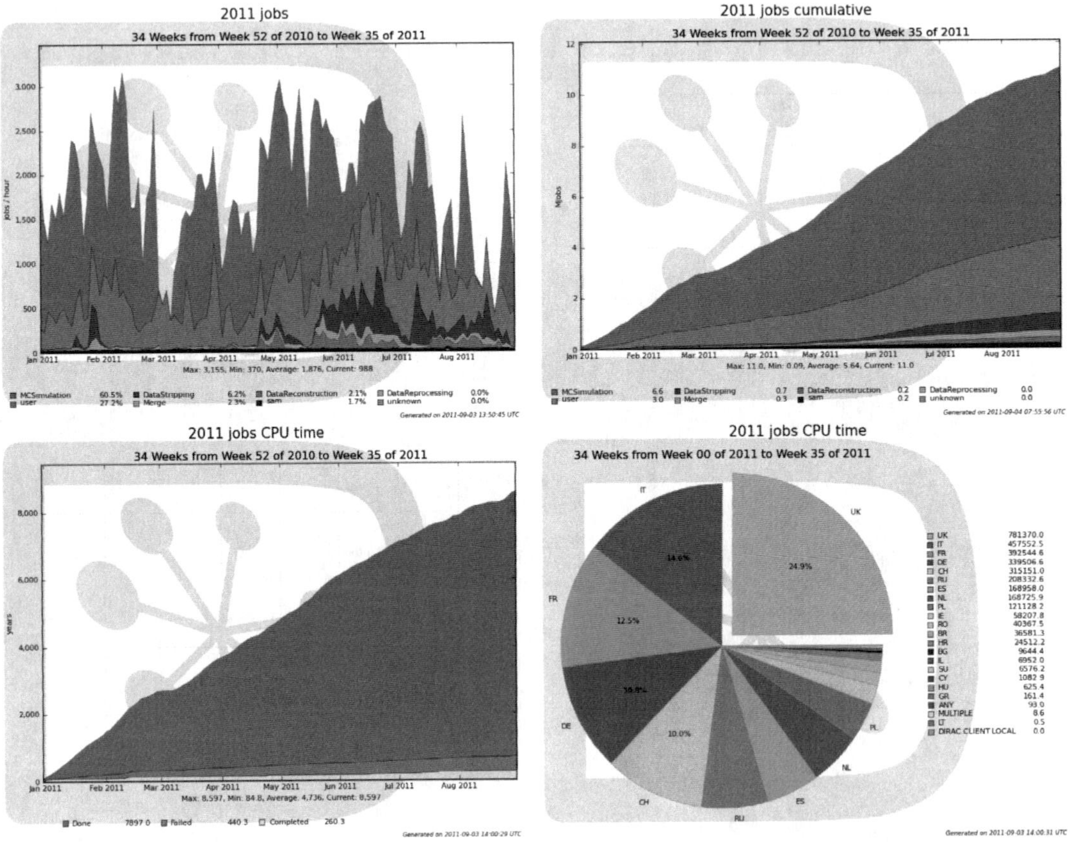

Figure 7. Some plots showing the results of data processing in 2011. Clockwise: the rate of jobs per hour in the system by job type, the cumulative number of jobs managed throughout the year, the percentage of CPU used by country by the totality of the jobs, and the cumulative number of jobs by their final major status.

6. Conclusions

DIRAC has proved to be a scalable, and extensible software. The production system is vital for a great part of LHCb distributed activities. The system first started to be used with simulation campaigns, and it is now used for Data Reconstruction, Reprocessing, Stripping, Calibration activities. It is also used for datasets manipulations like data replication or data removal. During the last years, we have seen how the number of jobs handled by the Production system steadily grew, and represents now more than half of the total number of jobs created. User analysis jobs still represent a good fraction of the jobs handled by the system, and since they have a priority over production jobs, when a large number of them are queued, they may represent an operational burden, slowing down if not completely blocking part of the production activities. For this reason, pressure has recently put on analysis working groups to integrate as much as possible their activities in the production framework. This would allow the operations team to better control the stress put on computing and storage resources, depending on the relative priorities of the various scheduled production activities. Working Groups will also gain advantages when coming to data provenance, recording and distribution of the produced data.

References

[1] Tsaregorodtsev A, Bargiotti M, Brook N, Ramo A C, Castellani G, Charpentier P, Cioffi C, Closier J, Diaz R G, Kuznetsov G, Li Y Y, Nandakumar R, Paterson S, Santinelli R, Smith A C, Miguelez M S and Jimenez S G 2008 *Journal of Physics: Conference Series* **119** 062048

[2] Diaz R G, Ramo A C, Agero A C, Fifield T and Sevior M 2011 *J. Grid Comput.* **9** 65–79

[3] Bonifazi F and al 2008 *J. Phys.: Conf. Ser.* **119** 042005

Journal of Physics: Conference Series **368** (2012) 012011 doi:10.1088/1742-6596/368/1/012011

Offloading peak processing to virtual farm by STAR experiment at RHIC.

Jan Balewski[1], Jerome Lauret [2], Doug Olson[3], Iwona Sakrejda[3], Dmitry Arkhipkin[2], John Bresnahan[4], Kate Keahey[4], Jeff Porter[3], Justin Stevens[5], and Matt Walker[1]

[1] Massachusetts Institute of Technology, Cambridge, MA 02139
[2] Brookhaven National Laboratory, Upton, New York 11973
[3] National Energy Research Scientific Computing Center, 1 Cyclotron Road Berkeley, CA 94720
[4] Argonne National Laboratory, Argonne, Illinois 60439
[5] Indiana University, Bloomington, Indiana 47408

E-mail: balewski@mit.edu

Abstract. The Virtual Machine framework was used to assemble the STAR-computing environment, validated once, deployed on over 100 8-core VMs at NERSC and Argonne National Lab, and used as a homogeneous Virtual Farm processing events acquired in real time by STAR detector located at Brookhaven National Lab. To provide time dependent calibration, a database snapshot scheme was devised. The two high capacity filesystems, localized at the opposite coasts of US and interconnected via Globus-Online protocol, were used in this setup, which resulted with a highly scalable Cloud-based extension of STAR computing resources. The system was in continuous operation for over 3 months.

1. Introduction

The Relativistic Heavy Ion Collider (RHIC), situated at Brookhaven National Laboratory, is the highest energy collider in the world that accelerates polarized proton beams in which the spin state of the majority of the protons will be aligned with the direction of the beam. This allows nuclear physicists to study spin properties of quarks and gluons - the constituents of the proton, schematically shown in inset a) of figure 1. The STAR experiment [1] records particles produced in the collision of two protons, shown in figure 1, which yields valuable information on the spin structure of the proton[6]. In particular, our understanding of the polarization (alignment of rotation axis) of the anti-quarks inside the proton is still very limited. By studying the relatively rare process where a quark and anti-quark annihilate during the collision of the two protons and produce a W boson (see figure 2), we can infer information about spin properties of the incident anti-quark. We measure the probability of such W production processes, by reconstructing W boson events shown in figure 2.a.

[6] Spin is central to a variety of scientific concepts and technologies, including Magnetic Resonance Imaging machines used for medical imaging in hospitals around the world.

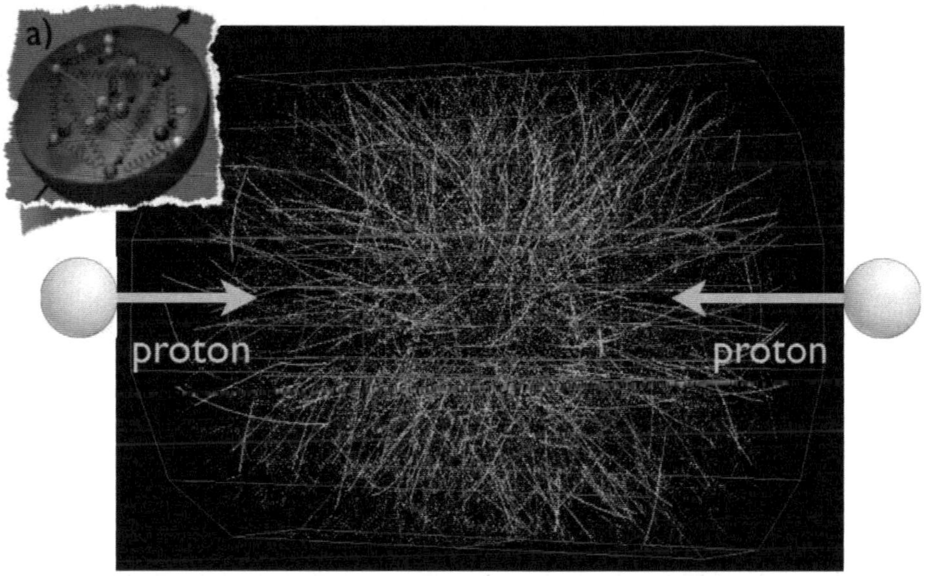

Figure 1. Example of ionization tracks recorded by STAR detector after the collision of two high energy protons. The track colors denote particles of different energy produced in the collision. The inset a) illustrates our current understanding of the structure of proton before collision: quarks, anti-quarks, and gluons interacting with each other.

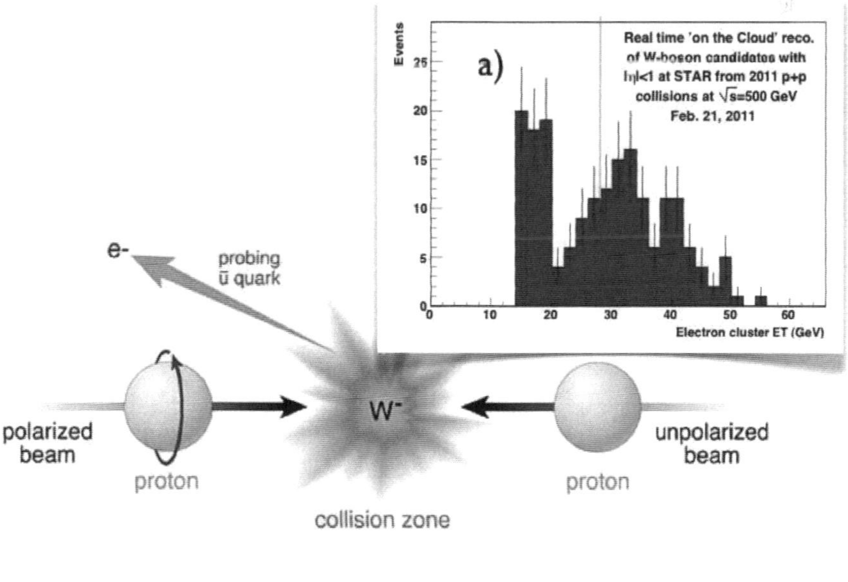

Figure 2. Illustration of the rare process of W boson production in the collision of two high energy protons. The W^- boson, shown here, decays into an electron, which if emitted in the opposite direction of the incident polarized proton, allows us to study polarization of anti-u quark inside the proton. The inset a) shows the energy distribution of electrons measured by STAR and reconstructed in almost real time using cloud resources - which are discussed later in this paper.

2. Computational challenge of STAR data mining

To study a very rare process, such as W boson production, one needs to give Nature many chances to produce it, hence RHIC is set up to provide a few million proton-proton collisions per second at the center of the STAR detector. We select and record on tape for several weeks only 1/1000-th of the most promising candidate events, still acquiring millions of events. The process of reconstructing recorded events involves sorting a hundred thousand space points, shown in figure 1 in to thousands of distinguishable particle tracks, and requires significant computing resources on the order of tens of CPU years. Furthermore, the measurement is completed in a broader sense, only after results are published. Therefore, one needs to fold in also the time needed to complete all the necessary aspects of data analysis, namely calibration of the STAR detector response, development and vetting of the W event reconstruction algorithm, discussion of the results within the STAR collaboration (consisting of ~500 members), and preparing a manuscript for publication.

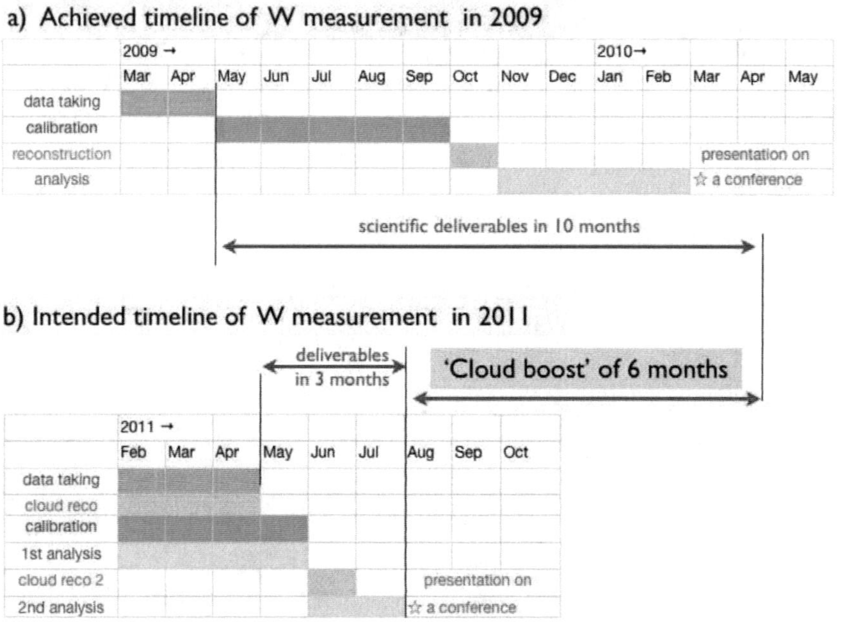

Figure 3. Comparison of the typical timeline from data collection to publication for traditional sequential data mining a) vs. parallel iterative approach benefiting from the cloud computing b).

2.1. Traditional sequential approach

The traditional approach at STAR calls for sequential accomplishing of required tasks, as illustrated by the diagram a) in figure 3. Prior to data taking the arrangement is made to reserve sufficient CPU power for the offline processing. After the data is collected on tapes, a subset of events is processed to establish detector calibrations. This calibration process takes several months. Next, all recorded events are processed on the RCF (RHIC Computing Facility) computing farm for about a month which reduces the footprint of the event files and allows for the development and tuning of the W reconstruction algorithm. Typically, it takes several months to identify and remediate problems with the data, reduce background, maximize efficiency, and vet intermediate results within STAR collaboration. In total, it is more than a year after experiment stops collecting data, when publication quality results are released by the STAR Collaboration to the public domain. This is a long time for a student or postdoc who often needs published results from their work to advance in their career.

2.2. 'Cloud boost' of data publication cycle

The advent of Virtual Machines (VM) [2, 3] combined with abundance of cloud computing resources, such as commercial Amazon EC2 or university based farms discussed later (see table 1), allows one to restructure the strategy of extracting scientific output from the raw measurements. The diagram in figure 3.b summarizes the changes leading to perhaps a half year shorter publication cycle. The key factor is continuous access to cloud computing resources of moderate magnitude, but powerful enough to process all events of interest in real time using a temporary calibration extrapolated from data processed the previous day. The availability of reconstructed events stimulates parallel work on the refinement of the W reconstruction algorithm and enables at the same time work on the quality assurance of data recorded on the previous day. Another positive side effect of these efforts is the continuous feedback to the ongoing experiment about the performance of various detector subsystems leading to a better utilization of the limited beam time. Finally, one should not forget about psychological impact of weekly updates on the number of recorded and reconstructed W events. It changes the attitude of the STAR shift crew if they can see that with the passing weeks of monotonous data taking there is an unambiguous measure of progress, e.g. a plot of reconstructed W candidates as function of measured energy, shown in figure 2.a.

Table 1. History of VM-based computation by STAR.

Date	Facility	Tools	Type of Task	# of VMs	# of Jobs per VM	Total CPU (days)	Calend. Days	Total Input (TB)	Total Output (TB)	Remarks
2009, March	Amazon EC2	Nimbus, Globus	simu	100	1	500	5	0	0.3	globus GK grid site
2009, Nov.	Amazon EC2	EC2	simu	10	1 or 2	1	1	0	0.1	use commercial interface
2010, Feb.	GLOW †)	Condor VM	simu	430	1	130	0.6	0	0.1	'call home' model .
2010, July	Clemson Uni	Ketrel KVM	simu	1000	1	17,000	20	0	7	short VM lifetime, no ssh to VM
2011, Spring	NERSC, ANL	Eucalyptus, Open-Stack	data reco	20-120	8	25,000+	120+	60	40	real time processing

† University of Wisconsin - Madison

3. STAR encounters with VMs

Over last 3 years STAR has ramped up its investment in the utilization of cloud computing resources [5, 6], as summarized in table 1.

At first STAR ran small batches of simulation jobs, which are easier to execute on a remote site since the input is identical for all jobs, negligible in size, and all compute nodes could be made identical. Also, produced events were transferred from the remote location to RCF which did not require ssh capability from RCF to compute nodes - the configuration preferred by IT at the remote site because of security concerns. The largest STAR simulation using cloud computing to date was completed in July of 2010. At it's peak, STAR was running on 1300 nodes, as shown in figure 4. We acquired a total of 47 CPU years over 3 calendar weeks and transferred back to RCF 7 TB of data. The practical benefit of this simulation was accelerating the work towards a Ph.D. for one of us (M.W.) by about a year.

Encouraged by this success we decided to increase the complexity and embark on the more ambitious task of almost real time processing of experimental data acquired at STAR in 2011. The challenges of two way data transfer, organization of asynchronous data processing, and updating calibrations will be discussed in the next section. Here, we only note this task lasted over 3 months, peaked at about 900 parallel running jobs, resulted in an integrated CPU time of above 70 CPU years over 3 calendar months, and transferred a total of 100 TB of data.

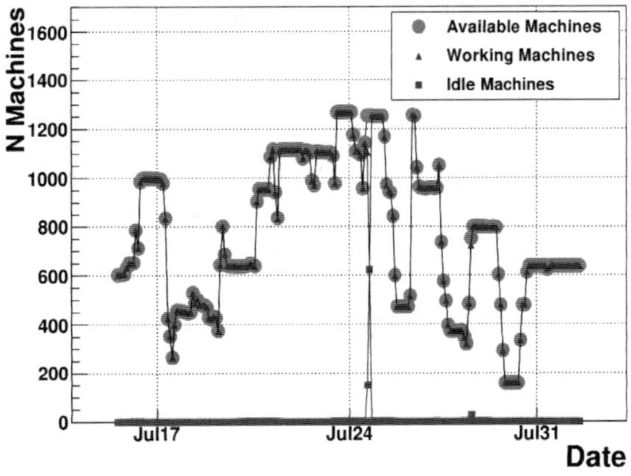

Figure 4. Load on GLOW cluster at University of Wisconsin - Madison, during 2 weeks of STAR simulations.

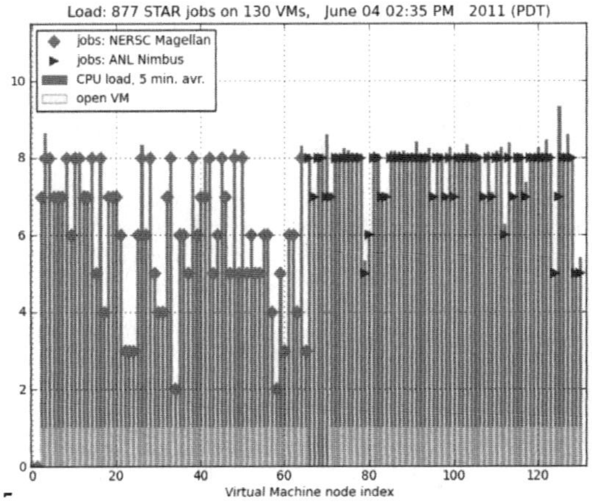

Figure 5. Load on Magellan cluster during real time STAR data processing, see text for details.

3.1. Advantages of VM

The use of Virtual Machines was an integral part of the STAR cloud computing exercises. The VM technology allowed STAR to separate details of the transient hardware/compiler features from persistent STAR code libraries. Some effort was necessary to understand the nuances between the installation of the STAR core code directly on a physical machine, and the assembling and remote deployment of a STAR VM at the computing facilities. However, the long term benefits of the use of VMs was of much larger value.

First of all it is sufficient to validate STAR VM only once and then re-use the same VM image multiple times on different machines located at different facilities. Additionally, the use of VMs has homogenized the STAR computing environment. The burden of accounting for the hardware differences at different computing facilities was shifted to the remote facility administrator assuring the VM hypervisor interacts properly with the local hardware and conforms with local security requirements. Finally, for a multi-decadal experiment like STAR it is essential to retain the ability to execute past analysis code at any time in the future. The typical obstacles are irreversible upgrades of the operating system and hardware (architecture) retirements. Archiving of the present STAR code in the form of a VM assures we will be able to execute it in the distant future on not yet defined architecture as long as it will support virtualization as we know it today.

4. Real time STAR data processing on the cloud

The resources of Magellan - the government-funded cloud computing testbed, have been made available to STAR. Magellan resources was split between two facilities. One, based at the National Energy Research Scientific Computing Center (NERSC) in Berkeley, California uses Eucalyptus, a widely used open source cloud platform. The second, based at Argonne National Laboratory near Chicago, Illinois, hosts two clouds; one runs the OpenStack software, while the other uses the Nimbus toolkit [7]. The availability of computing power was essential but not sufficient for the real time STAR data processing. We needed to also assure the input data and calibration were sent to the compute nodes, and then transported back the results. Finally, the STAR VMs themselves have been customized to continuously occupy all available cores.

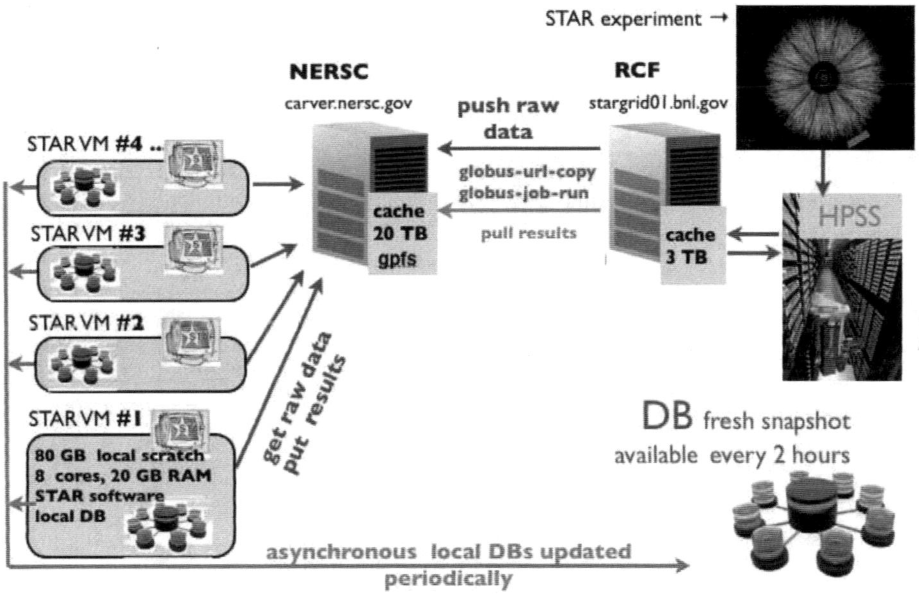

Figure 6. Example of the data flow for almost real time STAR processing on a single remote facility (NERSC) . The copies of input files produced by STAR experiment were retrieved from HPSS at RCF to a local scratch disc, sent via globus command to scratch disc at NERSC, and further using scp to individual VMs. The output files followed the same path in the reverse order.

4.1. Configuration of STAR VM

The general layout of the STAR data processing flow is illustrated in figure 6. Two independent processes check every half an hour for new files: one at RCF for input and another at NERSC for output. Both used Globus Online [8] to transfer the files they find to the scratch disc reserved at NERSC and RCF, respectively. The third process at RCF would archive output files at the local HPSS [9, 10] and free up disc space on both scratch discs to make room for new files. Every two hours, a fourth independent process takes a snapshot of the calibration data stored at Brookhaven, which changes much less rapidly than event data taking, and makes it available for download for all VMs.

At the remote computing facility, tens of eight-core VMs were running the STAR analysis software. Each VM could run up to eight jobs at a time to occupy all its cores. Each VM acted as an independent, greedy consumer of data by requesting a new input file from the scratch disc at NERSC whenever it determined it had available CPU to spend. The main challenge was to preserve independent, unsupervised data reconstruction on different VMs without processing the same input file multiple times. It was resolved by using the atomic rename operation of Linux, which uniquely renames a selected input event file and passes the new name to the VM that requested a new file. If multiple VMs try to 'lock' the same file at the same time, only one of the atomic rename processes will succeed. The remaining VMs would continue to request files periodically until their request succeeded; the result was that eventually, either all of the VMs will be analyzing data on all cores or the pool of input files on the scratch disc will be empty.

STAR data can only be processed on a machine if the time dependent calibration constants are valid up past the timestamp of the data. Once every 24 hours, at a fixed time chosen at random per VM, a cron job running on each VM pulled the most recent calibration snapshot from the cluster back in Brookhaven. Then the local copy of the calibration data on the VM was replaced. Since each VM initiated this process at a different time of a day, it ensured that some VMs always have a fairly recent copy of calibration and were ready to process fresh STAR data. A side benefit of this approach was the reduced load on the central machine distributing calibration constants.

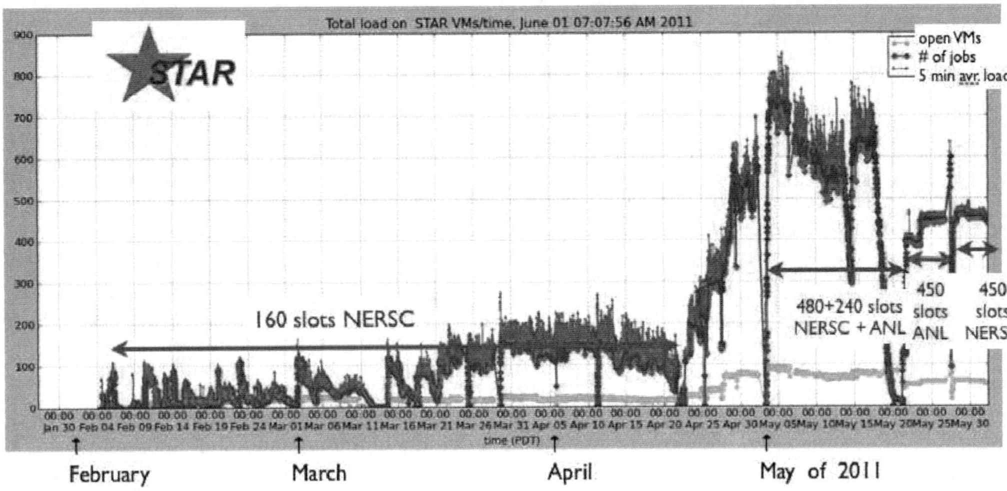

Figure 7. Total number of STAR jobs (red) and CPU load (green) for real time data processing over a period of 3 months. The total number of VMs used is shown in yellow. At different times we used resources from different facilities, as indicated by labels.

4.2. Monitoring of data processing

As with any real time process it was essential to deploy a short latency monitoring, allowing real time assessment of the performance of the data flow and pointing to possible bottle necks. A sequential survey of all responding VMs was conducted periodically and the aggregated performance summary plots automatically posted on the web. We have used NERSC portal for that purpose. An example plot showing instantaneous load of about 900 jobs running on 130 VMs is shown in figure 5. Each VM is represented on the x-axis as a single bin. The red diamonds (blue triangles) denote the number of jobs running on a given VM located at the NERSC (ANL) computing facility. Each VM had 8 cores, so the optimal utilization of the VM is achieved if 8 jobs are running on it. The green bar marks the CPU load on each VM. At the time of data transfer or during calibration updates the load on a VM would exceed the available number of cores. It was acceptable for short periods of times, but was a sign of a trouble if it persisted over more than an hour. A video clip available in the online version of this paper shows an animated graph of the load as a function of time.

We were improving the real time STAR data processing scheme over 3 months in the spring of 2011, reaching the sustained average load of 600 jobs over several weeks, as shown figure 7.

5. Future prospects

STAR has plans to expand the existing scheme and integrate all available resources as illustrated in figure 8. The goal is to refine our real-time cloud-based data processing scheme to function as a self-adjusting assembly line and handle variable rate of throughput. No human intervention, nor supervisor process that orchestrates the entire data flow should be needed. Every stage of the process is governed by local rules designed to handle time-outs and refusals from other elements of the system by waiting and starting over. Time will show if we will be be able to accomplish our plans as some of the computing facilities used so far are being re-tasked for other non-cloud related projects.

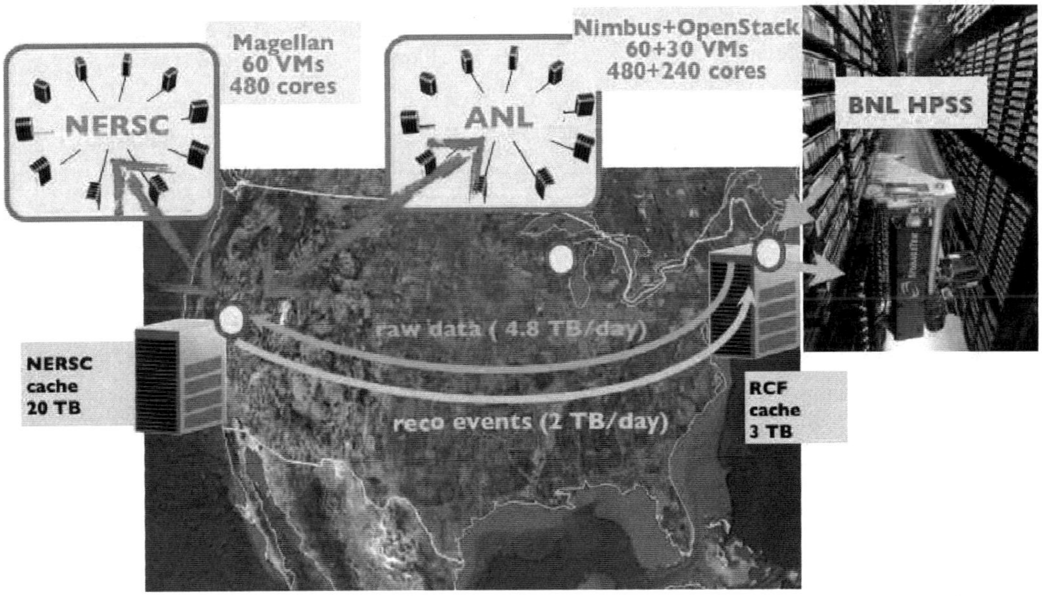

Figure 8. Target configuration of STAR data processing at multiple cloud computing facilities.

References

[1] Adams J et al. 2005 Experimental and theoretical challenges in the search for the quark gluon plasma: The STAR collaboration's critical assessment of the evidence from RHIC collisions *Nuclear Physics* A **757** 102

[2] Fenn M, Goasguen S and Lauret J 2010 Contextualization in Practice: The Clemson Experience *Proceedings of 13th International Workshop on Advanced Computing and Analysis Techniques in Physics Research* PoS(ACAT2010) 027

[3] Stour L, Murphy M and Goasgen S 2009 Kestrel: an XMPP-based framework for many task computing applications *Proceedings of the 2nd Workshop on Many-Task Computing on Grids and Supercomputers* (Portland, Oregon, November 16 - 16, 2009). MTAGS '09. ACM, New York, NY, 1- 6. DOI= http://doi.acm.org/10.1145/1646468.1646479

[4] Ramakrishnan L at al. 2011 Magellan: experiences from a science cloud *Proceeding ScienceCloud '11 Proceedings of the 2nd international workshop on Scientific cloud computing* (ACM New York, NY, USA 2011 ISBN: 978-1-4503-0699-7)

[5] Lauret J, Walker M, Goasguen S and Hajdu L 2010 From Grid to Cloud, the STAR experience *Scientific Discovery through Advanced Computing SciDAC 2010 Conference proceedings*

[6] Lauret J, Walker M , Goasguen S Stout L, Fenn M, Balewski J, Hajdu L and Keahey K 2010 When STAR meets the Clouds Virtualization & 'Cloudified' grids experience *CHEP 2010 proceedings, J. Phys. Conf. Ser.* **331** 062016 , iopscience.iop.org/1742-6596/331/6/062016

[7] Keahey K, Freeman T 2008 Science Clouds: Early Experiences in Cloud Computing for Scientific Applications, Cloud Computing and Its Applications *(CCA-08), Chicago, IL. October 2008* , see also http://www.nimbusproject.org/papers/

[8] Globus Online is a fast, reliable file transfer service that makes it easy for any user to move any data anywhere, for details see https://www.globusonline.org/about/

[9] LAURET J and YU D 2010 BNL Batch and DataCarousel systems at BNL: A tool and UI for efficient access to data on tape with faire-share policies capabilities *PoS(ACAT2010)* 023

[10] Yu D and Lauret J D 2010 Tape Storage Optimization at BNL *CHEP 2010 proceedings, J. Phys. Conf. Ser.* **331** 042045

Application of remote debugging techniques in user-centric job monitoring

T dos Santos[1], P Mättig[1], N Wulff[2], T Harenberg[1], F Volkmer[1], T Beermann[1], S Kalinin[1] and R Ahrens[1]

[1] Bergische Universität Wuppertal, Gaußstraße 20, 42119 Wuppertal, Germany
[2] Fachhochschule Münster, Hüfferstraße 27, 48149 Münster, Germany

E-mail: dos.santos@physik.uni-wuppertal.de

Abstract. With the Job Execution Monitor, a user-centric job monitoring software developed at the University of Wuppertal and integrated into the job brokerage systems of the WLCG, job progress and grid worker node health can be supervised in real time. Imminent error conditions can thus be detected early by the submitter and countermeasures can be taken. Grid site admins can access aggregated data of all monitored jobs to infer the site status and to detect job misbehaviour. To remove the last "blind spot" from this monitoring, a remote debugging technique based on the GNU C compiler suite was developed and integrated into the software; its design concept and architecture is described in this paper and its application discussed.

1. Introduction

In October 2009, the Large Hadron Collider (**LHC**) at the European Centre of Particle Physics Research, CERN, Geneva, has begun operation. In its four big experiments, **Alice**, **LHCb**, **CMS** and **ATLAS**, it generates massive amounts of data (about 47 PB) per year.

Readout, transfer, storage and - most importantly - analysis of this amount of data is a computationally intensive, challenging task. Not only single computers, but even whole computer clusters are undersized for such dimensions in computing and storage, and users from all over the world need fast access to the data. To address the enormous scope of this computational challenge, the world-wide LHC Computing Grid, **WLCG**, was created. Being a computing grid as defined by Foster, Kesselman et.al.[1], the WLCG is a distributed computing, data transfer and storage facility, which uses the grid middleware **gLite**[2] for operations.

The computational work is divided into fragments called **grid jobs**, of which large amounts can be managed and run in parallel. The jobs are executed on multiple computing sites called **Computing Elements** (CEs), each providing computing resources in form of clusters and batch systems with computers organized as **Worker Nodes** (WNs), and providing the needed data on storage sites named **Storage Elements** (SEs). In beforehand, the user has no knowledge of where his grid job will actually be executed[3].

[3] however, by specifying requirements, for example selecting CEs by available cores, memory, connectivity or available data fragments, he can take influence in the middleware's decision where to send his job

2. Motivation for a user centric grid job monitor

ATLAS grid jobs usually consist of at least five parts, implemented in several different programming languages and data formats.

(i) **An algorithm generating data or performing calculations on data**

This algorithm (= the "user algorithm") is implemented in C++ and utilizes library functions provided by the ATLAS software framework Athena[3] and analysis tools such as ROOT[4]. It is this code that does the actual analysis work: the generation of physics data (e.g. Monte Carlo or detector simulation), the conversion of this data into other formats for further processing or the analysis of such data. The binaries with these algorithms are either compiled locally on the user's machine and then submitted along the grid job, or they are sent in source code form and compiled directly on a worker node by a special form of grid job (the build job).

(ii) **Code setting up Athena and the environment**

The Athena environment is set up by a collection of shell- and python-scripts. Their main purpose is to set up binary search paths, determine platform specifics and other runtime information.

(iii) **Athena runtime code**

Athena itself consists of several modules written in Python which acquire and prepare the analysis[4] data and then call the user algorithm with the prepared data (e.g. physics event data). For this, the user algorithm is loaded before the job's main event loop execution as a shared library by Athena. The runtime furthermore provides services like data I/O, logging and error handling.

(iv) **A short launcher application, usually implemented as shell-script**

The grid job executable (that is, the executable launched by the grid middleware on the worker node) typically is a short launcher script that loads the Athena environment-setup scripts and triggers execution of Athena. It is either written by the user or automatically created by a job submission and management tool like PanDA[5] or Ganga[6].

(v) **The data that is being processed**

The data usually is stored on the SE located at the site the user job got scheduled to run at. According to the ATLAS computing model, jobs generally are only to be sent to sites providing the needed data locally. The data is accessed by the job by means of mass storage software developed specifically for ATLAS needs, for example **dCache**[7].

To summarize, ATLAS grid jobs are generated or user-written shell scripts running a framework of Python scripts that use algorithms from C++ shared libraries (figure 1), executed on arbitrary grid computing elements with varying available software- and hardware infrastructure, performing calculations on large, distributed datasets in multiple data formats.

The fact that this application model consists of several layers depending on each other, implemented by different groups of developers using different programming languages, already implies a number of possible failure conditions and errors. On top of that, one observes the usual errors contained in all human-written software (invalid data and memory access, floating point exceptions, logical errors, etc) and errors caused by the context, that is, the grid (submission errors, failures to query needed local and remote services, data lookup- and acquisition errors, environment errors and so on). Hence, there is a wealth of failure possibilities from which only a small subset is under control of the user.

[4] or simulation, data generation, conversion, etc.

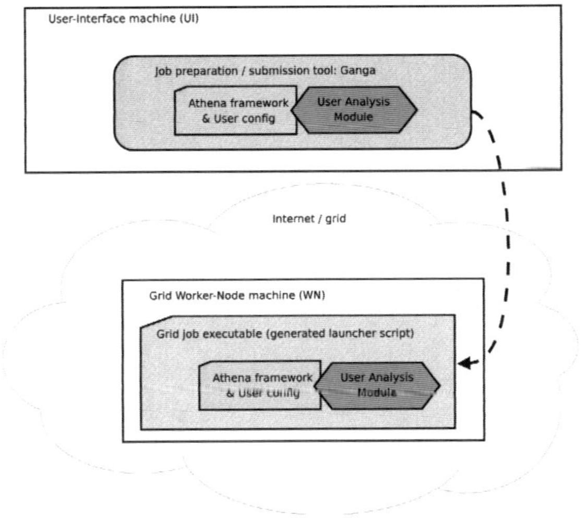

Figure 1: Submission of an Athena-analysis job to the grid using Ganga. The Athena-launcher is wrapped into a job script and uses the binary user analysis module.

To make things worse, a grid job's execution is an intransparent process for the user ("black box"), comparable to a **batch job** on a legacy batch computer: from the point of submission to the termination of the job (successful or not), one has no (or almost no) insight into the job's processing. Non-critical run time errors and invalid output data[5] can only be detected after the job finished. Likewise, the general progress of the job is not known during its run time. Finally, the health of the worker node the job is being executed on (resource usage, service status, etc) is not visible to the user.

The **Job Execution Monitor** (JEM), a user centric grid job monitoring software, has been developed to deal with these problems.

3. The job execution monitor

3.1. Overview

The Job Execution Monitor[8, 9, 10, 11] (JEM) is a distributed application developed to provide real-time worker node health- and job progress-monitoring. JEM provides job status-, job error- and environment data to the user in real-time. Its main module is submitted alongside user jobs to grid computing elements, and is executed in parallel to the user job on its worker node (see figure 2).

During the run of the user job, vital system information on the worker node the job was scheduled to run on, like CPU and memory usage and free disk space in the job's working directory and temporary directories, is gathered periodically. The scripts the user job consists of can be run in a supervised line-by-line mode, during which events like function and method calls, returns from functions and methods, exceptions being raised, or shell commands being executed are logged. Furthermore, log files can be watched and monitoring events can be created on key phrases (like progress annotations, error messages, etc).

All monitoring data gathered during the job run is transmitted in real-time to a messaging broker to allow for direct analysis by the user. The user can either subscribe directly to the

[5] invalid from the user's point of view / semantically invalid

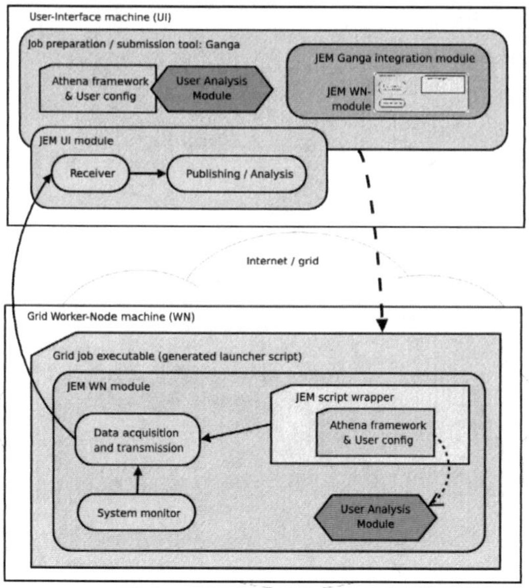

Figure 2: Workflow of job submission with JEM enabled. JEMs worker node module is run in addition to the Athena-launcher. At the same time, on the UI machine, the receiving module of JEM is run, receiving and publishing the monitoring data.

broker, thereby receiving all monitoring data of "his" job in real-time, or he can use JEM-provided summary web pages displaying aggregated monitoring data generated automatically from the data stream transferred via the broker. This analysis may lead to user decisions like aborting and re-submitting the job. "Post-mortem" job failure analysis even if all of the job's output was discarded by the grid middleware as a result of the failure is also possible by referring to the recorded monitoring data.

3.2. Limits of JEM without C tracing

Using the former version of JEM, the supervised line-by-line execution of the user job performed on the WN is performed for bash- and python scripts only. Hence, all of the job's initialization- and Athena management functions, like the environment setup and the staging of needed analysis data, is monitored. Errors in this part of the job run are discovered easily this way. The actual data analysis or generation code, however, cannot be monitored as it is implemented in C/C++ and shipped to the WN as object file or shared library. This user algorithm usually accounts for the biggest part of the job's run time.

So, great parts of the job's run - parts where logical errors might happen which lead to undesired and faulty job results - are only monitored in terms of system metric data. Because of that, such invalid job results are only discovered after the whole job has completed, possibly after several hours (or more) of job run time. If monitoring data would be available also during the execution of the user algorithm, the user might be alerted that the results differ from the expectations and so would have the possibility to abort the job run, correct the algorithm and re-submit it, saving time and increasing resource efficiency.

Similarily, job failures caused by infrastructure problems like depleted resources (disk space, memory), expired grid proxy certificates and the like can be more easily identified if monitoring information is provided during the execution of the user algorithm. Such problems often surface as algorithm crashes (null pointer references, floating point exceptions, segmentation faults etc.)

and are not distinguishable by itself from similar crashes caused by sloppy implementation, unless such monitoring data is available.

Finally, more often than not, grid jobs fail in a manner where the execution of the job just stalls in mid-processing, and the job eventually fails due to exceeded runtime. Usually, also this kind of failure occurs with the job's execution progress inside the binary user analysis, not inside the set-up scripts written in python. In such cases, when the job from a user's perspective "just doesn't continue", or "hangs", the user previously didn't have any possibility to handle the situation apart from just killing and resubmitting the job, hoping that this time the execution will succeed. By using a job execution monitor featuring a binary tracing module, the exact location in code or the exact data file the job stops on can be determined, allowing an analysis of the cause of the job failure.

4. C tracing library

4.1. Overview

To address the issue explained in the preceding section - the absence of monitoring data for the periods of the job's run inside the user algorithm - we developed an addition to JEM allowing to trace the progress of execution in compiled binaries (executables and/or shared libraries).

This library, named C-Tracer, logs every function- or method call and the corresponding return together with its code location, timestamp and associated data (function parameters, local scope variables). This data is then passed to JEM, where it gets inserted into the stream of other monitoring data transmitted by JEM to the UI machine (figure 3). In this way, the job's progress inside the user algorithm is no longer a blind spot for the job run analysis.

To use the C-Tracer, the target binaries must be prepared in a special way by the user, and additional configuration of the grid job is neccessary before submission. These are explained in section 4.6, after the functionality and implementation of the C-Tracer have been discussed in the following sections.

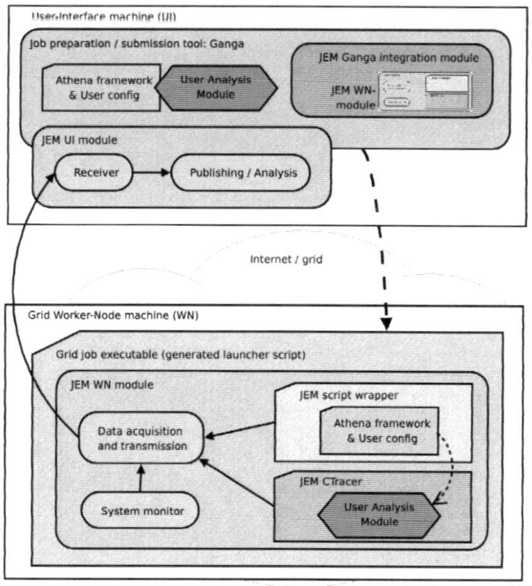

Figure 3: The C-Tracer is added to JEMs WN module and monitors the user analysis module (binary), transparently inserting the gathered monitoring data into JEMs publishing mechanism.

compile flag	description
`-g`	Include debug (DWARF) data in the object file.
`-finstrument-functions`	Add trace system call to each function call and return in the application.

Table 1: Compile flags needed to be passed to gcc when building the user application to prepare it for usage with the C-Tracer.

4.2. GNU C compiler trace hooks

Systems in the grid environment (CEs, WNs, UI machines, etc) almost exclusively run Linux as operating system (mostly Scientific Linux 5 [12], a Red Hat [13] derivate). Consequently, all binaries used in relation with Athena are compiled using the GNU C compiler [14] (GCC).

GCC offers the possibility at compile time (with the compile flag `-finstrument-functions`, see table 1) to instrument every function- or method call and every corresponding return in the built binary with an additional call to two callback functions (figure 4) which then separately must be implemented[15]. The implementation can be done directly within the built binary or, most usefully, in a library dynamically linked to the binary at run time. This is how our C-Tracer uses this feature of GCC.

To the callback functions, the addresses of the called and of the calling function are passed. The pointers to the stack frames of both functions can be determined using another of GCC's built-in functions (figure 5). These information are sufficient to resolve all wanted data about the called and calling functions (code location, arguments and local variables).

JEMs C-Tracer uses this callback system to log every[6] call/return and to publish it the same way as bash commands and python calls. To make the data useful, however, human-readable identifiers are needed, as well as further metadata like symbol types and line number information. These data can be gained by parsing the DWARF[16] debug data compiled into the binaries by GCC when the appropriate flag is appended to the compiler call.

```
void __cyg_profile_func_enter (void *func, void *callsite);
void __cyg_profile_func_exit (void *func, void *callsite);
```

Figure 4: Trace callback functions in binaries compiled by gcc.

```
void * __builtin_frame_address(unsigned int level);
```

Figure 5: Function used to resolve stack frame adresses.

[6] If the function/method called or returned from is contained in an include-list configurable for the C-Tracer to filter unwanted bloat data, e.g. calls inside system libraries like the STL

4.3. Resolving of symbol information

If an application is compiled with GCC set up to include debug symbols, DWARF data is written into the object file. This data contains information about every symbol (function/method, variable, etc) in the application's source code, including (but not limited to):

- The symbol's name
- Its location in source (file, line number)
- Its location in memory when the application is run (absolute address, offset from stack pointer or offset from structure start address for members)
- Its visibility and accessibility mode (for struct members), and its type

Furthermore, every type is listed in the debug data with its name, byte size and possibly linked types (e.g. "pointer to"-associations). Using this information, the monitoring data is augmented with the identifiers used in source code. This enables the user to follow the job's run through the algorithm. Furthermore, it is possible to interpret the algorithm's memory, displaying variable and parameter values and resolving whole structure contents. This allows the user to inspect the data the algorithm is working on, and possibly to discover logical errors caused by (or causing) invalid data in real time. One can call this operation mode of the C-Tracer as "remote debugging", although stopping and stepping-through the user algorithm is not (yet) supported.

4.4. Safe inspection of application memory

Because this data inspection takes place in the user application's working memory, one has to be aware of invalid or corrupt data and, most importantly, invalid or uninitialized pointers and references. To protect the C-Tracer - and with it the user application itself - from crashes caused by invalid memory access operations, special precautions must be taken.

To allow the C-Tracer to inspect user application memory and resolve pointers, a victim/watchdog system was implemented. Knowing that inspecting the user memory may very probably lead to an invalid access (segmentation violation), a dedicated victim thread that "is allowed to crash" is used to perform these accesses. The C-Tracer library, when loading and initalizing, starts a supervising "watchdog" thread, that in turn starts the "victim"-thread.

A piece of shared memory, protected from concurrent access by a mutex semaphore, is used to pass memory addresses to the victim and to return the inspected memory's contents back to the main thread of the C-Tracer monitoring the user application. The victim thread's main function just consists of an endless loop in that the victim waits for a memory address to try to access. If an address is written into the shared memory, the victim tries to copy the contents of the memory at this address into a buffer. If this copy attempt fails due to a protection fault (invalid memory access), a synchronous signal is sent to the victim thread by the Linux kernel, aborting this thread's execution without disturbing the other threads (main thread and watchdog).

The watchdog's task is to restart the victim-thread whenever it crashes. In this case, the watchdog informs the main thread about the failure. Instead of the value at the memory address referenced by the user pointer, a special value declaring the memory as invalid is then passed to JEM. Otherwise, if the memory content was successfully copied into the buffer, this content is published via JEM.

Memory inspected by this means includes the application's stack and heap, and pieces of code memory (static variables). By adding an offset to the base address of a structure (struct, class) in memory, also class members can be inspected. DWARF information, again, is used to resolve these offsets. The number of nested indirections inspected this way by the C-Tracer is limited; the exact number can be configured by the user. This way, endless loops are prevented when inspecting user data structures containing circular references. The structure of the victim/watchdog-system is displayed in figure 6.

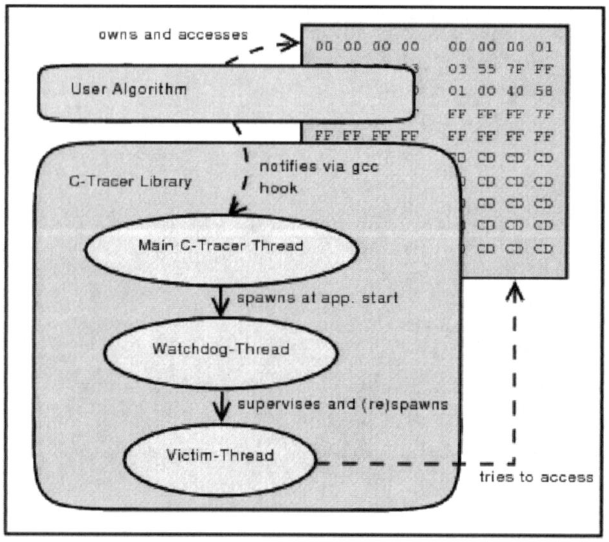

Figure 6: The Victim/Watchdog system's structure.

4.5. Summary of data gathered

Using the C-Tracer, as described in the preceding sections, rich monitoring information is gathered during the user algorithm's execution.

For each function (C-style function or class method, in the following just called "function") that is called, the function's location (file and line number), signature (name, return type and number, type and names of arguments - including an implicit this-pointer if existing) and the time of execution is recorded. If the C-Tracer is configured accordingly, for each argument, also the value is tried to be recorded.

The availability of those data is dependant on several factors. If the target binary has been compiled with optimizations, for example, not all variable data can be resolved. Even in non-optimized binaries, however, some variable content is not resolvable for the C-Tracer (variables assigned to registers instead of memory locations by the compiler). Those variables are denoted as "unresolvable" by the C-Tracer. If, otherwise, the value is available, it is read and inserted into the monitoring data.

From the caller (That is, the function containing the call to the monitored function), the location (again, file and line number), name as well as all local variables with their respective type, name and value are recorded (if available). The values of the local variables are resolved similarly to the function arguments as described above. For each return from a monitored function, the time of execution, the function's location, signature and its local variables are recorded similarly. For those events, as well, the availability of all variable contents is not guaranteed.

The data gained this way can be used to observe the parameters (sub)algorithms are called with, allowing to spot logical errors for example caused by specific input data. As the local variables of the calling method are recorded, the context of those calls is available, as well. This information also includes member variables and methods of user classes in C++ code.

4.6. Usage

In addition to the need to compile the user algorithm in a way providing the needed symbol information and callbacks for the C-Tracer by adding two special compile flags (see tab. 1) to the compiler call, the C-Tracer has to be configured before the user algorithm can be launched. This is done by means of several environment variables the C-Tracer reads and which have to be set beforehand by the user. To have the C-Tracer itself loaded, the dynamic library loader has to be advised to do so before the user application is launched. This, also, is configured just by setting an environment variable.

Since in most cases the user algorithm is used from inside of some launcher script, the environment variables can be set there. If the user job - and with it, JEM and the C-Tracer - is launched via a job submission tool like PanDA or Ganga, the configuration environment variables are set automatically according to the launcher-specific configuration of the job.

While the user job is executed on the foreign worker node, and if JEM has been set up to transmit data in real time, the monitoring data received by the UI-module of JEM is presented to the user in log-files and data files for further analysis; which files exactly are created depends on the local JEM configuration.

4.7. Data acquisition steps and performance impact

Using the C-Tracer adds a significant overhead to the user algorithm's run time. Measures are taken to allow for fine tuning the information gain against the performance impact of the C-Tracer. It is currently under investigation how much the performance of user algorithms degrades if the C-Tracer is used in different configurations and log levels. Already in its current version, the C-Tracer can be configured up to a certain degree, and very basic performance comparisons have already been taken. The performance impact is now explained by taking a detailled look into the processing of a single event (function call) in the C-Tracer.

Besides the natural possibility to skip the loading of the C-Tracer library completely by omitting the dynamic linker directive (`LD_PRELOAD`), it is possible to globally disable the C-Tracer for a user application run. This allowed us to measure the performance impact of just having the tracing library linked to the user application, without gathering any actual monitoring data. This also means no symbols are resolved, no user memory accesses are done and no data is passed to JEM. In this configuration, every call and every return inside the user algorithm causes a call of the tracing routines, with these routines then just immediately returning. This causes a performance degradation, for an empty user function, by a factor of 2.5, while a factor of roughly 3 could be expected for two additional function calls per original function call.

For actual user functions containing algorithm code, the factor is likely to improve, as the per-call overhead of the C-Tracer is constant.

If the C-Tracer was not globally disabled, the first action that is taken is to search the address of the called and calling functions in a whitelist (implemented as balanced red-black tree). If the address is not found there, the event is discarded by the C-Tracer. This mechanism was implemented to be able to aggressively filter unwanted events, such as all events occuring inside system- or thoroughly tested user-libraries, to focus the monitoring on specific user code and keep the performance costs as low as possible.

All other function calls and returns then are processed by looking up symbol-data stored in data structures optimized on fast by-address lookup. If the inspection of variable contents was enabled, this data also is used to discover the memory addresses that need to be accessed by the victim thread (plain memory adresses for heap-stored variables and offsets from the function's frame pointer for variables stored on the stack). The resolved data is written to memory and passed to JEMs data publishing process.

To give some absolute numbers (that are hardly comparable, due to being very test machine specific): Looking up all needed symbol data for a single function call event on average takes in

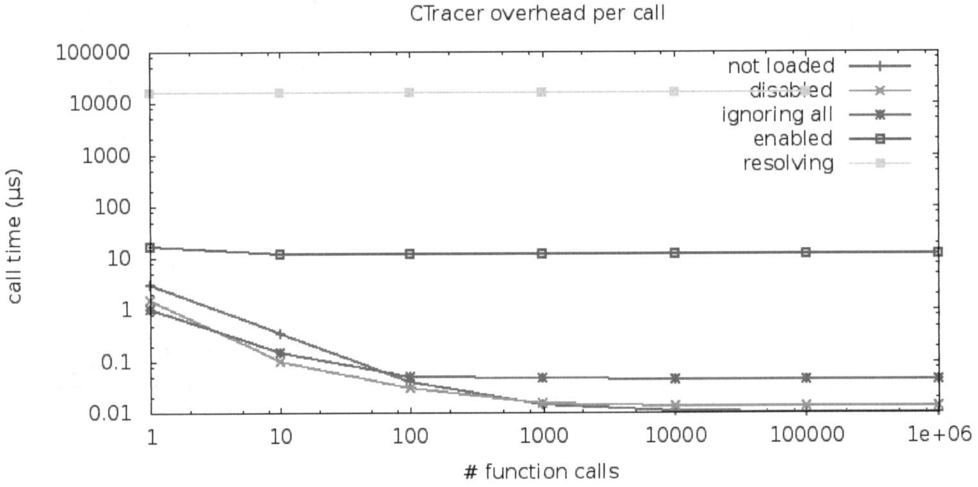

Figure 7: Per-Call overhead of the CTracer in several configurations.

the order of 50 ns. This delay is increased to up to 100 ns if variable values are to be resolved, and an extra 100 ns if a value couldn't be accessed, due to restarting of the victim thread. Similarly, each function return causes delay for monitoring data lookup and memory inspection.

The measurements on the test machine are shown in figure 7. One can see the basic overhead of loading and calling the C-Tracer library decreasing with increasing numbers of function calls. This is to be expected due to the (partial) static overhead nature of this basic functionality. Enabling the C-Tracer, but ignoring 100% of the events (thus causing one address lookup in the search tree per call/return) causes overhead in the order of one order of magniture for sufficiently large numbers of events.

Enabling the monitoring for all events raises the per-call overhead to three orders of magnitude; this, however, is an unlikely scenario, because in real application, only a small subset of the functions is expected to be white-listed. Resolving memory contents is a very expensive operation and will only be feasible for single key events, or on explicit user request (see the outlook in section 5).

Interestingly, the overhead of not loading the C-Tracer library at all is larger than the all-disabled overhead for low numbers of events. This effect is not yet fully understood, but may be a consequence of the Linux dynamic library loader unsuccessfully trying to find a suitable implementation for the trace-callbacks.

5. Conclusion and outlook

We have presented a binary trace application that is integrated into a user-space real-time Grid job monitoring software. This binary tracer, or C-Tracer, adds progress- and user data-monitoring capabilities to the monitoring software, and allows for extension of the transparent portion of the job's execution into compiled binary modules.

By protecting the C-Tracer's memory accesses using a novel victim-watchdog-architecture, tracer crashes affecting user job efficiency and reliability are prevented.

Provision of the data gathered by the C-Tracer into the job monitor's data stream allows for easy correlation with other monitoring data like system metrics, watched log file contents and job control script progress events.

Finally, the application of the C-Tracer to a user job is a straight-forward and uncomplicated operation if the monitoring software itself is already available and enabled. Integration into the two most common job control frameworks in ATLAS - PanDA and Ganga - has been implemented, allowing all ATLAS-based grid users to enable JEM by simply adding a configuration flag to their job submission.

Further development projects are being considered. Performance optimizations aside, several possible feature additions are worth to be considered for the C-Tracer. A user-defined rule base to narrow down the generated data to "interesting" events - by means of a "trigger"-like infrastructure - could increase the C-Tracer's applicability by a large amount.

Further, the possibility to control the C-Tracer's verbosity during the job's run time could be a helpful feature. Development is currently in progress to add a secure control channel from the user (or site admin, expert shifter, etc.) to the job's JEM instance; addition of the C-Tracer to this control channel is strongly being considered.

Such a control channel would allow the C-Tracer to provide in-depth "snapshot" information on demand to the user. This information could contain whole memory dumps, stack traces and call graphs, among other interesting options. By coupling these snapshot functionality to the aforementioned trigger-architecture, the C-Tracer would truly become a "remote debugger".

References

[1] Foster I and Kesselman C, 1999, *The Grid: Blueprint for a New Computing Infrastructure*, Morgan Kaufmann
[2] Laure E et al., 2006, *Programming the Grid with gLite*, in *Computational Methods in Science and Technology*
[3] The Atlas UK Collaboration, *The Athena software framework* http://www.atlas.ac.uk/comp.html
[4] Brun R, Rademakers F, 1996, *ROOT - An Object Oriented Data Analysis Framework*, (AIHENP conference, Laussanne) http://root.cern.ch
[5] The PanDA Production and Distributed Analysis System https://twiki.cern.ch/twiki/bin/view/Atlas/PanDA Accessed August 2010
[6] Harrison K et al., 2006, *Ganga: a Grid user interface for distributed analysis*, (Proc. Fifth UK e-Science All-Hands Meeting, Nottingham, UK, 18th-21st September, Ed. Simon J. Cox (National e-Science Centre), pp 518-525) http://ganga.web.cern.ch/ganga
[7] Ernst M, Fuhrmann P, Mkrtchyan T, Bakken J, Fisk I, Perelmutov T, and Petravick D, 2005 *Managed data storage and data access services for data grids*, (*Computing in High Engergy and Nuclear Physics*, 2004 (Aimar A, Harvey J, and Knoors N, eds.)), (Geneva), CERN
[8] Igdalov D, 2005, *Entwicklung eines Systems zur Analyse und Überwachung der Verarbeitung von Rechenanforderungen im LHC Computing-Grid*, Diploma thesis, Fachhochschule Niederrhein
[9] Hammad A, 2005, *Entwicklung eines Überwachungssystems für verteilte Prozesse im LHC-Computing Grid*, Master thesis, Fachhochschule Niederrhein
[10] München T et al., 2008, *Job centric monitoring for ATLAS jobs in the LHC Computing Grid*, (XII International Workshop on Advanced Computing and Analysis Techniques in Physics Research (ACAT) 2008, Ettore Majorana Foundation and Centre for Scientific Culture, Erice, Sicily)
[11] Igdalov D, Hammad A, Borovac S, Mechtel M, München T et.al., *JEM, The Job Execution Monitor* https://svn.grid.uni-wuppertal.de/trac/JEM
[12] Scientific Linux 5 Distribution http://www.scientificlinux.org
[13] RedHat Linux Distribution http://www.redhat.de
[14] GNU foundation *The GNU C Compiler*. http://gcc.gnu.org
[15] Stallman R et al., 2004, *Using the GNU Compiler Collection*, GNU GCC 3.4.6 reference manual http://gcc.gnu.org/onlinedocs/gcc-3.4.6/gcc.pdf
[16] The DWARF Standards Committee, 2006, *The DWARF Debugging Format Standard*, http://dwarfstd.org

Journal of Physics: Conference Series **368** (2012) 012013 doi:10.1088/1742-6596/368/1/012013

Monitoring the Grid at local, national, and global levels

P D Gronbech [1]

1 Department of Physics, University of Oxford, Denys Wilkinson Building, Keble Road, Oxford OX1 3RH, UK

Email: p.gronbech1@physics.ox.ac.uk

Abstract. The World-wide LHC Computing Grid is a global infrastructure set up to process the experimental data from the experiments at the Large Hadron Collider located at CERN. The UK component is provided by the GridPP project across 19 sites at the universities and Rutherford Lab. To ensure that these large computational resources are available and reliable requires many different monitoring systems, ranging from local site monitoring of individual components, through UK-wide monitoring of Grid functionality, to the worldwide monitoring of resource provision and usage. In this paper we describe the monitoring systems used for the many different aspects of the system, and how some of them are being integrated together.

1. Local site monitoring

Local GridPP [1] site monitoring covers cluster load, batch system utilization, network bandwidth monitoring and fault condition monitoring. Ganglia [2] is the most common software used to monitor a cluster figure 1. It is easily installed on clients and allows data to be collected on a master node and displayed via a web server.

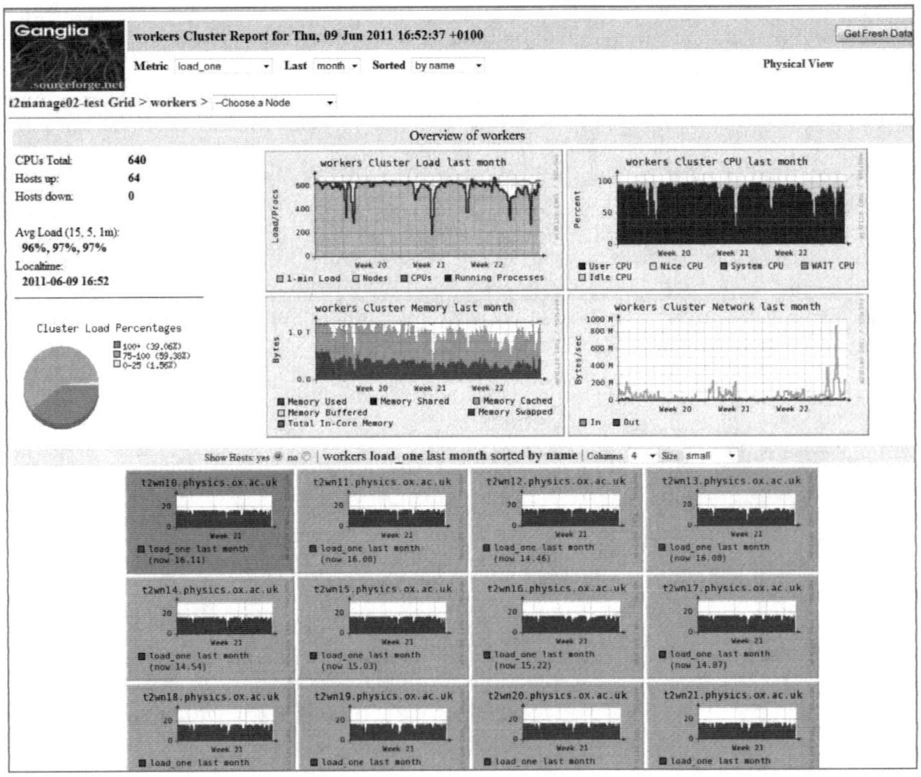

Figure 1. Ganglia monitoring showing a group of Worker Nodes.

Ganglia monitors and collects many parameters, such as load, memory used, network activity etc. The top part of the page shown in figure 1 shows an aggregate picture across the nodes in a particular logical group, such as worker nodes where the main priority is to ensuring they are busy. The web interface allows other aggregate parameters to be selected or the full details for any individual node can be displayed. The data are recorded, allowing time periods from an hour up to a year to be displayed. This can be extremely useful for spotting trends over time. Other parameters can be added to the Ganglia monitoring setup, but many sites are satisfied by the default setup.

Monitoring specific to the batch system used at a site is also typically used. Many GridPP sites use the torque batch system [3] (developed from PBS) monitored with pbswebmon [4], which provides a graphical way to monitor the occupancy of the cluster, and the job shares and efficiencies for each user figure 2. This package clearly shows which users jobs are running, something the ganglia programme does not do without modification. The efficiency of the job is also reported which can guide a systems administrator to investigate jobs which may be suffering through lack of I/O bandwidth to the disks or the network.

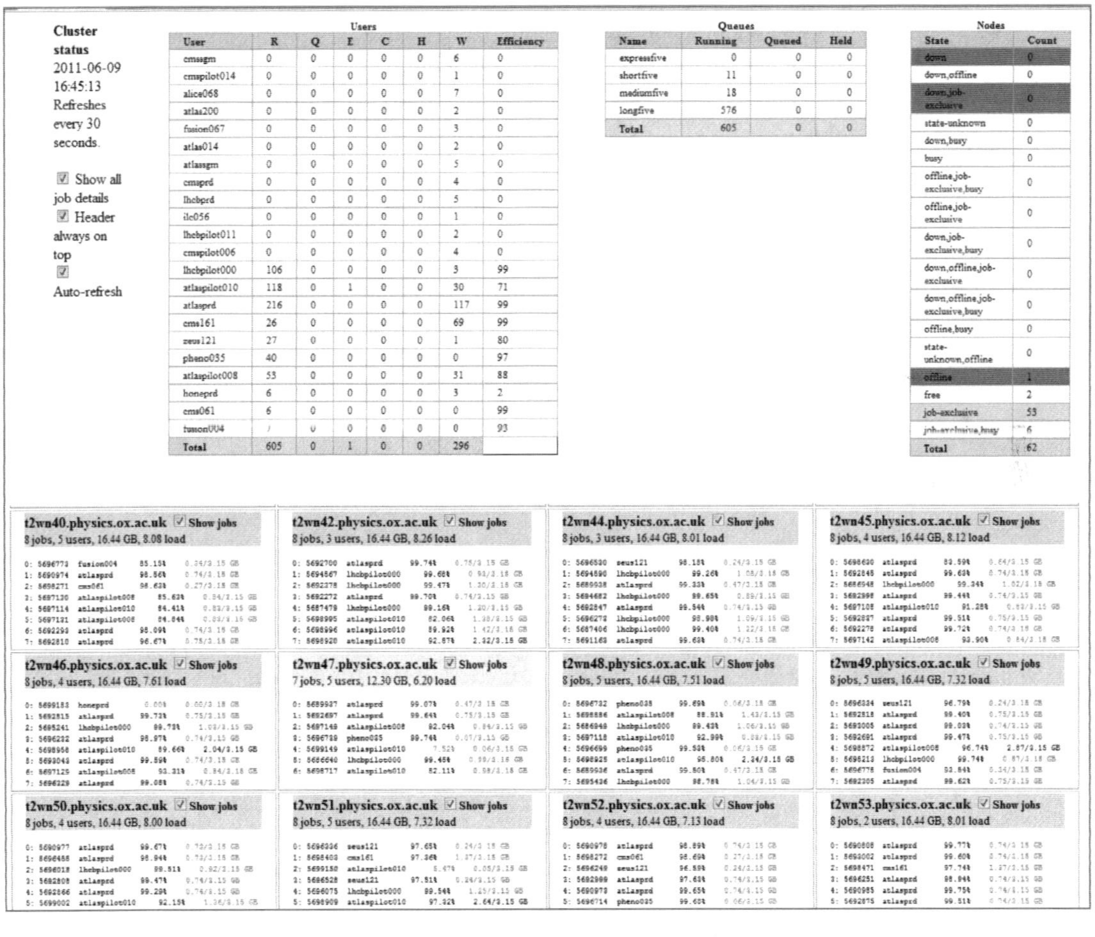

Figure 2. pbswebmon.

For security reasons it is important for sites to keep their systems package rpms up to date. Switching on automatic "yum" updates can help this, but for some critical packages this could break something so it is often not done. Similarly Kernel updates require a reboot and this is often delayed to avoid downtime. A useful way of seeing how many outstanding updates are required on a particular system is to run Pakiti [5]. This package, originally known as yumit, was developed by GridPP staff at the Rutherford Appleton Lab. On each client it effectively runs a "yum check-update" command to list any packages that have new updates available. The results are stored on a database on the Pakiti server and displayed on a web page figure 3. The graphical display allows systems administrators to spot systems that have failed to pick up expected updates, or are behaving differently to that expected.

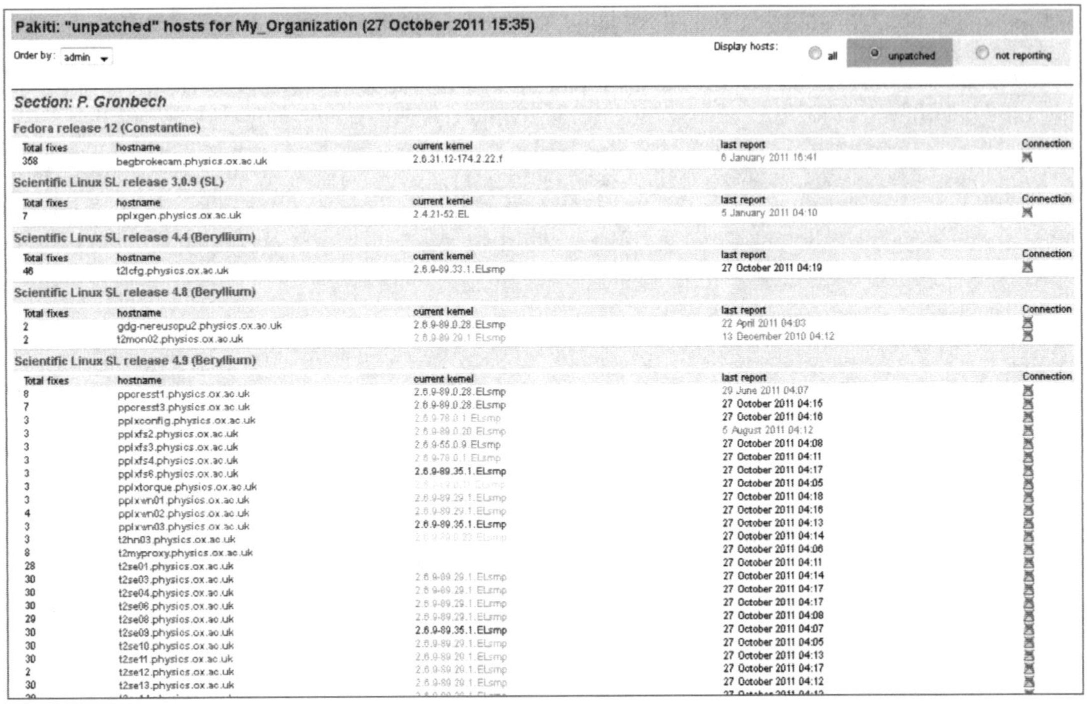

Figure 3. Pakiti.

The above three packages display results via a web interface that must be actively monitored. However, systems administrators want to be notified by an email or SMS message when systems start to fail for either hardware or software reasons. Nagios [6], which provides a very powerful framework, can be used to monitor the status of systems by running tests at specified intervals and then performing actions depending on the results. This could be emailing a warning message or running an event handler that takes remedial action to solve a problem. The advantage of Nagios is that it removes the need to monitor web pages when all is well but can provide notification via email, or SMS (as well as the web) when there is a problem. There are repositories of checking scripts that can be used to build up a customized array of tests for each system. If a specific check is not already available it is relatively simple to create it: For example to check the health of a particular hardware RAID array controller on a storage server or to check a particular piece of Grid middleware is running properly. Each time a new problem occurs the systems administrator can enhance the local Nagios monitoring, to watch for this new failure mode. Simple examples may be local disk partitions getting dangerously full, or a more sophisticated batch systems "black hole spotter". A black hole, in this context, describes a batch worker (CPU node), which develops a fault such that it accepts new jobs from the queue, then immediately fails them and then takes another job. In this situation all the jobs on the queue can very quickly be drained via this node and no further jobs will get run by the cluster, as that worker node is always apparently free.

Network health, usage and bandwidth should also be monitored at sites. Many sites use Cacti [7] and/ or Network Weathermap [8] to monitor the switches in and around their clusters. The switch connecting a cluster to the outside world is of particular relevance to Grid sites as almost all the work comes from other WLCG [9] sites around the world, and only a minority from the sites own local or Tier 3 facilities. Monitoring the usage of this link helps the site to know if they have enough bandwidth to feed their cluster. If it is continuously saturated for many days then thought should be given to increasing the site connection to JANET [10]. (UK sites are mostly connected at 1Gbps with a few sites at 2-5Gbps and a couple at 10Gbps. Over the next couple of years most sites are expected to move to approximately 2-5Gbps). Cacti can also be setup to monitor other systems that support the Simple Network Management Protocol (SNMP), such as the Power Distribution Units in each rack figure 4.

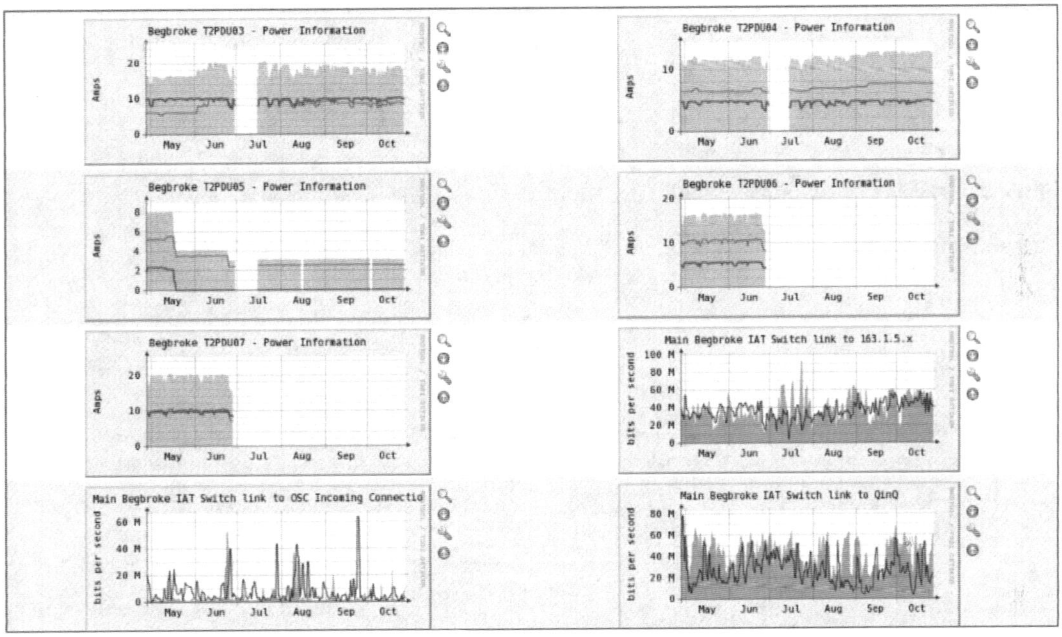

Figure 4. Cacti being used to monitor power and network switch traffic.

2. National testing

Although Cacti monitors the traffic actually flowing, it does not provide a measure of the available bandwidth between sites. The Gridmon [11] project was setup to address this problem and a dedicated 'Gridmon' test box is installed at each of the 19 GridPP sites to perform a matrix of iperf, udpmon and other network throughput tests. Each site performs tests against the other sites in its distributed Tier 2 and the Tier 1. The results are stored on a central database with a web frontend figure 5. This provides a useful history of the available bandwidth and can clearly show when the network gets degraded allowing problems to be diagnosed. A reduction in bandwidth can be easily seen which could be caused by a rate-cap that has been applied, or a faulty piece of networking equipment

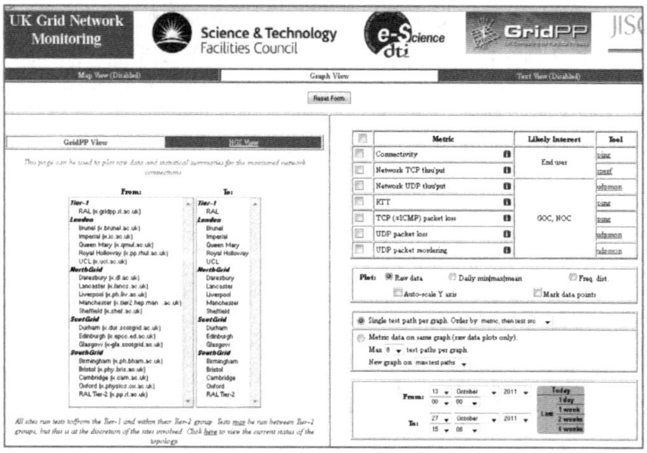
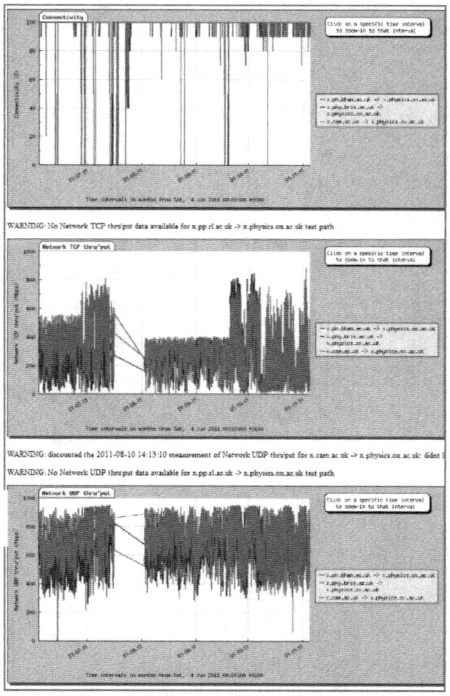

Figure 5. Gridmon Network Bandwidth monitoring.

Other UK wide testing includes a GridPP-developed summation of relevant WLCG tests coupled with dedicated UK tests [12] developed by Prof. S. Lloyd at QMUL figure 6. This started out by just summarizing the Global Service Availability Metrics for the UK sites, but has expanded to include experimental tests and customized tests required by the UK.

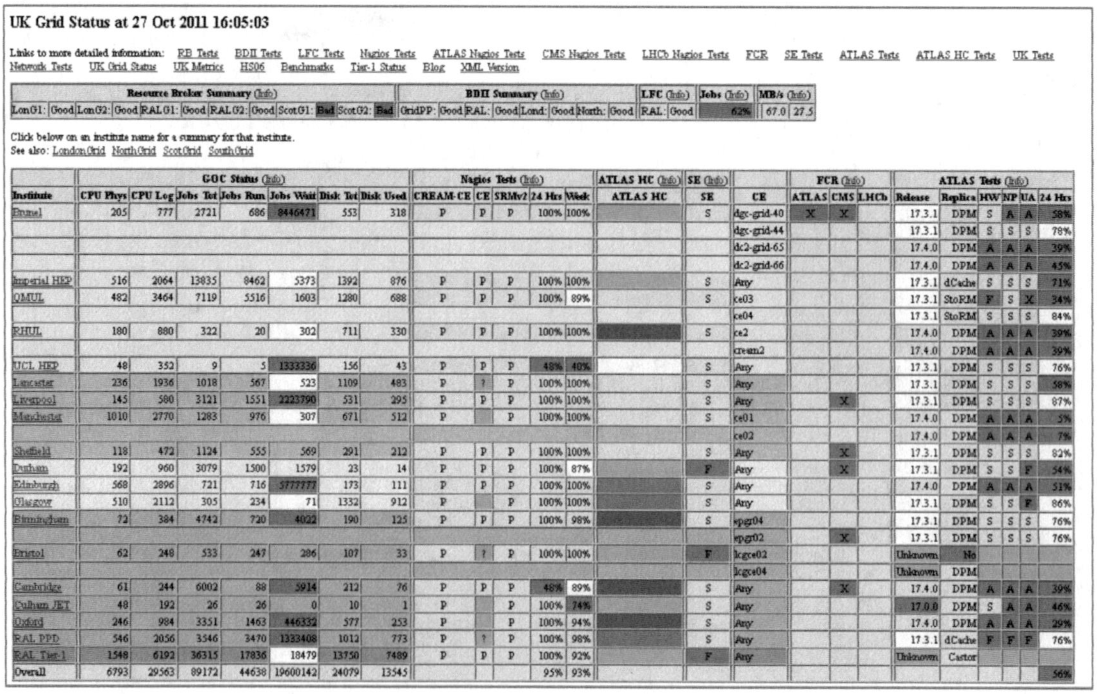

UK Grid Status at 27 Oct 2011 16:05:03

Links to more detailed information: RB Tests BDII Tests LFC Tests Nagios Tests ATLAS Nagios Tests CMS Nagios Tests LHCb Nagios Tests FCR SE Tests ATLAS Tests ATLAS HC Tests UK Tests Network Tests UK Grid Status UK Metrics HS06 Benchmarks Tier-1 Status Blog XML Version

Resource Broker Summary (Info)												BDII Summary (Info)								LFC (Info)		Jobs (Info)	MB/s (Info)	
LonG1:	Good	LonG2:	Good	RAL G1:	Good	RAL G2:	Good	ScotG1:	Bad	ScotG2:	Bad	GridPP:	Good	RAL:	Good	Lond:	Good	North:	Good	RAL:	Good	62%	67.0	27.5

Click below on an institute name for a summary for that institute.
See also: LondonGrid NorthGrid ScotGrid SouthGrid

	GOC Status (Info)							Nagios Tests (Info)					ATLAS HC (Info)	SE (Info)		FCR (Info)			ATLAS Tests (Info)					
Institute	CPU Phys	CPU Log	Jobs Tot	Jobs Run	Jobs Wait	Disk Tot	Disk Used	CREAM-CE	CE	SRMv2	24 Hrs	Week	ATLAS HC	SE	CE	ATLAS	CMS	LHCb	Release	Replica	HW	NP	UA	24 Hrs
Brunel	205	777	2721	686	8446471	553	318	P	P	P	100%	100%		S	dgc-grid-40	X	X		17.3.1	DPM	S	A	A	58%
															dgc-grid-44				17.3.1	DPM	S	S	S	78%
															dc2-grid-65				17.4.0	DPM	A	A	A	39%
															dc2-grid-66				17.4.0	DPM	A	A	A	45%
Imperial HEP	516	2064	13835	8462	5373	1392	876	P	P	P	100%	100%		S	Any				17.3.1	dCache	S	S	S	71%
QMUL	482	3464	7119	5516	1603	1280	688	P	P	P	100%	89%		S	ce03				17.3.1	StoRM	F	S	X	34%
															ce04				17.3.1	StoRM	S	S	S	84%
RHUL	180	880	322	20	302	711	330	P	P	P	100%	100%		S	ce2				17.4.0	DPM	A	A	A	39%
															cream2				17.4.0	DPM	A	A	A	39%
UCL HEP	48	352	9	5	1333336	156	43	P	P	P	48%	40%		S	Any				17.3.1	DPM	S	S	S	76%
Lancaster	236	1936	1018	567	523	1109	483	P	?	P	100%	100%		S	Any				17.3.1	DPM	S	S	S	58%
Liverpool	145	580	3121	1551	2223790	531	295	P	P	P	100%	100%		S	Any		X		17.3.1	DPM	S	S	S	87%
Manchester	1010	2770	1283	976	307	671	512	P		P	100%	100%		S	ce01				17.4.0	DPM	A	A	A	5%
															ce02				17.4.0	DPM	A	A	A	7%
Sheffield	118	472	1124	555	569	291	212	P	P	P	100%	100%		S	Any		X		17.3.1	DPM	S	S	S	82%
Durham	192	960	3079	1500	1579	23	14	P	P	P	100%	87%		F	Any		X		17.3.1	DPM	S	S	F	54%
Edinburgh	568	2896	721	716	5777777	173	111	P	P	P	100%	100%		S	Any				17.4.0	DPM	A	A	A	51%
Glasgow	510	2112	305	234	71	1332	912	P		P	100%	100%		S	Any				17.3.1	DPM	S	S	F	86%
Birmingham	72	384	4742	720	4022	190	125	P	P	P	100%	98%		S	epgr04				17.3.1	DPM	S	S	S	76%
															epgr02		X		17.3.1	DPM	S	S	S	76%
Bristol	62	248	533	247	286	107	33	P	?	P	100%	100%		F	lcgce02				Unknown	No				
															lcgce04				Unknown	DPM				
Cambridge	61	244	6002	88	5914	212	76	P	P	P	48%	89%		S	Any		X		17.4.0	DPM	A	A	A	39%
Culham JET	48	192	26	26	0	10	1	P		P	100%	74%		S	Any				17.0.0	DPM	S	A	A	46%
Oxford	246	984	3351	1463	446332	577	253	P		P	100%	94%		S	Any				17.4.0	DPM	A	A	A	29%
RAL PPD	546	2056	3546	3470	1333408	1012	773	P	?	P	100%	98%		S	Any				17.3.1	dCache	F	F	F	76%
RAL Tier-1	1548	6192	36315	17836	18479	13750	7489	P	P	P	100%	92%		F	Any				Unknown	Castor				
Overall	6793	29563	89172	44638	19600142	24079	13545				95%	93%												56%

Figure 6. UK Grid Status.

The UK regional Nagios based Service Availability Monitoring (SAM) [13] is run by Oxford University. This service queries a central database (GOCDB) and Grid information services to create a list of sites and systems to be tested. The services offered are tested and the results of the tests are sent via an active MQ message bus to the EGI Regional Operations Dashboard [14] figure 8. Each region has an operator on duty that can raise alarm tickets against sites that have failed critical tests. All EGI [15] and WLCG sites have to meet agreed SLA's, to respond to such tickets within a defined time depending on their tier status. The nagios web interface figure 7 can be used by Systems Administrators to investigate the error message resulting from the failed test. An alternative interface known as MyEGI is also available on the same server [16] that provides historical plots of availability.

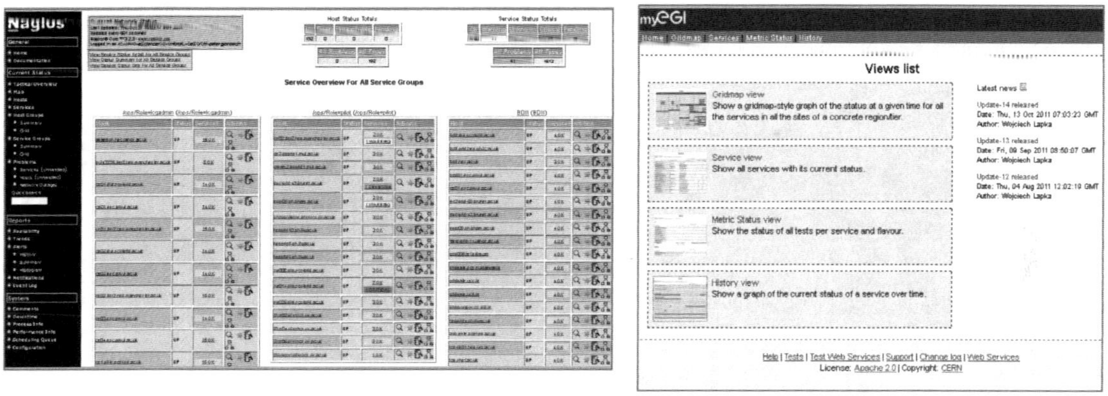

Figure 7. Gridppnagios and the MyEGI portal.

Figure 8. Regional Operations Portal.

3. Global monitoring

There are many monitoring pages that look at all the European Grid Infrastructure (EGI) grid, or focus particularly on the WLCG grid.

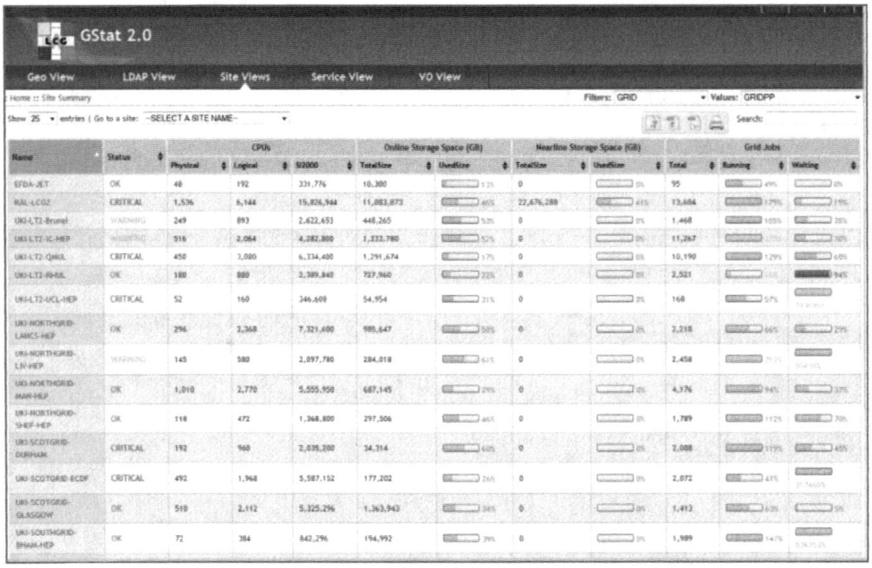

Figure 9. GSTAT.

The GOCDB (Grid Operations Central Database) [17] registers details of all sites, their production and certification status and the contacts and services hosted there. Each site publishes the same information plus additional real time attributes via a site-BDII using LDAP. This can be seen via LDAP queries or by using the GSTAT [18] web pages figure 9. The GOCDB lists what is expected to be there and the GSTAT shows what is actually there, if there is a difference there may be a problem and this would be shown up by the Regional monitoring. An extension to GSTAT is the WLCG REBUS (Resource, Balance and Usage) [19] systems that makes use of published capacities and compares them against the pledged resources to the LHC [20] experiments, over time. This can highlight sites that have failed to provide what they promised.

The actual work done at a site is measured by the Accounting system APEL [21], each night batch system logs are processed and correlated with the grid middleware logs to publish the number of jobs, and CPU time they took for each Virtual Organization. This is all collected together in a database and can be accessed by a very flexible web interface[22], which can show details of work carried out at individual sites, distributed Tier2s or by country, and can show the variations across time and users.

The main experiments at CERN submit jobs to the sites, with sophisticated job submission frameworks. The four LHC experiments dashboards now have a common frontend web page [23]. The dashboards track the jobs and report job submission, running, success or failure. Sites that have high failure rates are ticketed by the experiment team shifters, and can be automatically black listed from receiving new jobs.

4. Integrated dashboards

Systems Administrators are often overwhelmed by the number of different web sites and monitoring systems they should track. Attempts to integrate output from several systems into a site dashboard have been made at the Tier 1 figure 10 and some of the larger sites figure 11.

The RAL Tier1 Dashboard provides an overview of the Tier1 status. It was originally conceived as a dashboard to provide information externally (to the users). In practice it has proven an invaluable tool for internal monitoring as well. The dashboard pulls together a number of different sources of information. Several key ganglia plots provide trend information for batch and storage. Summaries of both the current SAM tests status and entries in the GOC DB are obtained via php scripts that request the information in XML and parse the result. A list of disk servers in intervention is provided by extract from the local database used to manage these systems. In addition the dashboard provides an area where Tier1 staff can display messages which are also copied to Twitter.

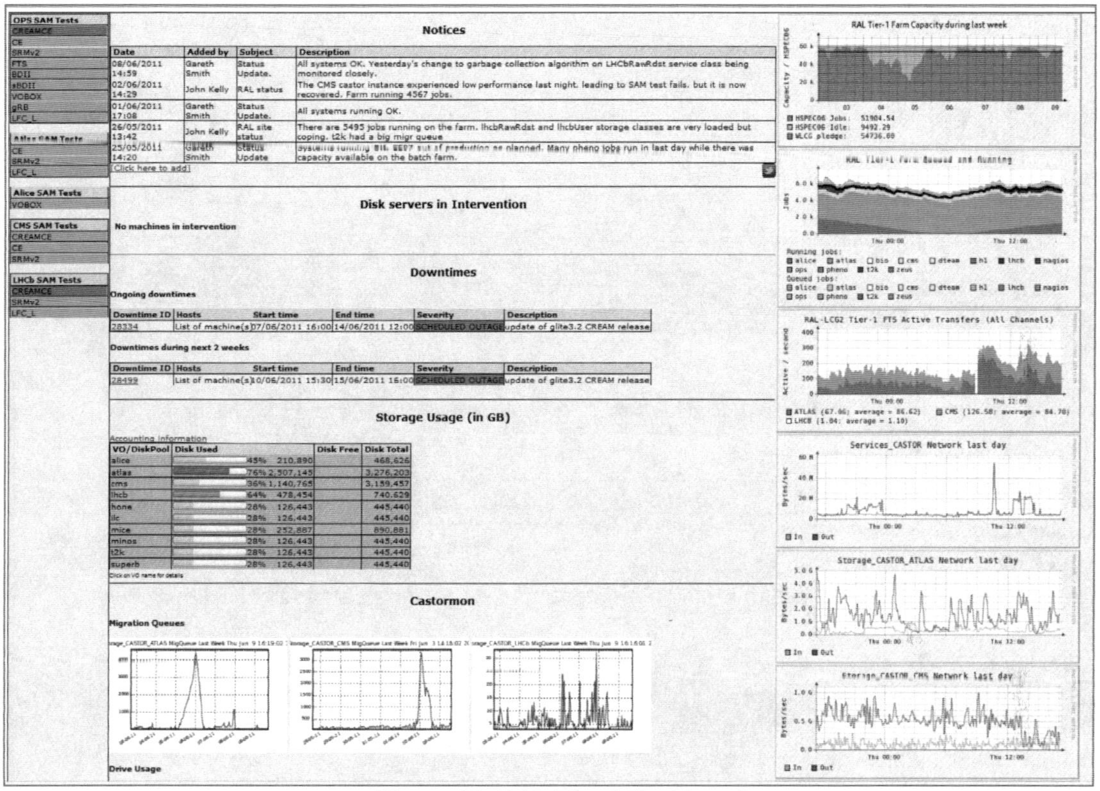

Figure 10. RAL Tier 1 Dashboard.

The Oxford dashboard is based on php code originally developed at Glasgow. It web scrapes the tables from Steve Lloyds SAM summary pages, the ganglia WN load graph, and some experiment dashboards. In addition as the computer room is in a remote location we find having a live camera feed useful. By showing the most relevant information on a single page that can be on display continuously in the support staff room, it can draw attention to faults at a glance.

5. Conclusions

Some of the many monitoring systems that are currently in use have been described. All areas; Fabric Monitoring, Grid Software; Experimental Job Submission and Accounting are continuously evolving. All providing views appropriate to each users particular interest. Sites running production resources have to take all of these into account. Systems administrators will use a combination of increasing the number of tests that local nagios monitoring systems check for and rectify automatically, coupled with use of site dashboards to reduce the time required to maintain an understanding of the health of their sites.

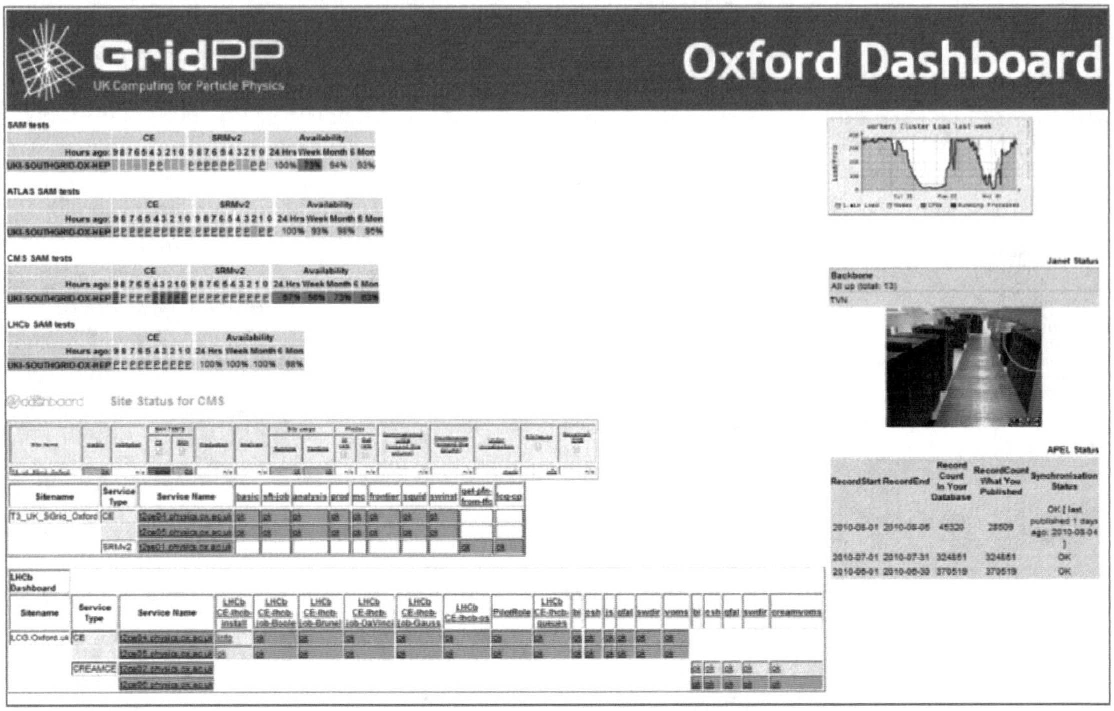

Figure 11. Oxford Site Dashboard.

Acknowledgements
This work is carried out on behalf of and with the support of the GridPP project [1].

References

[1] GridPP: the UK grid for particle physics D. Britton et al, UK e-Science All Hands Conference, Phil. Trans. R. Soc. A June 28, 2009 367:2447-2457; doi:10.1098/rsta.2009.0036
[2] Ganglia http://ganglia.sourceforge.net/
[3] Torque http://www.adaptivecomputing.com/products/torque.php
[4] Pbswebmon http://sourceforge.net/apps/trac/pbswebmon/wiki
[5] Pakiti http://pakiti.sourceforge.net/
[6] Nagios http://www.nagios.org/
[7] Cacti http://www.cacti.net/
[8] Network Weathermap http://www.network-weathermap.com/
[9] WLCG http://lcg.web.cern.ch/lcg/
[10] JANET http://www.ja.net/
[11] Gridmon http://gridmon.dl.ac.uk/gridmon/graph.html
[12] UKGrid Monitoring http://pprc.qmul.ac.uk/~lloyd/gridpp/ukgrid.html
[13] Distributed Nagios based SAM testing https://www.gridpp.ac.uk/wiki/UKI_WLCG_Regional_Nagios
[14] Regional Operations Dashboard https://operations-portal.in2p3.fr/dashboard
[15] EGI http://www.egi.eu/
[16] gridppnagios https://gridppnagios.physics.ox.ac.uk/nagios/ and https://gridppnagios.physics.ox.ac.uk/myegi
[17] GOCDB http://goc.egi.eu/
[18] GSTAT http://gstat-prod.cern.ch/gstat/summary/EGEE_ROC/UK/I/
[19] WLCG REBUS http://gstat-wlcg.cern.ch/apps/capacities/sites/
[20] The Large Hadron Collider http://lhc.web.cern.ch/lhc/

[21] APEL Accounting https://wiki.egi.eu/wiki/APEL
[22] Accounting Web Portal http://www4.egee.cesga.es/accounting/egee_view.html
[23] CERN Experimental Dashboards http://dashboard.cern.ch/

Journal of Physics: Conference Series **368** (2012) 012014 doi:10.1088/1742-6596/368/1/012014

Making distributed ALICE analysis simple using the GRID plug-in

A Gheata[1], M Gheata[1,2]
[1] Europeean Organization for Nuclear Research (CERN) - Geneva, Switzerland
[2] Institute for Space Sciences – Bucharest, Romania

E-mail: andrei.gheata@cern.ch

Abstract. We have developed an interface within the ALICE analysis framework that allows transparent usage of the experiment's distributed resources. This analysis plug-in makes it possible to configure back-end specific parameters from a single interface and to run with no change the same custom user analysis in many computing environments, from local workstations to PROOF clusters or GRID resources. The tool is used now extensively in the ALICE collaboration for both end-user analysis and large scale productions.

1. The ALICE computing model and data analysis

At a rate of up to *1.25 GBytes/second* data storage, the *ALICE* experiment has the biggest event size (tens of MBytes for Pb-Pb events) among the *LHC* experiments. This justifies the specific choices made for the *ALICE* computing model [1]. For scalability reasons, the reconstructed event summary data (*ESD*) is "democratically" replicated on the available distributed disk storages. The size of the data encouraged using a "pull" model in the *ALICE GRID* middleware, running jobs rather close to the data than on fixed sites and making *TIER 1* and *TIER 2* centers equivalent as role played in the data analysis chain. Moreover, the same resources are used for reconstruction, calibration and analysis tasks. All these facts, combined with a very ambitious physics program, are creating quite challenging scalability problems that had to be addressed by the analysis framework.

2. The analysis framework

The *ALICE* analysis framework was described in detail in a different paper [2]. In this section we just give a short overview of the components that are relevant for the way distributed analysis is steered. An *ALICE* analysis module is seen as an independent task that shares with others the main event loop steered by the framework. In fact, the event loop mechanism is provided by the *ROOT* framework [3] and the *ALICE analysis manager* is just a wrapper around the *ROOT TSelector* class. The advantage of this choice is that our framework can use directly the *ROOT* built-in tools for looping tree entries, including the formalism defined by the *TSelector* class [4] for the phases of data processing:

- Local initialization phase (corresponding to *TSelector::Init*) - allows initialization of the custom data structures for the analysis task. This phase is typically used to configure the analysis code.
- Remote initialization phase (corresponding to *TSelector::SlaveBegin*) – allows initializing the custom output objects produced by the analysis task (like booking histograms)

- Event processing phase (corresponding to *TSelector::Process*) - allows executing the custom task algorithm per event. This can be used as entry point for a specific user analysis framework
- Merging phase – This applies for all systems that allow parallel computing, where the workers produce only partial results that have to be merged. This phase is not directly defined by the *TSelector* interface but has to be handled by the framework.
- Post-processing phase (corresponding to *TSelector::Terminate* - allows any post-processing operation (like normalization, fitting) performed on the merged results

The analysis manager adds a specific model on top of the event loop, requiring that the user analysis follows the interface defined by a base analysis class (*AliAnalysisTask*). Using this interface, all analysis tasks registered to the manager are sequentially called in the different stages of processing. Analysis tasks are only allowed to communicate via output data, so that direct dependencies are avoided. This allows running together practically any combination of user analysis algorithms, taking profit from having the large-sized input data in memory. The analysis manager class acts here like a locomotive for a train having many analysis task wagons, while the deployment of the code, the data and merging services are provided.

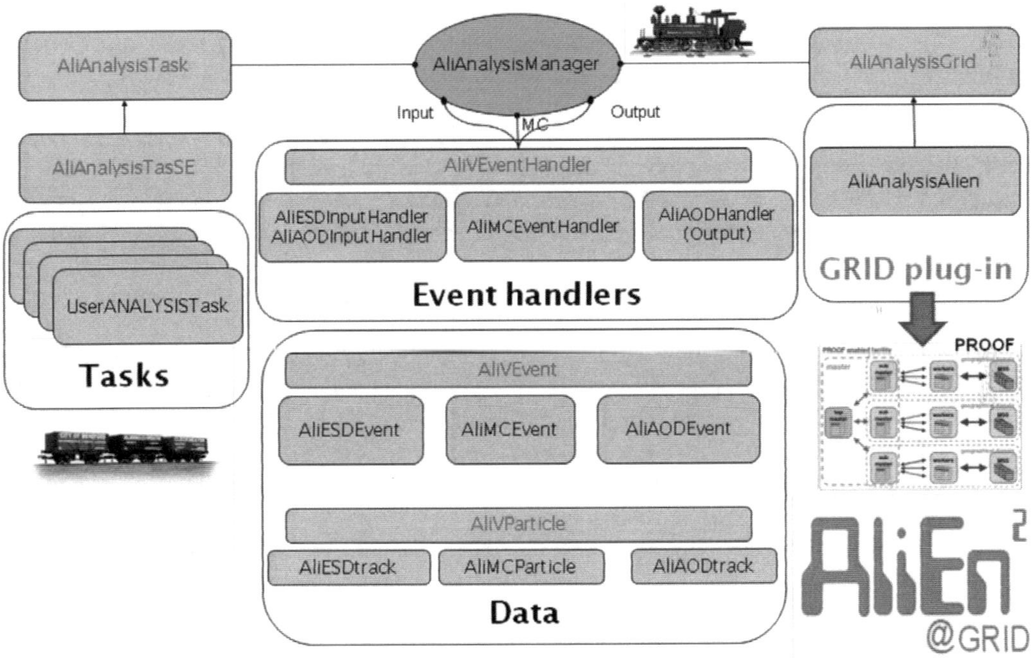

Figure 1. The *ALICE* analysis framework. The central role is taken by the analysis manager that steers the main event loop for a train of analysis tasks. The data services are provided via a base event handler class and specific data handlers. The distributed analysis is implemented as a plug-in that hides the complexity of the *PROOF* and *GRID* back-ends.

The reconstructed events are made available for analysis as event summary data, which in *ALICE* are rather large data structures containing information which is in many cases detector-specific. The *ESD* files can be skimmed or specialized to more compact formats (*AOD*, mini-*AOD*) for more specific analysis usage. The *ALICE* event model [5] derives from a base virtual class that makes the analysis transparent to the existing specific event formats. The same applies at the level of tracks, vertices or other structures. Similar approaches have been used by few other experiments (like STAR and RHIC) in order to make it possible to use generic data handlers to dispatch data to many analysis clients. The ALICE analysis framework [2] extended this concept in the sense that any analysis task can become at its turn a data producer and dispatch this newly computed data to an arbitrary number of clients. The train of tasks presented in figure 1 is just the trivial use case of the framework, which can propagate not only the input event, but also analysis results in a this data-driven tree of tasks.

One can include analysis wagons to an existing train (analysis manager) using custom ROOT macros. This is done via a simple standard procedure which is now a mandatory condition to be included in any train. The analysis trains were usually deployed using specific *GRID* or *PROOF* [6] setups. To make this operation simple and fully transparent for the user code we developed the so called **GRID plug-in**, described in detail in the following sections.

3. Distributed analysis in ALICE

There are two main use cases for distributed data analysis in *ALICE*. One can use dedicated *PROOF* clusters [7] to process a reasonable amount of data (like few millions events) with a fast turn-around cycle. This is quite useful to get to look into fresh new data, perform reconstruction and calibration tests or develop preliminary analysis cuts. A *PROOF* analysis session looks very much like a local one and uses the same analysis code. There are however specific settings that have to be made using the *TProof* interface in *ROOT* for: connecting to the *PROOF* cluster, staging/using datasets and configuring cluster parameters.

Once a certain analysis needs to run over a big part or the totality of the *ALICE* data in a period, it has to be deployed in the *ALICE GRID*. As described in the first section, distributed disk storage and computing resources are used here according to a "democratic" model. The system that allows running batch jobs on these resources is called **AliEn** (***Ali****ce* ***En****vironment*) [8]. It uses a specific job description language (*JDL*) to define: the software libraries to be used, the user executable code, the input data, the job splitting and other job parameters, together with the job output.

In practice, a user had to go through several steps before being able to run his analysis on the *GRID*: he/she needed to write the analysis code as an analysis task (as mentioned in the previous section), test and debug the code on local data, write the *JDL*, create a data set within the *AliEn* shell and copy all files required in the job input box. All this had to be done by keeping in sync the local and *GRID* software versions, monitoring the jobs and resubmitting in case of errors. The whole procedure (now obsolete) required a special training of rather qualified users. Gaining experience with the system was critical for the success rate.

A first step in making the system more user-friendly was the introduction of a web monitoring interface based on *Monalisa* [9]. This provides web interfaces for browsing and uploading files in the *AliEn* file catalog, monitoring the job status and user quotas, checking the job efficiency or inspecting central productions.

Efficiency is an important aspect for getting results out of a distributed system. The size of *ALICE* events encourages the use of CPU intensive analysis, which can be done by using big trains of analysis tasks, at the cost of extra (and sometimes limiting) memory requirements. Large productions have to be prioritized with respect to single user analysis. The life cycle of a given analysis passes through local and *PROOF* tests, fine tuning of cuts on the *GRID*, then entering when mature enough in a central production train conducted by one of the *ALICE* physics working groups. The whole procedure was not straightforward and required a certain effort and time to conclude, so development of central services and tools to shorten it became a priority.

4. The AliEn plug-in

To hide the complexity of the *AliEn* system from the user we have developed a utility class to describe in a general way the requirements of a given analysis job. The idea was to provide an *API* to allow defining in a simple key-value manner all the settings that are specific to the *GRID* job like: time to live (*TTL*), software versions, dataset splitting or run numbers to be processed. Based on these settings we can automatically generate and upload the needed datasets, *JDL* and analysis files. The system supports loading existing *AliRoot* [10] libraries or compiling the analysis code on the fly. Using the

GRID plug-in makes the analysis session look exactly like a local one, the only addition coming from the few extra plug-in configuration settings.

The implementation of this tool is based on the *TGrid* interface available in *ROOT*. This is specialized for the *AliEn GRID* via the *TAlien* class which allows executing shell commands in a programmatic way. This makes it possible to copy files directly to and from the *AliEn* file catalog, create datasets and most importantly, submit jobs and inspect their status. In practice, one has to connect in a plug-in fashion the utility ***AliAnalysisAlien*** to the analysis manager (as in figure 1). The way to configure it is straightforward and will be described in the next section. We wanted to provide a minimum amount of mandatory configuration settings (like software version and run numbers to be processed), while all the expert settings have default values.

While in a local session, the *AliEn* plug-in inserts into the analysis manager initialization stage without any change in the user calling sequence. The manager just detects its presence and delegates all *GRID*-specific tasks to it. To make the local analysis run on the *GRID*, the plug-in performs the following tasks:

- Connect to the *GRID* via the *TGrid::Connect("alien:")*. This uses the existing user token and is a mandatory step for being able to communicate with the *AliEn* system;
- Create the remote user working and output directories to hold the initial files and the future job output. A check for the possibility to copy files is performed by default at this stage;
- Check the availability of the requested input data and produce a dataset via the *AliEn "find"* command. The created dataset is checked to contain at least one entry and then copied to the *GRID* workspace.
- Write the initialized analysis manager from memory to a file, then copy it on grid storage
- Generate a steering analysis macro that is able to extract the analysis manager from file and launch a local analysis on a subset of the input data provided by the splitting procedure in *AliEn*. This macro is also copied in the input box and will be used to run the analysis on the remote batch nodes.
- Generate a *ROOT* macro to merge the outputs of the analysis jobs. This is an important step in distributed data analysis that otherwise has to be done separately.
- Generate validation scripts for both the analysis and merging jobs. These check after the job completion the presence of all the output files declared by the tasks connected to the manager. They also check the presence of a special end-of-job file marker named *outputs_valid* produced by the analysis manager in case of successful processing.
- Finally generate and upload the *JDL's* corresponding to the analysis and merging jobs. These keep the whole setup coherent and incorporate all the user-defined or default *GRID* configuration settings

An important part of this automatic procedure is the fact that all errors cause a well-documented execution abortion to prevent starting invalid code while on the GRID, which was very far from being the case before putting in place the plug-in. Besides that, handling all the information from a single point allows for a very powerful testing phase that was implemented in this tool from the very beginning.

Once the job input box and JDL are assembled as described above, the *AliEn* plug-in submits a configurable number of master jobs and starts automatically an *AliEn* shell. Note that all this happens from within the starting phase of a **local** analysis macro that hides completely the *GRID* API. The job submission is done gradually in case several runs are to be processed, to avoid overloading the job queue. Once the analysis jobs are submitted, the user has complete freedom to inspect the jobs status, resubmit the ones failing due to recoverable grid errors or leave the session.

An important feature of the plug-in is the ability to resume an interrupted session once a given phase of processing is completed. This first happens when the user finds out that all sub-jobs are in a final state and need to be merged. At this point the user needs to simply re-run his plug-in enabled local analysis macro, marking the start of the termination phase. The different analysis phases are described in the next section.

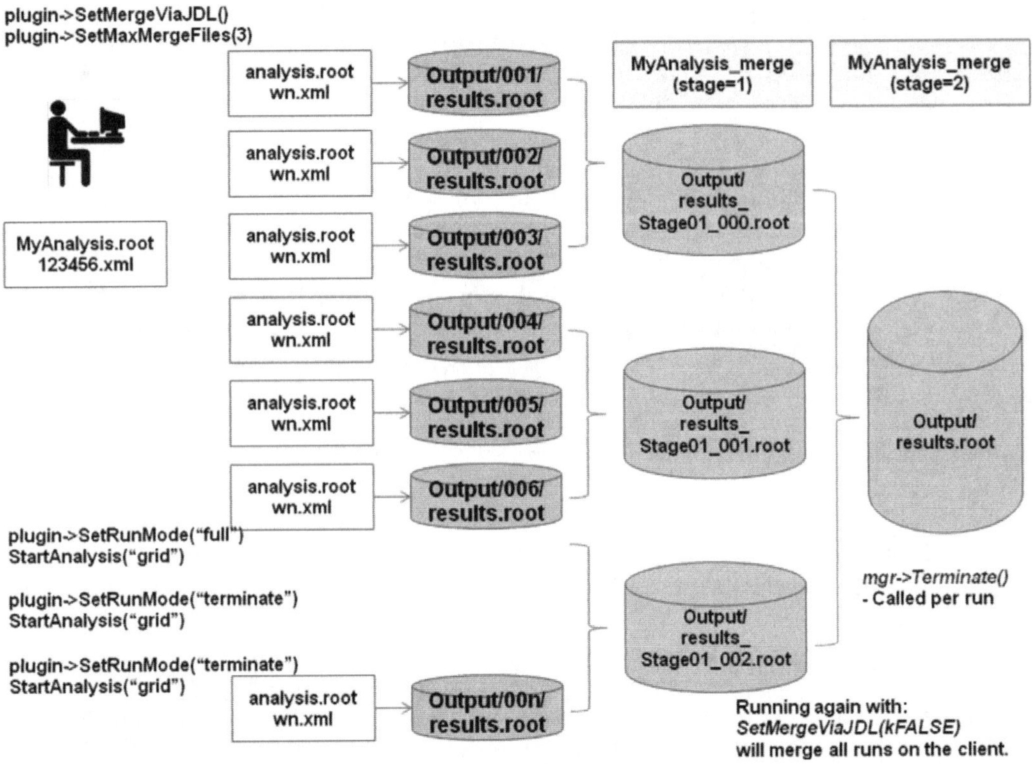

Figure 2. Merging scheme using the AliEn plug-in. The number of file per stage and the maximum number of stages are configurable.

After resuming the session, the tool first makes a new collection based on the output files available in the job output directories. This collection is used as input for the merging jobs that are subsequently submitted. We just mention here that merging is a specific phase in any distributed analysis, being a very complex problem in itself and subject to scalability issues. Merging is very sensitive to grid problems like load on storage and temporary network glitches, but also to memory limitations due to the type of the objects to be merged.

The *AliEn* plug-in alleviates this issue by providing a multi-stage iterative merging procedure that ends-up with a single file per master job (figure 2). The number of merging stages is configurable and the user has full control on (re)starting the next merging stage. The intermediate merging stages are run as independent grid jobs, taking as input a configurable fraction of the output files produced by the previous stage (or the analysis jobs in case of the first merging stage). The final merging stage is somewhat special since it runs as a single job, taking all previous outputs in one go. After performing the merging, the *Terminate()* phase of the analysis manager is automatically executed (executing post-processing analysis on the merged data) and the final merged files are copied to the grid output directory.

5. Run modes and configuration

As described in the previous section, the AliEn plug-in defines default values for most JDL parameters. These are however not always appropriate for a given analysis and need to be customized. The most important ones are described in this section.

The tool allows several different run modes to give a maximum of flexibility to the user analysis. The run mode has to be set as string via the method *SetRunMode()* of the plug-in. The supported run modes are described below:

- *"full"* – This mode is performing the full procedure described above, excluding the merging phase. If *GRID* merging is disabled or the number of stages is set to 0, merging is however executed locally. The plug-in makes a single collection over the job output files and starts a local file merger. This feature is however discouraged for more than 100 outputs. While doing local merging, the plug-in backs-up the last merged result so that in case of failure merging can be resumed. In case merging is done locally, the post-processing *Terminate()* phase is also executed locally.
- *"offline"* – This mode allows working with the plug-in without a *GRID* connection, skipping most tests and generating only the needed files. This allows modifying the intermediate analysis files or *JDL's*. This kind of configuration can be error prone and is discouraged, but in some cases it is very useful. Combined with the plug-in production mode it can be used for example to configure production jobs to be run by a different user.
- *"submit"* – This is used in correlation with the offline mode to submit the customized analysis files. It will first perform the remaining checks as in the full mode then it will submit the jobs
- *"terminate"* – This mode is used as trigger to pass to the next analysis phase. When the analysis jobs are done, one needs to put the plug-in in this mode and rerun the local analysis macro to start the merging phase. In case merging is done in stages, one needs to do the same to trigger the start of the next merging stage. As rule of thumb, it is important that all the previous jobs submitted by the plug-in are in a final state before running in this mode, otherwise the corresponding outputs will not be used.
- *"test"* – This mode performs a very powerful local test using the exact same files that are to be deployed and run on the *GRID*. The test first makes a collection over a configurable number of files from the requested input dataset. This is used to access the input data remotely via *xrootd* in the test session. The analysis is executed as on a *GRID* batch node and has to pass before running the full mode. The test mode is a unique feature of this tool and helps a lot to avoid most possible errors. Some of the features cannot be tested however, like the validity of the *JDL*, which in this case is not generated. Users have to make sure that the local software versions correspond to the ones that will be demanded on the *GRID*. To reproduce scalability issues it is recommended to set the number of test files the same as the maximum job split level. The test mode is available also in *LOCAL* or *PROOF* modes, as described in the next section.

Table 1. Most relevant plug-in configuration settings

AliEn plug-in method	Description
SetRunMode	Sets the plug-in run mode as described above
SetProductionMode	Produce files to be deployed and run by the production manager.
SetROOTVersion *SetAliROOTVersion* *SetAPIVersion*	Set requested package versions.
SetGridDataDir *SetGridWorkingDir* *SetGridOutputDir*	Define the grid directories
AddRunNumber *SetRunRange* *AddDataFile*	Define run numbers to be processed or custom collection files.
SetDataPattern	Set the pattern to be used for searching available

	input files (e.g. **AliESDs.root*)
SetAnalysisSource	Source code in case of dynamic compilation
SetExecutable *SetAnalysisMacro* *SetValidationScript* *SetJDLName*	Define custom names for the generated files
AddAdditionalLibrary *SetAdditionalLibs*	Instruct the plug-in to load these libraries (in the given order)
SetSplitMaxInputFileNumber	Maximum number of files to be allocated to a sub-job
SetDefaultOutputs *SetOutputFiles* *SetOutputArchive* *SetTerminateFiles*	Output file names as defined by analysis tasks Customize the output files Define how the outputs should be archived Extra files produced in the termination phase
SetMergeViaJDL *SetMergeExcludes*	Enable merging as *GRID* job (default) List of outputs to be ignored for merging
SetTTL *SetNtestFiles* *SetMaxInitFailed* *SetMasterResubmitThreshold*	Time to live for sub-jobs Number of files used by the test mode Number of failed jobs to trigger abortion Percentage threshold for automatic resubmission

6. Extension to local and PROOF modes

The plug-in has been recently upgraded to handle also the local and PROOF cases, so from the user perspective the computing infrastructure becomes completely transparent. The usage is straightforward, requiring the addition of few lines in the plug-in configuration. One just needs to have the plug-in connected to the analysis manager and start the analysis in the desired mode via: *AliAnalysisManager::StartAnalysis(const char *mode)*. The supported modes are: *"local"*, *"proof"* and of course *"grid"*. Table 2 presents the extra settings to be activated in case of PROOF.

Table 2. Plug-in configuration settings in PROOF mode

AliEn plug-in method	Description
SetProofCluster	Define the PROOF cluster to run the analysis
SetProofDataSet	Specify the dataset to be used. This has to be available via *gPROOF->ShowDatasets()*
SetProofReset	Reset the user session. Supported modes: *0-no reset, 1-soft, 2-hard*
SetNproofWorkers	Limit the number of workers (faster initialization)
SetNproofWorkersPerSlave	Define more than one worker per slave
SetRootVersionForProof	Request a specific *ROOT* version
SetAliRootMode	Load analysis libraries by default
SetClearPackages	Clear existing packages
SetProofConnectGrid	Connect to *GRID* on every worker
SetFileForTestMode	Define a file name containing pointers to data files stored locally, to be used for testing

A remarkable feature is that the test mode implemented initially for the *GRID* case was preserved also for the *PROOF* mode. This is activated in the same way, using the method *SetRunMode("test")* for the plug-in. In this case the file for the test mode must be defined. The test activates the *PROOF-lite* feature, which emulates a *PROOF* master on the local machine, using as many slaves as *CPU* cores. The analysis code is tested on the requested data without any connection to the *PROOF* cluster.

7. Future perspectives for organized analysis

The *AliEn* plug-in is just a simple utility that emerged from the need to make it easy for users to deploy analysis on large GRID data sets. Its usage increased very fast as well as the number of requests to support new features. This triggered several new developments and a tight integration with the GRID monitoring system, such that the tool became a key component of the analysis framework. A recent extension for *PROOF* support has made it even more useful, making all *ALICE* resources transparent and easy to use.

Besides end-user analysis, most central analysis productions are submitted based on this tool. We are currently making an effort to extend the plug-in functionality aiming to ease-up automatic generation and testing of custom analysis trains created via web forms. The idea is to provide tools that will allow for train "conductors" from the different *ALICE* physics working groups to assemble and manage trains using existing analysis modules from their group. This is now possible after standardizing the procedure of inclusion of an analysis task, and the *ALICE* analysis libraries contain already the analysis code in the necessary form. The gain in developing this new feature (already in an advanced beta-testing phase) is not only from a management perspective but also from a job efficiency one.

The implementation of this new feature required defining a new object type which describes the configuration needed to run a given analysis task. This is composed of:

- Location of the *ROOT* macro that adds the given task to an existing analysis manager. These macros already exist in the *AliRoot* library, so their path is defined with respect to the base installation location of this software.
- Required libraries to be loaded, in the right order.
- Supported input data types (*ESD*, *AOD*, *MC*). This is used to exclude incompatible analysis modules from a given train.
- Dependency analysis modules. Some analysis tasks represent general-purpose utilities (like centrality selection or physics selection) that are requested by most analysis modules even though there is no direct dependency between them. The analysis plug-in makes sure that dependencies are always included.
- Configuration for event handlers. Any train needs event handlers that may be differently configured.

This configuration can be imported from text files generated via the web forms. The configuration objects are internally looped and the macros adding each wagon are being executed. This finally creates the analysis train in memory and at this point a test can be generated by the plug-in. The generated test can be run on a test machine by the train administrator, who gets a detailed report on the CPU and memory footprint for EACH composing wagon. The idea is to create reference data to follow the evolution of a given analysis task. All errors coming out of the tests are pointing directly to the code to blame. The plug-in is also able to generate the final code to be run on the GRID, if all tests succeed.

Acknowledgements

The tools described in this paper were mainly developed by the authors, but there are many other people that contributed with ideas, developments or fixes. We would like to thank A. Morsch, J.F.Grosse-Oetringhaus, M.Vala, C.Grigoras for their direct or indirect participation to the development, together with all *ALICE* analysis users who were actively using this tool and giving their feedback.

References

[1] Carminati F *et al* 2004 *CERN-LHCC-2004-038*/G-086

Journal of Physics: Conference Series **368** (2012) 012014 doi:10.1088/1742-6596/368/1/012014

[2] Gheata A 2008 ALICE analysis framework *Proc. XII Int. Workshop on Adv. Comp. and Analysis Techniques in Phys. Research (Erice) PoS(ACAT08)* p 028
[3] http://root.cern.ch
[4] http://root.cern.ch/root/html/TSelector.html
[5] http://aliweb.cern.ch/secure/Offline/sites/aliweb.cern.ch.Offline/files/uploads/OfflineBible.pdf
[6] Ballintijn M *et al* 2003 *Proc. 2003 Conf. on Comp. in High-Energy and Nucl. Phys. (24-28 Mar La Jolla, California) arXiv:physics/0306110v1*
[7] Grosse-Oetringhaus J F 2008 *J. Phys.: Conf. Ser.* **119** 072017
[8] Bagnasco S *et al* 2008 *J. Phys.: Conf. Ser.* **119** 062012
[9] Legrand I *et al* 2009 *Computing Physics Communications* **180** Issue 12 December 2009 pp 2472–2498
[10] Brun R *et al* 2003 *Nuclear Instruments & Methods in Physics Research* A **502** pp 339–346

Journal of Physics: Conference Series **368** (2012) 012015 doi:10.1088/1742-6596/368/1/012015

Mass production of extensive air showers for the Pierre Auger Collaboration using Grid Technology

Julio Lozano Bahilo[1] for the Pierre Auger Collaboration[2]

[1] Universidad de Granada, Granada, Spain; [2] Pierre Auger Observatory, Malargüe, Mendoza, Argentina (Full author list: http://www.auger.org/archive/authors_2011_05.html)

e-mail address: julio@ugr.es

Abstract. When ultra-high energy cosmic rays enter the atmosphere they interact producing extensive air showers (EAS) which are the objects studied by the Pierre Auger Observatory. The number of particles involved in an EAS at these energies is of the order of billions and the generation of a single simulated EAS requires many hours of computing time with current processors. In addition, the storage space consumed by the output of one simulated EAS is very high. Therefore we have to make use of Grid resources to be able to generate sufficient quantities of showers for our physics studies in reasonable time periods. We have developed a set of highly automated scripts written in common software scripting languages in order to deal with the high number of jobs which we have to submit regularly to the Grid. In spite of the low number of sites supporting our Virtual Organization (VO) we have reached the top spot on CPU consumption among non LHC (Large Hadron Collider) VOs within EGI (European Grid Infrastructure).

1. Generation of extensive air showers (EAS) for the Pierre Auger Observatory

1.1. The Pierre Auger Observatory

The Pierre Auger Observatory [1] is a hybrid detector whose goal is to study ultra-high energy cosmic rays (UHECR). It makes use of fluorescence telescopes to collect fluorescence light produced as the cascade of secondary particles composing the EAS travels through the earth's atmosphere. It also uses water tanks which gather Cerenkov light emitted by the particles which remain at ground level after the partial absorption of the EAS in the air. Both detectors digitize light using photomultiplier tubes (PMT) and associated electronics.

The simulation of the events collected by the Pierre Auger detectors is done in two steps: first we generate the EAS with a given set of parameters and then we simulate the detector response and perform the reconstruction of the physics information of the shower. Those steps require different software packages with different requirements and thus it is more appropriate to write about both steps in separate sections. In this section we will describe briefly an EAS and discuss the software programs used in order to generate them.

1.2. EAS characteristics

The energies of the primary cosmic rays we are concerned with range roughly from 10^{16} to 10^{21} eV, thus being the most energetic particles studied by particle physics scientists. A single cosmic ray at those energies interacts soon after entering the earth's atmosphere producing an enormous cascade of secondary particles. It has a large electromagnetic component and a smaller hadronic component which have to be followed in order to characterize completely the development of the EAS. This means that the program to generate the shower must track billions of particles over tens of kilometers until the particles reach ground level. An example of such an EAS generated by a UHECR is depicted in figure 1 [2].

1.3. Simulation programs

There are two simulation packages widely used in our field in order to generate EAS: AIRES (AIRshower Extended Simulations) [3] and CORSIKA (Cosmic Ray SImulations for KAscade) [4] though we have only used CORSIKA in our Grid productions.

Both packages adopt high energy and low energy interaction models during the tracking of the secondary shower particles. Those models can be selected by the user and the executables are statically linked to the corresponding libraries when they are compiled. They are driven through input data files which contain parameters relevant to the set of EAS which we want to simulate: primary particle species, energy, zenith and azimuth ranges and other values changing the simulation conditions. We group these sets of events into *libraries*.

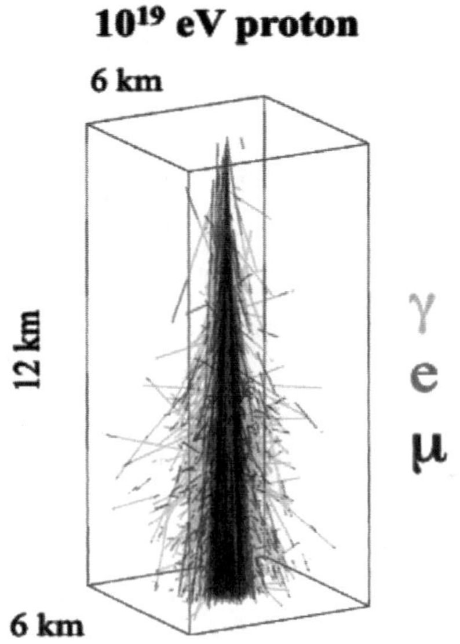

Figure 1. Image showing the development of a single EAS produced by a proton with an energy of 10^{19} eV. Most prominent particle types are shown in different colours [2].

The full simulation of a single EAS at the highest energies may take days, even months, of computing time and therefore we are forced to use statistical *thinning* techniques to avoid tracking individually each secondary particle while keeping an acceptable degree of precision. These techniques imply grouping a certain number of particles with similar characteristics when they fall below an energy value set by the user.

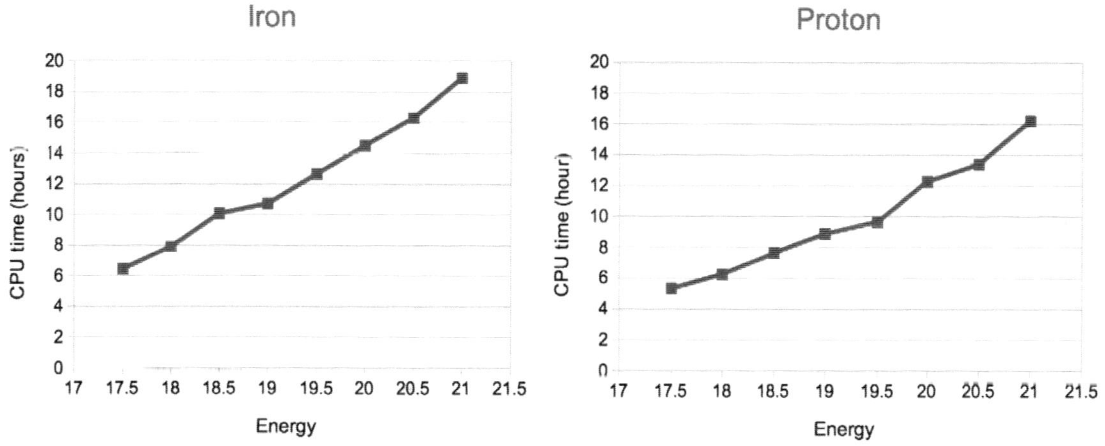

Figure 2. Left: CORSIKA average execution time as a function of the logarithm of the energy for iron primary cosmic rays. Right: same plot for proton cosmic rays

Usually, the applied *thinning* parameter depends directly on the energy of the cosmic ray. Even if the *thinning* is stronger for higher energies, the CPU consumption times, as can be seen in figure 2, increase almost linearly with the logarithm of the energy of the primary cosmic ray. There is also a dependence on the type of primary so that lighter nuclei primary consume lower amounts of time, and varying the models and parameters can make a difference. On average it takes several hours to complete a single simulation, although the spread is high.

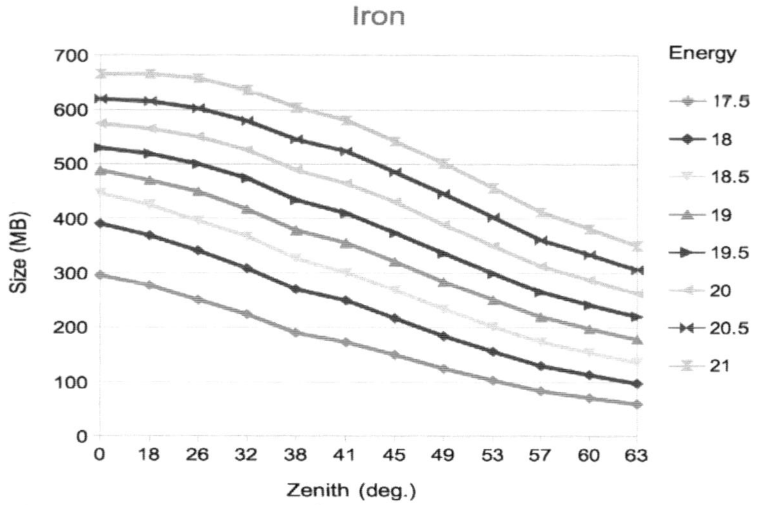

Figure 3. Size in MBs of the output files containing information on generated showers as a function of the zenith angle of the primary for iron primaries at diverse energies (expressed as log(E)).

The size of the output files change dramatically according to the primary energy, but they also have a strong dependence on the zenith angle as seen in figure 3. The values also depend on models and parameters so we present just the values obtained for our typical simulations which, on average, give sizes of tens to hundreds of MBs.

2. Detector simulation and reconstruction

The detector simulation package, referred to as Offline [5], is a software framework composed of independent modules which allows one to perform any step or set of steps within the simulation and reconstruction chain. XML files contain configurable parameters to modify the behavior of the modules. The tracking of the particles traversing the water tanks, which is done with the help of the GEANT4 [6] package, is the most time consuming process.

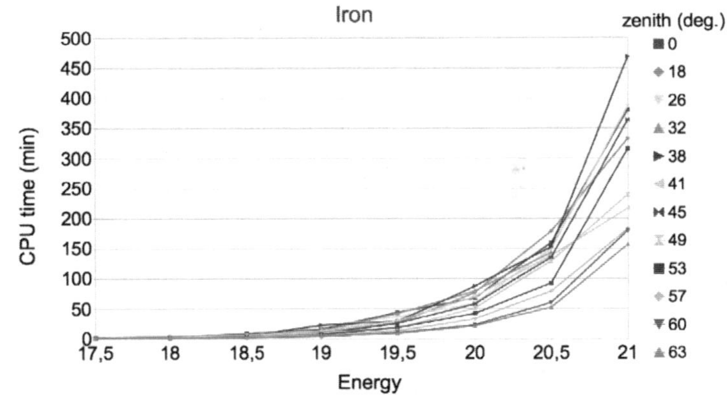

Figure 4. CPU time of an Offline execution as a function of the energy for various zenith angles.

The execution times of the Offline are smaller in general than those of the shower generation program. Whereas the generation of a shower requires several hours, the processing of one of those showers takes normally less than or about an hour except for the highest energies (see figure 4). The plot shows an approximately exponential rise.

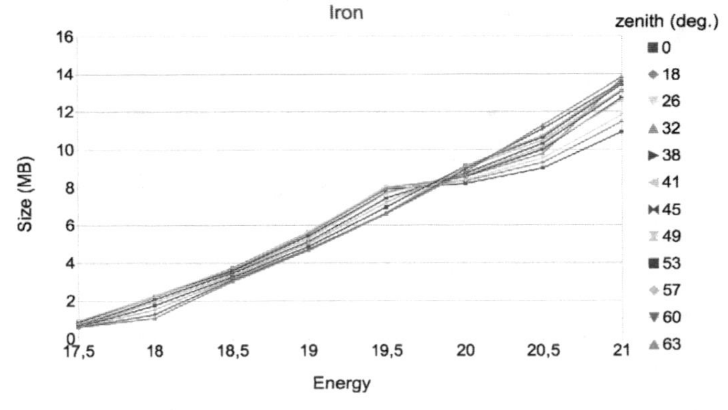

Figure 5. Total size of Offline output files as a function of the energy for various zenith angles.

The sizes of the resulting output files are much smaller (figure 5) since they just contain the information collected by the detectors, some processed information, and physical shower reconstructed parameters. They increase with the energy and do not depend very much on the zenith angle of the cosmic ray.

3. Grid computing model for official Auger simulations

The submission of jobs to Grid proceeds via a set of scripts written in Bash and Python which are running continuously at a User Interface (UI). They look for certain input files needed by our programs: CORSIKA needs input data files and the Offline requires a list of Logical File Names (LFN) associated to the input showers. The input files are represented by the squares within the light shaded areas in figure 7. In addition to the files mentioned previously, an execution script of the corresponding software package is also sent to the Grid within each individual job. This script is automatically generated from a common template. The Job Description Language (JDL) file required by the submission command is also created with the help of a template file. The presence of those files in certain directories is checked by the *Management scripts* (see also the same figure 6) and it triggers the submission of jobs to Grid Computing Elements (CE) as collections of jobs handled by the Workload Management System (WMS) services.

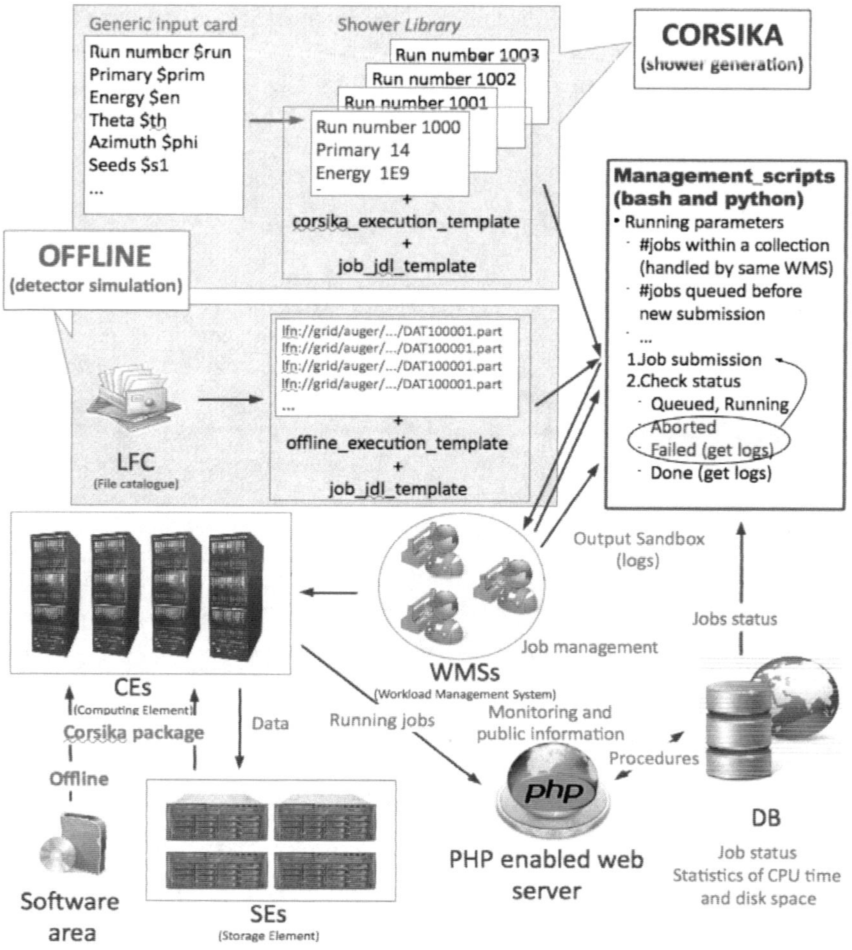

Figure 6. Workflow of the Auger simulation productions.

A set of parameters determines how the scripts behave:

- frequency of job submission depending on the ratio of running to queued jobs
- number of jobs submitted within a collection
- maximum number of jobs to be submitted in a submission cycle
- allowed number of aborted and failed jobs before interrupting the main script execution
- etc ...

The individual jobs are submitted together as a collection of jobs until the requested number of total jobs has been submitted. At this time, aborted and/or failed jobs may be resubmitted.

The scripts react according to the status of the previously submitted jobs and determine whether more jobs must be sent. The status is flagged and transmitted, along with some additional information, to a database which is coupled to a PHP-enabled web server. Thus we have appropriate means to monitor constantly the progress of the production and we can also set-up public web pages with updated information.

Detailed information on successfully executed jobs is obtained from the output logs and stored in the database in order to have detailed information of consumed CPU times, output files sizes, sites involved and other aspects of the job execution.

To reduce the load of the jobs on the network we copy our software packages in the Sofware Areas (SA) of the sites, though we can also fetch the CORSIKA package from Storage Elements (SE) where it has been previously placed. CORSIKA can be compiled statically and be copied in a single operational package. On the other hand the Offline software has many dependencies and needs a complicated and time consuming installation procedure; it is also heavier. In both cases, installation jobs are submitted before any new *library* is started. By *library* we mean the set of showers simulated with the same CORSIKA compilation options, high energy and low energy models in particular, and the same kind of primary cosmic ray.

4. Performance of our Grid production model

The Auger Virtual Organization is a relatively young VO with access to a limited number of EGI (European Grid Infrastructure) [7] sites, including some in latin America, and a low number of *prioritized* CPUs: just about a few hundred. In spite of the continuous and diverse problems which we have faced, the average production rate has been very good. As seen in figure 7, the daily CPU consumption presents remarkable fluctuations, though the tendency as shown in the cumulative mean (light shaded area) indicates a rise of the CPU consumption since the beginning of the year 2011 until we reached a stable average at about 850 CPU*days.

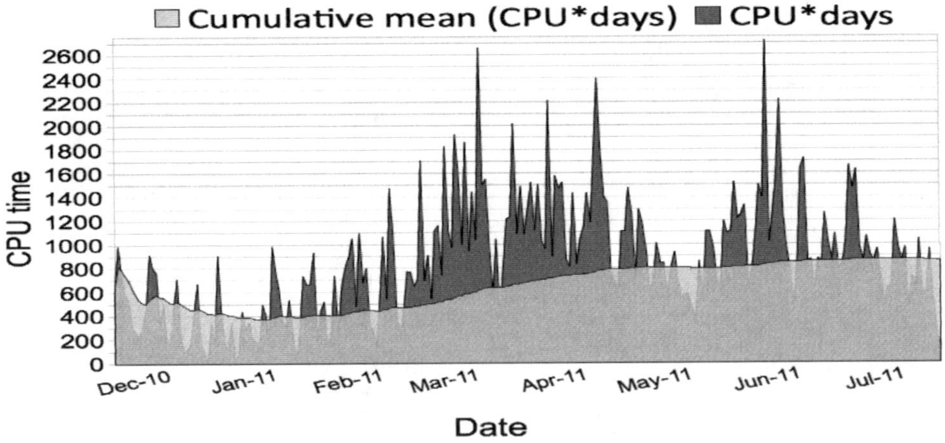

Figure 7. Daily CPU and cumulative mean CPU consumption of the Auger VO since end of 2010.

The continuous efforts to improve our set of automated scripts have paid off and we are being ranked consistently among the top 10 CPU consumers of the EGI Grid, only surpassed by the Large Hadron Collider (LHC) collaborations which dominate this statistic as shown in figure 8.

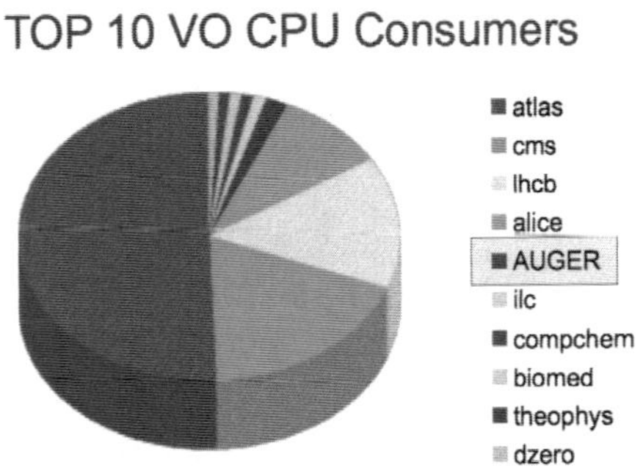

Figure 8. Percentage as pie chart of the top CPU consumers at EGI Grid. Auger ranks 5th.

Given the numbers presented for our consumed mean cumulative CPU time (figure 7) and the mean CPU time required for a shower with energy 10^{19} eV and zenith angle 38^0 (figure 2) we can calculate that the average daily number of iron showers of these characteristics we can produce is close to 2000.

5. Summary and conclusions

The Pierre Auger Observatory requires high numbers of simulated EAS which demand a very high level of computing time and storage space. Grid is an ideal technology for accessing the needed computing resources, although it suffers from the lack of stability. This can be specially the case for small Virtual Organizations where resources are concentrated on few sites.

We have developed a simple computing model based on scripts coupled to a web server and a database which make use of Grid middleware. The automation of all job related tasks and the development of monitoring tools have allowed us to produce hundreds of thousands of EAS to fulfill the needs of the Auger scientists. In spite of being a small team we have reached the 5th place in the CPU consumption ranking within the EGI Grid.

Acknowledgements

We gratefully acknowledge the contribution of the sites supporting the Auger Virtual Organization and the efforts of their administrators to maintain stable operation of their centers. The author also wants to thank the Auger Collaboration for its support to this project. He also wants to commend the remarkable effort put into this initiative by the members of the production teams at the University of Granada and at the FZU Institute in Prague.

References

[1] Abrahams J et al (Pierre Auger Collanoration) 2004 *Nucl. Instr. and Meth.* A **523**, pp 50-95.
[2] Pryke C 1998 *Phys. Dpt. Colloquium, Univ. of Wisconsin (Madison)* http://find.spa.umn.edu/~pryke/publ.html
[3] Sciutto SJ 2001 *astro-ph/0106044*, 27th ICRC Hamburg, August pp 7-15.
[4] Heck D Knapp J Capdevielle JN Schatz G Thouw T 1998 *Forschungszentrum Karlsruhe Report FZKA* 6019.
[5] Argiro S et al 2007 *Nucl. Instr. Meth.* A **580**, pp 1485–1496.
[6] Agostinelli S et al 2003 *Nucl. Instr. Meth.* A **506** pp 250-303.
[7] EGI, European Grid Infrastructure. http://www.egi.eu/

Do regions of ALICE matter? Social relationships and data exchanges in the Grid

E D Widmer[1], F Carminati[2], C Grigoras[2], G Viry[1] and G Galli Carminati[3]

[1] Department of Sociology, University of Geneva, Switzerland
[2] CERN, Geneva, Switzerland
[3] Units of Mental Development Psychiatry, Department of Mental Health and Psychiatry, University Hospitals of Geneva, Switzerland

E-mail: Federico.Carminati@cern.ch

Abstract. Following a previous publication [1], this study aims at investigating the impact of regional affiliations of centres on the organisation of collaboration within the Distributed Computing ALICE infrastructure, based on social networks methods. A self-administered questionnaire was sent to all centre managers about support, email interactions and wished collaborations in the infrastructure. Several additional measures, stemming from technical observations were produced, such as bandwidth, data transfers and Internet Round Trip Time (RTT) were also included. Information for 50 centres were considered (60% response rate). Empirical analysis shows that despite the centralisation on CERN, the network is highly organised by regions. The results are discussed in the light of policy and efficiency issues.

1. Introduction

This study aims at measuring the impact of regions on the structure of interactions between the centres of the Distributed Computing ALICE infrastructure, based on social network methods [2]. These centres are part of the Worldwide Large hadron collider Computing Grid (WLCG [3]) They form a large computational network where data and workload are exchanged, but also a large and complex social network with $\sim 3,000$ possible links. The operation experience of this Grid since ten years has shown that the coordination and collaborations between the different centres are as important as the material conditions to ensure the proper functioning of this complex system extended over different time-zones and continents.

In a previous article [1], we found that the centres of the ALICE Distributed Computing Infrastructure derive most of their support from CERN and that various types of interactions were centralised on CERN. However, we also found that there were signs of local organisation of data exchange and collaborations. In this paper, we estimate the impact of regional influences. The research issue considered is the extent to which ALICE is organised with reference to regional anchorage that may be related to cultural, historical, political or network connectivity issues.

2. Data

Information about collaborations were collected using a self-administered questionnaire which was filled by the technical manager of each ALICE centre in the Fall 2009 and early 2010.

Technical managers had to estimate by yes/no questions which centres of the Grid provided their centre with significant help in its work at least once a week (support). They also had to estimate which centres were in regular e-mail contact at least once a week with their own (interactions), and with which centres their centre would like to have more interaction in its work (wished collaborations). These indicators refer to self-reported exchanges. The questionnaire was sent via e-mail to the 73 centres of the ALICE Grid. Answers for 50 centres were received and considered in the empirical analysis (68% response rate).

Additional information were gathered by technical observation such as the theoretical capacity of the network linking the centres (bandwidth), the actual quantity of data exchanged and the Internet Round Trip Time (RTT). These indicators refer to observed exchanges, in contrast with self-reported exchanges provided by responses of technical managers in the self-administered on line questionnaire. Overall, this study focuses on support, interactions, wished collaborations, theoretical capacity of the links between any two centres (bandwidth), quantity of data exchanged, and Internet Round Trip Time (RTT). We investigate to what measure those interconnections are structured by regions.

Centres are dispersed in various regions. Overall, the sample includes 23 centres in Northern Europe, 9 in Southern Europe, 6 in Asia and 8 in Russia. In order to estimate the impact of regions, we could not include Africa (one centre), South-America (one centre) and north America (two centres) due to the very limited number of centres pertaining to each of these regions.

3. Measurements

A variety of measures exist in network analysis ([2, 4, 5]). We focus on density as a measure of cohesion, at the level of the full network and at the level of regions. Density in a directed graph is equal to the number of existing arcs (directed ties) divided by the total number of possible arcs. In addition, two measures of network centralisation were computed, which capture different conceptual dimensions of centrality. In-degree and out-degree centralisation express the degree of variability (or inequality) among centres in the number of ties pointing to and going from a specific centre. For instance, a network characterised by a small number of centres receiving many direct ties, and other centres receiving few ties, has a strong in-degree centralisation. These measures provide information on the local dimension of centralisation as a centre may have many connections within a rather isolated subgroup of centres. Quite distinctly, betweenness centralisation measures the degree of variability (or inequality) among centres in the proportion of interactions in the network captured by a centre. The network is said to be centralised if a small number of centres lie between all other centres´ chains of relationships. Finally, the number of cliques measures the extent to which the network is structured around multiple clusters of centres or, conversely, is composed of a small number of big clusters. Formally, a clique is a sub-set of a network including the maximum number of centres that have all possible ties present among themselves. For the present analyses, a clique includes at least four centres.

Using the software UCINET [6], we compute these parameters on the overall network, with and without CERN in it. This procedure is set up to control for the impact of CERN on the overall structure, as the centrality of CERN in the network was acknowledged [1] as it was to be expected since CERN is the largest laboratory, both source of the experimental data and of most of the software used on the WLCG Grid. Because CERN is by its institutional role lead to be central, it is necessary to estimate regional influences with and without it included in the network. Threshold values for observed exchanges were fixed to determine only the substantial interactions between the centres. Regarding bandwidth and the actual quantity of data exchanged, two centres are considered to be linked when the theoretical capacity and the data transfer from one centre to the other amount at least 100 and 0.1 megabyte per second respectively. Finally, RTT links were considered when the Internet Round Trip Time from one centre to the other was below 20 milliseconds. Table 1 presents various indices measuring the cohesion and the centralisation of the network. Indices are computed either with CERN included or CERN excluded, in order to estimate the impact of CERN on the overall organisation of the Network.

Table 1. Network indices (with and without CERN).

	Help	E-mail contacts	Wished collaborations	Bandwidth	Data transfers	RTT Capacity
			With CERN			
Density	4.3	4.9	3.6	16.2	8.4	18.2
Betweenness centralisation	36.2	31.3	26.3	6.2	8.0	21.0
Indegree centralisation (provided)	72.6	59.5	42.1	34.0	12.9	39.9
Outdegree centralisation (providing)	33.1	17.9	13.0	29.9	31.6	71.1
Number of cliques (min. size 3, sym. max)	24	19	23	38	24	67
			Without CERN			
Density	2.2	3.4	2.5	14.9	7.1	16.4
Betweenness centralisation	0.5	3.3	2.4	6.9	6.9	22.8
Indegree centralisation (provided)	12.7	11.5	6.0	33.9	11.8	34.0
Outdegree centralisation (providing)	10.5	9.3	12.4	29.6	26.7	72.3
Number of cliques (min. size 3, sym. max)	10	9	6	38	15	65

Results from the overall network, with CERN included, show a highly centralised network and a low level of density for all interactions. When CERN is excluded from the network, the centralisation becomes much weaker. The wished collaborations have somewhat a lower level of centralisation, thus showing that centre managers value to some extent the development of less centralised interactions within ALICE. For observed exchanges, there is also less centralisation. Getting CERN out of the network does not make such a great difference as in the case of self-reported exchanges.

4. Results

Results from the visual inspection of various graphs confirm that interactions, both self-reported and observed, depend to a large extent on regional influences. Figure 4 shows the interaction between regions and help provided. Figure 4 report responses given to the question about contacts by emails. The different regions are illustrated with the following colours: centres from Northern Europe (yellow, n=23), Southern Europe (green, n=9), Asia (white, n=6), Russia (orange, n=8), Africa (pink, n=1), South-America (blue, n=1) and north America (red, n=2). Both networks show a high level of centralisation on CERN as well as an organisation by regions. Figure 4 refer to the wishes of centre managers for collaborations. In that case, the network is much less centralised on CERN and the structuring by regions is less straightforward compared with current social interactions.

Are observed interactions among centres also significantly associated with their regional affiliations? As figure 4 shows, the theoretical capacity of the network is higher among the centres of the same region than among centres of distinct regions. Also, there are more exchanges

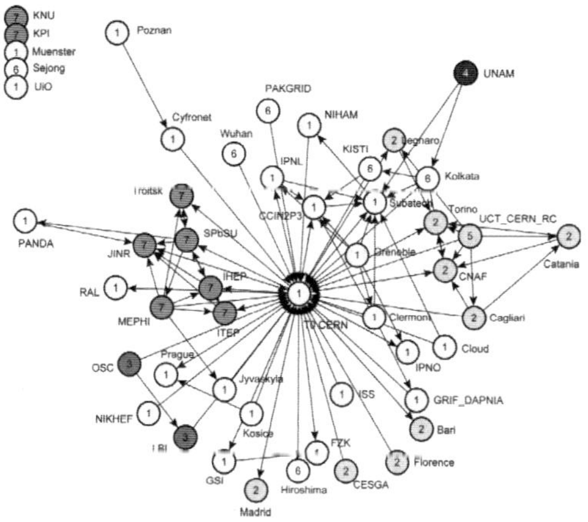

Figure 1. Help provided by regions.

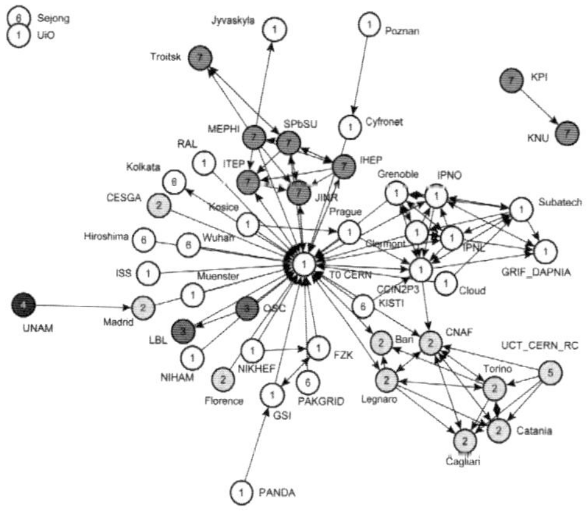

Figure 2. Exchanges of emails by regions.

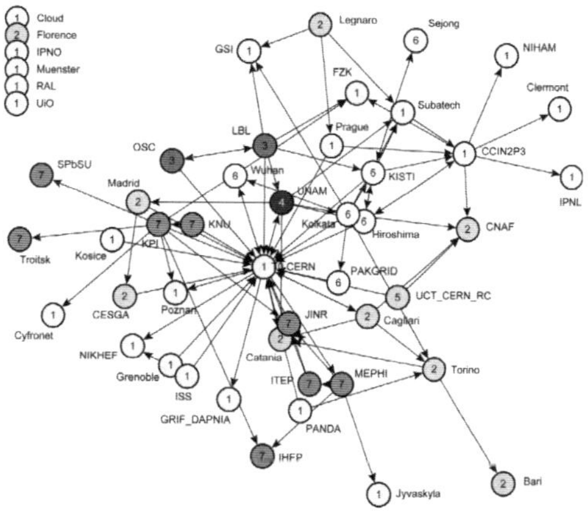

Figure 3. Wished interactions provided by regions.

among centres belonging to the same regions than among centres of distinct regions (figure 4), and the internet round-trip time (figure 4) is in their case lower.

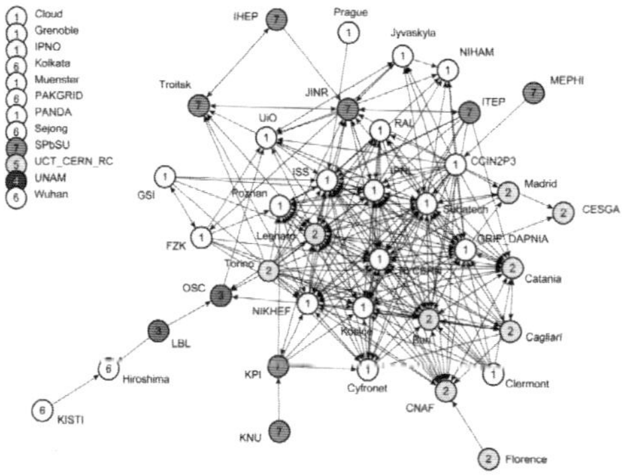

Figure 4. Bandwidth by regions.

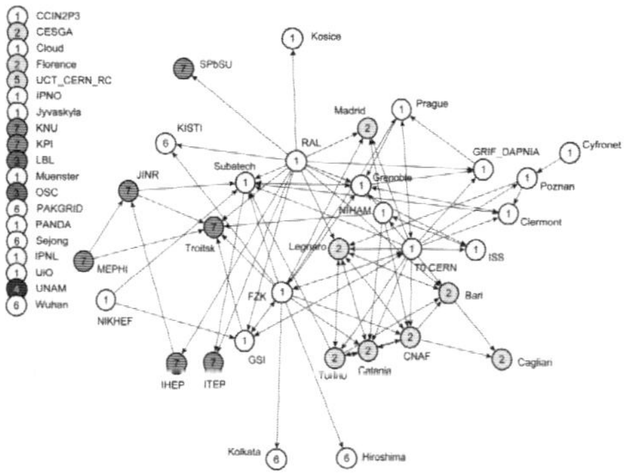

Figure 5. Data transfer by regions.

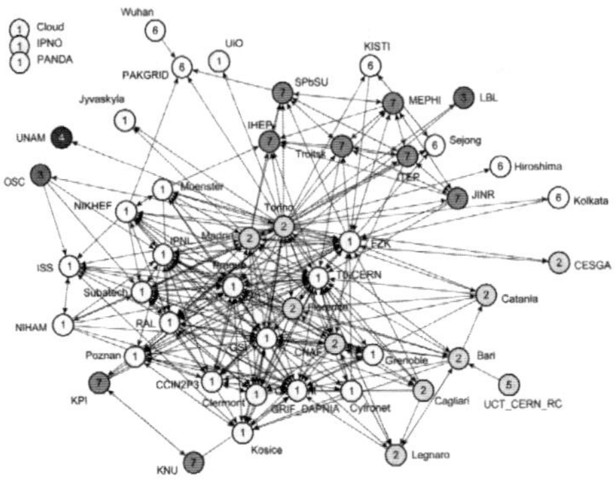

Figure 6. RTT Capacity by regions.

Are those graphical results confirmed by computational analysis? We first present the density within the main regions where most ALICE centres are located. Those regions are Northern Europe, Southern Europe, Russia and Asia. Indices for North America, South America and Africa were not computed as the number of countries associated with each regions is very small or even, in some cases, equals to 1. The density within Northern Europe was computed with and without CERN in it.

Table 2. Density within regions.

	Help	E-mail contacts	Wished collaborations	Bandwidth	Data transfers	RTT Capacity
North Europe (with CERN)	8.1	11.1	3.0	37.7	20.8	45.5
North Europe (without CERN)	3.5	8.2	1.5	35.3	16.8	41.1
South Europe	9.7	23.6	8.3	47.3	34.9	43.9
Asia	3.3	0.0	23.3	4.3	0.0	11.1
Russia	23.2	26.8	16.1	17.8	10.8	52.1
Total (with CERN)	4.3	4.9	3.6	16.2	8.4	18.2
Total (without CERN)	2.2	3.4	2.5	14.9	7.1	16.4

Table 2 shows that overall, the density on almost all indices based on self-reported exchanges is much higher within each region than in the full network. For instance, the density of contacts is twice higher in Northern or Southern Europe than in the full network. Similar patterns of collaboration within regions are found for email contacts. Interestingly, the observed exchanges are also frequently organised at the regional level. The only exception concerns wished collaboration, where there is a clear distinction between centres located in Northern Europe and centres in other regions. Centres in northern Europe are not especially seeking collaboration with other north European centres (average similar to the overall mean). Although the density of internal interaction is a bit higher, Southern Europe presents a quite similar pattern of results. Contrastingly, centres from Russia, and especially Asia, wish to increase their interactions at the regional level. The pattern of density of interactions among Asian centres is somewhat different from that in other regions. The density on self-reported and objective exchanges is low in comparison with other regions or even the full network, while the wishes of centre managers for collaborations are very high.

We now estimate the extent to which those results are statistically significant. In order to know whether the patterns revealed in table 2 are due to chance or reveal structures, we first build an hypothetical network perfectly structured around regions, i.e., in which a tie exists if and only if the centres are of the same region. We then run a set of permutation models, based on QAP (Quadratic Assignment procedures) [7]. The Quadratic assignment methods compute correlations between the entries of two square matrices and assess the frequency of random and true correlations between them, in order to test whether or not those dimensions are correlated beyond chance. The algorithm proceeds in two steps. In the first step, it computes a Pearson's correlation coefficient between corresponding cells of the two data matrices (corresponding to the observed and hypothetical network data). In the second step, it randomly permutes rows and columns (synchronously) of one matrix (the observed matrix, if the distinction is relevant) and recomputes the correlation. The second step is carried out hundreds of times in order to compute the proportion of times that a random correlation is larger than or equal to the observed correlation calculated in step 1. A low proportion (<.01) results in the rejection of the null hypothesis of independence (a network with permuted centres could have a correlation with the hypothetical network at least as high as the observed network) and suggests a strong relationship between the matrices that is unlikely to have occurred by chance ([6]). Here again, Pearson's correlation coefficients were computed with and without CERN.

Table 3. Qap testing of regional influences (Pearson's correlation coefficient).

	Help	E-mail contacts	Wished collaborations	Bandwidth	Data transfers	RTT Capacity
with CERN	0.15**	0.24**	0.07**	0.30**	0.27**	0.40**
without CERN	0.16**	0.27**	0.10**	0.30**	0.26**	0.39**
**p<.01						

Table 3 shows the average value of Pearson correlations between each empirical matrix and the theoretical matrix constructed on the assumption that ties only exist within regions.It also provides a p-value which reports the proportion of times in which the permuted empirical matrix shows a correlation with the theoretical matrix equal or higher than the original empirical matrix. The table reveals significant effects of regions for all indices considered in this research. Belonging to the same region makes the density of interactions among centres significantly increase. This holds true for both self-reported and observed exchanges. For some indices, the difference is important, as where it is marginal for others.

5. Discussion

The patterns emerging from this study help us understanding the underlying structure of the ALICE Grid within the Worldwide LHC Grid structure. Access to data and resources is actually ubiquitous within the ALICE Grid. Members of the ALICE collaboration can submit work requests (jobs) from any point in the world and this gets executed by any centre available with the only constraint of data locality for reconstruction jobs, while simulation workload gets assigned only on the basis of CPU availability.

The first remark is that the distribution of the links for the different relationships in the network is highly non-random, as it can be seen in table 3. This does not come as a surprise, as the Grid is a highly structured network with specific functionalities and connections. However this fact implies that the relations found in this paper are significant and can be interpreted as the effect of structuring factors.

If we look at the "physical" layer (RTT and bandwidth) we can "see" some regionalisation in figure 4 and figure 4. This is already a non trivial observation because the presence of large supranational initiatives like Géant in Europe establishing high-speed international networking.

Coming to the role of CERN, from table 1 we note that it does not have a particularly important role in the structure of the physical layer, as the different indicators for RTT and bandwidth show relatively little variation with and without it. Only the substantial reduction of the indegree centralisation in the RTT suggests the role of "hub" the CERN plays in the ALICE physical network. This role is confirmed by the opposite trend of the betweenness centralisation, indicating that the removal of CERN leaves a network where more centres play the role of relay between peers in a more "democratic" network.

If we now move to data transfer, we observe that CERN's role as a "hub" (Indegree centralisation) seems less prominent, while the network of data exchanges seems to be strongly hierarchised around CERN (Outdegree centralisation drops by 30% with CERN's removal). This is consistent with the fact that CERN is the ultimate "source" for the experiment data. This is mitigated by the fact that MonteCarlo data are generated at all the nodes of ALICE's Grid. A large change in the number of cliques also seems consistent with CERN's role of producer of data and receiver of elaborated data for custodial storage which makes of it a preferred vertex of fully connected subsets.

The above suggests that at the "physical" layer the ALICE Grid looks little hierarchised and rather homogeneous. The data transfer is characterised by CERN's special role as source of the data, but otherwise also shows little structure. Regional densities for the "physical layer" shown in table 2 differ largely for the different regions.

If we now turn to the "relational layer" we observe a largely different situation. The three categories Help, E-mail contacts and Wished collaborations are again shown in table 2. The density of the three categories show a large change with and without CERN. The nature of this change can be understood looking at the other parameters. The betweenness centralisation is reduced by an order of magnitude removing CERN and the effect is even more dramatic for the Help category, implying a structure where the largest number of links point to CERN, as can be easily seen in the pictures 4-4. The drop of Indegree centralisation is also very large, clearly indicating the "hub" role of CERN in providing help, email contacts and as a target for wished collaboration. Similar consideration for the number of cliques in wished collaborations. It seems that CERN is seen as the "missing vertex" for multilateral collaborations.

Interestingly enough the drop in outdegree centralisation, which measures the hierarchisation of the system, is less pronounced, and this may suggest that the hierarchical structure remains even if the "top" is removed. Looking at the graphs this hierarchisation appears very clearly to depend on geographical regions. The density of wished collaboration within regions seems to be inversely proportional to the perceived existing level of help and email exchange, which is an expected result.

It is interesting to compare these findings with the hierarchical organisation foreseen by the MONARC model [8] which has largely served as a blueprint for the WLCG Grid. The structure for the MONARC "physical layer" was influenced by the foreseen limitation on the network bandwidth and the projected need to optimise resource utilisation within this constraint. The actual evolution of the network capacity has been beyond all expectations. At the end of last century, when the MONARC proposal was put together, researchers were hoping to have 622 Mb/s links, while now links with a capacity one order of magnitude more are commonplace. As a consequence of this, at the "physical layer" this model has been replaced by a model evolving toward a more "democratic" cloud paradigm. In the case of ALICE, this move has been accompanied by precise architectural choices in the design of the middleware and in the operation of the Grid, with the aim of optimising the usage of resources. The fact that the data transfer follows this pattern is in this sense revealing that the architectural design is reflected in the actual dataflow.

At the "relational" layer a completely different picture emerges. Geographical regions play a large role in modelling the infrastructure, which looks highly hierarchised in a way similar to the one suggested by the MONARC model. While the machine-to-machine exchanges are designing a "cloud", the human relationships, both actual (email exchange), perceived (help) and wished (wished collaborations) follow a hierarchical schema strongly influenced by geographical and national / cultural elements. This layer looks like an example of dynamic "self organisation" capable to adjust itself in order to "optimise" the Grid usage from the "user's perspective" within the constraints coming from the "physical layer". Of course the "physical layer" itself also evolves as a result of the how the system is really used, aiming at improving efficiency and resource utilisation. Analysing the mutual influence between these two layers could be an interesting subject for future studies. Moreover, it seems that these two layers, although different, can work together in a complementary fashion.

We believe that this work is just the beginning of what could be a much deeper analysis of the complex human and technical relations that link the Grid nodes. We have built a large, highly non-linear system and, in some sense, we are just beginning to understand it. A massive amount of monitoring information is now accumulating, thanks also to the constant improvement of the monitoring tools. The analysis and understanding of these data will be essential to grasp the functioning of the Grid, and this paper can be considered part of this effort. However there is also another, even more exciting perspective that has to be considered, and this is the usage of these data to optimise the working of the Grid, making it more efficient and resilient. For the data that can be collected in real time, the tuning process can happen dynamically, making the Grid a self-adjusting system. Can "relational" data be used as part of this optimisation? This paper does not answer this question, but we hope it shows that these data can at least be collected and analysed meaningfully. In this sense a lot of interesting work lies ahead of us.

How much the results shown relate to the evolution of Internet as a whole outside HEP is also a very relevant question, and it would be important to compare our results to similar studies in different fields. At the same time it would of course be very interesting to repeat this study in the future to see how the situation evolves.

Acknowledgments

The authors wish to thanks Dr. Iosif Legrand for his insightful reading of the draft.

References

[1] E D Widmer F Carminati C G and Carminati G G 2010 How alice computing centres connect? an exploration of cooperative ties among computing centres. *13th International Workshop on Advanced Computing and Analysis Techniques in Physics Research. ACAT 2010* (Proceedings of Science)
[2] Wasserman S and Faust K 1994 *Social Network Analysis: Methods and Applications* (Cambridge University Press)
[3] Worldwide large hadron collider computing grid http://lcg.web.cern.ch/LCG/public last updated on 05-10
[4] Burt R 2001 *The social capital of structural holes* (New York: Russel Sage Foundation)
[5] Scott J 2000 *Social Network Analysis* (Newbury Park, CA: Sage Publications Ltd; 2nd edition)
[6] S Borgatti M G E and Freeman L C 2002 *Ucinet 6 for windows. software for social network analysis.* technical Report
[7] Hubert L J and Schultz J 1976 *British Journal of Mathematical and Statistical Psychology* 190–241
[8] The monarc project http://monarc.web.cern.ch/MONARC last updated on 13-01-2000

Journal of Physics: Conference Series **368** (2012) 012017 doi:10.1088/1742-6596/368/1/012017

Can Go address the multicore issues of today and the manycore problems of tomorrow?

Sébastien Binet

Laboratoire de l'Accélérateur Linéaire, Université Paris-Sud XI, 91898, Orsay, FR

E-mail: binet@cern.ch

Abstract. Current High Energy and Nuclear Physics (HENP) libraries and frameworks were written before multicore systems became widely deployed and used. From this environment, a 'single-thread' processing model naturally emerged but the implicit assumptions it encouraged are greatly impairing our abilities to scale in a multicore/manycore world. While parallel programming - still in an intensive phase of *R&D* despite the 30+ years of literature on the subject - is an obvious topic to consider, other issues (build scalability, code clarity, code deployment and ease of coding) are worth investigating when preparing for the manycore era. Moreover, if one wants to use another language than C++, a language better prepared and tailored for expressing concurrency, one also needs to ensure a good and easy reuse of already field-proven libraries. We present the work resulting from such investigations applied to the Go programming language. We first introduce the concurrent programming facilities Go is providing and how its module system addresses the build scalability and dependency hell issues. We then describe the process of leveraging the many (wo)man-years put into scientific FORTRAN/C/C++ libraries and making them available to the Go ecosystem. The ROOT data analysis framework, the C-BLAS library and the Herwig-6 MonteCarlo generator will be taken as examples. Finally, performances of the tools involved in a small analysis written in Go and using ROOT I/O library will be presented.

1. Introduction

The *"Free Lunch"* is over: Moore's law [1] can not be as easily leveraged as in the past, computer scientists and software writers have now to be familiar with Amdahl's law [2]. Indeed, computers are no longer getting faster: instead, they are growing more and more CPUs, each of which is no faster than the previous generation.

This increase in the number of cores evidently calls for more parallelism in HENP software. Fortunately, typical HENP applications (event reconstruction, event selection,...) are usually *embarrassingly parallel*, at least at the coarse-grained level: one "just" needs to call in parallel the portion of code which massages the events retrieved from the detector (the event loop) while still executing sequentially all the code processing each event.

However, this *'one event per core'* strategy may not scale up to hundreds or thousands of cores, thus requiring library implementers and code writers at large to deal with sub-event parallelism. Parallel programming in C++ has been done and is doable, but even if many libraries such as `Threading Building Blocks` [6] or `OpenMP` [7] exist to help the burden, such an endeavour is still quite tedious and error prone. Note that C++11 with lambda functions, `std::thread`, `std::future` and `std::promise` will likely improve the situation but at the price of sprinkling templates everywhere in the code, slowing down further the edit-compile-run cycle, and complicating further an already not so simple language. The recent gain in popularity of python [4] in HENP is probably a consequence of these two C++ defects, even if python is probably not the best language to tackle parallel high performance computing. At this point, it would seem reasonable to ask if using a new language more capable at leveraging multithreaded environments would be a better alternative.

This paper explores such a path. We first introduce some of Gomost relevant features with regard to concurrency and how its module system addresses the build scalability and dependency hell issues. We then describe the process of leveraging the many (wo)man-years put into scientific FORTRAN/C/C++ libraries and making them available to the Go ecosystem. The ROOT [8] data analysis framework, the C-BLAS [9] library and the Herwig-6 [10] MonteCarlo generator will be taken as examples. Finally, performances of a small analysis written in Go and using FORTRAN and C++ libraries will be discussed.

2. New languages

Since HENP and C++ met to produce (among other projects) GAUDI [11] and ROOT [8], the language landscape greatly changed. Many new languages appeared or became *"mainstream"* and, while closely following the language trend was not achieved, some adaptations were performed. For example, most of the GAUDI configuration and steering code is nowadays written in python and most, if not all, C++ components (from ROOT and GAUDI) are also available from python. But python (or more precisely CPython) has well-known scalability issues in a multithreaded environment because of its *Global Interpreter Lock (GIL)* which serializes access to python objects [1]. Moreover, even if this issue can be worked around by writing C extension modules, having an event loop in an interpreted language is not the best bet CPU-speed wise.

We investigated Go as a possible alternative to C++.

3. Elements of Go

Go [5] is a new open source language from Google, first released in November 2009. It is a compiled language with a garbage collector and builtin support for reflection, first-class functions, closures and object-oriented programming. The obligatory *"hello world"* program can be found in figure 1.

```
package main
import "fmt"
func main() {
    fmt.Println("Hello, world")
}
```

Figure 1. The obligatory *"hello world"* program, in Go.

Go is lauded to bring the best of both dynamic and static worlds:

- the feel of a dynamic language, thanks to its limited verbosity, its type inference system and its fast compile-edit-run cycle,
- the safety of a static type system,
- the speed of a machine compiled language. [2]

[1] Other python implementations (JPython, IronPython,...) do not present this limitation.
[2] the aim of the Go authors is to eventually bring the performances of a Go binary within 10% of C.

Moreover, `Go` support for interfaces which resembles the *duck-typing* motto of `python` fits nicely into `HENP` frameworks like GAUDI. Finally and more importantly, `Go` has language support for concurrency, following the *Communicating Sequential Processes (CSP)* [12] model: prefixing a method or function call with the keyword `go` will spawn off a `goroutine`: the function will be executed concurrently to other codepaths. `goroutines` are multiplexed onto multiple OS threads so blocked `goroutines` because of a non-finished I/O operation will not halt the execution of the others. Furthermore, `goroutines` are lightweight thanks to their variable stack size, starting small and growing as needed.

In `Go`, the typesafe mechanism to exchange data between `goroutines`, is a *channel*. Sending or receiving data on a channel is atomic and can thus be used as a synchronization mechanism. It should be noted that as of 2011, `Go` is lacking a few features which would probably make the implementation of typical `HENP` frameworks a bit easier, such as dynamic libraries and dynamic code loading. Another set of missing features more important for efficient scientific code is the lack of generics [3] and, for mathematical code clarity, the lack of operator overloading.

`Go` programs are constructed by linking together *packages*, whose properties allow efficient management of dependencies. Each package may in turn use facilities provided by other packages. From these inter-package dependencies, forming an acyclic dependency graph, a correct compilation order can be inferred. Within a package, all global variables, functions and types defined in that package are visible. From the outside, only the exported ones are available. In `Go`, the rule about visibility of information is simple: if a name (of a top-level type, function, method, constant or variable, or of a structure field or method) is capitalized, users of the package may see it. Otherwise, the name and hence the thing being named is visible only inside the package in which it is declared. This is more than a convention: the rule is enforced by the compiler. Each compiled package file imports transitive dependency informations. If `A.go` depends on `B.go` which itself depends on `C.go`, compiling the package `A` will result in first compiling `C.go`, `B.go` and then `A.go`. And to recompile `A.go`, the compiler will only have to read the object file `B.o` but not `C.o`. At scale, this can result in huge speedups: the improvements become exponential.

Third-party `Go` libraries and programs are also easily distributed *via* `goinstall`, an executable providing automatic package installation with dependencies' tracking. `goinstall` is the *de facto* standard to compile and distribute `Go` code [4]. An author wanting to distribute `Go` code just needs to publish it on some repository (`bitbucket`, `launchpad`, `github`, ...) and clients will be able to check out, compile and install it in one go, without special permissions, like in the shell session of figure 2.

```
shell> goinstall -v bitbucket.org/binet/go-croot/pkg/croot
goinstall: big: skipping standard library
goinstall: fmt: skipping standard library
[...]
goinstall: bitbucket.org/binet/go-ctypes/pkg/ctypes: gomake -f- install
[...]
goinstall: bitbucket.org/binet/go-croot/pkg/croot: gomake -f- install
```

Figure 2. Installing libraries or applications with `goinstall`. The `croot` `Go` package is being `goinstalled`. Note how `goinstall` automatically installed the `ctypes` package, a dependency of `croot`.

[3] also called templates in `C++`.
[4] `goinstall` is being refactored into a family of `go` binaries providing compilation and installation facilities in view of a frozen and long term supported `Go-1` version.

Then, clients can import these packages like any other, with the correct import path, as shown in figure 3.

```
package mypackage

import "bitbucket.org/binet/go-croot/pkg/croot"
// croot can now be used like any package:
// file := croot.OpenFile(...)
```

Figure 3. Importing and using third-party packages, using the correct import path.

Finally, a categorized list of all `goinstall`-able packages and commands is kept on the `godashboard` [13], providing the last piece for an effective `CPAN`-like facility.

4. Wrapping foreign language libraries

To paraphrase Tim Mattson, *"successful new languages build on existing languages and where possible, support legacy software. C++ grew out of C. Java grew out of C++."* `Go` is no exception and provides a standard way to access `C` libraries: `cgo`.

4.1. Wrapping `C` *libraries*

`cgo` allows to access and use entities defined in a `C` library with limited effort, as shown in figure 4.

```
package myclib
// #include <stdio.h>
// #include <stdlib.h>
import "C"
import "unsafe"

func foo(s string) {
  c_str := C.CString(s)  // create a C string from a Go one
  C.fputs(c_str, C.stdout)
  C.free(unsafe.Pointer(c_str))
}
```

Figure 4. Wrapping `C stdio` with `cgo`.

Wrapping bigger libraries, such as `C-BLAS`, is no different albeit quite repetitive and mechanical a task, which should probably be automatized, using `HENP` tools like `genreflex` or `rootcint`, or using a `SWIG` [14] interface.

4.2. Wrapping FORTRAN *libraries*

Wrapping FORTRAN 2003 code providing an `ISO C` interface is similar to the simple `C` library case. However, FORTRAN 77 libraries are more common in `HENP` and are dealt with the usual way:

- write a set of `C` wrappers to access the needed functionalities from the FORTRAN library,
- write `cgo` code to wrap the `C` wrappers.

```
package cblas

/*
 #include <complex.h>
 #cgo LDFLAGS: -lcblas
 #include "cblas.h"
*/
import "C"
import "unsafe"

/*
 C signature:
 float   cblas_sdsdot(const int N, const float alpha,
                      const float *X, const int incX,
                      const float *Y, const int incY);
 */
func Sdsdot(alpha float32, x, y []float32) float32 {
        if len(x) != len(y) {
                panic("slices' size differ")
        }

        c_N := C.int(len(x))
        c_alpha := C.float(alpha)

        c_X := (*C.float)(unsafe.Pointer(&x[0]))
        c_incX := C.int(1)

        c_Y := (*C.float)(unsafe.Pointer(&y[0]))
        c_incY := C.int(1)

        return float32(
                C.cblas_sdsdot(c_N, c_alpha,
                c_X, c_incX,
                c_Y, c_incY))
}
```

Figure 5. Wrapping C-BLAS with cgo.

Again, a tool automatizing writing this kind of repetitive and mechanical code would greatly ease the wrapping task. There is no technical showstopper to implement such a tool and evidence with f2py, a tool to automatically create python bindings for FORTRAN 77 code, shows it is indeed possible to generate the needed boilerplate code.

4.3. Wrapping C++ libraries

Wrapping C++ is a bit more involved a task. The current solution is to use SWIG and a SWIG interface file to generate cgo code. The Go backend of SWIG is sophisticated enough to handle and map C++ constructs not existing in Go, back to idiomatic Go code. For example, overloaded functions are dispatched from a common Go variadic function to each cgo-wrapped function; while multiple inheritance is handled by defining multiple Go interfaces; and virtual methods are implemented by methods on interfaces. And thanks to SWIG machinery, it is even possible to implement a C++ abstract class with a Go struct implementing the according interface.

```
/* c-hepevt/hepevt.h */

#ifdef __cplusplus
extern ''C'' {
#endif

/** event number
 */
int
hepevt_event_number();

#ifdef __cplusplus
}
#endif

/* c-hepevt/hepevt.cxx */
extern ''C'' {
  extern struct {
    char data[hepevt_bytes_allocation];
  } hepevt_;
}
#define hepevt hepevt_

static int
byte_num_to_int(unsigned int b)
{
  // consistency checks
  // ...
  if (s_sizeof_int == sizeof(short int)) {
    short int *si = (short int*)&hepevt.data[b];
    return (int)(*si);
  }
  // error handling
  // ...
  return 0;
}

/** event number
 */
int
hepevt_event_number()
{
  return byte_num_to_int(0);
}
```

Figure 6. Excerpts of the C wrapper around HEPEVT.

However, SWIG does not use a proper C++ compiler to parse the C++ header files and thus can not handle all of C++03 constructs. Attempts at wrapping ROOT's TObject class for this paper failed, even after trying numerous tricks like presenting a preprocessed TObject.h file to SWIG or injecting `#ifdef SWIG` preprocessor directives [5].

A more versatile tool, built on top of `gccxml` or `clang`, would resolve this issue. In the meantime, `Go` can still leverage `ROOT` libraries, resorting to the same recipe than in the FORTRAN77 case: write a `C-API` on top of the `C++` one and write `cgo` code to wrap the former.

5. Results

To test the performances and applicability of using `Go` with *"legacy"* `C++` code, the transliteration of the canonical `ROOT TTree` reading example from the `ROOT` tutorial has been performed. The code is provided on a `Mercurial` repository [15]. The same exercize has been done using the `C-API croot` calling the usual `ROOT C++` code to infer the overhead induced by this additional layer, and disentangle the contribution from the pure `cgo` overhead. Results are reported in figure 7.

```
shell> time ttree-ex-1
 29.04s user  1.03s system 86% cpu 34.83 total (C++)
 29.12s user  1.09s system 85% cpu 35.48 total (CRoot)
 44.83s user  1.24s system 87% cpu 54.36 total (go-croot)

shell> uname -a
Linux farnsworth 3.0-ARCH #1 SMP PREEMPT
x86_64 Intel(R) Core(TM)2 Duo
CPU T9400 @ 2.53GHz GenuineIntel GNU/Linux
```

Figure 7. Timings obtained for the `ROOT TTree` canonical example.

These timings show less than a factor two overhead with regard to the `C++` version. One should note that these timings were obtained with the 64bits version of the `gc` compiler (`weekly.2011-09-01 9619`). We should note that, while performances of the `gc` compiler are expected to improve, using the `gccgo` compiler would be a better match as far as pure number crunching and calling `C` functions are concerned. Indeed, `gccgo` can on one hand leverage the full set of optimizations that `GCC` can apply and on the other hand has smaller overhead calling `C` functions [6]. Nonetheless, we kept using `gc` as this is the compiler implementing all of the `Go` features.

6. Conclusions

We presented the work on investigating `Go` as a viable `C++` replacement for the multi-core era. `Go` has many builtin capabilities to ease the burden of exposing and leveraging concurrency in `HENP` applications as previously shown in [16]. The short development cycle together with the code distribution platform and tools provide `Go` with a short *'time-to-market'* length.

Wrapping foreign language libraries, and especially `C` libraries, is made easy thanks to `cgo`, a tool shipped with the `Go` distribution. Wrapping `C++` is possible - albeit cumbersome - with `SWIG` but is subject to `SWIG`'s limitations and does not seem to be applicable to large libraries such as `ROOT`. An automatized solution built on top of `gccxml` or `clang` would greatly improve the ability of `Go` to reuse *'legacy'* libraries.

[5] While this could theoretically be a solution for `SWIG`'s preprocessing troubles with some macros in `ROOT` code although not a particularly aesthetical one, this would not address `SWIG`'s limitations in handling `<` and `>` operators.

[6] `gc` has a different calling convention than `gcc`, so `cgo` needs to perform additional work when calling `C` functions. `gccgo`, for obvious reasons, has the same calling convention than `gcc`.

References

[1] Moore E 1965 Cramming more components onto integrated circuits *Electronics Magazine* vol 38 no 8 pp 114-117
[2] Amdahl G 1967 Validity of the Single Processor Approach to Achieving Large-Scale Computing Capabilities *AFIPS Conference Proceedings* pp 483-485
[3] The C++ programming language http://www.open-std.org/jtc1/sc22/wg21/docs/papers/2010/n3092.pdf
[4] The python programming language http://python.org
[5] The Go programming language http://golang.org/
[6] Intel Threading Building Blocks http://threadingbuildingblocks.org
[7] OpenMP (Open Multi-Processing) http://openmp.org
[8] The ROOT framework, http://root.cern.ch
[9] Lawson C L, Hanson R J, Kincaid D, and Krogh F T 1979 Basic Linear Algebra Subprograms for FORTRAN usage *ACM Trans. Math. Soft* pp 308-323 http://www.netlib.org
[10] Corcella G, Knowles I G, Marchesini G, Moretti S, Odagiri K, Richardson P, Seymour M H and Webber B R 2001 HERWIG 6.5 *JHEP 0101* (*Preprint* hep-ph/0011363)
[11] Mato P 1998 GAUDI-architecture design document *Tech. Rep. LHCb-98-064*
[12] CSP http://en.wikipedia.org/wiki/Communicating_sequential_processes
[13] The Go dashboard http://godashboard.appspot.com/project
[14] SWIG http://swig.org
[15] https://bitbucket.org/binet/go-io-benchmarks
[16] Binet S 2010 ng: what next-generation languages can teach us about HENP frameworks in the manycore era *J. Phys.: Conf. Ser.* **331** 042002

Multicore in production: advantages and limits of the multi-process approach in the ATLAS experiment

S Binet[1], P Calafiura[2], M K Jha[3], W Lavrijsen[2], C Leggett[2], D Lesny[4], H Severini[5], D Smith[6], S Snyder[7], M Tatarkhanov[2], V Tsulaia[2], P VanGemmeren[8] and A Washbrook[9]

[1] Laboratoire de l'Accélérateur Linéaire/IN2P3, 91898 Orsay Cédex, France
[2] Lawrence Berkeley National Laboratory, 1 Cyclotron Rd, Berkeley, CA 94720, USA
[3] INFN, CNAF & Bologna, Bologna 40127, Italy
[4] University of Illinois at Urbana-Champaign, 1110 W Green St, Urbana, IL 61801, USA
[5] University of Oklahoma, 440 W. Brooks Street, Norman, OK 73019, USA
[6] SLAC National Accelerator Laboratory, 2575 Sand Hill Rd, Menlo Park, CA 94025, USA
[7] Brookhaven National Laboratory, P.O. Box 5000, Upton, NY 11973, USA
[8] Argonne National Laboratory, 9700 S. Cass Avenue, Argonne, IL 60439, USA
[9] SUPA, School of Physics and Astronomy, The University of Edinburgh, James Clerk Maxwell Building, Mayfield Road, Edinburgh, EH9 3JZ, UK

E-mail: vtsulaia@lbl.gov

Abstract. The shared memory architecture of multicore CPUs provides HEP developers with the opportunity to reduce the memory footprint of their applications by sharing memory pages between the cores in a processor. ATLAS pioneered the multi-process approach to parallelize HEP applications. Using Linux fork() and the Copy On Write mechanism we implemented a simple event task farm, which allowed us to achieve sharing of almost 80% of memory pages among event worker processes for certain types of reconstruction jobs with negligible CPU overhead. By leaving the task of managing shared memory pages to the operating system, we have been able to parallelize large reconstruction and simulation applications originally written to be run in a single thread of execution with little to no change to the application code. The process of validating AthenaMP for production took ten months of concentrated effort and is expected to continue for several more months. Besides validating the software itself, an important and time-consuming aspect of running multicore applications in production was to configure the ATLAS distributed production system to handle multicore jobs. This entailed defining multicore batch queues, where the unit resource is not a core, but a whole computing node; monitoring the output of many event workers; and adapting the job definition layer to handle computing resources with different event throughputs. We will present scalability and memory usage studies, based on data gathered both on dedicated hardware and at the CERN Computer Center.

1. Introduction

Memory is a scarce resource for typical ATLAS reconstruction jobs and by running many individual jobs simultaneously we are likely to hit hardware memory limits on the production machines. Thus, in order to optimally exploit all available CPU cores on a given node, we need to have a parallel solution which allows us to share resources between processes or threads. The ATLAS experiment pioneered the process based parallelism in HEP applications. We have implemented a simple event task farm, which relies on the Linux kernel Copy On Write (COW) mechanism. The basic strategy of this approach is to go through the initialization stage of a job in a single master process, allocate and initialize as much of the memory that will be used by the job, then fork many worker processes and let them process their share of events of the job. Workers share a large fraction of process memory between each other and such memory sharing is solely provided by the COW mechanism. We have been able to run in parallel large reconstruction and simulation applications originally written to be run in a single thread of execution with little to no change to the application code.

Besides validating the software itself, an important and time-consuming aspect of running multicore applications in production was to configure the ATLAS distributed production system to handle multicore jobs. This entailed defining multicore batch queues, where the unit resource is not a core, but a whole computing node; monitoring the output of many event workers; and adapting the job definition layer to handle computing resources with very different event throughput (depending on the number of cores used).

This paper describes design and implementation concepts of AthenaMP [1] - the Multi-Process version of the ATLAS common reconstruction and analysis framework *Athena*. We present performance, scalability and memory usage studies based on data gathered both on dedicated hardware and at the CERN Computer Center.

In section 2 we go through details of AthenaMP implementation with the emphasis on minimized user code changes. Section 3 is devoted to running AthenaMP jobs on the Grid. Section 4 discusses AthenaMP validation process and also describes a series of large scale AthenaMP reconstruction tests performed over Summer 2011 at the CERN Computer Center ("Tier-0" of the LHC Computing Grid). In section 5 we present performance figures comparing AthenaMP reconstruction to N individual Athena processes run in parallel on the same machine. Finally, section 6 presents several ways to further improve AthenaMP performance, in particular by increasing reconstruction job size and by implementing specialized worker processes, which will be in charge of the event I/O operations.

2. AthenaMP

2.1. Sharing memory between processes using Copy On Write

It is trivial to achieve process-based parallelism on a multicore host. The simplest scenario, which *a priori* requires no code changes, is just to run N instances of the same application simultaneously. However, the obvious problem with this approach is the suboptimal usage of system resources. Nothing apart from dynamic shared objects in memory and files opened in read only mode is shared between these application instances. Such trivial multi-process architecture leads to increased memory requirements, which rapidly scale with number of individual processes.

One possible way to solve the memory duplication problem is to allocate and initialize as much memory as possible during the initialization phase of a single parent process and then fork() N child processes (workers). Upon invocation fork() clones the parent process including its entire address space. This may sound as counter-productive for the problem at hand, however modern operating systems use Copy On Write mechanism as an efficient optimization for such cases. The child is not given a private copy of the process memory, instead parent and child processes share physical memory up to the point when one of them writes to it. At that point, the affected memory page will be copied and become a 'property' of the writer process. Obviously the optimal strategy for such architecture is to fork() as late as possible in order to achieve maximal memory sharing between processes. The effect of such late forking can be quite dramatic and this is demonstrated by the performance figures in section 4 of this paper.

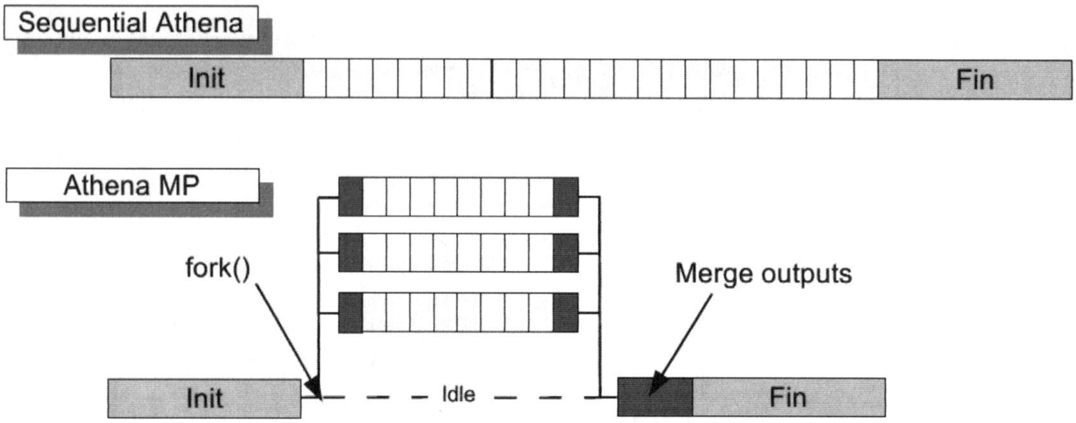

Figure 1. Athena MP versus sequential Athena job.

Another big advantage of the fork-based multi-process approach is that each child process can continue its work independently with little to no communication with either its parent process or other workers. As a result no changes are required in the user code for serializing access to shared data and to critical sections of the code under the assumption that all communications between processes will be handled by the framework itself.

Finally, relying on COW ensures that as much memory as possible is shared and automatically done so by the Linux kernel.

The main disadvantage of this approach is that once a memory page became unshared due to a write, it cannot be shared again later on. This may not be a real issue during event processing stage of the worker processes, assuming that workers don't need to allocate large chunks of memory individually. However, the implications of the memory un-sharing can be quite severe and we have observed huge spikes in overall memory consumption at the finalization stage of worker processes, which introduce considerable performance penalties when we hit machine memory limits. We believe that these memory spikes are caused by simultaneous calling of object destructors by worker processes. However the exact reason is yet to be understood.

2.2. Process management and output file handling

One of the constraints of parallelizing Athena is to avoid client code changes: AthenaMP fulfills this condition by hiding multi-process semantics inside the Athena/Gaudi [2] framework instead of publishing them as a layer atop.

Process steering in AthenaMP is implemented using python multiprocessing module [3]. A pool of forked sub-processes is created just before entering the event loop. The child processes are given a bootstrap function to handle I/O reinitialization (reopen file descriptors) and asynchronously scheduled to process all input events. The events are distributed between workers on a first come first served basis using a shared queue, which provides workers with event sequence numbers. Besides accessing common input files the event workers run independently in separate run directories, they don't communicate to one another and make their own output files. In the current implementation of AthenaMP workers process events one by one. For the future we plan to change this approach and let the workers process events in chunks in order to avoid reading the same portion of the input ROOT file by multiple processes and by this way improve overall performance of AthenaMP jobs.

When there are no more events to process the workers go through finalization stage and exit. Once all of the workers have terminated the control is passed back to the parent process, which then proceeds with merging workers' output files. This is probably the most problematic part of AthenaMP as POOL [4], the object persistency library used by ATLAS, does not support parallel I/O. Thus the parent process has to merge all types of outputs in serial and even after recent switch to the fast merger utility for POOL files, which improved merging performance by one order of magnitude, the output merging remains one of the critical areas which has a negative impact on the overall performance of the Athena MP.

3. AthenaMP on the Grid

In order to run AthenaMP reconstruction jobs on the ATLAS distributed production system it was necessary to define special multicore batch queues, where the unit resource is a whole computing node. As well as the whole-node production queue at CERN there are Tier-1 and Tier-2 Grid sites in the UK (RAL, Edinburgh, Glasgow), US (Illinois and OSCER - University of Oklahoma, OU Supercomputing Center for Education and Research) and Italy (INFN-T1, Naples) that have been configured to accept AthenaMP jobs submitted from the ATLAS production system. These candidate sites host a number of different batch system implementations (SGE, Torque/Maui, LSF and Condor) which are representative of the majority of Grid sites available to ATLAS.

In order for these computing resources to be made available for whole-node jobs it was necessary to apply configuration changes to the batch system scheduler at each participating site. At this time there is no prescriptive model on how to configure a Grid site for whole-node job execution. Rather, the choice of configuration was dependent on the existing resource allocation defined by each site.

AthenaMP use on the Grid poses a set of unique difficulties. In local use of AthenaMP, the user knows what hardware is being used, and is able to set the number of processes to match this hardware. But for jobs run on the Grid, this is not the case. The user will be able to submit jobs, but usually is not able to choose the resources where those jobs will run. So the number of processes in use for the job needs to be defined at run time, on the worker node.

The first method to answer this problem, is in the queue definitions on the Grid resources. The jobs will be assigned to certain sites, and from there run on sets of resources defined in a batch queue. In job management systems, there is added information about the queue, in the form of the number of CPU cores that exist in all resources on this queue. At run time, the job management system will use this information to define an environment variable, with the CPU core information. AthenaMP at run time can use the value of this variable to define the number of processes to use in running.

This creates some problems for the remote site and job management. The remote site usually has the CPU resources defined as one job per CPU core as a default. The queues for AthenaMP running have to be defined as all MP jobs, so there will be a certain number of CPUs per job defined. In practice sites should define the number of cores used per job as the number cores per machine in the queue. When a mixture of single core and multicore jobs compete for the same resources a non-optimal setup may impact on the overall job throughput. For example, a high priority whole-node job could block access to a compute node until all cores are available for its exclusive use. During this wait time it is feasible that single-core jobs could have been executed and completed on the host marked for use (a process known as backfilling) if the lifetime of the whole-node job was known in advance to the scheduler.

The only efficient use of the resources with low submit times is the use of one job per machine. When the jobs are finished, the whole node is then immediately free to accept the next in the queue, and submit times remain proportional to queue length. Although this was a relatively simple solution to implement it may not be the most suitable solution for every site, especially where the rate of whole-node jobs submitted is small. In low-usage periods any resources marked for whole-node use will remain idle and could have been used to process single-core jobs instead.

At this time several queues have been set up for AthenaMP use. Since AthenaMP is not in full data production the number of resources dedicated to these queues is relatively low. AthenaMP jobs have been run on each of these test queues and shown to be using the correct number of processes for the target worker nodes. Once run the status of the jobs is correctly determined, log files and the associated data produced by each job are copied to central storage resources. Job success and consequent data transfer is monitored by the ATLAS Panda job management system.

One difficulty with the above job submission procedure is that many sites need to manage large sets of non-homogeneous resources, and usually do not have the same number of CPU cores per machine in use throughout a single queue. Defining and managing special queues for MP use becomes difficult for sites in these cases and the queues will be defined on a set of machines with variable numbers of CPU cores.

In order to improve CPU efficiency of AthenaMP jobs it is necessary to adjust their size, i.e. the number of events to be processed by a given job, in accordance with number of CPU cores available on a given node. By running longer multi-process jobs we minimize the time the job spends in its sequential part and by this way we achieve better overall event throughput.

4. Validation of AthenaMP

The validation process of AthenaMP took several months of coordinated effort and which will now continue in step with the future Athena release cycle. At the first stage we had to make sure that AthenaMP can be used for running different types of reconstruction jobs, that there were no critical bugs and crashes and also the output produced by AthenaMP is sensible. After reaching a good level of stability we started to experiment with sending AthenaMP jobs to whole-node queues on the Grid and also we ran a series of large scale tests at the CERN Tier-0 cluster.

At the same time we started to perform more detailed comparisons of output files produced by AthenaMP and by serial Athena reconstruction jobs. For such comparison tests we usually ran reconstruction over small number of events and compared outputs bit by bit. After fixing all sources of differences seen between outputs of serial and multi-process jobs, we now plan to run AthenaMP reconstruction over large event samples and pass the results to ATLAS physics validation groups for detailed studies.

4.1. Large-scale tests of AthenaMP at CERN Tier-0 cluster

As the part of AthenaMP validation process we ran two series of large scale tests at the CERN Tier-0 cluster. The main goal of these tests was to ensure no disruptive failures are observed and to compare parallel and sequential reconstruction performance characteristics. In total we ran over 1.1K AthenaMP reconstruction jobs, which processed in total around 2M events. A few minor problems which were not known before were promptly fixed in advance of the second round of testing. The resulting performance characteristics were collected from LSF logs.

Both test rounds demonstrated that parallel reconstruction is considerably slower than its MJ counterpart. Total wall clock times for both modes of running are collected in table 1.

Table 1. Total wall clock times of sequential (MJ) and parallel (MP) Athena reconstruction jobs measured during two rounds of tests at CERN Tier-0 cluster.

	Athena MJ	AthenaMP	Athena MJ/AthenaMP
Round 1	173 core-days	360 core-days	0.48
Round 2	72 core-days	106 core-days	0.68

The speedup of AthenaMP reconstruction observed in the second round of testing was achieved by switching to the fast merger utility for POOL output files. The remaining difference between parallel and sequential reconstruction performance was due to both series of tests ran MP and MJ jobs of the same size, thus while every single parallel job ran faster than its sequential counterpart the overall observed performance degradation of parallel jobs was a direct consequence of Amdahl's law [5]. The only way to make AthenaMP competitive to the sequential reconstruction in terms of event throughput is to minimize the contribution of the serial part – initialization, merging, finalization – to the total wall clock time of the parallel job. In order to achieve that we need to extend AthenaMP job size, which means running over multiple input event files instead of just one for a single AthenaMP reconstruction job.

5. Performance studies

In order to study AthenaMP performance characteristics we ran a series of reconstruction jobs on dedicated hardware: dual quad-core 2.27 GHz Intel Xeon L5520 processors with Hyper-Threading enabled and with a total memory of 24GB. We ran either single AthenaMP job with N workers or N independent sequential Athena jobs.

Figure 2 shows maximal memory usage on the machine during the event processing loop. It does not take into account memory access spikes observed at finalization stage of worker processes. This figure demonstrates the effect of late forking on memory sharing optimization in AthenaMP jobs. In typical Athena reconstruction jobs a large memory chunk gets allocated after the initialization phase once the first event is processed. At this point the job loads and caches a large amount of data read from ATLAS conditions database. If we fork sub-processes after the first event then parent process memory is shared between AthenaMP workers, which results in a significant reduction in total memory footprint compared to running multiple instances of Athena. For certain types of reconstruction jobs we can share up to 80% memory pages among event worker processes.

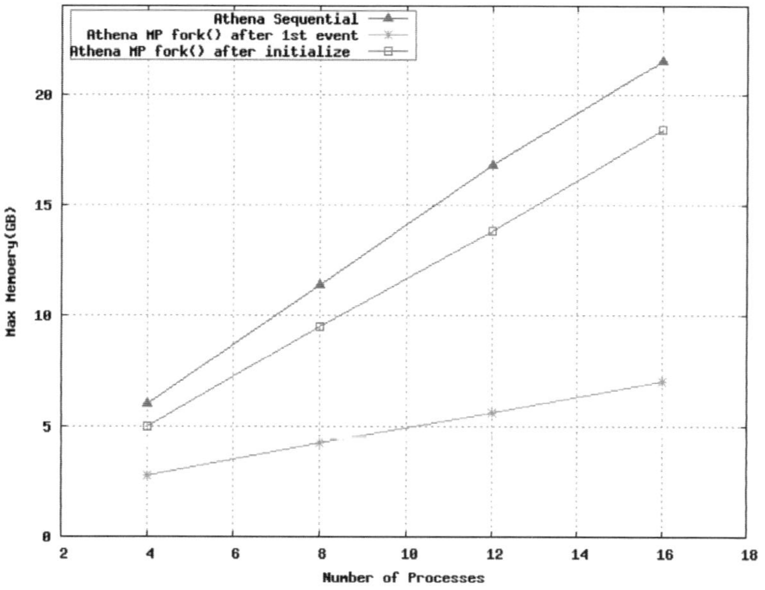

Figure 2. Maximal memory consumption seen on a dedicated machine during AthenaMP reconstruction event loop. Demonstrates the effect of late forking on memory sharing between worker processes.

6. Further improvements of AthenaMP performance

As already mentioned in section 4.1, in order to compete with event throughput of N individual sequential Athena reconstruction jobs executed in parallel, it is necessary to increase the size of AthenaMP jobs (i.e. increase the number of events to be processed by the jobs), because by doing this we minimize the fraction of total wall clock time spent in the sequential part of the job. Increasing the job size may seem to be fairly easy to achieve – one just needs to feed reconstruction with multiple input files instead of one, however in practice this will not be as simple when it comes to running such large jobs on the ATLAS distributed production system. In order to cope with this requirement the pilot process on a worker node will have to determine the number of processes to use depending on available system resources. Once determined the pilot will request a job of the appropriate size to be executed with that number of processes. Such mode of operation is currently under development.

An important area of AthenaMP, which requires long term development, is related to handling of I/O operations. The current implementation, when worker processes read and write files independently, does not scale with the number workers and in order to improve this situation we plan to develop two types of specialized I/O workers – Event Source (reader) and Event Sink (writer), which will communicate events with event workers via shared memory.

By introducing such processes we expect to reduce disk access and overall memory footprint, as, for example, all reading will be done in one single process and we will avoid duplicating ROOT [6] buffers. Also the combined event reader will reduce the number of open file handles. Specialized I/O processes will free event workers from dealing with I/O operations and by this way improve overall CPU performance of AthenaMP jobs. Finally, by introducing an Event Sink process we eliminate the need of output file merging. Events for writing will be provided to the Event Sink by event workers via dedicated shared memory segments and the actual writing to the file will be done only by a single process.

7. Summary

After initial development and validation phase AthenaMP is entering the stage when it can be used in large scale test campaigns, whose results will be handed over to ATLAS physics groups for further validation. By running reconstruction jobs in parallel using AthenaMP we can achieve much reduced overall memory footprint than by running the same number of independent serial jobs. The fast merger utility for output POOL file merging has greatly improved the event throughput of AthenaMP jobs, however in order to compete in CPU efficiency with serial jobs we need to increase AthenaMP job sizes by running them over multiple input files instead of just one. After passing the validation procedures we plan to start using AthenaMP widely on the ATLAS production system.

In order to be able to run AthenaMP jobs on the Grid special whole-node production queues have been configured at CERN and at several Tier-1 and Tier-2 sites. The necessary configuration changes have been applied to the batch system scheduler at each participating site to make computing resources available for whole-node jobs. AthenaMP use on the Grid poses a set of unique difficulties and there are several concerns related to job scheduling and efficiency, which are currently being addressed.

The optimal usage of available memory resources, good level of stability and good agreement between parallel and sequential reconstruction results makes AthenaMP a viable candidate to become a default mode for running large scale productions at CERN Tier-0 and on the Grid.

References

[1] Binet S, Calafiura P, Snyder S, Wiedenmann W and Winklmeier F 2010 Harnessing multicores: strategies and implementations in ATLAS *J. Phys.: Conf. Ser.* **219** 042002
[2] Mato P 1998 Gaudi - architecture design document *Tech. Rep.* LHCb-98-064 Geneva
[3] The multiprocessing module http://docs.python.org/library/multiprocessing.html
[4] POOL – persistency framework http://pool.cern.ch
[5] Amdahl G 1967 Validity of the single processor approach to achieving large-scale computing capabilities *AFIPS Conference Proceedings* (30) pp 483-85
[6] ROOT – data analysis framework http://root.cern.ch

PROOF on the Cloud for ALICE using PoD and OpenNebula

D Berzano[1,2], **S Bagnasco**[2], **R Brunetti**[2], **S Lusso**[2]

[1] Dip. di Fisica Sperimentale—Univ. di Torino—Via P. Giuria 1, 10125 Torino—Italy
[2] Istituto Nazionale di Fisica Nucleare—Via P. Giuria 1, 10125 Torino—Italy

E-mail: dario.berzano@to.infn.it

Abstract. In order to optimize the use and management of computing centres, their conversion to cloud facilities is becoming increasingly popular. In a medium to large cloud facility, many different virtual clusters may concur for the same resources: unused resources can be freed either by turning off idle virtual machines, or by lowering resources assigned to a virtual machine at runtime. PROOF, a ROOT-based *parallel* and *interactive* analysis framework, is officially endorsed in the computing model of the ALICE experiment as complementary to the Grid, and it has become very popular over the last three years. The locality of PROOF-based analysis facilities forces system administrators to scavenge resources, yet the chaotic nature of user analysis tasks deems them unstable and inconstantly used, making PROOF a typical use-case for HPC cloud computing. Currently, PoD dynamically and easily provides a PROOF-enabled cluster by submitting agents to a job scheduler. Unfortunately, a Tier-2 does not comfortably share the same queue between interactive and batch jobs, due to the very large average time to completion of the latter: an elastic cloud approach would enable interactive virtual machines to temporarily subtract resources to the batch ones, without a noticeable impact on them. In this work we describe our setup of a dynamic PROOF-based cloud analysis facility based on PoD and OpenNebula, orchestrated by a simple and lightweight control daemon that makes virtualization transparent for the user.

1. PROOF-based analysis facilities and the ALICE Experiment

PROOF (Parallel ROOT facility) [1] is a framework that enables ROOT [1] to perform parallel and interactive data processing in "embarrassingly parallel" use cases, such as typical HEP event-based activities. Events are processed in parallel by independent ROOT sessions (the *workers*) on multiple machines; results are collected and merged afterwards by a *master* session.

PROOF enables computing resources to be used *interactively*: all the requested PROOF resources are ideally available almost immediately. Workers can also be added or removed at runtime, and in case a worker crashes the job would proceed on the others.

A conventional static PROOF installation requires the *xproofd* daemon to listen for incoming connections on one machine: upon connection it will communicate to static *xproofd* daemons on the other machines to provide the user with workers.

Since the time required to complete the analysis of a single event depends on the available computing power and the complexity of the event itself, events are not evenly distributed to the workers; instead, the workers ask for new events when they are finished analyzing. This pull architecture, called the *dynamic packetizer* [2], leads to a uniform completion time.

A typical use case for a PROOF-based analysis facility in HEP is the optimization of user analysis: interactivity enables the physicist to immediately re-run the analysis after each small change in the code and cut parameters.

Those analyses run usually for a few minutes; moreover, since the tasks are run *interactively* by the user, there is a demand for PROOF resources only during working hours. Such discontinuous occupancy means that local PROOF resources may be largely underused.

PROOF is also available on larger scale deployments, where analyses on bigger samples or even full simulations or reconstructions can be run interactively.

Such facilities are normally used by a larger number of users, and fast network connectivity attracts non-local users. However, having a single static PROOF master process accepting connections for all the users is an approach that does not scale very well and can undermine the stability of the whole system.

At this point it should be clear that static PROOF deployments pose some issues when resources are either underused or heavily loaded: a *dynamic* PROOF deployment would rather make unused resources available for some other task in the first case, and would absorb peak loads in the latter.

PROOF is officially part of the ALICE computing model since 2006, when the CAF (Central Analysis Facility) was first set up at CERN [4]. In 2010 a set of guidelines for AAFs (ALICE Analysis Facilities) [5] were adopted to provide a uniform working environment throughout all ALICE PROOF clusters. AAF specifications cover: authentication via Grid proxy, software management via AliEn Packman [6][7], access to local data via a distributed disk pool aggregated and served via Scalla/xrootd [8] where pool disks are installed on the PROOF servers themselves, dataset management and staging [9] and monitoring via web through MonALISA [10].

The virtual analysis facility we are going to discuss is compliant to most of the AAF guidelines.

2. The tools: PROOF on Demand and OpenNebula

2.1. PROOF-on-Demand

PROOF on Demand (PoD) [11] is a set of tools that enables PROOF to be instantiated dynamically on an existing batch farm, making no particular assumptions on the underlying batch system. There are plugins for a wide range of resource management systems. Another independent plugin, *pod-ssh*, does not even need a batch system and uses standard passwordless SSH to access the worker nodes.

Its components act as wrappers around PROOF; the most important are:

- *pod-server*—launched only once by the user, it sets up the PROOF master and monitors the status of the PROOF cluster; it also terminates the cluster automatically in case of inactivity;
- *pod-agent*—it prepares the environment for PROOF and controls a single PROOF worker throughout its whole lifecycle, firing it up and cleaning up temporary files in the end; it is meant to be deployed by a batch system via *pod-submit*;
- *pod-submit*—queues the specified number of *pod-agent* jobs to the batch system, requesting for a certain number of PROOF workers: each *pod-agent*, once deployed, will communicate with the corresponding *pod-server*.

As soon as the first worker becomes available the user can start using PROOF; further workers will join the facility as soon as they are ready, or can be added to it at a later time.

Since each user has a different PROOF master, PoD helps addressing the scalability issues of a large PROOF deployment.

2.2. OpenNebula

OpenNebula [12] is an open source IaaS (Infrastructure as a Service) framework to set up, administrate and monitor public and private clouds. OpenNebula takes care of the setup and provisioning of the virtual resources, including storage areas and network connectivity. All of these virtual resources, and the virtual machines themselves, are described inside specific *templates*.

We favored OpenNebula over similar tools because it is widely used and almost every functionality is implemented via Bash or Ruby scripts, making it easy to add missing functionalities or to tailor the existing ones around our use case. OpenNebula refers to these scripts as MADs (Middleware Access Drivers): there exist MADs for transferring images, monitoring, handling the authentication and supporting different virtualization methods.

As of v3.0, the supported hypervisors [13] are Xen [14], KVM [15] and VMWare [16], all of them implementing basic deployment functionalities such as suspension, migration and even live migration. An interface to Amazon EC2 [17] is also available, as well as OCCI APIs for public cloud setups. Different hypervisors or cloud interfaces can be used together to set up a hybrid cloud.

Virtual machine images are registered inside a non-hierarchical repository: the *transfer manager* MAD is responsible of transferring the images from the image repository to the hypervisor *(cloning)* and vice-versa *(saving)*. Transfer managers supplied by default with OpenNebula include: the NFS transfer manager, for images on a filesystem shared between all the hypervisors; the LVM transfer manager, to run VM images written on LVM partitions instead of plain image files; the SSH transfer manager, to transfer the images using *scp*.

OpenNebula also features a web interface called Sunstone [18] that supports custom plugins.

3. PROOF on Demand and OpenNebula interplay

A prototype of a private cloud based on OpenNebula is being set up in the computing centre of INFN-Torino, a WLCG Tier-2 centre. The cloud runs a dynamic PROOF cluster based on PoD that uses a dedicated batch system whose worker nodes are virtual machines created and destroyed dynamically using the mechanisms supplied by OpenNebula.

There are many advantages in using a batch system over using *pod-ssh*. First of all, *pod-ssh* connects to machines that are already up and running, while through *pod-submit* we could place the *pod-agents* in a queue and the batch system will deploy them in order when the machines are up.

Secondly, the batch system knows the structure of the cluster and maintains a register of what is currently running on it, making tasks wait in a queue until the resources to run them become available—*pod-ssh* does not feature any queue and would be likely to overload the nodes.

Furthermore, *pod-ssh* needs a list of the valid PoD-enabled machines to connect to: making this list dynamically provided and making pod-ssh wait until the machines are up would require the reimplementation of features that are core functionalities of any batch system.

A batch system is also useful to customize the jobs distribution and to enforce policies (while no particular limitation can be enforced via *pod-ssh*): for instance, we could configure the batch system to keep jobs of different users separated on different virtual machines, or limit the maximum number of concurrent *pod-agents* or the maximum walltime.

We decided to use TORQUE v2.4 [19], based on PBS and simple to configure, but any of the batch systems supported by PoD would in principle work fine.

3.1. Farm configuration

3.1.1. Configuration of the access node. In our setup the user submits the jobs via pod-submit to a single access node (in this case it acts as a wrapper around TORQUE's qsub) which runs the TORQUE server (pbs_server) and the job scheduler (pbs_sched).

Given the very simple nature of our jobs, we were able to achieve an improved response time by configuring the schedule to reduce the default queue polling interval from 1 min to 7 s. This means that it takes TORQUE at most 7 s to dispatch the pod-agent to a freshly connected worker node.

Also, since *pod-agents* are disposable, we do not care to keep track of them in the queue when they are finished: the queue is configured to flush the agents as soon as they finish.

3.1.2. Configuration of the virtual dynamic worker nodes. Since worker nodes for PROOF are virtual machines, once the configuration is done on one machine the corresponding image can be cloned several times.

The virtual worker node runs the *pbs_mom* daemon that controls the running jobs and communicates with the server. Since the running worker nodes are created dynamically, the TORQUE server should be somehow made aware of the presence of the new node in order to start assigning jobs to it.

In our setup, virtual worker nodes add themselves automatically to the PoD queue by executing a command on the server via SSH right after the bootstrap: a private key is used for passwordless authentication, and for security reasons connections made with that key are limited to execute only one command, chosen to be the *qmgr* command that adds the node to the PoD queue.

This pull architecture is easier, less error-prone and quicker than adding nodes separately: in fact, if a node can successfully add itself to the queue it means that network connectivity works properly, and we avoid the cases where a virtual machine deployment fails, leaving a dangling node in "offline" state in the TORQUE nodes list.

Our operating system of choice for the virtual machines is CentOS 5.7, binary compatible with upstream RHEL releases, so that we can run precompiled ALICE software. The cloud environment currently runs on KVM hypervisors.

When running PROOF, the version of ROOT and AliRoot to use for the session is chosen upon connection: since this user decision happens way after virtual machine's boot, we cannot deploy a virtual machine self-containing only the requested software version; on the other side, a VM with all the ALICE software versions (tens of GB) would be both difficult to maintain and very clumsy to deploy.

ALICE software packages, as defined by AAF specifics, are exacty the same as the packages for the Grid worker nodes, so we just mount a shared ALICE Grid software area exported using the Lustre™ distributed filesystem [20], without the need for a software manager specific to PROOF.

In order to make the virtual machine image as small as possible, both the swap space and an empty scratch filesystem are created by OpenNebula upon the virtual machine deployment directly on the disks of the target hypervisor.

3.1.3. A physical PROOF cluster with a virtual elastic extension. It is worth noting that the configuration of the virtual worker nodes, apart from some configuration items specific to the virtualization (such as the dynamically-created disk partitions), is a perfectly valid configuration also for a physical machine. A virtual PROOF worker node and a physical one are, from the point of view of the batch system, nothing more than job slots, so we can actually use the batch system to deploy PROOF workers on a physical analysis facility with an elastic extension made out of virtual machines.

In this hybrid case the batch system can be, for instance, configured to favor physical over virtual worker nodes for performance reasons, a distinction that would not be possible to accomplish easily using *pod-ssh*.

3.1.4. Sharing virtual resources with other virtual clusters dynamically. The most important aspect of our PoD over OpenNebula cluster is the deployment swiftness: since interactive resources must be made available in a reasonably short amount of time, every configuration option has been tailored to gain precious seconds.

However, since we cannot afford to wait hours for some virtual resources to be freed by a certain virtual cluster, our PoD virtual machines can be configured to grab some CPU resources from other running virtual clusters without stopping them. This is currently supported on KVM by simply using *renice*, since virtual machines are treated as ordinary Unix processes.

Squeezing resources can not be done indiscriminately: for instance, it is agreed that an ALICE PROOF worker node can temporarily steal resources from a virtual Grid node running very long batch jobs from the same VO. The static prototype of the Virtual Analysis Facility we set up in Torino in 2008 [21][22] is now much easier to set up as a special use case of our OpenNebula cloud.

3.2. The load sensor

Turning the PROOF virtual machines on can be done manually by the system administrator or by the end user, but in a production facility it is much better to automatize such tasks. Moreover, since PoD is a well-established tool and very easy to learn, we would like to make the user completely unaware of the underlying virtualization layer, by not changing the standard PoD workflow (*pod-server* followed by *pod-submit*) and without patching PoD code.

The batch system can keep track of *pod-agents* waiting in the queue: if there are any, actions can be taken accordingly, such as spawning new virtual machines. On the other side, the batch system monitors the status of the nodes, so that we can turn off idle nodes, *i.e.* nodes that are haven't been executing PROOF workers for a certain amount of time.

The load sensor is a daemon (called *bananad*) that performs the two tasks above. It continuously checks the queue for waiting jobs and the nodes status for idle nodes. It also removes stale "offline" nodes hanging in the batch system for too long.

Continuously polling the status of the queue via the *qstat* command is a resources-consuming task, so wherever possible TORQUE logs are opened and read in a separate thread as more lines are appended to them. This strategy makes the daemon very lightweight.

All the *bananad* thresholds are configurable: the idle time before a virtual node is turned off, how long an "offline" node is left there before being removed, and how long a job is left waiting before asking for new virtual machines.

It is very important to point out that *bananad* is the only piece of software having direct access to the cloud infrastructure, concealing it completely both to the end user and the batch system.

3.3. OpenNebula tweaks

OpenNebula is the tool of our choice because it is easily customizable through simple scripting. The following OpenNebula configuration has been obtained only by modifying such scripts.

3.3.1. Virtual bridges: network isolation. Our OpenNebula infrastructure is not PROOF-exclusive, but it runs a certain number of virtual clusters: different virtual machines of the same cluster running on different hypervisors appear as being attached to the same virtual network bridge.

This can be obtained by configuring an OpenNebula *hook* (an event callback) executed before starting the VM that uses *ebtables* [23], an iptables-like kernel space tool that operates at Ethernet level to isolate the class of incoming and outgoing MAC addresses to a virtual machine. The hook is executed on the hypervisor running the virtual machine, and it has been modified to allow the specification of a list of extra authorized MAC addresses directly inside the OpenNebula VM template. This is needed to allow all the VMs of the same cluster to communicate with a service that resides outside the virtual bridge.

3.3.2. Quick deployment of virtual machines on LVM. In our setup, virtual machines are run on LVM partitions for performance reasons. This increases the performances with respect to running them from image files.

The LVM Transfer Manager MAD provided by OpenNebula has been heavily modified to drastically improve the virtual machine deployment time.

In the default LVM module, the image is first copied as a file via *scp* on the hypervisor, then it is dumped on the LVM partition using *dd*. To avoid this double transfer, plus the connection overhead introduced by SSH, the image repository resides on a shared GlusterFS [24] filesystem mounted on all hypervisors, and the image is "disk dumped" directly to the LVM partition. With this method, the transfer time of a 4 GB virtual machine has been reduced reduced from ~200 s of the standard LVM module (*scp* then *dd*) to the ~50 s of the modified LVM (direct *dd* over GlusterFS).

Table 1. Time needed to create on the fly an ext3 and a XFS filesystem: the results show how a XFS filesystem of any size may be created in near-real-time. These results were obtained on non-SDD disks.

Size	ext3	XFS
20 GB	13.1 s	1.5 s
40 GB	20.0 s	1.6 s

The second large improvement comes from the consideration that virtual machine images rarely change: it is therefore a waste of time to transfer a big image several times. Each virtual machine image is thus cached as a LVM partition, and the virtual machine instance is run on a snapshot partition of the image cache, using the *snapshot* feature built into LVM. Transferring the image to the hypervisor only happens if the image is not there yet, but further accesses only need to snapshot that partition, an operation that only takes a negligible amount of time (< 1 s).

Running VMs on snapshots of the original cache partition, without touching it, has some remarkable advantages. First of all, every potential damage made to a running virtual machine remains inside the snapshot and does not propagate to the original image. Moreover, multiple virtual machines can be run on the same hypervisor from multiple snapshots of the same LVM logical volume.

OpenNebula's LVM module also features functions to dynamically create disposable filesystems for the virtual machine. In particular, the swap creation module has been rewritten in order to create the swap partition on LVM too (OpenNebula's default module creates the swap on a file), and for the disposable scratch area XFS has been chosen because the filesystem creation is almost immediate with respect to some other well-established filesystems: as clearly shown in table 1 the creation of a XFS filesystem does not depend much on its size, while the creation of an ext3 filesystem is slow (compared to our interactive use case) and strongly dependent on the filesystem size. All of our hypervisors use regular non-SDD SATA disks.

3.3.3. Easier updates through contextualization. Contextualization is the process of running a set of functions to customize a general-purpose virtual machine image at boot time. By using contextualization, instead of hardcoding some modifications inside the virtual machine, we can keep our image repository smaller and easier to update. In the PROOF case, for instance, since all the software is on a shared partition and not within the virtual machine, the instantiated virtual machine is just a basic CentOS 5 installation, and all the customization is made with a contextualization script that takes < 2 s to complete, thus not negatively affecting the boot time.

3.4. SSH authentication using grid credentials and PoD connectivity

In this setup each user needs to login to the access node in order to run the PoD server: since v3.6 PoD features the *pod-remote* command, a local shell that issues remote PoD commands requiring only a SSH connection to the access node. The connection to PROOF is also tunneled through SSH, as depicted in figure 1, so that it is easy to configure a firewall allowing only *sshd* port (22/TCP) on the access node.

In order to allow ALICE users to transparently connect to SSH using their Grid credentials (a X.509 certificate and a private key), a two-step mechanism has been set up [25]. *Authentication* is performed via HTTPS using the certificate and the key: past this step, the user obtains a *temporary authorization* to connect via SSH with private key authentication by presenting his or her Grid private key, which happens to be in the same format (PEM) recognized by SSH.

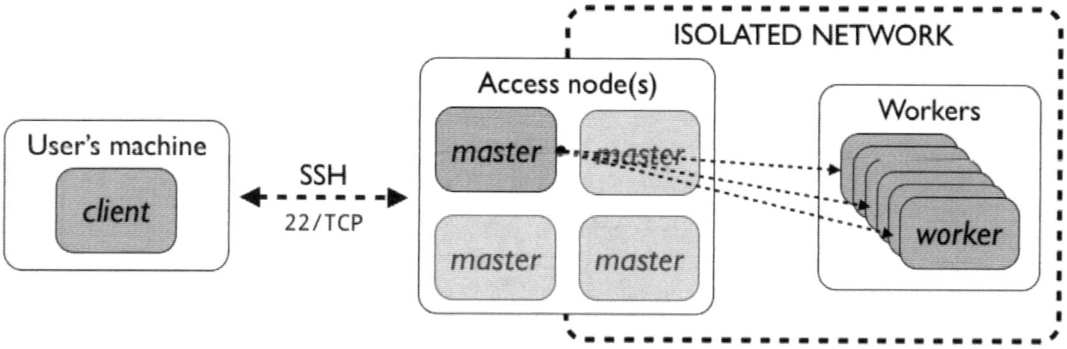

Figure 1. PoD connectivity schema: the only port opened for outbound connectivity is the SSH port, while both physical and virtual PROOF workers freely communicate in a private, isolated network. The access node, running the PROOF masters, acts as an interface between the two networks.

Certificate to Unix user mapping is performed using the ALICE LDAP user database, and home directories are created on the fly using the PAM *mkhomedir* module.

This approach has been made transparent to the user, which is only prompted for a passphrase to unlock the Grid private key.

4. Conclusions

This work shows how, using mainstream and widespread tools, it is possible to build a virtual elastic interactive analysis facility with a wise configuration of the single tools and minimal development effort.

Every piece has been configured by keeping in mind that interactive resources must be made available as soon as possible: the batch system has been configured accordingly, while the cloud infrastructure has been tweaked to avoid useless time consuming transfers, through a fast shared filesystem and local caching of the VM images. Once all the images have been cached, the maximum time a user waits to obtain 40 PROOF workers, even when the resources must be shared with other running virtual clusters, is under 1 min 30 s, perfectly acceptable for an interactive resource.

With this work we have shown that it is feasible to use an IaaS tool like OpenNebula to dynamically provide for interactive PROOF resources even on a small-sized computing facility such as a Grid Tier-2 centre.

References

[1] Ballintijn M, Brun R, Rademakers F and Roland G 2003 *Preprint* physics/0306110
[2] Brun R and Rademakers F 1997 *Nucl. Instr. and Meth. in Phys. Res.* A **389**
[3] Ganis G, Iwaszkiewicz J and Rademakers F 2007 *Proceedings of Science* PoS(ACAT)022
[4] Grosse-Oetringhaus J F 2008 *J. Phys.: Conf. Ser.* **119** 072017
[5] http://aaf.cern.ch/
[6] Bagnasco S, Betev L, Buncic P, Carminati F, Cirstoiu C, Grigoras C, Hayrapetyan A, Harutyunyan A, Peters A J and Saiz P 2008 *J. Phys.: Conf. Series* **119** 062012
[7] http://alien2.cern.ch/index.php?option=com_content&view=article&id=39&Itemid=86
[8] Furano F 2011 *Eur. Phys. J. Plus* **126** 12
[9] http://indico.cern.ch/getFile.py/access?contribId=66&sessionId=6&resId=0&materialId=slides &confId=138471
[10] Legrand I, Voicu R, Cirstoiu C, Grigoras C, Betev L and Costan A 2009 *Comm. of the ACM* **52** (9) 49
[11] Malzacher P and Manafov A 2010 *J. Phys.: Conf. Ser.* **219** 072009

[12] http://opennebula.org/
[13] http://opennebula.org/documentation:rel3.0:vmmg
[14] http://xen.org/
[15] http://www.linux-kvm.org/page/Main_Page
[16] http://www.vmware.com/
[17] http://aws.amazon.com/ec2/
[18] http://opennebula.org/documentation:rel3.0:sunstone
[19] http://www.adaptivecomputing.com/products/torque.php
[20] http://www.lustre.org/
[21] Berzano D, Bagnasco S, Lusso S and Masera M 2008 *Proceedings of Science* PoS(ACAT08)050
[22] Bagnasco S, Berzano D, Lusso S, Masera M 2009 *J. Phys.: Conf. Ser.* **219** 062033
[23] http://ebtables.sourceforge.net/
[24] http://www.gluster.org/
[25] Berzano D 2011 submitted to *INFN-CCR Internal Notes* CCR-42/2011/P

The PROOF benchmark suite measuring PROOF performance

S Ryu[1] **and G Ganis**[2]

[1] KISTI, 245 Daehak-ro Yuseong-gu, Daejeon 305-806, Korea
[2] CERN, Geneva, Switzerland

E-mail: sangsuryu@kisti.re.kr

Abstract. The PROOF benchmark suite is a new utility suite of PROOF to measure performance and scalability. The primary goal of the benchmark suite is to determine optimal configuration parameters for a set of machines to be used as PROOF cluster. The suite measures the performance of the cluster for a set of standard tasks as a function of the number of effective processes. Cluster administrators can use the suite to measure the performance of the cluster and find optimal configuration parameters. PROOF developers can also utilize the suite to help them measure, identify problems and improve their software. In this paper, the new tool is explained in detail and use cases are presented to illustrate the new tool.

1. Introduction

The Parallel ROOT Facility, PROOF, is an extension of the ROOT system [1] aimed at speeding-up analysis using multiple ROOT processes in parallel on a cluster of computers or a multi-core computer [2, 3].

PROOF implements multi-process parallelism to address specifically typical analysis problems encountered in HEP, which in most cases are either embarrassingly parallel or can be formulated as such. For systems like PROOF, scalability with respect to available resources is a natural metric of the efficiency of architecture design and implementation. The inherent scalability of a well-designed and efficient parallel system should be close-to-linear, with minimum overhead. When relevant, external factors, like the available I/O bandwidth, may change this picture [4]. Spotting, identifying and understanding departures of system scalability from linearity are important steps in system optimization process, and essential for any further improvement.

The new PROOF benchmark suite is a new module built into PROOF, providing a standard way to measure the scalability of PROOF system as a function of active processes in the cluster. It is a simple, easy to use, yet flexible enough tool with full support for user-specific customization.

2. PROOF benchmark suite

The new benchmark suite is a framework to perform scalability measurements on a PROOF cluster in a standard way. The primary goal of the new benchmark suite is to determine optimal configuration parameters for a set of machines to be used as PROOF cluster. The suite measures the performance of the cluster for a set of standard tasks as a function of the number of effective processes. From these results, indications about the optimal working point of the cluster – e.g. the maximum or optimal number of concurrent processes - should be derived. For large facilities, where a multi-tier architecture with a super-master supervising multiple sub-masters may help improve scalability and merging performance, the suite should also give indications about the optimal number of sub-masters into which the cluster should be partitioned.

The new suite should be beneficial to PROOF site administrators who wish to check their installation, find bottlenecks, and optimize configuration parameters. PROOF developers will find the suite useful in understanding and improving PROOF.

2.1. Design requirements

The new benchmark has been designed to be easy to use and flexible. This means that defaults are meaningful for most typical cases and straightforward to run, with only few or no settings from the user. At the same time the suite allows users to benchmark user-specific or experiment-specific cases, requiring dedicated datasets, selectors and software packages.

2.2. Analysis modes with PROOF

Typical analysis tasks in HEP application can be categorized into two types – cycle-driven analysis and data-driven analysis. Cycle-driven analysis work is typically CPU-intensive but it could also be I/O, network, or RAM intensive; generation of Monte Carlo events is a good example. For data-driven analysis, the unit of process is entries of a TTree fetched from distributed files. The analysis task reads in events from real data files from experiment or data files generated from simulation. This type of analysis task is typically disk I/O intensive but it could also be network, RAM, or CPU intensive. The new suite addresses both types of analysis tasks.

2.3. Key components

The suite consists of a set of client-side classes, of which essential classes are a steering class for user interface and a set of selectors for default tasks.

2.3.1. Steering class. TProofBench is a user interface class for the benchmark of the system. With the interface, the user makes a connection to PROOF cluster, prepares an output file for results of the test, uploads relevant selectors as PAR files [2], runs benchmark, and displays results. For typical use, this is the only class relevant to the user of the suite.

2.3.2. Selectors. The suite provides one default selector for each of analysis task types. TSelHist is a default selector class to be used for cycle-driven task and class TSelEvent for data-driven analysis task. These classes are uploaded to the cluster as PAR files at the beginning stage of benchmarking. The selector TSelHist intensively generates random numbers following normal distribution to fill 1-D, 2-D, or 3-D histograms, which will be merged at PROOF master and returned to client. 3-D histogram can be used to study the impact of merging large outputs on the scalability of the system. The selector TSelEvent reads in test events from files, which have been generated through TProofBench user interface class prior to benchmark, and fills in histogram with information from events read.

2.4. Modes of scan

To obtain the scalability of the system, the suite repeatedly performs measurements while enabling a certain number of additional workers at each step. In worker-scan mode, with default parameters, scan starts with one active worker, enabling another worker at each step until all available workers in the cluster are active. In core scan mode, with default parameters, measurement starts with one active worker on every node in the cluster, enabling additional worker on every node simultaneously at each step, until all available workers on nodes are active. The order workers are activated is determined by system configuration on master node. User can change how many active workers to start with, how many workers to activate each step, and the number of active workers to finish scanning at. With worker scan mode, overall behaviour of the system can be investigated, while core-scan mode is more effective for the investigation of the behaviour of the system inside a node. To minimize the statistical fluctuation of measurement and maximize measurement accuracy, measurements are repeated (four times, by default) for each step of the scan. For data-driven mode, file caches on worker nodes are cleaned after each measurement.

2.5. Generation of data files
For default data-driven benchmark, files with events should be generated on worker nodes, which can be done with user interface class TProofBench. The events are of type Event of ROOT[3]. Default is to generate two files with 30,000 events each for every worker in the system. All relevant parameters are configurable by user.

2.6. Output of benchmark
For cycle-driven test, average event rate (number of events processed on all active workers in unit time) and its RMS are calculated at each measurement point. For data-driven test, average I/O rate (MB read in on all active workers in unit time) and its RMS are calculated for each measurement point as well as average event rate and its spread. Event rate and I/O rate are plotted on display as a function of the number of active workers in the cluster and updated as progress is made, as well as saved to output files. Normalized event rate and normalized I/O rate (event rate and I/O rate divided by the number of active workers in the cluster, respectively) are also displayed to supplement the interpretation of system behaviour. Packet information from each active worker in the cluster such as CPU time, process time, and latency is saved to an output file for possible further investigation after the benchmark has finished.

2.7. Availability and documentation
The new benchmark suite is available from ROOT v5.29 and on. As it is a client-side module, it can be imported into previous ROOT versions. Additional documentation is available online at http://root.cern.ch/drupal/content/new-benchmark-framework-tproofbench.

3. Use cases
To illustrate the tool, some example results obtained from runs on PROOF facilities for ALICE experiment [5] and on a cloud facility [6] are shown. PROOF facilities that were used are summarized in table 1. All benchmark runs were performed with default parameters, if not stated otherwise.

Table 1. PROOF facilities used for use case study.

	ALICE CAF	KIAF	PoD on Frankfurt cloud
Nodes	1 master, 58 workers	1 master, 4 worker	50
CPU	2 / node, 4 cores/CPU (Intel Xeon L5520, 2.27 GHz)	2 / node, 6 cores/CPU (Intel Xeon X5650, 2.67 GHz)	2 / node, 12 cores / CPU (AMD Opteron 6172, 2.1 GHz)
RAM	24 GB / node	24 GB / node	64 GB / node
Storage	Local SATA disks	5 TB/node (NAS) 300 GB/node (SAS)	-

3.1. Cycle-driven benchmark on a homogeneous cluster
Figure 1 shows an example result of cycle-driven benchmark with default configuration on KIAF which is a homogeneous ALICE PROOF cluster. Total event rate divided by the number of active workers in the cluster as shown in figure 2 provides closer view on the scaling of the system.

[3] See the files test/Event.h and test/Event.C in any ROOT installation.

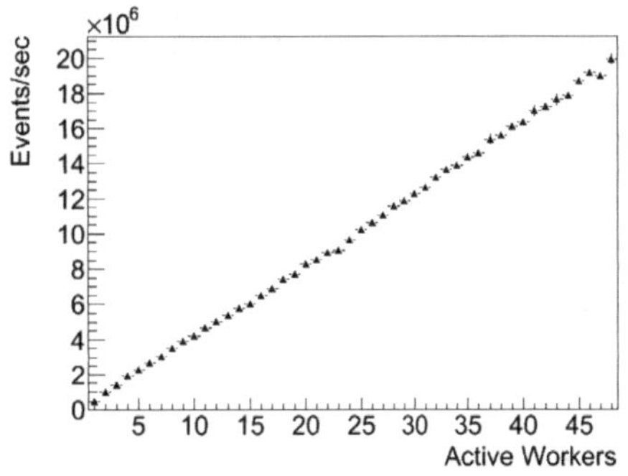

Figure 1. Event rate from a cycle-driven benchmark on KIAF.

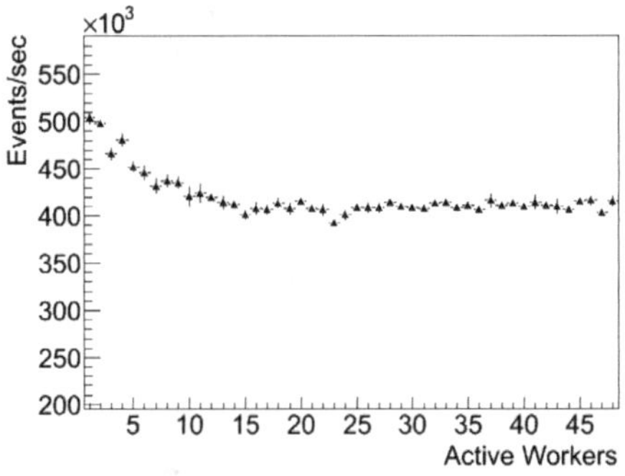

Figure 2. Events rate normalized by number of active workers from a cycle-driven benchmark on KIAF.

3.2. Cycle-driven benchmark on ALICE PROOF facility

Cycle-driven benchmark results on ALICE PROOF facility CAF [5] are shown in figure 3 and figure 4. The facility is currently actively utilized for analysis tasks by ALICE collaboration. The facility consists of 3 groups of computers with different performance, formed over time by adding new computers to existing cluster. The plots clearly show transition between the groups.

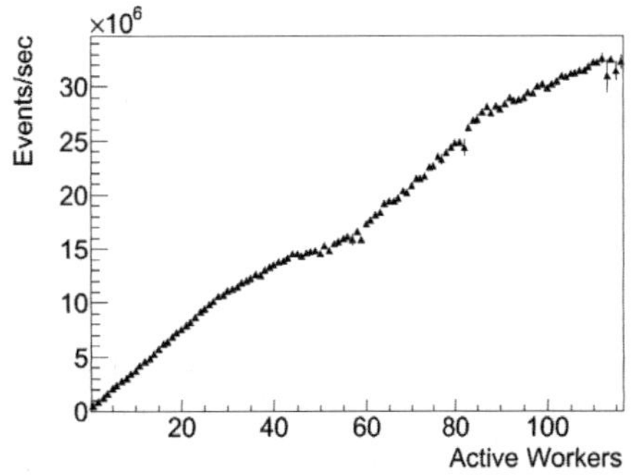

Figure 3. Event rate from a cycle-driven benchmark on ALICE CAF. Transition from one type of computers to another is clearly shown.

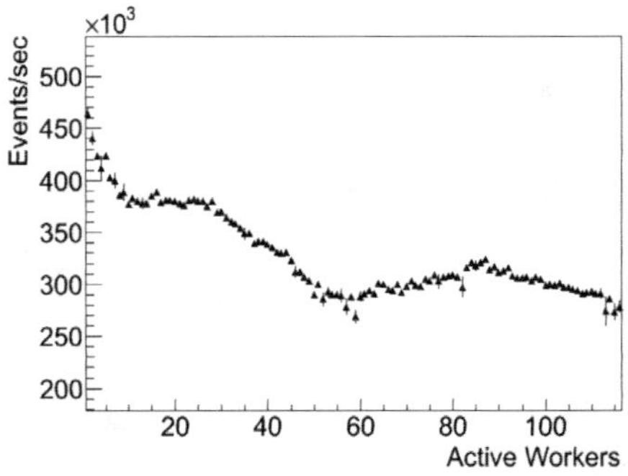

Figure 4. Normalized event rate from a cycle-driven benchmark on ALICE CAF.

3.3. Cycle-driven benchmark on PoD PROOF cluster

Figure 5 shows a cycle-driven benchmark result on PoD [7] cluster where PROOF cluster is dynamically set up on Frankfurt cloud facility [6]. We can see here that the system scales almost linearly up to around 300 workers, where the impact of the serial implementation of packet distribution by master becomes visible.

3.4. Data-driven benchmark of two storage systems on ALICE KIAF

Figure 6 shows results from a data-driven benchmark with test events on ALICE KIAF cluster [5]. KIAF had 2 storage systems available on the system – local SAS disks on every node and a NAS storage system mounted to all worker nodes via network. At 4~5 active workers per node, rate starts to saturate to the value representing the total amount of I/O that the device can provide [4]. The new benchmark suite is very effective in benchmarking and comparing performance of different hardware with PROOF.

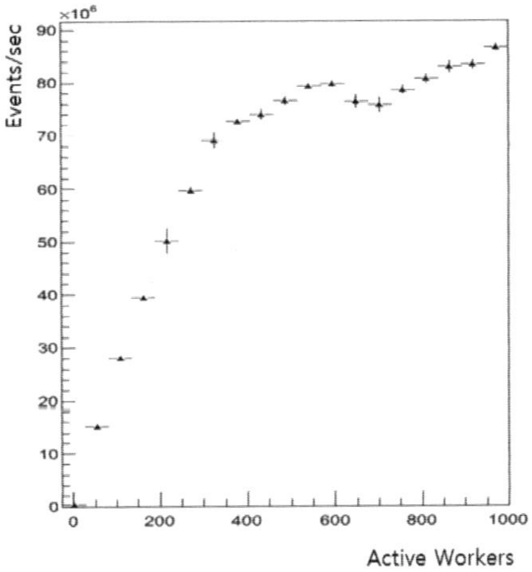

Figure 5. Event rate from a cycle-driven test on POD cluster on cloud system. Measurements were made for every 50 workers. Courtesy of A Manafov at GSI, Darmstadt.

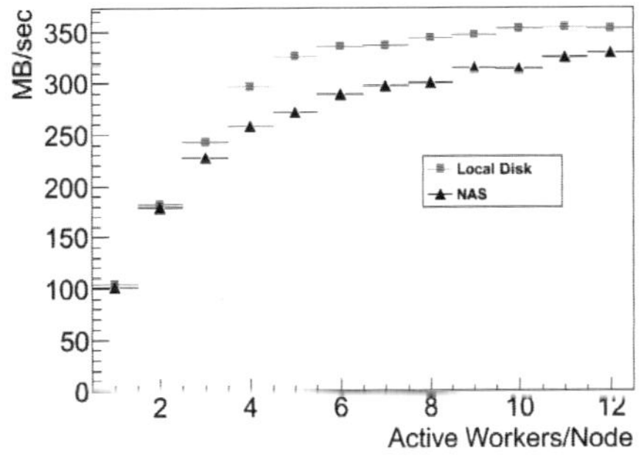

Figure 6. I/O rate from a data-driven benchmark on KIAF cluster. This is core scan result. Square points are with local SAS disks on every worker node, triangular points with a NAS system mounted to every worker node via network.

4. Conclusions

PROOF benchmark suite is a new module in PROOF system, which is easy to use, yet flexible enough to support user-specific or experiment-specific requirements. The new suite will help site administrator check their installation and optimize its configuration parameters. PROOF software developers can use the suite to spot, identify problems, and improve their software. The suite has been available from ROOT version 5.29/02 and on.

Acknowledgements

The authors gratefully acknowledge KIAF administrators at KISTI, Daejeon for PROOF facility and their support, A. Manafov at GSI, Darmstadt for the benchmark plots for PoD cluster on cloud system, and ALICE collaboration for ALICE PROOF facility.

References

[1] Antcheva I *et al*, ROOT – A C++ framework for petabyte data storage, statistical analysis and visualization *Comp. Phys. Comm.* **180**/12 (2009) 2499-2512. See also http://root.cern.ch

[2] Ballintijn M, Brun R, Rademakers F and Roland G 2003 The PROOF distributed parallel analysis framework based on ROOT *Proceedings of CHEP03 Intl. Conf. (La Jolla, California, US, March 2003*

[3] Ballintijn M, Brun R, Canal P, Gulbrandsen K, Roland G and Rademakers F 2005 Super scaling PROOF to very large clusters *Proceedings of CHEP04 Intl. Conf. (Interlaken, Switzerland , 27 September – 01 October 2004)*

[4] Aguado-Sanchez C, Blomer J, Buncic P, Charalampidis I, Ganis G, Nabozny M and Fademarkers F 2010, Studying ROOT I/O performance with PROOF-Lite *Proceedings of CHEP 2010 Intl. Conf. (Taipei, Taiwan , 11-22 October 2010)*

[5] ALICE analysis facilities http://aaf.cern.ch

[6] http://www.frankfurt-cloud.com/

[7] Malzacher P and Manafov A 2010 PROOF on Demand *J. Phys.: Conf. Ser.* **219** 072009

An exploration of SciDB in the context of emerging technologies for data stores in particle physics and cosmology

D Malon, P van Gemmeren and J Weinstein
Argonne National Laboratory, 9700 S Cass Ave, Lemont, IL 60439, USA

E-mail: malon@anl.gov

Abstract. Traditional relational databases have not always been well matched to the needs of data-intensive sciences, but efforts are underway within the database community to attempt to address many of the requirements of large-scale scientific data management. One such effort is the open-source project SciDB. Since its earliest incarnations, SciDB has been designed for scalability in parallel and distributed environments, with a particular emphasis upon native support for array constructs and operations. Such scalability is of course a requirement of any strategy for large-scale scientific data handling, and array constructs are certainly useful in many contexts, but these features alone do not suffice to qualify a database product as an appropriate technology for hosting particle physics or cosmology data. In what constitutes its 1.0 release in June 2011, SciDB has extended its feature set to address additional requirements of scientific data, with support for user-defined types and functions, for data versioning, and more. This paper describes an evaluation of the capabilities of SciDB for two very different kinds of physics data: event-level metadata records from proton collisions at the Large Hadron Collider (LHC), and the output of cosmological simulations run on very-large-scale supercomputers. This evaluation exercises the spectrum of SciDB capabilities in a suite of tests that aim to be representative and realistic, including, for example, definition of four-vector data types and natural operations thereon, and computational queries that match the natural use cases for these data.

1. Introduction

While the high energy physics community employs databases extensively for a wide range of purposes, the community has often chosen to develop its own in-house technologies to address many database functionality and persistence requirements that the computer science research community and the commercial marketplace have also endeavoured to address. There is, for example, a sizable literature on temporal databases spanning many decades, but the LHC experiments have developed their own "interval of validity" databases (cf. COOL) for managing time-varying conditions and calibrations, as did the high energy and nuclear physics experiments of earlier generations. Another example is the development of technologies for database content caching (cf. FroNTier), developed for Fermilab Tevatron experiments and now employed by LHC experiments. This tendency toward homegrown solutions is not restricted to databases and persistence technologies (nor indeed to high energy physics): high energy physics experiments have also developed their own configuration and release management tools, distributed data management infrastructure, workload management systems, and more.

This observation is not intended as a criticism—it is often true that available off-the-shelf technologies do not quite meet the requirements of high energy physics, or of other large-scale scientific enterprises—but it does suggest that from time to time it is appropriate to reassess the capabilities of current and emerging technologies and their potential utility to high energy physics.

Database technologies in particular have not traditionally been a good match for many large-scale data purposes in high energy physics. The LHC experiments process and analyze event data at a scale of many tens of petabytes (and soon, hundreds of petabytes), but even relatively small applications such as event-level metadata systems, at a scale of tens of terabytes, challenge the capabilities of relational databases. An event-level metadata infrastructure such as the ATLAS TAG database requires scalable support for variable-length structures (such as the properties of photons, electrons, muons, and jets, whose numbers vary from event to event), support for combinatorial queries (e.g., looking for pairs of oppositely-charged leptons with transverse energy above a threshold), and decoding (possibly bit-level decoding) of time-varying information (e.g., trigger decisions, where the trigger menus and hence the meanings of the bits may vary from run to run). All of these things are possible in principle and implemented in practice with relational databases, but they may be awkward to implement, or difficult to support in a scalable fashion. In this paper we take the opportunity to assess whether and how at least one state of the art development in scientific databases, SciDB, might be applicable to high energy physics data handling, doing so first in what is in some ways the simplest and smallest-scale event data setting in experimental particle physics, event-level metadata, then expanding these considerations to data produced by large-scale cosmological simulations.

2. Databases and SciDB: Some history

Even as the high energy physics (HEP) database "software stack" for LHC-era experiments has gradually emerged, the capabilities of commercial products and software from outside the community have been improving. Database technologies have evolved in a number of relevant ways. Column-oriented data storage, matching the access patterns of many kinds of analysis, is now available in several products. Temporal database technologies have grown in capability and range of applicability, and distributed caching is common in the web era. More importantly, high energy physics data scales are no longer rare: the data scales of Google and Amazon, and the data generated by the pervasive use of social media, now make HEP data challenges look tractable, though the access and analysis patterns are of course different in their demands. It is natural, then, to ask, whether there emerging database technologies that the HEP community should be tracking.

Motivated in part by the requirements emerging from the Large Synoptic Survey Telescope (LSST), the first XLDB (for extremely Large DataBase) workshop was convened in October 2007. At that meeting and at a followup meeting in March 2008, a consensus emerged that there are requirements common to a number of large-scale scientific enterprises that a collaboration between the academic and commercial database communities might be able to address. Shortcomings identified in current offerings included the lack of integrated analytical processing capabilities, the need for arrays, which are fundamental in science but difficult to emulate in relational databases, insufficient provenance and version control management, lack of internal support for uncertainty and error bars, an operations mismatch with SQL, and more. The commonality of such requirements led to the launch of the SciDB project, which seeks to address "the need for a scalable open-source data management with integrated analytics to support scientific research on massive datasets." In June 2011, the SciDB team released software that could be considered Version 1.0 of a product intended to address these common needs.

It is this product that is the subject of the explorations described in this paper.

3. Features of SciDB

Key features of SciDB are its array-oriented data model, its support for versioning and provenance, its shared-nothing architecture to allow massively parallel computations, its support for user-defined types and functions, and its native support for uncertainty.

3.1. Array data model

SciDB uses an array data model and supports multi-dimensional, nested arrays, which are often a natural representation for spatially- or temporally-ordered data. Array cells contain records with tuples of multiple attributes and/or one or more arrays. All cells in an array have the same attributes and data types. Arrays can have multiple named dimensions represented by contiguous integer values. SciDB supports unbounded dimensions in basic arrays. SciDB arrays may be sparsely populated and can have jagged edges, allowing each row/column to have a different dimensionality (see figure 1). SciDB supports two separate classes of "missing" information, and allows user-definable, context-sensitive treatment of these cell values.

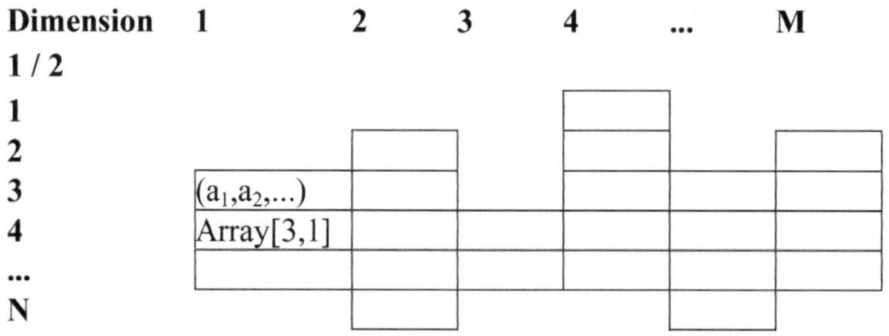

Figure 1. A two-dimensional SciDB array with jagged edges.

Data in a SciDB array is not updateable, but new array data can be appended to a SciDB database and the results of a SciDB query can be written back to storage. This behaviour matches the typical scientific use case in which data is read-only with updates being handled by versioning.

3.2. Preserving ordering helps complex computations

Unlike relational databases, which store their data in tables without row or column ordering, SciDB uses an array data model, with implicit ordering. This ordering allows for the notion of "adjacency," which is desirable for many scientific domains (see figure 2). In many science applications, the order of data is not simply given by its representation, but is fundamental to the semantics; e.g., for sensor data, cells that are adjacent in terms of cell indices are physically adjacent, and for collider data, event ordering may correspond to temporal sequencing during data taking.

Relational Database

I	J	Value
0	0	32.5
1	0	90.9
2	0	42.1
3	0	96.7
0	1	46.3
1	1	35.4
2	1	35.7
3	1	41.3
0	2	81.7
1	2	35.9
2	2	35.3
3	2	89.9
0	3	53.6
1	3	86.3
2	3	45.9
3	3	27.6

SciDB

I / J	0	1	2	3
0	32.5	46.3	81.7	53.6
1	90.9	35.4	35.9	86.3
2	42.1	35.7	35.3	45.9
3	96.7	41.3	89.9	27.6

Figure 2. In SciDB arrays adjacency of cells is preserved, whereas relational tables do not support ordering.

3.3. Data storage

In SciDB array dimensions can be "chunked" and data of each chunk can be stored on individual nodes. As each node has processing and storage capabilities, this partitioning allows shared-nothing operation. Chunk "overlaps" can be defined, so that certain operations involving neighboring cells are possible without communication among nodes (see figure 3).

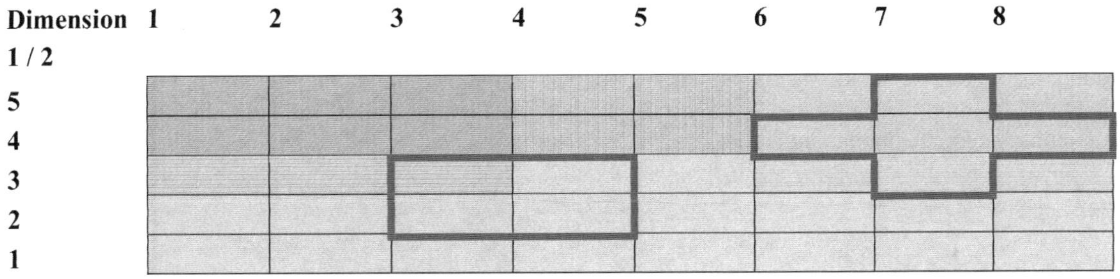

Figure 3. A two-dimensional SciDB array with chunking (indicated by the cell background color) and overlaps in both dimensions. Thanks to the overlap, both cell clusters (shown by the red line) can be processed by a single node without the exchange of data.

3.4. Architecture

The architectural concept of SciDB is built on a shared-nothing cluster of up to thousands of nodes on commodity hardware. A single runtime supervisor is used to dispatch queries and to coordinate the execution among the nodes' local executors and storage managers (see figure 4).

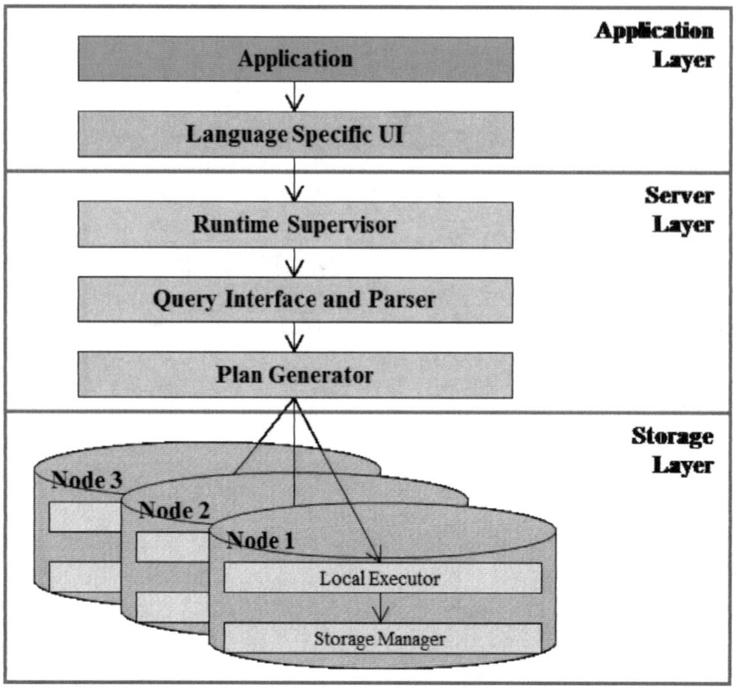

Figure 4. SciDB implements a shared-nothing architecture.

The array query language defined in SciDB refers to complete arrays agnostic to whether they are distributed or not. The SciDB plan generator optimizes queries for efficient data access and processing, taking into account the physical distribution of data among the nodes. The generated query plan is then run by a node's local executor, using its storage manager to access the data. The SciDB runtime supervisor coordinates the execution of the query.

4. Query language and operators

SciDB supports queries using either Array Query Language (AQL), a declarative SQL-like language extended for working with array data, or Array Functional Language (AFL), a functional language for internal implementation of operators.

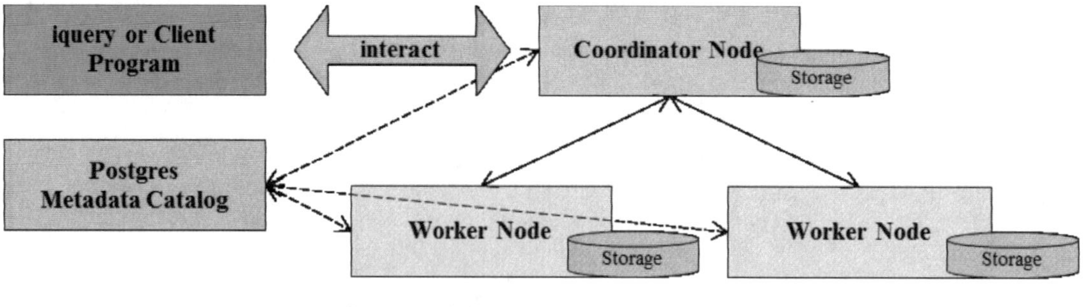

Figure 5. SciDB query plan execution.

Queries can be executed using the SciDB command line interface iquery as:

```
iquery -p <port> "<queryString>"
```

SciDB also provides a Python API to execute queries from within programs:

```
# Import scidb interface
import scidbapi as scidb
# Connect to the scidb database at host:port
db = scidb.connect(<host>, <port>)
# Execute a query string using AFL
result = db.executeQuery(<queryString>, 'afl')
```

SciDB provides a number of array-specific operators (e.g., Aggregate, Apply, Filter, Join, Regrid, Subsample), and linear algebraic and matrix operations (e.g., clustering, covariance, linear and logistic regression) and the language is extensible with Postgres-style user-defined functions. Interfaces to other packages (Matlab, R, …) will be provided in a future release.

4.1. Example query in AQL

The following AQL query creates a new one-dimensional array called AttributeList. Each cell will store tuples with RunNunber, EventNumber, DetFlags, IsSimulation and MissingET members, which are of uint32, bool or float data type. The only index of the array is called TAGID; the query specifies a chunk size of 1000 elements with an overlap of 10:

```
CREATE ARRAY AttributeList<RunNumber: uint32,
   EventNumber: uint32,
   DetFlags: uint32,
   IsSimulation: bool,
   MissingET: float> [TAGID=0:99999, 1000, 10]
```

The second example creates a two-dimensional Jet array using TAGID and JETID as indices, without chunking for JETID:

```
CREATE ARRAY Jet<EmFrac: float,
   Eta: float,
   Phi: float,
   Pt: float> [TAGID=0:99999, 10, 0, JETID=0:5, 6, 0]"
```

4.2. Array Functional Language (AFL)

The following AFL query filters out entries in the AttributeList array for which the EventNumber is greater than 12345:

```
filter (AttributeList, EventNumber > 12345)
```

The filter operator returns an array of the same dimension as the input array AttributeList, marking all cells in the input that do not satisfy the predicate expression as "empty."

5. SciDB and proton collider experiments?

Array-oriented operations with support for overlapping chunks, a major focus of initial SciDB development, is of inherent interest to a number of scientific disciplines, and in certain domains such features could be key. In proton collider experiments, though, such features are not determining factors in evaluating the suitability of an event data management product. The underlying parallel execution architecture, however, and a design explicitly intended for scalability to multi-petabyte stores, as well explicit support for sparse data, may be of genuine interest. A design with a view of transactions that is aware of the write-once nature of most scientific data is also relevant to efficient large-scale collider physics data handling. SciDB has plans in the near future to support *in situ* data, i.e., to allow access to and querying of data without translating and importing it into a SciDB-specific storage format, so that in principle bulk event data could be in a format known and proven within the community.

Initially, SciDB plans for *in situ* data support include HDF5 and possibly NetCDF plug-ins, but there is no reason that, say, a ROOT storage backend could not be accommodated. Since most collider physics experiments use ROOT as their principal event store technology for all but RAW data, such native support, with SciDB contributing a value-added layer, could be of great interest. Support for user-defined types and operators could be quite useful, as scientific data types and operations are often at best awkward to support in relational databases. An infrastructure for provenance tracking and management, though primitive in initial SciDB planning, is an important advantage. Support for uncertainty (e.g., error bars) in data is interesting in principle, though experimental particle physics has necessarily elaborate and sophisticated strategies for error estimation and bounds that may make such built-in features of academic interest at best.

SciDB's shared-nothing architecture on commodity clusters is definitely interesting by comparison with the purchase of special-purpose hardware for certain commercial database products. There are other non-database or alternative-database technologies and even relational databases that also take this approach, such as MySQL Cluster, which provides shared-nothing clustering capabilities for the MySQL database management system. There are also several products that provide array extensions to SQL—and while this is a worthwhile idea for some data, it may not be immediately apparent how this capability helps in the storage and management of bulk collider physics event data.

While event data stores may constitute the largest data stores in experimental particle physics, they are not the only ones, and it is also appropriate to consider whether technologies such as SciDB might be suitable for detector control system and conditions information (and such investigations are indeed underway), or for important subsets of an event data store, such as an event-level metadata system. The latter provides the principal setting for our investigations of SciDB.

6. SciDB for high energy physics data

The scope of the evaluation undertaken in the project described in this paper has been:

1. To install and configure first real release of SciDB, R11.06;
2. To provide machinery to upload data from multiple source types, including ROOT files;
3. To exercise the user interface and operators;
4. To explore tradeoffs in data organization;
5. To evaluate sparse versus dense data representations of identical data;
6. To provide user-defined types and functions in support of scientific data types.

6.1. Event-level metadata from proton collisions at LHC

ATLAS records key quantities describing properties of proton collision events as event-level metadata called TAGs, which are useful in event selection. TAG data include, for example, event identification information, triggers passed, and particle properties for variable numbers of jets, photons, electrons, muons, and taus, along with sufficient information to allow navigation to file-resident raw and derived event data. Variable-length array data members are not a natural fit to relational databases: while they can be supported they often have large performance penalties.

TAG data from 2011 ATLAS runs were chosen for early tests because the data structure is relatively simple, and event-level metadata records are already hosted both in files and in relational databases. The TAG format is almost identical to highly derived physics data used for analysis, so that results for testing SciDB on TAG data are applicable to analysis data as well.

6.2. Cosmological simulations

Data associated with cosmological simulations can challenge the capabilities of the largest leadership computing facilities. Still, these data often have a very simple structure, for example describing N-body problems with 8 floating point variables for each particle (position, velocity, mass, id):

x[Mpc], v_x[km/s], y[Mpc], v_y[km/s], z[Mpc], v_z[km/s], particle mass[M_sol], particle tag

Important derived data products are often no more complex. Dark matter halo catalogs, for example, store halos as:

halo_mass[10^10 M_sol], x[Mpc], y[Mpc], z[Mpc], vx[km/s], vy[km/s], vz[km/s]

Published reference datasets used in comparisons of cosmological simulations were uploaded in our tests. These simulations used 256^3 particles. Because of their structure, these data sets are natural candidates for exploiting three-vector constructs for position and velocity.

7. User-defined types and operators

A commonly-encountered drawback to the use of relational databases for scientific data is the lack of support for scientific data types and operators. The native array support in SciDB is an important step toward addressing this deficiency, but an important requirement for scientific data handling is the ability to support user-defined types and operators. Event selection in proton collider experiments, for example, requires identifying events with oppositely-charged leptons above an energy threshold, and this is a combinatorial operation on the variable number of muons and electrons produced in any given collision event. Spatial operations—for example, finding particles or jets whose trajectories have a particular geometric relationship (say, within a certain cone)—are another common requirement.

To explore the current capabilities of SciDB to support user-defined types and operators, several types and operators were defined and tested, including a user-defined three-vector type, a Lorentz vector, rotation operators, Lorentz transformations, and a number of related geometric operators.

The following example conveys the flavor of these constructions and these studies:

apply(AttributeList, Jet, threeVector(JetPt1, JetEta1, JetPhi1))
sort(AttributeList, angleBetween(Jet1, threeVector(1,1,1)))

The first line is a Jet constructor, building a Jet as a three-vector constructed from three separate floating point values in each AttributeList entry (transverse momentum plus the attributes to construct a directional vector {eta, phi}. The second line sorts the elements of AttributeList according to the angle between a contained Jet and a fixed vector (artificially chosen here as (1,1,1)).

As usual when introducing a new type into a system, the tedious work is not defining the data members (a struct of three doubles or three floats, say); rather it is in defining all the necessary supporting operators (constructors from values and from strings, assignment operators, input and output operators, and so on), and in registering the type so that it is recognized and its appropriate associated operators can be invoked by the system into which it is being introduced. The following figures illustrate how this happens in SciDB. The essential complexity of the task is no different than that of introducing a corresponding user-defined type in the underlying C or C++ language.

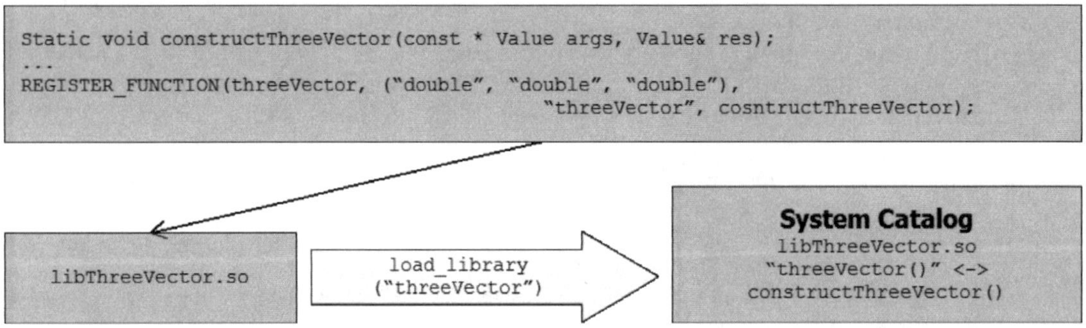

Figure 6. Machinery in support of user-defined types: Declaration and construction.

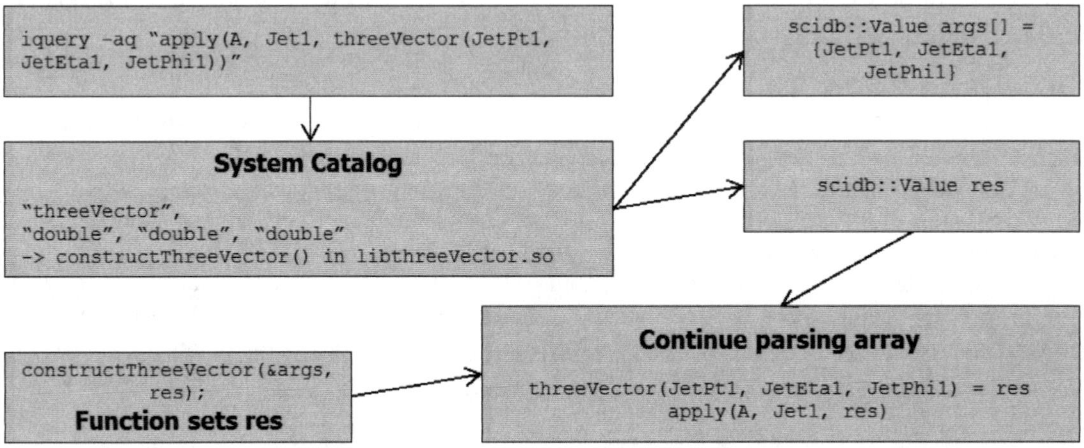

Figure 7. Machinery in support of user-defined types: Operator invocation.

8. Conclusion and outlook

SciDB has advanced considerably in functionality and in robustness since its earliest incarnations. Much of the early development in SciDB has been aimed at supporting array data models, both in storage and in operations, and most testable features are of this flavor. For many sciences, this focus is entirely appropriate: the product is designed to be well suited to image data, with LSST as a major target client. Support for array data structures, for array partitioning (chunking) across processing nodes, for overlap specification, and for a range of array operations, all have solid implementations. The advantages of SciDB for high energy physics data that do not have as natural an array organization as spatial data are less pronounced, and there are an increasing number of alternative data hosting technologies that also support efficient shared-nothing processing. For proton collision data and for cosmological simulation data, SciDB's support for user-defined data types and operators are of genuine interest, and significantly decrease the complexity of query formulation by physicists. Increasingly,other requirements of large-scale scientific data handling are being addressed, including improved support for user-defined types and operators, for *in situ* data, for provenance tracking, and more, making SciDB a technology worth tracking for the high energy physics computing community.

Acknowledgments
The submitted manuscript has been created by UChicago Argonne, LLC, Operator of Argonne National Laboratory ("Argonne"). Argonne, a U.S. Department of Energy Office of Science laboratory, is operated under Contract No. DE-AC02-06CH11357. The U.S. Government retains for itself, and others acting on its behalf, a paid-up nonexclusive, irrevocable worldwide license in said article to reproduce, prepare derivative works, distribute.

References

[1] Brown P et al 2010 *Proc. Int. Conf. on Management of Data SIGMOD 2010* (Indianapolis) pp 963-8
[2] Borne K, Becla J, Davidson I, Szalay A and Tyson J 2008 *AIP Conf. Proc.* **1082** pp 347-51
[3] http://www.scidb.org/about/history.php
[4] Malon D 2005 *Int. J. Mod. Phys.* A **20** pp 3871-73
[5] van Gemmeren P and Malon D 2009 *Proc. IEEE Int. Conf. on Cluster Computing and Workshops 2009* (New Orleans) pp 1-7
[6] Malon D, van Gemmeren P, Cranshaw J and Schaffer A 2006 *Proc. Computing in High Energy and Nuclear Physics 2006* (Mumbai: Macmillan) pp 312-5
[7] Malon D, Cranshaw J and Karr K 2006 *Proc. Computing in High Energy and Nuclear Physics 2006* (Mumbai: Macmillan) pp 316-9
[8] Cranshaw J, Goosens L, Malon D, McGlone H and Viegas F 2008 *J. Phys.: Conf. Series* **119** 072012
[9] van Gemmeren P and Malon D 2010 *J. Phys.: Conf. Series* **219** 042057

ACAT 2011 IOP Publishing
Journal of Physics: Conference Series **368** (2012) 012022 doi:10.1088/1742-6596/368/1/012022

One click dataset transfer: toward efficient coupling of distributed storage resources and CPUs.

Michal Zerola[1], Jérôme Lauret[2], Roman Barták[3], Michal Šumbera[1]

[1] Nuclear Physics Institute, Academy of Sciences of the Czech Republic, Czech Republic
[2] Brookhaven National Laboratory, Upton, USA
[3] Faculty of Mathematics and Physics, Charles University, Czech Republic

E-mail: `michal.zerola@ujf.cas.cz`, `jlauret@bnl.gov`, `roman.bartak@mff.cuni.cz`, `sumbera@ujf.cas.cz`

Abstract. The massive data processing in a multi-collaboration environment with geographically spread diverse facilities will be hardly "fair" to users and hardly using network bandwidth efficiently unless we address and deal with planning and reasoning related to data movement and placement. The needs for coordinated data resource sharing and efficient plans solving the data transfer paradigm in a dynamic way are being more required. We will present the work which purpose is to design and develop an automated planning system acting as a centralized decision making component with emphasis on optimization, coordination and load-balancing.

We will describe the most important optimization characteristic and modeling approach based on "constraints". Constraint-based approach allows for a natural declarative formulation of what must be satisfied, without expressing how. The architecture of the system, communication between components and execution of the plan by underlying data transfer tools will be shown. We will emphasize the separation of the planner from the "executors" and explain how to keep the proper balance between being deliberative and reactive. The extension of the model covering full coupling and reasoning about computing resources will be shown.

The system has been deployed within STAR experiment over several Tier sites and has been used for data movement in the favour of user analyses or production processing. We will present several real use-case scenario and performance of the system with a comparison to the "traditional" - solved by hands methods. The benefits in terms of indispensable shorter data delivery time due to leveraging available network paths and intermediate caches will be revealed. Finally, we will outline several possible enhancements and avenues for future work.

1. Introduction

Distributed computing offers large harvesting potential for computing power and brings multiple benefits as far as it is properly exploited. On the other hand it introduces several pitfalls including concurrent access, synchronization, communications scalability as well as specific challenges such as answering key questions like "how to parallelize a task?" knowing where my data and CPU power are located. In data intensive experiments, like the one from High Energy and Nuclear Physics (HENP) community and the STAR [4] [1] experiment, the problem

[4] Solenoidal Tracker at Relativistic Heavy Ion Collider is an experiment located at the Brookhaven National Laboratory (USA). See http://www.star.bnl.gov for more information.

is even more significant since the task usually involves processing and/or manipulation of large datasets.

For the colossal volumes of data being produced every year to be treatable, the technology and system must be manageable. In global collaboration, the needs for coordinated resource sharing and efficient plans solving the problem in a dynamic way are fundamental.

It is apparent that it will be hardly "fair" to users and hardly using network bandwidth efficiently unless we address and deal with planning and reasoning related to data movement and placement. We are addressing the paradigm of distributing the data focusing on one of the largest running physics experiment in the present. The article exploits and applies the solving techniques for designing and building the automated planning and transferring system; enabling scientists to reach fruits of their research leveraging the potential of their resources.

2. Related works and problem analysis

The needs of large-scale data intensive projects arising out of several fields such as bio-informatics (BIRN, BLAST), astronomy (SDSS) or HENP communities (STAR, ALICE) have been the brainteasers for computer scientists for years. Whilst the cost of storage space rapidly decreases and computational power allows scientists to analyze more and more acquired data, appetite for efficiency in Data Grids becomes even more of a prominent need.

Decoupling of job scheduling from data movement was studied by Ranganathan and Foster in [2]. Authors discussed combinations of replication strategies and scheduling algorithms, but not considering the performance of the network. The authors of [3] proposed and implemented improvements to the Condor, a popular cluster-based distributed computing system. The presented data management architecture is based on exploiting the workflow and utilizing data dependencies between jobs through study of related DAGs. Since the workflow in high-energy data analysis is typically simple and embarrassingly parallel without dependencies between jobs these techniques don't lead to a fundamental optimization in this field.

Sato et al. in [4] and authors of [5] tackled the question of replica placement strategies via mathematical constraints modeling an optimization problem in Grid environment. Solving approach in [4] is based on integer linear programming while [5] uses Lagrangian relaxation method [6]. The limitation of both models is a characterization of data transfers which neglects possible transfer paths and fetching data from a site in parallel via multiple links possibly leading to the better network utilization.

We focus on this missing component considering wide-area network data transfers pursuing more efficient data movement. An initial idea of our presented model originates from Simonis [7] and the proposed constraints for traffic placement problem were expanded primarily on links throughputs and consequently on follow-up transfer allocations in time. One of the immense advantages of the constrained based approach is a gentle augmentation of the model with additional real-life rules. Constraints identify the impossible and reduce the realm of possibilities to effectively focus on the possible, allowing for a natural declarative formulation of what must be satisfied, without expressing how.

Designing and implementing the automated planning system in a dynamic environment like Data Grid is, will not be fruitful unless we address the three main issues: *(1) Accuracy of estimation.* Estimating the computation cost of a job (w.r.t. either computing or transfer) is the key success factor, but at the same time the system can be hardly effective if it has to reason about all peculiarities from the environment. The proper abstraction of the real world is needed to provide balance between accuracy and complexity. *(2) Adaptation to dynamic environment.* Pure static approaches assume that resource and task set is given and fixed over time. Since this assumption is not always valid, the adaptation to the changing condition is needed. In other words, the system has to provide proper balance between being **deliberative** and **reactive** at the same time. *(3) Separation of planner from executor.* Fundamentally the first two issues are related to the lack of collaboration between planner and executor. Without a cooperation the planner cannot be aware of the grid environment change and cannot adapt to the more accurate estimations.

3. Architecture

It is important to pay close attention to the architecture of the system - the conceptual glue that holds every phase of a project together. In this section, we will describe the elements of the system, properties and relations between them. We introduce briefly each component following the work-flow (see Fig. 1 for illustration).

Let us start with explaining how requests are put into the system. End users (or stand-alone services) generate requests using the web interface, written in *PHP* following the *MVC* design pattern. There are two possible way how a request can be specified. **a)** either as an encapsulation of the meta-data query (as understood by STAR's File and Replica Catalogue), or **b)** providing the list of files using *filelist*. An example of the catalogue query is:

- **production**=P10ik,
- **filetype**=daq_reco_MuDst,
- **trgsetupname**=AuAu39_production

where we specified type of the production (data set), what type of files we are interested in and some trigger setup. This meta-data query covers about 220,000 files with a total size of 48TB. The population of the database with files belonging to the request is done asynchronously by separate component as we will see soon.

The second approach of entering the request is using a filelist, which has the following syntax:

```
SITE;STORAGE;PFN
```

Each line then describes exact location of the file given by its physical file name, the storage that holds it and finally the site where the storage is located.

The part of a request is also a desired destination for a data set (in the form of site + storage) which user selects using the web interface.

Afterwards, the request is stored in a *SQL* database (system supports *MySQL* and *PostgreSQL*) in a Catalog agnostic manner (any Catalog should work as far as they have a LFN/PFN concept our approach relies on) with the additional information like user name, group or date of the request.

Later, the component called *File Feeder* contacts the *File and Replica Catalogue* and makes the query for the requested meta-data. The output information is stored back to the database, including all possible locations for every file in a request. This is when population of the internal database with file repositories happens. Because of usually large volume of records that needs to be stored in a database, the *File Feeder* uses `LOAD DATA INFILE` syntax that provides high performance.

The main logic and reasoning about the plan happens in the brain of the system, a component called the *Planner*. It is the place where realization of the model is done and where the plan in iterations is computed. *Planner* takes a subset of all requests for files to be transferred according to the preferred fair-share function. It creates the plan (transfer paths, as we explained in the previous chapter) for the selected requests and stores the plan back to the database.

The individual file transfers are handled by the separate distributed component called *Data Mover*. As we specified at the beginning, we want to use existing point-to-point data transfer tools and use them as the back-end instruments. The *Data Mover* should serve as an intelligent wrapper on top of such tools handling the work. The role of these workers is to perform a point-to-point data transfer on a particular link following the computed plan. The results and

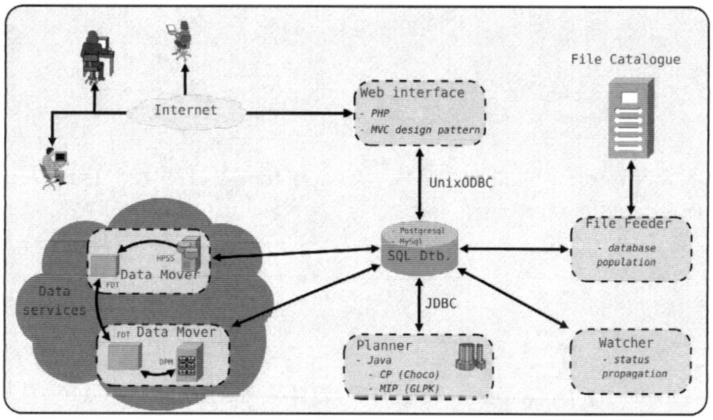

Figure 1: Architecture of the system.

intermediate status is continuously recorded in the database and user can check the progress at any time.

Because of the asynchronous nature of the communication between components, the system has to have well defined states and transitions from one state to another given by state diagrams. We will explain the flow in the following section. The *Watcher*, independent component running at each site, is responsible for changing states of the objects and cache management.

We can see that the whole mechanism is a combination of **deliberative** (assuring optimality) and **reactive** planning (assuring adaptability to the changing environment). Since this is crucial to the argument, in the next section we will describe the respective two components (*Planner* and *Data Mover*) serving up as a "reasoner" and a "worker".

4. Implementation of the planner

Linear programming is a method of minimizing a given linear function ($\min c^T x$) with respect to the system of linear inequalities ($Ax \leq c$). Vector x represents the variables to be determined. If all can be rational, the problem can be solved in polynomial time. However when some or all of the variables must be integer, corresponding to pure integer or Mixed Integer Programming (MIP) respectively, the problem becomes NP-complete (formally intractable). Several algorithms from operation research are widely used for solving integer programming instances in reasonable time. Reformulation of the problem into the set of linear inequalities often involves relaxation of several constraints. In the following text we introduce the formulation of the data transfer problem into the MIP syntax with involved approximations.

Since required datasets usually overlap together, we would like to minimize also the data movement of the common parts. In other words, if the same file is required by different users and the transfer paths share a link, we transfer the file on common link only once.

For this extension we have to slightly modify the constraint model, since the transfer path for a file can form a *forest* - using the terminology from the graph theory.

We denote the weight of an edge corresponding to the link bandwidth as $\mathbf{bw}(e)$ - bandwidth between two sites or average latency time for the storage elements (e.g. the time to stage the file from the tape system). The information about file's origins is a mapping of that file to a set of nodes where the file is available.

The input received from the users is a set of file names $\mathbf{F}$, where for every file $f \in \mathbf{F}$ we have a set of sources $\mathbf{orig}(f)$ - sites where the file f is already available and a set of destinations $\mathbf{dest}(f)$ - sites where the file f is supposed to be transferred.

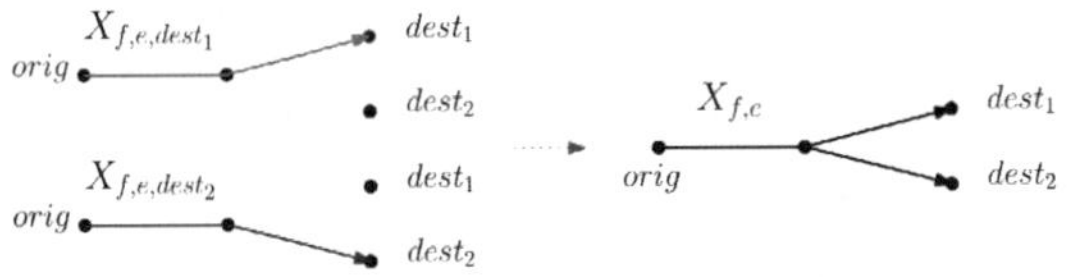

Figure 2: Two independent paths are *glued* together, so the file using their common links will be transferred only once (e.g. the file is staged only once, then transferred to two different destinations)

The essential idea is to use one decision variable for each file, its destination and edge in a graph. We will refer to this $\{0,1\}$ variable as X_{fed}, denoting whether file f is routed (value 1) over the edge e of the network or not (value 0) to its destination d. Mathematical constraints (1-3), ensuring that if all decision variables have assigned values the resulting configuration contains the independent transfer paths, are analogous to the Kirchhoff's circuit laws.

$$\forall f \in \mathbf{F},\ \forall d \in \mathbf{dest}(f):\quad \sum_{e \in \cup \mathbf{OUT}(n|n \in \mathbf{orig}(f))} X_{fed} = 1, \quad \sum_{e \in \cup \mathbf{IN}(n|n \in \mathbf{orig}(f))} X_{fed} = 0 \tag{1}$$

$$\forall f \in \mathbf{F},\ \forall d \in \mathbf{dest}(f): \sum_{e \in \mathbf{OUT}(d)} X_{fed} = 0, \ \sum_{e \in \mathbf{IN}(d)} X_{fed} = 1 \tag{2}$$

$$\forall f \in \mathbf{F},\ \forall d \in \mathbf{dest}(f),\ \forall n \notin \mathbf{orig}(f) \cup \{d\}: \quad \begin{array}{l} \sum_{e \in \mathbf{OUT}(n)} X_{fed} \le 1 \\ \sum_{e \in \mathbf{IN}(n)} X_{fed} \le 1 \end{array} \quad \sum_{e \in \mathbf{OUT}(n)} X_{fed} = \sum_{e \in \mathbf{IN}(n)} X_{fed} \tag{3}$$

Having generated all independent paths for a file to each of its destination, we need to *glue* them together. One can look at it as creating a *forest* using the terminology from the graph theory (Figure 2). We achieve it by defining new binary *two-index* variable X_{fe} stating whether file f uses link e (apart from reasoning about destinations).

$$\forall f \in \mathbf{F},\ \forall e \in \mathbf{E},\ \forall d \in \mathbf{dest}(f) : X_{fed} \le X_{fe} \tag{4}$$

$$\forall f \in \mathbf{F},\ \forall e \in \mathbf{E}: \sum_{d \in \mathbf{dest}(f)} X_{fed} \ge X_{fe} \tag{5}$$

$$\forall f \in \mathbf{F},\ \forall n \notin \mathbf{orig}(f) \cup \{d\}: \sum_{e \in \mathbf{IN}(n)} X_{fe} \le 1 \tag{6}$$

Finally, since we are minimizing the *makespan*, the time to transfer all files to the requested destinations, we define the constraints (7) for estimation of the completion time **T** variable and appropriate objective function: *minimize* T.

$$\forall e \in \mathbf{E}: \sum_{f \in \mathbf{F}} \frac{\mathbf{size}(f) \cdot X_{fe}}{\mathbf{bw}(e)} \le T \tag{7}$$

4.1. Performance comparison

In this section we will elaborate on the performance of the automated system under different environment configurations. The principal benefits of the solver depend on leveraging available data services and network links between sites. Let us focus first on performance comparison experienced by relying on data sources with diverse characteristic. All the following measurements were taken in real production environment while monitoring other exterior activities and requests for shared system resources. The monitoring included the control of *HPSS* usage over submitting large-size requests, extensive *WAN* third-party transfers between laboratories, etc. Therefore, the measurements spanning over several hours (and often repeated multiple times) are statistically stable and sound.

The performance of the system can be nicely described by comparing the makespan - the time it takes to bring requested files to the destination. It is important to see also the convergence, how fast were files appearing at the destination. Hence, we will display this tendency as a function of time in the following graphs. Figure 3 is displaying the comparison when the system alternates reasoning about data sources. The transfer was required between STAR *Tier-0* BNL laboratory and *Tier-1* center at LBNL, without involving any third site. We could concentrate thereby entirely on data sources and eliminate the influence of diverse network paths. We were comparing two data services which served as a source of the data. The services are in contrast by different characteristic:

- **HPSS** - holds all the files, works asynchronously. Usually involves waiting time at the beginning upon submission, then high throughput.
- **Xrootd** - holds only the portion of data, works synchronously. Usually provides low latency and high throughput.

The comparison consists of three several hours long transfers when solver was acting always in different mode. The blue line represents the mode when solver was using exclusively only *Xrootd* as the source service. We can see that files started to appear almost immediately at the destination and the slope of the line shows the fastest throughput. For better resolution the small plot in the same figure is displaying the zoomed region in the first 40 minutes. However, since *Xrootd* repository holds only a portion of all data, full data set could not be transferred. What portion of data the service holds usually depend on the age of files. Files from recent datasets are more likely available at *Xrootd* service. The red line represents the mode when system was relying only on HPSS. We can see that there was an initial small waiting time until files started to appear. More important is to realize also the "step-like" trend of the line. This is caused by the HPSS utilization. HPSS prioritizes requests from different users depending on tape locations, history, and other factors; and when there is our requests in turn, it serves the files usually fast. During several hours our request can be postponed and others are prioritized and we have to wait. Unfortunatelly, this "step" can often take several hours, depending on the current circumstances and load. The third black line is representing the last mode, when the system relied on both services concurrently. We can clearly see that the combination of both sources outperforms counting on the stable but not the fastest one (*HPSS*) even if *Xrootd* cannot provide all the files. Realizing which files where reside and coordinating the access to providing services is not something what users can efficiently do by themselves and hence, they often rely on the single one - stable but slow one. This is when automated solver can bring a significant benefit and increase the effectivity of their work.

Let us bring our attention now to the system's reasoning and utilizing diverse network paths. For the purpose of keeping the environment transparent and eliminating the effect of multiple data services, in this case we will concentrate on the sole *Xrootd* service as the source of all files. The slow latency and constant modest bandwidth of this data service allow us to concentrate on the influence of reasoning about network paths. The data transfer scheme is illustrated in Fig.4,

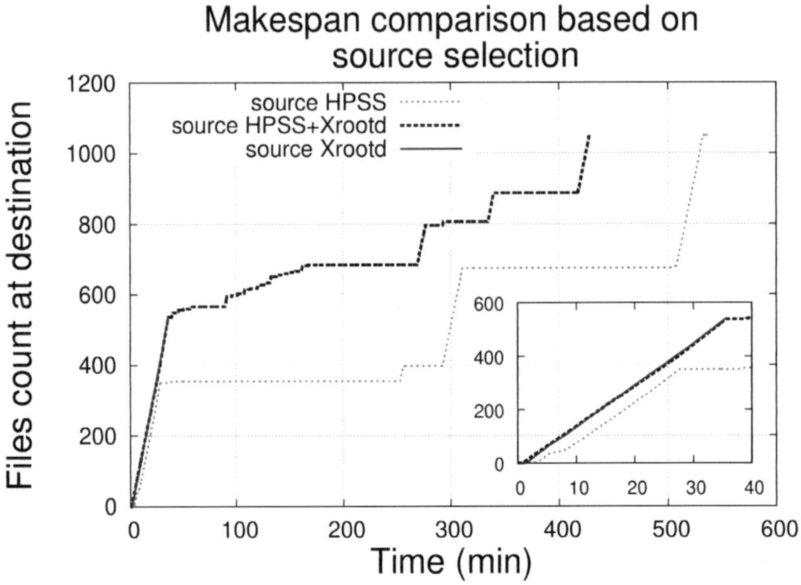

Figure 3: Graph displaying the trend how fast is the plan fulfilled concentrating on 3 different modes of source access. First one uses solely *Xrootd* service, second only the *HPSS* system, and finally third one uses combination of both.

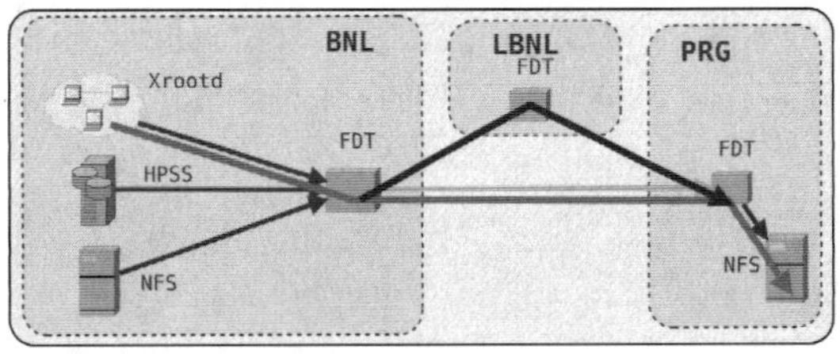

Figure 4: Transfer scheme for diverse network paths. Data movement relies on the *Xrootd* service as the sole source of files while leveraging also intermediate LBNL site.

where data are being moved from BNL laboratory (Tier-0 center) to Prague site (Tier-2 center). The system is allowed to reason also about intermediate site (LBNL laboratory, Tier-1 center) in order to increase the throughput. It is important to state that the connection between the BNL and Prague site is using static routing over dedicated link and is diverse from the path between BNL and LBNL as well as LBNL and Prague (using ESnet, Geant, and CESNET routing). We will again compare the speed how files appear at the destination service while looking at the impact of using intermediate site in parallel.

The graph in Fig.5 is exposing this comparison. The green line represents the mode where only direct BNL to Prague network path was used; while the red line the mode where also additional path through LBNL site was allowed. We can see the clear and very visible benefit in leveraging additional network path and routing part of the traffic via intermediate site. The overall gain in makespan (how less user waited for files) was almost one third of total time comparing to the direct and usual approach.

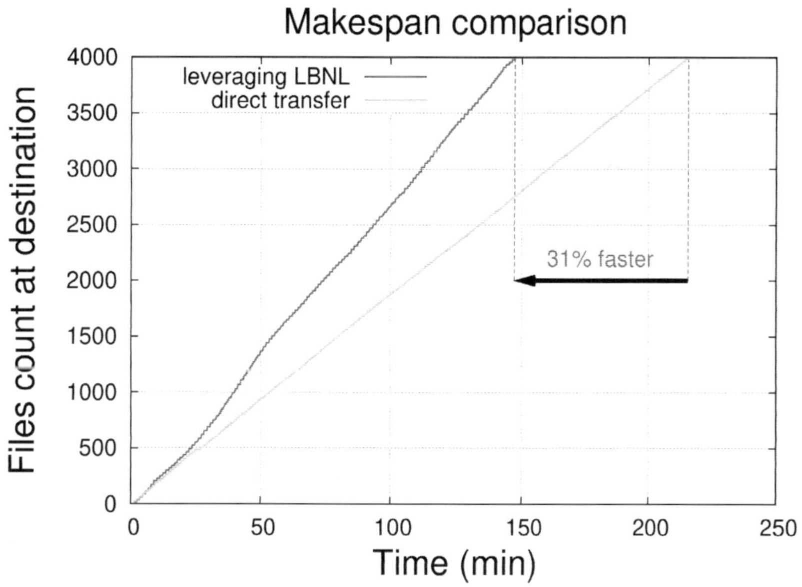

Figure 5: Graph displaying the trend how fast is the plan fulfilled comparing 2 different modes in network paths. Green line represents the mode when only direct transfer to Prague was allowed and red one the mode when part of the traffic was allowed to be routed via LBNL site.

5. Coupling with CPUs

In the previous sections we addressed and focused on the data transfer problem, where the task was to bring data sets to user specified locations. The role of the planner was to decide how to achieve it considering all constraints and having minimal makespan as an objective. However, very often the task is not finished by the time data are moved, but when data are analyzed. In other terms, the data movement itself only precedes the data processing.

Let us underline the benefit of CPU coupling by explaining the real case with production processing in STAR (see Fig. 6). Part of the production is being done at Argonne computing cloud (Chicago) together with PDSF/NERSC computing center (Berkeley). The workflow is following: files are continuously staged from BNL's HPSS system to the local 2TB cache. Since Argonne cloud is not equipped with any cache, the file can be transferred from BNL to Argonne only if there is a free CPU slot. To the contrary, PDSF site has sufficient 20TB cache and can hold data even if all the CPU slots are occupied. Therefore, and **this is being solved by hand**, it turns out that it is advantageous to feed Argonne's CPUs (when free) simultaneously from BNL and PDSF cache. If we had a system capable of dynamically solve this in automatic fashion the benefits could be clearly seen.

While in the pure file transfer problem the task was to locate and bring files to requested destinations, with CPU coupling the problem has to be reformulated. A user doesn't specify a single destination anymore, but a list of available processing sites along the full set of files. Each request R_i is therefore composed of set of files which need to be processed (F_{R_i}) together with a set of destinations - processing sites (D_{R_i}) where user is allowed to run jobs (equation 8). By saying allowed we mean that user has access and can run jobs on any of these sites.

$$\mathbf{R}_i = \{\underbrace{\{f_1, \ldots, f_{N_i}\}}_{\mathbf{F}_{R_i}}, \underbrace{\{d_1, \ldots, d_{M_i}\}}_{\mathbf{D}_{R_i}}\} \tag{8}$$

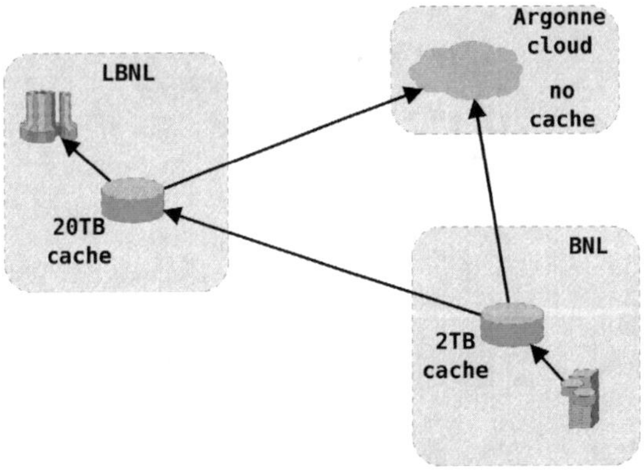

Figure 6: Representation of production processing using Argonne cloud with no cache space and PDSF farm with 20TB cache. Because of limited cache at BNL it turns out practical to feed Argonne cloud with data streaming from both coasts (PDSF and BNL).

The task of the planner is to find transfer paths for all files (every file has to appear in one of the destinations for each request it belongs to) considering the processing phase of the file at the computing site. The system may distribute files from a request among available destinations and execute job on the portion of data set at computing site A while on the other fraction of data set at computing site B.

The MIP model for sloving the pure file transfer problem can be reformulated and extended in such a way that it includes also the CPU reasoning. We will omit here the full set of mathematical constraints due to the length of the article. However, there is one difference in model's logic we would like to point out and explain.

If we look back into our initial motivation from Fig. 6, displaying the production processing schema in STAR, we can see that there is a substantial benefit of bringing files to the storage cache - closer to the processing sites, even if all the slots are used. The model, as it is defined above, reasons about bringing files to the processing sites since they are assigned as the only destinations. If the processing sites are occupied we would still like the solver to reason about bringing files to the storage space closer to the CPUs. In order to modify the reasoning we can create additional *dummy* links from the storages to the *dummy* destination vertex in the graph as represented in Fig. 7. The weights for these additional *dummy* edges need to be properly set so the solver will prioritize bringing files to the cache space in case the processing slots are taken. By setting the appropriate weight we can also control the preference between storage spaces.

6. Conclusions

The work presented in this paper deals and attacks the complex problem of efficient data movements on the network within a distributed environment. The problem itself arises from the real-life needs of the running nuclear physics experiment STAR and it's requirements for data storage and computational power.

To our best knowledge, it has been the first time in nuclear physics large scale experiments when automated planning approach was used for reasoning about data transfers and CPU allocations. We presented the underlying model using Mixed Integer Programming techniques and provided the inside view into the design and concept of the architecture. Implementation and coupling of components with standard STAR services proceeded with a real-case evaluation

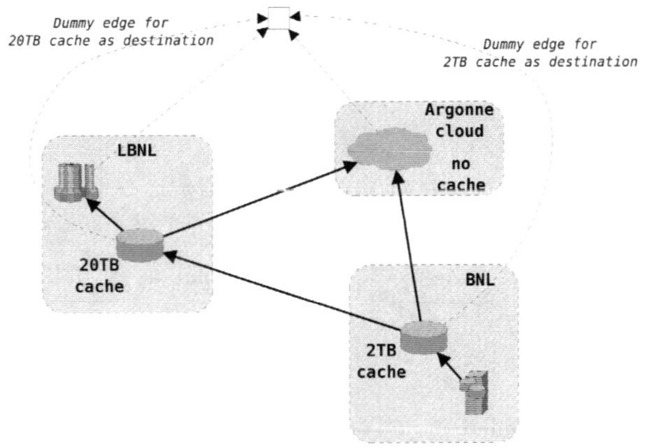

Figure 7: Representation of production processing using Argonne cloud as the graph input for the model. The storage cache nodes are also connected to the *dummy* destination in order to allow the model to bring files closer to the CPUs while they are taken by other jobs.

of the framework. We have demonstrated that this concept leads not only to the better comfort for physicists when dealing with data transfers but also brings fruits in the efficiency, load balancing and fail-over handling as was primarily requested.

With upcoming requirements for more frequent Cloud computing where data storage is constrained and the needs for prompt feeding CPUs by data is important, the automated data transfers and job allocation can greatly simplify the user's task. We have addressed this and proposed the extension of the model for reasoning about computational power as well.

The journey to the fully automated and intelligent system scheduling user's tasks considering all relevant constraints is certainly long and there are still numerous aspects that need to be addressed. We believe that the presented work contributed to this collaborated effort with several sound ideas, techniques and concepts.

Acknowledgments

The work has been supported in part by the Offices of NP and HEP within the U.S. DOE Office of Science and the U.S. NSF and by the grants LC07048 and LA09013 of the Ministry of Education of the Czech Republic.

References

[1] Adams S C J 2005 *Nuclear Physics A* **757** 102 URL `doi:10.1016/j.nuclphysa.2005.03.085`
[2] Ranganathan K and Foster I 2002 *11th IEEE Symposium on High-Performance Distributed Computing* vol 0 (Los Alamitos, CA, USA: IEEE Computer Society) pp 352–358 ISSN 1082-8907
[3] Shankar S and DeWitt D J 2007 *HPDC '07: Proceedings of the 16th international symposium on High performance distributed computing* (New York, NY, USA: ACM) pp 127–136 ISBN 978-1-59593-673-8
[4] Sato H, Matsuoka S, Endo T and Maruyama N 2008 *GRID* (IEEE) pp 250–257
[5] Rahman R M, Barker K and Alhajj R 2007 *CCGRID* (IEEE Computer Society) pp 171–178
[6] Fisher M L 1981 *Management Science* **27** 1–18
[7] Simonis H 2006 *Handbook of Constraint Programming* ed Rossi F, van Beek P and Walsh T (Elsevier) chap 25, pp 875–903

Evaluation of likelihood functions on CPU and GPU devices

Sverre Jarp, Alfio Lazzaro, Julien Leduc, Andrzej Nowak and Yngve Sneen Lindal

CERN openlab, Geneva, Switzerland

E-mail: sverre.jarp@cern.ch, alfio.lazzaro@cern.ch, julien.leduc@cern.ch, andrzej.nowak@cern.ch, yngve.sneen.lindal@cern.ch

Abstract. We describe parallel implementations of an algorithm used to evaluate the likelihood function used in data analysis. The implementations run, respectively, on CPU and GPU, and both devices cooperatively (hybrid). CPU and GPU implementations are based on OpenMP and OpenCL, respectively. The hybrid implementation allows the application to run also on multi-GPU systems (not necessarily of the same type). The hybrid case uses a scheduler so that the workload needed for the evaluation of function is split and balanced in corresponding sub-workloads to be executed in parallel on each device, i. e. CPU-GPU or multi-CPUs. We present the results of the scalability when running on CPU. Then we show the comparison of the performance of the GPU implementation on different hardware systems from different vendors, and the performance when running in the hybrid case. The tests are based on likelihood functions from real data analysis carried out in the high energy physics community.

1. Introduction

The evaluation of a likelihood function is required in several data analysis techniques, such as the maximum likelihood fitting procedure [1]. These techniques are largely used in high energy physics (HEP) experiments to analyze the data they collect. The data samples are a collection of N independent *events*, an event being the measurement of a set of O *observables* $\hat{x} = (x^1, \ldots, x^O)$ (energies, masses, spatial and angular variables...) recorded in a brief span of time by physics detectors. The events can be classified in S different *species*. Each observable x^j is distributed for the given species s with a probability distribution function (PDF) $\mathcal{P}_s^j(x^j;\hat{\theta}_s^j)$, where $\hat{\theta}_s^j$ are parameters of the PDF that can be related to the prediction obtained from physics models. If the observables are uncorrelated, then the total PDF for the species s is expressed by

$$\mathcal{P}_s(\hat{x};\hat{\theta}_s) = \prod_{j=1}^{O} \mathcal{P}_s^j(x^j;\hat{\theta}_s^j). \tag{1}$$

The PDFs are normalized over their observables, as function of their parameters, which implies an analytical or numerical evaluation of their integral. The *extended likelihood function* is

$$\mathcal{L} = \frac{e^{-\sum_{s=1}^{S} n_s}}{N!} \prod_{i=1}^{N} \sum_{s=1}^{S} n_s \mathcal{P}_s(\hat{x}_i;\hat{\theta}_s), \tag{2}$$

where n_s are the number of events belonging to each species. It is usual to evaluate the equivalent function $-\ln(\mathcal{L})$, the *negative log-likelihood* (*NLL*)[1]:

$$NLL = \sum_{s=1}^{S} n_s - \sum_{i=1}^{N} \left(\ln \sum_{s-1}^{S} n_s \mathcal{P}_s(\hat{x}_i; \hat{\theta}_s) \right), \tag{3}$$

that is a sum of logarithms. The terms of the sum can graphically be visualized as a tree, where the leaves are the PDFs $\mathcal{P}_s^j(x^j; \hat{\theta}_s^j)$ (basic PDFs), which are then linked to the corresponding product PDFs $\mathcal{P}_s(\hat{x}; \hat{\theta}_s)$, and finally the root that is $\sum_{s=1}^{S} n_s \mathcal{P}_s(\hat{x}_i; \hat{\theta}_s)$ (sum PDF). Product and sum PDFs are denoted as composite PDFs. Therefore, the root has S child nodes, each with O children, which means that in the tree there are $S \times (O+1) + 1$ nodes in total. The evaluation of the term in the sum of logarithms consists in traversing the entire tree, first evaluating the leaves up to the root. The time spent for this evaluation depends on the number of events and the complexity of the PDFs. Furthermore, the entire evaluation is performed several times during the data analysis for different sets of the PDF parameters. Hence, it becomes important to speed-up the evaluation to have fast data analyses.

The common software used in HEP community for the evaluation of the *NLL* is RooFit [2], which is part of the general data analysis framework ROOT [3]. Currently RooFit implements an algorithm for the *NLL* evaluation that cannot take full advantage of data vectorization and other code optimizations (like function inlining). A redesign of the algorithm and the corresponding implementation, with vectorization and several code optimizations, was described in [4] and [5]. This algorithm was also parallelized for CPU and GPU execution, based on OpenMP and CUDA, respectively. An improved version of the OpenMP implementation and a new implementation for GPU based on OpenCL are described in [6]. In the same reference, there is a description of a new implementation that allows the evaluations of the likelihood function to run cooperatively on a CPU and GPU belonging to the same computational node (hybrid evaluation). In this paper, the OpenMP and OpenCL implementations are briefly reported, while an improved hybrid implementation, which allows running on multi GPU systems (not necessarily of the same type) is introduced.

It is worthwhile to point out from the beginning that the implementations are specifically optimized for running on commodity systems, i.e. systems than can be considered, in terms of cost and power consumption, easily accessible to general data analysts (e.g. single socket desktop with a GPU whose main target is computer gaming). Data analysts can fully exploit their systems when performing the hybrid evaluation.

This paper is organized as follows: section 2 describes the OpenMP and OpenCL implementations and introduces the the new hybrid solution; section 3 reports the results obtained from tests done with a benchmark analysis on a testing system with NVIDIA and ATI GPUs; conclusion are given in section 4.

2. Parallel Implementations

This section describes the implementations for the evaluation on CPU, GPU, and the hybrid configuration. The code is implemented in C++ and all floating point operations are performed in double precision. Other details on these implementations can be found in [6].

2.1. OpenMP CPU Implementation

The parallelization on the CPU is based on OpenMP. Each type of PDF is implemented by a specific class. All inherit from a common interface that gives the public virtual method `getVal` and the protected pure virtual method `evaluate`. The former method computes the

[1] The $N!$ term in the expression is omitted, since it does not depend on the parameters.

normalization integral of the PDFs (in case of basic PDFs) and then it calls the latter method, which is overloaded inside each PDF class and it performs a loop for the evaluation of the function values. In total there are $S \times (O+1)+1$ loops that are implemented as `for` loops. There is only one parallel region for each evaluation, and this region starts at the root of the tree. This is implemented via a `#pragma omp parallel` directive. Therefore each loop executes a given subset of the N iterations. They are statically partitioned, i.e. each thread executes a fixed number of iterations. The partitioning is implemented in a way so that one thread can have at most one iteration of difference with respect to the other threads, to ensure an equal load-balancing. Eventually, the algorithm computes the logarithm of the final results, summing them up with a loop (reduction), which is also parallelized. To ensure the reproducibility of the results a specific algorithm for the reduction has been implemented. It preserves the order of the operations for a given number of threads and it reduces the rounding problem due to non-associative floating point arithmetic, using the double-double compensation algorithm 2Sum [7].

Each thread executes the entire evaluation from the root to the leaves within its own partition only and it accesses consecutive elements of the arrays of observables and results, allowing coalescing of memory accesses and data vectorization of the loops. These arrays are shared among the threads so that there is a negligible increase in the memory footprint of the application when running it in parallel. This implementation can lead to consequences that may be problematic in the case of a complex C++ code like RooFit. Indeed, the parallel region covers a larger portion of the execution, so it is crucial not to modify member variables of the object the method is running on, or global variables, without carefully assuring that race conditions are avoided.

Several arrays of temporary results and O arrays of observables have to be managed for each evaluation, each array composed by N double precision values. The amount of data to manage becomes significant in the case of complex models and large data samples, and it becomes crucial to have optimal treatment of the data inside the cache memories. Tests have proved that there is a significant penalty to the scalability when running with a high number of threads in a system where the largest cache memory is shared among the cores. Some different optimizations have been considered in order to reduce the load on memory. The composite PDFs, which have to combine arrays of results with just a simple operation, can send their results array "down" to the children, which then do their own evaluation and the corresponding combination directly on the array of value of the parent (results propagation). The main benefit of this change is that each basic PDF does not have to store its own results. Another optimization consists in splitting the data domain into blocks so that the entire procedure of evaluation is done one by one of them (block splitting). This optimization directly targets cache misses, since it increase locality and thereby cache efficiency. With these optimizations the number of total loops is $S \times (O+1)$ times the number of blocks and the number of arrays of results to $S+1$.

2.2. OpenCL GPU Implementation

In the OpenCL implementation the PDF loops are offloaded to be executed on a GPU. Each loop is thereby replaced by a corresponding OpenCL kernel which runs the N iterations using GPU threads. Also this implementation takes advantage of the results propagation as explained in section 2.1 for the OpenMP implementation. The block splitting is not considered because it does not fit with the way a GPU does computations. The results propagation for the composite PDFs is not feasible for OpenCL (it requires more coding with respect to the OpenMP implementation, e.g. because templates are not supported). There is an independent kernel to launch for each PDF, therefore $S \times (O+1)$ kernels, and a corresponding equal number of arrays of results to manage. The reduction is performed in parallel using a tree-based algorithm. It is deterministic and takes into account the non-associativity problem described in the OpenMP implementation.

The key reason for using OpenCL is the portability between vendors, which is particularly interesting for users using commodity systems. The implementation is fully performance-portable between NVIDIA and AMD GPUs. Tests have shown marginal improvements (less than 5%) when doing specific optimizations, e.g. using native vector types, which lead to different implementations for each device with respect to a common implementation.

The arrays of the observables are copied from the host to the GPU global memory using synchronous functions. These arrays are read-only during the entire execution of the application, so only one copy at the beginning is needed. Then they are used for all *NLL* evaluations. For each evaluation, the CPU traverses the *NLL* tree and it launches the corresponding kernels to be executed on the GPU. The arrays of results for each PDF can be kept resident in the GPU global memory. Eventually, the reduction is done on the final array of results and the value of the sum is copied back to the host memory for the finalization of the *NLL* value. The kernels are asynchronously executed, i.e. the evaluation of the tree is non-blocking and the kernels are just queued for execution on the GPU. The possibility of interleaving CPU and GPU computations reduces the impact of the operations that are not in loops, i.e. only executed by CPU, like the integral calculation for the normalization.

A very general procedure based on a heuristic approach has been used to decide the size of the workgroups. The rule is that if a kernel contains a transcendental function, the workgroup size is set to a "low" number. If the kernel does not contain transcendentals, but rather only basic arithmetics, the workgroup size is set to a "slightly higher" number. Tests have shown that 64 and 128, respectively, provided good results, with occupancy numbers ranging from 0.33 to 0.67 depending on the kernel. A comparison between using these numbers and the OpenCL default numbers gives about 14% improvement in performance. The number of blocks per kernel is then computed from the number of events divided by the number of threads per block, rounded to the greatest integer number.

2.3. *Hybrid Implementation*

The two implementations described in the previous sections make it possible to fully use the CPU and GPU computational devices independently. In the case of the GPU implementation, the CPU runs only a single thread, so a multicore CPU would be underutilized. Therefore, it is interesting to explore the possibility to fully load the CPU in an implementation-pleasant way. Another interesting possibility is to cooperatively use GPUs in a multi-GPU system. The hybrid solution described here allows to achieve these goals, running simultaneously the OpenMP and OpenCL implementations described in the previous sections.

In general a *compute device* is defined as some device capable of computation inside one physical computer. In the context of this work, it can be a CPU or a GPU. The physical computer has a single CPU and several possible GPUs. It then becomes important to have correct load balancing of the workload between such devices. Load balancing is a very fundamental and old problem, and much research has been done on it ([8, 9, 10] describe some examples of load balancing scenarios within computer science). The hybrid implementation described in this work uses a refining static load balancing for the evaluation of the *NLL*. The whole evaluation can be regarded as data-parallel, and therefore the whole domain is split into partitions so that each compute device is responsible for computing one and only one partition. A central concept is the importance of result determinism. It is *crucial* that each compute device has a fixed range before actual result computation starts, since different ranges in the same run would lead to different results when the final reduction is done. This means that the balancing must be an initial phase before a real evaluation is performed. The strategy used for implementing the heterogeneous data-parallel load-balancer in this work consists of:

- Starting an initial balancing phase by assigning to each compute device a range of length approximately N/K, where N is the total number of events, and K is the number of compute devices. This will lead to timing value t_k for each computational device, with

$k = 1, \ldots K$. The distribution of timing values will most probably be highly sub-optimal if the computational capabilities of the devices differ.

- Iterating and adjusting the partitioning based on these timings. Hopefully, this will converge within some time threshold, assuring a nearly optimal execution, i.e. a work partitioning that minimizes the difference between all timing values within some threshold.

Therefore, the determination of the optimal balancing is performed using *a priori* self-adjusting static load-balancing, so that the computation domain is first decomposed in such a way that each device will spent almost same amount of time to execute its own sub-workload. Then the decomposition is kept fixed (static) during all *NLL* evaluations. The downside of this strategy is that the entire subsequent evaluation will be based on this phase. If conditions in the system change, e.g. false timings during the initial phase or external load on the hardware during evaluation, the balancing can be sub-optimal.

The actual method used for dividing work between compute devices based on their respective execution times was originally described in [11] and was also used with seemingly good results in [12]. Let N_k^b denote the partition for compute device k (i.e. the number of events for that device) for a given balancing cycle b, so that the events are split in $\left[1, N_1^b\right], \left[1 + N_1^b, \sum_{k=1}^{2} N_k^b\right], \ldots \left[1 + \sum_{k=1}^{K-1} N_k^b, N\right]$ group for each device, respectively. Then each device evaluates its partition by mean of the specific implementation, i.e. OpenMP or OpenCL, spending the execution time t_k^b. The *relative power* for each device (i.e. its relative compute capability) is defined by:

$$RP_k^b = \frac{N_k^b}{t_k^b} \tag{4}$$

and the total power as

$$SRP^b = \sum_{k=0}^{K} RP_k^b. \tag{5}$$

Now each new partition can be computed by

$$N_k^{b+1} = N * \frac{RP_k^b}{SRP^b}. \tag{6}$$

The procedures ends when the maximum and the minimum of the distributions of t_k^b differ within some threshold. Each implementation runs the entire *NLL* tree evaluation. The measurement of the execution times of both CPU and GPU implementations is done using the `omp_get_wtime` OpenMP function. It is possible to consider averages from several *NLL* evaluations to reduce the fluctuations of these timings (the number of evaluation can be set by the users depending on the complexity of their *NLL*). The threshold value is increased when the condition is not satisfied after a given number of cycles of the load-balancer because of large fluctuations with respect to a small threshold. In any case, the minimum value of the threshold can be related to the time to execute a minimum number of events (granularity). Tests presented in this paper have shown that the load-balancer is able to reach convergence after only 3 cycles.

Once the best partition is found, the hybrid implementation consists in three different steps:

(i) Decomposition of the N iterations of the loops in groups of iterations N_k^{best} for each device k.

(ii) Each device runs the corresponding implementation for the entire *NLL* tree evaluation, considering its iterations. Then it runs the reduction.

(iii) The result of the reductions from each device are collected and summed together to finalize the *NLL* evaluation on the CPU.

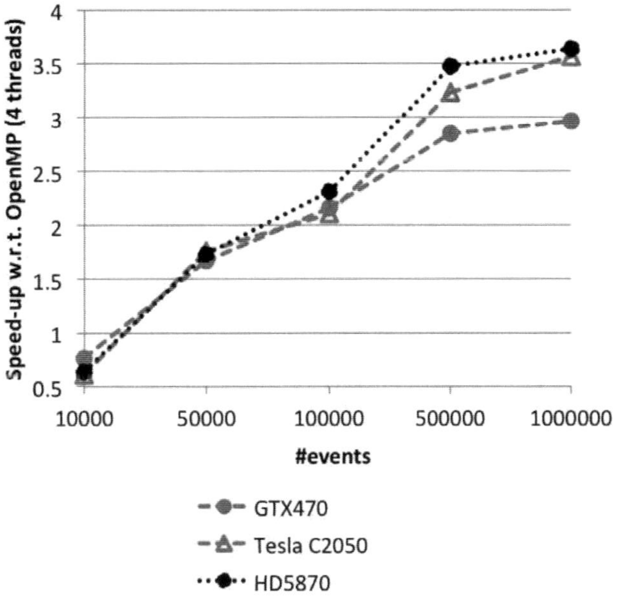

Figure 1. Comparison between OpenMP CPU and OpenCL GPU implementations. The reference is the OpenMP implementation with 4 threads.

In this strategy, the part that needs to be implemented is represented by the first step. The second step executes the OpenMP and OpenCL implementations already described on different ranges on the observables, and the last step is trivial.

The hybrid implementation is itself based on OpenMP. Essentially, the OpenMP implementation starts $P + K - 1$ threads, where P is the number of threads used for the NLL evaluation, while the other threads are used to run the GPU implementation for each GPU devices ($K - 1$ GPUs). This is possible since in the OpenMP implementation starts a single OpenMP parallel region at the root of the NLL tree, so that it is possible to branch between the two implementations at the beginning of the evaluation. Then the OpenMP standard guarantees an implicit synchronization at the end of the parallel region. It is important to remember that the OpenCL kernel calls are non-blocking and ideally consume minimal CPU time, i.e. the evaluation of the tree on the CPU imposes a minimal overhead. If this holds, then users can decide to run $P + K - 1$ threads on a CPU with P cores.

3. Tests

The following tests a statistical model based on the *BABAR* analysis for the branching fraction measurement of the B meson to $\eta' K$ decay [13]. There are 3 observables and 5 species. In total there are 21 PDFs: 7 Gaussians, 5 polynomials, 3 Argus functions, combined by 5 PDFs for multiplication and one for addition. All PDFs have an analytical integral. The number of events considered ranges between 10k and 1M. The model is implemented by a simplified version of the RooFit package. It is an easy-to-compile implementation which preserves the same programming interface of RooFit but without all the physics-related details. Tests are done considering 1000 NLL evaluations so that it is possible to compare the results for different number of events. The entire tests are repeated 5 times to achieve an accurate timing result.

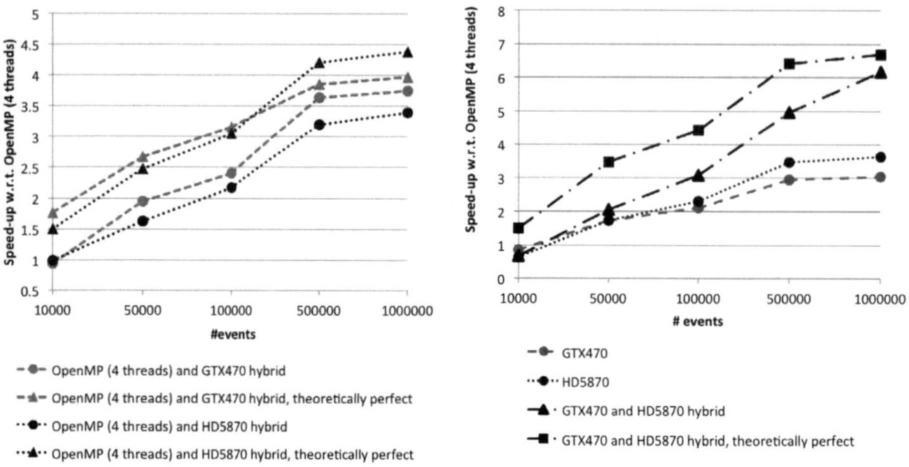

Figure 2. Comparison between OpenMP CPU, OpenCL GPU and the hybrid implementation (CPU + GPU on the left plot, multi-GPUs on the right plot). The reference is the OpenMP implementation with 4 threads. The "theoretically perfect" lines are obtained when summing the performance of the OpenMP and OpenCL implementation alone, i.e. no considering the overhead from the hybrid implementation.

The CPU system is an Intel Core i7 965 Nehalem running at 3.2 GHz, with 2GB DDR3 RAM. It is a quad-core CPU and supports SMT, which means that it has the ability to physically execute 8 threads simultaneously. Two GPUs are used: NVIDIA GeForce GTX470 and AMD Radeon HD5870. The code is compiled to run on the CPU using Intel Composer XE 2011 (icpc v12).

In the case of the OpenMP CPU implementation the speed-up is defined as the ratio between the execution times spent by the application running with a single thread and with a certain number of threads in parallel. The fraction of the execution time spent in code that can be executed in parallel is 96%. The results are: 1.8x with 2 threads and 3.6x with 4 threads. They are consistent with the theoretical speed-up values obtained by the Amdahl's law, i.e. 1.9x for 2 threads and 3.6x for 4 threads. Using the SMT threads, the overall speed-up becomes 4.7x with 8 threads, which means a benefit of 30% due to SMT. All results do not depend on the number of events used in the test.

In the case of the OpenCL GPU and the hybrid implementations the speed-up results are obtained from the ratio of the execution time of the exclusive OpenMP CPU implementation with 4 threads and the execution time of the configurations with GPUs. The reason of this choice is to see how fast we can go using the GPUs with respect to the parallel execution on the CPU only. The results for the OpenCL GPU implementation are shown in figure 1. For this case only, tests are also executed on a NVIDIA Tesla C2050. This implementation is not beneficial for a low number of events with respect to the OpenMP CPU implementation with 4 threads (speed-up < 1.0x), while it reaches a speed-up between 3.0x–3.6x for a high number of events. This is a direct consequence of the need to copy the final value over the PCI-Express bus. An interesting result is that neither the HD5870 nor the Tesla C2050 give any higher speedups than the GTX470, although theoretically they are almost 4x faster than the GTX470 when performing double-precision arithmetic. The reason is that the computation is completely memory-bound so all the ALUs on the cards are starved waiting for memory reads [14]. Figure 2 shows the results from the hybrid implementation. A general observation is that N must be large to gain anything since the overhead must be amortized. Balancing between the GTX470 and the CPU (with 4 threads in the OpenMP implementation) is very beneficial, achieving nearly perfect balancing for high workloads. In this case the speed-up increases from 3.0x when using the exclusive OpenCL GPU implementation to 3.8x of the hybrid case (4.0x is the theoretical maximum). Running with HD5870 and the CPU however leads to an almost negligible gain, i.e. the speed-up does not increase with respect to the exclusive OpenCL GPU implementation. This is due to an inefficiency of the AMD GPU driver for the OpenCL execution, which implies a significant overhead for the its execution on the CPU. The multi-GPUs configuration gives the best performance, reaching a speed-up of 6.1x, close to the theoretical maximum of 6.8x.

4. Conclusion

Two implementations for the *NLL* evaluation have been presented based on OpenMP and OpenCL and run on CPU and GPU, respectively. A novel approach has been presented that allows the hybrid evaluation for CPU and GPU computational devices including multi-GPUs systems. One can conclude that the GPU can be used for a sufficient number of events, and this is suitable since the need for computing power increases with N. Using the hybrid implementation further improves the performance; the multi-GPUs configuration gives the best performance. More details on this work can be found in [14].

Acknowledgments

The authors would like to thank Filippo Spiga of Irish Centre for High-End Computing for providing us the system with the NVIDIA Tesla card and Felice Pantaleo for his suggestions in the GPU implementation.

References

[1] Cowan G 1998 *Statistical Data Analysis* (Oxford: Clarendon Press)

[2] Verkerke W and Kirkby D 2006 *Proc. PHYSTAT05* (London: Imperial College Press) The RooFit Toolkit for Data Modeling

[3] Brun R and Rademakers F 1997 *Nucl. Instrum. Methods Phys. Res. Sect. A* **389**(1-2) 81

[4] Jarp S, Lazzaro A, Leduc J, Nowak A and Pantaleo F 2011 *2011 IEEE Int. Symp. Parallel Distributed Processing* 1349–1358 (*Preprint* CERN-IT-2011-010)

[5] Jarp S, Lazzaro A, Leduc J, Nowak A and Pantaleo F 2011 *J. Phys.: Conf. Series* **331** 032021 (*Preprint* CERN-IT-2011-009)

[6] Jarp S, Lazzaro A, Leduc J, Nowak A and Sneen-Lindal Y 2011 *Proc. Int. Conf. Parallel Computing* (Ghent, Belgium) Parallel Likelihood Function Evaluation on Heterogeneous Many-core Systems (*Preprint* CERN-IT-2011-012)

[7] He Y and Ding C H Q 2001 *J. Supercomputing* **18** 259–277

[8] Harchol-Balter M and Allen B D 1997 *ACM Trans. Computing Systems* **15**(3) 253–285

[9] Shirazi A B, Ali R H and Krishna M K 1995 *Scheduling and Load Balancing in Parallel and Distributed Systems* (IEEE Computer Society Press)

[10] Ferrer R *et al.* 2011 *Proc. 23rd Int. Conf. Languages and Compilers for Parallel Computing* pp 215–229 Optimizing the exploitation of multicore processors and GPUs with OpenMP and OpenCL

[11] Galindo I, Almeida F and Badia-Contelles J M 2008 *Lecture Notes in Computer Science* **5205** 64–74

[12] Acosta A, Corujo R, Blanco V and Almeida F 2010 *Proc. 2010 Int. Conf. on High Performance Computing and Simulation* pp 467–476 Dynamic load balancing on heterogeneous multicore/multiGPU systems

[13] Aubert B *et al.* (*BABAR* Collaboration) 2009 *Phys. Rev. D* **80** 112002

[14] Sneen-Lindal Y 2011 Optimizing a high energy physics toolkit on heterogeneous architectures (*Preprint* CERN-THESIS-2011-153)

Efficient pseudo-random number generation for monte-carlo simulations using graphic processors

Siddhant Mohanty[1,2], A K Mohanty[1,3] and F Carminati[1]

[1] CERN, CH-1211, Geneve 23, Switzerland
[2] Department of Computer Science, VNIT, Nagpur, India 440010
[3] Nuclear Physics Division, Bhabha Atomic Research Centre, Mumbai, India 400085

E-mail: siddhant9892@gmail.com, ajit@cern.ch, Federico.Carminati@cern.ch

Abstract. A hybrid approach based on the combination of three Tausworthe generators and one linear congruential generator for pseudo random number generation for GPU programing as suggested in NVIDIA-CUDA library has been used for MONTE-CARLO sampling. On each GPU thread, a random seed is generated on fly in a simple way using the quick and dirty algorithm where mod operation is not performed explicitly due to unsigned integer overflow. Using this hybrid generator, multivariate correlated sampling based on alias technique has been carried out using both CUDA and OpenCL languages.

1. Introduction

The future of high power computing is evolving towards the efficient use of highly parallel computing environment. The class of devices that has been designed having parallelism features in mind is the Graphics Processing Units (GPU) which are highly parallel, multi threaded computing devices. One application where the use of massive parallelism comes instinctively is Monte-Carlo simulations where a large number of independent events have to be simulated. At the core of the Monte-Carlo simulation lies the random number generators. For GPU programming, the random number generator should have (a) good statistical properties (b) high computational speed (c) low memory use, and (d) a large period . The most commonly used Mersenne Twister generator [1] has very good statistical properties with a long period of 2^{19937}, but not suitable for implementation in the GPU as it has a large state that must be updated serially. Each GPU thread must have an individual state in global RAM and requires multiple access per generator. The relatively large number of computation per generated number makes the generator too slow for GPU programming except in cases where the ultimate in quality is needed. In this paper, we have used a hybrid approach as used in NVIDIA-CUDA library [2]. The suggestion is to use a combination of three Tausworthe generators along with a simple Linear Congruential Generator (LCG) where the mod operation is not performed explicitly. The period of these combinations is quite high 2^{121} and has good statistical properties as the defects of one generator should hide those of the other. This hybrid generator requires four random seeds which can either be supplied using a CPU-side random number generator or through a simple quick and dirty type LCG generator which can be implemented on each GPU thread on fly. We have carried out alias Monte-Carlo sampling [3] from a two dimensional correlated distribution using this hybrid generator where each GPU thread is used to generate random variable in parallel.

This would mean each thread needs to be provided a random seed independently. In the present work, we have implemented alias sampling with NVIDIA GeForce GTX 480 GPU card using both CUDA and OpenCL kernels. It is noticed that the kernel execution in both cases is about 1000 times faster as compared to the CPU whereas the total code execution is only 10 times faster. This is due to the fact that memory copy from host to device or vice-versa is very slow. Therefore we try to minimize memory access time and implement a simple scheme to generate random seed per thread on fly from the formula $seed = 1099087573 * id$ where id is the thread index. This is known as quick and dirty LCG which has a period of 2^{32} and mod operation is not explicitly needed due to overflow of unsigned integer. It is shown that this hybrid generator which takes seed on fly is quite fast, reproduces the statistical properties reasonably well and can easily be implemented on each thread of GPU as well as CPU in an efficient way.

2. Random number generation:A review

A true random number generator provides truly unpredictable numbers which is difficult to achieve through computational implementation. The most common type of random number generators used are pseudo random in character as the implementation uses a deterministic software. However it is desirable to have the output of the generator practically indistinguishable and should not exhibit any correlations or patterns. Every deterministic generator must eventually loop but the goal is to make the loop period as long as possible. Poor generator quality can ruin the results of Monte-Carlo applications. In the following, we briefly mention various uniform PRNGS(Pseudo Random Number Generators) that are available for implementation [4].

It may be mentioned here that ROOT package [5] provides various classes for random number generators based on different algorithms. The TRandom1 class is based on the Ranlux engine originally developed by M. Lusher and implemented in Fortran by F. James [7]. The TRandom2 is based on the maximally equidistributed combined Tausworthe generator by L'Ecuyer [6] while the TRandom3 uses the Mersenne Twister algorithm [1].

2.1. Linear congruential generator

A Linear Congruential Generator (LCG) represents one of the oldest pseudo random number generator algorithms. It is a classical generator which uses a transition function of the form

$$X_{n+1} = (a\ X_n + c)\ mod\ m \tag{1}$$

The maximum period of the generator is m, which means that in 32-bit integer, the period can be at most 2^{32}. It is expected that c is relatively prime to m and $(a-1)$ is a multiple of p for every prime factor p of m. If $c = 0$, the generator is called multiplicative LCG. LCGs are fast and require minimal memory (typically 32 or 64 bit) to retain state. This makes them valuable for simulating multiple independent streams. Due to their statistical flaws, LCGs should not be used for applications where high-quality randomness is critical. For example it is not suitable for a Monte-Carlo simulation because of the serial correlation. A further problem of LCG is that the lower order bits of the generated sequence have a far shorter period than the sequence as a whole if m is set to a power of 2.

2.2. Multiplicative Recursive Generator

A derivative of the LCG is the multiple recursive generator (MRG), which additively combines two or more generators. If n generators with periods $m_1, m_2, ..., m_n$ are combined, then the resulting period is their LCM, thus the period can be at most $m_1\ m_2\ .....m_n$. These generators provide both good statistical quality and long periods, but the relatively prime moduli require complex algorithms using 32-bit multiplications and divisions, so they are not suitable for current GPU programing.

2.3. Mersenne Twister Generator

Mersenne Twister is one of the most respected methods for random number generation in software and has a period of 2^{19937} and provides a very high quality pseudo random numbers [1]. Even though it is good at generating high quality random numbers, it is not very elegant and is very complex to implement. This generator is not very sensitive to initialization and can take a long time to recover from a zero-excess initial state. It is not convenient for GPU implementation due to its large state that must be updated serially. Thus each GPU thread must have an individual state in global RAM and has to make multiple accesses per generator. In combination with the relatively large amount of computation per generated number, this requirement makes the generator too slow, except in cases where the ultimate in quality is needed.

2.4. Combined Tausworthe Generator

Linear Feedback Shift Register (LFSR) or the combined Tausworthe generator provides a fast implementation and make use of only exclusive-OR and shift operations. The following code shows the example of an one component combined Tausworthe generator,

```
------------------------------------------------------------------------------
 // S1, S2, S3, and M are all constants, and z is part of the
 // private per-thread generator state.
------------------------------------------------------------------------------
   unsigned TausStep(unsigned &z, int S1, int S2, int S3, unsigned M)
{
  unsigned b=(((z << S1) ^ z) >> S2);
  return z = (((z & M) << S3) ^ b);
}
------------------------------------------------------------------------------
```

A popular version of this procedure is the four component LFSR113 generator from L'Ecuyer 1999 which produces random stream with a period of approximately 2^{113} [6]. However, statistical tests show that even the four-component LFSR113 produces significant correlations across 5-tuples and 6-tuples for relatively small sample sizes. Also the periods are not sufficiently large enough. But in terms of speed, these are highly competitive to those which are currently available from software libraries.

3. A Hybrid generator for GPU programing

To increase the period length and improve the statistical behavior of a generator, it is always good to combine generators of different periods. If the periods of the individual generators are co-prime, then their product will be the period of the combined generators and also the combined generators will have less statistical flaws as compared to their individual counter parts. In the present work, we follow an approach suggested by NVIDIA [2] where three Tausworthe generator are combined with a LCG generators as given by,

```
--------------------------------------------------------------------------------
unsigned z1, z2, z3, z4;
float HybridTaus()
{
   // Combined period is lcm(p1,p2,p3,p4)~ 2^121
    return 2.3283064365387e-10 * (              // Periods
    TausStep(z1, 13, 19, 12, 4294967294UL) ^    // p1=2^31-1
    TausStep(z2, 2, 25, 4, 4294967288UL)   ^    // p2=2^30-1
    TausStep(z3, 3, 11, 17, 4294967280UL)  ^    // p3=2^28-1
    LCGStep(z4, 1664525, 1013904223UL)          // p4=2^32
   );
}
--------------------------------------------------------------------------------
```

The above combination provides an overall period of around 2^{121}. The LCGStep is a simple LCG generator of the type,

```
--------------------------------------------------------------------------------
 // A and C are constants
   unsigned LCGStep(unsigned &z, unsigned A, unsigned C)
{
  return z=(A*z+C);
}
--------------------------------------------------------------------------------
```

In the LCG, the mod operation is not explicitly required due to unsigned overflow and has a period of 2^{32} due to truncation in 32-bit arithematic. Since 2^{32} is also relatively prime to the periods of other three Tausworthe generators, the fourth combination with the LCG increases the period by a factor of 2^{32} without any heavy computation.

The above hybrid generator requires four random seeds $z1, z2, z3, z4$. One way is to generate them in the host machine using a CPU based program and then copying into the device. It is noticed that the memory copy from host to device or vice versa is very time consuming. Therefore, we use another LCG type generator for seed generation which requires only a multiplication on each GPU thread and given by,

```
--------------------------------------------------------------------------------
//Function giving seed for each thread

unsigned seed(int id)
{
return seed=id*1099087573UL;
}
--------------------------------------------------------------------------------
```

where the index *id* varies with the thread number of the GPU. This algorithm is known as quick and dirt algorithm and gives a random number due to same unsigned integer overflow. It is seen that this hybrid generator is quite fast as compared to the traditional Mersenne Twister algorithm and having nearly similar statistical performance. As an example, we have estimated the auto-correlation function defined as

$$C_m = (x[i] - \bar{x})\ (x[i+m] - \bar{x}) \qquad (2)$$

where $x[i]$ is an array of random numbers between 0 and 1, $\bar{x}$ is the average and m is the correlation distance. For $m = 0$, it represents the variance of the distribution. We have estimated C_m using both Mersenne Twister (TRandom3 of root package) and the present Hybrid algorithm. As shown in figure 1, the auto-correlation function (C_m/C_0) of TRandom3 and the present hybrid generator are comparable. It is one at $m = 0$ and decreases to zero quite fast as expected.

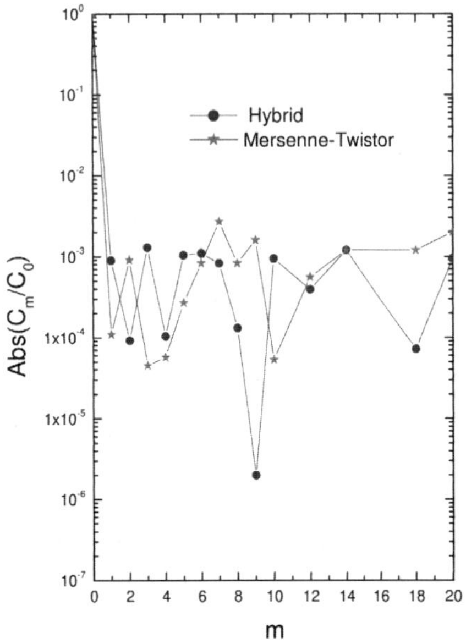

Figure 1. The comparison of auto-correlation function $Abs(C_m/C_0)$ as a function m for Mersenne-Twister (TRandom3) and Hybrid generator.

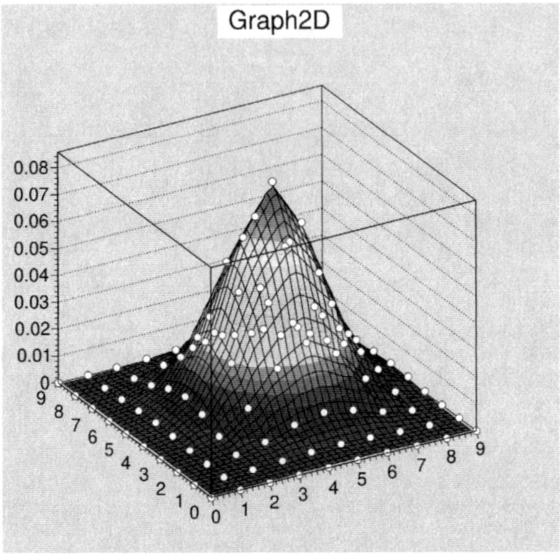

Figure 2. The comparison between actual 2D inputs (solid points) and generated points shown as continuous line obtained using alias technique.

4. Sampling algorithm and GPU implementation

In this conference, some of us have presented a methodology to carry out Monte-Carlo sampling from a multivariate correlated data set using alias technique. The motivation for this sampling is to develop a ROOT based low energy neutron transport code using ENDF nuclear data library where evaluated nuclear data like energy and angle of neutron emissions are highly correlated. In this work, we try to implement the alias algorithm using GPU to make the execution faster. The detailed of the algorithm can be found in [3, 8]. The alias method recasts the original non-alias outcome i and probability p_i into an alias outcome Λ_i and alias probability Π_i. An non-alias outcome i is chosen randomly between 0 and N with uniform probability i.e. $i = Nu_1$ where u_1 is an uniform random number between 0 and 1. The method then compares a second random number u_2 against the alias probability Π_i. If $u_2 \leq \Pi_i$, then non alias outcome i is chosen, otherwise the alias outcome Λ_i is chosen. Thus instead of the original probability array p_i, the alias method requires to store two arrays one for Λ_i and other for Π_i. This is in case of an one dimensional sampling. In case of two dimensional sampling, four tables are required to be computed and stored. Since these computations are one time, we generate them using the CPU based program in the host and copy to the device memory. We use a NVIDIA GPU (GeForce

GTX 480) which has compute capability 2.0, suitable for double precision arithematic. Other specifications are: it has total global memory of 1609760768 bytes, total constant memory 65536 bytes, multiprocessor count 15, shared memory per mp 49152 bytes, registers per mp 32768, threads in warp 32, maximum threads per block 1024 and maximum number of blocks 65535. The essential feature of GPU programing is the difference between the host and the device. The host program (CPU) executes the device kernel on each GPU thread. Although, we have carried out sampling using both OpenCL and CUDA, in the following, explanation is given as per the CUDA implementation. The host program invokes the device kernel (source code for CUDA kernel is listed at the end)

```
__global__ void cuda_kernel(int*c1,int*c2,double*fA,double*fR,double*fA1,double*fR1)
```

which is executed on each thread in parallel with thread index *id* that varies from 0 to $N = blockno * blocksize - 1$. For each thread, the kernel invokes two device routines *seedperthread* for random seed generation which is used in *TauStep* in step 1. For step 2 to step 3, the previous random number provides seed to the subsequent one. Thus, each thread generates four random numbers which are used for alias sampling. Since each thread is executed in parallel, this is an efficient and fast method to generate random number on each thread independently as *id* varies depending on the thread index. The sampling is carried out on each thread using the alias table fA, fR, $fA1$ and $fR1$ which are common to all threads and are copied from the host to device using the CUDA *cudaMemcpyHostToDevice* utility. Similarly, the results are copied back from device to host using *cudaMemcpyDeviceToHost* utility. As shown in the ode, each thread requires four random numbers $u1$, $u2$, $u3$ and $u4$ which are generated invoking the hybrid generator four times. For the first time, it takes the seed from the seed generator and use the previous number for the subsequent steps.

5. Results and conclusion

In the following, we carry out two dimensional sampling from a correlated Gaussian given by,

$$f(x, y) = e^{-[(x-x_0)^2+(y-y_0)^2+\alpha(x-x_0)(y-y_0)]} \tag{3}$$

which becomes uncorrelated when $\alpha = 0$ and can be written as the product of two independent Gaussians. We generate 100 discrete data points (as 10 by 10 matrix) from the above distribution and assign as probability $P[i][j]$ which is normalized to unity. We also fix $x_0 = y_0 = 5$ and $\alpha = 2$. This corresponds to a situation when the co-variance is non zero, although the choice of parameters are arbitrary. The sampling is carried out as follows. First we select a row J using alias technique from the distribution R_j where $R_j = \sum_i P[i][j]$ [3]. Once the row J is chosen, the corresponding column is fixed again using the same alias technique as per the distribution $P[i][J]$. Figure 2 shows typical plot of a 2D correlated Gaussian sampling where the solid points represent the actual distribution where as continuous curve is obtained through alias sampling. Note that the present implementation of alias sampling generates only the discrete data points which are connected smoothly as continuous lines using ROOT inbuilt algorithms. Although not shown here in detail, the sampling results obtained from the GPU are as accurate as what we would have got implementing the same algorithm on a CPU based program. On the other hand, the execution on GPU is about 1000 times faster as compared to the CPU. Table I shows the time taken to sample 10^6 events both on CPU and GPU. For GPU, we have implemented the algorithm both using CUDA and OpenCL kernels. The CUDA execution is faster than OpenCL as it has more overheads. It may be mentioned here that although GPU performance is much faster than CPU, the overall performance is only about 10 times faster. This is due to the large memory access time between host and device and vice versa. Work is in progress to develop better sampling algorithm and also to minimize on the memory access time so that over all execution can be made faster. Instead of transferring all the random numbers generated on GPU to host memory, it is possible to carry out Monte-Carlo simulation of interest on each GPU thread and finally transferring the desired results. This will make the overall code execution faster.

Table 1. The comparison of execution time.

	CPU	GPU-CUDA	GPU-OpenCL
Kernel Exec.	9.73 Sec	53 μ sec	220 μ sec
Total time	9.93 sec	0.92 sec	1.12 sec

Appendix

```
------------------------------------------------------------------------------
//Program in Cuda for 2D alias sampling
// Only the essential portions are given. This program
// performs a 2D correlated Gaussian sampling using alias technique
// for 100 data points given as 10 x 10 matrix.
------------------------------------------------------------------------------
#include<stdio.h>
#include<cuda.h>
#include<math.h>
#include<sys/time.h>

------------------------------------------------------------------------------
__device__ unsigned seed_per_thread(int id)
{
return id*1099087573UL;

}
------------------------------------------------------------------------------


__device__ unsigned TauStep(unsigned z, int s1, int s2, int s3, unsigned M)
{

unsigned b=(((z << s1) ^ z) >> s2);
return z = (((z & M) << s3) ^ b);
}

------------------------------------------------------------------------------


__global__ void cuda_kernel(int*c1,int*c2,double*fA,double*fR,double*fA1,double*fR1)
```

```
{
//fA, fR, fA1 and fA2 are alias tables which are computed in the host
//program and copied into the device using cuda memory copy utility.
//The final reasults which are stored in the device array c1 and c2 are also
//copied back to the host after the kernel execution is over.

int idx=blockIdx.x*blockDim.x+threadIdx.x;
float u1,u2,u3,u4;
int i1,i2,k1,k2;
unsigned b,z,z1,z2,z3,z4,M;
unsigned r0,r1,r2,r3;
unsigned seed;
//Generates four random numbers u1, u2, u3 and u4 in four steps.
//Only seed_per_thread is called in step 1 to privide the initial seed.

//STEP 1
seed=seed_per_thread(idx);
z1=TauStep(seed,13,19,12,429496729UL);
z2=TauStep(seed,2,25,4,4294967288UL);
z3=TauStep(seed,3,11,17,429496280UL);
z4=(1664525*seed+1013904223UL);
r0=(z1^z2^z3^z4);

//STEP 2
z1=TauStep(r0,13,19,12,429496729UL);
z2=TauStep(r0,2,25,4,4294967288UL);
z3=TauStep(r0,3,11,17,429496280UL);
z4=(1664525*r0+1013904223UL);
r1=(z1^z2^z3^z4);

//STEP 3
z1=TauStep(r1,13,19,12,429496729UL);
z2=TauStep(r1,2,25,4,4294967288UL);
z3=TauStep(r1,3,11,17,429496280UL);
z4=(1664525*r1+1013904223UL);
r2=(z1^z2^z3^z4);

//STEP 4
z1=TauStep(r2,13,19,12,429496729UL);
z2=TauStep(r2,2,25,4,4294967288UL);
z3=TauStep(r2,3,11,17,429496280UL);
z4=(1664525*r2+1013904223UL);
r3=(z1^z2^z3^z4);

--------------------------------------------------------------------------------
// u1, u2, u3 and u4 varies between 0 and 1.0

u1=r0*2.3283064365387e-10;
u2=r1*2.3283064365387e-10;
u3=r2*2.3283064365387e-10;
```

```
u4=r3*2.3283064365387e-10;


i1=10*u1;
 if(u2<=fR1[i1]) k1=i1;
 else k1=fA1[i1];
i2=10*u3;
 if(u4<=fR[k1*10+i2]) k2=i2;
 else k2=fA[k1*10+i2];
c1[idx]=k1;
c2[idx]=k2;
//sampling results are stored in two arrays of c1 and c2 with thread index idx.

}
```

References

[1] Matsumoto M M and Nishimura T 1998 *ACM Transactions on Modeling and Computer Simulations* vol 8 p 3-20
[2] http://developer.nvidia.com/GPUGems3
[3] Walker A J 1977 *ACM Transactions on Mathematical Software* vol 3 p 253
[4] Knuth D 1969 *The Art of Computer Programming* (V-2, Addison-Wesly)
[5] http://root.cern.ch
[6] L'Ecuyer 1996 *Mathematics of Computations* vol 65 p 203-213
[7] James F 1994 *Computer Physics Communication* vol 79 p 111
[8] Popescu L M 2000 *Journal of Computational Physics* vol 160 p 612

Journal of Physics: Conference Series **368** (2012) 012025 doi:10.1088/1742-6596/368/1/012025

Challenges in using GPUs for the reconstruction of digital hologram images

I D Reid, J J Nebrensky and P R Hobson

Centre for Sensors & Instrumentation, School of Engineering & Design, Brunel University, Uxbridge UB8 3PH, UK

E-mail: ivan.reid@cern.ch.

Abstract. In-line holographic imaging is used for small particulates, such as cloud or spray droplets, marine plankton, and alluvial sediments, and enables a true 3D object field to be recorded at high resolution over a considerable depth. To reconstruct a digital hologram a 2D FFT must be calculated for every depth slice desired in the replayed image volume. A typical in-line hologram of ~100 micrometre-sized particles over a depth of a few hundred millimetres will require O(1000) 2D FFT operations to be performed on an hologram of typically a few million pixels. In previous work we have reported on our experiences with reconstruction on a computational grid. In this paper we discuss the technical challenges in making efficient use of the NVIDIA Tesla and Fermi GPU systems and show how our reconstruction code was optimised for near real-time video slice reconstruction with holograms as large as 4K by 4K pixels. We also consider the implications for grid and cloud computing approaches to hologram replay, and the extent to which a GPU can replace these approaches, when the important step of locating focussed objects within a reconstructed volume is included.

1. Introduction

In the last decade digital in-line holography has replaced silver-halide based techniques for the 3D recording of volumes of small particles. It has applications in a number of areas of science and engineering, for example cloud or spray droplets [1], marine plankton [2] and sediments [3]. Digital holography replaces the older technique, which primarily used high-resolution silver-halide emulsions, with direct recording onto a solid-state pixel sensor such as a CCD or CMOS array. There are a number of different mathematical techniques that can be used to calculate the replayed image volume; ours is based upon the convolution approach [4]. To replay the image volume encoded by a digital hologram requires that the two-dimensional discrete Fourier transform of the hologram is multiplied by a phase factor related to the recording beam and the distance of the replaced depth slice and then the image is created by taking the inverse discrete Fourier Transform and calculating its magnitude. A typical CCD or CMOS sensor used in digital holography has a few million pixels, although individual sensors with over eighty million pixels have been made. To fully reconstruct the volume recorded on a digital hologram requires 2D depth-slices at sub-millimeter intervals to be reconstructed over the entire recorded depth which is often a few hundred millimeters in extent. Thus for a single hologram of order one thousand individual 2D images will be replayed. The computational task, even with modern multi-core processors and efficient discrete Fourier transform algorithms such as FFTW [5] is considerable. In previous work [6], [7] we have exploited the trivially parallel nature of the hologram replay to enable us to use computational grid resources. Unfortunately the stochastic nature of submission to grid resources, via a broker, and the possibility of being queued until the validity of a job's X.509 proxy certificate expires means that the initial significant speed advantage over a single local computational node is rarely maintained and quite often not all of a volume is reconstructed. In earlier work [8] a number of different jobs each reconstructing ten, fifty or one hundred sequential depth slices were submitted to nodes on the EGEE grid. The total number of depth planes required to fully reconstruct the volume was 2200. Figure 1 illustrates the initial large advantage over a single processor but the rate in all cases dramatically slowed and only a fraction of all the jobs completed. This behavior is typical of current scientific production grids.

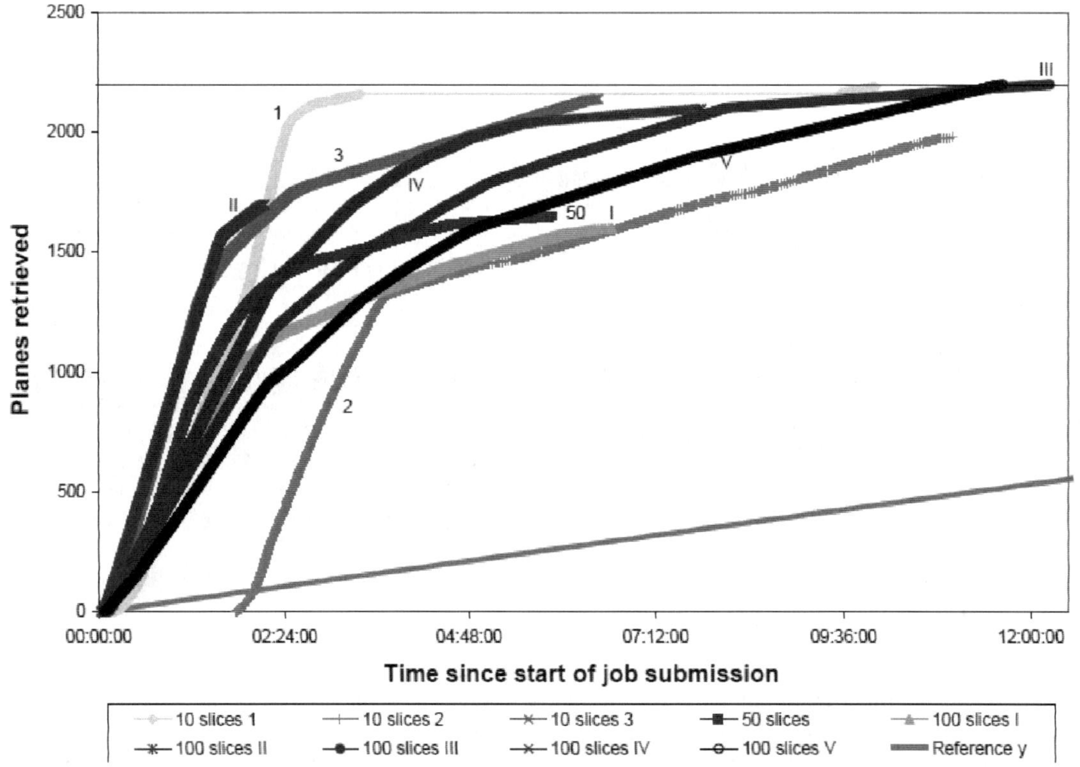

Figure 1. Overall rate of grid job completion (times in hh:mm:ss format after submission). The straight line labelled "Reference y" refers to serial replay on a single processor (AMD Athlon XP model 10 2600+ (1920 MHz), 1.5 GB memory) [8].

With the advent in recent years of powerful and increasingly inexpensive GPU chips considerable effort from ourselves [9] and many other researchers (see for example [10], [11], [12]) has been expended in porting, or developing, digital hologram replay code to run efficiently on these systems. In this paper we describe our latest work using NVIDIA Tesla and Fermi class GPU systems, the optimization techniques that we have deployed to achieve very significant replay speed, and the prospects for further improvement. Finally we speculate on whether a hybrid approach of GPU and distributed (grid or cloud) based computing may be optimum when the task of both replay and object extraction is considered.

2. Recording and replaying a digital hologram

We consider in this paper only the recording and replay of in-line (Fraunhofer) holograms recorded digitally on a pixel area sensor. Other forms of digital hologram, for example those recorded in off-axis holography, do not differ in the essential principles of replay.

2.1. Hologram recording

We recorded in-line (Fraunhofer) holograms of objects in water using an 8 megapixel camera (Atmel Camelia 8M, 2300 by 3500 pixels with 12-bit depth) with a collimated beam from a c.w. HeNe laser (λ=633 nm, 1 mW). As in earlier work [9] we recorded a sample volume consisting of cenospheres mostly of 100-300 μm dia. (Fillite Trelleborg Specialty Grade (High Alumina) SGHA 500) dispersed in a water tank (figure 2). A variety of other subjects, for example air bubbles in water have also been used as test objects.

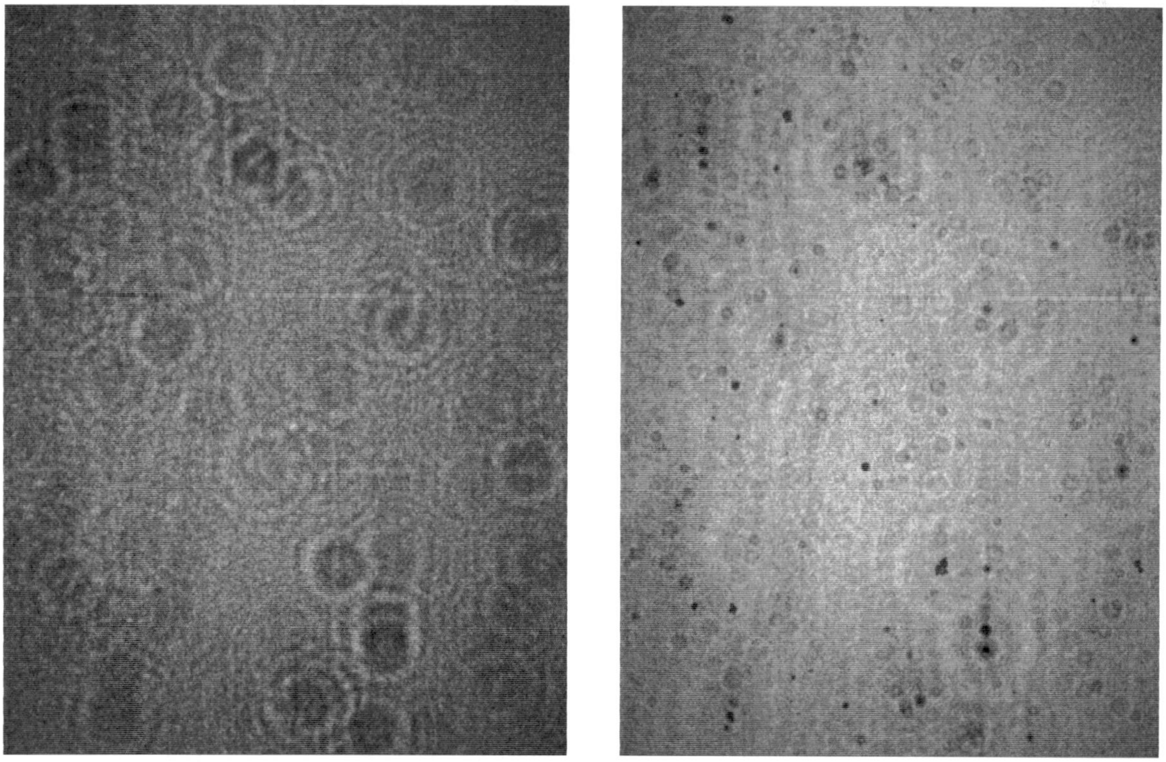

Figure 2. (Left): An extract (10 mm wide) from a digital hologram of cenospheres in water. (Right): The region of a numerically-reconstructed arbitrary plane through the sample volume. The smallest particles imaged here are approximately 30 micrometres in size [8].

2.2. Hologram replay

Our general approach is based on the convolution method [4] and can be summarised as follows:

- Numerically multiply the (zero-padded) hologram by a complex amplitude reference wave (in the case of collimated beam in-line holography the reference wave is 1+0i).
- Perform a 2D discrete Fast Fourier transform (FFT); real-valued data is transformed to complex-valued data.
- Multiply the result by a transfer function in the frequency domain based on the Rayleigh-Sommerfeld equation. The transfer function depends on the wavelength and the distance of the desired depth plane from the recording plane.
- Perform an inverse 2D FFT to generate the reconstructed image for a single depth plane. Extract the magnitudes, remove the zero-padding and normalise intensity.

This process is computationally heavy as it requires multiple 2D complex FFT to be computed. Regarding only the reconstruction of a digital hologram, and ignoring the challenging and similarly computationally intensive task of locating "regions of interest" (ROI) within the replayed volume, it can be seen that every depth slice may be calculated in parallel. Since the initial Fourier transformed hologram is common to any depth slice calculation there are clearly potential advantages in doing this step once and making the transformed data available to all processors. With the recent availability of very powerful GPU processors and the porting of the FFTW algorithm to the NVIDIA CUDA library, most recent work in the digital holography field has made use of such GPU based systems. In the next section we describe in detail our work on porting our existing replay code *HoloReco* [13] to the NVIDA CUDA environment, the refactoring needed to generate significant replay speeds on commodity GPUs and some analysis of the remaining bottlenecks and the prospects for further improvement.

3. Code refactoring for the NVIDIA CUDA environment.

The original *HoloReco* code was written to carry out all of the four steps outlines in subsection 2.2 for every depth slice. Carrying out the first FFT and then using it subsequently for a series of different depth slices eliminates all but one of the time intensive serial reads from host memory to GPU memory. The most significant improvement to the processing speed was finding a way to parallelise the transfer function (unique to each depth plane). After many code rearrangements, a way to parallelise the inner loops of the transfer function was realised. This allowed the results of the forward FFT to be kept in the GPU memory for the transfer function step and reverse FFT, thus no data movement is needed until the output stage. Some further optimization of the input file format (PGM) and the power spectrum and intensity normalization steps resulted in a very significant improvement in computational speed compared to a conventional processor.

3.1. Tesla based system with no video display

Using a system containing three NVIDIA Tesla S1070 (240 thread processors per GPU core, shaders clocked at 1.30 GHz) we were able to process our hologram at 13.6 frames/second. This should be compared to our original code running on an eight-core Xeon E5420 @ 2.50GHz based system where a speed of only 0.057 frames/second was achieved. Thus code refactoring and using three GPUs provided a factor of nearly 240 improvement in reconstruction speed. These results used V2.2 of the FFTW libraries for CUDA. It might be worth noting that at the time this work was done (2009) the relative speed per pound Sterling spent on the hardware was in favour of the GPU based system by only about a factor of six.

3.2. Tesla and Fermi based systems with video display

There are two clearly different user cases for hologram data visualisation:

- An expert user (e.g. a Marine Biologist) searches through a video stream of replayed hologram depth slices looking for “interesting” objects. This is a very good use of the expert *until they become bored!*
- An image-processing system searches through replayed hologram image slices looking for “objects”. The set of detected objects is then sent to classifier.

A better use of the human expert is for them to locate quickly holograms that contain some objects of interest, for example a zooplankton species, and then to create a selected list of holograms to be systematically replayed in full and processed by an image-processing system and object classifier. Alternatively the expert user could look, in context, at those regions of the replayed volume in which an automated system had previously located and classified objects. Some Tesla and many Fermi-class GPUs have a video output and an OpenGL compatibility library is available. For human visualisation these are ideal since one can efficiently copy (using **cudaMemcpy()**) the replayed hologram to the video buffer and the overall performance is not compromised. The sequence of events for each replayed frame is shown in figure 3.

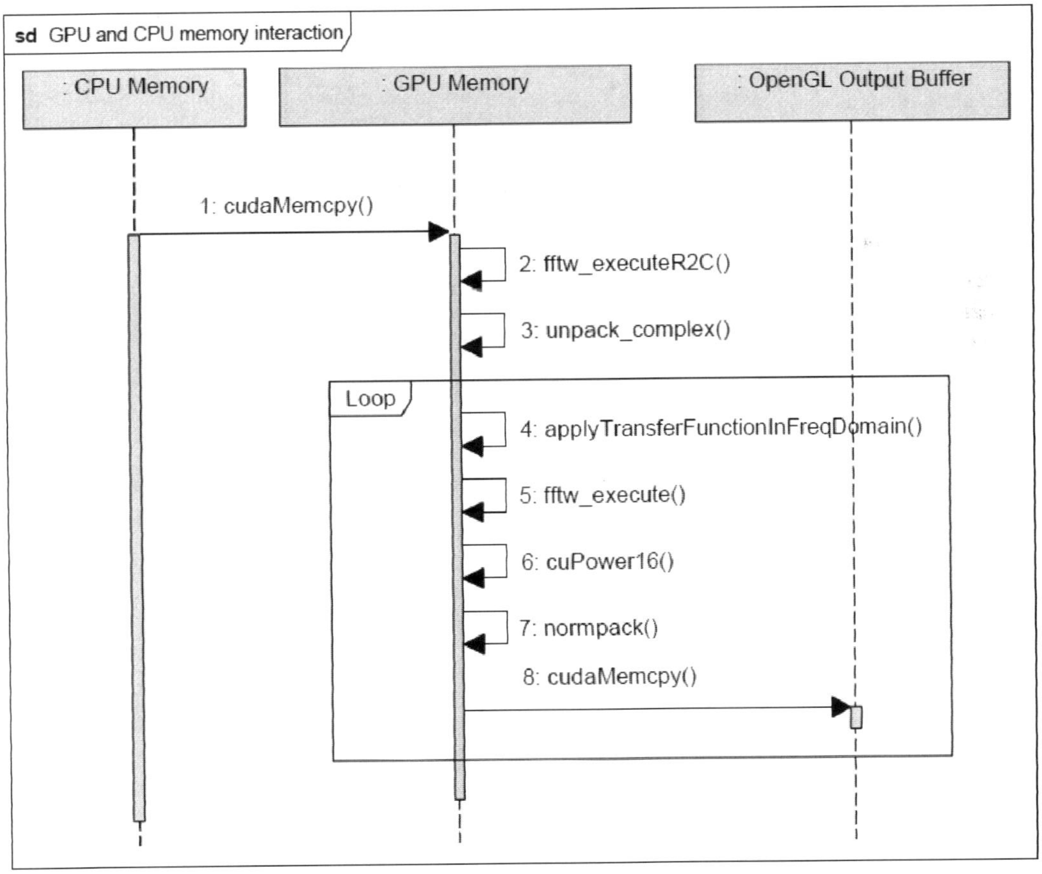

Figure 3. A UML representation of the sequence of key events in our software that occur during the processing of a digital hologram. If a number of frames from the same hologram are replayed to video then the sequence numbers 4 through 8 are repeated as often as required.

Table 1 lists the video rates we achieved for two different sizes of hologram using a Tesla C2070 (448 thread processors in one GPU core, shaders clocked at 1.15 GHz); CUDA 4.0; Driver 285.03; in a Xeon 5620 @ 2.40 GHz, Ubuntu 11.04.

Table 1. Frame rates for two different hologram sizes and an NVIDIA C2070 GPU.

Hologram size	Padded size	Video replay rate (frames/s)
2300×3500	4096×4096	15.8
2048×1389	2048×2048	41.2

3.3. Profiling the code on a Tesla GPU

The CUDA SDK also includes a profiler for mapping GPU usage and this is a powerful tool for locating and understanding which operations dominate the GPU processor time. Unfortunately the GUI version turned out to be temperamental on our systems; because of this we ran the CLI version to get a timestamp snapshot of the programme running for 15 seconds on a 4-Core Xeon computer running Ubuntu 11.04 and equipped with a Tesla C2070 GPU. This gave the start-up and 50 iterations of the display loop (at ~4.4 frames/s; without the profiler it ran at ~16 frames/s). The hologram was 3500x2300 and zero-padded to 4096x4096 for processing. Table 2 shows information on GPU usage for this run.

Table 2. (a) GPU and CPU procedure call timings and GPU occupancy for the initialisation step common to all slices. (b) GPU and CPU timings and GPU occupancy, averaged over 50 iterations, of the procedure calls that are executed for each depth slice calculated and then displayed. Note that there are two occupancy figures for the FFT because there are two separate kernels executed for each FFT; one executes once, the other twice.

(a) GPU kernel	Number of calls	GPU time (μs)	CPU time (μs)	GPU occupancy	Comment
memset32_post	3	6.7	27	0.042	
memcpyHtoD	1	12203	12313		Copy data from Host to Device
Forward FFT operations	3	5011	36	0.667/0.417	
Post-process	1	2379	5	1.000	
memcpyHtoD	1	2.0	7		Parameters for unpacking
copy_kernel	4095	11123	18524	1.000	cudaBlas unpacking FFT
insert_conj	1	15609	4	0.167	Reconstruct 2nd half of FFT
setUpX22_kernel	1	421.1	9	0.021	Calculate a lookup table for transfer

(b) GPU kernel	Number of calls/loop	Average GPU time/loop (μs)	Average CPU time/loop (μs)	Average GPU occupancy	Comment
doInnerLoop	4096	20780	19891	0.167	Parallelized Transfer Function

Reverse FFT operations	3	9898.9	26.0	0.667/0.333	
cuPower16_kernel	1	22333	4.6	0.500	Extract amplitudes from complex & remove zero-padding
memcpyDtoH	1	2.6	40.1		Return subroutine results
memcpyDtoH	1	2.1	33		Return subroutine results
memcpyHtoD	1	1.6	5.4		Parameters for minmax to GPU
thrust::minmax_element	2	516.0	11.7	0.500	Find min & max
memcpyDtoH	1	2.0	539.0		Return subroutine results
setminmax	1	4.3	6.5	0.021	Place scaling factors in device
packfloat	1	5128.4	4.4	0.167	Create scaled, packed pixels
memcopyDtoD	1	272.4	7.8		Copy to OpenGL buffer

The timing data show that, once the initial file is read, the majority of the GPU time is spent manipulating data – performing the transfer function to prepare for the reverse FFT and converting back from the complex transform to real magnitudes. While we have profiled our code, we have yet to use these profiling data in our code optimisation; for example the GPU warp occupancy figures in table 2 imply that further refactoring to increase the occupancy could improve performance.

4. Discussion

With our refactored code achieving a replayed frame rate approaching standard video for holograms as large as 4096×4096 pixels and significantly exceeding it for 2048×2048 pixel holograms, there is no doubt that for visualisation a GPU is far superior to a conventional CPU or a computational grid. Recent advances (for example the GTX460 and GTX560 based systems) in inexpensive "desk-top Fermi" processor GPU systems, developed primarily for the high-end computer gaming community, now enable a very impressive price-to-performance ratio to be obtained for a modest monetary outlay. It is clear from our work that considerable refactoring of existing code is needed to obtain the significant performance increases that a GPU can provide and further optimisation, with an attendant increase in replay speed, is likely if we can improve the occupancy figures. However the picture is less clear when considering the overall task of replaying a digital hologram and then searching in 3D for in-focus images of "interesting" objects, extracting these and then classifying them.

Much existing code, including extensive image processing and classification libraries, would need to be ported to a GPU to generate a significant speed increase. The challenge lies not in the porting but in the code refactoring needed to make efficient use of the GPU. In our earlier work [9] the naive approach of recompiling our unmodified *HoloReco* code with the CUDA FFTW libraries resulted in only a factor of six improvement in processing speed and only extensive code refactoring coupled with detailed profiling has enabled the current factor of 240 to be achieved. In addition the task of locating objects in a replayed 3D volume is not a trivially parallel one. Several correlated depth slices need to be considered in order to locate, using some appropriate focusing metric [14], the best one so that each object may be extracted for subsequent classification. We hypothesise that a hybrid approach, with replay using a GPU and subsequent processing of the datasets using grid or commodity cloud computing may be the most appropriate combination of the two different architectures.

References

[1] Yang Yan, Kang Bo-seon 2011 *Opt.Lasers Eng* **49** 1254–63
[2] Sun H, Benzie P W, Burns N, Hendry D C, Player M A and Watson J 2008 *Phil. Trans. R. Soc. A* **366** 1789–806
[3] Graham G W, Smith W and Nimmo A M 2010 *Limnol.Oceanogr.Meth.* **82** 1–15
[4] Kreis T M, Adams M and Juptner W P O 1997 *Proc. of SPIE* **3098** 224–33
[5] Frigo M and Johnson S G 2005 *Proc. IEEE* **93** 216–31
[6] Nebrensky J J, Hobson P R and Fryer P C 2005 *Proc. of SPIE* **5775** 285–96
[7] Nebrensky J J and Hobson P R 2006 *Proc. of SPIE.* **6252** 62521I.1–6
[8] Nebrensky J J and Hobson P R 2009 Replay of Digitally-Recorded Holograms Using a Computational Grid [available] http://bura.brunel.ac.uk/handle/2438/3443
[9] Nebrensky J J, Hobson P R and Reid I D 2009 Digitally-Recorded Hologram Replay - a Comparison Between a Computational Grid and a GPU Based System, EOS Topical Meeting on Blue Photonics, Aberdeen, UK 18-19 August 2009
[10] Pandey N, Kelly D P, Naughton T J and Hennelly B M, 2009 *Proc. of SPIE* **7442** 744205–1
[11] Zulfiqar A, Hyun-Eui Kim, Dongbiao Han, Jae-Hyeung Park and Nam Kim 2011 *3D Res.* **2** 1–5
[12] Zhuqing Zhu, Min Sun, Heping Ding, Shaotong Feng and Shouping Nie 2009 Fast numerical reconstruction of digital holography based on graphic processing unit [Available] http://dx.doi.org/10.1109/CLEOPR.2009.5292249
[13] [Available] http://holoreco.sourceforge.net/
[14] Meinecke T, Sabitov N and Sinzinger S 2010 *Appl.Opt.* **49** 2446–55

Data preservation in High Energy Physics

R Kogler[1], D M South[2] and M Steder[3]
on behalf of the ICFA DPHEP Study Group

Deutsches Elektronen Synchrotron, Notkestraße 85, 22607 Hamburg, Germany

E-mail: roman.kogler@desy.de[1], david.south@desy.de[2], michael.steder@desy.de[3]

Abstract. Data from high-energy physics experiments are collected with significant financial and human effort and are mostly unique. However, until recently no coherent strategy existed for data preservation and re-use, and many important and complex data sets have simply been lost. While the current focus is on the LHC at CERN, in the current period several important and unique experimental programs at other facilities are coming to an end, including those at HERA, b-factories and the Tevatron. To address this issue, an inter-experimental study group on HEP data preservation and long-term analysis (DPHEP) was convened at the end of 2008. The group now aims to publish a full and detailed review of the present status of data preservation in high energy physics. This contribution summarises the results of the DPHEP study group, describing the challenges of data preservation in high energy physics and the group's first conclusions and recommendations. The physics motivation for data preservation, generic computing and preservation models, technological expectations and governance aspects at local and international levels are examined.

1. Introduction

Particle physics experiments are designed to probe the structure of matter, the character of fundamental interactions and to ultimately extend our understanding of nature. Since the advent of collider experiments the available range in energy and intensity has been enlarged by many orders of magnitude [1]. However, the development, building and commissioning of colliders and the experimental detectors takes considerable human, technological and financial effort. Most colliders and their associated scientific programmes have been unique in energy range, process dynamics or experimental techniques. The data preservation effort aims to ensure long-term availability of these data after the end of the experimental collaborations.

Nowadays the focus of the particle physics community is on the Large Hadron Collider (LHC) at CERN, which operates mainly as a *pp* collider, currently at a centre–of–mass energy of 7 TeV. At the same time, a generation of other high energy physics (HEP) experiments have concluded their data taking and the experimental collaborations are finishing their physics programmes. These include the HERA, LEP, KEK and PEP experiments [2], as well as those at the Tevatron, where data taking has recently ended [3]. Data collected from these experiments continue to be crucial to our understanding of particle physics, ranging from precision measurements [4, 5] to searches for new signatures beyond the Standard Model [6]. Besides the ongoing analyses that remain to be completed, these data may also provide important future scientific opportunities [7]. Moreover, the data from these experiments are often unique in terms of the initial state particles and are unlikely to be superseded anytime soon, even considering such future projects as the LHeC [8]. The issue of HEP data becoming inaccessible due to technological advances in operating systems, storage hardware or data loss becomes imminent already at the LHC, where the lower–energy and lower–luminosity data collected at 0.9, 2.36 and 7 TeV in 2010 and 2011 will not be repeated. Such datasets, which are unique in terms of energy regime and experimental conditions should be secured in order to ensure re-use and future accessibility [9]. It is therefore prudent to envisage some form of conservation of the respective data sets. However, HEP has little or no tradition or clear current model of long term preservation of data in a meaningful and useful way. The preservation of and supported long term access to the data is generally not part of the planning, software design or budget of a HEP experiment [10].

In the following, the physics case for data preservation is examined, followed by an overview of the study group on Data Preservation and Long Term Analysis in High Energy Physics (DPHEP). Different aspects of HEP data are reviewed and a summary is given of the different models for data preservation identified by the DPHEP study group. Current inter–experimental data preservation initiatives are then presented, followed by some words on governance and structures, before finally concluding with an outlook and summary of future working directions.

2. The physics case for data preservation

The motivation behind data preservation in HEP should have its roots in physics. A detailed report on physics cases for data preservation is given elsewhere [11], only a brief review is presented in these proceedings. One of the main assumptions concerning experimental HEP data is that older data will always be superseded by that from the next generation experiment. However, this is not always the case and several scenarios exist where the preservation of experimental HEP data would be of benefit to the particle physics community:

- An extension of the existing physics programme may be necessary to ensure the long term completion of ongoing analysis. In particular, precision analyses continue long after the end of data taking, making use of the full statistical power and the best knowledge of the systematic uncertainties. This scenario has proven to be true for LEP [12] and a similar tail in the publication timeline is expected for H1 and BaBar [13].
- It may be favourable to re-do previous measurements to achieve an increased precision via new and improved theoretical calculations (MC models) or newly developed analysis techniques. A re-analysis of the JADE (1979–1986) data has lead to a significant improvement in the determination of the strong coupling $\alpha_s(M_Z)$, as shown in figure 1 (a), in an energy range that is still unique [4, 5]. Figure 1 (b) shows a variety of $\alpha_s(M_Z)$ measurements, as well as the current world average, where it can be seen that for the latest H1 measurements the theoretical uncertainty dominates the error [14]. In a situation that mirrors this JADE analysis, it is hoped that the uncertainty on $\alpha_s(M_Z)$ will be further reduced at some point in the future by re–analysing the very accurate HERA data once improved theoretical predictions become available.
- Combining similar measurements from different experiments increases the statistical significance and reduces systematic uncertainties via cross–calibration techniques to arrive at a more precise result. In figure 1 (c) individual and combined H1 and ZEUS measurements of the reduced neutral current cross section are shown. The combined cross sections serve as the sole input for a QCD analysis at NLO which determines a new set of parton distributions, HERAPDF1.0, with reduced experimental uncertainties [15].
- If new phenomena are found by analysing data recorded at the LHC or some other future collider, it may be useful or even mandatory to go back, if possible, and verify such results using older data.

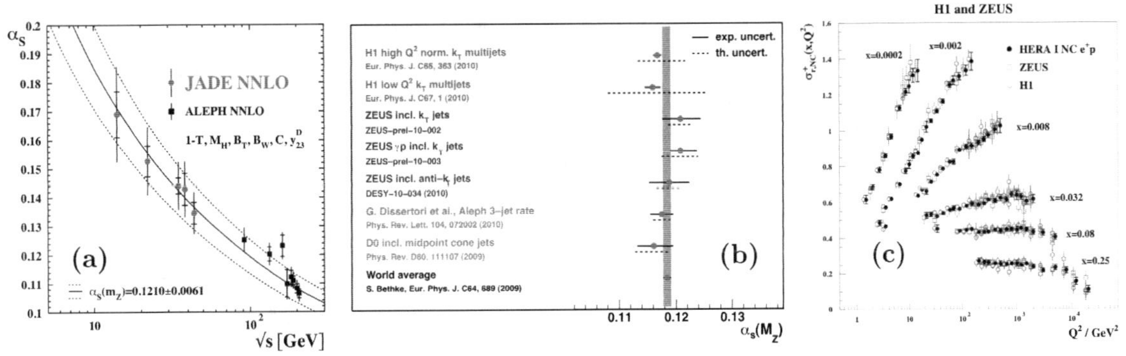

Figure 1. (a) Measurements of the strong coupling, α_s from an event shape analysis of JADE data at various centre–of–mass energies, $\sqrt{s}$. The full and dashed lines indicate the result from the JADE NNLO analysis [5]; (b) Recent determinations of the strong coupling $\alpha_s(M_Z)$ from a variety of experiments compared to the 2009 world average [14]; (c) HERA combined NC e^+p reduced cross sections as a function of Q^2 for different x bins [15].

3. DPHEP

To address the issue of data preservation in a systematic way, the DPHEP study group was formed at the end of 2008 [16]. The aims of the study group include to confront the data models, clarify the concepts, set a common language and investigate the technical aspects of data preservation in HEP. The experiments BaBar, Belle, BES-III, CLAS, CLEO, CDF, DØ, H1 and ZEUS are all represented in DPHEP, with representatives from the LHC experiments ALICE, ATLAS, CMS and LHCb having joined the study group recently. The associated computing centres at CERN (Switzerland/France), DESY (Germany), Fermilab (USA), IHEP (China), JLAB (USA), KEK (Japan) and SLAC (USA) are all represented in DPHEP. A series of workshops [16] have taken place over the last three years, beginning at DESY in January 2009 and most recently at Fermilab in May 2011. The study group is officially endorsed with a mandate by the International Committee for Future Accelerators (ICFA) [17] and the first DPHEP recommendations were published in 2009, summarising the initial findings and setting out future working directions [16]. The role of the DPHEP study group is to provide international coordination of data preservation efforts in high energy physics and to provide a set of recommendations for past, present and future HEP experiments. Participants of the fifth DPHEP workshop at Fermilab are shown in figure 2.

4. What is HEP data?

In order to identify different models of data preservation, it is necessary to survey the different aspects of HEP data which contain important information for physics analyses.

The collision data recorded by the experiments are usually structured in different layers, each successive layer containing decreasing complexity as the detector based quantities such as measured charges or track hits are consecutively replaced by reconstructed and calibrated tracks, particle candidates and jets. Hence the top–layer data have the full potential for physics analyses, but improvements for example of the reconstruction algorithms or calibrations can only be incorporated if the first layer, typically referred to as the raw data, is available, thus retaining full flexibility. However, since these data usually have a total volume of a few PB[4], today's computing centres are able to store them without significant additional costs.

[4] The collisions recorded by the LHC experiments result in 10's of TB of data per day, or up to 15 PB per year.

Figure 2. Participants of the fifth DPHEP workshop at Fermilab, May 2011.

In addition to the raw and reconstructed data the various software, such as simulation, reconstruction and analysis software need to be considered. If the experimental software is not available the possibility to study new observables or to incorporate new reconstruction algorithms, detector simulations or event generators is lost. Without a well defined and understood software environment the scientific potential of the data may be limited.

In every experimental collaboration different levels of documentation are available. Publications of data analyses or detector studies may be in journals, on SPIRES or on arXiv, with the physics results usually stored in a database like HEPDATA [18]. Detailed information about the analyses is usually only available in Ph.D. or masters' theses or internal notes, where usually not all of them are electronically available. Many types of internal meta–data may also exist, such as the detailed detector layout and performance, hardware replacements, manuals or the documentation of meetings.

The implementation of a data preservation model as early as possible in the lifetime of an experiment may greatly increase the chance that the data will be available in the long term, and may also simplify the data analysis in the final years of the collaboration.

The unique expertise of collaboration members is also at risk, as the person power associated to the experiment decreases. By planning a transition of the collaboration structure to something more suited to an archival mode, this particular loss may be minimised (see section 7).

5. Models of data preservation

The resurrection of the JADE analysis chain to perform the analyses described above, proved to be an eventful exercise and often a subject of luck rather than careful planning [7]. The general status of the LEP data, which was recorded as recently as the year 2000, is a concern, despite the continued paper output. A recent review of the status of the data of the four experiments identified that efforts are needed to ensure long term access [19].

The different data preservation models established by DPHEP are summarised in table 1, organised in levels of increasing benefit, which comes with increasing complexity and cost. Each level is associated with use cases, and the preservation model adopted by an experiment should reflect the level of analysis expected to be available in the future. More details on each of the preservation levels is given in the first DPHEP publication [11].

Although a level 1 preservation model, to provide additional documentation, is considered the simplest, this still requires some, often substantial, activity by the experiment. The HERA collaborations, as well as BaBar, are all currently involved in dedicated efforts to safeguard and

Table 1. Data preservation models, listed in order of increasing complexity, cost and benefits.

Preservation Model	Use Case
1. Provide additional documentation	Publication-related information search
2. Preserve the data in a simplified format	Outreach, simple training analyses
3. Preserve the analysis level software and the data format	Full scientific analysis based on existing reconstruction
4. Preserve the reconstruction and simulation software and basic level data	Full potential of the experimental data

streamline the available documentation concerning their respective experiments (see section 6.2). A level 2 preservation, to conserve the experimental data in simplified format, is considered to be unsuitable for high level analyses, lacking the depth to allow, for example, detailed systematic studies to be performed. However, such a format is ideal for education and outreach purposes, which many experiments in the study group are also actively interested in (see section 6.3). Past experiences with old HEP data indicate that new analyses and complete re–analyses are only possible if all the necessary ingredients to retrieve, reconstruct and understand the data are accounted for. Only with the full flexibility does the full potential of the data remain, equivalent to the DPHEP level 4 data preservation. Accordingly, the majority of participating experiments in the study group plan for a level 4 preservation programme, although different approaches are employed concerning how this goal can be achieved.

6. Common data preservation projects

Since the formation of DPHEP, activities and models of the experiments have aligned to a certain degree and joint initiatives have been launched, related to all four data preservation levels as described in the following.

6.1. A generic validation suite

In the case of HEP, data preservation means also preserving the software and environment employed to analyse the data, equivalent to level 3 or 4 of the preservation model. Experience has shown that freezing the software in the latest state would sustain analysis capability for only a limited amount of time. Moreover, frozen simulation and reconstruction software would prevent the implementation of improved reconstruction or calibration algorithms or new MC simulations. It is therefore beneficial to keep the experimental software running and keeping it up–to–date with changes of the system architecture. For this purpose it is advantageous to have a framework available to automatically test and validate the software and data against changes and upgrades to the environment, as well as changes to the experimental software itself. As such a framework would examine many facets common to several current HEP experiments interested in a more complete data preservation model, the development of a generic validation

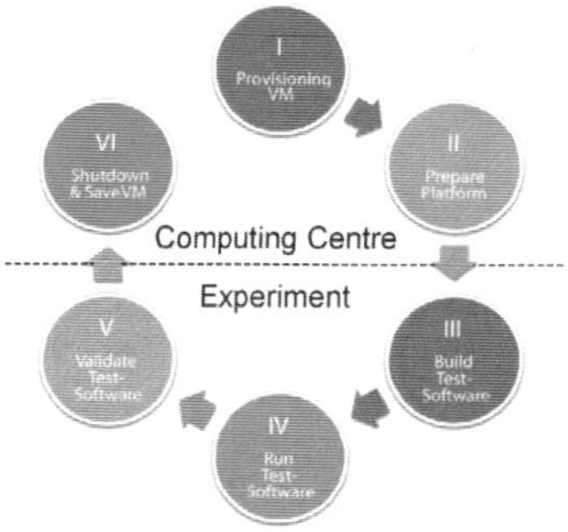

Figure 3. A sketch of the proposed software validation scheme under development at DESY.

suite is favourable. A full version of such a suite is realised at DESY-IT, in co-operation with the HERA experiments. The scheme, illustrated in figure 3, makes use of a virtual environment capable of hosting an arbitrary number of different virtual machine images and includes automated software build tools and data validation. Such a framework is by design expandable and able to host and validate the requirements from multiple experiments.

6.2. Global documentation initiatives
As well as the aforementioned individual documentation efforts, global information infrastructures in HEP may be beneficial to the data preservation project. INSPIRE [20], the successor to SPIRES, is an existing third-party information system for HEP, and is thus ideally situated to provide external management of experimental documentation. The INSPIRE project is preparing for the ingestion of much more high–level information in addition to the scientific papers themselves. These additions range from simple, documented information from the experiments about a given analysis, through wikis and news–forums, to even the data themselves, in a storage model where controlled access is possible. The additional information may, if desired, be only visible to collaboration members, which allows internal information such as notes to be stored, a project already in place between INSPIRE and the HERA collaborations. Another major advantage of such a scheme is that the responsibility of hosting the information passes from a defunct experiment to an active environment.

6.3. HEP data for outreach, education and open access
The development of a HEP data format for outreach and education is an attractive proposition. For most collaborations such a project would run in parallel to preserving the full re–analysis potential. In recent years there is a notably increased global effort to improve the overall level of education in particle physics and to provide access to HEP to more people than ever before. Tutorials using a simplified format of real HEP data with associated pedagogical exercises might help further the public understanding of science. First projects have started within the BaBar [21] and Belle [22] collaborations. The challenge of releasing such formats to the public domain is to keep the balance between a useful open access of HEP data on the one hand and control of the data, correctness and reputation of the experiment on the other hand.

7. Resources, governance and structures
The transition to a data preservation model should be planned in advance of the anticipated end date of an experiment. An early preparation is needed and sufficient resources should be provided in order to maintain the capability to re–investigate older data samples and collect all necessary documentation. However, the additional resources are estimated to be rather small in comparison to the person power allocated to the planning, construction and running of an experiment. Typically, a surge of 2–3 FTEs for 2–3 years, followed by steady 0.5–1.0 FTE per year per experiment is required for the implementation of a level 3 or 4 preservation model, which should be compared to 300–500 FTEs per experiment for many years. Therefore, the data preservation cost estimates represent typically much less than 1% of the original investment, for a potential 5–10% increase in physics output [11].

The transition to a new operational model of a collaboration should be considered by HEP experiments before the end of their lifetime. If the change to a long term analysis model is begun too late the experimental collaboration risks being left in an undefined state. In particular, the scientific supervision of the data preservation process and decisions regarding authorship, access to data and supervision of physics output after the collaboration's lifetime may benefit from a restructuring of the collaboration towards the final years. The presence and influence of DPHEP may facilitate this transition by providing common standards and global solutions within the HEP community.

8. Conclusions and outlook

The collection of high energy physics data represents a significant investment and physics cases can be made to demonstrate the potential for scientific results beyond the lifetime of a collaboration. However, until recently no coherent strategy existed regarding long term access of HEP data and an international study group, DPHEP, was formed to address this issue in a systematic way. Given the current experimental situation, data preservation efforts in HEP are timely, and large laboratories should define and install data preservation projects in order to avoid catastrophic loss of data once major collaborations come to an end. The preservation of the full analysis capability of experiments, including the reconstruction and simulation software, is recommended in order to achieve a flexible and meaningful preservation model. Such a project requires a strategy and well–identified resources, but provides additional research at relatively low cost, enhancing the return on the initial investment in the experimental facilities.

Data preservation efforts are best performed within the common organisation at the international level DPHEP, through which there is a unique opportunity to build a coherent structure for the future. Common requirements on data preservation are now evolving via DPHEP into inter–experimental R&D projects, optimising the development effort and potentially improving the degree of standardisation in HEP computing in the longer term. A second publication from the DPHEP group is expected shortly, describing the current projects in more detail and providing recommendations and guidelines for future HEP experiments.

References

[1] Panofsky W 1997 *Beam Line* **27** 1 ed Donaldson R, Kirk B and Riordan M p36
[2] South D M 2011 *Proc. of International Conference on Computing and High Energy Physics 2010, CHEP 2010 (Taipei) J. Phys.: Conf. Series* **331** (2011) 012005 (*Preprint* arXiv:1101.3186)
[3] Reich E S 2011 *Nature* **477** 379 http://www.nature.com/news/2011/110921/full/477379a.html
[4] Bethke S et al. (JADE) 2009 *Eur. Phys. J. C* **60** 181 Erratum 2009 *Eur. Phys. J. C* **62** 451 (*Preprint* arXiv:0810.2933)
[5] Bethke S et al. (JADE) 2009 *Eur. Phys. J. C* **64** 351 (*Preprint* arXiv:0810.1389)
[6] Schael S et al. (ALEPH) 2010 *J. High Energy Phys.* **1005** 049 (*Preprint* arXiv:1003.0705)
[7] Bethke S 2010 *Proc. of 15th International QCD Conference, QCD 10 (Montpellier) Preprint* arXiv:1009.3763
[8] Klein M and Newman P 2009 *CERN Courier* **49** 3 ed Sutton C p22 (*Preprint* arXiv:0910.0388)
[9] Malik S and Lassila-Perini K 2011 *Preprint* arXiv:1110.0355
[10] Reich E S 2011 *Nature* **474** 16 http://www.nature.com/news/2011/110527/full/474016a.html
[11] Asner D et al. (ICFA DPHEP Study Group) 2009 *Preprint* arXiv:0912.0255
[12] Brooks T 2009 *Pres. talk at 2nd Workshop on Data Preservation and Long Term Analysis (SLAC)* http://indico.cern.ch/conferenceDisplay.py?confId=55584
[13] Smith D 2010 *Pres. talk at International Conference on Computing in High Energy and Nuclear Physics, CHEP 2010 (Taipei)* http://event.twgrid.org/chep2010
[14] Kogler R 2010 *Proc. of 18th International Workshop On Deep Inelastic Scattering And Related Subjects, DIS 2010 (Florence) Preprint* arXiv:1006.4184
[15] Aaron F D et al. (H1 and ZEUS) 2010 *J. High Energy Phys.* **1001** 109 (*Preprint* arXiv:0911.0884)
[16] ICFA Study Group on Data Preservation in High Energy Physics (DPHEP) http://dphep.org
[17] The International Committee for Future Accelerators (ICFA) http://www.fnal.gov/directorate/icfa
[18] The Durham HepData Project http://durpdg.dur.ac.uk
[19] Holzner A G et al. 2009 *Preprint* arXiv:0912.1803
[20] INSPIRE http://inspirehep.net
[21] Bellis M HEP Outreach Efforts at SLAC http://stanford.edu/group/burchat/cgi-bin/bellis mediawiki/index.php/HEP_Outreach_efforts
[22] B–Lab http://belle.kek.jp/b-lab/b-lab-english

A validation system for data preservation in HEP

Yves Kemp[1], **Marco Strutz**[2], **Hermann Hessling**[2]
[1]DESY, Notkestrasse 85, D-22607 Hamburg
[2]HTW Berlin, Wilhelminenhofstrasse 75A, D-12459 Berlin

E-mail: yves.kemp@desy.de

Abstract. Preserving data from past experiments and preserving the ability to perform analysis with old data is of growing importance in many domains of science, including High Energy Physics (HEP). A study group on this issue, DPHEP, has been established in this field to provide guidelines and a structure for international collaboration on data preservation projects in HEP. This contribution presents a framework that allows experimentalists to validate their software against a previously defined set of tests in an automated way. The framework has been designed with a special focus for longevity, as it makes use of open protocols, has a modular design and is based on simple communication mechanisms. On the fabrics side, tests are carried out in a virtual environment using a cloud infrastructure. Within the framework, it is easy to run validation tests on different hardware platforms, or different major or minor versions of operating systems. Experts from IT or the experiments can automatically detect failures in the test procedure by the help of reporting tools. Hence, appropriate actions can be taken in a timely manner. The design and important implementation aspects of the framework are shown and first experiences from early–bird–users will be presented.

1. Introduction

Data preservation is a topic of increasing importance. This is true in general for any scientific data in the digital era, and it is especially true for High Energy Physics data. This has been recognized by ICFA, the International Committee for Future Accelerator, as they endorse since August 2009 the "Study Group for Data Preservation and Long Term Analysis in High Energy Physics", DPHEP[1]. As the DPHEP initiative is presented elsewhere in these proceedings, we simply summarize the main objectives of the initiative:

- Review and document the physics objectives of the data persistency in HEP.
- Exchange information concerning the analysis model: abstraction, software, documentation etc. and identify coherence points.
- Address the hardware and software persistency status.
- Review possible funding programs and other related international initiatives.
- Converge to a common set of specifications in a document that will constitute the basis for future collaborations.

In the context of the DPHEP initiative, many different aspects of data preservation and adjacent topics are discussed[2, 3]. In this contribution, we focus on the question how to preserve best the ability for performing analysis on preserved data. We start with some general considerations from the perspective of the HERA experiments at DESY and, then, present a framework for preserving software developed at DESY.

2. Conserving analysis capability

When faced with the question "How long should we be able to analyze old data?", a universal answer cannot be given. There are of course scientific rules imposed by Science bodies or journals, but these are usually rather lose, do not specify the detail level of reproducibility, do not ask for new analyses to be done with old data, and sometimes do not extend to more than five years. In discussions among different experiments, we can find two categories:

- Experiments for which a superseding experiment is on the horizon
- Experiments for which no superseding experiment is on the horizon.

It is clear that in the first case, the interest in long term analysis is reduced: Often people want to be able to analyze data of their old experiment until new and potentially better data is available from the new experiment. Eventually, comparisons will be made between old and new data, but after some period of overlap, the old data and its analysis is felt obsolete.
In the second case, long term analysis is crucial: The collected dataset is unique, it should be preserved as well as the knowledge how to analyze it – potentially for eternity. This is clearly the case for the HERA data: No e^-p or e^+p collider is currently running and there are no serious dicussions for planning such a collider. Probably the most tangible plan would be LHeC, for which a "Conceptual Design Report" is being written.
It is, therefore, clear that the strategy for HERA concerning analysis preservation must differ from other experiments.
We propose an analogy with the real world: "Pizza preservation".

2.1. Discussion strategies: "Pizza preservation"

If the task is to preserve a pizza, the timeline plays a role. If we want to preserve it for a short time (in this case, some days up to some months), a freezer or deep freezer is a solution known to work. For example, the BaBar collaboration has set up an elaborated system for archival, more details can be found in several presentations[4].
If the task is to preserve a pizza for a longer time, conserving the pizza as a material entity is not a good solution. The timescale is simply too long to ensure integrity and constance of the result. Instead, it is advisable to preserve the recipe of the pizza. To ensure integrity and constance, it is important to bake the same pizza repeatedly, check for deviations - and potentially fix them. Along this line, we would propose a similar scheme for preserving the HERA software: Experiments would not preserve ready–made executables or OS images, but rather recipes how to get to these. Together with this recipe, one or more test suites would be provided that would check the integrity of the result.
All this would be build against the newest OS flavors or other dependencies. The build and test process would be performed automatically at regular intervals. In case of problems occurring at any step, expert intervention would be triggered in a timely manner for a single problem.
We will shortly discuss the advantages (+) and disadvantages(−) of such an approach, compared to a scenario in which software is not compiled and OS images are frozen. These issues are discussed in more details elsewhere[5].

− At first glance, the continuous rebuilding looks like a higher effort. It clearly is in the beginning, if the experiment does not already have an automated build and testing infrastructure. If there is, the task would be to maintain this infrastructure after the end of the active analysis period. An experiment-independent framework maintained elsewhere (e.g. in an IT department) would be helpful

− Some kind of organization would still be necessary even after the end of the collaboration: Experts needs to be named that are able to address potentially occurring problems.

+ Runability and correctness of the code is guaranteed at any given moment.

+ Changes between each test run are small if the frequency is high enough. This eases problem finding.
+ Necessary changes are potentially small.
+ As the software "lives", adaptations to new technologies (network, protocols, data analysis, MC generators, ...) can be made with reasonable effort.

2.2. Towards a DESY wide solution: A generic recipe
We have tried to identify a basic, generic recipe for all steps of the different HERA experiments–the atomic life–cycle of a validation test. This comprises six steps that involve either IT or experiments at some stage. Based on this, we have designed and implemented a framework in which such steps can be executed [6].

3. The generic recipe
3.1. Concept and design
The prototype of our software preservation system is based on some general concepts. The modular design of the framework allows to replace components easily and to add new components to extend the functionality. New and unforeseen requirements by future environments may be included by minor changes of the structure of the code. Each configuration is formatted in plain text, human readable and self-explaining. This ensures longevity of the information both for operators and users. Open standards are preferred against proprietary ones. The communication between the components is mostly based on HTTP.

Data preservation covers a broad view of different fields. The focus is set rather on the software layer than on data archival or access. The standard *Open Archival Information System* (OAIS) [8] provides a general reference model for long–term preservation of digital data. It is identical to ISO 14721:2003. Our prototype is OAIS oriented but not OAIS conform as the standard has shortcomings with respect to software preservation and validation.

The validation system is composed of six different components as summarized in figure 3.1.

- The **controller** is the central component and manges all other components of our software preservation system. Incoming messages are processed and generate new events or tasks which, then, are transmitted to appropriate other components.
- The **ingest manager** executes all necessary steps needed to integrate a new test collection into the system. It extends a test specification with additional information, downloads external software packages and stores previous test collections.
- Most parts of the tests are executed inside virtual machines. The **workflow engine** provides mechanisms for mapping a life–cycle (see figure 3.1) inside a virtual machine.
- **IaaS cloud service.** The Infrastructure as a Service (IaaS) provider is responsible for managing and hosting virtual machines that can be instantiated dynamically and automatically for the software preservation system. Therefore, it is crucial to define a set of common interfaces being provided by the IaaS system. Providers of cloud computing differ from each other concerning the offered interfaces and protocols. Using only one of these interfaces may lead to a problematic dependency. In our prototype only open interfaces are used which are already standardized or at least well established such as the Open Cloud Computing Interface (OCCI[9]) driven by the Open Grid Forum[12]. OCCI is used as a protocol and as an API for any management tasks where the focus is set on portability, integration, innovation and interoperability.
- The **cloud infrastructure manager** comprises all feature related to IaaS interaction with the software preservation system. It is a management layer between controller and IaaS cloud service which decouples specific cloud protocols or function from the controller. This

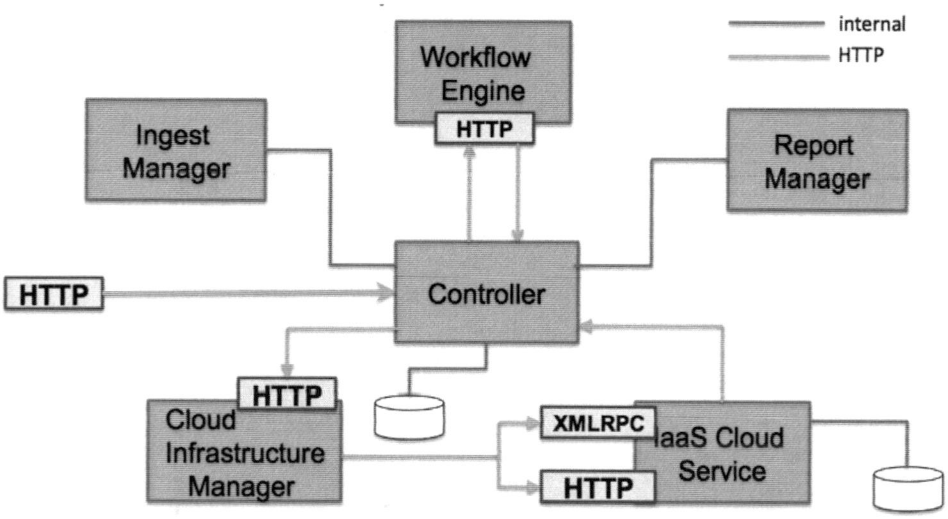

Figure 1. Components and their communication protocols.

enables an exchange of the cloud service without touching the controller but requires only a modification in the cloud infrastructure manager itself.

o The **report manager** provides statistics and reports about test runs and the software preservation system itself. The output can be rendered as a JSON message or presented as HTML.

During the life–cycle of the system different types of resources such as virtual machines, software packages, scripts, system state values, log files, test results, configuration files and test packages are created. Most of them are used by different components and during different time periods. Some need to be available only temporarily while others must be stored permanently. Our prototype uses Mercurial[11] as a source code management system and SQLite[10] as a database management system. The configuration data are stored in an ext3 file system.

A major pillar of the software preservation system is the integration of vitalization and cloud services to enable continuous testing of software packages. OpenNebula [7] is used for providing a cloud infrastructure as a service (IaaS). It can easily be adapted to special needs and is open–source.

Each operation accessing the cloud, for instance if a virtual machine image is requested, is handled at first by the controller. It receives messages and creates new actions. This may result in sending requests to the cloud–infrastructure manager. The cloud service is hidden from other components since only the controller is allowed to communicate with the cloud–infrastructure manager. OpenNebula can easily be replaced by a different cloud system, only the cloud–infrastructure manager needs to be modified.

3.2. Prototype implementation

Each test collection consists of a set of configuration files. They are collected in a tar–gzip archive and sent to the software preservation system via a HTTP–POST request.

The JavaScript Object Notation (JSON) is used as common notation for all configuration files. The basic idea is to use only key–value pairs, each element has the datatype string. By this structure it is not difficult to interpret the content in other programming languages. Additionally

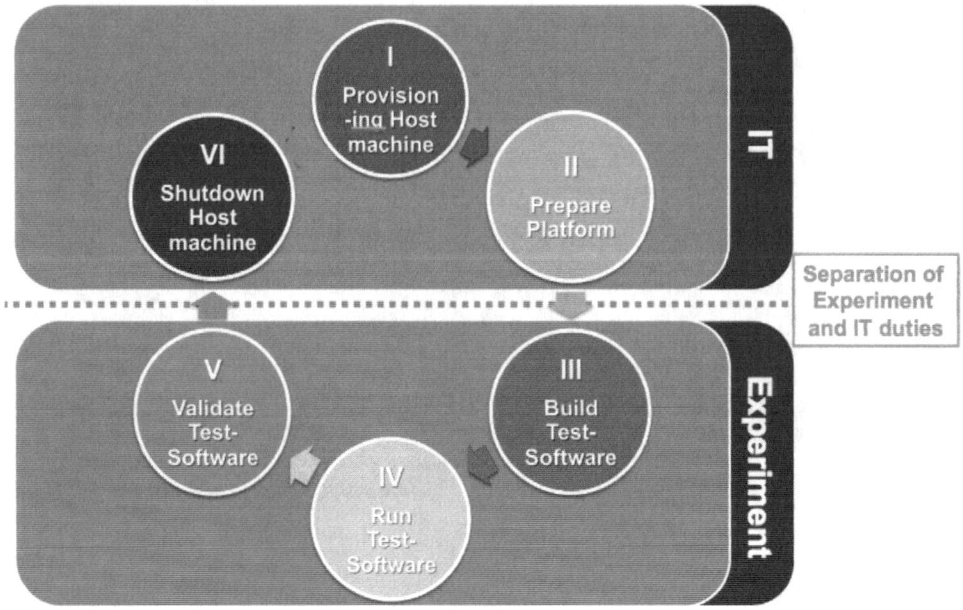

Figure 2. Life–cycle of a validation test. The responsibilities for the individual steps are separated between the IT group and the users.

the key–value notation is easily extendable. New attributes can be added without affecting with the old ones.

- software.txt
  This file describes the analysis software to be validated by the software preservation system. The key *archive* characterizes a single archive that contains the full analysis software. If the software needs to be compiled the key builder describes the file that can be used for building the software. The key *executable* represents a script executing the analysis software. Both builder and executable have to be part of the compressed software archive.
- validator.txt
  Validator is a program that can validate results of an analysis software. The result and log files of an analysis software are passed to the validator which then returns the test result, for instance SUCCESS or FAIL.
- rpm.txt
  This file contains a list of key–value pairs describing the packages the analysis software is depending on. For example, the key *zip* may be the name of a package and its metadata are stored as a value. The package name must be available for the package manager which is installed in the virtual machine template. For each virtual machine package managers are defined in their metadata.
- vm.txt
  Details about the virtual machine used to run the analysis software on, are stored here. The software preservation system provides a list of virtual machine templates. Each template can be contextualized and instantiated.
- contextualization.txt
  Each test has different requirements, therefore virtual machines must be prepared

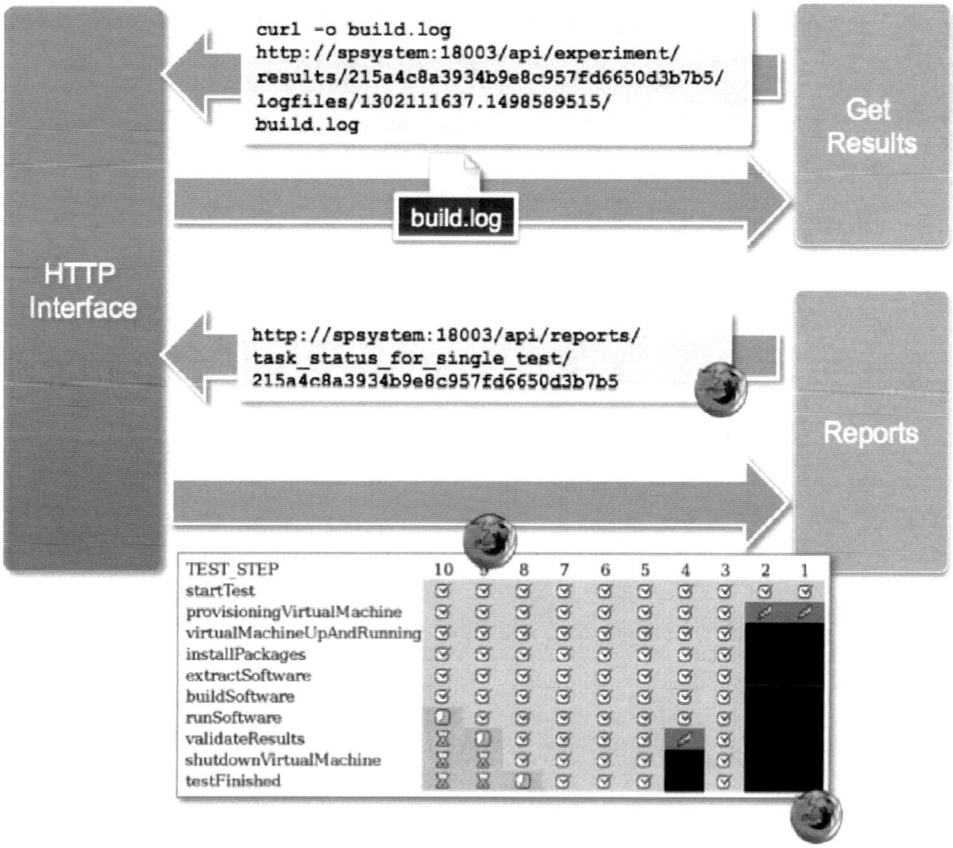

Figure 3. Monitoring.

individually to fit the needs. In the current system this is achieved by the contextualization mechanisms of OpenNebula. The first time a virtual machines boots a shell script is executed to prepare the system (e.g. init.sh).

o configuration.txt
All remaining information about the test collection are part of this configuration file.

As soon as a valid archive is created it can be registered to the software preservation system by sending a HTTP–POST request to a webservice, for example via the command-line tool curl. Then, the software preservation system provides a reference number, the TESTCOLLECTION UUID. This id is then used to trigger a test run for the referenced test collection which will generate a new TESTRUN UUID.

The execution of each test step produces a log file. A user can download these files by sending an HTTP–GET request to the software preservation system, specify an id for the test and its test run as illustrated by figure 3.2. Furthermore the software preservation system provides graphical reports that can be rendered by a web browser as indicated at the bottom of figure 3.2.

4. Experiences with the generic recipe and the prototype implementation

HERA users have generally well received the idea of keeping all or parts of their code alive. Some concern has been raised that migrating to newer operating systems might be difficult or impossible in the future. The framework could also be used to test binaries compiled on an

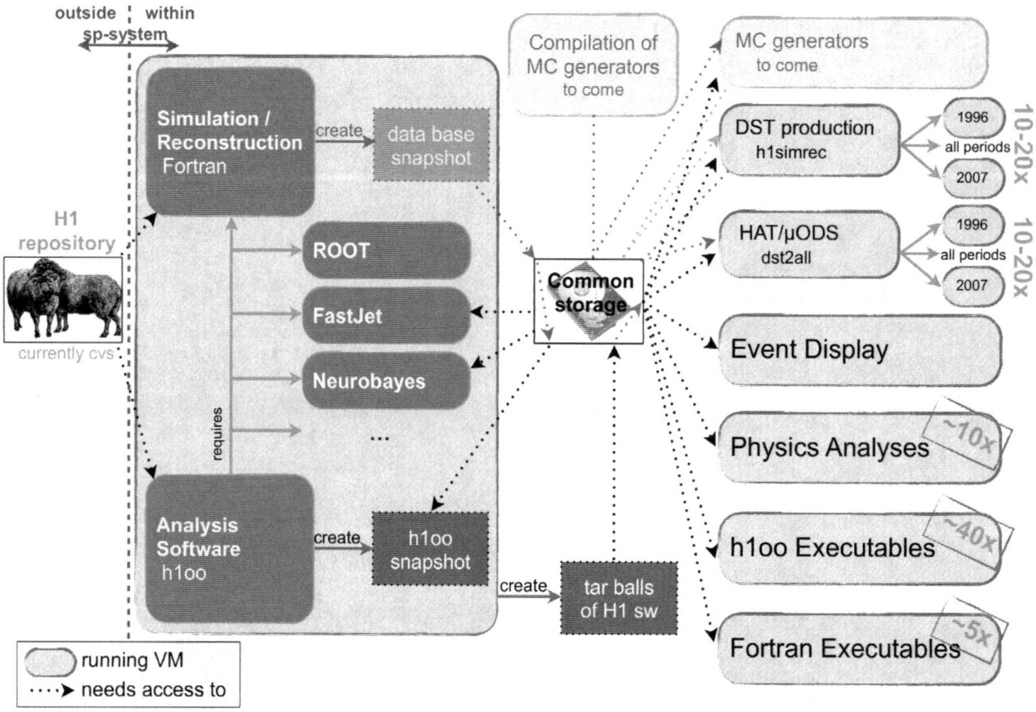

Figure 4. An example of complex workflows within the compilation, running and testing procedure and their interdependencies. (courtesy Michael Steder, H1).

old OS which are executed in an new OS. This is sometimes possible to some extend using compatibility libraries.
Users have started writing build, run and evaluation scripts in parallel to the prototype implementation. At a rather late stage of the implementation, these scripts were run on the system. We identify two main lessons:

(i) Users tend to recreate the environment they are used elsewhere. It seems to be more work to disentangle the dependencies against e.g. AFS and software being available there then to recreate such environments. We think that for the long term success of the system, users must be helped in understanding such dependencies and solving them in a clear way.

(ii) It has turned out that the workflows of the experiments are more complicated than initially thought and need information exchange between different test scenarios. It is for example envisageable to separate the compilation of the whole framework into smaller pieces (e.g. modules) which are also tested independently. Some modules might however depend on the compilation of previous modules, so the modules from previous compilations should be available in the framework. Figure 4 shows a sketch of the H1 plans for such a structure. This was not foreseen in the initial design. Any implementation that would go into real production should be able to handle such scenarios.

5. Conclusion and outlook

In this contribution, we discuss different scenarios of preserving the ability to analyze data.
We have successfully planned and implemented a proof of concept of one scenario, which allows physicists to constantly and automatically validate their code and help migrating to new OS or hardware.
Experience of HERA physicists working with this prototype has shown that it is useful to some extend for them. It has turned out that their workflows are more complex that initially thought. All of the steps – compilation, running, testing – should be divided into smaller sub–steps with inter–dependencies.
To achieve production readiness, more resources and work are needed to plan and implement a framework which is aware of such complex workflows and inter–dependencies.

References

[1] ICFA Study Group on Data Preservation and Long Term Analysis in High Energy Physics `http://www.dphep.org`
[2] DPHEP Study Group 2009 Data preservation in high energy physics *Preprint* arXiv:0912.0255 [hep-ex]
[3] South D 2010 Data preservation in high energy physics *Preprint* arXiv:1101.3186v1 [hep-ex]
[4] Neal H and Cartaro T 2011 Data preservation at BaBar/SLAC *5th DPHEP Workshop FNAL 2011* `http://indico.cern.ch/conferenceDisplay.py?confId=116485`
[5] Kemp Y 2009 Use of virtualization techniques for data preservation and long term analysis *DPHEP workshop DESY 2009*
`http://www.dphep.org/sites/site_dphep/content/e20/e36625/e43750/dlpta.pdf`
[6] Strutz M 2011 Virtualisation technologies and cloud computing in data preservation for high energy physics *Master's thesis Berlin University of Applied Sciences (HTW) Berlin* (available in university library).
[7] OpenNebula Open Source Toolkit for Cloud Computing `http://opennebula.org`
[8] Iso 14721:2003 Space data and information transfer systems - open archival information system - reference model OAIS 2003
[9] OCCI Open Cloud Computing Interface `http://occi-wg.org/`
[10] SQlite A SQL database engine `http://www.sqlite.org/`
[11] Mercurial *A Distributed Source Control Management Tool* `http://mercurial.selenic.com/`
[12] OGF - Open Grid Forum. `http://www.ogf.org/`

ACAT 2011 IOP Publishing
Journal of Physics: Conference Series **368** (2012) 012028 doi:10.1088/1742-6596/368/1/012028

Advanced event reweighting using multivariate analysis

D Martschei, M Feindt, S Honc, J Wagner-Kuhr

Institut für Experimentelle Kernphysik - Karlsruhe Institute of Technology KIT, DE

E-mail: martschei@ekp.uni-karlsruhe.de

Abstract. Multivariate analysis (MVA) methods, especially discrimination techniques such as neural networks, are key ingredients in modern data analysis and play an important role in high energy physics. They are usually trained on simulated Monte Carlo (MC) samples to discriminate so called "signal" from "background" events and are then applied to data to select real events of signal type. We here address procedures that improve this work flow. This will be the enhancement of data / MC agreement by reweighting MC samples on a per event basis. Then training MVAs on real data using the $_s$Plot technique will be discussed. Finally we will address the construction of MVAs whose discriminator is independent of a certain control variable, i.e. cuts on this variable will not change the discriminator shape.

1. Introduction

In order to have a well trained MVA which gives a calibrated discriminator, i.e. a discriminator which is linearly dependent on the signal probability of the given event, it is necessary to have a training sample which is as similar to data as possible. This requirement can not fully be met by generated events in most cases. Some of the possible reasons are:

- Cross sections and branching ratios of different processes have yet to be measured
- Hard process simulation is not good enough (e.g. it is only calculated to some finite precision by theorists)
- Detector simulation is not yet optimal (material budget, dead channels, ...)
- Physics "behaves" differently than initially expected in the new energy range.

Usually some of these effects, which appear as differences between data and MC, are tried to be eliminated by introducing scale factors. These scale factors are usually just constants or functions of some measured variable like p_T or η of the leading jet. This can already reduce differences between data and MC a lot, but the agreement is usually not yet perfect in all variable distributions and can even make certain distributions worse than they were without weights. As will be shown in section 2, it is possible to use MVA methods to derive individual per event weights, which will outweight differences in all input variables simultaneously.

In section 3 the $_s$Plot method will be introduced, which can be used to train on measured data events only, and not depend on any generated MC samples. The third section 4 will focus on the issue of biasing the distribution of a control variable through discriminator cuts. This happens if the discriminator is correlated to the variable in question, which might be the case if the control variable is able to discriminate signal and background itself. Also in this case using appropriate weights can solve the issue by outweighting the correlation but still keeping all the discrimination power which is orthogonal to the control variable.

2. Multivariate reweighting to correct for differences between generated Monte Carlo and real data events

In physics experiments there are usually differences between the simulated MC events and measured data. These arise from limited knowledge of the underlying physics processes, detector mismodeling and other issues.

In order to still be able to use the MC ensemble as a whole qualitatively, weights may be applied to make certain groups of events more pronounced and make observable distributions agree better between MC and data.

Instead of using scale factors in bins of certain measured values, we propose to use MVA methods to derive individual weights per event.

2.1. Task description

The reweighting procedure has to emphasize those events in the MC sample which have a high probability to be found in data, based on a certain phase space under inspection. This can be achieved by performing a training in which real data events are used as target/signal events and MC is fed to the MVA as background events. The MVA will then pick up which MC events can be separated from data. Optimally, if data/MC agreement was perfect, the MVA would not be able to learn anything in this training. Everything it learns is due to mismodeling in MC.

In the next step the MVA is applied to MC events and the discriminator output is transformed into an event weight.

2.2. Derivation of transformation for MVA output to event-weight

If the MVA was trained to discriminate data from MC, the MVA will yield the probability p for an event to be data when selecting from the whole sample: $p = \frac{data}{data+MC}$. What we need is the probability w for the MC event to be data when selecting from the MC sample only $w = \frac{data}{MC}$. In the following the derivation for NeuroBayes output NB_{out} is given, which is linearly dependent on p in the range [-1,1].

$$p = \frac{NB_{out}+1}{2} = \frac{data}{data+MC} \quad \Rightarrow \quad w = \frac{data}{MC} = \frac{1+NB_{out}}{1-NB_{out}} \tag{1}$$

2.3. Illustrative example

For illustration, the above steps have been applied on simulated toy samples. Two MC-samples with 9 observables were generated from arbitrary analytical distributions. Some correlations among the observables were introduced in order to have a more realistic example, e.g. var3 and var6 have a correlation coefficient of 0.2. Two of these observables have different distributions in the two samples, while the other 7 observables are distributed equally. These distributions are the black and blue ones in figure 1.

In this example the differences were exaggerated to be visible with the naked eye and to show the power of this method. As a first step towards individual event weights, a NeuroBayes training to discriminate between pseudo data and MC has been performed. The discriminator distribution can be seen in figure 2(a).

Since the two input variables shown in figure 1 have quite different shapes for the two samples, the NeuroBayes output is able to discriminate the two sample types quite well. The weights w for the MC events are derived using formula (1). The weight distribution is shown in figure 2(b)

As can be seen, most events will get weights smaller than 1, while some will get huge weights of up to 45. This is mainly due to the small overlap for the distributions of var3 and the

Journal of Physics: Conference Series **368** (2012) 012028 doi:10.1088/1742-6596/368/1/012028

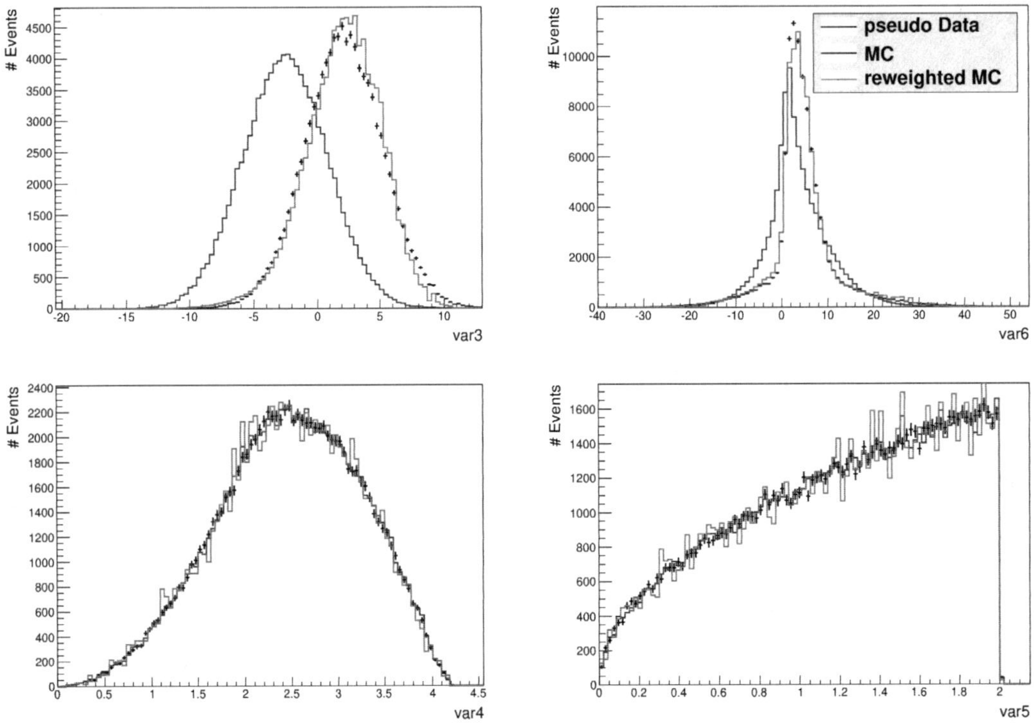

Figure 1. Distributions for 4 input variables. The MC (blue) was reweighted (red) with the weights from NeuroBayes to become comparable to the pseudo data.

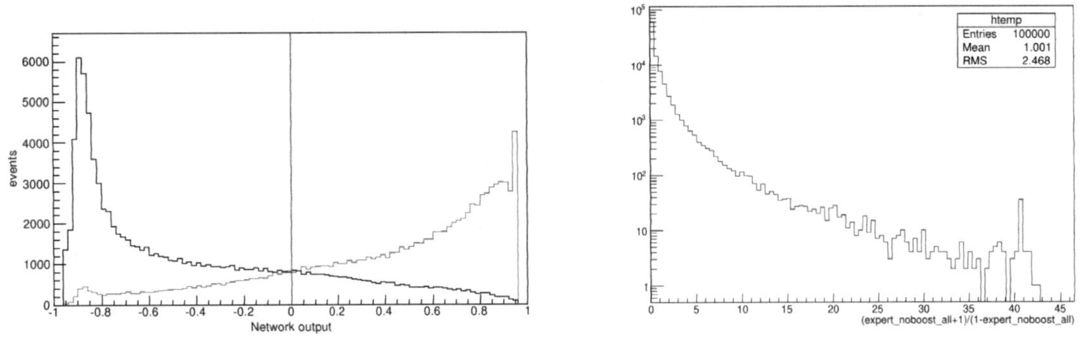

(a) (b)

Figure 2. Histograms for the MC to pseudo data reweighting. On the left the distribution of NeuroBayes discriminator for pseudo data (red) and MC events (black) is given. On the right the histogram shows the distribution of event weights for MC events derived from NeuroBayesExpert output.

big height difference in some regions. Applying the weights to the MC events leads to the distributions shown in figure 1. The method does not work well if there are no MC events at all in regions where real data events exist since then the weights diverge. Obviously the reweighted distribution agrees quite well in those variables which were differing without weights, while those variables which already matched are still agreeing. This would not necessarily be the case if simple histogram based weights were applied, since they don't take correlations into account.

3. Training on data without MC: side band subtractions and the $_s$Plot technique

3.1. Side band subtraction

Another way of getting around the issue with the disagreement between data and MC would be to directly train on data. However this is usually not possible because one cannot find absolutely pure samples of signal events. Still in some cases, one can at least define regions in phase space which are signal enriched and where a signal peak is already visible over some background. This peak can be used to estimate the yield of signal events and perform a side-band subtracted training. An overview of the different regions needed to perform such a side-band subtraction is shown in figure 3

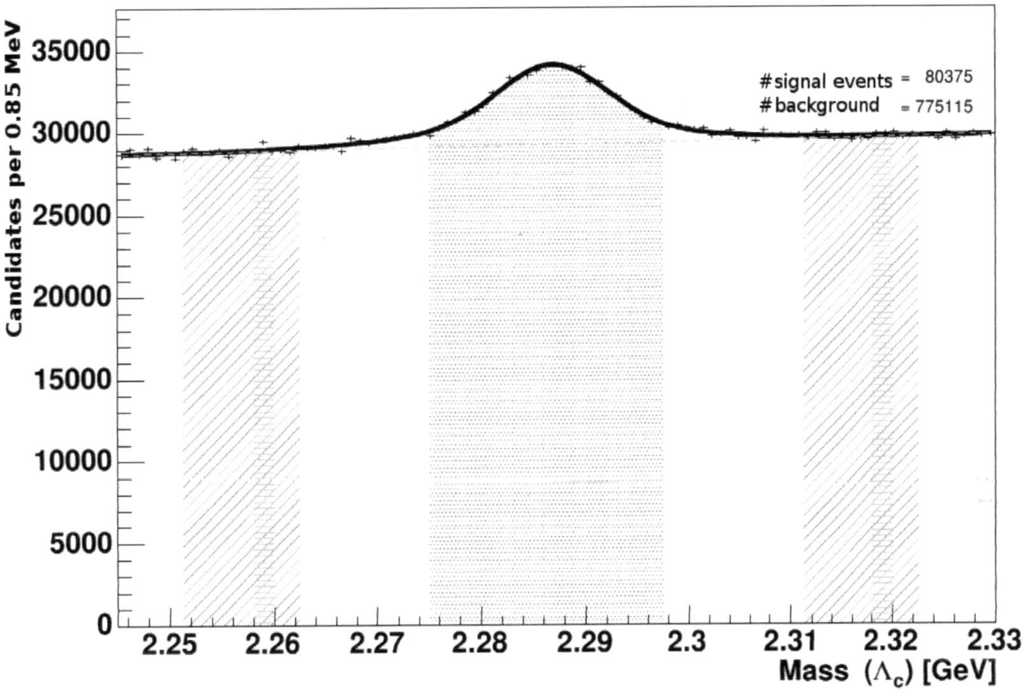

Figure 3. Mass distribution of Λ_c candidates. Red is the signal region, while blue and green are the side-bands [3] .

Three regions are defined and the events are used during the training in the following way:

region	amount of events	target type	weight
red	$N_{sig} + N_{back}$	signal	1
green	N_{sig}	background	1
blue	N_{back}	signal	-1

N_{sig} is the estimated yield of signal events in the red region (in the peak above the background), while N_{back} is the estimated amount of background events in this region. With the weight[1] and target definitions given above, the MVA will get a training sample which (after

[1] A short introduction on how weights are used during a MVA training can be found in Appendix A.

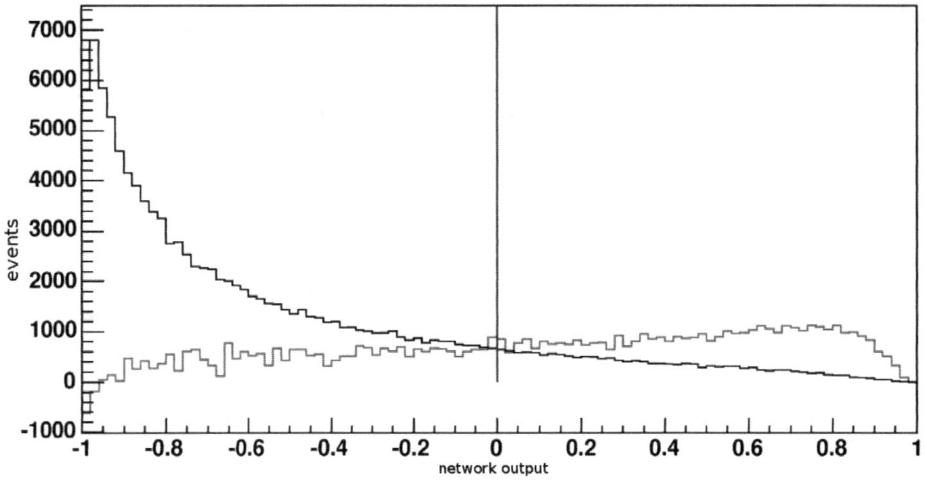

Figure 4. NeuroBayes discriminator distribution for the sideband training [3].

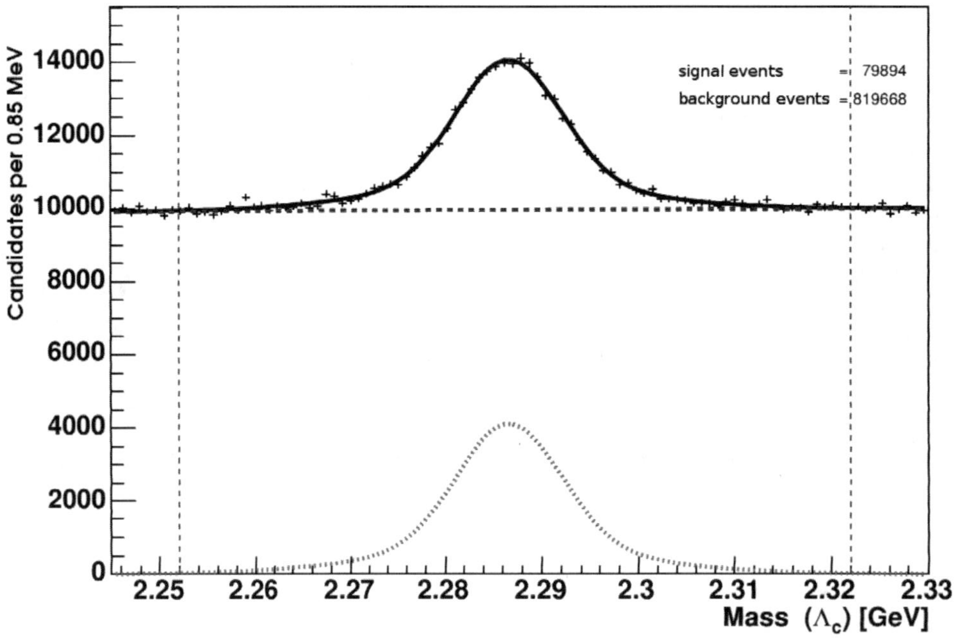

Figure 5. Mass distribution for Λ_c candidates after a discriminator cut of 0.5 [3].

subtraction) consists of an equal amount of signal and background events. One has to make sure that the MVA is not able to learn the dependency of the variable (mass in this case) one uses to estimate the signal and background yields, e.g. by excluding input variables that can distinguish between left and right side-band or the method in section 4. In this example mainly variables which parametrize the identification quality of the Λ_c decay particles kaons and protons and their kinematics are used, but not their full momentum or energy since these are too closely related to the Λ_c mass.

3.2. The $_s$Plot formalism: advanced side-band subtraction

An advanced version of the side-band subtraction is the $_s$Plot formalism. It exploits the complete probability densities of signal and background. Also a pure background region is not necessary, it is enough that the PDFs differ measurably.

A full derivation of the $_s$Plot weights can be found at [4]. They result is:

$$\begin{pmatrix} w_s(x) \\ w_{bg}(x) \end{pmatrix} = \frac{1}{f(x)} \cdot \mathbf{V} \cdot \begin{pmatrix} f_s(x) \\ f_{bg}(x) \end{pmatrix} \tag{2}$$

with:

$$f(x) = N_s \cdot f_s(x) + N_{bg} \cdot f_{bg}(x) \tag{3}$$

Here the following variable definitions apply:

- w_s the $_s$Plot signal weight, w_{bg} the background weight, with $w_s + w_{bg} = 1$. w_s and w_{bg} may be negative or larger than 1.
- $f_{s,bg}$ are the probability density functions of signal and background.
- $\mathbf{V}$ covariance matrix of f_s and f_{bg}.

What the $_s$Plot formalism will do, is the following. One has to deduce the probability density function for the two event types signal and background, e.g. using function fits to a mass peak. Using the given formula above, these pdfs can be transformed into event weights. By applying the signal/background $_s$Plot weights to data, we will get the pdf of any variable independent of the one (mass) we used to deduce the pdfs in the first place. In order to actually use these weights for a MVA training one has to use each event twice[2], once as signal with weight w_s and once as background with weight w_{bg}. During the training the MVA will then learn that there are certain events in the trainings sample which have a high signal weight and a lower background weight, and reside in a different phase space than the rest of the sample and will learn their properties to distinguish them.

3.3. Example for the application of the $_s$Plot formalism

An example is taken from the PhD thesis of Claudia Marino [2]. The $_s$Plot formalism was used there to train NeuroBayes on the $\Upsilon(1s)$ resonance in CDF data. The goal was to reduce the background under the Υ 1s,2s,3s resonances for further studies. The result can be seen in figure 6. The background could be reduced drastically while barely loosing signal events. Again this was done by training on real, measured data events and not using any kind of generated events. Another successful application of this method was the excited charm baryon analysis by the CDF collaboration [5].

4. Orthogonal discriminator: preventing distribution bias through event selection

In some analyses the final result (e.g. signal yield or certain properties) should be deduced by fitting a template to a certain control variable distribution. We will refer to this variable as varX in the following. This variable is chosen such that it will show a clear shape difference between the distribution of signal and background events. This is essential for the template fit to converge. Usually the amount of background has to be reduced drastically to have a decent signal/background ratio, which is done by cutting on several other variables and/or with a cut on a MVA discriminator. Especially the latter may introduce a bias in the control variable varX if the MVA learned the dependence of the signal probability on varX. This might even happen

[2] For NeuroBayes trainings this is not necessary any more, since it provides a special $_s$Plot training mode. It will automatically use each event as signal and background event if given the **$_s$Plot weight as target** and will account for the *real* given statistics and the correlation of the two entries of a given event.

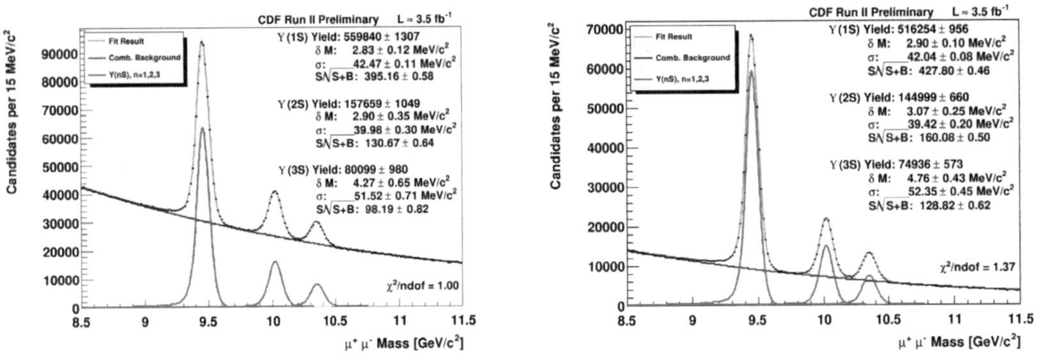

Figure 6. Υ resonances before and after cutting on a $_s$Plot trained NeuroBayes network. Extracted from Phd thesis Claudia Marino figure 5.8

if varX is not one of the input variables because the other variables may be correlated to varX. In any case what might happen is illustrated in figure 7.

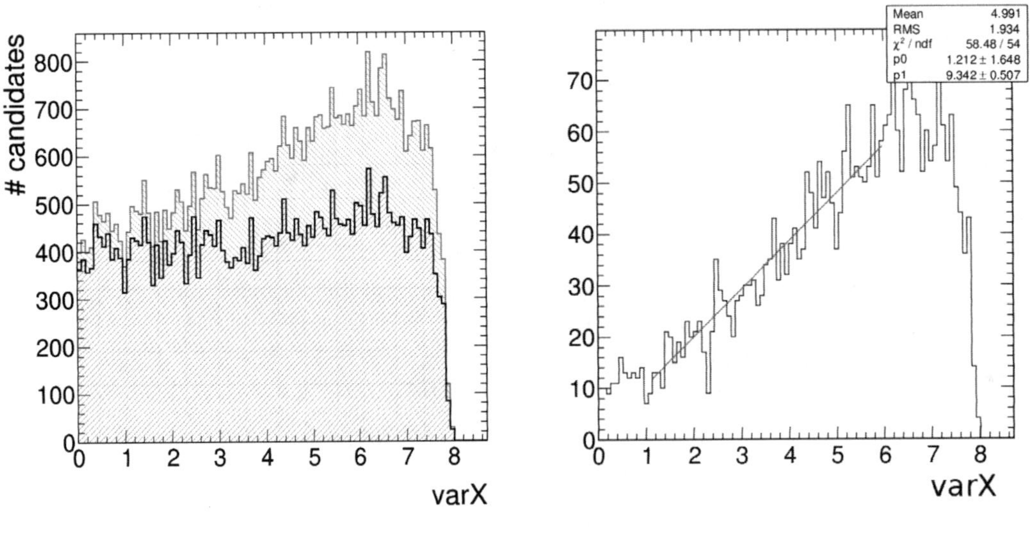

Figure 7. The left, stacked plot shows the original varX distribution for signal(red) and background(black). The background distribution is relatively flat and has only a small slope, while the signal events are more prominent at higher values of varX. Due to the rising of the signal distribution in varX, a cut on a discriminator which uses varX as one of its inputs, will result in a background distribution which is shown on the right.

Obviously the background distribution gets tilted by the discriminator cut, which is an issue in some analysis and we will therefore present a way to prevent this.

4.1. Task description

The task is to find a way to train the MVA such that its discriminator value is independent of varX and that cuts will equally remove background events from all bins in varX.

The easiest way would be to eliminate those variables which carry information about varX from the set of input variables. On the one hand this would remove the discriminator's dependence on varX but it will probably also remove some discrimination power from the MVA which is orthogonal to varX and should be saved.

Since the problem mainly arises from the different dependence of the signal and background on varX, another possible solution is to reweight the training sample such that this dependence vanishes. For this, the ratio between the desired distribution and the current height of the bin in each histogram is used as weight for events in this bin. Any distribution would be possible of which two have been studied here.

- Signal and background are both reweighted to be flat in varX. This is shown in figure 8(a).
- Both distributions are reweighted to the shape of the inclusive distribution, as shown in figure 8(b). This has the advantage over the flat shape that especially in the low varX region for the signal distribution the weights are not as large. Large weights should be avoided during MVA trainings as they will make the internal error calculation more difficult.

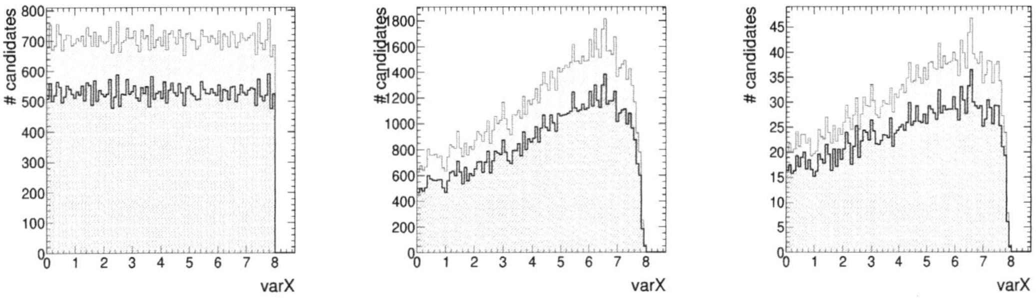

Figure 8. Reweighted shapes of varX with the three different weight types for signal(red) and background(black). left: signal and background reweighted to be flat in varX. middle: both distributions are reweighted such, that each of them has the same shape as the original inclusive distribution. right: Events are reweighted with individual weight-network values.

The weight calculation based on the varX histograms for signal and background has several shortcomings:

- The differences between the signal and background are only taken into account in bins and not continuously or individually per event.
- Only reweighting based on varX does not take complex correlations between varX and the other input variables into account. Thus some of the unwanted discriminator dependence might still be present.

The third approach we have studied is more elaborate. The basic idea is to train a MVA to discriminate between events which have high or low values of varX. This network will carry all information about varX which is contained in the input variables. The output of this MVA can then be translated into a weight for the actual signal/background training, which will be insensitive to varX. In the following we will give a more detailed explanation on how this MVA based weights are derived.

4.2. MVA based weight derivation

In order to deduce individual per event weights a more complex approach was developed. Here we use a chain of NeuroBayesExperts to calculate the weights for the final signal against background training. In the following we have to distinguish between those NeuroBayes trainings which are needed for the weight derivation (called weight-network) and the one which does the actual signal against background discrimination (called discriminator-network).

The first step in developing the weight-network is based on the idea, of letting NeuroBayes learn the information whether varX is high or low in an event based on the input variables which will be used in the discriminator-network. So the target definition for the weight-network training is "Is varX higher than a certain threshold?“. The steps to get to the definition of what is target (T) and non-target (!T) during the weight-network training is illustrated in figure 9(a) and are as follows:

(i) Draft of signal and background in a stacked plot.

(ii) We do not really care of what is signal or background for the time being, but we divide the whole dataset in the middle of varX (4 in this case).

(iii) Only defining the high values as target and the low values as non-target is not quite sufficient, because then the weight-network will implicitly already learn what signal and background look like. Therefore the target and non-target definition is opposite between signal and background for the weight-network training in one half of the dataset.

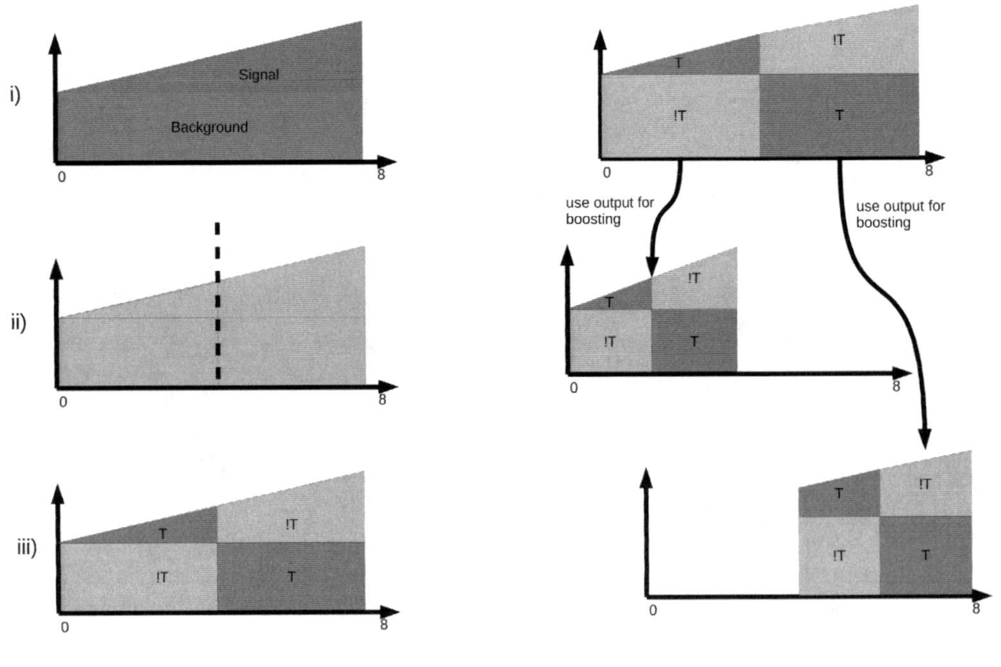

(a) (b)

Figure 9. Structure illustration of the weight-network boost chain. The output from the first network is used as boost weight in the successive layers weight-networks. On the left the defining regions of target(T) and non-target (!T) events for the weight-network training are depicted. In the right column is an llustration of the first boost step. One divides the sample into two subsamples and runs a boost training on each sample separately.

One weight-network will learn the differences between events with very high and very low values of varX but it will not be very sensitive to differences in events which are closer in varX.

Thus in order to refine the weight quality the dataset is split in a Haar-wavelet expansion. In each step, the previous training sample is split in the middle of varX into two subsamples and on both a boosted weight-network is trained with the adapted target definition. An illustration of the first step of this procedure is shown in figure 9(b). It is repeated several times (3 times in our example) until the following boost networks are not able to learn new information.

A "boosted network“ in this case refers to a special training mode, in which the output from one network is used as weight for the trainings in the next layer of networks. The weight-network boost weight is derived by transforming the NeuroBayes output into the probability $p = \frac{\text{NBout}+1}{2}$. This is already the weight if the event was a non-target (!T) event in the previous training, otherwise the weight is $1 - p$. In general this choice of weights leads to a enhancement of events which have been misclassified by the previous network, since these will get higher weights.

For the example shown here, a tree with 4 layers was used, which consists of 1+2+4+8=15 NeuroBayes networks which were trained successively on different subsamples as boosted networks (except the first one of course). In order to calculate the discriminator-training weight for each event one has to multiply the returned probability of the four corresponding weight-networks depending on its value of varX.

The weighted distribution of varX is shown in figure 8(c). It looks similar to the ones reweighted to the inclusive distribution (sum of signal and background), but it yields better performance as we will see next.

4.3. Performance comparison

In summary six different NeuroBayes trainings with different weights were performed. The full list of all trainings and whether they used varX as input or not can be seen in table 1.

Table 1. List of trainings, showing which variables were left out and which weights are applied

training title	variables left out of training inputs	weights
with varX	none	none
w/o varX	varX	none
w/o varX and corr. vars	varX and variables highly correlated	none
reweighted to sum	varX	histogram weights for flat shapes
reweighted flat	varX	histogram weights for inclusive shape
reweighted with NB boost chain	varX	NeuroBayes boost chain derive weights

There are 31 variables available in the trainings sample, including varX. In most trainings all but varX are used. For one training also varX is included and in another varX and those variables which are highly correlated to it are excluded.

The fitness of MVAs is usually compared by means of discrimination power. In order to deduce whether one MVA is better than the other, so called ROC (receiver operating characteristic) curves are compared. In figure 10(a) the background efficiency versus the signal efficiency is shown for the different NeuroBayesExperts (i.e. the part of NeuroBayes[1] which is used after the training for the actual classification).

As can be seen the NeuroBayesExpert which was trained without varX and all correlated variables yields the worst performance, as expected. The network which yields slightly better

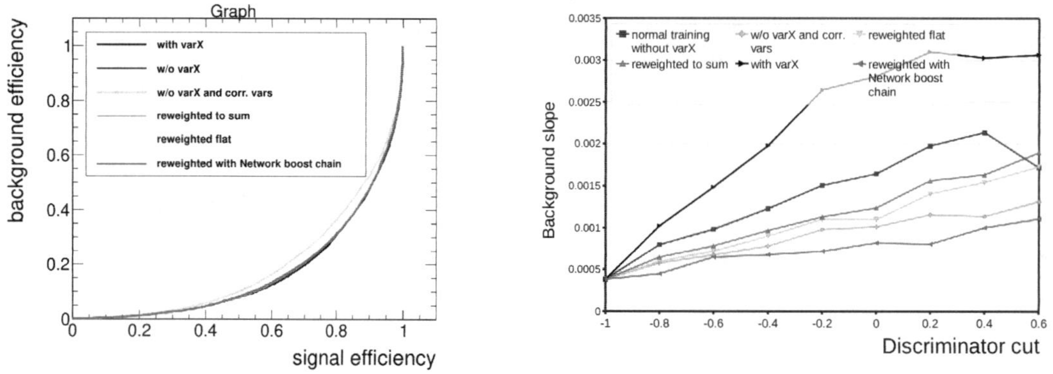

(a) (b)

Figure 10. Benchmark results for different NeuroBayes trainings concerning discrimination power and background flatness after discriminator cuts. The left figure shows performance curves for the different NeuroBayesExperts, while on the right background slopes after discriminator cuts for all studied NeuroBayesExperts are drawn.

performance than the others is the one with all variables included and no weights applied but all the other four are close by.

Since a good separation power is not the only figure of merit in this task, but we also want the background distribution to remain flat after discriminator cuts, we defined a second benchmark to evaluate which training type is suited best for our needs. The goal there is to measure the insensitivity of the background distribution of varX to discriminator cuts. Therefore straight lines were fitted to the central range of the distribution after several discriminator cuts and the slopes were compared. Such an example of a fitted line is shown in figure 7 in the right plot. The slope values for all the background distributions after several cuts on each of the NeuroBayesExperts are shown in figure 10(b).

Cuts were made in steps of 0.2 in the discriminator range [-1,1]. The results of a cut at 0.8 are not shown, since the fits did not yield reasonable results due to the small statistics in the remaining distribution.

The black line illustrates the behaviour of the ordinarily trained NeuroBayes network including varX in the inputs. With harder cuts the slope of the distribution of the remaining events gets steeper, which is the unwanted feature we are trying to avoid. The second worst slope behaviour is returned by the network which was trained without varX and without any weights applied. This already gives quite an improvement, but due to the variables correlated to varX the slopes still rise. Leaving out also the correlated variables leads to the light blue points, which is the second best in this comparison, but yielded the worst discrimination performance. In this benchmark the slope remains the flattest when cutting on the network which was trained with NeuroBayes boost chain weights (since it is even better than the one without varX and correlated variables, there still seems to be some correlation left). This means the effort of using 15 NeuroBayesExperts for the weight calculation is worthwhile, since it yields a good discriminator while keeping the background distribution of the remaining events flat.

5. Summary and Conclusion

We've presented three use cases in which NeuroBayes was either used to calculate individual weights for each event and/or weights were used to train NeuroBayes in a special way. These are not just theoretical concepts, but we have shown, that they are applicable in practice.

(i) We have used NeuroBayes to calculate individual MC event weights which were able to correct significant differences in the distribution of two variables between pseudo data and MC, while keeping the already matching ones untouched. The latter one is an advantage of MVA deduced weights over most commonly calculated scale factors since they usually are not able to take all correlations fully into account.

(ii) the ${}_s$Plot method was presented which has already been successfully used in real physics analysis to train NeuroBayes on real measured events without the need for any generated MC events. The only prerequisite is that a signal peak is visible and can be modelled in order to deduce a probability density function for signal and background for calculating the ${}_s$Plot weights.

(iii) The third use case of advanced event weights we have shown will train a discriminating network in such a way, that the shape of the distribution of a certain control variable is barely influenced by making selection cuts on the discriminator. This is achieved by training a tree like chain of NeuroBayes networks in order to absorb any information about the control variable into an event weight, which will make the discriminating-network blind to any control variable information.

Appendix A. Training of MVAs and where weights are taken into account

From the mathematical point of view, a well trained discriminating MVA is a non-linear transformation

$$f(\overrightarrow{x} \in \mathbb{R}^n) \to y \in \mathbb{R} \tag{A.1}$$

such that a loss function

$$\mathrm{err}(y) = \sum_i g(y_i, t_i) \qquad (\ g(y) \text{ could be } \chi^2 \text{ for example}) \tag{A.2}$$

is minimal with respect to the target value t (usually $t = 1$ for signal and $t = 0$ or -1 for background).

Introducing weights leads to

$$\mathrm{err}(y) = \sum_i w_i \cdot g(y_i, t_i) \tag{A.3}$$

So according to their weights, events are taken into account differently when minimizing the loss function. It can be understood as emphasizing certain events and suppressing others during the training procedure. This can be used to achieve several effects during a training.

References

[1] Feindt M 2004 A Neural Bayesian Estimator for Conditional Probability Densities *Preprint* arXiv:physics/0402093v1
[2] Marino C 2009, PhD thesis, IEKP-KA/2009-29
[3] Wick F 2009 Diploma thesis, IEKP-KA/2009-2
[4] Pivk M 2006 sPlot: A Quick Introduction, *Preprint*' arXiv:physics/0602023v1 [physics.data-an]
[5] Wick F 2011 Measurements of the properties of Lambda_c(2595), Lambda_c(2625), Sigma_c(2455), and Sigma_c(2520) baryons *Preprint* arXiv:1105.5995v2 [hep-ex]

Journal of Physics: Conference Series **368** (2012) 012029 doi:10.1088/1742-6596/368/1/012029

Tau identification using multivariate techniques in ATLAS

D C O'Neil on behalf of the ATLAS collaboration

Simon Fraser University

E-mail: doneil@sfu.ca

Abstract. Tau leptons play an important role in the physics program of the LHC. They are being used in electroweak measurements, in detector related studies and in searches for new phenomena like the Higgs boson or Supersymmetry. In the detector, tau leptons are reconstructed as collimated jets with low track multiplicity. Due to the background from QCD multijet processes, efficient tau identification techniques with large fake rejection are essential. Since single variable criteria are not enough to efficiently separate them from jets and electrons, modern multivariate techniques are used. In ATLAS, several advanced algorithms are applied to identify taus, including a projective likelihood estimator and boosted decision trees. All multivariate methods applied to the ATLAS simulated data perform better than the baseline cut analysis. Their performance is shown using high energy data collected at the ATLAS experiment. The improvement ranges from a factor of 2 to 5 in rejection for the same efficiency, depending on the selected efficiency operating point and the number of prongs in the tau decay. The strengths and weaknesses of each technique are also discussed.

1. Introduction

Tau leptons (τ) play an important role in the physics program of the LHC. In addition to measurements of standard model processes involving τ, they provide an important signature of several types of new physics. Notably, both low mass Standard Model Higgs boson searches and several new particle searches in Supersymmetric models involve decays to τ. Figure 1 shows the branching ratio for Higgs boson decay in the Standard Model as a function of Higgs mass [1]. The two-τ channel is an important signature for Higgs masses below about 140 GeV.

The challenge in exploiting τ signatures in the LHC environment is separation from background. The τ lifetime is too short to observe them directly. Instead, the τ decay products are observed. Figure 2 shows a pie-chart of the decay branching ratio for τ. The leptonic decays, representing about 35% of the total, are usually considered to be indistinguishable from prompt lepton production and are therefore included in other signal channels. For this reason, τ identification concentrates on identifying the hadronic decay modes, representing about 65% of the total. The difficulty with the hadronic modes is distinguishing them from an overwhelming background of quark and gluon jets at the LHC. To overcome this difficulty, a multivariate approach is required.

This paper presents the status of τ identification in the ATLAS experiment at the time of the 14th International Workshop on Advanced Computing and Analysis Techniques in Physics Research (ACAT 2011).

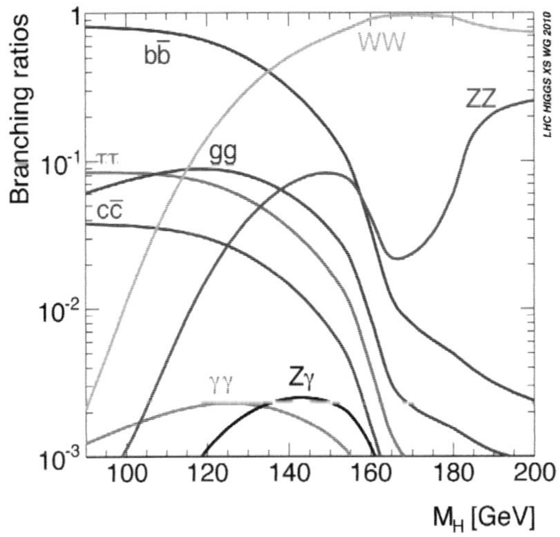

Figure 1: Standard Model Higgs branching ratios as a function of the mass of the Higgs [1].

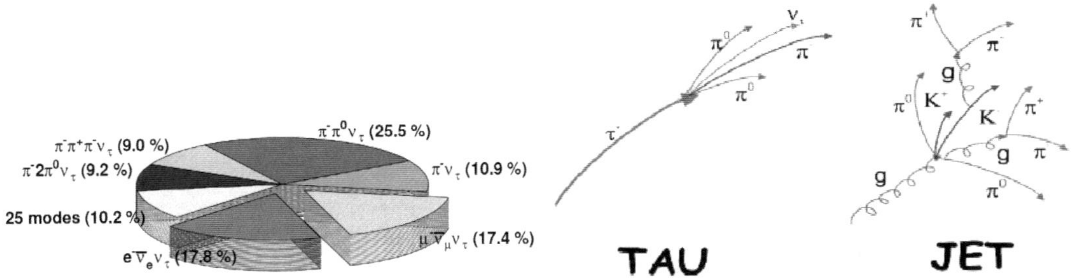

Figure 2: Left: pie-chart of hadronic tau decay branching ratios. Right: artist rendering of a τ jet and a gluon jet.

2. τ reconstruction and identification

The reconstruction and identification of hadronic τ in ATLAS are described in detail in [2]. Separation of hadronic τ decays from jet backgrounds happens in two steps. The first is to reconstruct a list of τ candidates in each event. Jets reconstructed with the anti-K_T algorithm [3] with distance parameter $R = 0.4$ are used to "seed" the reconstruction. Tracks are then associated to each seed and variables useful in distinguishing τ from quark and gluon jets are calculated and stored for each candidate. Very little separation between jets and τ is gained by this reconstruction step. An incomplete list of distinguishing variables includes:

- Track Radius (R_{track}): the p_T-weighted average track radius measured from the tau candidate axis.
- Electromagnetic Radius (R_{EM}): the transverse energy weighted average calorimeter radius measured from the tau candidate axis. Only cells in the EM calorimeter with $\Delta R < 0.4$ are used[1].
- Core Energy Fraction (f_{core}): the fraction of transverse energy in the core of the τ candidate. The core is defined as being within $\Delta R < 0.1$ of the tau axis.
- Cluster Mass ($m_{clusters}$): Invariant mass computed from the constituent clusters of the jet seed.

[1] $\Delta R = \sqrt{(\Delta\eta)^2 + (\Delta\phi)^2}$

- Track Mass (m_{track}): Invariant mass computed from the constituent tracks of the τ candidate.
- Transverse Flight Path Significance (S_T^{flight}): The decay-length significance of the secondary vertex for multi-track τ candidates in the transverse plane.

Figure 3 shows these distinguishing variables for both simulated candidates and real candidates extracted from the 2010 ATLAS dataset. Most of these variables exploit the narrower energy deposits and lower particle multiplicity in τ decays compared to quark or gluon jets.

The second step is to use the distinguishing variables calculated during reconstruction to separate τ from jets. ATLAS has used three different approaches for this step: simple cuts, log likelihood (LLH), and boosted decision trees (BDT). The latter two will be described herein.

Both the LLH and BDT techniques were "trained" using monte carlo generated events ($Z \to \tau\tau$) as signal and dijet ATLAS data as background. The two techniques are then compared on common samples using common definitions of signal efficiency and background rejection. They are trained and evaluated separately for 1-track (1-prong) and multi-track (3-prong) τ candidates.

2.1. LLH ID technique

The likelihood function is defined as the product of the distributions of the identification variables:

$$L_{S(B)} = \prod_{i=1}^{N} p_i^{S(B)}(x_i)$$

where $S(B)$ refers to signal (background), and $p_i^{S(B)}$ is the signal (background) probability density function of variable x_i. The likelihood function neglects correlations between variables and represents the joint probability distribution for the input variables.

A discriminant is built from the log-likelihood ratio between signal and background:

$$d = \ln\left(\frac{L_S}{L_B}\right) = \sum_{i=1}^{N} \ln\left(\frac{p_i^S(x_i)}{p_i^B(x_i)}\right).$$

A representative discriminant distribution is shown in figure 4. Good separation between signal (τ) and background (jets) is seen.

2.2. BDT ID technique

Boosted decision trees [4] turn a simple cut-based approach into an advanced multivariate technique. A simple decision tree makes a series of selections (cuts) to classify each τ candidate as either signal or background. However, candidates which fail each cut are not discarded, rather they are further processed by the algorithm. By successively applying the best selection on the best available variable to achieve optimal classification of each candidate, a simple decision tree can effectively make a piecewise continuous cut through the multidimensional space defined by the input variables, as illustrated in figure 5. The process of "boosting" then creates a second decision tree optimized to work well on the candidates which were misclassified by the first tree. A large number of trees can be created in this way, each optimized to work well on the misclassified candidates in the previous tree. Boosting improves the stability of the training and the overall performance of the signal and background classification. At the end of the process, the final BDT "score" is determined by a weighted average of the scores from each of the trees in the BDT set.

In this case, the BDT was trained using the TMVA [5] program and 50 trees were used. The boosting was performed by the adaBoost [4] algorithm with boosting strength=0.2. Figure 5

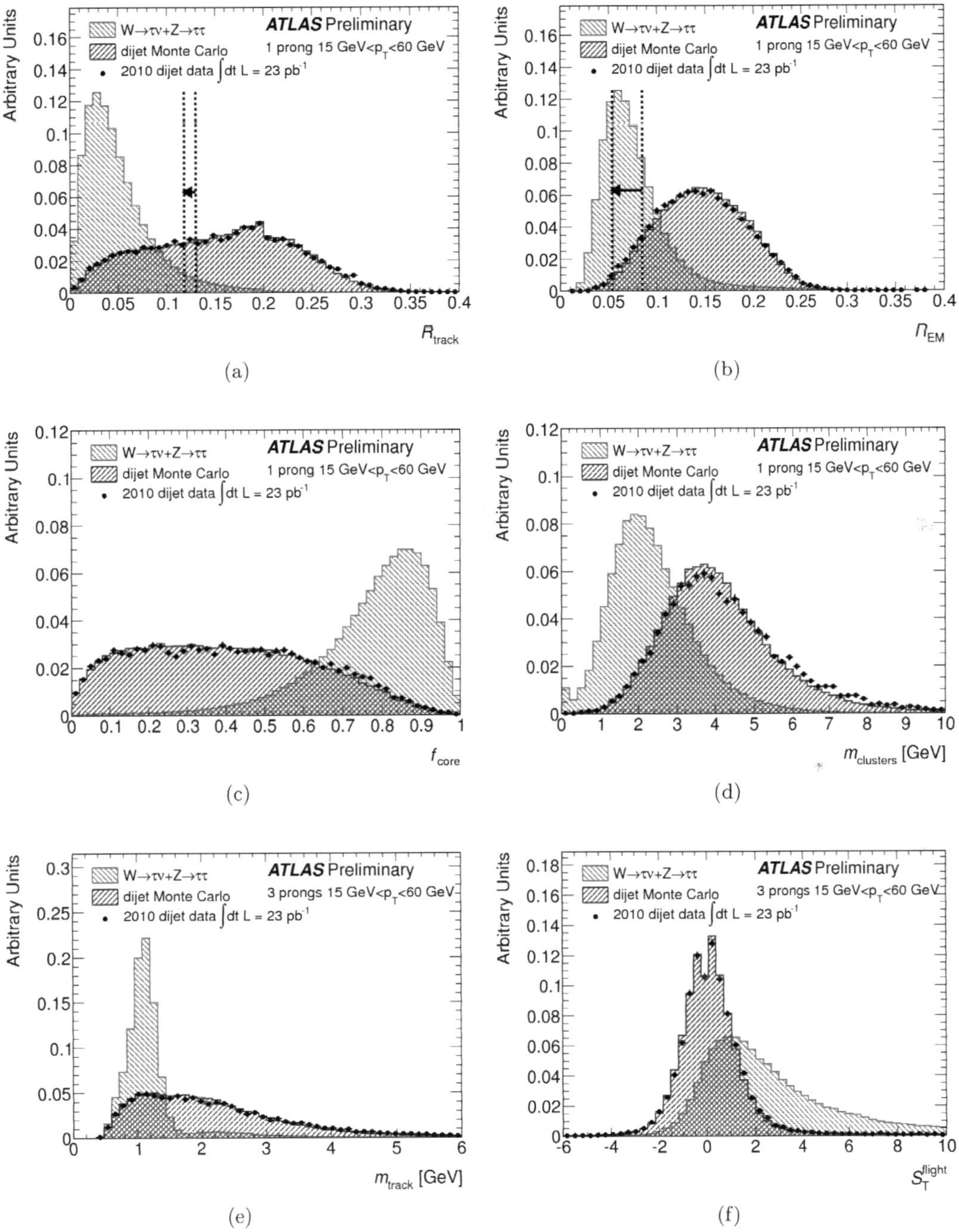

Figure 3: Variable distributions used to separate hadronic τ from quark and gluon jets [2]. (a) Track radius (b) Electromagnetic radius (c) Core energy fraction (d) Cluster mass (e) Track mass (f) Transverse flight path significance.

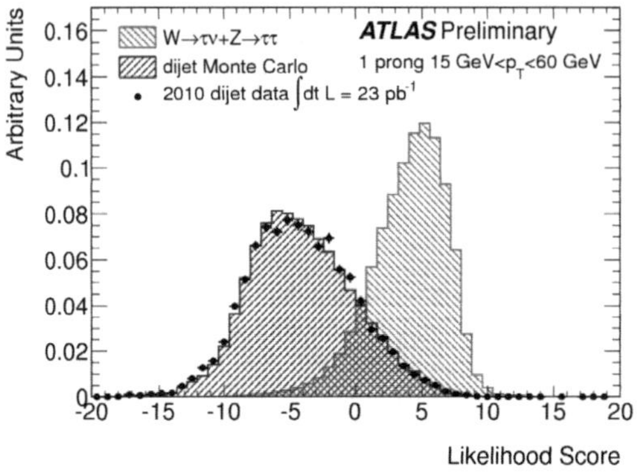

(a)

Figure 4: The log-likelihood-ratio for 1-prong τ candidates [2].

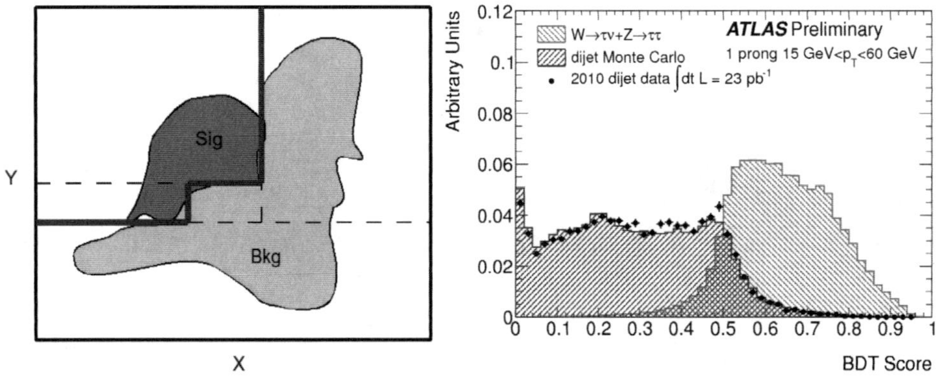

(a) (b)

Figure 5: (a) Illustration of a piecewise continuous cut through a 2D plane made by a simple decision tree. Signal is in red, background in blue. (b) BDT discriminant output scores for signal and background for 1-prong τ candidates [2].

shows a representative discriminant distribution for BDT using signal Monte Carlo and dijet ATLAS data. Good separation between signal (τ) and background (jets) is seen.

In addition to its good performance, the BDT technique has several distinct advantages over other approaches. Correlated variables are handled easily by the technique. At each decision point, the best variable is chosen to separate signal from background. Adding a well-modeled variable does not worsen the performance of the algorithm, regardless of the correlation with existing variables. The training and optimization of BDT is straightforward and fast, with very few tunable parameters to optimize. A disadvantage of BDT compared to simple cuts or LLH is the number of candidates needed in the training sample. The power of the technique makes it necessary to use larger training samples.

2.3. Performance of MV techniques

Figure 6 shows the final performance of three τ identification techniques in ATLAS: simple cuts, LLH and BDT. The figure shows inverse background efficiency (rejection) vs. signal efficiency.

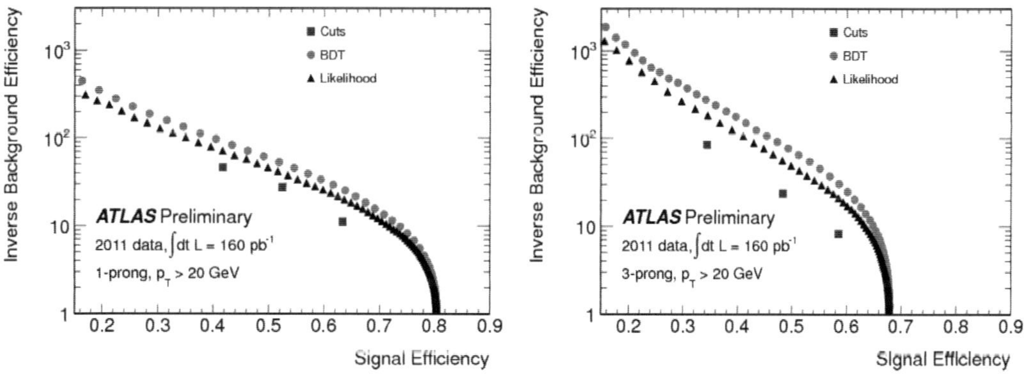

(a) (b)

Figure 6: Performance summary for three different techniques to separate τ from jet backgrounds: cuts, LLH and BDT. (a) 1-prong candidates (b) 3-prong candidates. Taken from [6].

As expected, the least performant solution is the simple cuts approach, the next solution is the LLH and the best is BDT.

3. Physics Applications

The ATLAS τ identification algorithms have been applied in several public physics results in 2010 and 2011. In this section, two such results are described. Each of these results uses the BDT τ identification technique.

3.1. $W \rightarrow \tau\nu$ cross section

Rediscovery of the W boson and measurement of its cross section was one of the most important milestones acheived in the first months of LHC running. Rediscovery of the τ channel in the same dataset was challenging. The first paper published in this channel at the LHC is the ATLAS result using the BDT τ identification [7]. In this analysis, a cut was made on the BDT output score to purify the sample. This cut accepted roughly 30% of signal events while applying a jet rejection factor of about 100 for 1-prong candidates. For 3-prong candidates, the signal efficiency was 35% for a jet rejection factor of about 300.

Figure 7 shows the number of charged tracks distribution after application of all selection cuts, including the BDT τ identification. The supression of the 2-track bin relative to the 3-track bin is a strong indication of a signal. The final inclusive cross section measurement is

$$\sigma^{tot}_{W\rightarrow\tau\nu} = 11.1 \pm 0.3(stat) \pm 1.7(syst) \pm 0.4(lumi)nb.$$

3.2. $t\bar{t} \rightarrow \mu + \tau + X$ cross section

Measurement of the branching ratio of top quarks to τ is of interest in probing for physics beyond the Standard Model. As an example, if the charged Higgs boson of Supersymmetry has a mass less than the top quark mass minus the b-quark mass, the top quark can decay to it. If this decay is possible, the charged Higgs will then favour decays to $\tau + \nu$ over a wide range of SUSY parameter space. Therefore, a measurement of an unexpectedly high cross section in the $t\bar{t} \rightarrow \tau + X$ channel would be an indication of new physics.

Journal of Physics: Conference Series **368** (2012) 012029 doi:10.1088/1742-6596/368/1/012029

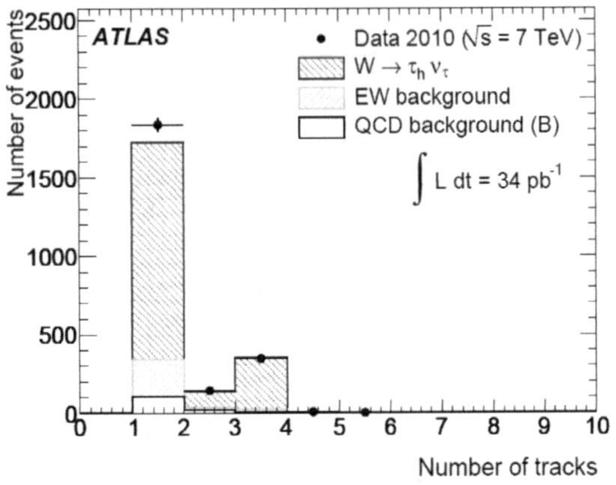

Figure 7: The number of charged tracks reconstructed as part of the τ candidate. True τ should have an odd number of associated tracks (charge conservation) [7].

In the summer of 2011, the most precise measurement of the $t\bar{t} \to \tau$ cross section came from the ATLAS experiment in the $\mu + \tau$ channel [8]. After preselection, the dominant background to this channel is $t\bar{t} \to \mu + jets$ with one of the jets faking a τ. The most powerful remaining feature to exploit to remove this background is BDT τ identification.

This analysis used BDT τ identification in a different way than the W analysis described earlier. Rather than cutting on the BDT output score, the full score distribution was used. Since the jets faking τ are from $t\bar{t}$ events, they are dominantly quark jets. The BDT score distribution for a quark-rich data sample was obtained from $W + jets$ data. Figure 8 shows two different quark-rich templates, one extracted from data in which the muon and the jet had the opposite sign charge (OS) and another from the same sign (SS) sample. It also shows a true τ template extracted from simulation. Figure 9 shows the result of fitting these templates to Monte Carlo simulation (a) and data (b). The results in simulation (MC) agree well with expectation. The results in data yield

$$\sigma_{t\bar{t}}^{tot} = 142 \pm 21(stat) \pm^{20}_{16} (syst) \pm 5(lumi)pb$$

in agreement with the Standard Model.

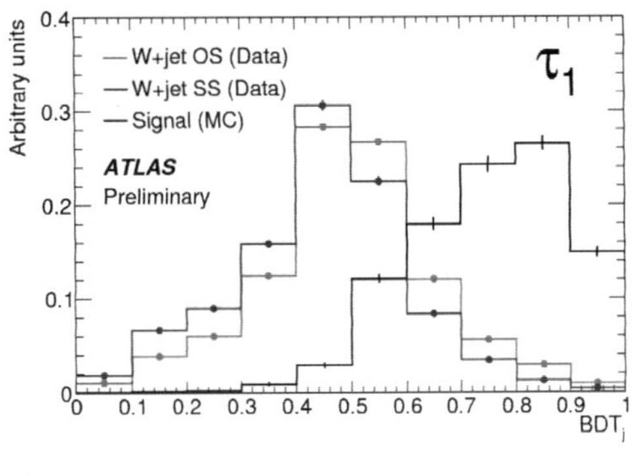

Figure 8: Templates used to extract signal cross section in $t\bar{t}$ analysis [8].

4. Summary and Conclusions

The ATLAS τ identification relies on multivariate techniques to separate signal from quark and gluon jet backgrounds. The perfomance gain compared to a simple cut-based approach is substantial (factor of 2-5 in background rejection for the same signal efficiency). These techniques has been applied to physics measurements in challenging channels, allowing cross section measurements in τ channels for the first time at the LHC.

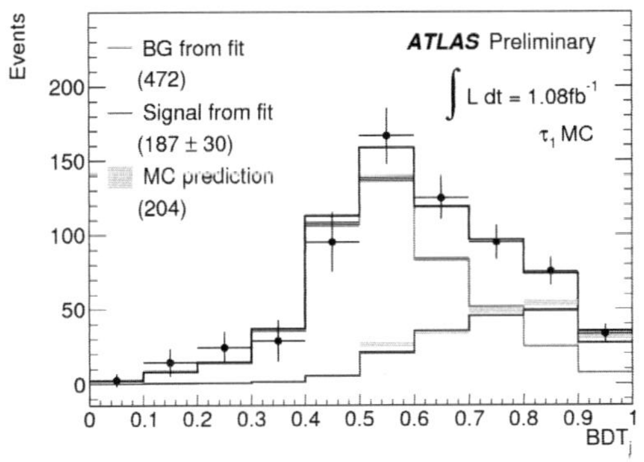

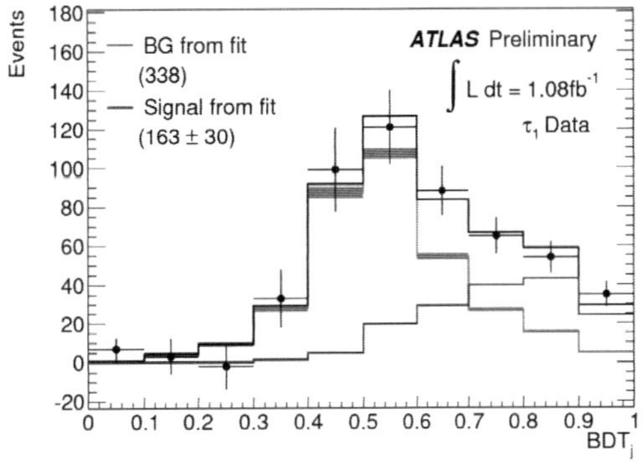

Figure 9: Results of template fits from the $t\bar{t} \to \mu + \tau + X$ analysis. (a) The fits performed on simulation only, (b) The fits performed on data [8].

References

[1] Dittmaier S, Mariotti C, Passarino G and Tanaka R (editors) 2011 CERN-2011-002, *Preprint* arXiv:1101.0593v3

[2] The ATLAS Collaboration 2011 ATLAS-CONF-2011-077, https://cdsweb.cern.ch/record/1353226

[3] Cacciari M, Salam G P, and Soyez G 2008 *Journal of High Energy Physics* **04**, 063

[4] Breiman L, Friedman J, Stone C, and Olshen R 1984 *Classification and Regression Trees* (Chapman and Hall)
Freund Y and Shapire R 1996 *Proceedings 13th International Conference on Machine Learning* (July 3-6, 1996, Bari, Italy)

[5] *Toolkit for Multivariate Data Analysis* 2011 http://tmva.sourceforge.net/, Jan, 2011. version 4.0.4 (available as part of ROOT version 5.26).

[6] The ATLAS Collaboration 2011
Tau Public Results https://twiki.cern.ch/twiki/bin/view/AtlasPublic/TauPublicCollisionResults

[7] The ATLAS Collaboration 2011 Accepted by *Physics Letters B.*, CERN-PH-EP-2011-122, (*Preprint* arXiv:1108.4101v1)

[8] The ATLAS Collaboration 2011 ATLAS-CONF-2011-119 https://cdsweb.cern.ch/record/1376411

Online particle detection with Neural Networks based on topological calorimetry information

T Ciodaro[1], D Deva[1], JM de Seixas[1], D Damazio[2]
[1]Federal University of Rio de Janeiro
[2]Brookhaven National Laboratory

E-mail: ciodaro@cern.ch, dhiana.deva@cern.ch, seixas@lps.ufrj.br, damazio@cern.ch

Abstract. This paper presents the latest results from the Ringer algorithm, which is based on artificial neural networks for the electron identification at the online filtering system of the ATLAS particle detector, in the context of the LHC experiment at CERN. The algorithm performs topological feature extraction using the ATLAS calorimetry information (energy measurements). The extracted information is presented to a neural network classifier. Studies showed that the Ringer algorithm achieves high detection efficiency, while keeping the false alarm rate low. Optimizations, guided by detailed analysis, reduced the algorithm execution time by 59%. Also, the total memory necessary to store the Ringer algorithm information represents less than 6.2 percent of the total filtering system amount.

1. Introduction

Pattern recognition based on artificial neural networks is widely used in many scientific areas achieving good performance [1]. For example, neural networks can be found in several high-energy physics experiments, for particle identification [2], particle tracking [3], and other related issues [4]. The experiments in the Large Hadron Collider experiment (LHC), in CERN [5], have used artificial neural networks widely, e.g. [6, 7].

The LHC accelerates and collides proton beams at an event rate of 40 MHz, with up to 14 TeV center-of-mass energy (currently, it operates at 7 TeV). At this unprecedent energy, interesting physics processes never seen before, such as the Higgs boson, can be searched for.

ATLAS is the largest LHC particle detector [8]. In order to provide detailed information about the collision subproducts, ATLAS is split into three main subdetectors: the inner detector, which is responsible for tracking particles; the calorimeters, responsible for measuring their energy; and the muon chambers, which aim at tracking and measuring the momentum of muons. figure 1 shows the ATLAS detector.

Altogether, ATLAS produces around 1.5 MB of information per collision event [9], which must be stored and analyzed. Considering the LHC collision rate, the total ATLAS data flow is around 60 TB/s. As much of the events don't contribute to the new physics which is the goal of the experiment, a three level online filtering system (trigger) was designed to select interesting events [9].

This article presents the latest results of the Ringer algorithm [10], which combines both feature extraction, from the ATLAS calorimetry system, and neural networks to select electrons from the background at the ATLAS second level trigger system. Optimizations are applied

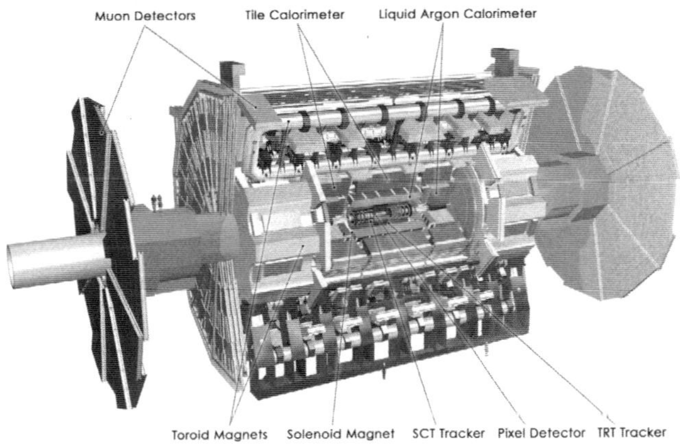

Figure 1. The ATLAS detector.

for improving the processing speed and the whole detection performance is evaluated through algorithm implementation at the trigger framework using simulated data, which include pileup effects.

The article is divided as follows. Section 2 describes the ATLAS calorimeters and section 3 details the ATLAS trigger system. The Ringer algorithm implementation is presented in section 4. Section 5 discusses the methodology and the results achieved, while conclusions are derived in section 6.

2. Calorimetry

The ATLAS calorimetry system [11] is divided into two sections: electromagnetic and hadronic, as can be seen on figure 2(a). The calorimeters are segmented in-depth into layers. The granularity (number of cells per layer) is layer dependent and it was designed to efficiently characterize the desired particles.

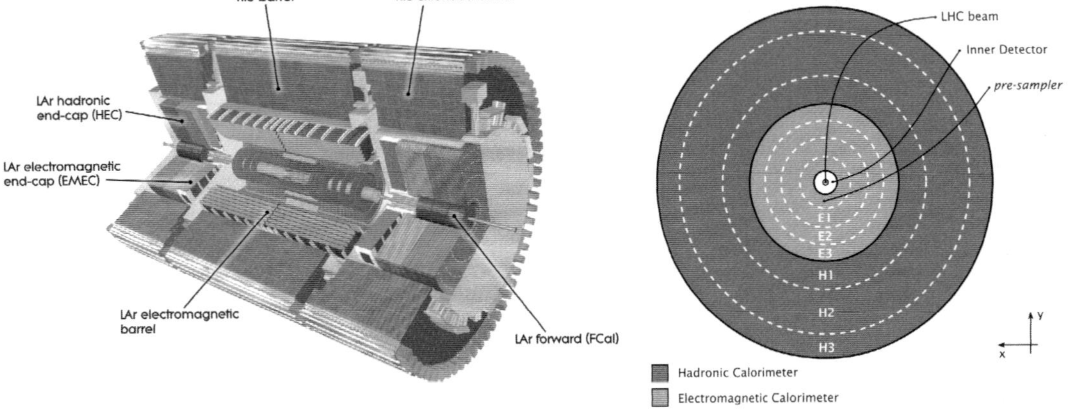

(a) (b)

Figure 2. The ATLAS calorimeters (a) and the schematic cross-sectional view of the ATLAS detector (b).

The electromagnetic calorimeter uses liquid Argon technology [12] and is divided into two parts: the barrel and the end-caps. The electromagnetic calorimeter also has an additional pre-sampler layer, which is used to correct for energy lost in dead materials in front of the calorimeter (solenoid, cryostat, ...), totalizing four layers. The electromagnetic section comprises, approximately, 180,000 readout cells.

The hadronic calorimeter, however, uses two technologies: scintillating tiles (TileCal) [13] in the barrel region, and liquid Argon in the hadronic end-cap regions. The TileCal is segmented into three layers. The electromagnetic and hadronic calorimeters cover a region of $-\pi < \varphi < \pi$ and $-2.5 < \eta < 2.5$. The region $|\eta| > 2.5$ is covered by the forward calorimeter, which, however, is not used in this work. Also, the calorimeter has a gap (crack region) in $|\eta| \approx 1.4$, which permits services and infrastructures to reach the most inner parts of ATLAS.

3. Trigger system

The trigger system is split into three cascaded levels, each with its own maximum event rate and latency requirements [9]. Eventually, if the event is accepted by all three levels, data from the entire ATLAS detector are stored for offline analysis [9]. Figure 3 shows the ATLAS trigger system diagram.

3.1. The L1 trigger

The first level of the ATLAS trigger system (L1) is based on highly programmable devices and its main objective is to reduce the 40 MHz event rate to no more than 75 kHz. The L1 identifies the regions at the detector with possibly interesting physics (RoIs, Regions of Interest) [14]. The time budget for the L1 trigger decision (latency) is 2.5μs.

In order to achieve the highest event rate reduction, in the shortest time possible, L1 has access to compacted information from the detector. Only calorimetry and muon information is used. In the calorimeters, trigger tower regions are defined in the $\eta \times \varphi$ plane (0.1×0.1). The energy of the cells inside a trigger tower is summed up, for both electromagnetic and hadronic sections. For the muon chambers, low precision muon transverse momentum information is used. Electrons and other electromagnetic objects are the most frequent signature accepted by L1 [14], and this is the reason for their study in this work.

3.2. The High Level Trigger

The ATLAS High Level Trigger (HLT) is formed by the second level trigger (L2) and the event filter (EF) and it is implemented in software [15]. The HLT operates on the RoIs selected by L1, accessing the detector information in its full resolution.

Large computer farms are used to support the HLT framework. The event processing is split into several processing units according to their usage. For the L2, the mean processing time available is 40 ms, and for the EF, it is 4 s.

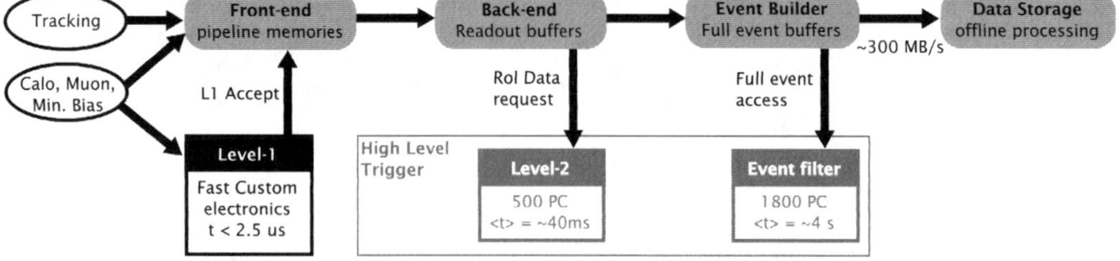

Figure 3. The ATLAS trigger system.

Differently from L1, where only a rough classification is possible, the L2 runs specialized algorithms, which use information from the different subdetectors (calorimeters, tracking and muon chambers). The algorithm's complexity is further increased at the EF, where adapted versions of the offline reconstruction algorithms are executed. Accessing the full event data, the EF can better exploit the physics properties of each event.

3.2.1. Calorimeter HLT: The calorimeter High Level Trigger (HLTCalo) is a set of algorithms that selects particles by performing different clustering methods at the cell level. An optimized infrastructure, common for L2 and EF, for data retrieval is available for all trigger algorithms.

The HLTCalo algorithms are split into two parts: feature extraction (FeX) and hypothesis testing (Hypo). The FeX part accesses the detector data, through the common HLTCalo infrastructure, and calculates relevant variables needed for particle identification. These variables are used by the Hypo part to classify the RoI and, eventually, to accept or reject event.

3.3. Athena

All levels of processing of ATLAS data, from high level trigger to event simulation, reconstruction and analysis, take place within the Athena framework [16]. Athena is fully implemented in objected oriented language (C++ and python) and it is structured into projects and packages.

The HLT community maintains a set of algorithms to emulate the trigger system, which runs after the L1 event acceptance. Also, common tools are provided for analyzing the trigger performance for the associated physics. New algorithms can be developed for running at the HLT, as long as they are implemented within the Athena framework, following some standard procedures and common methods.

3.4. Electron identification

According to the expected event rates in L1, the detection of electromagnetic particles (electrons and photons) is extremely important for the identification of interesting physics channels. A slice of the trigger system is fully dedicated to the detection of such particles (electron/photon slice) [17].

In order to control the electron trigger flow, linear thresholds are applied to the quantities extracted from electromagnetic showers in HLT. These thresholds can also change during the experiment, in order to allow both detection efficiency and false alarm rate to be controlled. If less background is desired, tighter criteria for electron identification are applied, but also more electrons are discarded. Otherwise, loose criteria are selected and more electrons (and also fake electrons from the background), are triggered. Through such selection, the experiment can tune the trigger rate and other physics demands optimizing the trade-off between particle identification efficiency and speed [15].

Offline algorithms extrapolate tracks from the inner detector to the electromagnetic calorimeter, improving the characterization of the electromagnetic showers. The standard quantities extracted can be used as guidelines for trigger performance evaluation. In particular, the R_{core} quantity extracted offline is used in this study, which is defined as follows:

- R_{core} $\rightarrow$ characterizes the shower spread at the second electromagnetic calorimeter layer: $R_{core} = E_{3\times7}/E_{7\times7}$, where $E_{m\times n}$ is the energy deposited in a region of $m \times n$ cells in the $\eta \times \varphi$ plane around the extrapolated point of impact in the calorimeter of the associated track.

Journal of Physics: Conference Series **368** (2012) 012030 doi:10.1088/1742-6596/368/1/012030

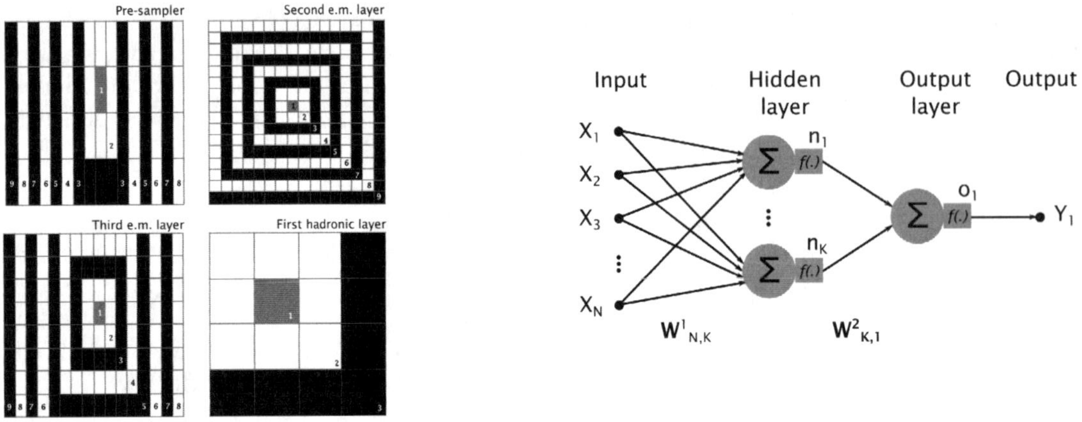

(a) (b)

Figure 4. Illustration of possible rings from the Ringer feature extraction (a) and an example of a feedforward neural network for the Hypothesis testing (b).

4. The Ringer algorithm

The Ringer algorithm is an alternative L2 algorithm for electron identification [10]. Feature extraction is performed through data description by means of concentric rings extracted from seven calorimeter layers of a selected L1 RoI. Hypothesis testing is applied through an artificial neural network [18] operating on the extracted calorimetry features (ring sums).

4.1. Feature extraction

The energy sum over all the cells in each of the rings in an RoI are calculated. In each calorimeter layer, the algorithm searches for the most energetic cell, which defines the first energy sum. The energies of the outer cells with respect to the previous ring are summed up to form the next ring energy sum. This procedure is implemented until the RoI is covered, or a maximum number of rings is reached.

Figure 4(a) shows some ring examples. It can be seen that, in some cases, the ring is completely conceptual, as some are not closed rings. For RoI description, a total of 100 rings are used and each layer contributes with a fixed number of rings (see table 1).

4.2. Hypothesis testing

The extracted rings are presented to a three-layer feedforward neural network (Multi-Layer Perceptron, MLP), whose architecture is illustrated in figure 4(b). The network is trained offline and its parameters are set before the start of the experiment. Several normalizations have been tested to improve the discrimination performance [10]. The present work implemented an event normalization, which consists of dividing each ring sum component by the total event energy.

Table 1. Number of rings per calorimeter layer.

Layer	Electromagnetic Calorimeter				Hadronic Calorimeter		
	pre-sampler	EM 1	EM 2	EM 3	HAD 1	HAD 2	HAD 3
# Rings	8	64	8	8	4	4	4

4.3. Implementation
The Ringer algorithm is implemented on the Athena framework, which allows the algorithm to run on both real and simulated data. Several optimizations have been performed to ensure that the Ringer implementation respects the time constraints of the L2 trigger. Also, the total amount of data required to store the Ringer information must be reasonable. Results from tests on such requirements are discussed in the following sections.

5. Performance
The Ringer algorithm performance is first evaluated using 2010 14 TeV Monte Carlo simulations of electrons and particle jets (background). Electrons are from filtered simulated $Z \rightarrow ee$ events. In both simulations, pileup is considered (the current event might have information from previous events), which tends to deteriorate the trigger performance.

Time performance is evaluated through detailed analysis using the Valgrind tool [19], which is a framework specially designed to analyze code implementation. Through this framework, it is possible to reduce code bottlenecks, optimize function calls and spot unnecessary memory consumption and memory leaks. Athena has a patch for Valgrind analysis, such that the Ringer implementation can be analyzed within the ATLAS software framework.

For the hypothesis testing, (trained) neural network parameters (weights and biases) are exported and uploaded to the Athena framework.

5.1. Methodology
In Z→ee events there will be detector activity registered which is not related to the electrons (for instance it could originate from jets occurring elsewhere in the event). In order to consider only RoIs that correspond to electrons, we use only RoIs (from simulated events) that match the true direction of an electron (defined as $\sqrt{\Delta\eta^2 + \Delta\varphi^2} < 0.05$ between the directions of the RoI and of the actual electron). For the jet simulation, all fake electron RoIs (accepted by a simulated L1 trigger) are considered background. In total, nearly 7k events are selected for electrons and 5k for jets, which are split into half, one forming the training dataset, and the other forming the testing dataset. While the training dataset is used to adapt the MLP parameters, the testing dataset is used to evaluate the MLP generalization (capacity to generalize what was learnt from the training dataset). Also, the testing dataset is used as validation set to avoid overtraining [18].

The resilient back-propagation algorithm [20] is used for training the MLP in batch mode, because of its fast convergence. Also, the MLP is built with one hidden layer only, in order to keep its architecture simple (interesting for online operation, because of its speed). The hyperbolic tangent is used as activation function in both hidden and output layers, where electrons and jets are targeted to the output values '1' and '-1', respectively. The training cost function is the mean squared error between the network output and the desired target. Previous studies have shown that 10 neurons at the hidden layer gives the best trade-off between performance and architecture complexity. In order to avoid local minima, the training algorithm was initialized randomly 100 times and only the network with the best performance was considered.

The SP index is used to evaluate the network performance:

$$SP = \sqrt{\sqrt{P_e \times (1 - F_j)} \times \left(\frac{P_e + (1 - F_j)}{2}\right)} \quad (1)$$

where P_e is the electron detection efficiency and F_j is the jet false alarm rate. Previous studies concluded that this index is the best compromise between electron detection and jet rejection, when compared to their average value, for instance [21]. The SP index results in '1' if all electrons are accepted and all jets are rejected. If all electrons are rejected, or all jets are accepted, it results in '0'. The MLP output is compared to a threshold in order to finally classify the input as electron or jet. The discrimination threshold is selected in order to maximize the SP index.

5.2. *Relevance mapping*

A relevance mapping [22] can reduce the MLP complexity by selecting only the input variables (rings) which are most relevant for the discrimination process. Here, the relevance R_i of a given input variable i is measured from the relative difference between the MLP SP index and the SP index achieved when the input variable is fixed to its mean value:

$$R_i = \frac{SP_{MLP} - SP_i}{SP_{MLP}} \tag{2}$$

Rings having relevance values below a given threshold are removed from the training and test sets and a new MLP is designed. Note that, if negative relevant values are found, it means that the given input variable confuses the MLP, and its removal may increase performance.

5.3. *Tag-and-probe*

In real ATLAS data (as opposed to simulated data) it is not known if a given trigger RoI is truly from an electron or a jet. Thus, a data-driven method for efficiency estimation must be employed. The so-called Tag-and-probe method is often used, as it is extremely suitable for physics processes characterized by a double-object final state signature (like the $Z \rightarrow ee$ process) [23].

The method consists in evaluating the efficiency of detecting 'probe' objects within a properly 'tagged' sample of events. For the L2 online trigger, it consists of selecting possible electron pair candidates with invariant mass consistent with the Z boson mass and which have been accepted by the electron offline algorithm. The event is tagged if one of the electrons (at least) is accepted by a L2 tight electron criteria. The probe object is selected according to the Ringer algorithm hypothesis testing, considering the discriminator threshold that maximizes the SP index. Only the test set is used to evaluate the Ringer performance with the Tag-and-Probe method.

5.4. *Results*

The Ringer algorithm implementation was evaluated with respect to the total execution time and amount of data needed to save its information (ring sums and neural network output). The Valgrind tool did not report any memory leak after several events. The total time needed to run the entire L2 electron algorithm was measured. After a total of ~17k RoI's, the new Ringer implementation (after the Valgrind analysis) is estimated to take, on average, 1.075 ms to fully execute (feature extraction and hypothesis testing), while before the Valgrind analysis it took, on average, 1.445 ms. The detailed analysis led to a 59% faster execution time. Also, the new Ringer implementation adds an overhead of ~0.3 ms per RoI when compared to a standard L2 algorithm. With respect to the ring energy sums and neural network output, the Ringer information represents 6.2% of the total amount of stored data used by electron/photon algorithms.

The neural network performance for the training set indicates a 5.18% false alarm probability (jets misclassified as electrons), for a fixed electron detection probability of 98.00%. Figure 5 shows the neural network output for the training set. Notice that most of the electrons were correctly mapped to the output value '1', while jets were mapped to '-1'. The threshold needed to maximize the SP index is also shown, together with the threshold needed to achieve 98.00% electron detection. The maximum SP index translates into a 97.28% electron detection probability, against a 4.00% false alarm probability.

In order to reduce the network complexity, the input variables (ring energy sums) were ranked according to their relevance. Figure 6(a) shows the relevance for each ring. The full vertical lines indicate the end of a calorimeter layer and the beginning of the next one. By changing

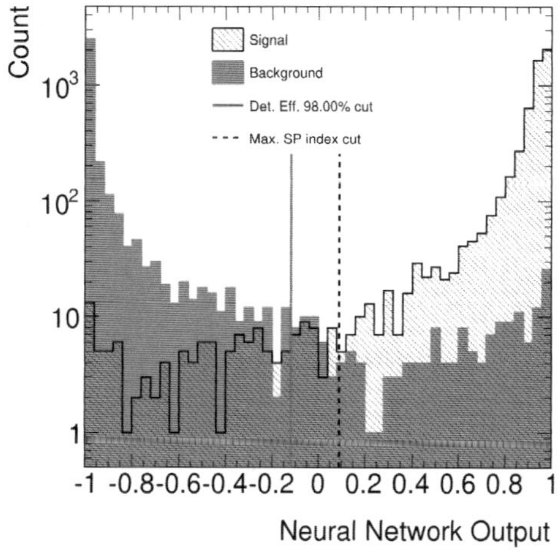

Figure 5. Neural network output for the training set over the MLP trained with full ring description.

the relevance threshold, the number of rings considered relevant for the discrimination can be adjusted. In order to achieve a reduction of ~70% of the total amount of rings, for instance, a relevance threshold of 0.05% is required (dashed horizontal line). Note that some rings have negative relevance, meaning that they confuse the particle classification, instead of assisting it. In terms of performance, figure 6(b) shows the ROC (Receiver Operating Characteristic) [24] curve achieved after training an MLP using only the relevant rings as inputs (again initializing 100 times to avoid local minima). The ROC curve for the MLP using the entire ring set is

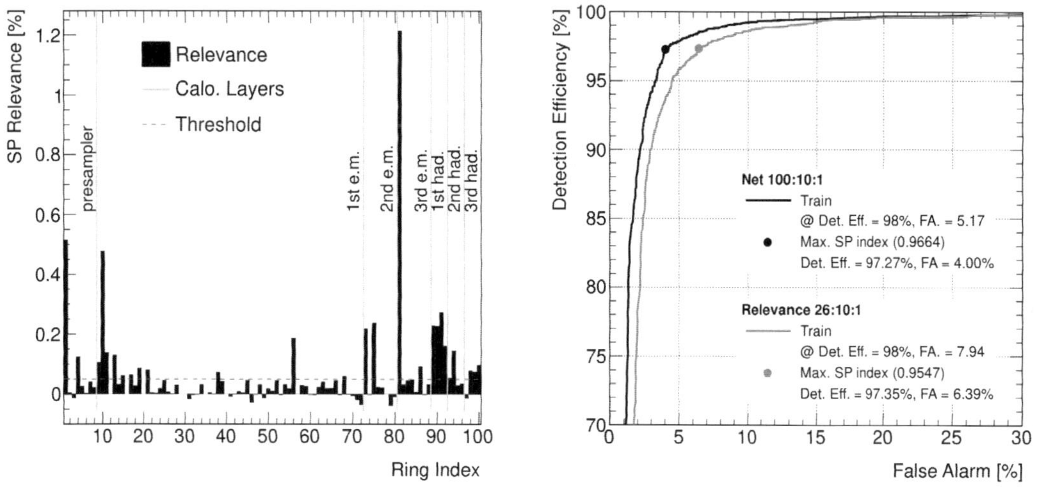

(a) (b)

Figure 6. Relevance mapping using the training dataset (a) and the resulting ROC curve for MLPs using only relevant rings, and using full ring information (b).

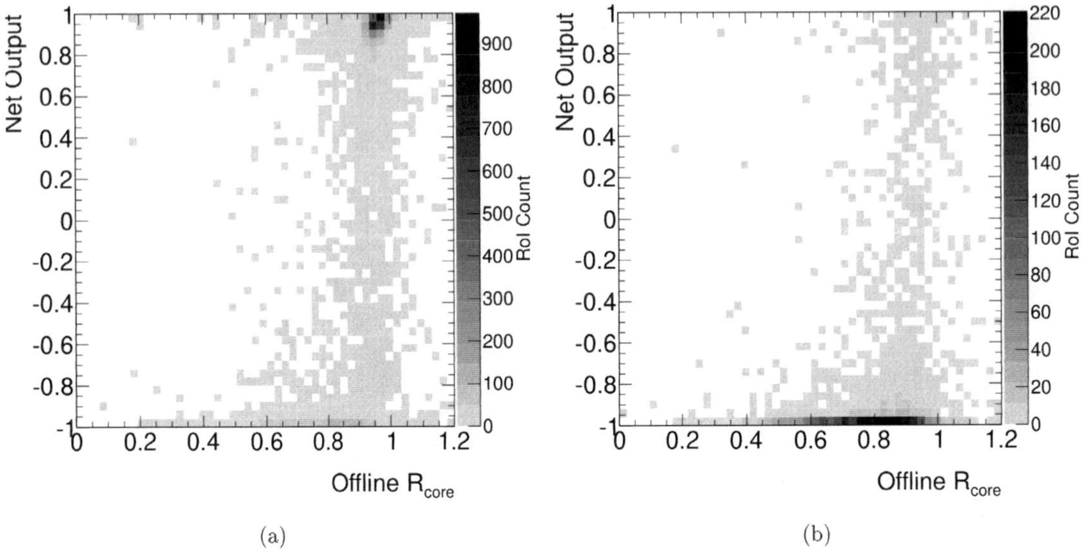

Figure 7. MLP output as a function of the electron offline R_{core} discriminant variable for electrons (a) and jets (b).

also shown for the sake of comparison. It can be seen that the ~70% information reduction increased the false alarm rate by less than 3% (fixing the electron detection efficiency to 98%), while the SP index is slightly reduced (~0.1). The relevance threshold value can be used as an adjustable parameter to regulate the MLP performance in a compromise with the Ringer hypothesis test speed and the total false alarm rate.

In order to extract physics conclusions, the neural network output can be compared to some key offline electron variables for a given RoI. Figure 7 shows the MLP trained without the relevance cut (full ring information) with respect to the R_{core} offline variable (see section 3.4), for both electrons and jets. The results are considering the test set, in order to evaluate the MLP generalization. The R_{core} measures the electromagnetic shower spread. Electrons produce narrower showers, leading to an R_{core} value close to unity, while jets tend to form wider showers, thus with a R_{core} value far from unity. It can be seen from figure 7(a) that the R_{core} value is close to unity for most of the electrons that pass the MLP cut. This is particularly interesting and useful when real data is considered, as the actual RoI classification is unknown. The same result can be seen for the rejection of jets, on figure 7(b), where most of the jet RoI's are correctly classified by the Ringer hypothesis test, as their R_{core} value is mostly around 0.8.

Finally, the performance with respect to the Tag-and-Probe method was evaluated. As described before (see section 5.3), the Tag-and-Probe is a data-driven method for efficiency estimation, and it is an important performance estimator used in real data. Figure 8(a) shows the Ringer performance and false alarm rate as a function of the RoI transverse energy. RoIs near the crack region ($|\eta| \approx 1.5$) and the end-cap ($|\eta| \approx 2.5$) are excluded. It can be seen that the Ringer performance in the $Z \to ee$ channel is very good for most of the energy range. On average, for the transverse energy range 20 GeV< E_t <50 GeV, the Ringer algorithm achieves an electron detection efficiency of 98.24±2.54%, for a 15.78±10.16% false alarm probability. More statistics is needed to increase the measurement accuracy.

Similarly, the performance for the electron detection based on the Tag-and-Probe method and the false alarm rate as a function of η is shown on figure 8(b). Only RoI's whose respective transverse energy is greater than 20 GeV are considered. It can be seen that the performance

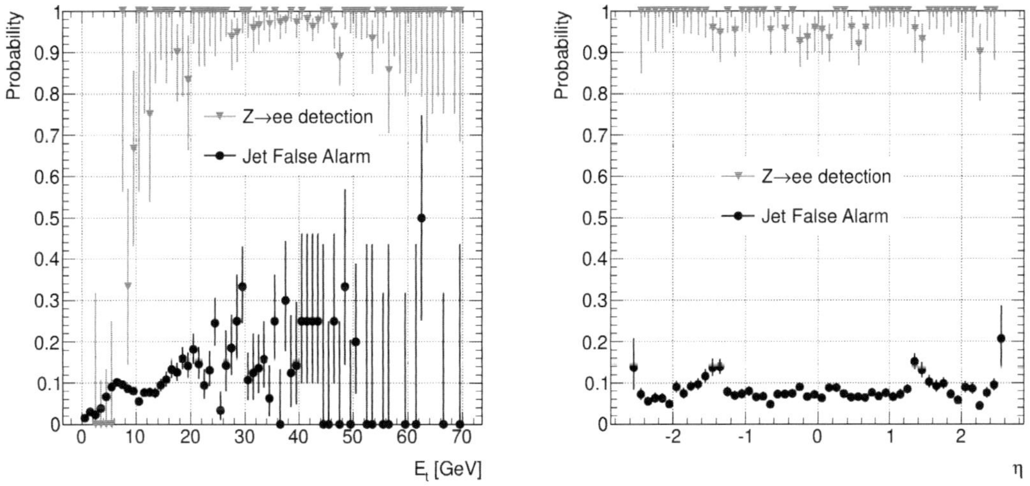

(a) (b)

Figure 8. MLP performance using the and the Tag-and-Probe method as a function of the RoI offline transverse energy (a) and η (b).

over the $Z \rightarrow ee$ channel is almost uniform in η (97.34±2.82% electron detection efficiency against 8.08±2.46% false alarm probability, considering $|\eta| < 1.5$). For the false alarm rejection, the regions around the calorimeter crack ($|\eta| \approx 1.5$) and at the far end of the calorimeter suffer from a higher false alarm probability. This is expected due to the missing information and limited energy resolution in these regions.

6. Conclusions

Results from the ATLAS HLT Ringer algorithm performance were shown, with respect to the implementation in the ATLAS Athena framework, and considering the detection performance in the $Z \rightarrow ee$ channel, including pileup. In addition, a methodology for reducing the neural network complexity, by ranking the input variables according to their relevance, was studied. Finally, the relevance study reduced the MLP input dimension in ~70%, with a small impact on detection performance.

The detailed Valgrind analysis led to an optimized implementation, which became 59% faster, while the total amount of data needed to store the Ringer information was kept low (6.2% of the total used by the electron trigger slice).

The detection performance was evaluated from 2010 14 TeV Monte Carlo simulations. The Ringer algorithm showed high performance also when a data-driven method for performance estimation is considered (Tag-and-Probe). Crosscheck with the offline R_{core} discriminant variable shows good agreement with the MLP response for electrons and jets.

The results presented here show that it is promising to use the Ringer algorithm for electron identification in the trigger system for collider experiments.

Acknowledgements

The authors would like to thank the Brazilian funding agencies CNPq, CAPES, FINEP, FAPERJ and RENAFAE, as well as CERN and the European Union, for the support to this work. We also thank our colleagues from the ATLAS/TDAQ collaboration for the interesting discussions concerning this work.

References

[1] Bishop C M 1994 *Review of Scientific Instruments* **65** 1803
[2] Abramowicz H *et al.* 1995 *Neural network based electron identification in the ZEUS calorimeter* vol 1 p 763
[3] Badalà A, Barbera R *et al.* 2003 *Nucl. Instrum. Methods Phys. Res.* A **502** 503
[4] Bortolotto C, Angelis A D, Groot N D and Seixas J 1992 *Int. J. Mod. Phys.* C **3** 733
[5] Evans L *et al.* 2008 *Journal of Instrumentation* **3** S08001
[6] Prevotet J C, Denby B *et al.* 2003 *Nucl. Instrum. Methods Phys. Res.* A **502** 511
[7] McGlone H M 2009 *Neural Network Analysis in Higgs Search using* $t\bar{t}H, H \rightarrow b\bar{b}$ *and TAG Database Development for ATLAS,* Ph.D. thesis , University of Glasgow (*Preprint* `http://cdsweb.cern.ch/record/1287877/files/CERN-THESIS-2005-062.pdf`)
[8] The ATLAS Collaboration 2008 *Journal of Instrumentation* **3** 1
[9] Kordas K *et al.* 2007 *Nuclear Physics B - Proceedings Supplements* **172** 178
[10] dos Anjos A R, Torres R C, J M de Seixas B C F and Xavier T C 2006 *Nucl. Instrum. Methods Phys. Res.* A **559** 134
[11] Wigmans R 2000 *Calorimetry: energy measurement in particle physics* (Oxford: Clarendon Press)
[12] Zhang H *et al.* 2010 *The ATLAS Liquid Argon Calorimeter: Overview and Performance* vol 1 pp 1–6
[13] Ariztizabal F *et al.* 1994 *Nucl. Instrum. Methods Phys. Res.* A **349** 384
[14] Berge D, Aielli G, Andrei V *et al.* 2007 *Nucl. Instrum. Methods Phys. Res.* A **581** 476
[15] Stelzer J *et al.* 2011 *Journal of Physics: Conference Series* **331** 022026
[16] Cranmer K 2008 *Nuclear Physics B, Proceedings Supplements* **177** 126
[17] The ATLAS Collaboration 2010 Performance of the electron and photon trigger in p-p collisions at a centre of mass energy of 900 GeV Tech. Rep. ATLAS-CONF-2010-022 CERN Geneva
[18] Haykin S 2008 *Neural Networks and Learning Machines* (Prentice Hall)
[19] Nethercote N and Seward J 2007 *Valgrind: A Framework for Heavyweight Dynamic Binary Instrumentation* vol 1 pp 1–12
[20] Riedmiller M and Braun H 1993 *A Direct Adaptive Method for Faster Backpropagation Learning: The RPROP Algorithm* pp 586–591
[21] Torres R C, de Lima D E F, Simas Filho E and Seixas J M 2009 *IEEE Nuclear Science Symposium Conference Record* **1** 530–536
[22] Gruber A *et al.* 1994 *New Computing Techniques in Physics Research* **1** 429
[23] Straessner A and Schott M 2010 *Journal of Physics: Conference Series* **219** 032023
[24] van Trees H 2001 *Detection, estimation, and modulation theory part I* (John Wiley & Sons)

A population-based approach to background discrimination in particle physics

Federico Colecchia
Department of Physics and Astronomy, University College London, Gower Street, London WC1E 6BT, UNITED KINGDOM
Now at Brunel University, Kingston Lane, Uxbridge, Middlesex UB8 3PH, UNITED KINGDOM
E-mail: federico.colecchia@brunel.ac.uk

Abstract. Background properties in experimental particle physics are typically estimated from control samples corresponding to large numbers of events. This can provide precise knowledge of average background distributions, but typically does not take into account statistical fluctuations in a data set of interest. A novel approach based on mixture model decomposition is presented, as a way to extract additional information about statistical fluctuations from a given data set with a view to improving on knowledge of background distributions obtained from control samples. Events are treated as heterogeneous populations comprising particles originating from different processes, and individual particles are mapped to a process of interest on a probabilistic basis. The proposed approach makes it possible to estimate features of the background distributions from the data, and to extract information about statistical fluctuations that would otherwise be lost using traditional supervised classifiers trained on high-statistics control samples. A feasibility study on Monte Carlo is presented, together with a comparison with existing techniques. Finally, the prospects for the development of tools for intensive offline analysis of individual interesting events at the Large Hadron Collider are discussed.

1. Introduction

Background discrimination in particle physics is usually performed by identifing events that are more likely to contain a physics process of interest, the primary goal being rejection of contributions from uninteresting processes that mimic the signal and thus make its extraction and measurement more complicated. Traditional approaches achieve this goal by focussing on entire events, comparing kinematic and topological properties with reference distributions usually obtained from control samples.

This article presents a novel approach that builds on a population-based view of particle physics events, which are treated as mixtures of subpopulations comprising particles originating from different physics processes such as a hard scattering of interest as opposed to background. The main goal is to decompose an input data set by assigning individual particles a probability for them to originate from a given process based on particle-level information.

This is achieved by adapting and applying mixture decomposition techniques [1] that are well established in statistics and that have been used in other disciplines to solve formally-similar problems. In this formulation, events are treated as heterogeneous statistical populations comprising particles whose kinematics reflects the process they originated from.

This contribution describes an initial investigation of the possibility to use mixture model decomposition techniques for background discrimination at the Large Hadron Collider (LHC). The study is based on a sampling algorithm inspired by the Gibbs sampler [2] and by Expectation Maximization (EM) [3] whose goal is to decompose an input data set into collections of particles originating from a hard scattering of interest as opposed to background, mapping individual particles to signal or background on a probabilistic basis. A number of well-established methods and results set a context for this investigation in addition to the Gibbs sampler and to EM, namely (i) other simulation-based methods such as [4], (ii) a more general use of Markov Chain Monte Carlo (MCMC) techniques, recently applied to the study of the Cosmic Microwave Background radiation [5], (iii) a recent renewed interest in Bayesian numerical methods for data analysis in particle physics [6] [7] [8], in addition to (iv) the use of MCMC with reference to specific optimization problems in the field [9].

In this study, the proposed sampling algorithm was used to classify individual particles into signal and background. Results obtained on a collection of ~ 600 simulated particles from a hard scattering and from background are presented and discussed in this article, together with cross-checks on toy Monte Carlo as described in the appendix.

In general, different events in particle physics can look very different from one another even when the underlying physics processes are the same, and statistical fluctuations can be non-negligible in low-statistics data sets. When classification is performed using traditional supervised algorithms, fluctuations are usually not taken into account, since training typically relies on high-statistics control samples. On the other hand, the algorithm presented in this article can estimate properties of signal and background probability density functions (PDFs) from the data: in principle, this makes it possible to use information obtained from a data set of interest to improve on the description of background PDFs obtained from control samples, which do not normally take statistical fluctuations into account.

From a broader perspective, this contribution illustrates a new population-based approach that aims to improve on the description of background PDFs obtained from a high-statistics control sample by using information about statistical fluctuations extracted from a lower-statistics data set: this is done by assigning individual particles a probability for them to originate from signal or background, i.e. by decomposing an input collection of particles into signal and background-associated subpopulations.

2. The sampling algorithm

This approach to background discrimination is presented with reference to the general problem of decomposing a collection of particles from high-energy particle collisions into subpopulations associated with different underlying physics processes and described in terms of different PDFs.

The input data set consists of a mixture of particles, some of which originated from a hard scattering of interest, others from background. Provided that the corresponding subpopulations can be characterized sufficiently well in terms of their kinematic or topological properties, it is in principle possible to ask, for each particle, what the probability is for it to originate from signal as opposed to background. In particular, the proposed algorithm estimates such probabilities by iteratively sampling from subpopulation PDFs.

As opposed to classical mixture models, which typically rely on a parametric formulation requiring the shapes of the subpopulation PDFs to be known a priori, our formulation is based on a more general mixture of the form

$$\sum_{j=1}^{K} \alpha_j f_j(x) \tag{1}$$

where the PDFs f_j satisfy a set of constraints associated with a histogram regularization procedure as outlined in section 3. Subpopulation fractions α_j ("mixture weights") are required to sum to unity, i.e. $\sum_{j=1}^{K}\alpha_j = 1$.

The variable x can correspond to particle pseudorapidity η, a kinematic variable related to the particle polar angle θ in the laboratory frame in terms of $\eta = -\ln(\tan\theta/2)$, or p_T i.e. the transverse momentum of the particle with respect to the beam direction. The subpopulation PDFs f_j are defined in terms of regularized histograms of x, as described in section 3, where the associated constraints imposed on the PDFs are detailed. The symbol φ_j will be used to denote the estimate of the generic subpopulation PDF f_j throughout the text.

The choice of (1) was driven by our previous studies, where assuming a predefined PDF functional form led to significant bias on mixture weight estimates. That bias ultimately related to assuming that PDFs obtained from high-statistics control samples were also appropriate to describe the corresponding probability distributions in a lower-statistics data set. However, statistical fluctuations are sometimes appreciable, and for this reason it is necessary for the model to provide more flexibility if fluctuations in the data set of interest are to be described. While a rigorous treatment may call for the use of nonparametric Bayesian methods [10], which can be used to provide an additional dimension of flexibility to statistical models, it was decided to adopt a simplified intuition-driven approach for this study, in order to avoid introducing additional complications not related to the algorithm itself in this phase of the development.

The histogram regularization procedure described in section 3 can be seen as a simplified version of established methods such as Tikhonov regularization [11], which can be used to impose smoothness constraints to a likelihood maximization problem. From a conceptual point of view, an alternative way of interpreting the model used in this study is as a simplified version of established kernel or wavelet-based techniques, where regularized histograms effectively play the role of a set of basis functions. In the absence of any constraints to the PDFs in the mixture, the statistical model (1) would instead not be well defined, so this is an essential ingredient. Additional remarks about existence and uniqueness of a stationary distribution for the Markov Chain associated with the algorithm in the configuration used for this study will be provided in section 3 after the discussion of the Monte Carlo analysis.

Given the mixture of probability distributions (1) and a set of observations $\{x_i\}_{i=1,...,N}$, the problem of clustering the latter into K groups by probabilistically associating each of them with a distribution of origin has been solved numerically in a Bayesian framework using MCMC techniques. In particular, the Gibbs sampler [1], which directly inspired this work, has been used for this purpose in different disciplines.

The basic pseudocode of the proposed sampling algorithm is reported below. The value of variable v at iteration t is indicated with $v^{(t)}$ throughout.

(i) **Initialization:** Choose $\underline{\alpha}^{(0)} = \{\alpha_j^{(0)}\}_j$ and $f_j^{(0)}=\varphi_j^{(0)}$, $j = 1, ..., K$ as described in section 3.

(ii) **Iteration t:**

  (a) Generate the allocation variables $z_{ij}^{(t)}$, $i = 1, ..., N$, $j = 1, ..., K$ based on probabilities $P(z_{ij}^{(t)} = 1|\alpha_j^{(t-1)}, \varphi_j^{(t-1)}, x_i)$ proportional to $\alpha_j^{(t-1)} f(x_i|\varphi_j^{(t-1)})$. The quantity $z_{ij}^{(t)}$ equals 1 when observation i is mapped to distribution j at iteration t, and 0 otherwise. In general, the variables $z_{ij}^{(t)}$ depend both on the mixture weights α_j and on the estimates φ_j of the subpopulation PDFs from the previous iteration.

  (b) Generate $\underline{\alpha}^{(t)}$ from the probability density function of $\underline{\alpha}$ given $\underline{z}^{(t-1)} = \{z_{ij}^{(t-1)}\}_{ij}$, $\rho(\underline{\alpha}|\underline{z}^{(t-1)})$. Knowledge of which particles are mapped to process j at iteration $t-1$ makes it possible to generate the subpopulation fractions $\underline{\alpha}$ at iteration t.

  (c) Obtain an updated estimate of the subpopulation PDFs from the data $\underline{x}$ based on knowledge of which particles are mapped to subpopulation j at iteration $t-1$. Details are provided in section 3.

A specific choice for the function ρ and a way to obtain updated estimates of the subpopulation PDFs f_j are described in section 3 with reference to the Monte Carlo study.

The central idea of the algorithm is the following: the better the observations $\{x_i\}_i$ are mapped to the subpopulations $j = 1, ..., K$, the more accurate the estimates of $\rho(\underline{\alpha}|\underline{z})$ and of the subpopulation PDFs φ_j. Once some correct values of z_{ij} are found, $\rho(\underline{\alpha}|\underline{z})$ and φ_j begin to roughly reflect the correct distributions, which in turn leads to additional correct mappings z_{ij} to be found at subsequent iterations.

The above pseudocode corresponds exactly to the Gibbs sampler, where updated estimates of subpopulation PDFs are obtained at each iteration, as indicated at step (c). On the other hand, when step (c) is removed from the pseudocode, particles are mapped to signal and background based on the subpopulation PDFs provided at initialization, and the algorithm is then more akin to EM. Throughout the paper we will refer to the two versions of the algorithm with step (c) included or not in the pseudocode as "unconstrained sampler" and "constrained sampler", respectively.

The primary objective of this article is to study the use of the proposed sampling technique in different configurations in order to:

(i) Obtain estimates φ_j of the subpopulation PDFs from the input data set.

(ii) Estimate the subpopulation weights α_j. In the context of this study, this corresponds to estimating the fractions of background and signal particles contained in the input data set.

(iii) Assign individual particles a probability for them to originate from a given process based on the subpopulation PDFs estimated at step (i) as opposed to relying exclusively on high-statistics templates. In the context of this study, this allows classification of individual particles into signal and background.

3. Monte Carlo study

The algorithm was applied to a Monte Carlo data set generated using Pythia 8.140 [12] [13], obtained superimposing $gg \to t\bar{t}$ signal events from pp interactions at $\sqrt{s} = 14$ TeV to soft QCD interactions, so called Minimum Bias events, in order to simulate background. The signal process was chosen in order to illustrate the use of the algorithm for background discrimination at the particle level. Further studies will be needed in order to extend these results beyond the initial investigation presented in this article, and to assess the potential of population-based techniques for background discrimination in the context of specific analyses at the LHC.

The sampler was run over a collection of charged particles with 2 GeV/c $< p_T <$ 5 GeV/c, and individual particles were assigned a probability for them to originate from signal as opposed to background based on their η and p_T values.

The pseudocode of the algorithm used for this application is shown below. Subscripts sig and bkg relate to signal and background, respectively.

(i) **Initialization:** Set $\alpha_{bkg} = \alpha_{bkg}^{(0)} = 0.5$, $f_j = \varphi_j^{(0)}$. Initial conditions for the estimates $\varphi_j^{(0)}$ of the subpopulation PDFs $f_j^{(0)}$ are given by regularized η and p_T distributions from the high-statistics control sample, as described in section 3.1.

(ii) **Iteration** t**:**

  (a) Generate $z_{ij}^{(t)}$ for all particles ($i = 1, ..., N$) and distributions ($j = 1, 2$ corresponding to background and signal, respectively) according to $P(z_{ij}^{(t)} = 1|\alpha_j^{(t-1)}, \varphi_j^{(t-1)}, x_i) \propto \alpha_j^{(t-1)} f_j(x_i|\varphi_j^{(t-1)})$, where $\alpha_1 = \alpha_{bkg}$, $\alpha_2 = 1 - \alpha_{bkg}$.

  (b) Set $\alpha_j^{(t)} = \sum_i z_{ij}^{(t-1)}/N$, $\forall j$. This corresponds to the simplest choice of setting $\rho(\alpha_j|\underline{z}^{(t-1)}) = \delta(\alpha_j - \sum_i z_{ij}^{(t-1)}/N)$ for the probability density function of $\underline{\alpha}$ given $\underline{z}$.

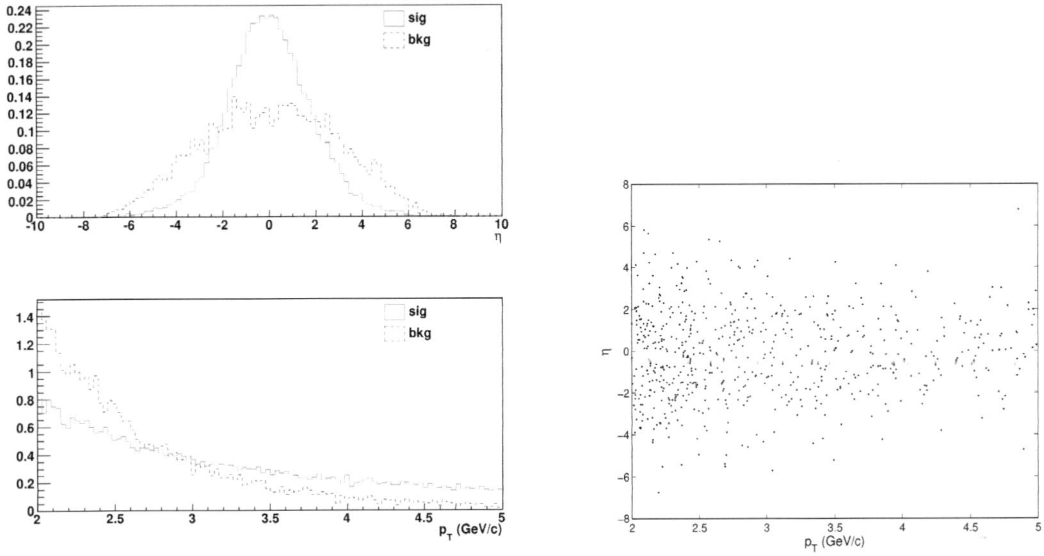

(a) (b)

Figure 1. (a) Generator-level η and p_T distributions for signal (solid green histograms) and background particles (dashed red histograms) with 2 GeV/c $< p_T <$ 5 GeV/c from the high-statistics control sample. The distributions correspond to a total number of $\sim 33,000$ particles and are normalized to unit area. (b) The corresponding two-dimensional distribution.

(c) Obtain updated estimates of the subpopulation PDFs by regularizing the η and p_T distributions corresponding to particles mapped to the relevant subpopulation at iteration $t-1$, i.e. based on $z_{ij}^{(t-1)}$.

In general, the functions f_j are the joint PDFs for η and p_T corresponding to background ($j = 1$) and signal ($j = 2$) particles. This study is restricted to charged particles with 2 GeV/c $< p_T <$ 5 GeV/c, which makes it possible to neglect the correlation between η and p_T as a first approximation. For this reason, the joint PDFs take the form $f_{sig/bkg} = f^{(\eta)}_{sig/bkg} f^{(p_T)}_{sig/bkg}$, and obtaining updated estimates of the subpopulation PDFs reduces to regularization of one-dimensional histograms, as described in the following.

As for the number of iterations to be used with the algorithm, no rule is documented in the statistics literature with reference to related techniques, and the choice is generally problem-dependent. The number of iterations was set to 1,000 in this study, and probabilities were averaged over the last 100 iterations. Runs were also performed letting the sampler run for a longer time: the algorithm exhibited a relatively fast convergence on the data set analyzed, and no gain was found in choosing a higher number of iterations. Moreover, multiple runs were performed under different initial conditions in order to make sure that the algorithm converged. In particular, the initial conditions for the subpopulation PDFs were perturbed by using different initial conditions for the fits to the high-statistics distributions from the control sample. Similarly, the generation parameters in the toy Monte Carlo study were varied around their nominal values by $\pm 10\%$, with no appreciable difference in the results.

In order to obtain initial conditions $\varphi_j^{(0)}$ for the subpopulation PDFs, a Monte Carlo data set was used containing a total of about 33,000 charged particles in the kinematic range 2 GeV/c $< p_T <$ 5 GeV/c. In addition to estimation of $\varphi_j^{(0)}$, this high-statistics control sample was also used to guide the histogram regularization procedure as described in the following. Figure 1 (a) shows the η and p_T distributions for signal and background particles (solid green and dashed red histograms, respectively).

As anticipated, one of the goals of the sampler is to estimate the background PDFs from the input collection of particles. In other words, the algorithm classifies particles into signal and background without relying exclusively on predefined background templates: the background PDFs that are estimated by the algorithm are thus expected to reflect the specific background conditions in the input data set, which can be different from the average conditions obtained from a high-statistics control sample.

The algorithm basically tries to uncover a signal and a background subpopulation in the input collection of particles based on the data and on initial conditions on the subpopulation PDFs. The results presented in this article relate to an input data set comprising 636 charged particles in the kinematic region 2 GeV/c $< p_T <$ 5 GeV/c, out of which 481 originate from a signal hard process and 155 from background, corresponding to a fraction of background particles of $\sim 24\%$. The total number of particles in the input data set is in line with typical charged particle multiplicities at the LHC as of July 2011.

The signal and background η and p_T distributions corresponding to the Monte Carlo input data set used in this study are shown in figure 2. The solid green (dashed red) histograms in the upper panel display the signal (background) η distributions, normalized to unit area. The corresponding p_T distributions are given in the lower panel. It is worth noticing that some of these distributions are appreciably different from the corresponding ones obtained from the control sample due to statistical fluctuations, as expected. In particular, the background η distribution exhibits two modes that are shifted with respect to zero, while the corresponding distribution from the control sample is centered around zero.

In order to illustrate the histogram regularization procedure used in this study, figure 3 shows an example of the η distribution of particles mapped to the signal subpopulation at a given iteration of the algorithm. As opposed to assuming a functional form for the PDF and fitting a function to the histogram, the histogram is regularized i.e. the subpopulation PDF is obtained by means of spline interpolation of the histogram contents, as further discussed in the following. The superimposed curve on the figure corresponds to the regularized histogram, and is used by the algorithm as an estimate of the corresponding subpopulation PDF.

As anticipated, this approach gives the algorithm more flexibility in terms of estimating the subpopulation PDFs from the input data set with respect to our previous attempts relying on a predefined PDF functional form, while still leading to a well-defined target distribution for the associated Markov Chain.

3.1. Regularization

Step (c) in the pseudocode shown in section 3 requires iterative PDF updates based on current mapping of individual particles to different subpopulations. This operation is performed when the algorithm is operated in unconstrained mode, as discussed below.

As anticipated, it was decided to adopt a simplified statistical model for the purpose of this investigation, while at the same time providing enough flexibility for the algorithm to be able to describe fluctuations. In the context of this study, this was done by performing spline interpolation of one-dimensional η and p_T histograms. As previously mentioned, this can be seen as a simplified version of established regularization techniques, for instance as a way to use a priori information about the underlying distributions in order to get rid of spurious oscillatory components (such methods have been analyzed in detail in particle physics in order to develop unfolding procedures, with a view to "removing" detector effects from observed distributions, see e.g. [14]).

The complexity of the histogram regularization procedure that was used in this study was

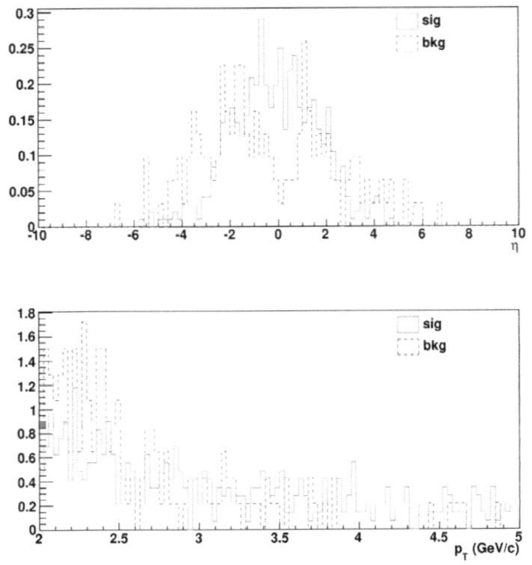

Figure 2. Particle η and p_T distributions from the Monte Carlo input data set used in this study. Solid green and dashed red histograms correspond to signal and background, respectively. Distributions are normalized to unit area.

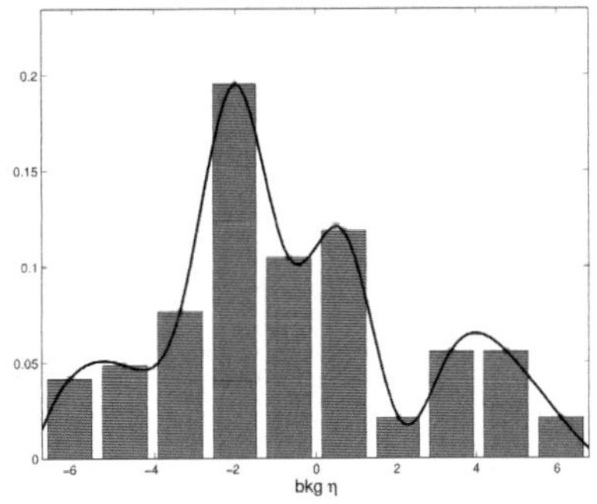

Figure 3. Example of the pseudorapidity η distribution of particles mapped to the background subpopulation at a given iteration of the algorithm. The superimposed curve is the result of the regularization procedure described in the text, and is used by the algorithm as an estimate of the corresponding subpopulation PDF.

intentionally kept minimal in order to avoid the introduction of additional complications that might obscure the response of the algorithm at this stage of the development. Further studies will be needed in order to understand in detail how results are affected by the regularization procedure.

In the context of this investigation, a priori information about signal and background PDFs was obtained from the high-statistics control sample. When subpopulation PDFs are updated iteratively during the execution of the sampler, i.e. when the algorithm is operated in unconstrained mode, a "regularization window" is applied to the η histograms in order to get rid of outliers: in other words, a spline interpolation of the histogram contents is obtained using only the part of the histogram that lies between a minimum and a maximum η value, which leads to extreme statistical fluctuations on the tails of the distribution to be excluded. It is worth noticing that assuming a functional form for the PDF and fitting it to the histogram would effectively produce a similar result, i.e. it would reduce the impact of outliers on the estimated PDF. However, as anticipated, that approach was observed to introduce significant bias in previous studies, and was thus abandoned in favor of the statistical model presented in (1), where subpopulation PDFs are defined as the output of a histogram regularization procedure without reference to any predefined functional form, but still subject to regularization constraints. Figure 4 shows signal (solid green) and background (dashed red) η distributions from the high-statistics control sample, with superimposed arrows indicating the regularization window. The maximum $|\eta|$ value was set to $|\eta| = 5$ ($|\eta| = 7$) for signal (background).

On top of this, again with a view to getting rid of extreme statistical fluctuations when regularizing histograms, boundary conditions were introduced on the η and p_T PDFs, constraining the value of f_j to points chosen based on control sample distributions: in particular, the signal (background) η PDF was constrained to 0 when $|\eta| > 5$ ($|\eta| > 7$), and signal (background) p_T PDFs were constrained at 2 GeV/c and 5 GeV/c to 0.7 (1.2) and 0.1 (0)

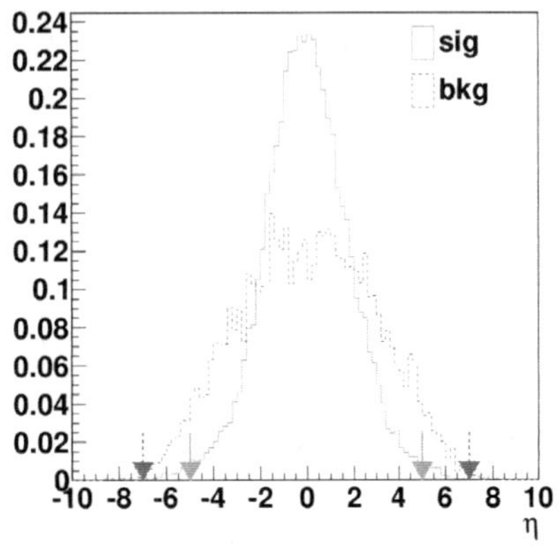

Figure 4. Illustration of the regularization window used in this study. The histograms correspond to signal (solid green) and background (dashed red) η distributions from the high-statistics control sample. The position of the solid green and dashed red arrows correspond to the regularization window: at each iteration of the algorithm, only the part of the distribution that lies between the arrows is used for spline interpolation, which makes the results robust against outliers. Additional details are given in the text.

(see figure 1(a)).

Results were found to be stable with respect to reasonable changes to the above regularization contraints.

3.2. Choice of configuration

As anticipated, the algorithm can be operated in unconstrained or constrained mode, depending on whether step (c) in the pseudocode given in section 3 is included or not.

As already pointed out in section 2, the algorithm can process an input collection of particles in order to obtain one or more of the following results:

(i) Estimate the subpopulation PDFs from the input data set.

(ii) Estimate the fraction of particles associated with a given process in the input data set, e.g. the fraction of background particles.

(iii) Assign individual particles a probability for them to originate from a given process, such as a hard scattering of interest as opposed to background, based on subpopulation PDFs estimated at step (i) as opposed to relying on predefined templates that reflect average background conditions.

Depending on the objective, it may be appropriate to run the algorithm in different modes.

For instance, the histogram regularization procedure that is used here to obtain iterative estimates of the subpopulation PDFs when the algorithm is operated in unconstrained mode inherently leads to bias on mixture weights, because imposing a regularization window changes the number of particles that are mapped to signal or background at a given iteration. For this reason, it may be more appropriate to use a different approach in order to estimate the fraction of background particles.

One option is described below:

(a) The constrained sampler is first used to estimate the mixture weights. In the two-subpopulation scenario described in this study, goal (ii) above corresponds to estimating the fraction of background particles contained in the input data set. The initial conditions for the mixture weights are $\alpha_1^{(0)} = \alpha_2^{(0)} = 0.5$, corresponding to no prior knowledge about the fraction of background particles in the input sample. The subpopulation PDFs are kept fixed at the estimates provided by the high-statistics control sample. The corresponding results are described in section 3.2.1.

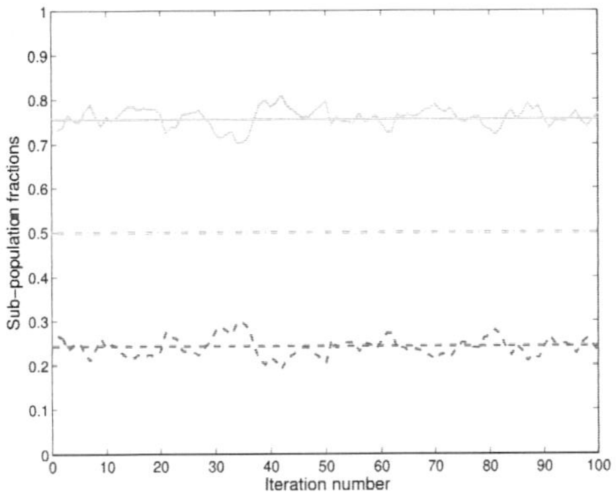

Figure 5. Mixture weights obtained running the constrained sampler on the Monte Carlo input data set. Results from the last 100 iterations are shown. The solid green (dashed red) curve denotes the estimated fraction of signal (background) particles. The solid green (dashed red) horizontal line indicates the true value for signal (background) from the simulation, and the dash-dot line corresponds to the initial conditions for the mixture weights.

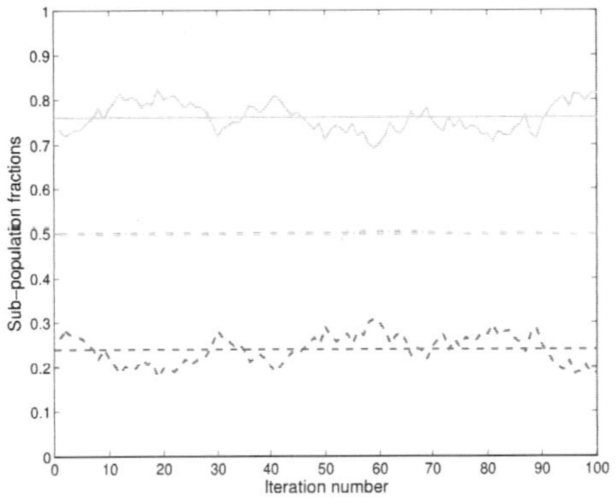

Figure 6. Mixture weights obtained running the constrained sampler on a toy Monte Carlo data set, as described in the text. Results from the last 100 iterations are shown. The solid green (dashed red) curve corresponds to the estimated fraction of signal (background) particles. The solid green (dashed red) horizontal line indicates the true value for signal (background) from the toy Monte Carlo, and the dash-dot line corresponds to the initial conditions for the mixture weights.

(b) The algorithm is then run again on the input data set in unconstrained mode, i.e. subpopulation PDFs are now updated at each iteration, starting from initial conditions corresponding to regularized distributions from the high-statistics control sample. However, mixture weights are kept fixed at the results from the previous step.

It is worth noticing that the algorithm differs from a proper Gibbs sampler in both cases.

As for assigning individual particles in the input data set a probability for them to originate from signal as opposed to background, the most appropriate approach may again depend on the specific application. In general, probabilities may be assigned directly using the unconstrained sampler at step (b) above, as done in this study, or an additional run of the algorithm in constrained mode may alternatively be added after the previous two, with fixed PDFs given by the estimates from step (b). Further studies will be necessary in order to better understand the classification performance of the algorithm in different configurations and to guide this choice.

Results obtained running the constrained an unconstrained sampler as described above on the Monte Carlo input data set used in this study are reported and discussed in the following sections.

3.2.1. Constrained sampler As anticipated, the algorithm in constrained mode was primarily used in this study in order to estimate mixture weights, i.e. the fraction of background particles in the input data set. Figure 5 shows the corresponding estimates over the last 100 iterations. The solid green and dashed red curves correspond to the estimated fractions of signal and background particles, respectively. The solid green (dashed red) horizontal line indicates the signal (background) true value from the simulation, while the dash-dot line corresponds to the initial conditions for the mixture weights.

Additional runs on toy Monte Carlo samples were performed as a cross-check, as described in the appendix. In particular, figure 6 displays the estimated mixture weights obtained by running the constrained sampler on a toy Monte Carlo data set with subpopulation PDFs kept fixed at truth information.

3.2.2. Unconstrained sampler The unconstrained sampler was used in this study in order to estimate the signal and background PDFs from the input data set, while keeping the mixture weights fixed at the results obtained from the previous run of the algorithm in constrained mode.

Figure 7 shows the subpopulation PDFs estimated by the algorithm on the Monte Carlo input data set. The curves correspond to the output of the histogram regularization procedure averaged over the last 100 iterations, superimposed to the true distributions (histograms). The η (p_T) distributions are displayed in the top (bottom) plots, figures on the left-hand (right-hand) side correponding to background (signal). All distributions are normalized to unit area. The bottom panel in each figure shows the corresponding ratio between the relevant subpopulation PDF estimated by the algorithm and truth information.

The figure illustrates a distinctive characteristic of the proposed algorithm as opposed to well-established techniques. As already pointed out, the background η distribution in the Monte Carlo data set used in this study differs appreciably from the corresponding distribution obtained from the control sample, as shown by the two modes around $\eta \simeq -2$ and $\eta \simeq 1$ in the figure, as opposed to the symmetric distribution centered around $\eta \simeq 0$ that is obtained from the high-statistics data set. As it can be seen, the sampler was able to identify with reasonable performance the presence of such deviations with respect to the control sample templates. Such properties of the background PDFs are specific to the data set under investigation, and could not have been extracted using traditional supervised classification techniques, since those would have relied on predefined background templates those features would have been absent from in the first place.

In conclusion, although this study shows that traditional techniques can in some cases outperform this algorithm in terms of classification performance as discussed in the following, the primary objective of the proposed approach is not to improve on existing methods in terms of classification performance, but rather to extract information about statistical fluctuations from a data set of interest.

In addition to obtaining data-driven estimates of the subpopulation PDFs in unconstrained mode, one of the goals of the algorithm in this application is to assign individual particles a probability for them to originate from a given process, such as a hard scattering of interest as opposed to background. In this study, the latter probabilities were obtained from the same unconstrained run of the algorithm that provided the PDF estimates shown in figure 7. In general, other choices are possible, such as performing an additional run of the algorithm in constrained mode with subpopulation PDFs kept fixed at the estimates shown in figure 7, as previously mentioned. Detailed studies will be necessary in order to understand the implications of different choices before population-based tools for background discrimination can be applied to physics analysis at the LHC.

An initial comparison of the classification performance of the algorithm in the configuration chosen for this study with the corresponding performance of existing techniques is described in section 3.3.

The probabilities returned by the algorithm were validated by comparing the true kinematic distributions with the corresponding ones for particles with $P_{sig} > 0.5$, P_{sig} being the estimated probability for a given particle to originate from the signal process, averaged over the last 100 iterations. Results are shown in figure 8, where histogram bars indicate the η distribution for particles with $P_{sig} > 0.5$ and stars correspond to true distributions.

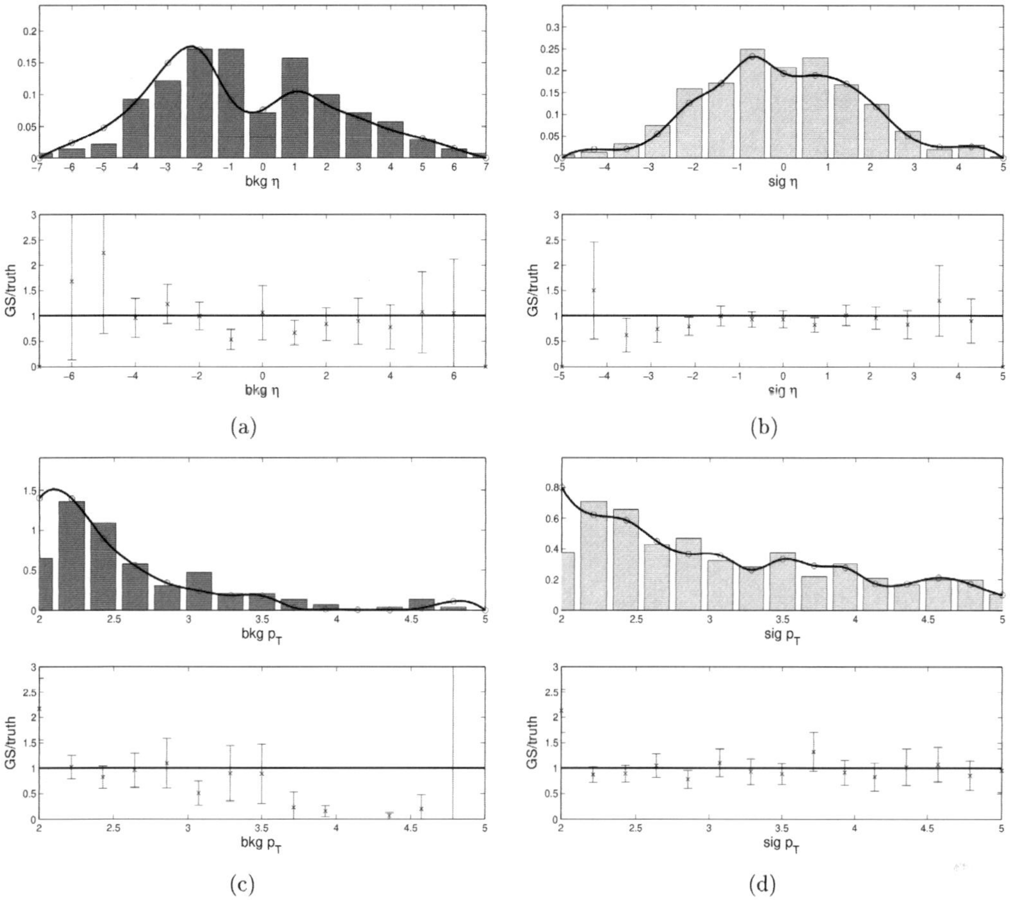

Figure 7. Subpopulation PDFs estimated by the unconstrained sampler on the Monte Carlo input data set used in this study, averaged over the last 100 iterations. (a) Background η. (b) Signal η. (c) Background p_T. (d) Signal p_T. In each subfigure, the upper panel shows truth information (histogram bars) superimposed to the result of the regularization procedure averaged over the last 100 iterations (curve). The lower panels display the ratio between subpopulation PDFs estimated by the algorithm and the corresponding truth-level information.

Table 1. Average numbers of pile-up particles (second column) expected at different LHC instantaneous luminosities (first column) [17]. A 25 ns bunch crossing is assumed. The third column reports the corresponding ratios between the number of background and signal particles observed in the kinematic region considered in this study. These estimates were used to generate the curves shown in figure 10, as described in the text.

$L(\mathrm{cm}^{-2}\mathrm{s}^{-1})$	$\langle n_{PU} \rangle$	N_{bkg}/N_{sig}
10^{33}	2.3	0.1
10^{34}	23.0	0.9
10^{35}	230.0	4.4

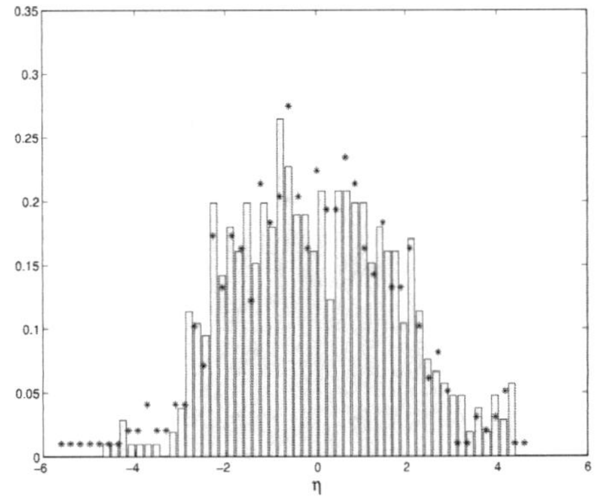

Figure 8. Comparison between the η distribution for particles with $P_{sig} > 0.5$ (histogram bars) and the corresponding distribution from truth (stars). Additional information is given in the text.

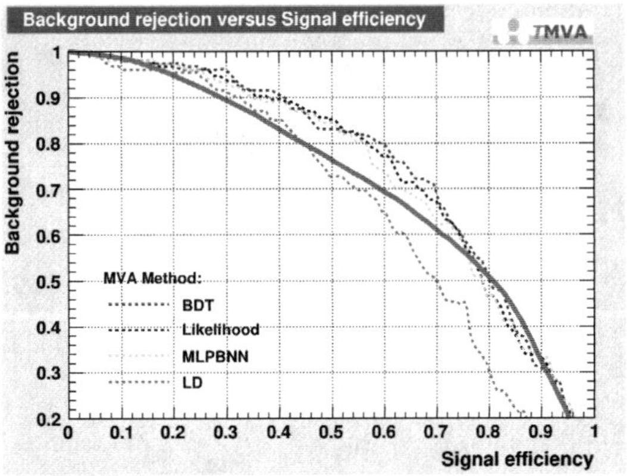

Figure 9. Comparison between the ROC curve obtained using the unconstrained sampler, shown here as background rejection rate as a function of signal efficiency, and the corresponding curves obtained using existing supervised classification techniques from TMVA, as described in the text. The solid red line is the curve from the unconstrained sampler corresponding to the last 100 iterations. The other curves correspond to TMVA algorithms, namely Boosted Decision Trees (dashed blue), Naive Bayes classification (dashed black), the Neural Network-based classifier MLPBNN (dashed green), and Linear Discriminant (dashed red). Additional information is given in the text.

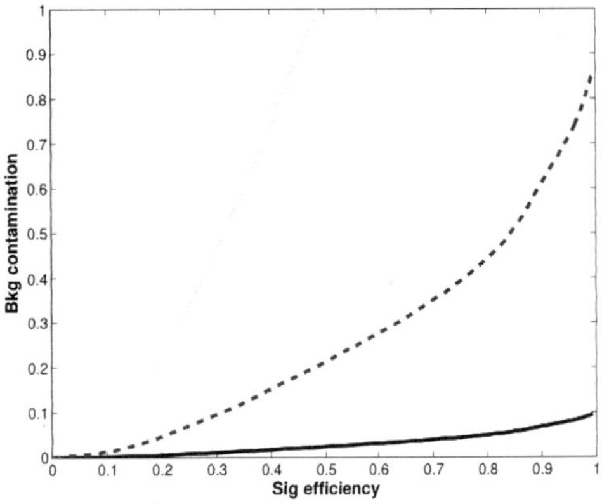

Figure 10. Background contamination as a function of signal efficiency at different LHC instantaneous luminosities (10^{33} cm^{-2}s^{-1} solid black, 10^{34} cm^{-2}s^{-1} dashed blue, 10^{35} cm^{-2}s^{-1} dotted green). Background contamination is defined as number of misclassified background particles normalized to number of signal particles. Misclassification probabilities correspond to the ROC curve from the sampler in figure 9. Additional information is given in the text.

3.3. Classification performance

Operating the sampler as presented in this article is equivalent to using it as a binary classifier. Its performance can thus be quantified using the Receiver Operating Characteristic (ROC) curve, which displays true-positive as a function of false-positive probability. The area under the curve is a number between 0 and 1: the higher its value, the better the classifier is able to discriminate between the two categories (signal and background in this case). The ROC curve of a random classifier would be a straight line along the main diagonal on the true-positive vs false-positive plane ("chance diagonal" [15]).

Figure 9 shows a comparison between the ROC curve obtained using the unconstrained sampler on the Monte Carlo input data set used for this study and the corresponding curves from different supervised multivariate classification methods using TMVA [16] V04-01-00. The curves are displayed using an equivalent representation in terms of background rejection rate as a function of signal efficiency[1] . The dashed lines refer to different TMVA methods[2], namely Boosted Decision Trees (dashed blue), Naive Bayes classification (dashed black), the Neural Network-based classifier MLPBNN (dashed green), and Linear Discriminant (dashed red).

The solid red line corresponds to the proposed algorithm. The figure suggests that classification performance of the sampler is similar to that of existing supervised methods, although other methods perform better in terms of ROC curve on the data set used in this study. However, the advantage of the proposed sampling algorithm with respect to existing methods is not in terms of improved classification performance, but instead relates to estimating features of the background distributions that reflect statistical fluctuations in the data, which is generally not possible using established supervised classifiers trained on control samples.

It may also be useful to provide a more precise idea of the background rejections and signal efficiencies that can be achieved using the proposed algorithm corresponding to different LHC instantaneous luminosities[3]. Figure 10 shows estimates of background contamination as a function of signal efficiency at three different LHC instantaneous luminosities. Background contamination is defined as the number of misclassified background particles normalized to the number of signal particles, and is calculated by rescaling the abscissa of the ROC curve by the ratio between the number of background and signal particles in the kinematic region considered in this study, as given in table 1[4]. The abscissa of the ROC curve in fact corresponds to false-positive rate, i.e. to the probability for a background particle to be misclassified as signal, and multiplying it by the ratio between the number of background and signal particles provides the desired result. The three curves in figure 10 correspond to instantaneous luminosities of 10^{33} cm^{-2}s^{-1} (solid black), 10^{34} cm^{-2}s^{-1} (dashed blue), and 10^{35} cm^{-2}s^{-1} (dotted green).

3.4. Convergence issues

A remark is necessary with regards to the convergence properties of the Markov Chain associated with the proposed sampling algorithm in the form presented in this article. The proposed technique is here justified primarily based on the results it provides, and based on its ability to extract additional information related to statistical fluctuations from a data set of interest. This is to be compared with the description of background distributions obtained using control samples, which, despite its level of precision, usually only reflects average background conditions and does not take statistical fluctuations into account.

Although the statistical model (1) may be questioned from a theoretical point of view and a more rigorous approach based on Bayesian nonparametric methods may be required, the model presented here in practice leads to a well-defined target distribution for the algorithm to sample

[1] Based on our previous terminology, "background rejection rate" is equivalent to $1 - P_{bkg\to sig}$, and "signal efficiency" corresponds to $P_{sig\to sig}$, where $P_{bkg\to sig}$ ($P_{sig\to sig}$) is the probability for a background (signal) particle to be mapped to the signal subpopulation.

[2] The algorithms were run using the high-statistics control sample for training and the same collection of particles the sampler was run on for testing.

[3] The expected average number of pile-up interactions, i.e. the expected average number of primary vertices in the events, is here taken as a measure of background activity for illustrative purposes.

[4] The average numbers of pile-up interactions at different LHC instantaneous luminosities are taken from [17], and correspond to a 25 ns bunch crossing.

from. As anticipated, this is primarily due to the constraints associated with the histogram regularization procedure adopted in this study, which effectively restricts the search space and leads to the existence of a well-defined stationary distribution for the Markov Chain. This was also verified explicitly by using flat distributions as initial conditions for the subpopulation PDFs, making sure that reasonable estimates of the PDFs were still obtained by the sampler.

3.5. Dependence on initial conditionus

One more issue that is worth discussing is dependence of results on initial conditions. The ability to reach the equilibrium distribution regardless of the starting point is a defining feature of Markov Chains. Throughout this study, it has been verified that the initial conditions on the subpopulation PDFs can be perturbed without altering the final resuls. Results are actually independent of the PDF initial conditions well beyond the deviations that are normally expected given the high level of precision with which initial PDF estimates are generally obtained from control samples.

The similarity of the PDFs estimated by the sampler with the PDF initial conditions obtained from the control sample should not be mistaken for a limitation of the proposed method, but should rather be seen as a defining feature. Although incorporating a more rigorous formulation of the statistical model will be important for the method to be developed further, it should be noticed that one of the goals of the algorithm is to improve on knowledge of background PDFs obtained from high-statistics control samples by extracting additional information about statistical fluctuations from a data set under study. For this reason, the PDFs estimated by the sampler will normally be similar to the initial PDFs, and the associated Markov Chain will generally exhibit a relatively-fast convergence by construction.

3.6. Concluding remarks

The possibility to estimate features of signal and background distributions from a data set under study is a distinctive characteristic of the proposed method as compared to existing multivariate approaches such as those available in ROOT [18] with TMVA. Although established techniques in some cases provide better classification performance on the data set analyzed in this study, as shown by the comparison in figure 9, existing methods are in general unable to describe features of the background distributions that are not already encoded in the training sample. And since the latter typically corresponds to a high-statistics control sample, this usually leads to statistical fluctuations in the input data set being neglected.

This technique has been investigated with the prospective goal of developing novel methods for intensive offline analysis of individual interesting events at the LHC, and more generally in particle physics. Data analysis in the field in fact often results in a number of candidate events that may contain a signal process of interest. Traditional methods perform background subtraction based on fixed templates that typically provide a precise description of average background properties. However, this approach normally leads to neglecting features of background distributions due to statistical fluctuations that may be present in the candidate events of interest even though those features cannot be spotted from background templates obtained from control samples. Developing dedicated tools for background subtraction based on event-level templates taking fluctuations into account may then lead to improved background subtraction and to lower systematic uncertainties. This aspect will be the subject of future studies, as will quantification of the impact of the algorithm in a realistic analysis environment.

It is also worth noticing that, from a conceptual point of view, the proposed population-based approach is in a sense based on a similar phylosophy as particle flow analysis, which has been increasingly used in particle physics [19], in that the focus is on individual particles inside events. However, the prospective objective of the proposed technique is different, and concentrates on extracting from the data event-level background templates that take statistical fluctuations into account.

Efforts to eliminate noise in event-by-event analysis of high-energy multiparticle production are reported in the literature, most notably with reference to the study of dynamical fluctuations in heavy-ion collisions, where the notion of "event-by-event fluctuations" was introduced [20], e.g. for mean transverse momentum or mean transverse energy measurement. In the context of such studies, the focus is e.g. on analytically obtaining moments that can be used to eliminate statistical fluctuations from the data with a view to extracting information about the underlying dynamics [21]. Although those studies are conceptually related to the prospective goal of the approach presented in this article in that they aim to subtract noise from individual events, they are fundamentally different. First of all, [21] requires fluctuations to be Poissonian, while this method works under more general conditions. Moreover, one of the novel aspects of this work is the idea of concentrating on individual particles inside events, reformulating background discrimination in terms of a classification problem at the particle level. In other words, the emphasis of this work on a new population-based view of particle physics events is an important aspect that distinguishes the proposed approach from previous efforts.

As a concluding remark, it should also be noted that the iterative nature of the algorithm leads to a disadvantage with respect to established multivariate algorithms in terms of execution time. However, the running time of the sampler corresponding to 1,000 iterations on the Monte Carlo input data set used in this study was $\sim$ 20 s on a 2 GHz Intel Processor with 1 GB RAM, so still reasonable for offline use. In any case, given the parallelization potential of the sampler, which is a consequence of a similar property of the Gibbs sampler as pointed out in [2], improvements may be possible in this respect, for example using commodity Graphics Processing Units (GPUs) that have been used extensively both in particle physics and in other disciplines for compute-intensive applications.

4. Conclusions and outlook

This contribution has presented an initial investigation of a novel approach to background discrimination in particle physics that builds on a population-based view of events from high-energy particle collisions. Collections of particles are treated as mixtures of subpopulations associated with different physics processes, and sampling techniques related to statistical mixture decomposition models are used to assign individual particles a probability for them to originate from a hard scattering of interest as opposed to background. This application of the proposed sampling algorithm to a classification problem at the particle level has been pursued with the prospective goal of developing a suite of tools for extraction of background properties from individual interesting events at the LHC, and more generally in particle physics. For instance, a major objective is to obtain estimates of PDF shapes from the data without relying exclusively on templates from high-statistics control samples and without assuming predefined functional forms.

This study has highlighted strengths and limitations of the algorithm operated in different configurations. In general, systematic uncertainties associated with the use of the algorithm will have to be evaluated in the context of a given analysis.

Detailed understanding of how classification performance in different configurations compares to existing techniques will also require further study, as will the possible development of subsequent versions optimized in terms of execution time, building on the inherent parallelizability of the algorithm.

As anticipated, the total number of particles in the Monte Carlo input data set used in this study is in line with typical charged particle multiplicities at the LHC corresponding to operating conditions as of July 2011. For this reason, the results presented in this article are a promising starting point for futher development, with a view to building dedicated software tools for offline analysis of individual interesting events at the LHC.

Acknowledgments

The author wishes to thank the High Energy Physics Group at the Department of Physics and Astronomy, University College London, for their kind hospitality that created the conditions for this study to be completed. Particular gratitude goes to Prof. Jonathan M. Butterworth for his precious comments on this project. The author also wishes to thank Prof. Trevor Sweeting at the Department of Statistical Science, University College London, for his feedback, and Dr. Alexandros Beskos at the same department for fruitful discussions. Particular gratitude also goes to the Department of Astronomy and Theoretical Physics at Lund University for their kind hospitality when this project was started, and particularly to Prof. Carsten Peterson and to Prof. Leif Lönnblad for their advice and for fruitful discussions.

Appendix: Toy Monte Carlo studies

Results from the Monte Carlo study described in section 3 were cross-checked on toy Monte Carlo data sets. Samples of ~ 600 signal and background particles were generated according to η and p_T distributions similar to those obtained using Pythia. Particle η and p_T were generated independently: Gaussian PDFs centered at zero with standard deviations comparable to those observed in Monte Carlo were used for η, and p_T values were generated based on polynomial PDFs in the range 2 GeV/c $< p_T <$ 5 GeV/c parametrizing the corresponding Monte Carlo distributions.

Additional cross-checks were performed by varying toy Monte Carlo generation parameters by $\pm 10\%$ with respect to their nominal values, in order to make sure that results did not depend on a specific parameter choice. The algorithm was also run on different numbers of particles in the input data set, with no appreciable changes to the results.

Journal of Physics: Conference Series **368** (2012) 012031 doi:10.1088/1742-6596/368/1/012031

References

[1] Marin J M 2005 *Handbook of Statistics* eds Dey D and Rao C R (Elsevier North-Holland) chapter 16, pp 459-503
[2] Geman S and Geman D 1984 *IEEE T. Pattern Anal.* **6** 721
[3] Dempster A P, Laird N M and Rubin D B 1977 *J. Roy. Statist. Soc. Ser. B* **39**(1):1-38
[4] Tanner M A and Wong W H 1987 *J. Amer. Statistical Assoc.* **82** (398):528-540
[5] Groeneiboom N E 2009 A self-contained guide to the CMB Gibbs sampler (*Preprint* astro-ph.CO:0905.3823)
[6] D'Agostini G 1999 *Bayesian Reasoning in High Energy Physics - Principles and Applications* CERN Yellow Report 99-03
[7] D'Agostini G 2003 *Bayesian Reasoning in Data Analysis: A Critical Introduction* (Singapore:World Scientific Publishing)
[8] Ciuchini M *et al.* 2001 *J. High Energy Phys.* JHEP07(2001)013
[9] Buckley A *et al.* 2006 (Preprint hep-ph/0605048)
[10] Ferguson T 1973 *Ann Stat* **1**(2) 209-30
[11] Tikhonov A N 1943 *Doklady Akademii Nauk SSSR* **39**(5):195-8
[12] Sjöstrand T, Mrenna S and Skands P 2006 *J. High Energy Phys.* JHEP05(2006)026
[13] Sjöstrand T, Mrenna S and Skands P 2008 *Comput. Phys. Comm.* **178**
[14] Hoecker A and Kartvelishvili V 1996 *Nucl. Instrum. Methods* A **372**(3) 469-81 (*Preprint* hep-ph/9509307)
[15] Krzanowski W J and Hand D J 2009 *ROC Curves for Continuous Data* (Boca Raton:Chapman & Hall/CRC Monographs on Statistics & Applied Probability)
[16] Hoecker A *et al.* 2007 PoS(ACAT) **040**
[17] Lockman W 2009 SLUO/LHC Workshop
[18] Brun R and Rademakers F 1996 *Proc. Fifth International Workshop on New Computing Techniques in Physics Research (Lausanne)*
Brun R and Rademakers F 1997 *Nucl. Instrum. Methods* A **389** 81-86
http://root.cern.ch
[19] The ALEPH Collaboration 1995 *Nucl. Inst. Meth.* A **360** 481-506
[20] Voloshin S A, Koch V and Ritter H G 1999 *Phys Rev* C **60** 024901
[21] Jinghua F and Lianshou L 2003 *Phys Rev* C **68** 064904

Semi-supervised anomaly detection – towards model-independent searches of new physics

Mikael Kuusela[1,2], Tommi Vatanen[1,2], Eric Malmi[1,2], Tapani Raiko[2], Timo Aaltonen[1] and Yoshikazu Nagai[3]

[1] Helsinki Institute of Physics, PO Box 64, FI-00014 University of Helsinki, Finland
[2] Aalto University, Department of Information and Computer Science, PO Box 15400, FI-00076 Aalto, Finland
[3] University of Tsukuba, 1-1-1 Tennodai, Tsukuba, Ibaraki 305-8577, Japan

E-mail: mikael.kuusela@aalto.fi, tommi.vatanen@aalto.fi, eric.malmi@aalto.fi

Abstract. Most classification algorithms used in high energy physics fall under the category of supervised machine learning. Such methods require a training set containing both signal and background events and are prone to classification errors should this training data be systematically inaccurate for example due to the assumed MC model. To complement such model-dependent searches, we propose an algorithm based on semi-supervised anomaly detection techniques, which does not require a MC training sample for the signal data. We first model the background using a multivariate Gaussian mixture model. We then search for deviations from this model by fitting to the observations a mixture of the background model and a number of additional Gaussians. This allows us to perform pattern recognition of any anomalous excess over the background. We show by a comparison to neural network classifiers that such an approach is a lot more robust against misspecification of the signal MC than supervised classification. In cases where there is an unexpected signal, a neural network might fail to correctly identify it, while anomaly detection does not suffer from such a limitation. On the other hand, when there are no systematic errors in the training data, both methods perform comparably.

1. Introduction

Machine learning techniques have been extensively used in high energy physics data analysis over the past few decades [1]. Especially supervised, multivariate classification algorithms, such as neural networks and boosted decision trees, have become commonplace in state-of-the-art physics analyses and have proven to be invaluable tools in increasing the signal-to-background ratio in searches of tiny signals of new physics. Such methods work under the assumption that there exists labeled data sets of signal and background events which can be used to train the classification algorithm. These training samples are usually obtained from Monte Carlo (MC) simulation of the new physics process and the corresponding Standard Model background.

Supervised machine learning algorithms are the tools of model-dependent new physics searches. In the beginning of the analysis, one decides to focus on a particular variant of a certain new physics model. This could be for example a certain SUSY parametrization. Events corresponding to this choice are then generated using a MC generator followed by training of a classification algorithm to optimize the yield of such events. In the best case scenario, applying this classifier to real observed data would produce a statistically significant excess over

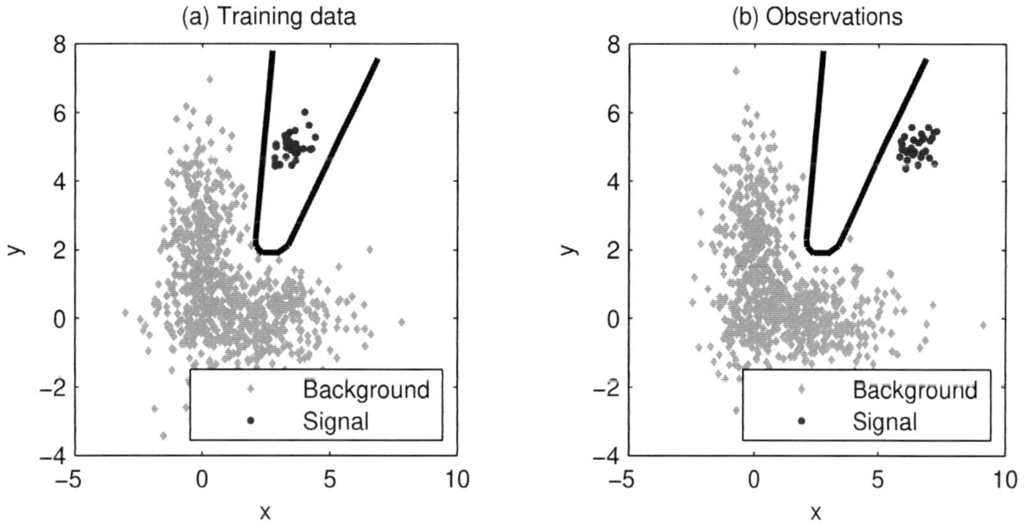

Figure 1. Demonstration of the limitations of model-dependent searches of new physics. When a neural network is trained with the data of figure (a), the classification decision boundary is given by the black curve. But if the training data is systematically inaccurate, the new physics signal could be completely missed in the analysis. For example, all the observations of figure (b) would be classified as background.

the Standard Model background and result in a discovery of new beyond the Standard Model physics.

Unfortunately, this process could go wrong on several different levels. Firstly, there are no guarantees that nature actually obeys any of the existing theoretical new physics models. The solution to the existing problems with the Standard Model could be something that no one has even thought of yet. Even if one of the existing theories is the right way forward, such models often have a large amount of free parameters. Exploring the whole high-dimensional parameter space one combination of values at a time could be a very laborious, time-consuming and error-prone process. Lastly, MC simulation of such processes often requires simplifications and approximations and could be a major source of systematic errors in its own right.

This kind of uncertainties and systematic errors are not well-tolerated by supervised classification algorithms. Figure 1 illustrates the problem. When a neural network was trained using the signal and background data of figure 1(a), we obtained the black decision boundary to separate signal events from background. But what if due to one of the above mentioned problems, the real new physics signal looked like the one in figure 1(b)? The algorithm trained with wrong type of training patterns would completely miss such a signal. The signal events would all be regarded as background even though the data clearly contains a signature of an interesting anomalous process.

While it is possible to alleviate this type of problems up to some extent by tweaking the training data and the classification algorithm of the supervised classifier, a more principled solution is provided by performing the new physics search in a model-independent mode where, instead of trying to provide an experimental validation of a particular beyond the Standard Model physics model, one is simply looking for deviations from known Standard Model physics. The goal of model-independent new physics searches is to be sensitive to any deviations from known physics, whether or not they are described by one of the existing theoretical models. This kind of ideas have been put forward both at the Tevatron and at the LHC. For example, at the

CDF experiment, a combination of algorithms called Vista and Sleuth were used to perform a global model-independent search of new physics with early Tevatron Run II data [2], while an algorithm called MUSiC is currently being used to scan the data of the CMS experiment at the LHC for such deviations [3].

In this paper, we present an alternative algorithm for model-independent searches of new physics based on semi-supervised anomaly detection techniques. As opposed to existing methods, our aim is not only to detect deviations from known Standard Model physics in a model-independent manner but also to produce a model for the observations which can be used to further analyze the properties of the potential new physics signal. Our method is also inherently multivariate and binning-free, while the existing methods are based on exploiting the properties of one-dimensional histograms.

2. Semi-supervised detection of collective anomalies

2.1. Motivation

Our data analysis problem can be formulated as follows: given a labeled sample of known physics expected to be seen in the experiment and the empirical observations, how can we find out if there is a previously unknown contribution among the measurements and, if there is, how can we study its properties? In statistics, such a problem is solved using semi-supervised anomaly detection algorithms [4]. These are algorithms designed to look for deviations from a labeled sample of normal data. The neural network based background encapsulator [5] of the trigger of the H1 experiment at HERA can, for instance, be regarded as an example of such an algorithm.

Unfortunately, existing semi-supervised anomaly detection algorithms can rarely be directly applied to solve the model-independent search problem. This is because they are designed to classify observations as anomalies should they fall in regions of the data space where there is a small density of normal observations. The problem is that in many theoretical models the excess of events corresponding to the new physics signal appears among the bulk of the background distribution instead of the tail regions considered by standard anomaly detection algorithms. We solve the problem by exploiting the fact that if there is a new physics signal, it should occur as a collection of several events which, only when considered together, constitute an anomalous deviation from the Standard Model.

2.2. Fixed-background model for anomaly detection

We start by fitting a parametric density estimate $p_{\mathrm{B}}(\boldsymbol{x})$ to a labeled background sample. This density estimate, which we call the *background model*, serves as a reference of normal data. We conduct a search for collective deviations from the background model by fitting to the observations a mixture of $p_{\mathrm{B}}(\boldsymbol{x})$ and an additional *anomaly model* $p_{\mathrm{A}}(\boldsymbol{x})$:

$$p_{\mathrm{FB}}(\boldsymbol{x}) = (1-\lambda)p_{\mathrm{B}}(\boldsymbol{x}) + \lambda p_{\mathrm{A}}(\boldsymbol{x}). \tag{1}$$

We call this mixture model the *fixed-background model* to emphasize that when modeling the observations, $p_{\mathrm{B}}(\boldsymbol{x})$ is kept fixed which allows $p_{\mathrm{A}}(\boldsymbol{x})$ to capture any unmodeled anomalous contributions in the data. Fitting of both $p_{\mathrm{B}}(\boldsymbol{x})$ and $p_{\mathrm{FB}}(\boldsymbol{x})$ to the data is done by maximizing the likelihood of the model parameters.

Pattern recognition of the anomalies with the fixed-background model enables a variety of data analysis tasks:

(i) *Classification*: Observations can be classified as anomalies using the posterior probability as a discriminant function

$$p(\text{anomaly}|\boldsymbol{x}) = \frac{\lambda p_{\mathrm{A}}(\boldsymbol{x})}{(1-\lambda)p_{\mathrm{B}}(\boldsymbol{x}) + \lambda p_{\mathrm{A}}(\boldsymbol{x})} =: \mathcal{D}(\boldsymbol{x}). \tag{2}$$

An observation $\boldsymbol{x}$ is then classified as an anomaly if $\mathcal{D}(\boldsymbol{x}) \geq T$ for some threshold $T \in [0,1]$.

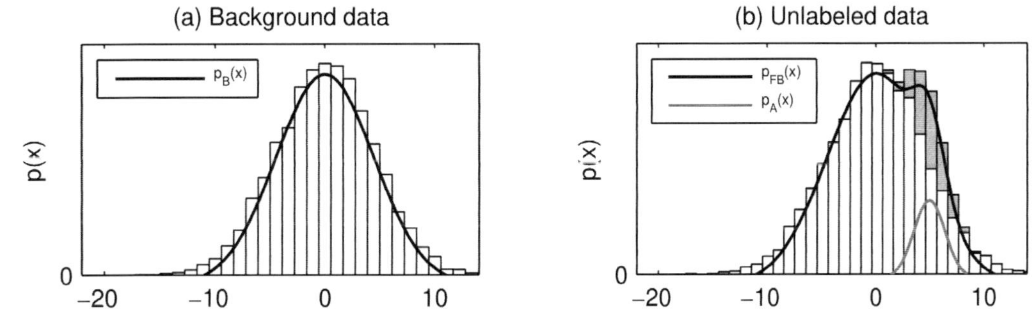

Figure 2. (a) A histogram of background data from a univariate Gaussian distribution and an estimated background model $p_{\mathrm{B}}(x)$. (b) An illustration of the fixed-background model in a univariate case. The histogram shows the unlabeled observations (the gray excess in the histogram denotes the anomalous contribution). The estimated fixed-background model $p_{\mathrm{FB}}(x)$ is shown with a black line and the anomaly model $p_{\mathrm{A}}(x)$ with a gray line.

(ii) *Proportion of anomalies*: The mixing proportion of the anomaly model λ directly gives us an estimate for the proportion of anomalies in the observations. This could be further used to derive a cross section estimate for the anomalous physics process.

(iii) *Statistical significance*: The statistical significance of the anomaly model can be estimated by performing a statistical test for the background-only null hypothesis $\lambda = 0$ using the likelihood ratio test statistic Λ [6]. This enables us to discriminate between statistical fluctuations of the background and real anomalous processes. Following [7], we obtain the distribution of the test statistic using nonparametric bootstrapping. That is, we sample with replacement observations from the background data, fit $p_{\mathrm{FB}}(\boldsymbol{x})$ to this new sample and compute the corresponding value of Λ. This allows us to recover the distribution of Λ under the background-only null hypothesis and hence to compute the significance of the observed collective anomaly.

As a simple illustration of the fixed-background model, figure 2(a) shows a univariate data set of background data generated from a Gaussian distribution and a maximum likelihood Gaussian density $p_{\mathrm{B}}(x)$ estimated using the data set. Figure 2(b) shows a very simple anomalous pattern that can be modeled with a single additional univariate Gaussian. Given a sample contaminated with these anomalies, our goal is to find an optimal combination of the parameters of the anomaly model (μ_{A}, σ_{A}^2) and the mixing proportion λ. The resulting model $p_{\mathrm{FB}}(x)$ is shown with a black line and the anomaly model $p_{\mathrm{A}}(x)$ with a gray line in figure 2(b).

2.3. *Algorithmic details*

We model all the densities in equation (1) using a multivariate Gaussian mixture model [8]

$$p(\boldsymbol{x}|\boldsymbol{\theta}) = \sum_{j=1}^{J} \pi_j \mathcal{N}(\boldsymbol{x}|\boldsymbol{\mu}_j, \boldsymbol{\Sigma}_j). \tag{3}$$

Here J is the number of Gaussian components in the model. The components have means $\boldsymbol{\mu}_j$ and covariances $\boldsymbol{\Sigma}_j$, while π_j are the mixing proportions which satisfy $\sum_j \pi_j = 1$.

Fitting a Gaussian mixture model with J components to the background sample by maximizing its log-likelihood is a standard problem in computational statistics and can be carried out using the iterative expectation-maximization (EM) algorithm [9]. The algorithm proceeds in two steps. In the *expectation step* (E-step), we compute the posterior probabilities for each data point $\boldsymbol{x}_i$ to have been generated by the jth Gaussian component

$$p(z_{ij} = 1|\boldsymbol{x}_i, \boldsymbol{\theta}^k) = \frac{\pi_j^k \mathcal{N}(\boldsymbol{x}_i|\boldsymbol{\mu}_j^k, \boldsymbol{\Sigma}_j^k)}{\sum_{j'=1}^{J} \pi_{j'}^k \mathcal{N}(\boldsymbol{x}_i|\boldsymbol{\mu}_{j'}^k, \boldsymbol{\Sigma}_{j'}^k)} =: \gamma_{ij}^k. \tag{4}$$

Here, $\boldsymbol{\theta}^k$ contains the parameter estimates at the kth iteration and $\boldsymbol{z}_i$ indicates which component generated the ith observation. In the subsequent *maximization step* (M-step), the parameter values are updated according to the following equations:

$$\pi_j^{k+1} = \frac{1}{N}\sum_{i=1}^{N} \gamma_{ij}^k, \quad \boldsymbol{\mu}_j^{k+1} = \frac{\sum_{i=1}^{N} \gamma_{ij}^k \boldsymbol{x}_i}{\sum_{i=1}^{N} \gamma_{ij}^k}, \quad \boldsymbol{\Sigma}_j^{k+1} = \frac{\sum_{i=1}^{N} \gamma_{ij}^k (\boldsymbol{x}_i - \boldsymbol{\mu}_j^{k+1})(\boldsymbol{x}_i - \boldsymbol{\mu}_j^{k+1})^{\mathrm{T}}}{\sum_{i=1}^{N} \gamma_{ij}^k}. \tag{5}$$

The algorithm alternates between these two steps until the log-likelihood is not improving anymore. This gives us an estimate of the background model $p_{\mathrm{B}}(\boldsymbol{x})$.

The next step of the anomaly detection algorithm is to fit to the unlabeled observations the fixed-background model (1) where the anomaly model is a Gaussian mixture model with Q components. When we keep the background model fixed, the free parameters of the full model are the parameters of the anomalous Gaussians, their mixing proportions and the global mixing proportion λ and we seek to find such values of these parameter that their log-likelihood is maximized. This can be done easily by noting that as a mixture of two Gaussian mixtures, the fixed-background model itself is a Gaussian mixture with $J+Q$ components. Hence, we can use the same EM update equations as above to update the free parameters of the model and simply skip the updates of the fixed parameters. The complete update equations based on this idea can be found in the technical report [10], where we also describe a number of heuristics implemented in the algorithm to overcome the issues related to choosing the model complexities J and Q.

3. Demonstration: Search for the Higgs boson

We demonstrate the application potential of the proposed anomaly detection framework by using a data set from the Higgs boson analysis at the CDF detector. It should be stressed that the results presented here should not be regarded as a realistic Higgs analysis. Instead, the goal is to merely demonstrate the performance and the potential benefits of the proposed algorithm.

3.1. Description of the data set

We consider a data set produced by the CDF collaboration [11] containing background events and MC simulated Higgs events where the Higgs is produced in association with the W boson and decays into two bottom quarks, $q\bar{q} \to WH \to l\nu b\bar{b}$. In the data space, this signal looks slightly different for different Higgs masses m_{H}. The goal is to show that semi-supervised anomaly detection is able to identify such a signal without a priori knowledge of m_{H}. More generally, this could be any set of free parameters in the new physics theory under consideration.

Each observation in the data set corresponds to a single simulated collision event in the CDF detector at the Tevatron proton-antiproton collider. We follow the event selection and choice of variables described in [11] for double SECVTX tagged collision events. We also consider an additional neural network based flavor separator from [12], giving us a total of 8 variables to describe each event. To facilitate density estimation and visualization of the results, the dimensionality of the logarithmically normalized data was reduced to 2 using principal component analysis (PCA) [13].

We used 3406 data points to train the background model which was then used to detect signals of 400 data points for masses $m_{\mathrm{H}} = 100, 115, 135, 150$ GeV among another sample of

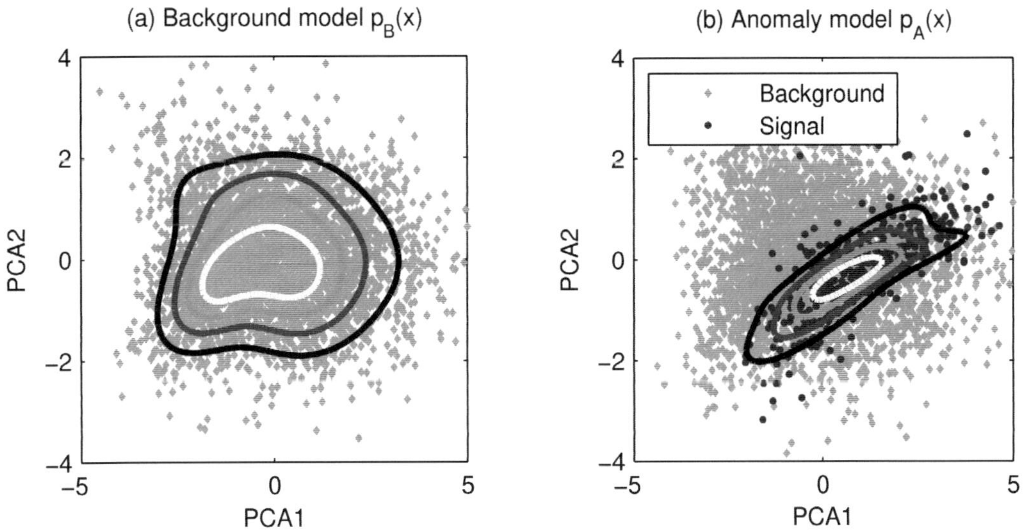

Figure 3. (a) A projection of the WH background data into its two-dimensional principal subspace. The solid lines show contours of the estimated 5-component Gaussian mixture model for the background. (b) A projection of the $m_{\mathrm{H}} = 150$ GeV test data set into the two-dimensional principal subspace. The solid lines show contours of the estimated 2-component Gaussian mixture model for the signal.

3406 observations of background data. Hence, the unlabeled sample contained 10.5 % of signal events. In reality, the expected signal is roughly 5 to 50 times weaker than this, but due to the limited number of background events available, the signal had to be amplified for this demonstration. Based on experiments with artificial toy data reported in [10], we expect to be able to detect signals which contribute only a few percent to the unlabeled sample when we have two orders of magnitude larger background statistics.

3.2. Modeling the Higgs data

We used the cross-validation-based information criterion (CVIC) [14] to select a suitable number of components J for the background model. When a 5-fold cross-validation was performed, the evaluation log-likelihood was maximized with $J = 5$. Figure 3(a) shows contours of the resulting five-component background model in the two-dimensional principal subspace.

We then learned the fixed-background models for the signals with different masses starting with $Q = 3$ anomalous components. The algorithm converged with one anomalous component for $m_{\mathrm{H}} = 100$ GeV and two components for the rest of the masses. The resulting anomaly model for $m_{\mathrm{H}} = 150$ GeV is shown in figure 3(b).

3.3. Anomaly detection results

The statistical significances of the anomaly models were evaluated using the bootstrap technique with 50 000 resamplings. The significances, starting from the lowest mass, were 1.8σ, 2.8σ, 3.1σ and 3.3σ. Hence, in our toy analysis, all the signals would have been significant enough to draw closer attention.

Figure 4(a) shows the receiver operating characteristic (ROC) curves for anomaly detection with different Higgs masses. One can see that regardless of the mass, the algorithm is able to identify the signal with a relatively constant accuracy. The classification results are slightly

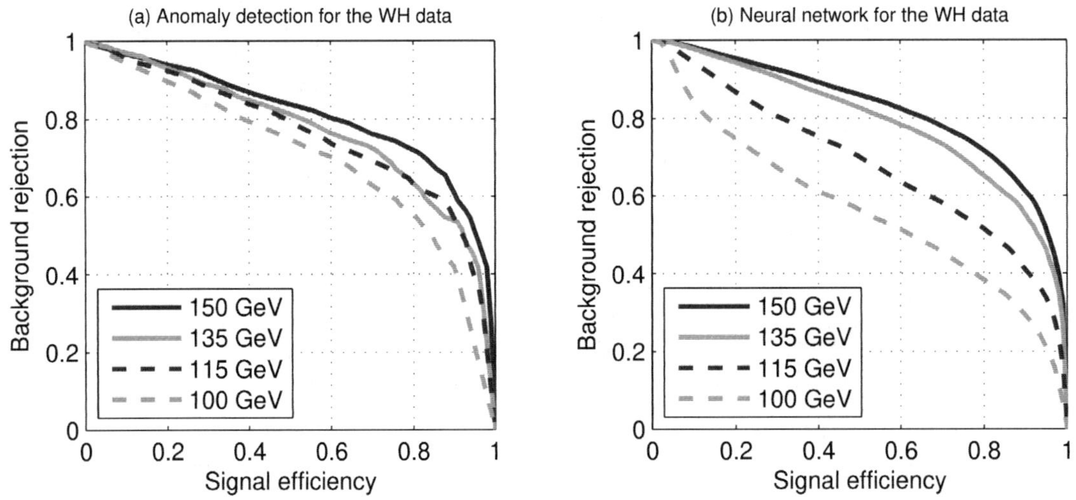

Figure 4. (a) ROC curves for the WH data with different Higgs masses m_H with semi-supervised anomaly detection. The method is able to identify the signal without a priori knowledge of the mass. (b) ROC curves for the WH data for a neural network classifier trained with the 150 GeV signal. The neural network is able to efficiently identify only the signal it has been trained for.

better for the high-mass signal because it lies on a region of the data space with a lower background density. Starting from the lowest mass, the estimated anomaly proportions are $\lambda = 0.100$, 0.121, 0.118, 0.122, which are all in agreement with the correct proportion of 0.105.

We also trained a supervised MLP neural network for each of the mass points to act as an "optimal" reference classifier and compared the ROC curves to the ones obtained with anomaly detection. Figure 4(b) shows the ROC curves for a neural network trained with the 150 GeV signal and tested with all the mass points. When the neural network was tested and trained with the same mass, the ROC curve was comparable to anomaly detection, which shows that the proposed model-independent framework is able to achieve similar performance as model-dependent supervised classification. However, when the neural network was tested with a mass different from its training mass, its performance started to decline. For example the 150 GeV neural network would be likely to miss the 100 GeV signal. On the other hand, anomaly detection is able to successfully identify the signal regardless of the mass. In other words, this experiment shows that supervised classifiers are able to efficiently identify only the signals they have been trained for with potentially severe consequences, while model-independent approaches, such as semi-supervised anomaly detection, do not suffer from such a limitation.

4. Discussion

The semi-supervised anomaly detection algorithm could be used to scan the measurements for new physics signals by focusing on some particular final state which is thought to be especially sensitive to new physics. One could then use the framework to look for deviations from the expected Standard Model background in this final state. If a statistically significant anomaly is found, one could use the fixed-background model to study its properties. Since the anomalous events are likely to lie within the bulk of the background, the best way to reconstruct their properties would be a soft classification based approach [15] where the contribution of a single event to some physical observable is weighted by the posterior class probability of the event. Reconstructing a number of physical spectra for the anomaly should allow one to produce a physics interpretation for the observed deviation. It is likely that most observed anomalies correspond detector effects and background mismodeling. If this is determined to be the case, new cuts could be introduced to isolate such regions or the background estimate could be corrected to account for the anomaly. The analysis could then be repeated iteratively until all anomalies are understood. If at some stage we encounter a significant anomaly which cannot be explained by just adjusting the background estimate, it could be a sign of new physics and should be studied further to see if there is a plausible new physics interpretation for it.

The computational experiments of this paper were carried out using well-understood standard techniques from computational statistics in order to convey the basic idea of semi-supervised anomaly detection as clearly as possible. It is likely that some of the shortcomings of the algorithm could be alleviated by using more advanced computational tools. One of the obvious limitations of the algorithm is the curse of dimensionality. With a reasonable sample size, the algorithm seems to perform relatively well up to three dimensions, but beyond that, the number of observations required to estimate the parameters of the Gaussian mixture models becomes prohibitively large. We demonstrated with the Higgs example that one possible way of solving the problem is dimensionality reduction with PCA or some other dimensionality reduction algorithm. Another possibility would be to consider parsimonious Gaussian mixture models, where the number of parameters is reduced by constraining the structure of the covariance matrices [16]. The current algorithm is also only able to handle anomalies which manifest themselves as an excess over the background. That is, a deficit with respect to the background estimate is not treated properly, although it might be possible to circumvent this restriction by allowing λ to take negative values.

5. Conclusions

We presented a novel and self-consistent framework for model-independent searches of new physics based on a semi-supervised anomaly detection algorithm. We showed using a Higgs boson data set that the method can be successfully applied to searches of new physics and demonstrated the potential benefits of the approach with respect to conventional analyses. To make sure that no new physics signals have been missed by the current model-dependent searches, it would be important to complement them by scanning the collision data with model-independent techniques, one example of which is the proposed anomaly detection framework. We hope that the work presented here helps to revive interest in such techniques among the HEP community by showing that model-independent new physics searches can be conducted in a feasible and practical manner.

Acknowledgments

The authors are grateful to the CDF collaboration for providing access to the Higgs signal and background Monte Carlo samples, to the Academy of Finland for financial support and to Matti Pöllä, Timo Honkela and Risto Orava for valuable advice.

References

[1] Bhat P C 2011 *Annu. Rev. Nucl. Part. Sci.* **61** 281–309
[2] Aaltonen T *et al.* CDF Collaboration 2008 *Phys. Rev. D* **78**(1) 012002
[3] Chatrchyan S *et al.* CMS Collaboration 2008 MUSiC – an automated scan for deviations between data and Monte Carlo simulation CMS-PAS-EXO-08-005 CERN URL `http://cds.cern.ch/record/1152572`
[4] Chandola V, Banerjee A and Kumar V 2009 *ACM Comput. Surv.* **41**(3) 15:1–15:58
[5] Ribarics P *et al.* 1994 *New Computing Techniques in Physics Research III: Proc. 3rd Int. Workshop on Software Engineering, Artificial Intelligence and Expert Systems for High Energy and Nuclear Physics* ed Becks K H and Perret-Gallix D (Oberammergau, October 4-8, 1993: World Scientific) pp 423–428
[6] Knight K 2000 *Mathematical Statistics* (Chapman and Hall)
[7] Wang S, Woodward W A, Gray H L, Wiechecki S and Sain S R 1997 *J. Comput. Graph. Stat.* **6** 285–299
[8] McLachlan G J and Peel D 2000 *Finite Mixture Models* (Wiley-Interscience)

[9] McLachlan G J and Krishnan T 2008 *The EM Algorithm and Extensions* 2nd ed (Wiley-Interscience)
[10] Vatanen T, Kuusela M, Malmi E, Raiko T, Aaltonen T and Nagai Y 2011 Fixed-background EM algorithm for semi-supervised anomaly detection Tech. rep. Aalto University School of Science
[11] Aaltonen T *et al.* CDF Collaboration 2009 Search for the Standard Model Higgs boson production in association with a W boson using 4.3/fb CDF/PUB/EXOTIC/PUBLIC/9997 Fermilab
[12] Chwalek T, Feindt M, Hirschbühl D, Müller T, Renz M, Richter S and Wagner W 2007 Update of the neural network b tagger for single-top analyses CDF/ANAL/TOP/CDFR/8903 Fermilab
[13] Jolliffe I T 2002 *Principal Component Analysis* 2nd ed (Springer)
[14] Smyth P 2000 *Stat. Comput.* **10**(1) 63–72
[15] Kuusela M, Malmi E, Orava R and Vatanen T 2011 *Proc. Diffraction 2010: Int. Workshop on Diffraction in High Energy Physics* ed Capua M *et al.* (Otranto, September 10-15, 2010: AIP) pp 111–114
[16] Fraley C and Raftery A E 2002 *J. Am. Stat. Assoc.* **97** 611–631

Journal of Physics: Conference Series **368** (2012) 012033 doi:10.1088/1742-6596/368/1/012033

Real time algorithms in the ATLAS tau trigger system at 7 TeV center of mass energy

P Kadlečík on behalf of the ATLAS Collaboration

Department of High Energy Physics, Niels Bohr Institute, Copenhagen University, Blegdamsvej 17, 2200 Copenhagen, Denmark

E-mail: kadlecik@nbi.dk

Abstract. The ATLAS hadronic tau trigger plays an important role in many analyses. Among these analyses are searches for H^0, $H^{\pm}$, W' and Z' in the tau decay channel. In order to achieve the needed sensitivity in these measurement it is important to reduce the QCD background, but at the same time to keep the signal efficiency high. Furthermore it is important to understand the trigger efficiency in real data. This paper summarizes the performance of the tau trigger in data collected by the ATLAS detector in 2011.

1. Introduction

The ATLAS tau trigger system is designed to select hadronically decaying taus while keeping the QCD jets at acceptable rates. The hadronic decays of the taus are identified by the track multiplicity as 1-prong (1 track, 76.5% of hadronic decays) and 3-prong (3 tracks, 23.5% of the hadronic decays). Unlike the QCD jets which have in general a high track multiplicity, the track multiplicity of the tau is low. Other characteristics of the taus are their narrowness and isolation. All the characteristics of the taus, narrowness, low track multiplicity and isolation are exploited by the tau trigger.

2. The ATLAS trigger

The Large Hadron Collider (LHC) provides proton-proton collisions at 40 MHz bunch crossing rate. The trigger system is designed to reduce the initial 40 MHz rate to an acceptable value which can be handled by the data acquisition system, by preselecting only the "interesting" events. The ATLAS trigger is a three level trigger system. The first level (L1) is a hardware-based trigger, while the Level 2 (L2) and Event Filter (EF) are software-based triggers. The L2 and EF triggers are often referred to as the Higher Level Trigger (HLT).

At L1, regions with high detector activity, the so called regions of interest (RoI's), are selected, which then seed the HLT. The trigger decision is based on the information contained in the RoI. The goal of L1 trigger is to reduce the initial 40 MHz rate to ~75 kHz. The average processing time at L1 is ~2 μs, which means that a decision on whether or not an event will be passed to the HLT, must be made within this time scale.

At L2 and at the EF sophisticated algorithms are used. The goal at L2 is to reduce the rate from ~75 kHz to ~5 kHz. The mean processing time at L2 is ~40 ms. At L2, full granularity data inside the RoI is analyzed. The final rate reduction from 5 kHz to ~200 Hz is done at the

EF. The events that have passed the L2 are analyzed at the EF using similar algorithms as run offline. The mean processing time at EF is $\sim$4 s.

The trigger bandwidths (L1 = 75 kHz, L2 = 5 kHz, EF = 200 Hz) at every trigger level are shared by a number of trigger items (for example electron, tau, muon, missing energy, calibration and detector performance trigger items). The rate of the trigger items can be controlled by the tightness of the trigger cuts, or by trigger prescales.

3. L1 tau trigger

The L1 tau trigger uses electromagnetic (EM) and hadronic (HAD) trigger towers of the size $\Delta\eta \times \Delta\phi = 0.1\times0.1$ for identifying the tau candidates, where η is the pseudorapidity[1] and ϕ is the azimuthal angle in the ATLAS detector.

The tau object at L1 is selected based on the requirements on the energy in the trigger towers of the size 2×2 in the calorimeter, the so called core region. The core region is surrounded by an isolation ring of 12 trigger towers, allowing to use isolation quantities at L1 if needed. The energy of the L1 tau object is calculated from trigger towers with the size 2×1 in the electromagnetic calorimeter[2], and the energy in the trigger towers with size 2×2 (full core region) in the hadronic calorimeter. The schematic view of the trigger towers can be seen in figure 1.

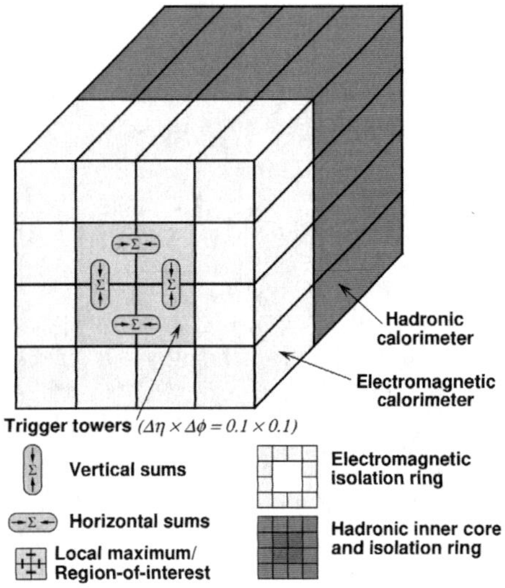

Figure 1. Simplified picture of the EM and HAD trigger towers showing the isolation (yellow and pink color) and the core region (green). The individual EM and HAD trigger towers are corresponding to the EM and HAD calorimeter cells (granularity: $\Delta\eta \times \Delta\phi = 0.1\times0.1$).

The rate of the L1 tau trigger items depends on the energy thresholds applied, and on the instantaneous luminosity. Some L1 items that have been deployed during the 2011 data taking are the L1_TAU6, L1_TAU8, L1_TAU11 and L1_TAU50. The number in the trigger name corresponds to the energy threshold of the L1 tau candidate in GeV. The rates of some of the L1 tau items, as a function of the instantaneous luminosity, are shown in figure 2. The tau trigger items presented in this paper did not require isolation at L1.

4. L2 tau trigger

The rate reduction at L2 is performed by defining and applying a selection based on variables, that are sensitive to the specific characteristics of the taus. The L2 tau candidate is reconstructed from calorimeter cells in the $\Delta\eta \times \Delta\phi$ =0.8×0.8 (Nor) region around the refined L1 direction. One of the most important variables, that uses the feature of the tau being narrower than the

[1] $\eta = -\ln\tan\frac{\theta}{2}$, where θ is the polar angle measured from the beam axis

[2] The electromagnetic calorimeter covers the layers 0 (presampler), 1, 2 and 3 of the ATLAS calorimeter [1].

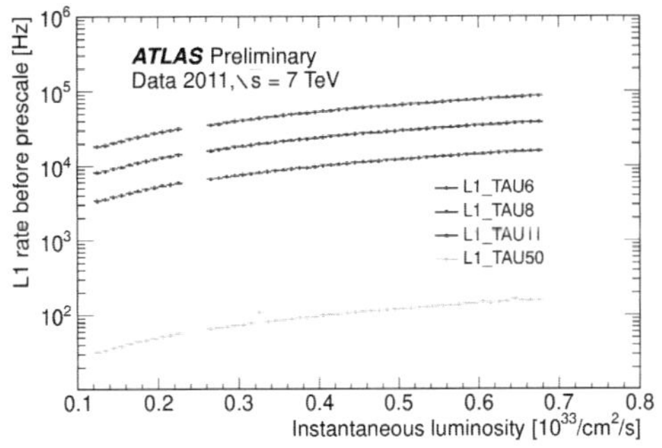

Figure 2. L1 rates before prescale versus the instantaneous luminosity measured by ATLAS for four different L1 tau items that are feeding primary HLT tau chains. The last digits in the L1 item name correspond to the transverse energy, E_{T}, requirement, e.g. a L1 $E_{\mathrm{T}} > 11$ GeV for L1_TAU11. The data has been collected in 14 runs at a centre-of-mass energy of 7 TeV in spring 2011.

typical QCD jet, is the electromagnetic radius, R_{EM}. It is an energy weighted radius in the Nor region calculated from cells in the EM calorimeter:

$$R_{\mathrm{EM}} = \frac{\sum_{Nor} E_{\mathrm{cell}} \Delta R_{\mathrm{cell}}}{\sum_{Nor} E_{\mathrm{cell}}}, \tag{1}$$

where $\Delta R_{\mathrm{cell}} = \sqrt{\Delta\eta^2 + \Delta\phi^2}$. $\Delta\eta$ and $\Delta\phi$ are the distances in the η and ϕ coordinates of the calorimeter cell with respect to the direction of the L2 tau candidate. E_{cell} is the energy of the calorimeter cell in the EM calorimeter.

Another variable that is used at L2 to reduce the rate is the transverse energy, E_{T}, of the L2 tau candidate. The L2 tau E_{T} is calculated from calorimeter cells inside the Nor region, and the suppression against the electronic and pile-up noise is applied.

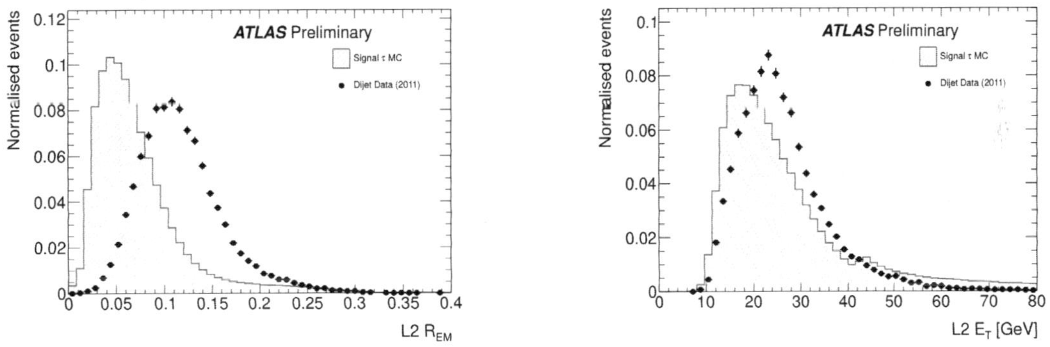

(a) (b)

Figure 3. Distributions of R_{EM} (a), and E_{T} (b) in a cone of radius 0.4 around the tau direction, at L2. The hatched histogram represents the combined contributions from $Z \to \tau\tau$, $W \to \tau\nu$ and $Z' \to \tau\tau$ signal Monte Carlo samples while the black points represent the data. A dijet selection has been applied to select the events in data.

The L2 distributions of R_{EM} and E_{T} for the QCD dijet events estimated from data and taus obtained from the Monte Carlo simulation are shown in figure 3(a) and (b). For the signal Monte Carlo distribution in figure 3, the L2 tau trigger candidates require to be associated with an offline tau candidate with $E_{\mathrm{T}} > 15$ GeV. For the dijet data in figure 3, the EF tau candidate is associated to an offline leading tau candidate with $E_{\mathrm{T}} > 30$ GeV, which is matched to a L1 tau RoI, and furthermore, the event must contain an additional subleading tau candidate with $E_{\mathrm{T}} > 15$ GeV.

5. EF tau trigger

At the EF we define various selection variables that take advantage of e.g. better tracking at the EF than at L2, or a better knowledge about the position of the trigger tau candidate. The EF tau candidate direction is reconstructed from calorimeter cells in the rectangular RoI, $\Delta\eta \times \Delta\phi = 0.8\times0.8$, centered around the L2 tau candidate direction. The EF tau candidate E_{T} is calculated from calorimeter cells within the RoI, centered around the refined EF tau candidate direction, applying electronic and pile-up noise suppression, and hadronic and tau specific energy calibration.

Tracks that are reconstructed at EF from inner detector hits by using a fast Kalman filter [2] are associated to the EF tau candidate within the region $\Delta\eta \times \Delta\phi = 0.4\times0.4$, centered around the L2 tau direction. For a track to be associated to the tau, the track must pass loose quality criteria including the cut on transverse momentum, $p_{\mathrm{T}} > 1$ GeV.

The selection that provides the rate reduction is parametrized as a function of transverse momentum of the EF tau candidate. Depending on the multiplicity of the tracks associated to the EF tau candidate, the selection is optimized for 1-prong (one associated track) and 3-prong (three associated tracks) taus separately.

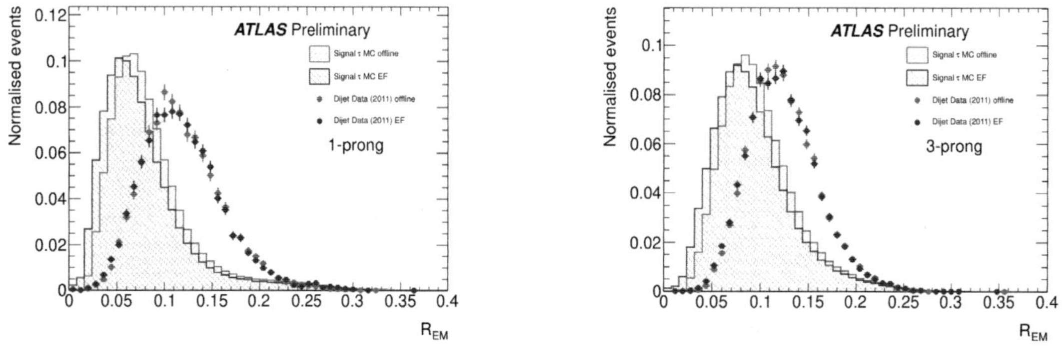

(a) (b)

Figure 4. In figures (a) and (b) are the distributions of radius of energy deposits in electromagnetic calorimeters, R_{EM}, for EF and offline reconstructed tau candidates with exactly 1 (a) and 3 (b) associated track(s). The hatched histogram represents the combined contributions from $Z \to \tau\tau$, $W \to \tau\nu$ and $Z^{'} \to \tau\tau$ signal Monte Carlo samples while the points represent the data. A dijet selection has been applied to select the events in data. A less precise energy calibration applied at EF causes a shift on this distribution with respect to offline.

Calorimeter and track based variables are used to accept tau candidates. The most important calorimeter based variable is R_{EM}, which is defined at EF similarly as at L2, but takes advantage of the refined position of the EF tau candidate. The distributions of R_{EM} for 1-prong and 3-prong tau candidates at the EF are shown in figure 4, for signal Monte Carlo (which is the combination of $Z \to \tau\tau$, $W \to \tau\nu$ and $Z' \to \tau\tau$) and for dijet sample selected from data. In figure 4, the EF tau candidates in data and in Monte Carlo have to be associated with the offline tau candidates with the same requirements as mentioned in the discussion to figure 3. A good agreement between EF and offline variables is observed in both Monte Carlo and in data.

6. Tau trigger performance and analyses using the tau trigger in 2011

The tau trigger in combination with missing E_{T} trigger has been successfully used in 2010 for the observation of the $W \to \tau\nu$ events [3] and in 2011 for the $W \to \tau\nu$ cross section measurement

[4]. In 2011, the tau items have been used in combinations with other trigger items such as with muon trigger (shortly denoted as "mu"), electron trigger ("e"), missing E_T trigger ("xe"), or jet trigger ("j"), in various physics analyses.

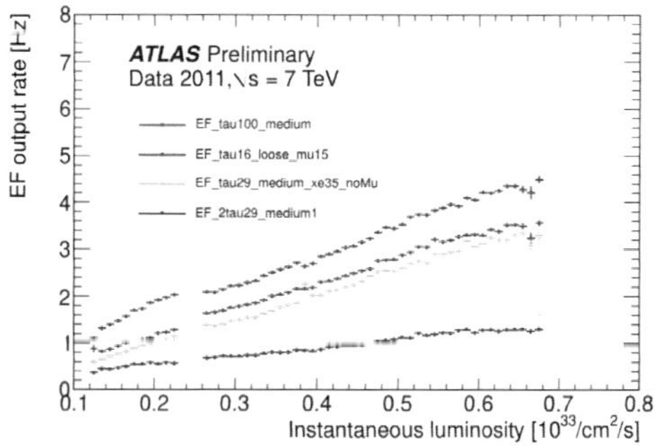

Figure 5. EF output rates versus the instantaneous luminosity of four different trigger items, each including at least one tau trigger.

Figure 5 shows the rates at EF as a function of the instantaneous luminosity in early 2011 runs of four tau (and combined) trigger items: EF_TAU100_MEDIUM, which is a single trigger item that requires at EF a tau with $E_T > 100$ GeV and medium identification requirements; EF_TAU16_LOOSE_MU15, which is a combined trigger that requires at EF a tau with $E_T > 16$ GeV and with loose identification requirements, and a muon with $p_T > 15$ GeV; EF_TAU29_MEDIUM_XE35_NOMU, which is a combined trigger requiring at EF a tau with $E_T > 29$ GeV and medium identification requirements, and missing E_T greater than 35 GeV; 2EF_TAU29_MEDIUM1, which is a combined trigger requiring two taus at EF, both with E_T greater than 29 GeV and both passing medium identification requirements with strict requirement on the number of tracks associated to the tau trigger object. The identification requirements (loose and medium) at the trigger level are similar to the offline tau identification requirements, but applying looser cut values due to the resolution effects.

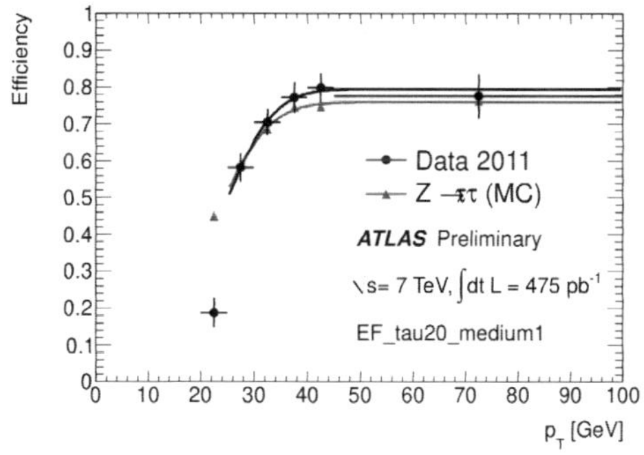

(a)

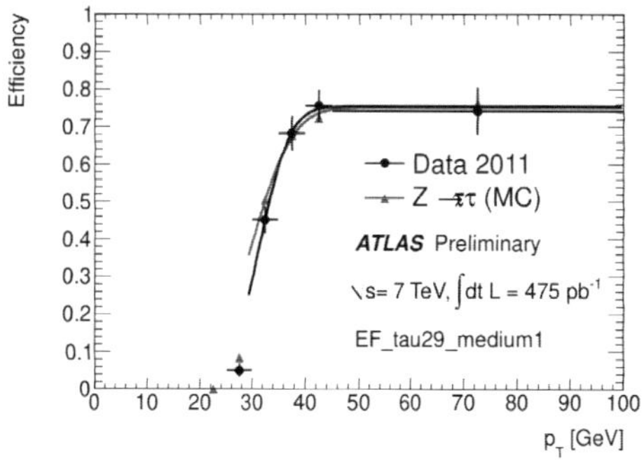

(b)

Figure 6. Efficiency of the EF_TAU20_MEDIUM1 (a) and EF_TAU29_MEDIUM1 (b) trigger with respect to offline reconstructed tau candidates, as a function of the offline p_T. The tau candidates are required to pass medium identification criteria and include a strict requirement on the number of tracks associated to the trigger object.

The measurement of the trigger efficiency in the 2011 data has been done using a tag and probe analysis with $Z \to \tau\tau \to \tau_\mu \tau_{had}$ events. A single lepton trigger has been used to tag the events in data, while measuring the tau trigger efficiency of the probed τ_{had}. The event selection followed closely the selection used in the $Z \rightarrow \tau\tau$ cross section measurement described in [5]. In this way it was possible to measure the trigger efficiency of some tau items in a data driven way. The comparison of the trigger turn on curves in data and in $Z \rightarrow \tau\tau$ Monte Carlo is shown in figure 6(a) for EF_TAU20_MEDIUM1, and in (b) for EF_TAU29_MEDIUM1 item. A good agreement of data and Monte Carlo has been observed.

7. Conclusions

In 2010 the tau trigger has been successfully commissioned and fully operational in 2011. During the 2011 data taking more than 5 fb^{-1} of real data has been collected by ATLAS at $\sqrt{s} = 7$ TeV. With this amount of data we will be able to provide many exciting analyses and the tau trigger in many of those will play an important role. The results we have obtained so far have given confidence about the performance of the tau trigger. The future improvements of the tau trigger involve using a selection more robust against the pile-up and multivariate methods at the EF.

References

[1] The ATLAS Collaboration *et al* 2008 The ATLAS Experiment at the CERN Large Hadron Collider *JINST* **3** S08003.

[2] Fruhwirth R 1987 Application of Kalman filtering to track and vertex fitting. *Nucl. Instrum. Methods* A 262 444

[3] The ATLAS Collaboration *et al* 2010 Observation of $W \rightarrow \tau\nu$ Decays with the ATLAS Experiment *Preprint* ATLAS-CONF-2010-097

[4] The ATLAS Collaboration *et al* 2012 Measurement of the $W \rightarrow \tau\nu$ Cross Section in pp Collisions at $\sqrt{s} = 7$ TeV with the ATLAS experiment *Phys. Lett.* B **706** 276-294

[5] The ATLAS Collaboration *et al* 2011 Measurement of the $Z \rightarrow \tau\tau$ Cross-Section with the ATLAS Detector *Phys. Rev.* D **84** 112006

Track reconstruction and b-jet identification for the ATLAS trigger system

A Coccaro on behalf of the ATLAS Collaboration
Department of Physics, University of Genoa and I.N.F.N., Genoa, Italy
E-mail: andrea.coccaro@ge.infn.it

Abstract. A sophisticated trigger system, capable of real-time track reconstruction, is used in the ATLAS experiment to select interesting events in the proton-proton collisions at the Large Hadron Collider at CERN. A set of *b*-jet triggers was activated in ATLAS for the entire 2011 data-taking campaign and successfully selected events enriched in jets arising from heavy-flavour quarks. Such triggers were demonstrated to be crucial for the selection of events with no lepton signature and a large jet multiplicity. An overview of the track reconstruction and online *b*-jet selection with performance estimates from data is presented in these proceedings.

1. Introduction

The Large Hadron Collider (LHC) has just finished a tremendously successful year of proton-proton collisions with a total integrated luminosity in 2011 of nearly $6\,\mathrm{fb}^{-1}$ and a peak instantaneous luminosity constantly increasing over the year and reaching its maximum at $3.65 \times 10^{33}\,\mathrm{cm}^{-2}\,\mathrm{s}^{-1}$. The enormous collision rate required an efficient and flexible trigger system to select only a few hundred of events per second and to evolve following the continuous improvement of the LHC performance during the year.

The ability to distinguish between jets arising from heavy-flavour quarks (*b*- and *c*-jets) and light jets (jets from *u*-, *d*-, *s*- and gluon jets) already at the trigger level offers significant improvement and increases the flexibility in the triggering strategy to help maximize the output for physics analyses. The strategy adopted in ATLAS is reviewed in these proceedings. A description of the *b*-jet trigger selection together with details on performance and a summary of the trigger menu is given. A set of dedicated triggers for the commissioning and calibration of the online *b*-jet selection is also discussed.

2. The ATLAS detector and its trigger system

The general-purpose ATLAS detector [1] is built for probing the proton-proton collisions at the LHC. ATLAS satisfies all the requirements needed to successfully operate in the challenging LHC environment and consists of several highly granular and hermetic sub-detectors arranged in concentric layers oriented coaxially with respect to the beam line and centered around the nominal interaction point.

The Inner Detector (ID) provides measurements for high precision tracking and vertexing in the central volume of ATLAS. The ID consists of three independent subsystems immersed in a 2 T magnetic field generated by a central solenoid. Closest to the beam line is the pixel detector [2] with approximately 80 million silicon pixels of size $50 \times 400\,\mu\mathrm{m}^2$ arranged in three layers. A silicon micro-strip detector arranged in four nested cylindrical layers surrounds the pixel detector with a stereoscopic geometry for three-dimensional hit position measurements. The outermost part of the ID consists of layers of straw tubes with 2 mm radius providing $R-\phi$ measurements out to a radius of about 111 cm from the beam line. This detector contributes significantly to the track momentum determination and also provides electron identification capabilities through transition radiation measurements.

The ID is designed to reconstruct tracks with a precision of $\mathcal{O}(10\,\mu\mathrm{m})$ in the transverse plane, to reconstruct masses of unstable particles, to identify the primary vertex among multiple pile-up interactions and to find secondary vertices and displaced tracks to tag jets originating from b-quarks (b-tagging).

The ATLAS trigger performs the online event selection in three stages of increasing complexity. The Level-1 trigger (LVL1) is a hardware-based system using coarse information from the calorimeter and dedicated muon trigger chambers. The Level-2 (LVL2) and the final Event Filter (EF) trigger levels are together referred to as the High Level Trigger (HLT). The event selection at both LVL2 and EF is based on software algorithms running on large farms of commercial processors.

LVL2 is the first trigger stage where information from the ID is available. The reconstruction at LVL2 uses specialized fast algorithms to process events in only a few tens of milliseconds on average. At EF, the available processing time is on average a few seconds which enables the use of algorithms developed for the offline reconstruction, but adapted for running in the HLT software framework and tuned for faster execution.

The trigger at LVL1 identifies so called Regions of Interest (RoIs) which are portions of the detector associated to a specific type of physics signature (electron/photon, muon, τ and jet). RoIs are widely used in the subsequent trigger levels to restrict the amount of data read from the detector readout buffers while accessing the most important part of the event in full granularity.

3. Online b-tagging strategy

The identification of a jet arising from b-quark is experimentally possible thanks to several properties: the fragmentation process is hard and the B hadron retains about 70% of the original b-quark momentum, the B hadron mass is relatively high ($> 5\,\mathrm{GeV}$) and the long B-hadron lifetime ($\sim 1.5\,\mathrm{ps}$) corresponds to a measurable decay length of $\sim 3\,\mathrm{mm}$ for a B hadron of $\approx 50\,\mathrm{GeV}$. Thus, impact parameters of charged tracks and properties of a secondary vertex provide good discriminant power and are widely used as input variables for the majority of the b-tagging algorithms.

The reconstruction at HLT starts from the RoIs selected by the LVL1 items and the b-tagging signatures are activated starting from a jet RoI with dimensions equal to 0.8 in η and ϕ. The LVL1 jet trigger is a fixed-size sliding window algorithm that sums energy in projective towers of size $\Delta\eta \times \Delta\Phi = 0.4 \times 0.4$. At the HLT, the sequence of algorithms for b-jet triggers consists of two major steps: the calorimeter jet reconstruction and the b-jet identification. This is a major difference compared to the b-jet trigger configuration in 2010 which was exclusively relying on tracking information at the HLT [3] (b-jet triggers were not actively selecting events in 2010 due to

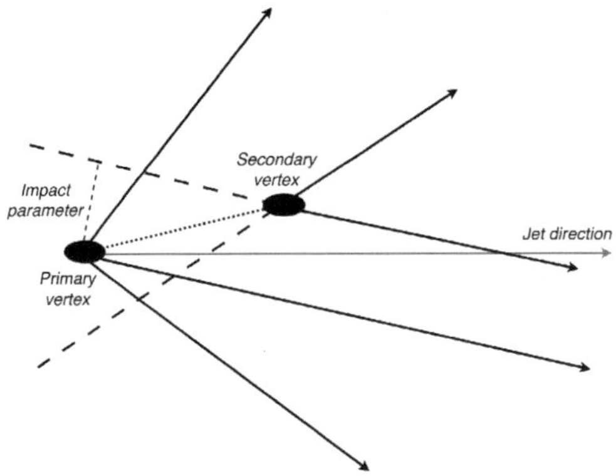

Figure 1: Schematic sketch (not to scale) of a displaced vertex coming from a b-jet with high impact parameter tracks.

the relatively low LHC instantaneous luminosity). Each step includes a set of reconstruction algorithms ending with an hypothesis algorithm to stop the trigger execution as soon as it is clear that it doesn't fulfill the trigger requirement anymore.

Three main features must be evaluated to perform the b-jet selection: reconstruct tracks of charged particles, measure the primary interaction vertex and estimate the b-tagging discriminant variables. This scheme is repeated at both LVL2 and EF.

3.1. Track reconstruction

Online track reconstruction is essential in various trigger signatures (electron, muon, τ and b-jets) and for the online determination of the luminous region in ATLAS. As the various signatures have different requirements, flexible configurations of the pattern recognition are used for ad-hoc optimization purposes.

At LVL2, two fast custom-built algorithms based on conceptually different techniques, a combinatorial and an histogramming approach, are implemented. In both cases, track candidates are fitted using a fast Kalman filter algorithm. Online operations proved that both approaches at LVL2 can be successfully applied to reconstruct tracks in the harsh tracking environment at the LHC [3]. Based on slightly different tunings, the primary algorithm adopted for high-multiplicity signatures such as b-jets was the combinatorial approach while for high-momentum particles, such as muons, the histogramming approach was the default one.

At EF, track reconstruction benefits from sharing the majority of the reconstruction software with the offline but slightly adapted to fulfill the trigger requirements of the average event processing time.

In order to reduce the processing time, the tracking in b-jet triggers is performed in a smaller RoI, half the size in η and ϕ compared to the corresponding LVL1 jet RoI. A dedicated tuning optimized to efficiently reconstruct tracks with a transverse momentum larger than 1 GeV and with a transverse impact parameter up to 3 mm was used in 2011. The track reconstruction efficiency was found to be stable in presence of pile-up and close to 100% if compared to offline track reconstruction in a wide region of the phase space, as shown in figure 2.

3.2. Primary vertex reconstruction

The knowledge of the primary vertex position is an essential ingredient to identify b-jets. A fast algorithm, exploiting a simple histogramming method based on a sliding window, is found to be highly efficient in finding the true primary vertex position along the beam line with a resolution of about 120 μm (100 μm) at LVL2 (EF).

There is no direct attempt in determining the primary vertex position in the transverse plane as a consequence of reconstructing tracks only in particular RoIs (the primary vertex resolution is strongly correlated with the track multiplicity of the vertex which is relatively low in each RoI, as shown in figure 4b). The adopted solution for 2011 was to exploit the online measurement of the luminous region, also called the beamspot, which is estimated in the trigger framework using ad-hoc prescaled triggers. The per-event vertex position is integrated to measure the average position and shape of the beamspot in intervals as short as a few minutes [5]. An example of the distribution of primary vertices used in the extraction of the beamspot parameters is shown in figure 3.

The commissioning and evaluation of a three-dimensional primary vertex finding algorithm for online b-tagging is ongoing and considered for 2012.

3.3. Discriminating variable estimation

The online b-tagging in 2011 is based uniquely on the transverse impact parameter of the reconstructed tracks. The signed impact parameter significance is calculated for each track fulfilling specific b-tagging selection criteria. The sign is defined using the angle between the jet

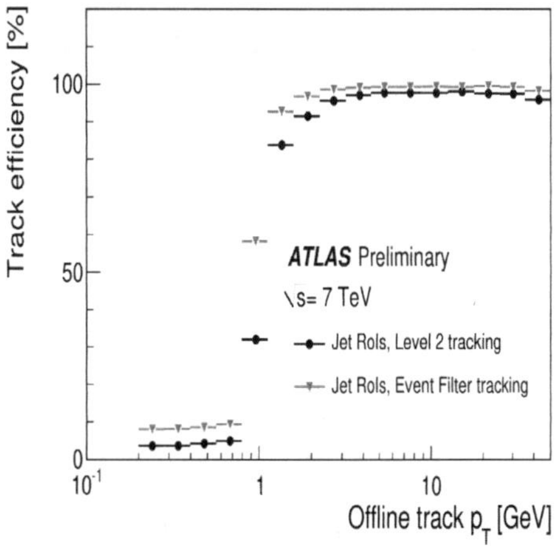

Figure 2: Track reconstruction efficiency at LVL2 and EF as a function of the offline track transverse momentum. Online and offline tracks are matched with a geometrical requirement $\Delta R < 0.1$ and must fulfill basic quality requirements.

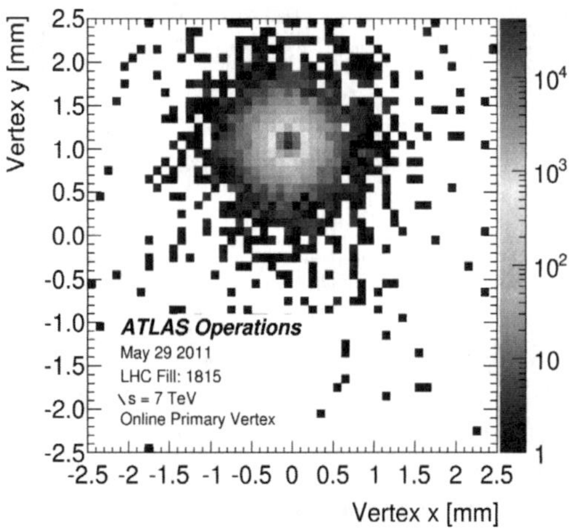

Figure 3: The transverse distribution of primary vertices used to derive the luminous region parameters in the trigger framework. The distribution corresponds to 1 minute of data taking and contains $\mathcal{O}(10^5)$ vertices.

axis and the line between the primary vertex and the point of closest approach of the track, such that it becomes negative when the track appears to originate behind the primary vertex. As a consequence, most of the tracks produced from decays of particles with a long lifetime, such as B hadrons, are signed as positive, as can be seen in figure 4c.

In the significance calculation the transverse impact parameter (d_0) is corrected using the beamspot information and its uncertainty using the measured transverse width of the beamspot,

$$\sigma = \sqrt{(\sigma_{d_0}^2 + \sigma_{BS}^2)} \,. \tag{1}$$

The jet probability method [4], also called JetProb, is the technique adopted in ATLAS for the online event selection in 2011. It computes the probability for a jet to originate from the primary vertex using the signed transverse impact parameter significance of tracks associated to the jet. A probability, P, is assigned to each selected track

$$P = \int_{-\infty}^{-|d_0/\sigma(d_0)|} \mathcal{R}(x)dx \,, \tag{2}$$

where $\mathcal{R}$ is the resolution function for prompt tracks which can be easily measured in data using the negative side of the impact parameter distribution in a light-jet dominated sample. Figure 4 shows the signed transverse impact parameters and other track variables at the EF. A good agreement between data and simulation is obtained. In particular, the b-jet enrichment in the positive tail of the transverse impact parameter distribution is evident. This enrichment is also enhanced thanks to the employment of a track-jet algorithm which greatly improves the jet direction resolution compared to the jet sliding-window approach at LVL1.

The individual track probabilities, P_i, are then combined into a per-jet quantity

$$P_{jet} = P_0 \sum_{i=0}^{N_{tracks}-1} \frac{(-\ln P_0)^i}{i!} \,, \tag{3}$$

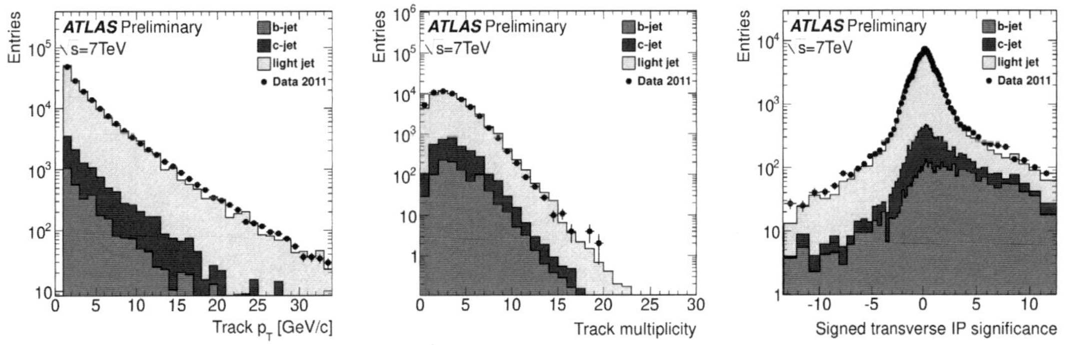

(a) (b) (c)

Figure 4: (a) Track transverse momentum (b) track multiplicity and (c) signed transverse impact parameter significance of reconstructed tracks at the EF trigger level. Tracks are reconstructed starting from a low transverse momentum jet identified by the LVL1 and are requested to fulfill online b-tagging criteria.

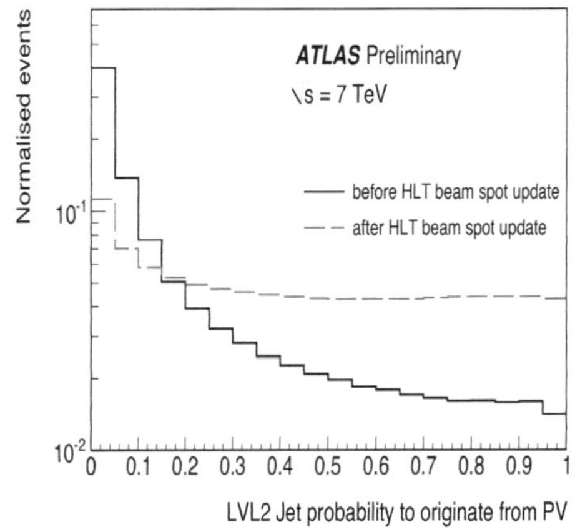

Figure 5: Probability for the LVL2 tracks to originate from the primary vertex in case of updated or outdated knowledge of the beam profile in the transverse plane.

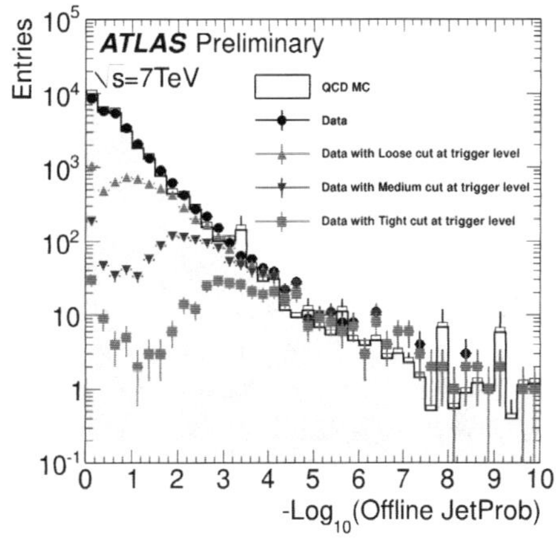

Figure 6: The JetProb distribution at the reconstruction level for all jets and jets b-tagged at the trigger level with three different operating points.

where $P_0 = \prod_i P_i$. The distribution of P_{jet} is uniform between zero and one for jets containing tracks compatible with the resolution function $\mathcal{R}$ and is peaked at zero when highly-displaced tracks are reconstructed in the RoI.

An example of P_{jet} is shown in figure 5 in case of an updated or outdated knowledge of the beamspot parameters. These beamspot parameters are updated for the b-jet trigger if the position along any of the three axes changes by more than 10% or the width along any of the three axes changes by more than 10% or there is a reduction of 50% in the statistical uncertainty of any of the values [5]. These criteria ensure a flat JetProb distribution for light jets and only jets with JetProb values close to zero, the exact cut being dependent on the algorithm instance, will be contributing in filling the bandwidth allocated to b-jet triggers.

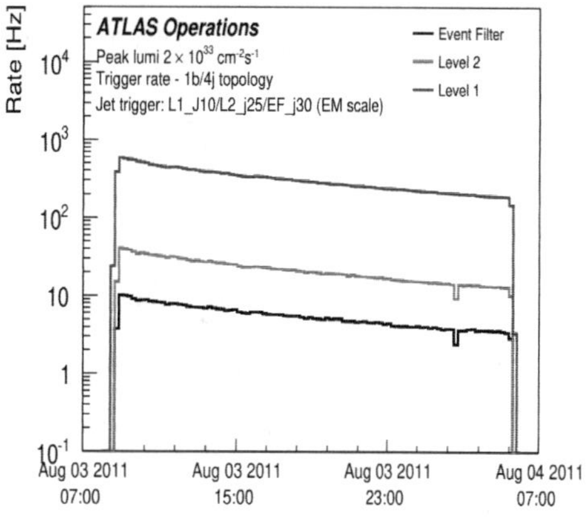

Figure 7: Rate of a b-jet trigger requiring at least four jets in the event and at least one b-tagged jet. The jet thresholds correspond to 10, 25 and 30 GeV at LVL1, LVL2 and EF with energies measured at the electromagnetic scale. The b-jet requirement is applied at LVL2 and EF and is tuned to give 60% efficiency on truth b-jets.

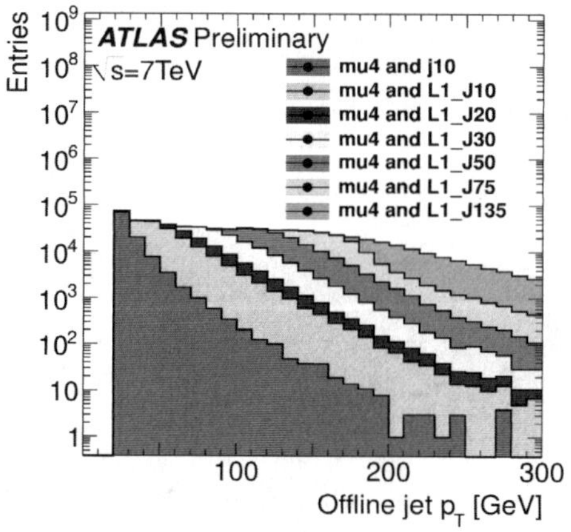

Figure 8: Muon-in-jet trigger distribution as a function of the jet transverse momentum in offline muon-in-jet candidates. Several triggers with various jet thresholds select events in order to collect a sample of muon-in-jet candidates in the entire jet transverse momentum spectrum.

Likelihood-ratio methods based on impact parameter significances are also implemented in the trigger domain but did not contribute to the event selection in 2011. The decision to use the JetProb algorithm was driven by its intrinsic robustness, relying only on the knowledge of the negative transverse impact parameter distribution of prompt tracks. Likelihood-ratio methods exploiting both track parameters and secondary vertex properties are being commissioned and considered for 2012.

4. Trigger menu and impact on offline quantities

The primary goal of b-jet triggers is to lower jet energy thresholds while keeping the output rate at a manageable level. For 2011, three working points were defined corresponding to 70 %, 50 % and 40 % b-jet efficiency on a simulated top quark sample, called the loose, medium and tight instances of the b-tagging algorithm. The light-jet rejection for the medium instance was found to be ~ 40 in the same top quark sample. The medium instance was the most widely used in b-jet triggers for physics analyses.

Figure 6 shows the bias on the offline JetProb weight distribution coming from the online b-tagging. Data and simulation are overlaid in case of no b-jet trigger requirement and the effect on data when a b-jet trigger requirement is included produces a bias on the offline distribution. Offline b-jet candidates are however recorded by the trigger system, a clear indication of the correlation between the online and offline taggers.

The b-jet trigger rejection is evident from figure 7 where the rate at the output of the three trigger levels is shown for a particular b-jet trigger in a run with peak instantaneous luminosity of $2 \times 10^{33}\,\mathrm{cm}^{-2}\mathrm{s}^{-1}$. This particular trigger was designed for the all-hadronic $t\bar{t}$ final state. Other b-jet triggers were implemented and running online for benchmark channels such as $H \to b\bar{b}$, especially useful for the low Higgs mass region still to be fully explored, and for supersymmetric models with no leptons in the final state.

The combination of the online plus offline b-tagging efficiency and mis-tag rate is a key ingredient to fully integrate b-jet triggers into physics analyses. Since detailed tracking information is not perfectly reproduced by the simulation, these measurements are currently being derived with data-driven techniques and are carried out using analyses already employed in ATLAS for the offline b-tagging [6]. These methods mainly explore muon properties from heavy-flavour decays or explicit reconstruction of B hadron decay chains. To successfully apply these methods a set of dedicated triggers collected muons close to jets in 2011, obtaining a sample enriched in B hadrons with a muon in the decay chain. Triggers with different jet thresholds were used to obtain a rather flat jet transverse momentum spectrum, as clearly shown in figure 8. In the second part of 2011, an additional trigger was deployed online to enable the b-jet efficiency measurement using techniques based on the reconstruction of the top quark.

5. Conclusions and outlook

Exploiting b-jet identification at the trigger level allows for lowering jet trigger thresholds. b-jet triggers are a reality in ATLAS since the beginning of 2011. Various triggers were used to collect events for a set of benchmarks channels and contributed to maintain an acceptable output rate. Online tracking is an essential ingredient for the b-jet selection and has shown high performance and robustness over a large range of working conditions.

Physics analyses relying on b-jet triggers are in progress and will benefit from the ongoing measurements of the online and offline b-tagging efficiency and mis-tag rate using data-driven techniques.

References

[1] The ATLAS Collaboration 2008 *JINST* **3** S08003.
[2] Aad G et al 2008 *JINST* **3** P07007.
[3] The ATLAS Collaboration 2012 *Eur. Phys. J.* **C72** 849.
[4] ALEPH Collaboration 1993 *Phys. Lett. B* **313** 535.
[5] Strauss E, These proceedings.
[6] The ATLAS Collaboration 2011 *ATLAS CONF-2011-089*.

Alignment of the ATLAS inner detector

Jike Wang on behalf of the ATLAS Collaboration
Institute of Physics-Academia Sinica
E-mail: jike.wang@cern.ch

Abstract. In order to reach the track parameter accuracy motivated by the physics goals of the experiment, the ATLAS tracking system needs to determine accurately its almost 700,000 degrees of freedom. The demanded precision for the alignment of the silicon sensors is below 10 μm. The implementation of the track based alignment within the ATLAS software framework unifies different alignment approaches and allows the alignment of all tracking subsystems together. The alignment software relies on the tracking information (track-hit residuals) but also includes the capability to set constraints on the beam-spot and primary vertex as well as the momentum measured by the Muon System or the E/p using the calorimetry information. The alignment chain starts at the trigger level where a stream of high p_T isolated tracks is selected online. Also a cosmic ray trigger is enabled while ATLAS is recording collision data, thus a stream of cosmic-ray tracks is recorded exactly with the same detector operating conditions as the normal collision tracks. We will present results of the alignment of the ATLAS tracker using the 2011 collision data. The validation of the alignment is performed using track-hit residuals as well as using more advanced physics observables. The results of the alignment with real data reveals that the attained precision for the alignment parameters is approximately 5 μm.

1. Introduction

The ATLAS experiment [1] is a general purpose experiments at the Large Hadron Collider (LHC) at CERN. It records proton-proton collisions at the center-of-mass energy up to 14 TeV. The Inner Detector (ID) [2] is the most important tracking system occupying a cylindrical volume of 2.1 m in diameter and 6.2 m in length around the interaction point that surrounds the beam-pipe. It is enclosed inside the central superconducting solenoid providing a 2 T axial field. The ID consists of two silicon subsystems, the Pixel Detector and the Semiconductor Tracker (SCT), complemented by the Transition Radiation Tracker (TRT) composed of straw tubes. Figure 1 shows an overview of the ID.

The Pixel detector consists of a barrel region with three cylindrical layers and two symmetric end-caps each containing three disks for tracking in the forward region. All pixel modules (1744 in total) are identical, with a sensor segmented in 50 μm $\times$ 400 μm pixels providing a 2D readout with a resolution of 10 μm and 115 μm in the $r\phi$ and rz coordinates, respectively. The SCT is made of four layers in the barrel region and nine disks in each of the two end-caps. Different types of modules have been installed in the SCT, all with the same components but differing in geometry. Each one of the 4088 modules is composed of two pairs of single-sided silicon micro-strip detectors glued back-to-back with a relative stereo angle of 40 *mrad*. The strip pitch is 80 μm for the barrel and varying from $\sim$ 55 μm to $\sim$ 90 μm for end-cap modules due to their fan-out geometry. The intrinsic resolution in the $r\phi$ and rz coordinates is 17 μm and 580 μm, respectively. The TRT is the outermost and largest of the ID sub-detectors, which is made of

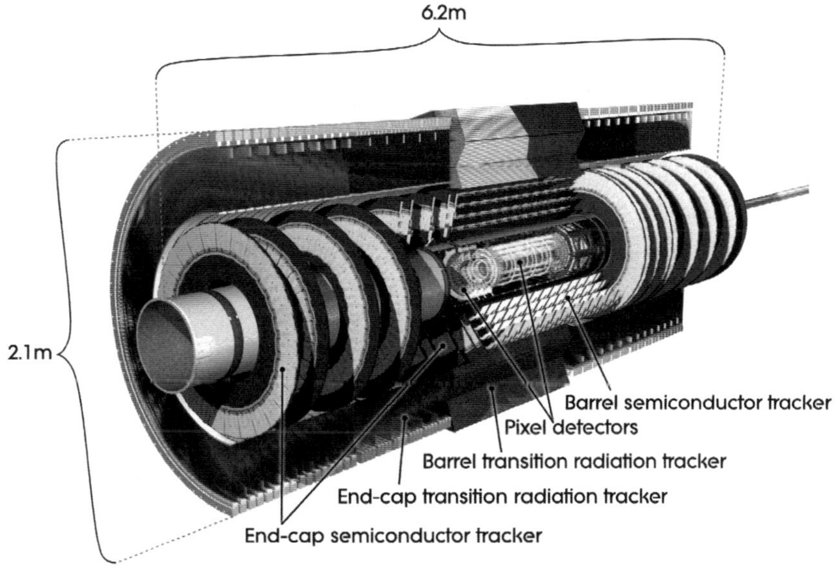

Figure 1. Overview of the ATLAS Inner Detector.

straw drift tubes which have a single hit resolution of 130 μm in the $(r\phi)$ coordinates. The straw tubes are arranged in 32 modules in each of the three barrel layers and 2 $\times$ 40 end-cap wheels (176 modules in total).

1.1. Alignment goals

After the assembly of the detector, the position of the individual modules is known with much worse accuracy than their intrinsic resolution. Therefore the alignment procedures have to be applied. The baseline goal of the alignment is to determine the position and orientation of the modules with such precision that the track parameters' determination is not worsened by more than 20% with respect to that expected from the perfectly aligned detector. This is crucial for efficient track reconstruction, precise momentum measurement and vertex reconstruction.

1.2. Alignment challenges

The alignment challenges come from huge numbers of DoFs to be aligned. For each module 6 alignment degrees-of-freedom (DoF) can be defined (if ignoring the module level deformation): 3 translations and 3 rotations. Thanks to the strict assembly tolerance of the sensors in the structure of the SCT modules [3][4], alignment corrections at the level of the individual single-sided sensor are not computed. This means for Pixel and SCT, considering 6 DoFs per module, in total ~35,000 DoFs; For TRT, considering 2 DoFs per Straw, in total ~700,000 DoFs, should be aligned respectively. In order to ensure the quality of the alignment, huge statistics such as ten million high quality tracks are needed which is rather computing intensive. Several hundred or even thousand CPUs are simultaneously used in every iteration.

2. Alignment strategies and algorithms

Two independent algorithms have been developed and validated in the ATLAS offline software framework [5]. All are iterative and make use of the residuals of the reconstructed hits on tracks (See the explanation in figure 2).

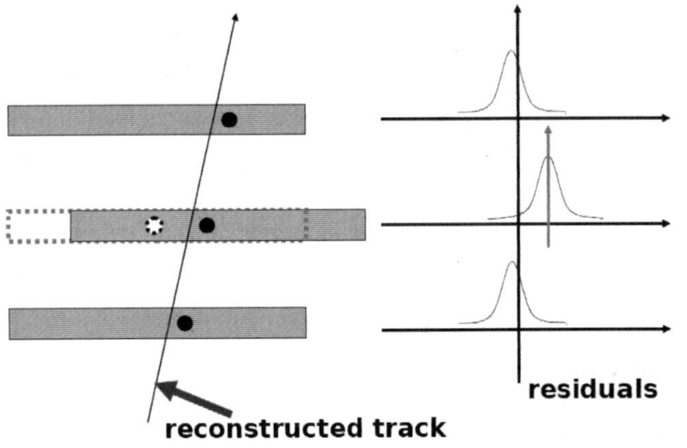

Figure 2. The middle module is shifted away from its design position. Real positions (green filled) are not known, thus the reconstruction uses the nominal position (red dashed). Consequently track fit quality is degraded. The residual distribution of the displaced module (and the others) will be biased away from zero. This bias measures the shift of the module and is the basis for the alignment.

The main idea is that the sum of the residuals over a large number of reconstructed tracks should be minimal for the aligned geometry. This can be formulated as a large χ^2 minimization problem [6][7] :
Let $\mathbf{r}(\mathbf{a}, \tau)$ be the vector of the residuals of the track hits. Residuals depend on both the alignment parameters $(\mathbf{a})$ and the track parameters (τ). V is the covariance matrix of the hit measurements. The alignment χ^2 is built as follows :

$$\chi^2 = \sum_{tracks} [\mathbf{r}(\mathbf{a}, \tau)]^T V^{-1} \mathbf{r}(\mathbf{a}, \tau) \tag{1}$$

The alignment corrections $(\delta\mathbf{a})$ are obtained by applying the minimization condition to the χ^2 and by making use of the linear expansion of the residuals around $\mathbf{r}_0$ (their initial estimates). This requires solving linear systems of the size equal to the number of the alignment DoFs as follows:

$$\frac{d\chi^2}{d\mathbf{a}} = 0 \implies \delta\mathbf{a} = -\left[\sum_{tracks} \left(\frac{d\mathbf{r}}{d\mathbf{a}}\right)^T V^{-1} \left(\frac{d\mathbf{r}}{d\mathbf{a}}\right)\right]^{-1} \cdot \left[\sum_{tracks} \left(\frac{d\mathbf{r}}{d\mathbf{a}}\right)^T V^{-1} \mathbf{r}_0\right] \tag{2}$$

2.1. Global χ^2 alignment
In the Global χ^2 approach, the solution given by equation 2 is calculated by using the total derivatives of residuals with respect to alignment parameters :

$$\frac{d\mathbf{r}}{d\mathbf{a}} = \frac{\partial\mathbf{r}}{\partial\mathbf{a}} + \frac{\partial\mathbf{r}}{\partial\tau}\frac{d\tau}{d\mathbf{a}} \tag{3}$$

which includes all correlations between the modules. This allows it to converge quasi-instantly. The calculation is numerically very challenging as it needs an inversion of a large matrix. The Global χ^2 approach is currently the baseline alignment algorithm used for the ID.

2.2. Local χ^2 alignment
In the Local χ^2 approach, the solution is also calculated according to equation 2, but partial derivatives with respect to alignment parameters are used instead of the total ones :

$$\frac{d\mathbf{r}}{d\mathbf{a}} = \frac{\partial\mathbf{r}}{\partial\mathbf{a}} \tag{4}$$

This way the correlations between different detector modules are neglected and the alignment matrix is reduced to a series of 6×6 matrices (2×2 for TRT), which can be inverted quickly, as opposed to the Global χ^2 approach. The inter-module correlations are restored by performing a certain number of iterations, updating the geometry of the detector and then re-fitting the tracks.

3. Computing Model

3.1. Alignment level

To address the realistic misalignments of the detector, the alignment is done at different levels of granularity motivated by the mechanical structure of the ID. First, the largest structures, subdetector barrels and end-caps, are aligned (L1). A second level of alignment (L2) treats barrel layers and end-cap disks or wheels as separate alignable objects. At this level, more structures, and consequently more degrees-of-freedom, are aligned. The final alignment level consists of the module-by-module (Pixel, SCT) or wire-by-wire (TRT) alignment. This third alignment level has the most degrees-of-freedom and requires the greatest statistics.

3.2. Each iteration

The first part is the track accumulation, in which for each selected track the first and the second order derivatives of χ^2 with respect to the alignment parameters are calculated and added to equation 2. The accumulation is split into several hundred parallel subjobs.

The second part is the solving for the alignment corrections ($\delta\mathbf{a}$). Matrices and vectors resulting from the accumulation jobs are first merged together. Based on equation 2, very large linear matrix equation is solved to get the corrections. Different matrix solvers have been implemented, some are based on the direct diagonalization libraries (like CLHEP [10] and LAPACK [8]), the others are based on the fast-solving techniques (like MA27 [9]) which typically exploit the special properties of the large matrix to be inverted, such as sparseness and symmetry. Even though the fast-solvers don't provide the errors for the solution, their speed allows them to be used for the problems of very large size.

3.3. The full chain

Figure 3 shows the flow diagram of the full alignment iterations chain. The alignment is mainly performed on data from a special trigger stream, the ID Calibration stream. However in some cases it is also run over high quality tracks in physics streams.
Track reconstruction is followed by the alignment monitoring which checks for the quality of the current constants. Afterwards the Silicon and TRT alignment is performed sequentially as described in the flow diagram. The obtained constants are fed into the next iteration. The final alignment constants are uploaded into the database after the iteration process has been completed.

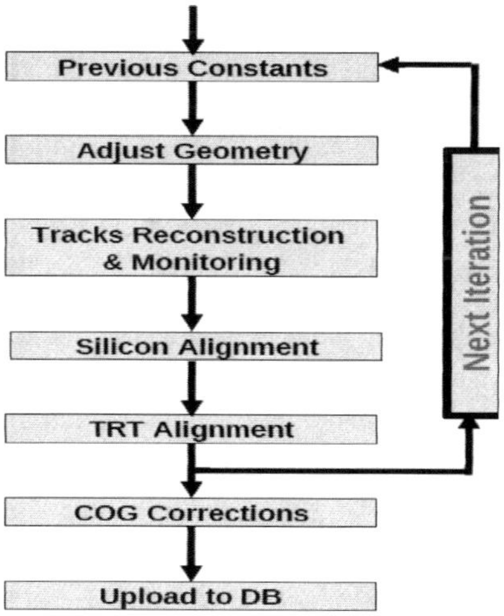

Figure 3. Flow diagram showing the full alignment iterations chain.

3.4. Run on Grid and automatic alignment

An infrastructure for running the alignment on Grid has been developed and is extensively used. Traditionally, the ID alignment is run on the CERN batch system (CAF). With the increasing

LHC data statistics the CPU and storage resources for ID alignment are no longer sufficient. The datasets accessible on CAF are quite limited too. The Grid environment offers access to nearly all ATLAS datasets and massively larger CPU and disk storage resources. Another advantage is that there are more powerful and convenient tools and interfaces for job management.

Last year the detectors were found to be slowly moving, hence decided to perform the alignment run by run to promptly detect the movements. Thus a mechanism that can automatically run the L1/L2 alignment in the calibration loop has been implemented. The shifters watch the constants changes, if significant changes are observed then some detailed study will be triggered and an updated set of constants will be produced.

4. Weak modes and constraints

The minimization of track residuals is necessary but not sufficient. The global distortions which preserve the helical trajectory of tracks and leave the χ^2 unchanged while systematically biasing the track parameters are known as weak modes. These kind of distortions are difficult to be removed by the minimization of the residuals, on the other hand they are very dangerous to physics results. Figure 4 shows a few example weak modes which may potentially have big impact on physics.

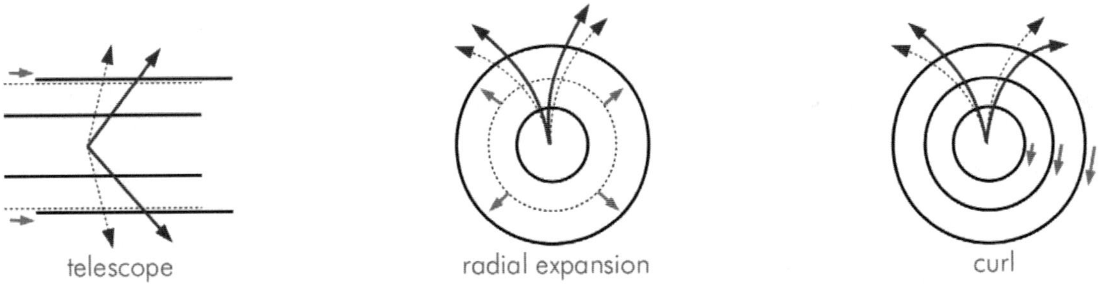

Figure 4. Selected possible systematic distortions for cylindrical structures.

Two main kinds of methods can deal with the weak modes. The first is the usage of different track topologies such as cosmics, beam halo and beam gas. The second is using constraints. Several constraint tools have been implemented and work quite effectively: These include beam-spot and vertex constraint, constraint on invariant masses of well known resonant decays, constraint on momentum from other systems (such as the constraint from Muon Spectrometer standalone momentum measurement, E/p constraint from the curvature asymmetry between e^+ and e^-).

5. The alignment results

5.1. Overall Residuals

The currently achieved alignment delivers nearly perfect residuals (see in figure 5 the exemplary plots of SCT local x residuals. In the Pixel and the SCT the local x and local y axes are in the detector plane with the x-axis pointing along the most precise measurement direction), which indicates the algorithms work correctly [6]. Now we are focusing on effects that are beyond just getting the residuals correct.

5.2. Residuals maps

There are also huge improvements in the detailed 2-D maps of the mean of the residual after the alignment for every sub-detector. Exemple plots are shown in figure 6 and 7. They are detailed

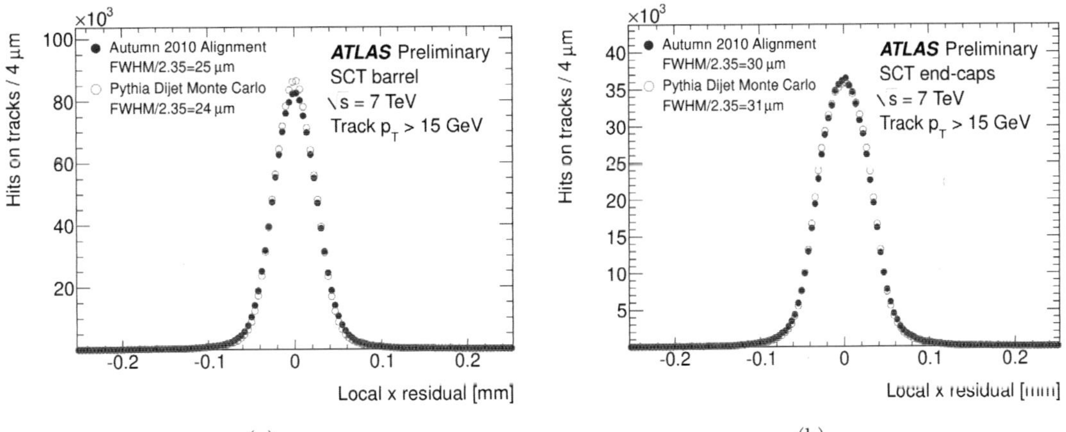

(a) (b)

Figure 5. The SCT local x residual distributions for the jet trigger data sample reconstructed with the Autumn 2010 Alignment (full circles), compared with the dijet MC simulation sample (open circles). The distributions are integrated over all hits-on-tracks in barrel modules (left) and end-cap modules (right). Tracks are required to have $p_T > 15$ GeV.

2-D maps of the mean of the residual for the Pixel and the TRT before and after the alignment, clearly can see most of the average Pixel local x residuals after module level alignment are within 5 μm.

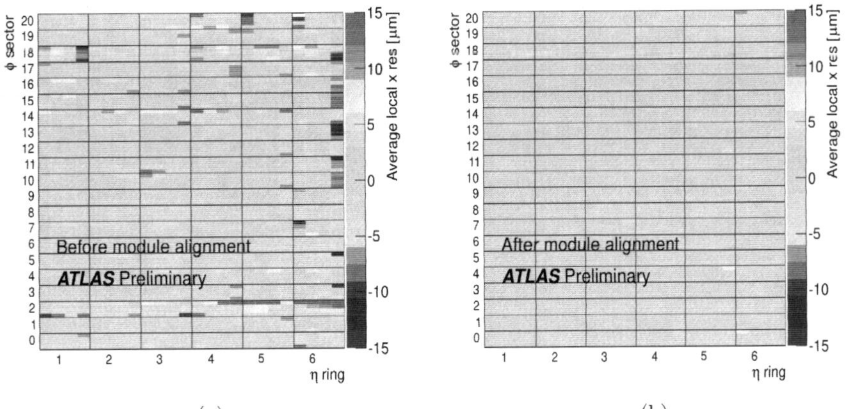

(a) (b)

Figure 6. Detailed residual maps of the Barrel pixel modules. Each pixel module has been split in a 4×4 grid and the average residual in each cell is plotted. Each module is identified by its position in the layer. This is given by its "η ring" and "ϕ sector" indices. Figure 6(a) shows the average local x residual before the module level alignment; figure 6(b) shows the average local x residual after module level alignment (including pixel module distortions). Only a subset of the pixel modules of the intermediate pixel barrel layer is shown.

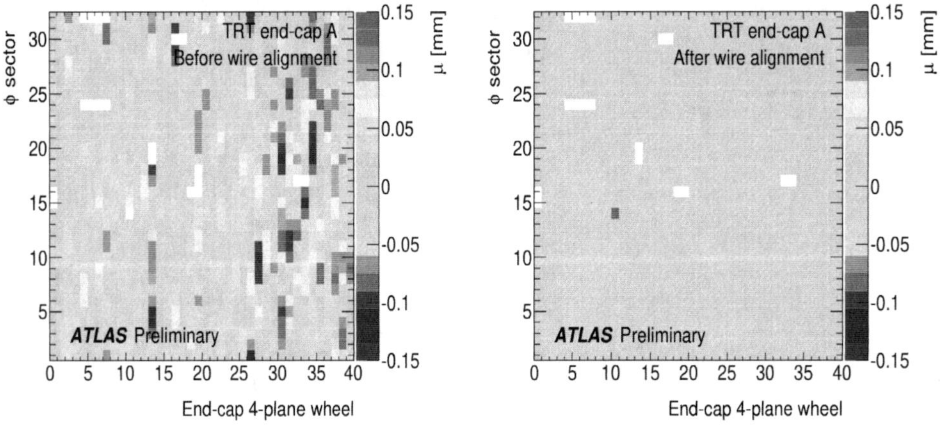

(a) (b)

Figure 7. Mean of the Gaussian fit to TRT residuals vs ϕ-sector and wheel before, left, and after, right, the wire-by-wire alignment. The plots illustrate the end-cap A results. The white bins are due to dead channels.

5.3. Detector stability

Figure 8 shows an example of the alignment corrections for translations in the global x direction, which are performed on a run by run basis starting from a common set of alignment constants [11]. Sizable movements ($\mathcal{O}(10\ \mu\text{m})$) coincide with sudden environmental changes. In other periods a generally good stability ($\mathcal{O}(1\ \mu\text{m})$) is observed.

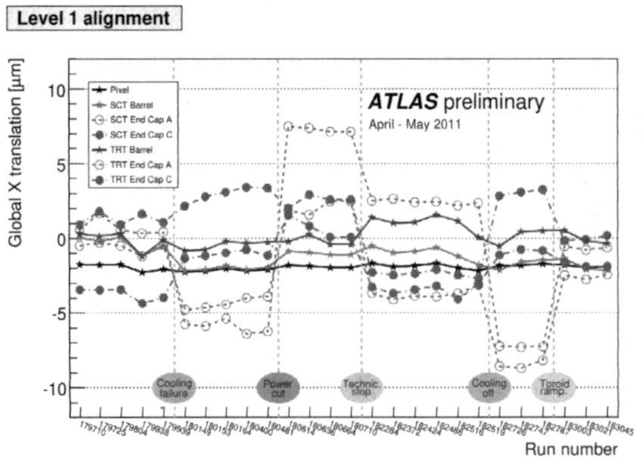

Figure 8. Level 1 alignment corrections for translations in the global x direction vs. the Run number.

5.4. $Z \to \mu^+\mu^-$ performance

For the ATLAS 2011 Summer reprocessing, a new set of constants was derived, for which the E/p constraint was employed for the first time. Preliminary results indicate significant performance improvement especially in the endcap regions. The mass resolution of the $Z \to \mu^+\mu^-$ resonance decay very nearly reproduces the one expected from the perfectly aligned MC [11]. Figure 9 shows the Z mass resolution in different ID regions; the black circles are for the Spring 2011 alignment, the red circles for the Summer 2011 alignment, shadow area for the MC perfect geometry.

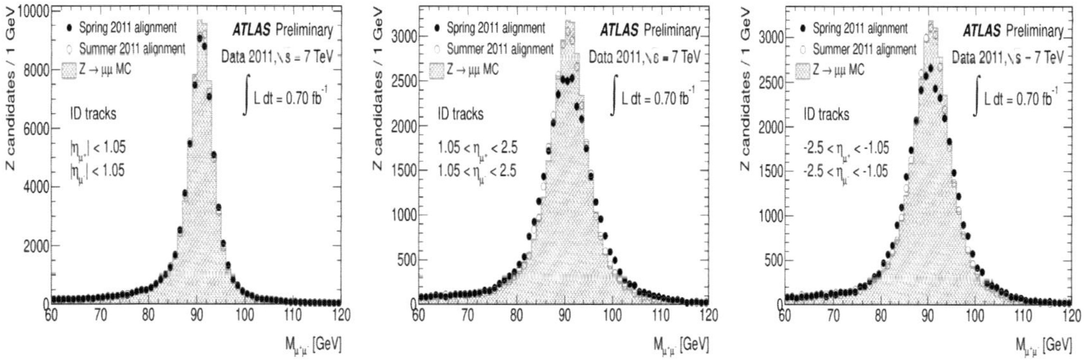

Figure 9. Invariant mass distribution of $Z \rightarrow \mu^+\mu^-$ decays, where the mass is reconstructed using track parameters from the ID track of the combined muons only, using about 702 pb^{-1} of data collected during Spring 2011. Performance in case of no misalignment based on Monte Carlo is compared to observed performance of data processed with the Spring 2011 alignment and data processed with the Summer 2011 alignment constants.
The left figure requires both tracks in the barrel region $|\eta|$ <1.5; the middle require at least one track in the end-cap A region 1.05< η <2.5; the right requires at least one track in the end-cap C region -2.5< η <-1.05.

The performance improvement of the latest alignment constants can be attributed to the fact that they eliminate most of the momentum related weak modes. Figure 10 shows the Z invariant mass vs. ϕ for μ^+ and μ^- [11]. Clearly visible sinusoidal structures in the Spring 2011 alignment indicate weak distortions of the geometry. Most of the structures disappear in the Summer 2011 alignment.

6. Summary

Alignment of the Inner detector of ATLAS is a very complex and challenging task. After a fast learning phase, already now the resolution on the reconstructed track parameters very closely approached the MC expectation for a perfect geometry. The future plans include further understanding of the alignment systematics, better description of time dependent changes and of the internal module distortions. The ID Alignment is making a significant contribution to ATLAS precision measurements and new physics searches.

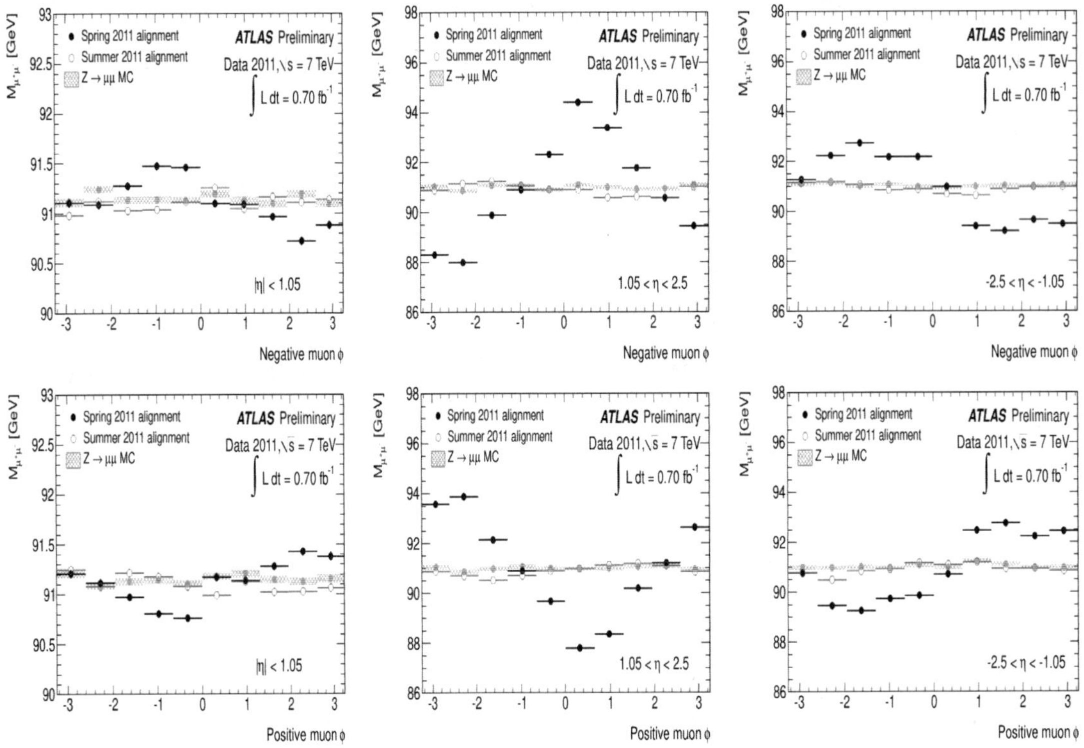

Figure 10. Mean Z invariant mass vs. ϕ for positive and negative muons, respectively. The mass distributions are fitted in RooFit using an unbinned maximum likelihood fit of a Breit-Wigner distribution, describing the intrinsic Z width, convolved with a Crystal Ball function as resolution function. Only Z candidates within the mass range [71, 111] GeV are used. Performance in case of no misalignment based on Monte Carlo is compared to observed performance of data processed with the Spring 2011 alignment and data processed with the Summer 2011 alignment constants.
The left figure requires both tracks in the barrel region $|\eta|$ <1.5; the middle requires both tracks in the end-cap A region 1.05< η <2.5; the right requires at least one track in the end-cap C region -2.5< η <-1.05.

References

[1] ATLAS Collaboration 2009 *Journal of Instrumentation* **3** S08003
[2] ATLAS Collaboration 1997 CERN-LHCC-97-16/17
[3] A. Abdesselam et al 2006 *Nucl. Inst. Meth. A* **568** 642-671
[4] A. Abdesselam et al 2007 *Nucl. Inst. Meth. A* **575** 98-118
[5] ATLAS Collaboration 2010 ATLAS Note ATL-COM-PHYS-2010-402
[6] ATLAS Collaboration 2011 ATLAS Note ATLAS-CONF-2011-012
[7] Brückman de Renstrom P, Hicheur A and Haywood S 2005 ATLAS Note ATL-INDET-PUB-2005-002
[8] http://www.netlib.org/lapack/
[9] http://www.hsl.rl.ac.uk/
[10] http://proj-clhep.web.cern.ch/proj-clhep/
[11] https://twiki.cern.ch/twiki/bin/view/AtlasPublic/InDetTrackingPerformanceApprovedPlots#Figures

Alignment of the CMS silicon tracker

Gero Flucke on behalf of the CMS Collaboration
Deutsches Elektronen-Synchrotron, Hamburg, Germany
E-mail: gero.flucke@desy.de

Abstract. The CMS all-silicon tracker consists of 16 588 modules with 25 684 sensors in total. In 2010 it has been successfully aligned using tracks from cosmic rays and proton-proton collisions, following the time dependent movements of its innermost pixel layers. In 2011, ultimate local precision is achieved by determining sensor curvatures in addition to module shifts and rotations, challenging the alignment procedure to determine about 200 000 parameters. This is achieved in a global fit approach using MILLEPEDE II with the General Broken Lines track model. Remaining alignment uncertainties are dominated by systematic effects that bias track parameters by an amount relevant for physics analyses. These effects are controlled by including information about the Z boson mass in the fit.

1. Introduction

The inner tracker of the CMS detector [1] at the LHC is completely based on silicon sensor technology. The tracker is placed inside a large solenoid providing a magnetic field of 3.8 T. The CMS coordinate system is defined such that the z-axis coincides with the nominal beam line. The radial distance from the z-axis is denoted as r. The polar angle θ is measured from the positive z-axis and the azimuthal angle ϕ is measured in the plane perpendicular to it. The innermost part of the tracker ($r < 20$ cm) consists of 1440 pixel modules. The barrel (BPIX) and forward (FPIX) pixel sub-detectors are surrounded by 15 148 strip modules up to $r < 110$ cm, arranged in four sub-detectors: the Tracker Inner and Outer Barrels (TIB and TOB), the Tracker Inner Disks (TID), and the Tracker End-caps (TEC). All sub-detectors are concentrically arranged around the z-axis. Modules at $r > 60$ cm contain two daisy-chained sensors, leading to 24 244 strip sensors in total.

To fully exploit the single hit resolution of down to 9 μm for pixel and of 23 to 60 μm for strip sensors, the positions of the sensors must be known to a precision of a few micrometers. This can best be achieved by track-based alignment algorithms. For the CMS tracker these have been applied using tracks from cosmic ray events [2], from a mixture of cosmic ray and first LHC collision events [3], and from complete 2010 data taking [4]. Based on the experience with these data and having collected a large sample of high momentum tracks in 2011, alignment development focuses now on systematic effects biasing reconstructed track parameters.

This article first recalls the principle of track-based alignment and explains the technological improvements needed to face the large amount and the required precision of the alignment parameters that are needed at CMS. Then the main results of the 2011 alignment campaign and their validation are presented, followed by a summary.

2. Track-based alignment

2.1. Principle

Track-hit residual distributions are generally broadened if the assumed positions and orientations of the silicon modules differ from the true ones. Following the least squares approach, alignment algorithms minimise the squares of normalised residuals, summing over many tracks. If the (hit or virtual) measurements m_{ij} with uncertainties σ_{ij} are independent, the minimised function is

$$\chi^2(\mathbf{p}, \mathbf{q}) = \sum_{j}^{tracks} \sum_{i}^{measurements} \left(\frac{m_{ij} - f_{ij}(\mathbf{p}, \mathbf{q}_j)}{\sigma_{ij}} \right)^2 \tag{1}$$

where f_{ij} is the track model prediction at the position of the measurement, depending on the alignment ($\mathbf{p}$) and track ($\mathbf{q}_j$) parameters. In the global fit approach as implemented in the MILLEPEDE II program [5], $\chi^2(\mathbf{p}, \mathbf{q})$ is minimised after linearising f_{ij}. This leads to the normal equations $\mathbf{Ca} = \mathbf{b}$ with $\mathbf{a}^T = (\mathbf{p}, \mathbf{q})$, i. e. all alignment parameters $\mathbf{p}$ and all parameters of all n used tracks $\mathbf{q}^T = (\mathbf{q}_1, \ldots, \mathbf{q}_n)$. Due to the special structure of $\mathbf{C}$, block matrix algebra facilitates the reduction of this large system of equations to a smaller one for the alignment parameters only,

$$\mathbf{C}'\mathbf{p} = \mathbf{b}'. \tag{2}$$

$\mathbf{C}'$ and $\mathbf{b}'$ sum contributions from all tracks. For each track a matrix equation $\mathbf{C}_j\mathbf{q}_j = \mathbf{b}_j$ has to be solved and $\mathbf{C}_j^{-1}$ needs to be calculated.

2.2. The general broken lines fit

Since multiple scattering effects are relevant in the CMS tracker, they have to be taken into account in the track model. This can be achieved by increasing the number of parameters for a charged particle in the magnetic field to $n_{par} = 5 + 2n_{scat}$, e. g. adding two deflection angles for each of the n_{scat} thin scatterers.[1] Since the expectation value of these scattering angles β is zero, virtual measurements $\beta = 0 \pm \sigma_\beta$ are added for each of them with σ_β according to the track momentum and the traversed material. For cosmic ray tracks this often leads to $n_{par} > 50$. Since in the general case the effort to calculate $\mathbf{C}_j^{-1}$ is proportional to n_{par}^3, a significant amount of computing time would be spent to calculate $\mathbf{C}_j^{-1}$ and thus $\mathbf{C}'$ and $\mathbf{b}'$. The progressive Kalman filter fit as used in CMS track reconstruction avoids the n_{par}^3 scaling, but cannot be used with MILLEPEDE II that requires the complete covariance matrix $\mathbf{C}_j^{-1}$.

The General Broken Lines track refit [6] provides $\mathbf{C}_j^{-1}$, but avoids the n_{par}^3 scaling by defining a track parametrisation with local meaning. These parameters are $\mathbf{q}_j = (\Delta\frac{q}{p}, \mathbf{u}_1, \ldots, \mathbf{u}_{(n_{scat}+2)})$, where $\Delta\frac{q}{p}$ is the change of the charged inverse momentum and $\mathbf{u}_i$ are the two-dimensional offsets to a reference trajectory in local systems at each scatterer and at the first and last measurement (figure 1). Predictions $\mathbf{u_{int}}$ for hit measurements are derived by interpolating between scatterers and the multiple scattering angles are determined from triplets of adjacent scatterers. This locality of all track parameters (except $\Delta\frac{q}{p}$) results in $\mathbf{C}_j$ being a bordered band matrix with band width $m \leq 5$ and border size $b = 1$. Using root free Cholesky decomposition, the effort to calculate $\mathbf{C}_j^{-1}$ and $\mathbf{q}_j$ is reduced to $\propto \left(n_{par}^2 \cdot (m+b)\right)$ and $\propto \left(n_{par} \cdot (m+b)^2\right)$, respectively.

The MILLEPEDE II program detects bordered band matrices while constructing $\mathbf{C}'$ and $\mathbf{b}'$, saving a factor 7 in CPU time for refitting cosmic ray tracks. The equivalence of Kalman filter and General Broken Line fits is shown for simulated muon tracks in figure 2 [6].

[1] For thin scatterers as used in CMS track reconstruction, the trajectory offsets induced by multiple scattering can be ignored. If a scatterer is thick, it can be treated as two thin scatterers.

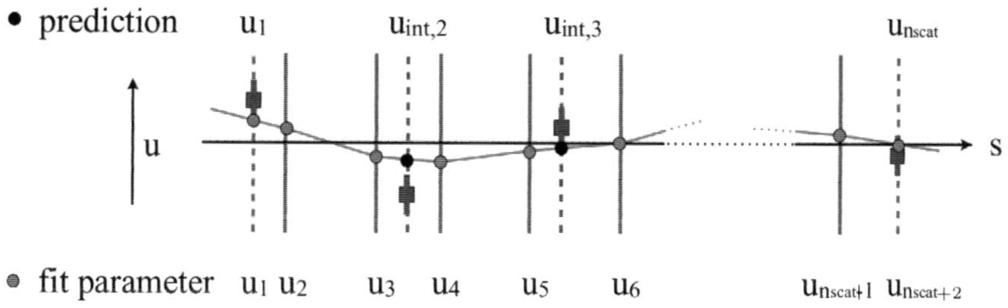

Figure 1. General Broken Lines track parametrisation along the path s of a reference trajectory. Dashed lines indicate the measurement planes, solid lines the thin scatterers, blue squares the measurements, black circles the trajectory predictions and red circles the fitted track parameters.

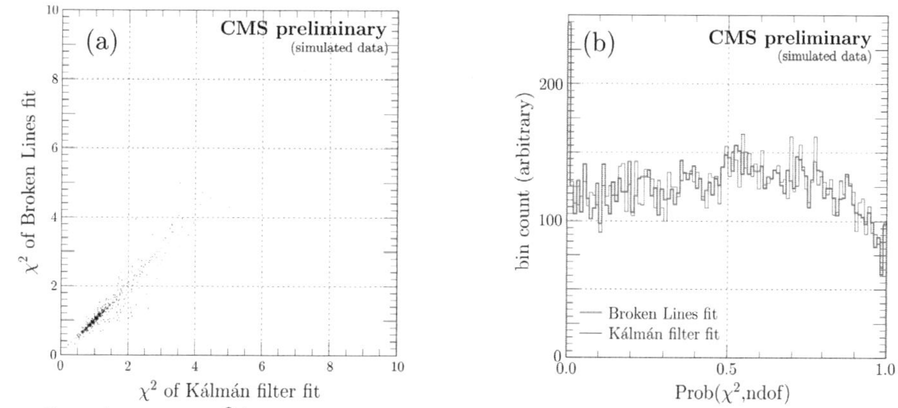

Figure 2. Correlation of χ^2/ndf values of single tracks (a), and probability value distributions (b), comparing the General Broken Lines with the Kalman filter fit.

2.3. Choice of alignment parameters

Previous alignment approaches treated the silicon modules as flat surfaces. Up to 5 (6) degrees of freedom per strip (pixel) module had been determined. But tracks with large angles relative to the module normal, e. g. tracks from the interaction point with small angles relative to the beam line, and cosmic ray tracks, are sensitive to small flatness deviations. These variations arise from silicon sensors being curved and, for strip modules with two sensors in a chain, from their relative misalignment. Therefore the vector of alignment parameters **p** has been extended to up to 9 degrees of freedom per sensor (instead of 6 per module). The sensor shape is parametrised as a sum of modified (orthogonal) Legendre polynomials up to order 2 where up to order 1 is equivalent to the rigid body parameters (figure 3).

Further alignment parameters are used for time dependent shifts and rotations of BPIX layers and FPIX half disks while the modules are kept fixed within these structures using constraints. Nine periods of time with relevant movements have been determined using the track-vertex residual monitoring introduced in section 3.2. In total, more than 200 000 alignment parameters have been determined in a common fit.

2.4. Momentum changing weak modes

Minimising residuals is not sufficient to find the alignment corrections **p**. Systematic movements of silicon sensors and changes of track parameters can balance each other without changing the overall χ^2, biasing the track parameters. Avoiding these so-called weak modes is the major alignment challenge once local precision is reached using many tracks from collisions. The best

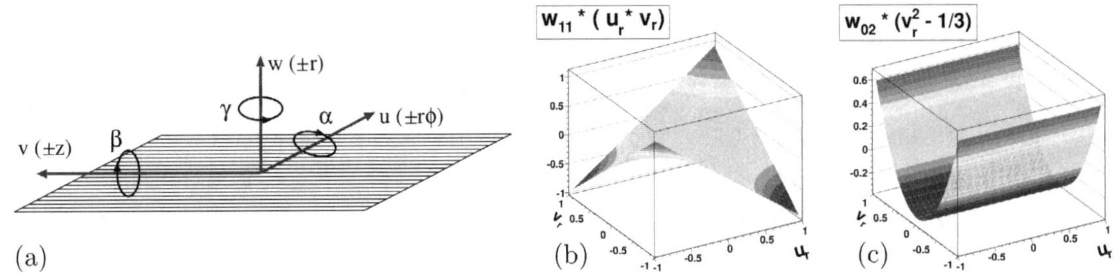

Figure 3. Sensor alignment parametrisation: The Cartesian coordinate system uvw and the rotations $\alpha\beta\gamma$ for the six rigid body parameters, where u and v lie in the sensor plane and u is the more precisely measured direction (a). Two of the three polynomials to describe deviations from the flat plane, where $u_r = \frac{2u}{l_u}$ and $v_r = \frac{2v}{l_v}$ are the relative positions on the plane with length l_u and width l_v (b and c).

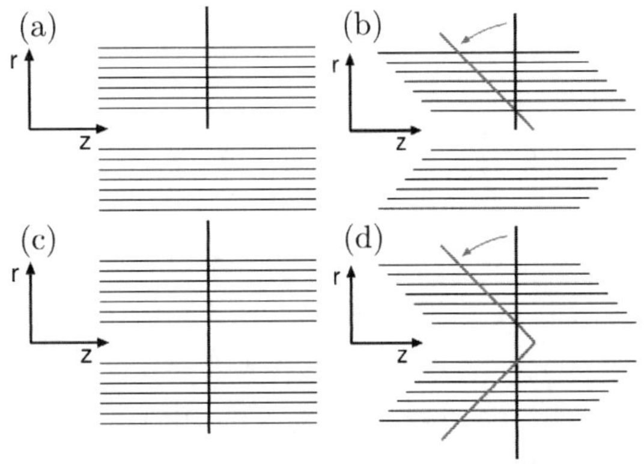

Figure 4. Using cosmic ray tracks avoids the telescope weak mode as explained in the text.

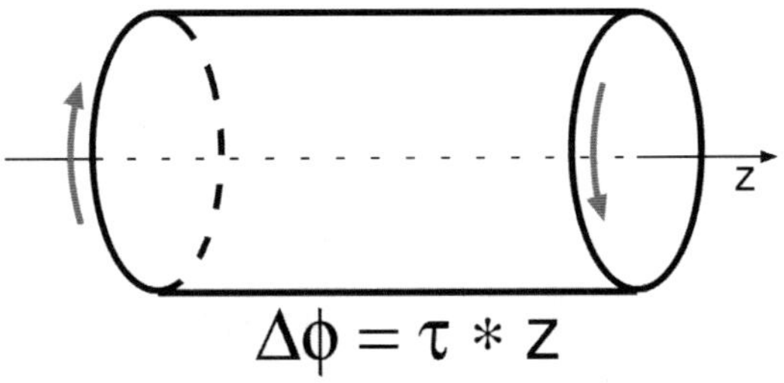

Figure 5. The twist weak mode.

handle to control these weak modes is their dependence on the data used in equation (1).

Making use of cosmic ray tracks is a key ingredient, controlling several distortions that are weak modes if using only tracks originating from the interaction region. Such a track is sketched in figure 4(a) together with a cut through the tracker along the z-axis. A so-called telescope distortion (z-position of the module biased proportionally to its radial coordinate, $\Delta z \propto r$) does not change the track-hit residuals, but clearly biases the track direction as illustrated in figure 4(b). A cosmic track connects the upper and lower part of the tracker as shown in figure 4(c). This prevents the telescope distortion, since either the track would be kinked, which is prevented by the track parametrisation, or the residuals would increase as can be seen from figure 4(d).

A systematic distortion $\Delta\phi \propto z$ (figure 5) is only partially controlled using cosmic ray tracks. This so-called twist biases the curvature of high momentum tracks. The transverse momentum measurement is oppositely affected as a function of the pseudorapidity $\eta = -\ln(\tan\theta/2))$ for tracks of positively and negatively charged particles. The effect on the reconstructed mass of a resonance decaying into two muons depends on the charge and η of the decay products. The sensitivity on the twist depends on the difference of the longitudinal momenta of the muons, i.e. is largest for large differences in η. Thus the bias is more severe for heavy resonances produced with smaller boosts, like the Z boson.

To avoid this twist effect, information about the mass of the Z boson is used in the alignment fit as follows. A common parametrisation for the two trajectories of the decay tracks is found [7]. Instead of 2×5 parameters (plus those accounting for multiple scattering), the nine common parameters are the position of the decay vertex, the momentum of the Z candidate, two angles defining the direction of the muons in the rest-frame of the Z boson, and its mass. The nominal Z mass is added as a virtual measurement with an uncertainty corresponding to the standard deviation of the Z lineshape around the peak position. In the sum on the right hand side of equation (1) the two individual tracks are replaced by the common fit object. The corresponding $\mathbf{C}_j$ has a border size $b = 9$. It should be noted that this approach implies a rigorous implementation of a vertex constraint as well since the parametrisation forces the tracks to a common vertex.

2.5. Optimisation of memory usage and speed

The task of the MILLEPEDE II program is to setup matrix equation (2) and to solve it to get the alignment parameters $\mathbf{p}$.

The matrix $\mathbf{C}'$ is symmetric and its elements need to be stored in double precision for accuracy. For 200 000 alignment parameters this would require 160 GB of RAM that are not easily affordable. But the matrix is rather sparse, so only non-zero elements need to be stored, reducing the storage need. Since these elements are usually close to each other, further compression is reached by bit-packed addressing of non-zero blocks in a row. Finally, some matrix elements sum contributions of only a few tracks, e. g. cosmic ray tracks from rare directions. For these elements single precision storage is sufficient.

The most important ingredients that speed up MILLEPEDE II are the following. The iterative MINRES algorithm [8] is used to solve equation (2). Besides the speed gain compared to invert $\mathbf{C}'$ this makes the sparse storage mentioned above possible. As already explained in section 2.2, bordered band matrices $\mathbf{C}_j$ are automatically detected when setting up equation (2). Finally, CPU intense parts as MINRES and track fits are parallelised using OpenMP® [9]. Reading data from local disk and memory access are the remaining bottlenecks. To reduce the time needed for reading, MILLEPEDE II can read compressed input and caches the information of many tracks to reduce the number of disk accesses.

3. CMS tracker alignment 2011

The alignment of the CMS tracker in 2011 was performed based on data collected until end of June, using about 15 million loosely selected isolated muon tracks, 3 million low momentum tracks, 3.6 million cosmic ray tracks (collected between LHC fills, during collisions and before collision data taking) and 375 thousand muon track pairs from Z boson decays. Within a few hours, 246 parallel jobs produced the compressed input files of in total 46.5 GB, containing residuals, uncertainties and derivatives for the linearised track model f. More than 200 000 alignment parameters were determined in the final minimisation program. The constructed matrix $\mathbf{C}'$ contained 31% non-zero off-diagonal elements. With a compression ratio of 40% this fitted well into 32 GB of memory. The input files were read 13 times: twice to setup the addressing for the reduced memory usage, four times to run MINRES with subsequently tighter rejection of bad tracks, and seven times for one dimensional line searches after the MINRES calls. Using eight threads on an Intel® Xeon® L5520 with 2.27 GHz, the CPU usage was 44.5 h with a wall clock time of only 9:50 h.

3.1. Modules with curved sensors

The parametrisation with polynomials describes the sensors very well. This can be seen in figure 6 where the offsets dw are calculated from the residuals in u and the track angle ψ from the sensor normal in the uw plane. For an alignment with the flat sensor assumption, a parabolic shape is seen that vanishes taking into account the additional parameters.

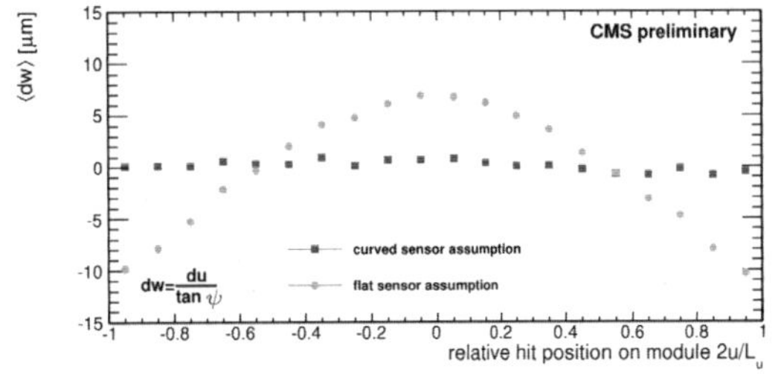

Figure 6. Average track-hit offsets perpendicular to the sensor plane across TIB modules.

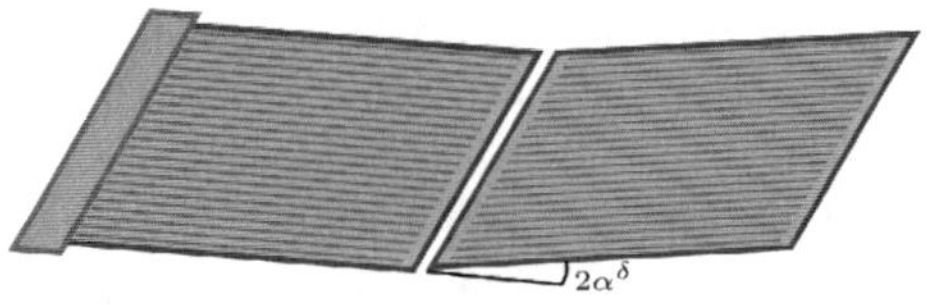

Figure 7. Sketch of a strip module with two sensors in a chain.

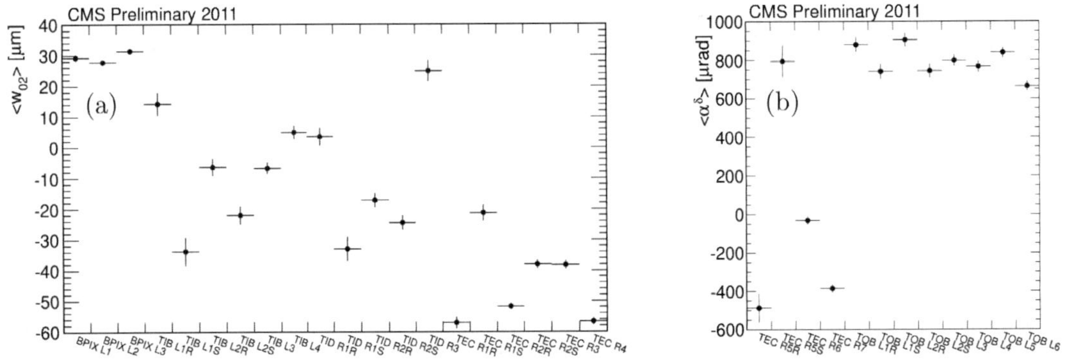

Figure 8. Curvature along the local v coordinate for single sensor modules (a) and half difference of the rotation around local u for double sensor modules (b), averaged for all modules of same type and similar mounting. The module radii increase with layer (ring) number L (R).

Figure 8 shows average values of some of the determined parameters describing deviations from flat modules. A curvature along the local v axis of $w_{02} = 30$ μm in the BPIX means that the sensor position in w at the edges ($v_r = \pm 1$) differs by 20 μm from the average. Modules in the first layer with largest $|z|$ are crossed by tracks from the interaction point under large angles relative to the sensor normal. Therefore hit position corrections of around 100 μm have to be applied. These are quite relevant compared to the hit resolution of down to 9 μm.

For TOB modules, the largest deviations from flatness result from the kink between the two sensors (figure 7). A size of $\alpha^\delta = 0.8$ mrad results in a deviation of up to 40 μm in w.

While figure 8 shows average values to demonstrate the module and mounting type dependence, the variances of these parameters are relevant. For w_{02} they range from 10 μm in the BPIX to 50 μm in the TIB and for α^δ from 0.8 mrad in the TOB to 2 mrad in the TEC. Strip sensor curvatures are expected due to the single sided processing and are requested to be below 100 μm [1]. This requirement is fulfilled for the majority of the modules, but not for the tails of the distributions.

High momentum tracks from the interaction region cross the strip modules under small angles relative to the module normal. Therefore sensor curvatures have only a small effect. This is different for cosmic ray tracks that cross the tracker with a large closest distance to the beam line, d_0. This is demonstrated in figure 9: The larger d_0, the larger the average track angle from the module normal, leading to degraded fit results for the flat module assumption. If curvature parameters on sensor level are determined, the average fit probability is almost flat as

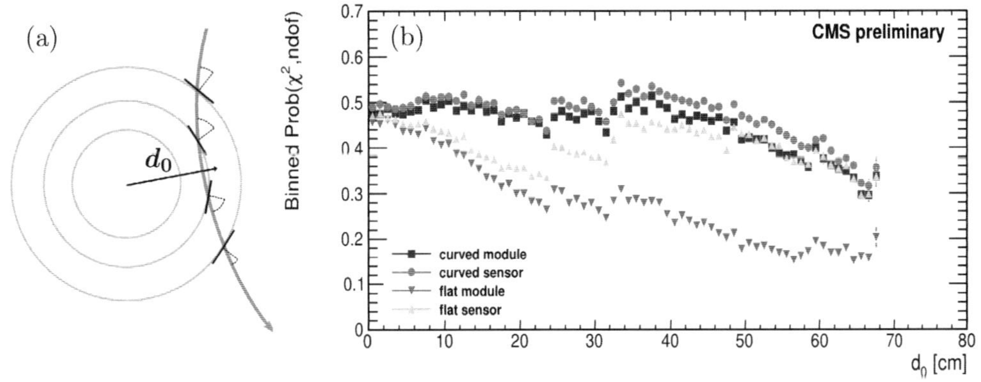

Figure 9. Illustration of a cosmic ray track passing the tracker with large d_0 (a) and the average fit probability as a function of d_0 for different module shape parametrisations (b).

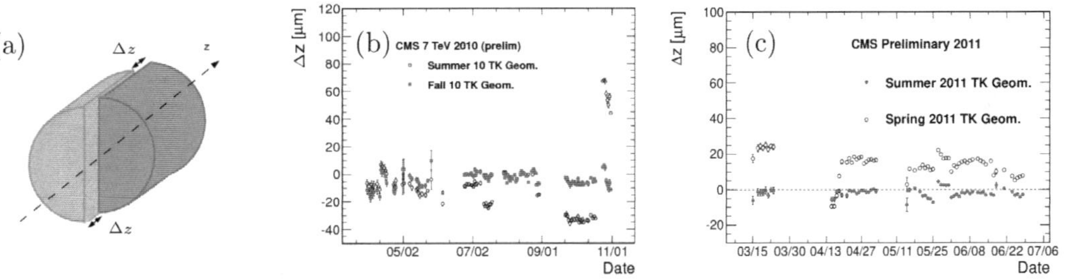

Figure 10. The relative misalignment of the two pixel half barrels along the beam line (Δz, a) is monitored versus time using track-vertex residuals for 2010 (b) and 2011 (c). Black open circles (red dots) refer to the situation before (after) re-alignment, respectively.

a function of d_0 up to 50 cm, thus improving substantially the consistency between tracks from the interaction point and cosmic rays. The latter are invaluable to control weak modes in the alignment procedure.

3.2. Time dependence of large structures of the pixel detector

Unbiased track-vertex residuals are used to monitor the position of the two pixel half barrels relative to each other. Each primary vertex is refitted after removal of one of its tracks. This is repeated for each track of the vertex. The track-vertex residuals Δz along the beam line are averaged as a function of the polar angle ϕ of the track. A difference of the mean values for tracks stemming from the one half barrel or the other indicates a relative misplacement. Jumps of up to 100 μm are seen before alignment in figure 10. After the alignment with time dependent parameters for the positions and orientations of large pixel structures, the remaining half barrel separations are well below 10 μm, a value that has no effect on the alignment sensitive b-tagging algorithms.

3.3. Control of momentum changing weak modes

Systematic momentum biases for tracks with high momenta have been investigated using events with a Z boson decaying into oppositely charged muons. The dependence of the position of the mass peak as a function of the pseudorapidity η and the polar angle ϕ of the positively charged muon is shown in figure 11. Without using the Z-mass information in the alignment, a large

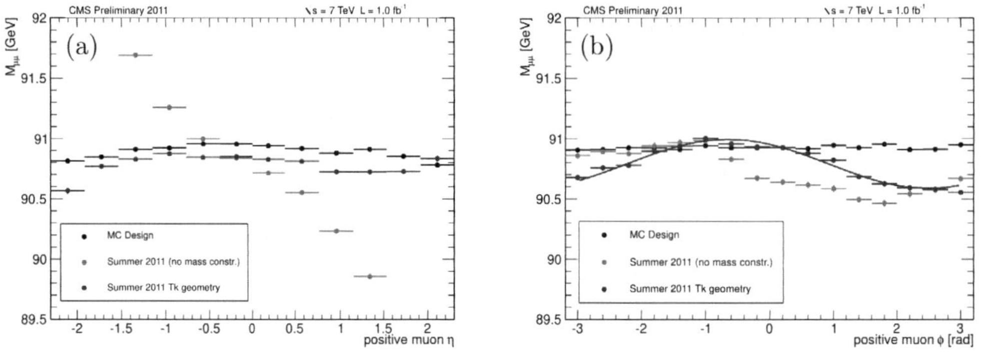

Figure 11. Reconstructed mass peak position for $Z \to \mu^+\mu^-$ decays as a function of η (a) and ϕ (b) of the μ^+, for detector simulations with perfect alignment (black) and for 2011 data aligned with and without special treatment of the Z events (blue and red, respectively).

η-dependence is observed. It can be related to a twist in the alignment result. In contrast, using this information in the alignment fit as described in section 2.4, the remaining spread of Z-mass peak values is almost as small as in the detector simulation with perfect alignment. The reason for the small overall offset between simulation and data and for the sine wave-like dependence on ϕ are currently under investigation.

Investigations of the ratio of the energy reconstructed in the hadronic calorimeter and the track momentum for well isolated positively and negatively charged hadron tracks have confirmed the results obtained using Z events, and thus serve as an independent cross check.

4. Summary

Recent alignment results for the CMS silicon tracker have been determined using the global fit approach of track-based alignment with the MILLEPEDE II program. The use of the General Broken Lines track model makes a rigorous treatment of multiple scattering effects possible and substantially reduces the computing time to set up the large matrix equation of the alignment procedure. Parallelisation of the CPU intense parts of MILLEPEDE II and sophisticated storage of the large sparse matrix are further key ingredients to facilitate the determination of more than 200 000 alignment parameters in 9:50 h wall clock time for the final fit, using eight threads and less than 32 GB memory.

This large number of parameters is needed to describe deviations from the flat module assumption and therefore significantly improves the consistency between tracks from the interaction point and cosmic rays. Also tracks passing pixel modules under large angles profit from taking into account the sensor curvature. Movements of larger structures of the pixel detector are taken care of by a time dependence of their alignment parameters, validated using residuals between tracks and their primary vertex.

Inclusion of cosmic ray tracks in the alignment procedure is not sufficient to control all weak modes biasing the curvatures of high momentum tracks. Making use of the mass of the Z boson as an external information reduces the variation of the position of the reconstructed $Z \to \mu^+\mu^-$ mass peak as a function of the direction and charge of its decay products to well below 500 MeV. This remaining dependence is currently being studied in detail, profiting from a fast turnaround due to the speed and modest memory requirements of the alignment procedure using MILLEPEDE II with the General Broken Lines track model.

Journal of Physics: Conference Series **368** (2012) 012036 doi:10.1088/1742-6596/368/1/012036

References

[1] CMS Collaboration 2008 *JINST* **3** S08004

[2] CMS Collaboration 2010 *JINST* **5** T03009

[3] Lampén T 2011 *J. Phys.: Conf. Ser.* **331** 032043

[4] Draeger J 2011 *Track based alignment of the CMS silicon tracker and its implication on physics performance* Ph.D. thesis Universität Hamburg (available from `http://www.physnet.uni-hamburg.de`)

[5] Blobel V 2006 *Nucl. Instr. and Methods* **A566** 5–13

[6] Blobel V, Kleinwort C and Meier F 2011 *Comput. Phys. Commun.* **182** 1760–3

[7] Widl E and Frühwirth R 2007 *Representation and estimation of trajectories from two-body decays* CERN-CMS-NOTE-2007-032

[8] Paige C and Saunders M 1975 *SIAM Journal on Numerical Analysis* **12**(4) 617–29

[9] `http://openmp.org`

An alternative method for the TileCal signal detection and amplitude estimation

B Sotto-Maior Peralva on behalf of the ATLAS Collaboration
Universidade Federal de Juiz de Fora, Brazil
E-mail: bernardo@cern.ch

Abstract. The Tile Barrel Calorimeter (TileCal) is the central section of the hadronic calorimeter of ATLAS. It is a key detector for the reconstruction of hadrons, jets, taus and missing transverse energy and it assists the muon measurements due to a low signal-to-noise ratio. The energy deposited in each cell is read out by two electronic channels for redundancy and is estimated by reconstructing the amplitude of the digitized signal pulse sampled every 25 ns. This work presents an alternative approach for TileCal signal detection and amplitude estimation under low signal-to-noise ratio (SNR) conditions, exploring the applicability of a Matched Filter. The proposed method is compared to the Optimal Filter algorithm, that is currently being used at TileCal for energy reconstruction. The results for a simulated data set showed that for conditions where the signal pedestal could be considered stationary, the proposed method achieves a better SNR performance than the Optimal Filter technique.

1. Introduction

Experimental high energy physics often faces problems related to signal detection and reconstruction in harsh conditions. Therefore, advanced techniques of digital signal processing should be used in order to reconstruct signatures with low signal-to-noise ratio (SNR). The ATLAS (A Toroidal LHC ApparatuS) is a general purpose detector built for probing proton-proton collisions and heavy ion collisions at the Large Hadron Collider (LHC). It is composed of six different subsystems: The Inner Detector [1], the Solenoidal Magnet that surrounds the inner detector, the Electromagnetic [2] and Hadronic [3] calorimeters, the Toroid Magnets [4] and the Muon Spectrometer [5], as illustrated in figure 1. The detector has a total length of 42 m and radius of 11 m.

The ATLAS detectors comprise a large number of channels that should be read out when a valid event is selected by the complex online trigger system [6]. Besides selecting the valid events, those systems should perform substantial operations such as estimating the correct energy and time of the events. This process should be carried out online, and for this reason, their algorithms must be fast. It is worth mentioning that many channels are likely to have low energy deposition which impacts on the SNR level. Due to this fact, they could be discarded during event reconstruction. The correct channel selection selection impacts on the trigger system and the offline analysis.

In this paper, a Matched Filter (MF) to perform both the signal detection and the amplitude estimation for the ATLAS Hadronic calorimeter (TileCal) is presented. The Matched Filter technique is known to be the optimal signal detector in terms of the SNR for cases where the received signal could be described as a known signal (deterministic) corrupted by Additive White Gaussian (AWG) noise [7]. The main goal of the proposed technique is the selection of the calorimeter cells (around 5,000) that should be used for energy reconstruction. As each calorimeter cell has two associated readout channels, this work also

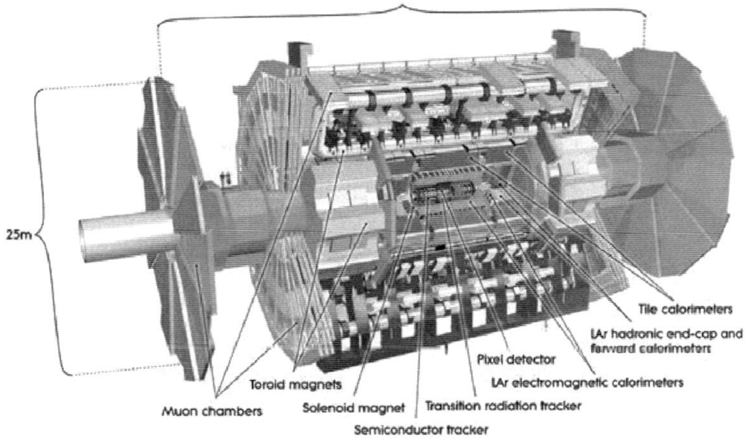

Figure 1. The ATLAS detector and its subsystems (extracted from CERN Document Server CERN-GE-0803012 01).

analyzes the impact on signal detection when these signals are summed before the energy estimation. The performance achieved by each of the methods is evaluated using experimental noise acquired on the ATLAS experiment and simulated TileCal pulses based on its reference pulse shape corrupted by the experimental noise.

Next section describes the TileCal environment and section 3 presents the alternative method for the TileCal signal detection and amplitude estimation. Section 4 presents the achieved performance for both methods and a comparison between them is carried out. Finally, in section 5, some discussions and conclusions are outlined.

2. The TileCal calorimeter

The ATLAS Hadronic Tile Calorimeter (TileCal) is a sampling calorimeter made of steel and plastic scintillating tiles. It comprises three cylindrical parts: one central barrel and two extended barrels. Each cylindrical part is composed of 64 modules. Each central (extended) barrel module is divided into 23 (16) cells with double readout, resulting in almost 10,000 channels (electronic signals). The TileCal structure can be seen in figure 2 where the ATLAS calorimeter system is shown.

The deposited energy in a particular TileCal cell is sampled by a group of scintillating tiles and transmitted through wavelength shifting fibers, connected on both sides of the scintillating tiles, up to the photo-multipliers where the light is converted into an electrical signal. Therefore, for each TileCal cell two electronic signals (channels) are available.

The fast electronic pulse generated at the photo-multiplier is conditioned by a shaper circuit which provides an electronic pulse for each channel, with 50 ns Full Width at Half Height (FWHH) and the amplitude proportional to the deposited energy. Therefore, the TileCal pulse shape can be considered almost invariant from channel to channel and the energy deposited by the particle at a given cell can be retrieved by the corrected estimation of the TileCal pulse amplitude. Figure 3 illustrates the analog TileCal reference pulse.

The TileCal analog signals are converted to digital signals with 40 MHz sampling frequency at the digitizer board and a window with 7 samples (150 ns) is enough represent the entire pulse. Then, the seven samples from each TileCal channel are transmitted through optical fibers up to the readout drivers (ROD), where the energy estimation is performed online.

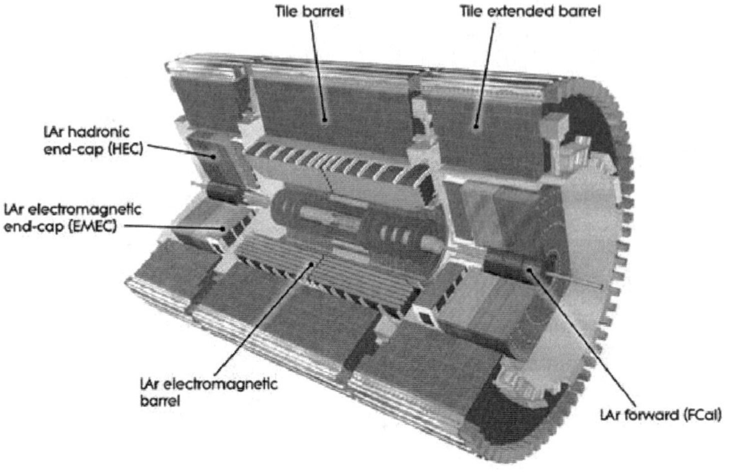

Figure 2. The ATLAS calorimetry system (extracted from CERN Document Server CERN-GE-0803015 01).

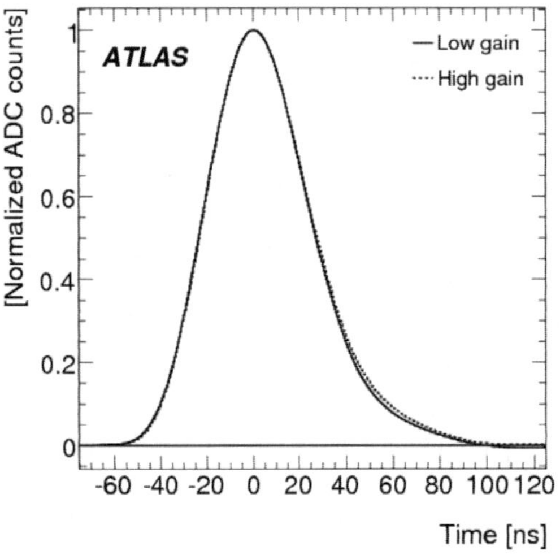

Figure 3. TileCal reference pulse shape.

The Optimal Filtering (OF) method [8] is the currently used algorithm to reconstruct the energy and the time at the ROD level during detector operation. It makes use of a weighted linear combination of the signal samples to obtain the amplitude, time and pedestal (baseline offset) of the pulse.

In this paper, an alternative study of signal detection and amplitude estimation for TileCal is presented and its performance is compared to the current method used in TileCal.

3. An alternative method for TileCal signal detection and amplitude estimation

This section briefly introduces the theory of signal detection and describes the proposed method to perform the signal detection and amplitude estimation for TileCal.

3.1. Signal detection

For a given discrete signal $s[k]$, where k represents the samples, which is transmitted through a channel that introduces an additive noise $n[k]$, with a received signal expressed by $r[k] = s[k]+n[k]$, it is desired to detect, with maximum efficiency, the presence of $s[k]$ in $r[k]$. In other words, a processing should be performed on the received signal in order to optimize the decision among two possible hypothesis:

- Hypothesis H_0: Only noise.
- Hypothesis H_1: Signal plus noise.

Thus, the detection problem can be described as a function of the received signal, as follows:

$$\begin{aligned} H_0 : r[k] &= n[k] \qquad & k = 0, 1, 2, ..., N-1 \\ H_1 : r[k] &= s[k] + n[k] \qquad & k = 0, 1, 2, ..., N-1 \end{aligned} \tag{1}$$

By representing the sequence $r[k]$ as an array $\mathbf{r}$ and considering $\mathbf{R}$ a given outcome at the receiver, it can be demonstrated [7] that the relationship which maximizes the detection efficiency is given by the maximum likelihood ratio, as follows:

$$L(\mathbf{r};\mathbf{R}) = \frac{f_{\mathbf{R}|H_1}(\mathbf{r}|H_1)}{f_{\mathbf{R}|H_0}(\mathbf{r}|H_0)} \underset{H_0}{\overset{H_1}{\gtrless}} \gamma \tag{2}$$

The terms $f_{\mathbf{R}|H_0}$ and $f_{\mathbf{R}|H_1}$ are the probability density functions of the received signal $\mathbf{R}$, given that H_0 (noise) and H_1 (signal plus noise) occurred, respectively, and γ is the detection threshold. The idea is to decide in favor of hypothesis H_0 if the probability of the received signal given that noise was transmitted is greater than the probability of the received signal given that signal was transmitted and in favor of H_1 if the opposite occurs. In such way, the maximum likelihood ratio maximizes the detection probability and minimizes the detection error probability.

The probability density functions are usually unknown for the majority of the real detection problems, therefore, the detection algorithms consist mainly in estimating $f_{\mathbf{R}|H_0}$ and $f_{\mathbf{R}|H_1}$ from a set of data where each event is known in advance.

3.2. The matched filter method

In the case where the probability density functions of the noise samples are Gaussian with covariance matrix $\mathbf{C}$ and the signal of interest $s[k]$ is deterministic, the maximum likelihood ratio reduces to

$$L(\mathbf{r}) = \frac{\frac{1}{\sqrt{2\pi\det(\mathbf{C})}} exp\{-\frac{(\mathbf{r}-\mathbf{s})^T\mathbf{C}^{-1}(\mathbf{r}-\mathbf{s})}{2}\}}{\frac{1}{\sqrt{2\pi\det(\mathbf{C})}} exp\{-\frac{\mathbf{r}^T\mathbf{C}^{-1}\mathbf{r}}{2}\}} \underset{H_0}{\overset{H_1}{\gtrless}} \gamma \tag{3}$$

By performing some mathematical operations on Equation (3), the optimum detector results in:

$$\sum_{k=0}^{N-1} r_k s_k^{'} \underset{H_0}{\overset{H_1}{\gtrless}} \gamma' \tag{4}$$

where $\mathbf{s}^{'} = \mathbf{C}^{-1}\mathbf{s}$.

The new detection threshold γ' relates to the old threshold γ according to

$$\gamma' = ln(\gamma) + \frac{\mathbf{s}\mathbf{C}^{-1}\mathbf{s}^T}{2} \tag{5}$$

Hence, Equation (4) shows that the decision between two hypothesis H_0 and H_1 is optimized through the inner product between the received signal and a replica of the deterministic signal of interest $s[k]$ after pre-whitening. Therefore, the decision between two classes can be optimized by filtering the received signal by its matched filter, whose impulse response $h[k]$ is given by [7]:

$$h[k] = s^{'}[-k + N] \tag{6}$$

For TileCal, the electronic noise readout during nominal TileCal operation has been shown to be Gaussian-like [9]. The pulse shape of an incoming signal is rather similar to the TileCal reference pulse shape as the latter was acquired and studied over an intense calibration period [10]. Therefore, even working under sub-optimum conditions, the MF approach may achieve good results for TileCal. For this, optimal TileCal signal detection is approximately performed by using the reference signal pulse shape. Since this corresponds to using seven samples, MF implements a fast finite impulse response (FIR) [11] filter according to Equation 6, which is appropriate for online applications.

3.3. Amplitude estimation through the matched filter

The acquired TileCal signal could be described by the components shown in Equation (7), where $r[k]$ is the received signal, ped is the baseline offset, $n[k]$ is the electronic noise, A is the amplitude and $s_{ref}[k]$ is the TileCal reference pulse:

$$r[k] = ped + n[k] + A \cdot s_{ref}[k] \tag{7}$$

By performing the inner product between the received signal and the TileCal reference pulse after pre-whitening, the amplitude of the incoming signal can be estimated. Equation (8) shows the inner product operation and Equation (9) shows the expression for the estimated amplitude:

$$y = \sum_{k=0}^{N-1} r[k]s^{'}[k] \tag{8}$$

$$A = \frac{y - ped\sum_{k=0}^{N-1} s^{'}[k]}{\sum_{k=0}^{N-1} s_{ref}[k]s^{'}[k]} \tag{9}$$

4. Results

Both MF and OF methods were implemented in software and their detection efficiencies under low SNR conditions are evaluated below. Two different analysis are performed, at single channel and cell level.

4.1. Database

The database used comprises two classes with 5,016 events each: the noise data set, which is composed of noisy signals taken from a pedestal run during nominal TileCal operation, and the signal data set, which is constructed from TileCal reference pulse shape with added noise. In order to simulate realistic low SNR conditions, both amplitude and time shifting distributions were taken into account to generate the signal data set.

4.2. Amplitude estimation

In order to evaluate the energy estimated by both methods, the amplitude of the signals were estimated. The amplitude distributions estimated by the MF and OF methods are shown in figures 4(a) and 4(b), respectively. The constant ped in Equation (9) was estimated by taking the mean value of the first sample of the incoming signals.

As was mentioned in section 2, the TileCal has double readout, therefore, Two channels are associated with each TileCal cell. Currently in TileCal, the energy is estimated per channel and summed up to form the final cell energy used for offline analysis.

Supposing that the noise between channels of the same cell is uncorrelated and the signals are summed before estimating the amplitude, the noise standard deviation of the resultant signal would be increased

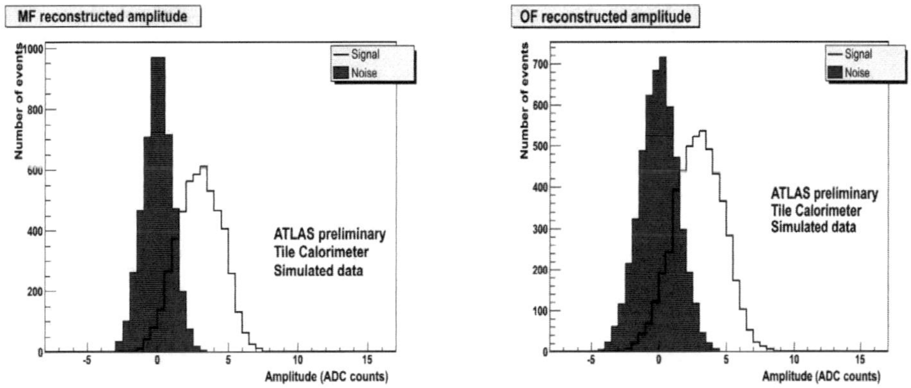

(a) (b)

Figure 4. Amplitude reconstructed by MF (a) and OF (b) for TileCal single channel.

approximately by a factor of $\sqrt{2}$ [12]. Figures 5(a) and 5(b) show the distributions of the estimated amplitude for the MF and OF approaches, respectively, considering the sum of the TileCal channels for a given cell.

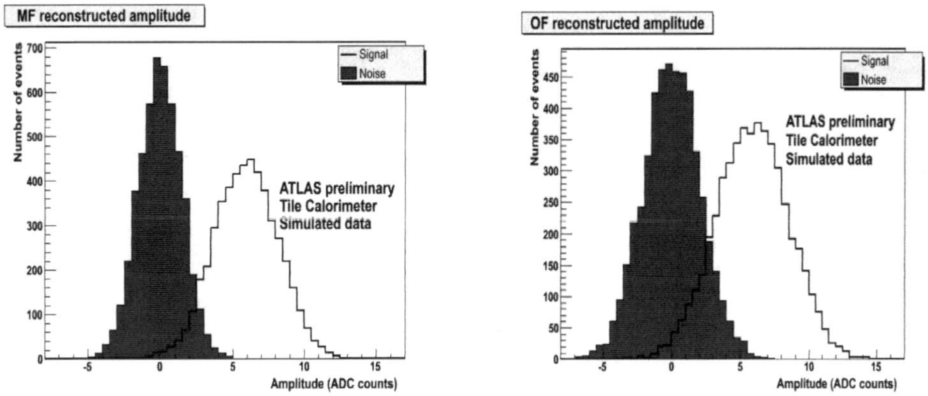

(a) (b)

Figure 5. Amplitude reconstructed by MF (a) and OF (b) for TileCal cell.

On both approaches, it can be seen that the MF method shows a better separation between signal and noise.

Since the standard deviation of the noise distribution is the error estimate in the measurement of the signal amplitude, table 1 summarizes this result for both types of analysis performed.

Table 1. Error estimate normalised to the mean signal ADC count.

	Single channel	Cell
OF	1.4149	1.0417
MF	1.0116	0.7458

For the purpose of illustrating the estimation quality of both methods, figures 6(a) and 6(b) show the relative error estimation with respect to the reference amplitude for single channel and cell analysis, respectively. In addition to that, it can be seen that both methods have similar performance in estimating the signal amplitude.

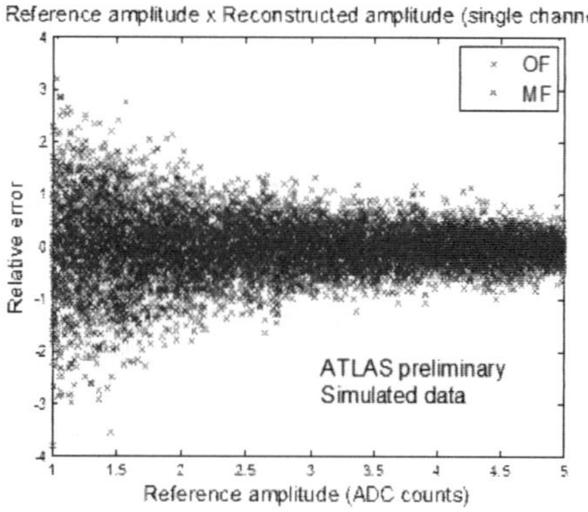

(a)

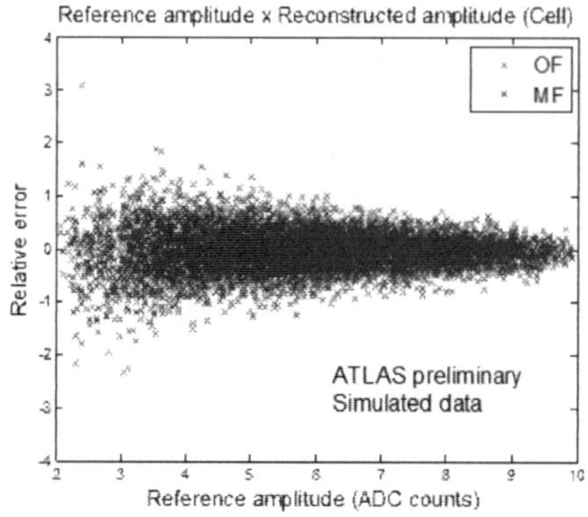

(b)

Figure 6. Relative error for single channel (a) and cell (b) analysis.

4.3. Signal detection

In order to evaluate the performance of each method for signal detection, the associated threshold (eg γ' for the MF method) were varied and the Receiver Operating Curve (ROC) [13] could be constructed for each scenario. Such curves can be seen in figure 7. It can be noticed that the best detection efficiency is achieved by the MF method with summed signals.

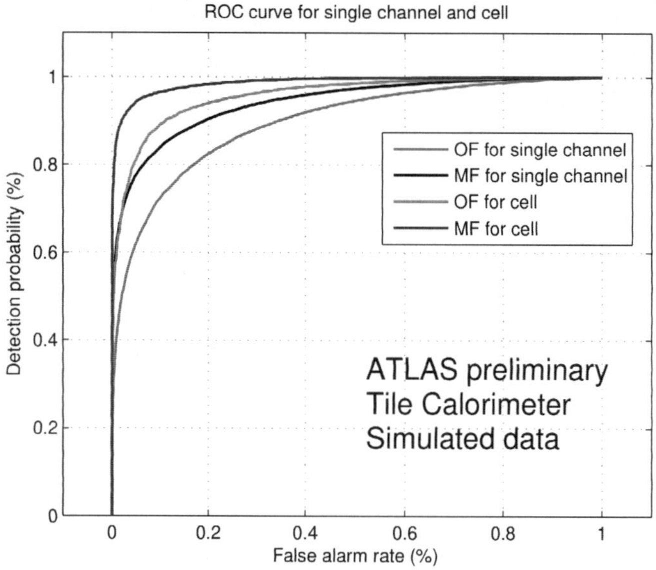

Figure 7. Detection efficiency against false alarm rate for all four analysis.

In order to quantify the results shown in figure 7, for a false alarm rate of 0.01%, the MF achieved 66.23% and 86.30% of probability of detection for single channel and cell analysis, respectively. For the same false alarm rate, the OF method attained 43.36% and 60.55% of probability of detection for single channel and cell analysis, respectively.

These results can be reproduced for experimental data in scenarios where the pedestal and the signal shape remain constant or with low variation rate, which is the case for the current LHC data taking setup.

5. Conclusions

In this paper, an alternative method for TileCal signal detection and amplitude estimation was presented. Differently from the current algorithm that performs this task, the proposed method makes use of an estimate of the signal pedestal whereas this parameter does not affect the performance of the current method. In conditions where the pedestal can be considered stationary, the MF approach has been shown to surpass the OF method in terms of detection efficiency. Besides that, since the MF method implements a FIR filter, it could be simply employed in the TileCal digital signal processors since these devices are suitable to carry out such fast multiply-accumulates (MACs) operations.

Concerning future studies, a pre-processing step has been implemented aiming to decorrelate the signals of the channels from the same cell before adding them up. In addition to that, other approaches to improve the pedestal estimation, such as adaptive filters, are under consideration. In scenarios with high pile-up (high luminosity) such methods could be very useful.

Acknowledgements

We are thankful to CNPq, CAPES and FAPEMIG (Brazil) for the financial support to this work. We also would like to thank many colleagues from the ATLAS collaboration for the fruitful discussions concerning this work.

References

[1] Ros E 2003 *ATLAS inner detector*, Nuclear Physics B Proceedings Supplements, 120:235-238.
[2] Perrodo P 2002 *The ATLAS Liquid Argon Calorimeter System*, Proceedings of ICHEP, pp 909-912.
[3] ATLAS Tile Calorimeter Technical Design Report 1996 ATLAS TDR 3, CERN/LHCC/96-42, CERN, Geneva, Switzerland.
[4] The ATLAS Collaboration 2008 *The ATLAS Experiment at the CERN Large Hadron Collider*, Journal of Instrumentation, JINST 3 S08003.
[5] Palestini S 2003 *The Muon Spectrometer of the ATLAS Experiment*, Nuclear Physics B, 125:337-345.
[6] The ATLAS Collaboration 2003 *ATLAS high-level trigger*, DAQ and DCS Technical Design Report.
[7] Trees H L V 2001 *Detection, Estimation and Modulation Theory, Part I*, Wiley.
[8] Valero A *et al* 2011 *Implementation and performance of the signal reconstruction in the ATLAS Hadronic Tile Calorimeter*, ATLAS Note ATL-TILECAL-PROC-2011-008 (URL: http://cdsweb.cern.ch/record/1381535).
[9] Ciodaro T *et al* 2011 *Optimal Signal Discrimination in a Low Signal-to-Noise Ratio Environment*, ATLAS Note ATL-TILECAL-PROC-2011-002 (URL: http://cdsweb.cern.ch/record/1330363).
[10] Tylmad M *et al* 2009 *Proc. Real Time Conference RT'09 (Beijing). 16th IEEE-NPSS*, p 543.
[11] Mitra S K 1998 *Digital Signal Processing: A Computer Based Approach*, McGraw-Hill.
[12] Papoulis S 1991, *Probability, Random Variable and Stochastic Processes*, McGraw Hill.
[13] Shamugan K S, Breipohl A M 2001 *Random Signals detection, estimation and data analysis*, Wiley.

Fractal dimension analysis in a highly granular calorimeter

M Ruan, V Boudry, J-C Brient, D Jeans, H Videau
Laboratoire Leprince-Ringuet,
École polytechnique, CNRS/IN2P3, Palaiseau, France
E-mail: Manqi.Ruan@llr.in2p3.fr

Abstract. The concept of "particle flow" has been developed to optimise the jet energy resolution by distinguishing the different jet components. A highly granular calorimeter designed for the particle flow algorithm provides an unprecedented level of detail for the reconstruction of calorimeter showers and enables new approaches to shower analysis. In this paper the measurement and use of the fractal dimension of showers is described. The fractal dimension is a characteristic number that measures the global compactness of the shower. It is highly dependent on the primary particle type and energy. Its application in identifying particles and estimating their energy is described in the context of a calorimeter designed for the International Linear Collider.

1. Introduction

1.1. Linear Collider and a highly granular calorimeter

In recent years, the concept of a highly granular calorimeter has been studied in the context of a proposed e^+e^- collider, either the International Linear Collider (ILC)[1] or the Compact Linear Collider (CLIC)[2]. At these colliders, many interesting physics channels will involve multi-boson production, and a good separation between the Z, W and Higgs bosons will be crucial. As many of their decay channels contain jets in the final states, a detector providing good jet energy resolution is required. For example, to separate different hadronic final states when looking at WW scattering (i.e, the $ZZ\nu\nu$ and $WW\nu\nu$ events), a jet energy resolution better than 3% (sometimes referred to as $30\%/\sqrt{E_J/\mathrm{GeV}}$) is required. This is roughly a factor of two better than previous experiments (e.g, ALEPH), and is one of the biggest challenges for the detector design. This goal has now been achieved at the full simulation level by using a Particle Flow Algorithm (PFA)[3]. The key idea is to measure the energy of each jet particle individually in the best suited subdetector: charged particles with tracker devices, photons with the electromagnetic calorimeter (ECAL) and neutral hadrons with the ECAL and hadron calorimeter (HCAL).

The most crucial task for a PFA is to separate showers from different particles in the calorimeter, in order to avoid losses and double counting. On average, 65% of the jet energy is in charged particles, 26% in photons (γ) and 9% in neutral hadrons (h^0). Charged particles and photons can be measured to much higher precision in the detector than neutral hadrons. For example, at the International Large Detector (ILD, one of the proposed ILC detectors), the momenta of charged particles can be measured to a precision of $\delta(1/P_T) = 2.0 \times 10^{-5}\mathrm{GeV}^{-1}$,

Published under licence by IOP Publishing Ltd

the energy of photon measured to $\delta(E_\gamma)/E_\gamma \sim 15\%/\sqrt{E_\gamma/\text{GeV}}$, and neutral hadrons measured to $\delta(E_{h^0})/E_{h^0} \sim 50\%/\sqrt{E_{h^0}/\text{GeV}}$ [4]. Therefore, for a perfect PFA with no confusion between these components, the ultimate jet energy resolution can reach $\delta E_{Jet}/E_{Jet} \sim 14\%/\sqrt{E_{Jet}/\text{GeV}}$ at typical precision of subdetectors [5].

The confusion contribution dominates the jet energy resolution, and to reduce it, the calorimeter granularity is more important than the intrinsic single particle energy resolution. The calorimeters designed for PFA typically have of order 10^8 readout channels with a volume of $1\,\text{cm}^3$ each. A sampling structure is used to provide a high granularity in the longitudinal direction, and each transverse sensor layer is divided into numerous cells. Calorimeters with such high granularity are referred to as PFA oriented or imaging calorimeters.

1.2. Shower Fractal Dimension

Many objects in nature have a self-similar pattern, or in other words "a rough or fragmented geometric shape that can be split into parts, each of which is, or at least approximately, a reduced-size copy of the whole" [6]. A self-similar set is also known as a fractal. The Fractal Dimension (FD) is a characteristic property of a fractal. It is defined as:

$$FD = -\lim_{\epsilon \to 0} \frac{\log N_\epsilon}{\log(\epsilon)}, \tag{1}$$

where N_ϵ represents the number of subsets at a scale ϵ.

A particle shower is a cascade of particles produced when an energetic primary particle interacts with dense matter, producing multiple daughter particles with less energy. Each of these then interacts in a similar way. This process stops when particles are absorbed in matter. Because each shower particle with sufficient energy can produce multiple daughter particles, the shower has a self-similar pattern, and can be regarded as a fractal. Thus a shower can be described by its fractal dimension.

Based on the type of interactions, showers can be classified into three different kinds: the minimum ionization particle (MIP) shower/track, the electromagnetic shower and the hadronic shower. Electromagnetic showers (composed of e^+, e^- and photons) are developed by pair-production and bremsstrahlung, whose large cross sections ensure the compactness of electromagnetic showers. Hadronic showers are governed by the production of hadrons (pions, kaons, protons, neutrons, etc.) and ions in the hard collisions with nuclei. Therefore they are composed of long travelling hadrons and compact clusters from nuclear fragments and electromagnetic showers stemming from π^0 decays.

2. Measurement of shower fractal dimension in a highly granular calorimeter

Showers of different particle types (π^+, μ^+, e^+, K_L^0, proton and neutron, see figure 1) with energies from 2 GeV to 40 GeV were simulated with Mokka[7], the official GEANT4 [8] based full simulation software for ILD. The simulated detector was only the Digital Hadron Calorimeter (DHCAL) in a solenoidal magnetic field of 3.5 T. The DHCAL is a sampling Resistive Plate Chamber(RPC)-iron HCAL with 48 longitudinal layers, each 26.5 mm thick with a 20 mm iron absorber and a 6.5 mm RPC sensor layer. In the simulation, each RPC layer is divided into $1 \times 1\,\text{mm}^2$ cells (modified from the original design of $10 \times 10\,\text{mm}^2$). This cell size is chosen to record detailed information about shower development information. In the following analysis, only digital information has been used, meaning only one bit is used for each channel to record whether this channel was hit or not.

The effective cell size can be varied by grouping $\alpha \times \alpha$ nearby cells. α then defines the scale at which the shower is analyzed, see figure 2. The shower fractal dimension can be estimated

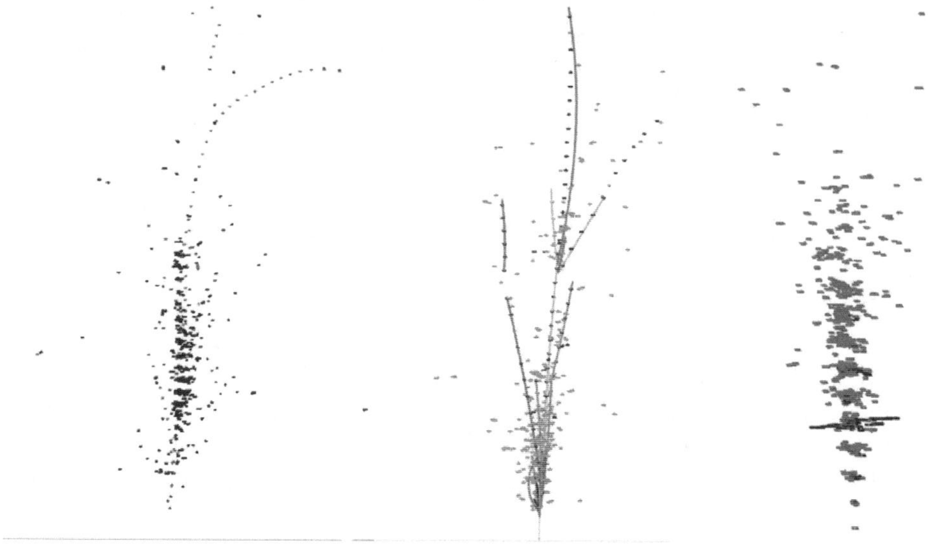

Figure 1. From left to right: simulated shower of 40 GeV π^+, K_L^0 and e^+.

by exploring the variation of the number of hits with the scale. Defining N_α as the number of hits at a scale α, the ratio of hit counts at different scales can be written as:

$$R_{\alpha,\beta} = N_\beta / N_\alpha. \tag{2}$$

At a fixed β, the ratio $R_{\alpha,\beta}$ is a function of α. In the following discussing β will be called the initial scale. To make full use of the recorded information, β is usually fixed at the initial cell size. Figure 3 shows the correlation between $R_{\alpha,1}$ and the scale α for different samples. An approximate linear correlation with α is observed in a double logarithmic scale (especially for the e^+ and π^+ samples), thus we define the shower fractal dimension as the average slope + 1, where +1 is added for the longitudinal degree of freedom:

$$FD_\beta = \left\langle \frac{\log(R_{\alpha,\beta})}{\log(\alpha)} \right\rangle + 1. \tag{3}$$

The shower fractal dimension depends on the type of the primary particle. A MIP has a trajectory close to a helix, and its fractal dimension is close to 1 (in the ideal case, a MIP track has only one hit per layer). On the other hand, electromagnetic showers (e^+ samples) have the largest fractal dimension because of their compactness. Hadronic showers (for example π^+ samples) are composed of MIP tracks from long travelling charged hadrons and compact clusters from electromagnetic interactions and heavy nuclear fragments, thus their fractal dimension lies in between.

Typical values for the shower fractal dimension for 40 GeV particles are (see figure 6):

$$FD_{1mm}(\mu) = 1.2 \tag{4}$$

$$FD_{1mm}(\pi^+) < FD_{1mm}(K_L^0) = 1.5 \tag{5}$$

$$FD_{1mm}(e^+) = 1.75 \tag{6}$$

At the same particle energy, the fractal dimension of the K_L^0 showers is slightly larger than that of the π^+ showers, as the K_L^0 showers don't have an initial MIP track before the first interaction.

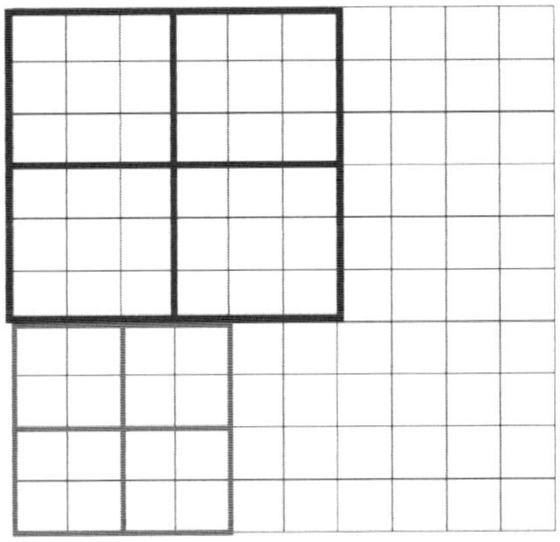

Figure 2. Basic method: group nearby cells.

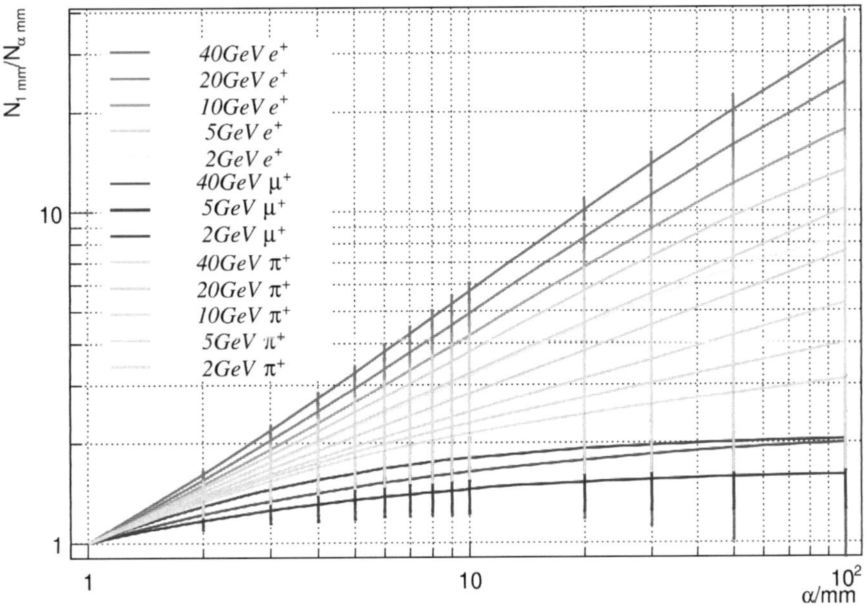

Figure 3. Dependence of hit count ratio and scale for different samples (the error bar scaled to rms, 1k events for each sample).

The shower fractal dimension also depends on the energy of the primary particle. The volume of a shower increases logarithmically with the primary particle energy while, to first order, the number of hits is proportional to the particle energy (ignoring the effects of saturation). A higher energy shower is therefore more compact, and has a larger fractal dimension (except for a MIP). The fractal dimension of electromagnetic and hadronic showers follows approximately (see figure 4):

$$FD_{em,1\mathrm{mm}}(E) = 1.41 + 0.21 \times \log_{10}(E/\mathrm{GeV}), \tag{7}$$

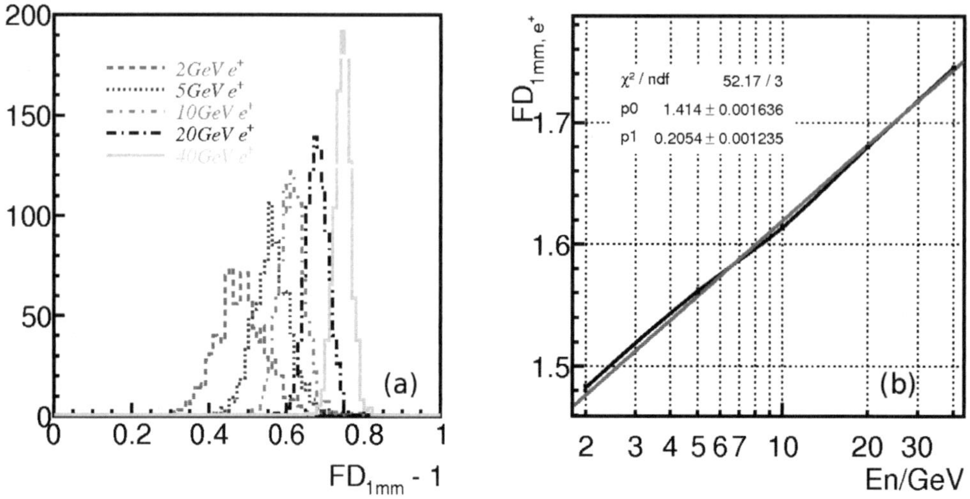

Figure 4. (a) Fractal dimension for e^+ samples at different energies; (b) Correlation between fractal dimension and particle energy for e^+ samples.

$$FD_{had,1\text{mm}}(E) = 1.24 + 0.15 \times \log_{10}(E/\text{GeV}). \tag{8}$$

3. Application of shower fractal dimension

The fact that the shower fractal dimension is sensitive to the particle type and energy makes it a potential tool for particle identification and shower energy estimation.

3.1. Particle identification

A particle identification technique purely based on the calorimeter information has been developed. It makes use of two input variables: the measured fractal dimension of the shower and the number of hits. This technique has been tested on different data samples at an energy of 40 GeV, see figure 5. To study its performance at different cell size, these two observables have been measured at three different initial scales: 1 mm, 10 mm and 30 mm.

A clear separation between muons, hadrons and positrons is observed at each initial scale. Using some simple cuts (for example, at initial scale of 1 mm, showers with $FD \geqslant 0.68$ are identified as electromagnetic showers, showers with $FD < 0.68$ and $N_{hits} > 500$ are identified as hadronic showers, and all the rest are identified as MIPs), the tables in figure 5 show the output of this particle identification, where the rows present input samples and columns present the output for 1000 events. In fact, the separation is clear at each initial scale and similar performance is achieved at different scales.

These results are achieved with single particles at a fixed energy, thus in a full event, the performance of this method will be affected by the performance of the PFA to identify individual showers and the measurement of the shower energy. However, for well isolated particles, especially charged particles whose track momentum can be taken as a measurement of their energy, this performance should be easily obtained.

The evolution of the fractal dimension with the initial scale can provide more information. Figure 6 shows the measured fractal dimension for the same data samples at three different initial scales. Because of the compactness of electromagnetic showers, the e^+ samples have a narrower

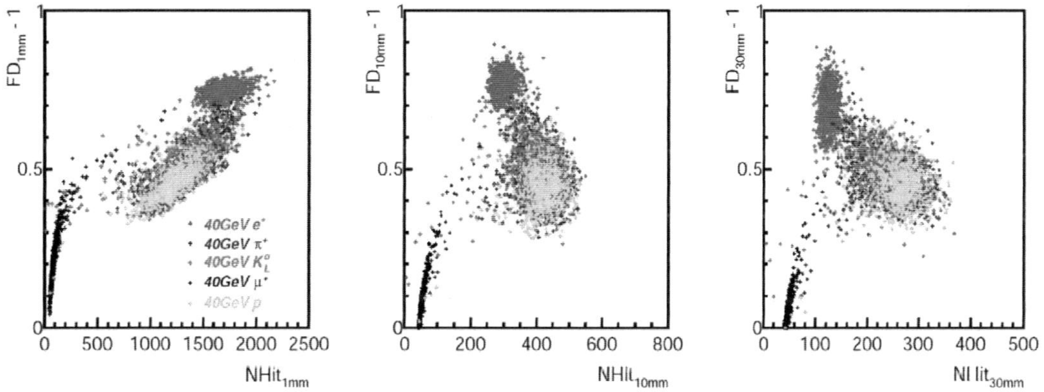

1mm	e+	u	h
e+	998	0	2
u	1	994	5
h	15	14	971

10mm	e+	u	h
e+	1000	0	0
u	0	995	5
h	17	14	969

30mm	e+	u	h
e+	1000	0	0
u	0	996	4
h	18	11	971

Figure 5. Particle identification with the fractal dimension of shower and the number of hits.

fractal dimension distribution at smaller initial scale, and get significantly smeared at larger initial scales. On the other hand, with increasing cell size, the fractal dimension distribution of the μ^+ sample becomes narrower, providing a better muon-hadron separation. The reason is that at large cell size, the fractal dimension is less sensitive to soft bremsstrahlung photons and noise (especially the noise caused by low momentum charged particles trapped in the gas gap by the magnetic field), see figure 7.

At any scale, the fractal dimension of π^+ samples is a continuous distribution which lies between those of μ^+ and e^+ of the same energy. The pion events with comparable fractal dimension to muons are due to pion decay ($\pi^+ \to \mu^+ + \nu_\mu$) before reaching the calorimeter, resulting in a MIP cluster with small fractal dimension. Events with large fractal dimension are cases in which the majority of energy is deposited electromagnetically: for example, a charged pion can convert into π^0 through isospin exchange ($\pi^+ + n \to \pi^0 + p$). This can be distinguished from e^+, e^- or γ by tagging the position of the first interaction point, or by measuring the fractal dimension at different depths in the calorimeter, see figure 7.

In the μ^+ sample, the fractal dimension of the shower can be as large as that of the electromagnetic shower through hard bremsstrahlung. More interestingly, in viewing the fractal dimension measured at different scales, some of the muon events have small fractal dimension at large scale, while having a large fractal dimension at small scale. As explained above, these showers have some micro structure, which is not resolved at large scale, see figure 7.

The fractal dimension depends strongly on the type of interaction, and is therefore a promising tool for particle identification. It also has the capability to tag some of the details of shower development.

3.2. Energy Estimation

A strong anti-correlation is observed between the fractal dimension measured at small cell size and the number of hits at a large scale. For example, the left plot in figure 8 shows the distribution of the fractal dimension measured from 10 mm cells and the number of hits at

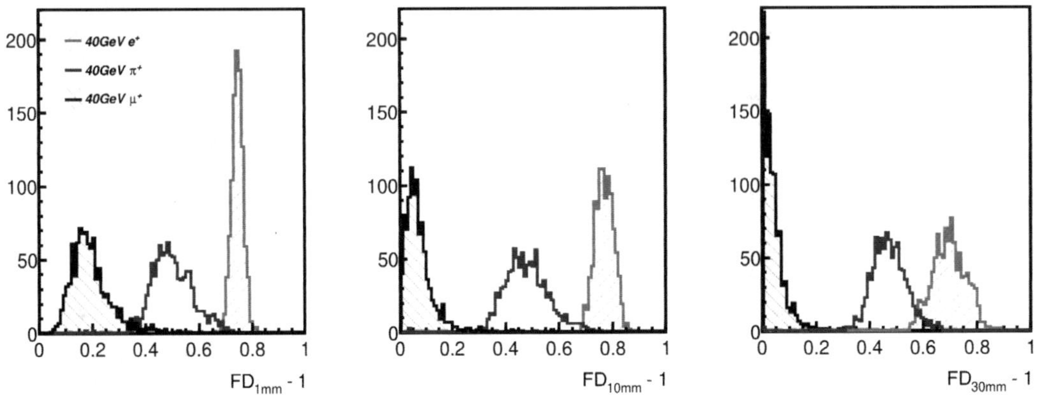

Figure 6. Fractal dimension measured from 1, 10 and 30 mm cell for μ^+, e^+ and π^+ samples.

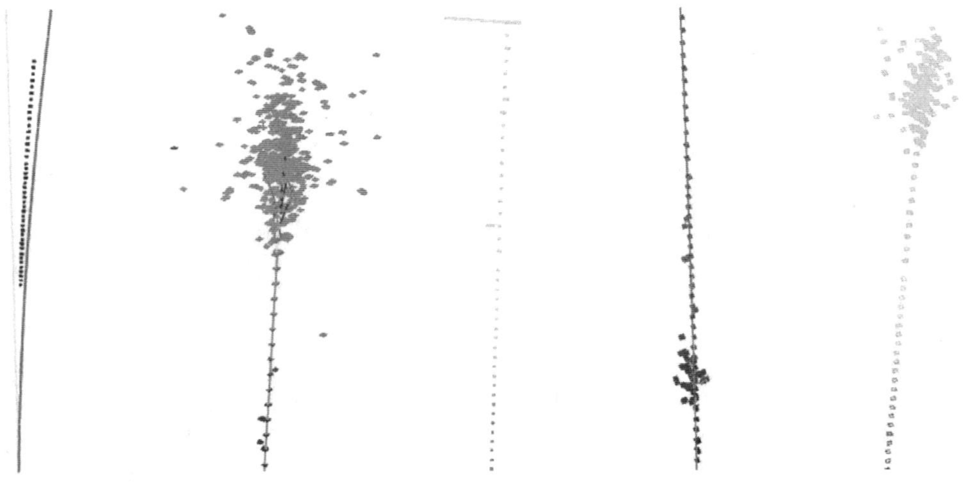

Figure 7. Extreme cases for π^+ and μ^+ shower, from left to right: pion decay, isospin exchange, noise, soft and hard bremsstrahlung radiation of μ^+.

30 mm cells for the 40 GeV μ^+, π^+ and e^+ samples. This correlation allows a linear combination of these two observables to be used as an energy estimator:

$$E = N_1 \times (Nhit_{30\text{mm}} + c_1 \times FD_{\beta\text{mm}}), \beta < 30, \tag{9}$$

where N_1 is a calibration constant, $Nhit_{30\text{mm}}$ is the number of hits at 30 mm cell size, and $FD_{\beta\text{mm}}$ is the shower fractal dimension measured at the cell size of β mm.

The right plot of figure 8 shows the distribution of measured energy with different estimators for 40 GeV π^+ sample (with mean value of the distribution normalized to 1). Using the energy estimator shown in equation 9, the shower energy resolution can be improved by roughly a factor of two comparing to hit counting at any scale. Unfortunately, the correlation coefficient c_1 depends on the particle energy. This energy resolution can only be obtained if the energy of the primary particle is known. Therefore, the application of the fractal dimension on energy estimation is two-fold. First, for charged particles, this phenomenon can be used to improve the

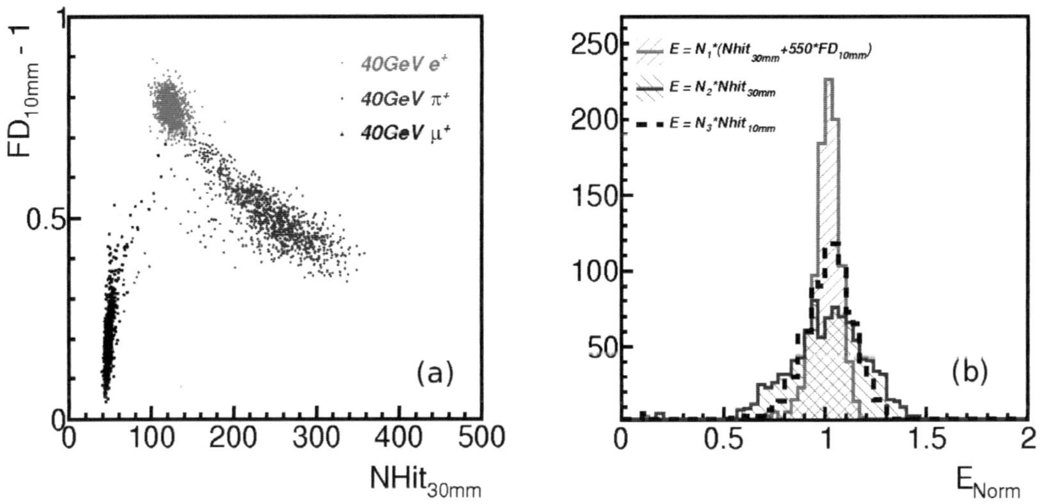

Figure 8. (a) Correlation between $FD_{10\,\mathrm{mm}} - 1$ and $Nhit_{30\mathrm{mm}}$ for different samples. (b) Energy measurement with different estimators for 40 GeV π^+ sample.

track-cluster matching in the PFA, which means the comparison of the position and momentum of tracks to a given cluster, to check if the cluster should be linked to a track and reconstructed as a charged particle. In this case, the track momentum can be taken as a reference of the injected particle energy, with which the correlation coefficient can be determined, and a better energy estimation can be achieved. Secondly, to first order, the correlation coefficient c_1 is proportional to the particle energy. Equation 9 can be written as:

$$E = N_1 \times (Nhit_{30\mathrm{mm}} + \delta \times E \times FD_{\beta\mathrm{mm}}), c_1 = \delta \times E. \tag{10}$$

From which the energy of the shower can be resolved:

$$\begin{aligned} E &= N_1 \times (Nhit_{30\mathrm{mm}}/(1 - \delta \times N_1 \times FD_{10\mathrm{mm}})) \\ &= N_1 \times (Nhit_{30\mathrm{mm}}/(1 - c_2 \times FD_{10\mathrm{mm}})), c_2 = \delta \times N_1, \end{aligned} \tag{11}$$

where c_2 is a constant. The right hand side of equation 11 is independent of the primary particle energy, therefore it can be used for the neutral particle energy measurement.

The performances of the different estimators tested on K_L^0 samples have been compared. The distributions of the measured energy have been fitted with a Gaussian function. On the left plot of figure 9, the resolution is expressed as the ratio between the width and mean of the Gaussian distributions (σ/M). The right plot shows the linearity in terms of the ratio between the measured energy and true particle energy (E/E_{exp}), fixing this ratio to 1 at 80 GeV.

As the digital readout can not distinguish if a hit is induced by a single or multiple particles, at higher energy, the curve for number of hits at 10 mm has a significant saturation effect and a degraded resolution. However, this can be corrected using the fractal dimension. With a roughly tuned coefficient c_2, the resolution significantly improves at energies above 40 GeV and below 2 GeV, while it is worse in the range 10 − 20 GeV. The linearity simultaneously improves to a level comparable to that of 5 mm.

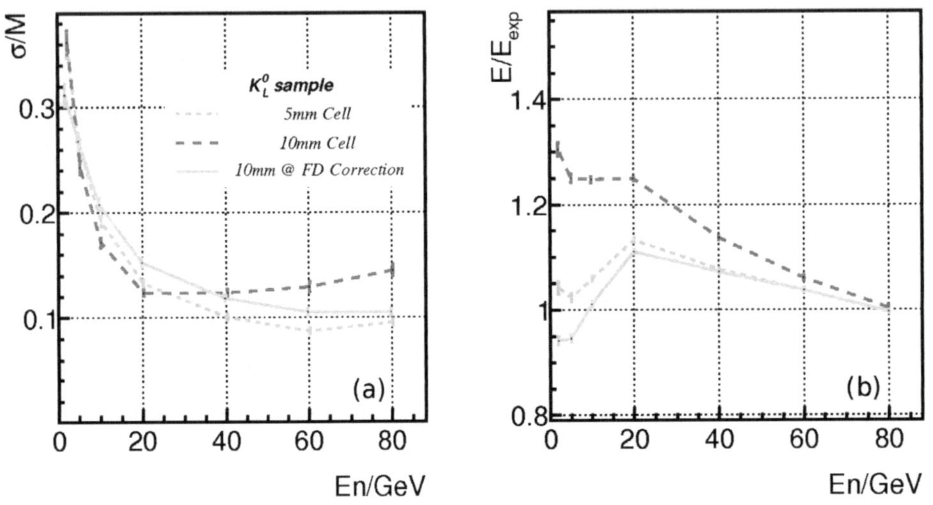

Figure 9. (a) σ/M and (b) linearity of different energy estimators. Pink curve: hit counts at 10 mm. Light blue curve: hit counts at 5 mm. Green curve: estimator of equation 11.

4. Summary

Showers created by particle interactions in material are fractal objects, to which a fractal dimension can be assigned. This shower fractal dimension can be measured with a highly granular calorimeter such as the ones designed for a future linear collider. The shower fractal dimension depends strongly on the type and energy of the incident particle, making it a potential tool for particle identification and shower energy estimation. Preliminary results from full simulation of the detector show that the fractal dimension can provide particle identification with excellent efficiency and purity on positrons, muons and pions at 40 GeV using only calorimetric information. In using the fractal dimension measured at different scales, this method can tag more detailed information of the shower development.

A strong anti-correlation between the fractal dimension at small scale and the number of hits at large scale is observed. For charged particles with a known track momentum, this correlation enables an energy resolution improved by a factor of up to two compared to hit counting, and can be used to enhance the track-cluster linking in the PFA. For neutral showers, the fractal dimension can improve the energy measurement in terms of linearity and resolution, especially at large energy where the saturation effects limit the performance of a digitally readout calorimeter with large or medium cell sizes.

Acknowledgments

The research leading to these results has received funding from the European Commission under the FP7 Research Infrastructures project AIDA, grant agreement no. 262025. We are grateful to the CALICE collaboration for their constructive comments, and to the linear collider software team, especially to G. Musat from LLR, for the development of many tools used in this study.

Reference

[1] Brau J *et al.* 2007 International linear collider reference design report *CERN Report* CERN-2007-006 ILC-REPORT-2007-001

[2] Battaglia M *et al.* 2004 Physics at the CLIC Multi-TeV Linear Collider: report of the CLIC Physics Working Group *CERN Report* CERN-2004-005 (*Preprint* hep-ph/0412251)
[3] Thomson M A 2009 *Particle Flow Calorimetry and the PandoraPFA Algorithm* Nucl. Instrum. Meth. A **610**, 25-40
[4] Abe T *et al.* 2010 The International Large Detector: Letter of Intent *DESY Report* DESY-2009-87
[5] Brient J-C 2009 *Particle Flow Algorithm and calorimeter design J. Phys.: Conf. Series* **160** 012025
[6] Mandelbrot B B 1982 *The Fractal Geometry of Nature. W.H. Freeman and Company* ISBN 0-7167-1186-9
[7] http://polzope.in2p3.fr:8081/MOKKA
[8] Agostinelli S *et al.* 2003 *Geant4 a simulation toolkit* Nucl. Instrum. Meth. A **506**, 250-303

Visual physics analysis – from desktop to physics analysis at your fingertips

H-P Bretz, M Erdmann, R Fischer, A Hinzmann, D Klingebiel, M Komm, J Lingemann, M Rieger, G Müller, J Steggemann, T Winchen

RWTH Aachen University, III. Physikalisches Institut A, 52062 Aachen, Germany

E-mail: `rfischer@physik.rwth-aachen.de`

Abstract. Visual Physics Analysis (VISPA) is an analysis environment with applications in high energy and astroparticle physics. Based on a data-flow-driven paradigm, it allows users to combine graphical steering with self-written C++ and Python modules. This contribution presents new concepts integrated in VISPA: layers, convenient analysis execution, and web-based physics analysis. While the convenient execution offers full flexibility to vary settings for the execution phase of an analysis, layers allow to create different views of the analysis already during its design phase. Thus, one application of layers is to define different stages of an analysis (e.g. event selection and statistical analysis). However, there are other use cases such as to independently optimize settings for different types of input data in order to guide all data through the same analysis flow. The new execution feature makes job submission to local clusters as well as the LHC Computing Grid possible directly from VISPA. Web-based physics analysis is realized in the VISPA@Web project, which represents a whole new way to design and execute analyses via a standard web browser.

1. Introduction

Visual Physics Analysis (VISPA) is an integrated analysis development environment for high-energy and astroparticle physics [1, 2]. The goal is to provide one environment that addresses the typical cycle of designing, executing, and verifying of analyses.

In VISPA, analyses are designed and represented visually as an analysis flow. This flow consists of individual modules and connections among them via a variable number of input and output ports. During the execution, data sets are guided through the flow on an event by event base. Depending on internal criteria, user programmed modules decide to which output port the current event will be propagated. Along with the attached connection to the respective port, this also determines the next module to handle the event. Once an event is not forwarded to another module anymore, the analysis loop continues with the next event in the data set.

The modular structure simplifies the exchange of whole analyses or even parts of it between physicists due to the clear module interface and the fact that modules are independent from each other. In order to make the design process more flexible, VISPA supports C++ and Python modules. While C++ offers better runtime performance, Python modules are faster to prototype due to advantages of interpreted code compared to compiled code. Part of the flexibility is also the option to include arbitrary external packages.

Analyses in VISPA employ three main paradigms, see figure 1. Within modules, physics objects are accessible via the physics library PXL [3,4] in an object-oriented manner. The analysis flow based on the module system implements a data-flow driven paradigm. Visual programming and steering are provided within one integrated graphical user interface (GUI).

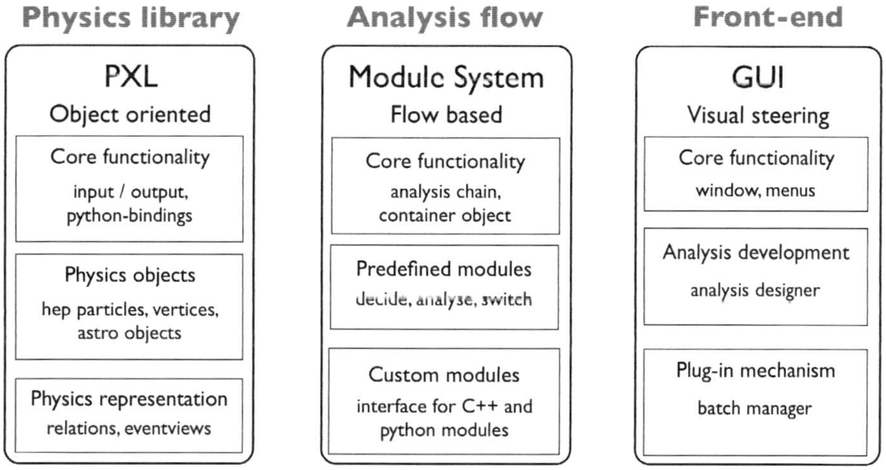

Figure 1. VISPA is based on three main components. The physics library PXL, the module system, and the GUI are summarized along with some of their core features.

In addition to the the physics objects, PXL also manages relations among these objects and handles input / output operations. Part of the module system are a container object and the analysis chain itself, through which the container is guided. It also contains predefined modules, e.g. a simple switch module, which can be used to split a data stream. User interfaces to C++ and Python build the foundation for user-written modules. The graphical front-end is written in Python [5] and uses the Qt framework [6] with the Python-bindings PyQt [7]. While the core part only provides common GUI elements, such as menus, toolbars and dialogs, the user interacts with plug-ins that are loaded into the GUI. Subsequently, two plug-ins, which provide core functionality to the analyst, are presented, see figure 2.

The analysis designer is the tool used to create and configure the analysis flow.

The PXL data browser visualizes data files event by event in order to verify them.

VISPA has been successfully used for published analyses in high-energy and astroparticle physics [8, 9].

As outlined so far, VISPA integrates the analysis design and verification steps of the development cycle. In the following section, the project's scope is extended to further improve analysis handling and comfortably integrate the execution step. Finally, with VISPA@Web a new perspective to visual physics analysis via a web-interface is opened.

2. New developments

2.1. Improved handling of complex analysis designs

Analyses often contain steps that the analyst would like to execute independently. There might also be parts of the analysis that should run several times with only slightly different configuration, e.g. for different types of data sets. These are two examples of complex problems during the analysis process that could be addressed by creating separate analysis configurations. In order to integrate such cases into one design, we introduce a new layering technique into VISPA.

The idea is to create one analysis flow as usual, however, containing all steps. This analysis flow is regarded as the base layer. New layers can be created upon the base layer and will automatically inherit all settings from the underlying layer. Within each layer, module options can be changed and modules

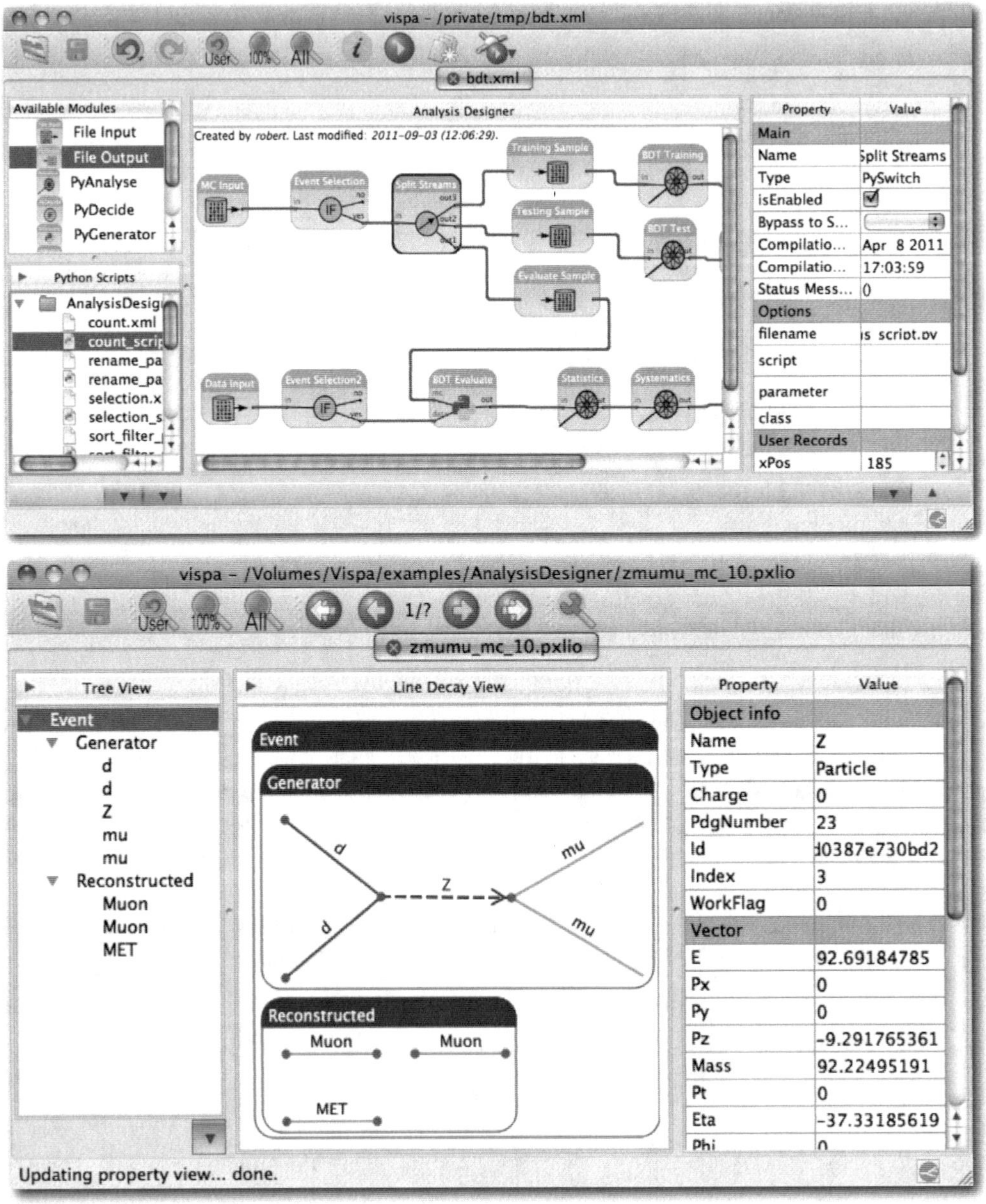

Figure 2. Top: The picture shows the analysis designer, which consists from left to right of a list of all available analysis modules, the workspace where the actual analysis flow is designed and visualized, and a property view to configure options of the selected module. Bottom: Here, the PXL data browser is depicted showing a physics event. On the left-hand side there is a tree view representation, while the center shows a Feynman-like diagram. The property view shows detailed information about a selected physics object.

can be enabled, disabled, and bypassed. By just leaving a subset of all modules active, while all other modules are deactivated, the first example is covered. In the case of the second example, certain module options could be changed to reflect the requirements of a certain data set. For the execution, the analyst selects the layer that should be run. Figure 3 shows an example analysis flow using layers.

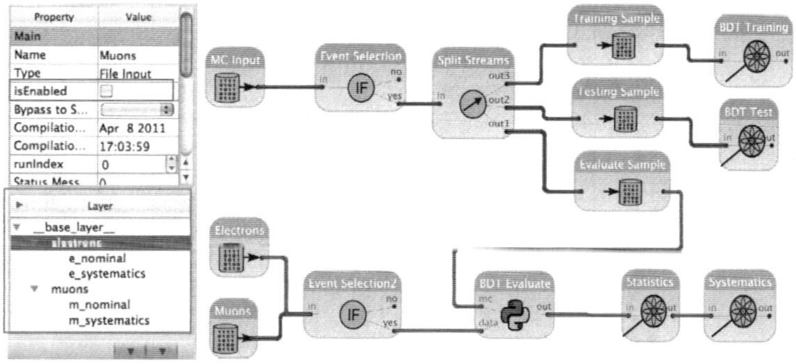

Figure 3. Left: The property view of the analysis designer is extended by a layer control box in the bottom. Right: An analysis flow with activated and deactivated (greyscale) modules. The currently active analysis path is realized as a layer.

2.2. *Convenient analysis execution*

Analyses are executed in different computing environments. While basic tests may be run locally on a desktop or laptop computer, there might be resource requirements, e.g. computing power or storage needs, that can only be covered by batch systems. VISPA provides a batch system plug-in making execution in these different environments transparently accessible. Based on an analysis developed with the analysis designer, a batch job is created. At this stage it is possible to override module options, e.g. to scan over ranges or a list of parameters. Use cases are for instance to try different cut values or to give a list of input files.

Once the job is configured, it is submitted to one of VISPA's batch managers. For each supported back-end batch system, a separate batch manager exists. These managers make sure to configure the batch job correctly for the corresponding back-end. Currently, a local batch manager, a condor batch manager and a grid batch manager are included. The local batch manager executes the analysis on the local machine running VISPA. The condor batch manager supports the condor batch system [10], while the grid batch manager submits jobs to the world wide computing grid [11]. The mechanism described allows the user to design the analysis and the batch job independent from the desired back-end. Submitting the job is the only part of the mechanism in which the back-end becomes important.

2.3. *Web-based physics analysis*

Software packages need to be installed and maintained. This has two consequences. First of all, the packages are available only on computers where they are installed. Secondly, the software has to be kept up-to-date on each machine. A complete new perspective of accessing physics analysis is VISPA@Web, which is the realization of the VISPA concept accessible via a web browser. Since VISPA@Web follows a server-client approach, the software has to be installed once only to serve several users. At the same time, each analysis is stored on the server and is, therefore, accessible from any computer connected to the internet without special requirements on the client side. While it is already fairly easy to exchange analysis modules with VISPA, this is even easier for analysts who use the same server. Thus, collaboration efforts are further enhanced.

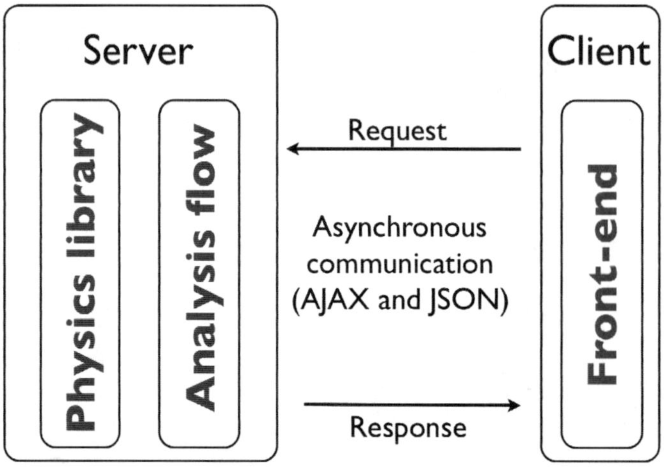

Figure 4. VISPA@Web consists of the same components as the desktop application as seen in figure 1. Here these components are devided between server and client side.

VISPA@Web employs the same components as the desktop version, see figure 4. However, while the front-end is the web interface, the physics library and analysis flow run on the server. Server and client communicate in an asynchronous manner using the AJAX technique [12]. Therefore, the web page does not have to be completely reloaded each time the user interacts with the front-end. The interactions are rather transferred to the server in the background in the JSON format [13] and are immediately applied to the analysis design. The user can just shut down his browser and reconnect, even from a different computer, and will find all of his analyses in the same state as when he quit the session before.

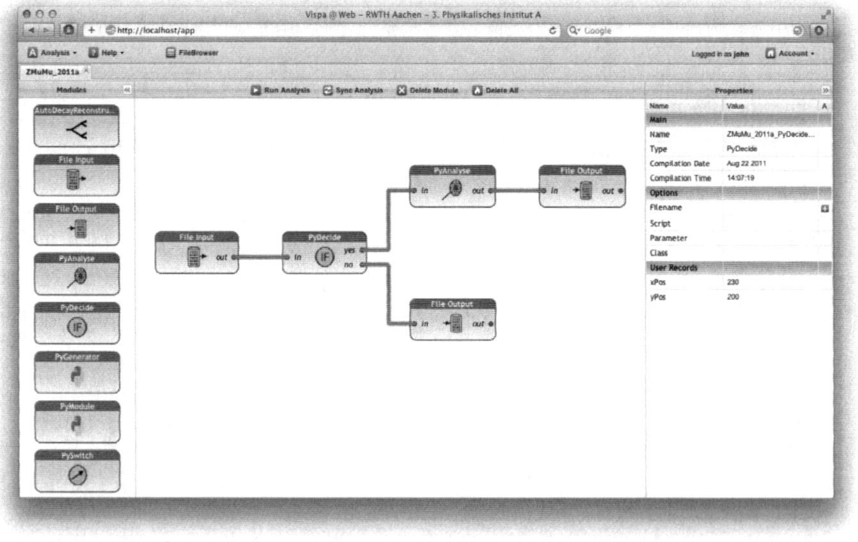

Figure 5. VISPA@Web screenshot. The interface is designed to resemble the desktop application in order to enable users familiar with VISPA to get started immediately.

Besides AJAX, which is provided by the ExtJS package [14], VISPA@Web also uses other up-to-date web technologies. The front-end uses HTML5 [15] and CSS [16].

Currently, VISPA@Web is in a beta phase. Basic analyses can be designed and executed. The next steps planned are to perform a multi-user field test and to integrate the aforementioned batch system plug-in in order to allow the system to scale with the number of users.

3. Summary

VISPA is a well-tested analysis development environment with applications in high-energy and astroparticle physics. Within the project, new technologies are explored. Layers unify and simplify the handling of likewise analysis steps while they increase the flexibility at the same time. The integrated batch system plug-in makes job submission to multiple back-ends effortless. Both layers and the batch system are consequent improvements based on the foundations of VISPA. VISPA@Web represents a new perspective and demonstrates that web-based physics analysis is possible.

References

[1] Actis O, Erdmann M, Fischer R, Hinzmann A, Kirsch M, Klimkovich T, Müller G, Plum M and Steggemann J 2008 (*Preprint* 0810.3609v1) URL http://arxiv.org/abs/0810.3609v1
[2] Brodski M *et al.* 2010 *Proceedings ACAT2010, Jaipur, India, PoS(ACAT2010)* **064**
[3] 2011 Physics eXtension Library (PXL) URL http://pxl.sourceforge.net
[4] 2010 PXL Class Documentation URL http://pxl.sourceforge.net/doxygen3a
[5] 2011 Python Programming Language URL http://www.python.org
[6] 2008 Qt - A cross-platform application and UI framework URL http://www.qtsoftware.com
[7] 2010 Pyqt URL http://www.riverbankcomputing.co.uk
[8] Erdmann M and Schiffer P 2010 *Astropart. Phys.* **33** 201 (*Preprint* 0904.4888)
[9] 2011 PARametrized Simulation Engine for Cosmic rays (PARSEC) URL http://www.physik.rwth-aachen.de/parsec
[10] Thain D, Tannenbaum T and Livny M 2005 *Concurrency - Practice and Experience* **17** 323–356
[11] Donno F and Litmaath M 2008 Data management in WLCG and EGEE. Worldwide LHC Computing Grid Tech. Rep. CERN-IT-Note-2008-002 CERN Geneva
[12] Garrett J J 2005 Ajax: A new approach to web applications URL http://www.adaptivepath.com/ideas/ajax-new-approach-web-applications
[13] Crockford D The application/json media type for javascript object notation (json) URL http://tools.ietf.org/html/rfc4627
[14] Sencha Inc 2011 JavaScript Framework for Rich Apps in Every Browser URL http://www.sencha.com/products/extjs/
[15] Hickson I *et al.* Html5 - a vocabulary and associated apis for html and xhtml URL http://www.w3.org/TR/html5/
[16] Bos B, Lie H and Lilley C The text/css media type URL http://tools.ietf.org/html/rfc2318

Druid, displaying root module used for linear collider detectors

M Ruan

Laboratoire Leprince-Ringuet,
École polytechnique, CNRS/IN2P3, 91128 Palaiseau, France

E-mail: `Manqi.ruan@llr.in2p3.fr`

Abstract. Based on the ROOT TEve/TGeo classes and the standard linear collider data structure, a dedicated linear collider event display has been developed. It supports the latest detector models for both International Linear Collider and Compact Linear Collider as well as the CALICE test beam prototypes. It can be used to visualise event information at the generation, simulation and reconstruction levels. Many options are provided in an intuitive interface. It has been heavily employed in a variety of analyses.

1. Introduction

Visualization plays an important role in various aspect of experimental particle physics, including the verification of detector geometry, debugging of simulation/reconstruction codes, physics analysis and online detector monitoring. Druid has been developed as a dedicated event display for the linear collider.

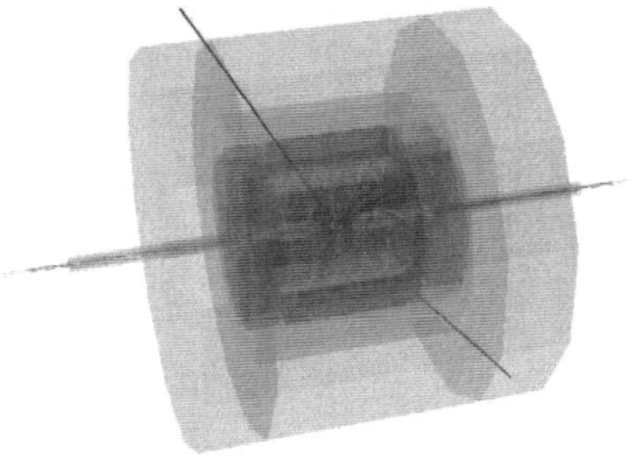

Figure 1. 230GeV ZH event at ILD detector simulation.

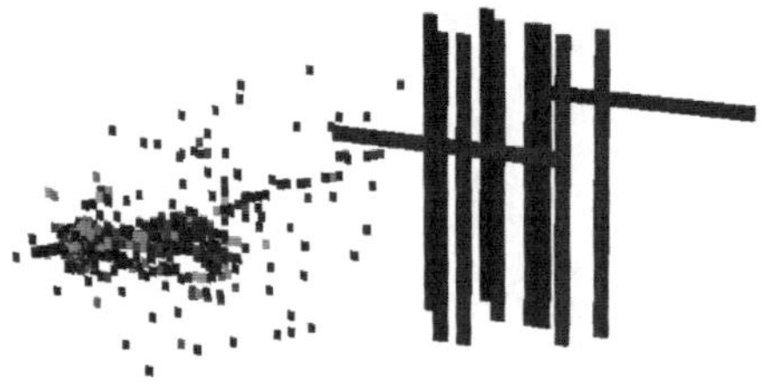

Figure 2. CALICE test beam event.

In this paper we present the dependency, development history, objects, display patterns and options for Druid, as well as an example on its utilization for debugging the reconstruction software.

Published under licence by IOP Publishing Ltd

2. TEve, gdml and Development history of Druid

Reading LCIO[1] data files and gdml[2] geometry files, Druid can visualize both event data and the detector geometry. For the event, Druid generates ROOT[3] TEve[4] objects according to the LCIO data file. It has been intensively tested on CALICE[5] test beam data and simulated/reconstructed full events, see figure 1, 2. For the detector geometry, Druid visualizes the geometry to different level of details according to the gdml file and the user's setting. The gdml file can be written by the official GEANT4[6] based simulation software Mokka[7]. The later release of Druid includes the gdml files for 5 most recent full detector concepts and CALICE test beam prototypes.

TEve is a framework for object management, providing hierarchical data organization, object interaction, and visualization through ROOT GUI and OpenGL. It can visualize detector geometry, simulated and reconstructed data such as hits, tracks, clusters. TEve is heavily used in LHC event displays. More details can be found in [4].

LCIO is the standard data format for the linear collider. The detector geometry is described with gear [8] or gdml file. Gear is a geometry description toolkit for ILC reconstruction software. Gear file records the large scale detector geometries, for example the radius and the half length of the TPC and the calorimeters. Gdml is an application-independent geometry description format based on XML. It can be used as the primary geometry implementation language and it provides a geometry data exchange format. In the Monte-Carlo study for linear collider, the gdml file is used to record all geometry information known to the simulation: the size, material, orientation and shape of every volume.

Druid has been optimized to have minimal dependencies: only ROOT (with version 5.28.00 or higher) and LCIO is requested. Druid has been integrated into the standard ilcsoft, which allows an automatic installation using ilcinstall [9].

The development of Druid was launched at the summer of 2009. The first prototype is a root macro, reading root files converted from LCIO files with the event information. Detector geometries was hard coded into cylinders and polygons with tunable parameters.

Soon Druid was modularized into an independent application compiled with LCIO and gear. This leads to the first release of Druid in November 2009.

In May 2010, after discussing with American colleagues working on the Silicon Detector (SiD), gdml support was added to Druid. This option allows the visualization of detector geometry up to the simulation level.

At November 2010, multiple windows option was enabled in Druid, which allows a synchronized display with R-phi and YZ plane projection (see figure 3). To minimize package dependency, gear support was dropped. Druid was stabilized since then, with minimal upgrading from time to time.

Testing on Dell E6500 (CPU: Intel Core 2 Duo T9600, 2.8GHz) with linux system, it takes about 5 second to start Druid on any data file. The time to display a event depends on the event size, for example, it takes about 3 seconds to display a 1 TeV ttH event at Compact Linear Collider (CLIC) SiD detector (see figure 3).

3. Objects

LCIO files contain event information at different levels and organize them into different collections. Druid accordingly organizes the event information into different groups.

Monte Carlo truth level:

MCParticle: displayed as helix.
SimCalorimeterHit: displayed as cuboid with different size, orientation and colors.
SimTrackerHit: displayed as points and colored according to track charge.

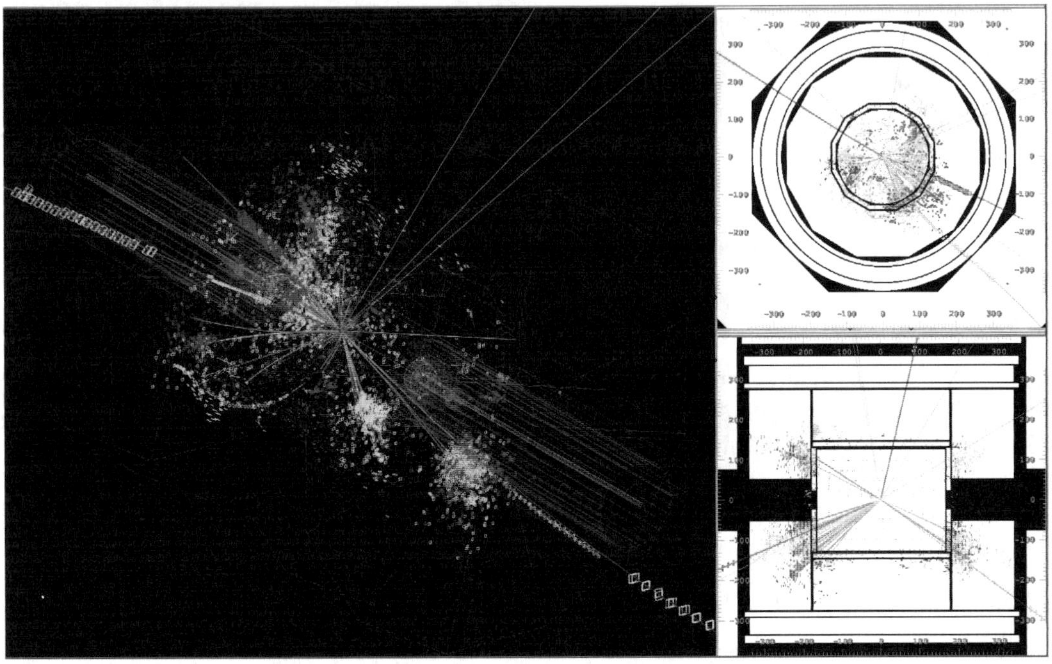

Figure 3. 1 TeV ttH event at the CLIC SID with projections.

Digitized level:

CalorimeterHit: displayed as cuboid with different size, orientation and color.
TrackerHit: displayed as point with uniform green color.

Reconstruction level:

Intermediate reconstructed objects: TrackAssignedHit, Vertex, ClusterHit. Displayed as points or cuboid and colored according to their index (the order in corresponding collection).
ReconstructedParticle: displayed as helix, colored according to the type of particle.

MCParticle and ReconstructedParticle collections are divided into subgroups according to the particle type/energy, while the detector hits are divided into subgroups according to the subdetectors.

For the detector geometry, each volume is displayed as a polyhedron, whose color and transparency is defined according to the material or to the user's setting. The full geometry information is usually too dense for a display focus on the event information, Druid therefore uses two options to reduce the workload of geometry display. First, for any "unwanted" volume, Druid allows not to display it. Secondly, a tunable parameter called display depth is employed, which refers to the level of hierarchy in the geometry organization. More details will be displayed with larger display depth. For the event display focus at event topology and debugging reconstruction code, staying at the default display depth, at which the contour of sub detectors are displayed, should be sufficient, see figure 4, 5.

4. Display Options

4.1. Options inherited from TEve and ROOT

From TEve, Druid inherits many display options with hotkey access, for example zoom, rotation, drawn back to the original orientation and scale, and switch the color of the background.

Journal of Physics: Conference Series **368** (2012) 012040 doi:10.1088/1742-6596/368/1/012040

Figure 4. ILD detector with Yoke, Coil and part of Calo dismounted.

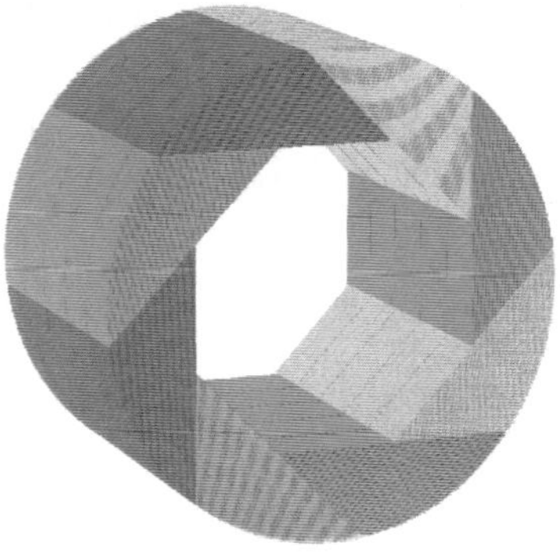

Figure 5. ILD HCAL Barrel.

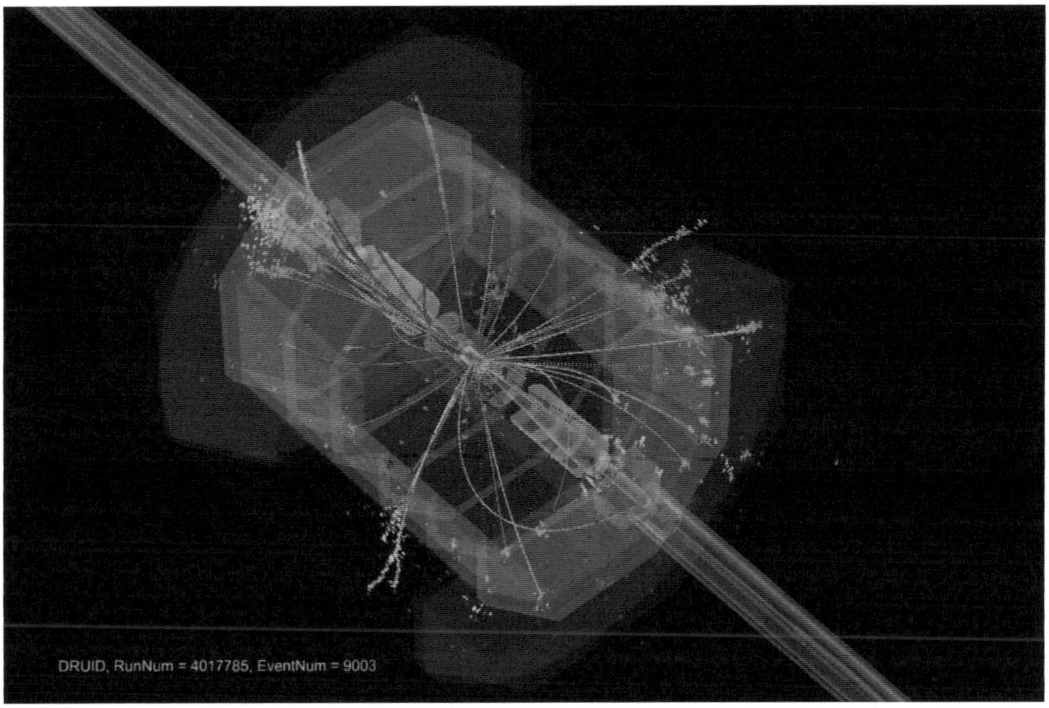

Figure 6. 500GeV ttbar event at ILD.

TEve allows a cut away view of the event display, which removes part of the display as defined by a box or a plane with interactively tunable parameters. An example is given in figure 6, where one eighth of the detector is removed.

Another useful function provided by TEve is the text attachement for each object. By picking up an object with the cursor, the attached text will be printed in the display, see figure 7.

Besides these options, there are lots of interactive actions at the ROOT GUI interface. The GUI interface is divided into three pages: the Eve page and file page inherited from TEve, as well as the option panel page defined by Druid. The file page browses the file directory while the Eve page contains a browser to all the displayed/hidden objects. By checking the appropriate boxes, any object can be hidden/displayed individually or by groups. The hide/display status will be

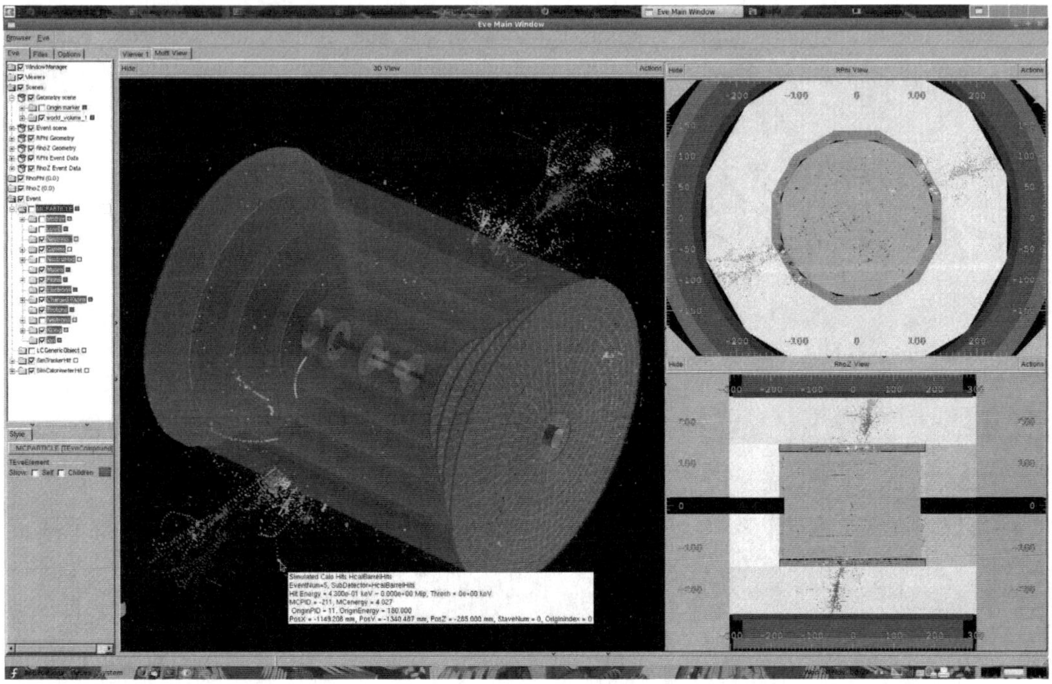

Figure 7. Z Threshold uds event at SID, with part of detector dismounted.

kept (by collection name) when navigating to a new event. Other options such as changing background color and illumination setting, tuning geometry display depth, setting reference point/frame are also available.

4.2. Event navigation and cuts

Many practical options are defined in the Druid option panel, for example:

Event navigation: to display the previous/next event, or specify the event number.

Druid defined options with buttons, there are (from left to right):

(1), Select rotation center.

(2), Regenerate the color: generate another color according to the object index. Here index means the order of a given object in its collection, for example the clusters. This option allows to a straightforward visual separation of different objects in one collection.

(3), Switch between scenarios: the minimal scenario which ignores all the intermediate reconstructed objects (as digitized hits, tracks and clusters) and the maximal scenario that displays every possible collection. To give fastest performance, the minimal scenario is set as the default.

Cuts:

The MCParticle list can be very long for the simulated event. Display all of the MCParticle may froze the display. a cut on the minimal transverse momentum (with default value 1.5GeV) is therefore applied.

A default energy cut of 0.2 mip (the most probable value of energy deposited by a minimal ionization particle in a certain volume) has been applied to both simulated and digitized calorimeter hits. The low energy hits can be shown (with default grey color) by clicking on

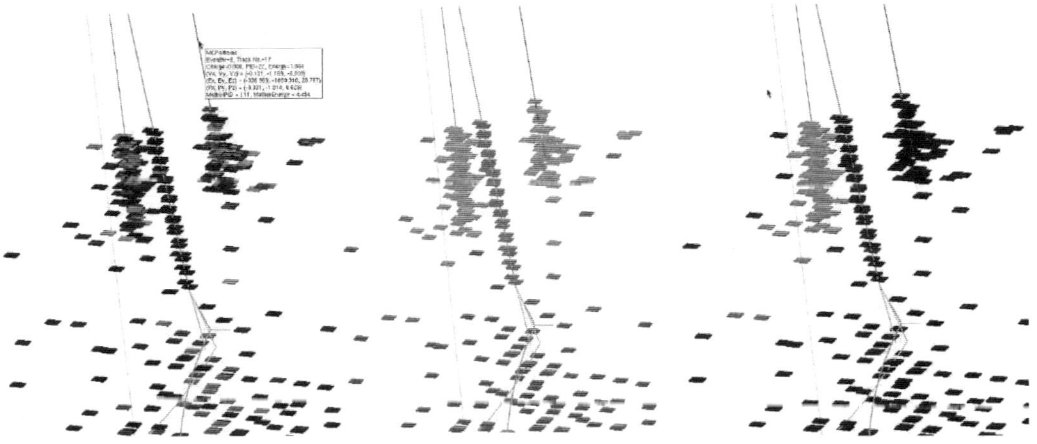

Figure 8. Tau jet $\tau \rightarrow \nu + \pi^0 + \pi^+$ with different color options: from left to right, energy, track ID and random index on track.

the box marked with 'show lowE cell'. The value of hit energy threshold can be interactively adjusted. Each time the color or size option is changed, the names and statistics of affected collections are printed on the terminal.

4.3. Color/size option for calorimeter hits

There may be three different types of calorimeter hit collections in the lcio file: the simulated, digitized and clustered calorimeter hits. Colors and sizes for these hits can be specified in the option panel according to different information.

Simulated hits can be colored according to the energy (calibrated to mips), the type or index of the particle (direct or origin) that creates this hit, the time, or a uniform color. figure 8 shows a τ jet displayed with different color option. The digitized calorimeter hits are colored to the energy calibrated to mips. A global color factor could be tuned when colored with the energy or time. Clustered hits are colored according to the cluster index.

The default size and orientation of calorimeter hits has been tuned to the cell sizes corresponding to different models and subdetectors. The hit size can be adjusted independently for simulated, digitized and clustered hits on the option panel.

5. Example application: debugging reconstruction code

One of the most important applications for Druid is the debugging of reconstruction code. Here we explain how it works with one example.

Figure 9 shows the same τ jet at simulation and reconstruction level. The left view is the reconstructed particle tracks with assigned calorimeter hits, where you can see two photons and one pion is reconstructed. The middle view overlays the MCParticle information, thus we know that reconstruction module successfully found the gammas and pions, some secondary particles generated in the shower and the neutrino are also displayed. The right view overlays the simulated calorimeter hits: some simulated calorimeter hits are dropped during the digitization/reconstruction.

6. Summary

Druid, a dedicated event display for linear collider has been developed. It has been heavily used in the geometry verification, the data analysis and the reconstruction algorithm developments.

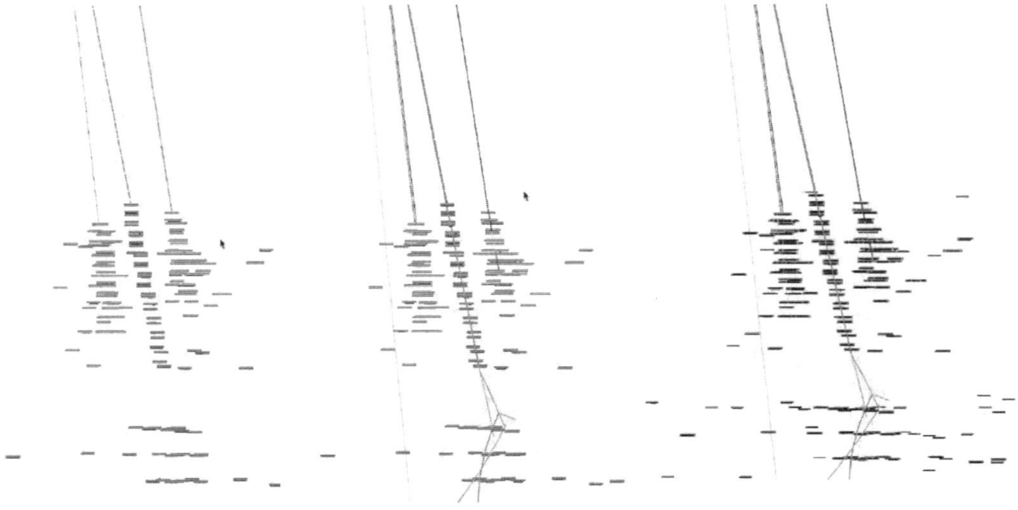

Figure 9. Tau jet $\tau \rightarrow \nu + \pi^0 + \pi^+$ at MCTruth and Reconstruction level.

Acknowledgments

I am grateful to Gabriel Musat, Daniel Jeans for the polishing of the codes, to Vincent Boudry, Jean-Claude Brient and Henri Videau for the valuable discussion and their continuous support. Thanks to Jayant Pande for his hard work in the early developing phase.

Special Thanks goes to Matevz and Alja, for all the nice discussions and beer we had.

The research leading to these results has received funding from the European Commission under the FP7 Research Infrastructures project AIDA, grant agreement no. 262025.

References

[1] Gaede F, *et al.* 2003 LCIO-A persistency framework for linear collider simulation studies *LC-TOOL-2003-053*

[2] Chytracek R, McCormick J, Pokorski W, Santin G 2006 Geometry Description Markup Language for Physics Simulation and Analysis Applications *IEEE Trans. Nucl. Sci.*, Vol.**53**, Issue: 5, Part 2, 2892-2896

[3] Brun R and Rademakers F 1997 ROOT - An Object Oriented Data Analysis Framework, Proceedings AIHENP'96 Workshop, Lausanne, Sep. 1996, *Nucl. Inst. Meth. in Phys. Res. A* **389** 81-86

[4] Tadel M 2010 Overview of Eve - the Event Visualization Environment of the ROOT *J. Phys.: Conf. Series.* **219** 042055

[5] The CALICE collaboration 2008 Design and electronics commissioning of the physics prototype of a Si-W electromagnetic calorimeter for the International Linear Collider *Journal of Instrumentation* **3(08)**: P08001, arXiv:0805.4833v1 [physics.ins-det].

[6] Agostinelli S, *et al.* 2003 Geant4 - a simulation toolkit *Nucl. Instrum. Meth.* A **506**, 250-303

[7] http://polzope.in2p3.fr:8081/MOKKA

[8] http://ilcsoft.desy.de/portal/software_packages/gear

[9] http://ilcsoft.desy.de/portal/software_packages/ilcinstall

Journal of Physics: Conference Series **368** (2012) 012041 doi:10.1088/1742-6596/368/1/012041

Green's function based unparameterised multi-dimensional kernel density and likelihood ratio estimator

P Kövesárki, I C Brock and A E Nuncio Quiroz

Physics Institute, University of Bonn

E-mail: kovesarki@physik.uni-bonn.de, brock@physik.uni-bonn.de, nuncio@physik.uni-bonn.de

Abstract. This paper introduces a probability density estimator based on Green's function identities. A density model is constructed under the sole assumption that the probability density is differentiable. The method is implemented as a binary likelihood estimator for classification purposes, so issues such as mis-modeling and overtraining are also discussed. The identity behind the density estimator can be interpreted as a real-valued, non-scalar kernel method which is able to reconstruct differentiable density functions.

1. Theory of density estimation

Generally, a density estimator is an identity operator for the probability density function. If it is a linear operator, then it is an approximation of the Dirac-delta function:

$$g_{\text{est}}(x) = \int \delta_{\text{approx}}(x - x')g(x')\,\mathrm{d}x\,. \tag{1}$$

Fixed kernel methods usually estimate the density with a bias, since the kernel does not converge to the delta function. The same can be seen on histograms with fixed bin sizes, and to make them unbiased certain adaptivity is necessary. Different kernels usually maintain different features of the g_{est}, such as differentiability or easy recognizability of peaks [1]. Non-adaptive methods are the estimation from the moments via Fourier integrals [2] or the estimations that target the integral probability density. A problem with the methods using Fourier integrals, is that they may require computationally expensive volume integrals, and the result is not necessarily a density function until sufficient number of moments are known.

To avoid the complexity of finding the global minimum of the error function, most of the overtraining problems and heavy volume integrals, we tried to define a successive adaptive kernel method to estimate the probability densities, using a kernel that is suppressed at large distances. The starting point is, that functionals like linear operators are preferred, as they can be successfully approximated with a sample of a distribution, because they act like expectation values of random variables.

$$\lim_{N\to\infty} \sum_{i=1}^{N} Lp = \int L(x, x')p(x')\,\mathrm{d}\mu(x')\,, \tag{2}$$

where L is a linear operator, p is a parameter vector and $\mu(x)$ is a probability measure. To have such an estimator for the probability density itself, an identity operator is needed:

$$\mu(x) = I(\mu) = P\left(\int L(x, x')p(x')\,\mathrm{d}\mu(x')\right), \tag{3}$$

where P is a necessary post-processing to convert the linear operator's output to a scalar. Identity operators can be constructed using Green functions of operators. A particularly interesting one is the Green function of the Laplace operator in $n \geq 3$ dimensions[1].

$$\Delta G(x, x') = \delta(x - x') \tag{4}$$

$$G(x, x') = -\frac{1}{S_n}\frac{1}{|x - x'|^{n-2}}\,. \tag{5}$$

Here S_n is a constant, the surface of a unit sphere in n dimensions. For twice differentiable functions $c(x)$ that disappear at the boundaries, $\lim_{x\to\infty} c(x) = 0$, the identity operator can be constructed:

$$c(x) = -\frac{1}{S_n}\int_{\mathbb{R}^n} \frac{\Delta c(x')}{|x - x'|^{n-2}}\,\mathrm{d}^n x' = \frac{1}{S_n}\int_{\mathbb{R}^n} \partial_{\mu'}\frac{1}{|x - x'|^{n-2}}\cdot\partial_{\mu'}c(x')\,\mathrm{d}^n x'\,. \tag{6}$$

This can not be used directly for density estimations, since it is linear in $\partial_{\mu'}c(x')$ instead of $c(x')$, but an identity can be constructed for the vector field of the first derivative:

$$\partial_\mu c(x) = \frac{1}{S_n}\int_{\mathbb{R}^n} \partial_\mu\partial_{\mu'}\frac{1}{|x - x'|^{n-2}}\cdot\partial_{\mu'}c(x')\,\mathrm{d}^n x'\,. \tag{7}$$

The amplitude of $\partial_\mu c(x)$ is still a scalar function; hence for those probability measures, $g(x)$, where a $c(x)$ exists[2] with $|\partial_\mu c(x)| = g(x)$, there must be at least one $\phi_\mu(x)$ vector field consisting of $\phi_\mu\phi^\mu = 1$ unit vectors, for which the following identity holds:

$$g(x)\phi_\mu(x) = \frac{1}{S_n}\int_{\mathbb{R}^n} \partial_\mu\partial_{\mu'}\frac{1}{|x - x'|^{n-2}}\cdot\phi_{\mu'}(x')g(x')\,\mathrm{d}^n x'\,. \tag{8}$$

Since the integral operator was derived from the Laplace operator, in three dimensions it is similar to the *dipole* interaction operator in electromagnetism, hence ϕ_μ might be called a dipole field. What is important is that this $\phi_\mu(x)$ field can be found without the knowledge of the true $g(x)$; only the possibility of calculating the expectation values on the left-hand side of equation (8) is required. In this case it is possible to calculate the left-hand side for any configuration of $\phi_{\mu'}(x')$, until the expectation value $E_\mu(x)$ has the same direction as $\phi_\mu(x)$ at every x:

$$\phi_\mu(x) \parallel E_\mu(x) = \frac{1}{S_n}\int_{\mathbb{R}^n} \partial_\mu\partial_{\mu'}\frac{1}{|x - x'|^{n-2}}\cdot\phi_{\mu'}(x')g(x')\,\mathrm{d}^n x'\,. \tag{9}$$

A way to show that $E_\mu(x) \approx g(x)\phi_\mu(x)$ is to do an integration by parts first:

[1] Although the Green's function is different for $n = 2$, the same procedure can be repeated and it gives the same form of functions at the end, making method valid in 2 dimensions.
[2] See the subsection targeted at the existence of such a $c(x)$.

$$E_\mu(x) = \frac{1}{S_n}\int\limits_{\mathbb{R}^n} \partial_\mu \partial_{\mu'} \frac{1}{|x-x'|^{n-2}} \cdot \phi_{\mu'}(x')g(x')\,\mathrm{d}^n x' \qquad (10)$$
$$= -\frac{1}{S_n}\int\limits_{\mathbb{R}^n} \partial_\mu \frac{1}{|x-x'|^{n-2}} \cdot \partial_{\mu'}\left(\phi_{\mu'}(x')g(x')\right)\,\mathrm{d}^n x' ,$$

where the integral on the boundaries is zero, since the derivatives of the probability measures are also suppressed at infinity. It is possible to do an external differentiation now, calculating the divergence of $E_\mu(x)$, and using equation (4):

$$\partial_\mu E_\mu(x) = \int\limits_{\mathbb{R}^n} \underbrace{-\frac{1}{S_n}\partial_\mu\partial_\mu \frac{1}{|x-x'|^{n-2}}}_{\Delta G(x,x')=\delta(x-x')} \cdot \partial_{\mu'}\left(\phi_{\mu'}(x')g(x')\right)\cdot \mathrm{d}^n x' = \partial_\mu\left(\phi_\mu(x)g(x)\right) . \qquad (11)$$

For an arbitrary ϕ_μ, E_μ and $g\phi_\mu$ can only differ by a divergence-free vector field. In order to show $|E| \approx g$, a certain measure is needed for the difference. With the square of the difference it becomes:

$$\int\limits_{\mathbb{R}^n} \left(g(x)\phi_\mu(x) - E_\mu(x)\right)^2\,\mathrm{d}^n x \geq 0 . \qquad (12)$$

Expanding the square, it will have three terms:

$$\int\limits_{\mathbb{R}^n} \mathrm{d}^n x \underbrace{g^2}_{\text{invariant}} -2 \underbrace{g\phi_\mu E_\mu}_{\text{dipole energy}} + \underbrace{E^2}_{\text{field energy}} . \qquad (13)$$

The field energy can be expressed in terms of the dipole energy. Explicitly writing it out:

$$\int\limits_{\mathbb{R}^n} \mathrm{d}^n x E^2 = \frac{1}{S_n^2}\int\limits_{\mathbb{R}^{2n}} \mathrm{d}^n x\,\mathrm{d}^n y\left[g^x\phi_\mu^x g^y\phi_\nu^y\right] \underbrace{\int\limits_{\mathbb{R}^n} \mathrm{d}^n z \partial_{x_\mu}\partial_{z_\eta}\frac{1}{|x-z|^{n-2}}\partial_{y_\nu}\partial_{z_\eta}\frac{1}{|y-z|^{n-2}}}_{I_{\text{int}}} . \qquad (14)$$

Performing integration by parts on the I_{int} term, $\Delta\frac{1}{|x-z|^{n-2}}$ appears. Due to equation (4) this can be substituted with a delta function. The remaining part has the derivative of the delta function, which can be easily evaluated:

$$I_{\text{int.}} = -S_n \int\limits_{\mathbb{R}^n} \mathrm{d}^n z \partial_{x_\mu}\delta(x-z)\partial_{y_\nu}\frac{1}{|y-z|^{n-2}} = S_n \partial_{y_\nu}\partial_{x_\mu}\frac{1}{|x-y|^{n-2}} . \qquad (15)$$

As a result the field energy equals the dipole energy

$$\int\limits_{\mathbb{R}^n} \mathrm{d}^n x E^2 = \int\limits_{\mathbb{R}^n} \mathrm{d}^n x g\phi_\mu E_\mu , \qquad (16)$$

so the equation (12) can be modified accordingly:

$$\int\limits_{\mathbb{R}^n} \left(g(x)\phi_\mu(x) - E_\mu(x)\right)^2\,\mathrm{d}^n x = \int\limits_{\mathbb{R}^n} \mathrm{d}^n x g^2 - g\phi_\mu E_\mu \geq 0 . \qquad (17)$$

The minimum is not necessarily zero, but it can be reached by finding the energy minimum of the dipole system.

1.1. Existence of the optimum

It was assumed for equation (8), that an identity operator exists for a given $g(x)$ probability density with a special $\phi_\mu(x)$ configuration:

$$\partial_\mu c(x) = g(x)\phi_\mu(x)\,. \tag{18}$$

In other words, such a $c(x)$ has to exist which has a gradient with specified absolute value $|\partial_\mu c(x)| = g(x)$. This only exists if $g(x)\phi_\mu(x)$ can be made curl free, and all possible loop integrals should result in zero:

$$\oint_L g(x)\phi_\mu(x)\,\mathrm{d}l_\mu = 0\,. \tag{19}$$

As an n-dimensional cube has $n(n-1)$ faces, from which $n(n-1)/2$ share the same corner, $n(n-1)/2$ linearly independent infinitesimal loops can be imagined. So the curl in n-dimensions may have that many independent variables, which has to be zeroed out by a ϕ_ν vector field with only $n-1$ independent variables. That seems to be non-trivial, but for differentiable $g(x)$, it turns out it can be done easily. The curl in question is:

$$[\mathrm{curl}\ g(x)\phi_\mu(x)]_{\nu\eta} = \partial_{(\nu)}(g\phi_\mu)e_\mu^{(\nu)} - \partial_{(\eta)}(g\phi_\mu)e_\mu^{(\eta)} \stackrel{?}{=} 0\,, \tag{20}$$

where $e_\mu^{(\nu)}$ denotes the unit basis vector in the ν direction, and the bracketed indices are *not* summed up. In the basis $e_\mu^{(\nu)} = \delta_{\nu\mu}$, the curl is simplified:

$$[\mathrm{curl}\ f(x)\phi_\mu(x)]_{\nu\eta} = \partial_{(\nu)}(g\phi_{(\nu)}) - \partial_{(\eta)}(g\phi_{(\eta)}) \stackrel{?}{=} 0\,. \tag{21}$$

As only certain derivatives of ϕ_μ appear, one possibility to anull the curl is to set every subcomponent to zero:

$$\partial_{(\nu)}(g\phi_{(\nu)}) = 0 \tag{22}$$

$$\frac{\partial_{(\nu)}\phi_{(\nu)}}{\phi_{(\nu)}} = -\frac{\partial_{(\nu)}g}{g}\,. \tag{23}$$

In this differential equation only the relative change of ϕ_ν appears, so setting $\phi_\mu\phi_\mu = 1$ at every x is still possible. As a consequence, the ϕ_μ vector field exists for any differentiable g, making $g\phi_\mu$ curl-free, and a $c(x)$ function exists with the property $\partial_\mu c(x) = g(x)\phi_\mu(x)$, making equation (8) an identity.

2. Density estimation from a finite sample

As long as a suitable $\phi_\mu(x)$ field is found that satisfies equation (9), it can be used to estimate any differentiable probability density function from a sample $\{x_i\}$:

$$g(x) = \lim_{N\to\infty}\left|\frac{1}{NS_n}\sum_{i=1}^{N}\partial_\mu\partial_\nu\frac{1}{|x-x_i|^{n-2}}\cdot\phi_\nu(x_i)\right|\,. \tag{24}$$

However, since the Green's function has divergences, a finite sample may produce unnecessarily large fluctuations. Fortunately, volumes around most of the divergences can be neglected, as their integral is zero. A homogeneous $g(x) = g$, homotropic $\phi(x)_\mu = \phi_\mu$ sphere surface around a point x gives no contribution:

$$\int\limits_{S_n} \partial_\mu \partial_{\mu'} \frac{1}{|x-x'|^{n-2}} g\phi_{\mu'} \, \mathrm{d}^n x' = g\phi_{\mu'}\partial_\mu \int\limits_{S_n} \partial_{\mu'} \frac{1}{|x-x'|^{n-2}} \, \mathrm{d}^n x' \tag{25}$$

$$= g\phi_\nu \partial_\mu \underbrace{\int\limits_{S_n} \underbrace{-\frac{1}{|x-x'|^{n-1}}}_{const.} \hat{r}_{\mu'} \, \mathrm{d}S_n \, \mathrm{d}r'}_{\int\limits_{S_n} \hat{r}'_\mu \, \mathrm{d}S_n = 0} = 0\,.$$

As a result, a small sphere of radius ΔR around the point of interest x can be excluded from the summation in equation (2), since the error will go as $\partial_\mu(g(x)\phi_\nu(x))O(\Delta R)$ for any differentiable $g(x)$ and $\phi(x)$. The only exception where the error is not $O(\Delta R)$ is for δ functions, so these points have to be treated independently.

In order to have control on the fluctuations, one has to choose the ΔR radius of the exclusion volume in such a way that a δr thick shell around this sphere contains enough sample points to estimate the integral belonging to that area. In this case the points in this layer will have a dominant contribution to the sum between $\{\frac{1}{\Delta R^n}, \frac{1}{(\Delta R+\delta r)^n}\}$, since points at larger distances will have decreasing values. To have the contribution in the shell to be of the same order of magnitude, δr has to be around $\Delta R/n$. This approximation is derived from the behaviour of the $(1+{}^1/_n)^{-n}$ function, which becomes nearly flat in n, with a value around ${}^1/_2$. From this it follows that if one needs N_{large} number of points in this stable layer, approximately $n \cdot N_{\text{large}}$ points have to be discriminated, which eventually determines ΔR. In practice one may choose a different N_{large} for the iterative search of the $\phi_\nu(x_i)$ parameters and for the evaluation of the density estimation.

2.1. Finding the parameters

With a suitable N_{discr}, the set of $\{\phi_{\mu i} = \phi_\mu(x_i)\}$ that can be used for density estimations can be found with iterative search. The stopping criterium can be formulated with discretising equation (9):

$$\phi_{\mu i} \parallel E_\mu(x_i) = \frac{1}{NS_n} \sum_{\substack{j \in \{j:|x_i-x_j|>R^i_{\text{discr}}\} \\ |\{k:|x_i-x_k|<R^i_{\text{discr}}\}|=N_{\text{discr}}}} \partial_\mu \partial_{\mu'} \frac{1}{|x_i - x'_j|} \phi_{\mu' j}\,. \tag{26}$$

As this requirement is similar to looking for the ground state of the dipole system, a way to find it is the gradient descent on the energy function. With $E_{i\mu} = E_\mu(x_i)$ this direction is:

$$\mathrm{d}\phi_{i\eta} = (1 - \phi_{i\eta}\phi_{i\mu})E_{i\mu}\,, \tag{27}$$

which is a rotation with the E_μ field in the directions perpendicular to ϕ_μ. A gradient descent method typically contains a line search part, but it is not necessary here. It is enough to set a limit for the angular change of the $\phi_{i\mu}$, an order or magnitude smaller than the maximal possible, $|\mathrm{d}\phi_{\max}| = 0.1 < \pi$, seems to be a good choice, which ensures that the E_μ field won't change significantly during the iteration, keeping it a valid gradient. Further speed increase can be achieved, while taking care not to overshoot the angle between $E_{i\mu}$ and $\phi_{i\mu}$, since this is what minimises the energy. However, there might be several local energy minima, but the identity in equation (8) is only true if $g\phi$ is curl free. Following from the inequality in equation (12) it means that the non-global minima solutions are not satisfying the identity and they are not curl free. These solutions will hold a bias in the density estimations, which can be relatively big in the low density regions. This is because the non-optimal minima have a remnant polarisation, for example the destructive interferences do not fully cancel at zero density regions.

As can be seen on figure 1, the algorithm actually performs as a density estimator. A sample with a size of two thousand points from a two dimensional Gaussian distribution was fed into the parameter search, and the density estimation was evaluated in the described manner, at the position of sample points. The figure shows the average of the estimated densities for a given radius, where the theoretical value is constant, because of the rotational symmetry of the chosen Gaussian distribution. The spread was calculated from the variance of the estimations on the same radius. The uncertainty can be directly compared with the simplest, kernel-less nearest k-neighbour method [1] in figure 2. It can be seen, that the uncertainty for the discussed density estimator in figure 1 is approximately half of that from the nearest k-neighbour method with flat kernel in figure 2. The resolution were chosen to be the same for the two figures, given that the exclusion radii, defined in section 2, were the same as the radii that was used for volume calculation in the nearest k-neighbour method.

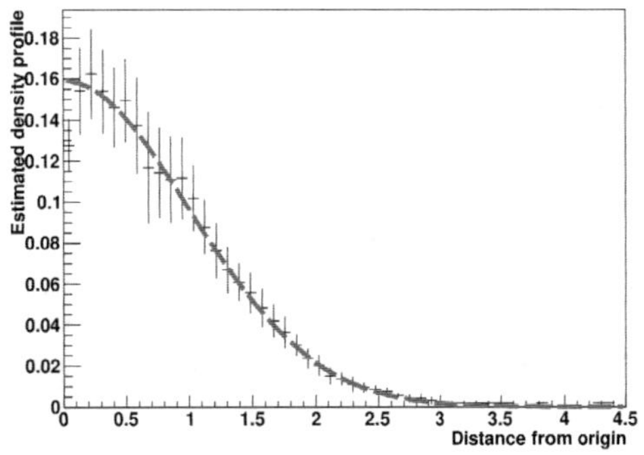

Figure 1. Radial density estimate profile of a Gaussian distribution placed at the origin. The crosses show the spread of the Green's function density estimate compared to the true value shown as the dashed red curve.

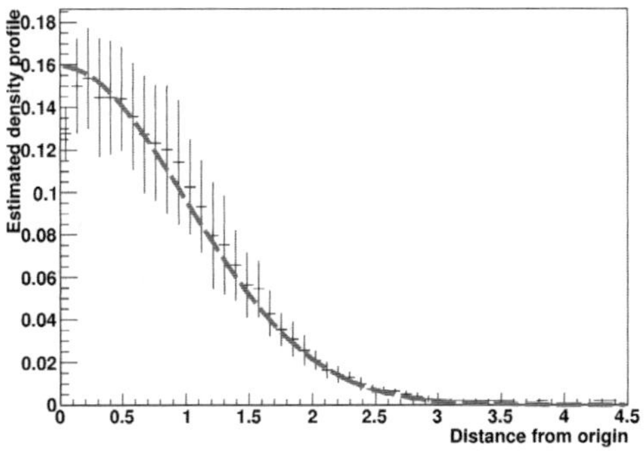

Figure 2. Profile plot for the density estimate made for the same Gaussian sample as in figure 1 made with the nearest k-neighbour method with a flat kernel.

3. Classification and overtraining

Although binary classification can be done with binary regression models, their results typically depend only on the fraction of the signal $s(x)$ and background density $b(x)$, which is optimal due to the Neyman-Pearson Lemma [3]. This is also true for an optimally trained neural network with a quadratic loss function, where the response is usually the following [1][4]:

$$r_{\rm NN} = \frac{s(x)}{s(x) + b(x)} \,. \tag{28}$$

In case the network is not optimally trained, the error of the response $r_{\rm NN}$ can be related to the uncertainty of the $s(x)/b(x)$ fraction. The same fraction can be calculated from a density estimator, giving a likelihood estimation. The uncertainty in the latter case is much better under control, since the calculation of the signal and background densities are made independently. For the method described in this article, a density estimation for a single point comes from the superposition of the kernels of the most dominant $N_{\rm large}$ sample points, hence by construction the likelihood estimation can not come from a single point, as happens during overtraining in regressions.

Figure 3 and figure 4 show the response distribution for a network trained on a signal consisting of twelve non-overlapping Gaussians and a flat background [5]. Unlike in many regression models, here there is no need for an independent test sample for an early stopping criterium. Figure 3 is evaluated on the full training sample, while figure 4 was made on an independent sample. The shape of the two responses are essentially same, the slight difference is at the part where either the signal or the background distribution density is very low, since these areas have the largest uncertainty in density estimators.

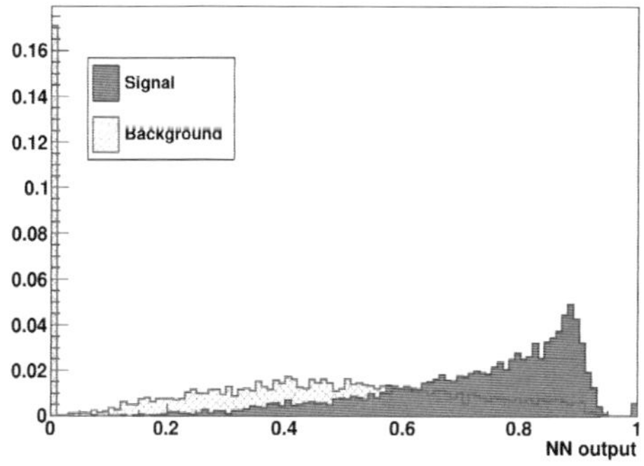

Figure 3. Distribution of the response function on the 10^4 training sample.

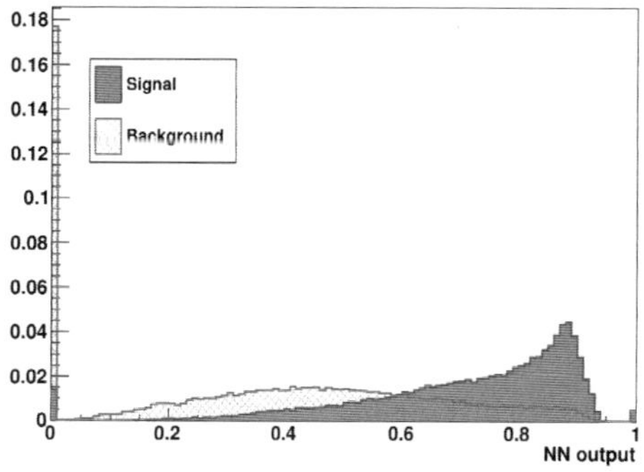

Figure 4. Distribution of the response on an independent 10^5 test sample.

4. Conclusions

The article presents a novel class of density estimators, based on Green's function identities. The investigated one is derived from the Green's function of the Laplace operator which provides meaningful results for once differentiable probability density functions. It is possible to construct density estimators from other linear operators, but they are expected to require either higher order differentiability or other functional constraints on the investigated probability density. As it was shown, such a density estimator can be used for binary classification. In such case, due to the handling of the fluctuations, overtraining is not so prominent as it is in binary regression models. Therefore there is less need for independent test samples.

Acknowledgements

We would like to thank Eckhard von Törne for giving encouragement, for pointing out the important topics needed to be touched and for the sample he prepared and gave us to use for testing the classification power of the algorithm.

References

[1] Bishop C M 2006 *Pattern Recognition and Machine Learning* (Secaucus, NJ, USA: Springer-Verlag New York, Inc.) ISBN 0387310738 iSBN 0-387-31073-8

[2] James F 2006 *Statistical methods in experimental physics* (5 Tuh Tuck Link, Singapore 596224: World Scientific Publishing Co. Pte. Ltd.) ISBN 9789812567956 iSBN 981-256-795-X

[3] Neyman J and Pearson E S 1933 On the problem of the most efficient tests of statistical hypotheses *Philosophical Transactions of the Royal Society of London. Series* A **231** 289–337 URL `http://rsta.royalsocietypublishing.org/content/231/694-706/289.short`

[4] Feindt M 2004 A neural bayesian estimator for conditional probability densities *ArXiv Physics e-prints* (*Preprint* `arXiv:physics/0402093`)

[5] von Toerne E private communications

ACAT 2011 IOP Publishing
Journal of Physics: Conference Series **368** (2012) 012042 doi:10.1088/1742-6596/368/1/012042

Continuous simulation of hypothetical physics processes with multiple free parameters

J Zhong[1,2,3] and S-C Lee[2]

[1] School of Physics, Nanjing University, Nanjing, CN - Jiangsu 210093, China
[2] Institute of Physics, Academia Sinica, TW - Taipei 11529, Taiwan
[3] Physics Department, University of Oxford, UK - Oxford OX13RH, United Kingdom

E-mail: `Jiahang.Zhong@cern.ch`, `phsclee@phys.sinica.edu.tw`

Abstract. We present a new approach to simulate Beyond-Standard-Model (BSM) processes which are defined by multiple parameters. In contrast to the traditional grid-scan method where a large number of events are simulated at each point of a sparse grid in the parameter space, this new approach simulates only a few events at each of a selected number of points distributed randomly over the whole parameter space. In subsequent analysis, we rely on the fitting by the Bayesian Neural Network (BNN) technique to obtain accurate estimation of the acceptance distribution. With this new approach, the signal yield can be estimated continuously, while the required number of simulation events is greatly reduced.

1. Introduction

Many Beyond Standard Model (BSM) physics are extensively searched in collider experiments, such as Large Hadron Collider (LHC). As the properties of such new physics are unknown, they are often defined by multiple free parameters, including the masses of hypothetical particles and the related coupling constants. These free parameters can largely affect the kinematics and event topology, as well as cross-sections and branching ratios of these processes. It's important to understand such dependency, in order to constrain the phase space of possible new physics.

The conventional approach to study such dependency is the so-called "grid-scan" approach. A discrete grid with regular space is defined within the concerned parameter space, and a Monte-Carlo (MC) sample is generated at each point of the grid. There are several disadvantage of this approach. Firstly, it suffers the "curse of dimensionality", i.e. the number of points on the grid increases exponentially with the number of parameters. The grid often has to be reduced to two-dimensions, with all the other parameters fixed to certain values. Secondly, as the MC samples at different points are treated independently, a decent statistics is need for each sample in order to evaluate the acceptance precisely. Thirdly, the discreteness of the grid requires interpolation between points, which may leads to non-negligible systematic uncertainties when the spacing is large.

In this presentation, we demonstrated a new method to provide prediction of signal yield continuously over the multivariate parameter space, with much less MC simulation required.

2. Continuous simulation

Given a set of free parameters $\mathbf{x}$, the signal yield is defined as

$$N = \mathcal{L}\sigma(\mathbf{x})\epsilon(\mathbf{x}) \quad (1)$$

$\mathcal{L}$ is the integrated luminosity, which is irrelevant to the BSM parameters. σ is the cross-section times branching ratio of the concerned process. Although it is affected by the parameters, it can be calculated easily at generator level. ϵ represents the detector acceptance and reconstruction efficiency. It needs to be estimated with many events of detector simulation, which is generally the bottleneck of computing power. In this new method, we focused on facilitate an easier estimation of ϵ as a function of the parameters.

We first pick a large number of space points stochastically from the concerned parameter space. Then only one or a few events are simulated at each point $\mathbf{x}$. After that, we process all the events through the reconstruction and analysis procedures. A target value t of 1 or 0 is assigned to each event, depending on whether it passed the selection criteria of the analysis. A Bayesian Neural Network (BNN) [1, 2] is then used to perform an unbinned fitting with all these events, in order to approximate the probability distribution $\epsilon(\mathbf{x})$ by its output distribution $y(\mathbf{x})$. This is achieved by minimizing the cross-entropy cost function

$$\mathrm{CE} = \sum_k (-t_k \log y(\mathbf{x_k}) - (1 - t_k)\log(1 - y(\mathbf{x_k}))) \quad (2)$$

k loops over all simulated events. Minimizing this cost function is equivalent to maximizing the Bernoulli likelihood, which naturally gives $y(\mathbf{x})$ the meaning of success probability.

The Neural Network (NN) has been extensively used in Partticle Physics in the past decade, mostly as a multivariate discrimnant for signal/background separation [3, 4, 5]. In recent years, its usage as a universal function approximator [6, 7] has also been exploited by physicists [8, 9, 10].

There are several benefits to use the NN algorithm for this fitting. Firstly, it's a non-parametric model which does not require *a prior* knowledge of the distribution. Secondly, this algorithm assumes smoothness in the target distribution, which generally is true over the BSM parameter space. Thirdly, as an unbinned fitting, the information of input events can be fully exploited. Fourthly, NN is well-known as a multivariate tool. It can handle higher dimensional distributions easier, even if the dependency of $\epsilon(\mathbf{x})$ on each BSM parameters are correlated.

In addition, the Bayesian implementation of the BNN algorithm also allows us to estimate the uncertainty of each predicted $\epsilon(\mathbf{x})$ based on Bayesian inference framework. It provides us a handle on both the statistical fluctuation of the generated events, as well as the goodness of BNN fitting.

3. LRSM acceptance study

To demonstrate this new technique, we studied the acceptance $\epsilon(\mathbf{x})$ of di-lepton final state, for a hypothetical Left-Right Symmetrical Models (LRSM) processes [11]. In this process, a right-handed W boson (W_R) is produced, which subsequently decays into a heavy Majorana neutrino (N_R), as shown in figure 1.

The masses of the hypothetical W_R and N_R are the free parameters of this model. We considered the parameter space region with W_R mass between 0.5 TeV and 1.5 TeV, and N_R mass between 0 GeV and $M(W_R)$. The Pythia generator [12] is used to simulate this process. We randomly pick 100,000 points form the concerned parameter space, and simulate one event at each point. For simplicity, we only studied the acceptance at generator level, with 100% fiducial efficiency assumed for particle identification. Two leptons (e,μ) are required in each event, with transverse momentum greater than 20 GeV, and pseudo-rapidity η between -2.5 and 2.5. A

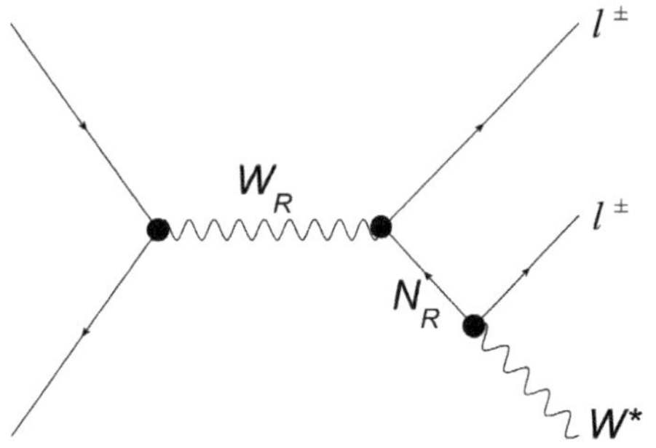

Figure 1. The Feynman diagram of W_R production in the LRSM model. The right-handed W boson will couple with a lepton and a hypothetical Majorana neutrino, which subsequently decays into another lepton and two jets.

typical isolation cut for QCD suppression is used, requiring the transverse energy flow of all interactive (non-neutrino) particles around the lepton, within a $\Delta R = \sqrt{(\Delta\eta)^2 + (\Delta\phi)^2} \leq 0.2$ cone, to be less than 10% of the lepton's transverse momentum.

Then all the simulated events are passed as input into a BNN. Target values t were assigned based on whether the event passed the above-mentioned selections. The BNN has a topology of 20 neurons in the first hidden layer and 10 neurons in the second hidden layer. The probability function of fitted $\epsilon(\mathbf{x})$ is shown in figure 2(a).

To demonstrate the performance, we also simulated another set of grid-scan samples. In this typical 2D grid, the values of $M(W_R)$ are from 600 GeV to 1400 GeV with 100 GeV steps, and the values of $M(N_R)$ are from 50 GeV to $(M(W_R) - 50)$ GeV with 50 GeV steps. The $\epsilon(\mathbf{x})$ evaluated from these samples can be seen in figure 2(b).

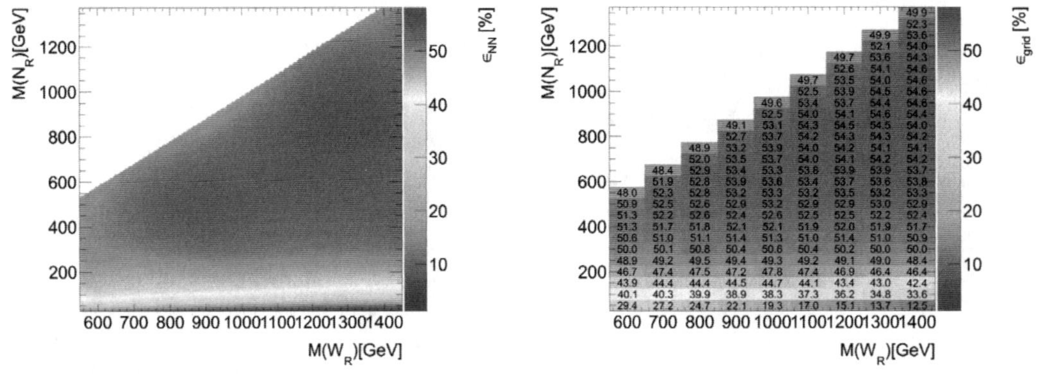

(a) (b)

Figure 2. The acceptance distribution fitted by BNN (a) and estimated by the grid-scan samples (b).

The relative difference between the BNN fitted distribution and the reference values estimated from the grid-scan samples are shown in figure 3. For most points, the deviation is quite small, except the region where $M(N_R)$ approaches 0. In this edge region, the accuracy is worse because that $\epsilon(\mathbf{x})$ is relatively smaller, also the statistics of training sample is poorer due to the lack of neighbouring points.

With the Bayesian uncertainty estimation, the above mentioned poorer accuracy is actually predictable. Figure 3 shows the uncertainty estimated by the BNN, based on the input statistics and goodness of fitting. We denote this uncertainty as $\sigma_{NN}(\mathbf{x})$. Further comparing this

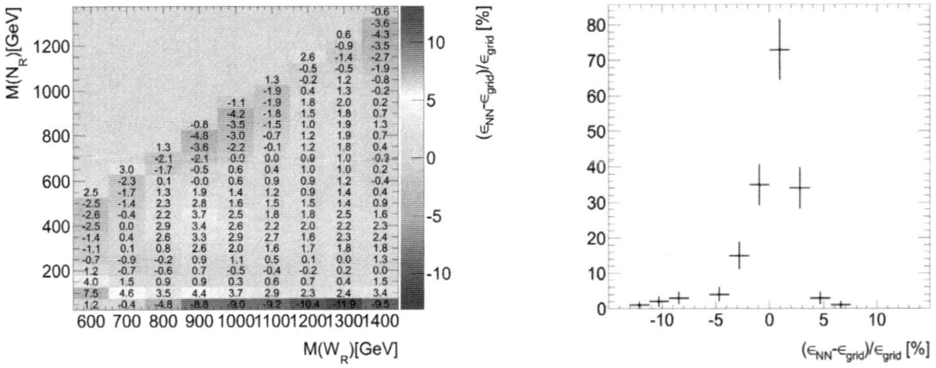

(a) (b)

Figure 3. The relative deviation of the BNN predictions from the estimations by grid-scan, at the selected parameter space points (a), and the distribution of these deviations (b).

uncertainty to the actual deviation between the fitted values and the reference values (figure 3), we can see good consistency over the whole parameter space. This gives us the confidence on the BNN's uncertainty, allowing future application to evaluate the precision of BNN estimation without generating grid-scan samples for reference.

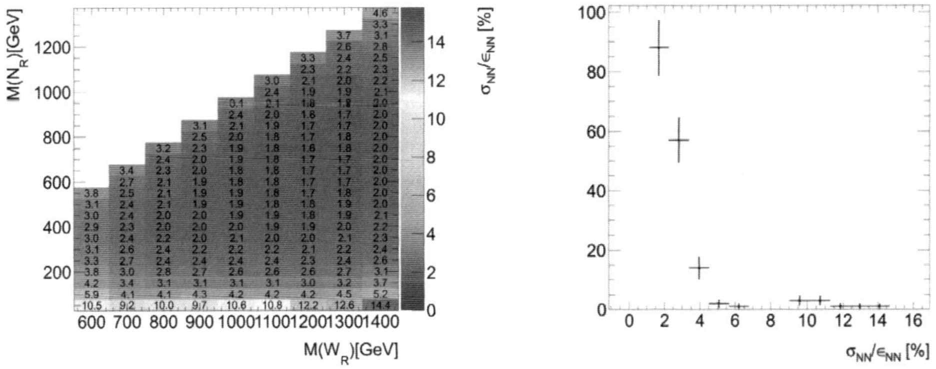

(a) (b)

Figure 4. The uncertainty estimated by the BNN, at the selected parameter space points (a), and the distribution of these values (b).

4. Summary

In this work, we introduced a novel technique to simulate and analyse BSM processes defined by multiple parameters. A BNN algorithm is used for the unbinned fitting of the acceptance, which could help to suppress the required computing load greatly, while providing continuous estimation over the parameter space. The exemplar demonstration with LRSM model indicated that the technique can provide good approximation of the acceptance distribution, as well as precise uncertainties associated to the fitted distributions.

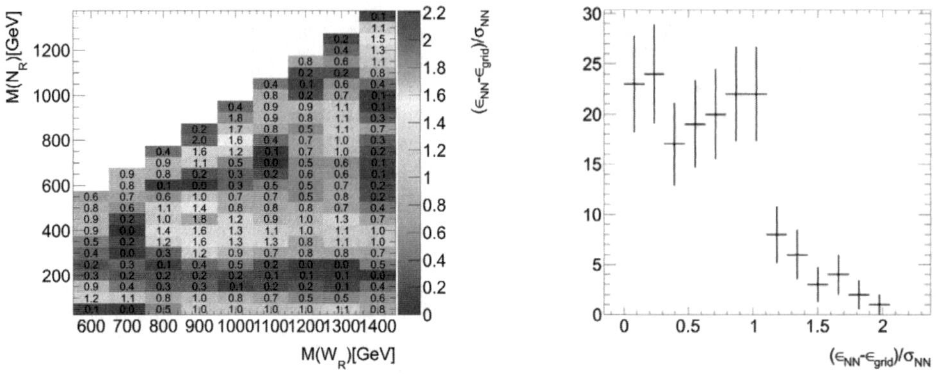

(a) (b)

Figure 5. The deviation between the BNN predictions and the estimations by grid-scan, divided by the BNN uncertainties, at the selected parameter space points (a), and the distribution of these ratios (b).

Acknowledgments
We thank Song-Ming Wang, Rachid Mazini for helpful discussions. Jiahang Zhong and Shih-Chang Lee are partially supported by the National Science Council, Taiwan under the contract number NSC99-2119-M-001-015.

References
[1] MacKay D J C 1995 *Network: Computation in Neural Systems* **6** 469–505
[2] Zhong J, Huang R S and Lee S C 2011 *Computer Physics Communications* **182** 2655 – 2660 ISSN 0010-4655
[3] Abazov V M *et al.* (D0) 2001 *Phys. Lett.* **B517** 282–294
[4] Chiappetta P, Colangelo P, De Felice P, Nardulli G and Pasquariello G 1994 *Phys. Lett.* **B322** 219–223
[5] Abramowicz H, Caldwell A and Sinkus R 1995 *Nucl. Instrum. Meth.* **A365** 508–517
[6] Cybenko G 1989 *Mathematics of Control, Signals and Systems* **2** 303–314
[7] Hornik K, Stinchcombe M and White H 1989 *Neural Networks* **2** 359 – 366 ISSN 0893-6080
[8] Forte S, Garrido L, Latorre J I and Piccione A 2002 *Journal of High Energy Physics* **2002** 062
[9] Rojo J 2006 *Journal of High Energy Physics* **2006** 040
[10] Graczyk K M, Plonski P and Sulej R 2010 *JHEP* **09** 053 (*Preprint* 1006.0342)
[11] Ferrari A, Collot J, Andrieux M L, Belhorma B, de Saintignon P, Hostachy J Y, Martin P and Wielers M 2000 *Phys. Rev. D* **62** 013001
[12] Sjostrand T, Mrenna S and Skands P 2006 *Journal of High Energy Physics* **2006** 026

Journal of Physics: Conference Series **368** (2012) 012043 doi:10.1088/1742-6596/368/1/012043

A linear iterative unfolding method

András László
Wigner RCP, P.O.Box 49, H-1525 Budapest, Hungary
CERN, CH-1211 Geneve 23, Switzerland
E-mail: laszlo.andras@wigner.mta.hu

Abstract. A frequently faced task in experimental physics is to measure the probability distribution of some quantity. Often this quantity to be measured is smeared by a non-ideal detector response or by some physical process. The procedure of removing this smearing effect from the measured distribution is called unfolding, and is a delicate problem in signal processing, due to the well-known numerical ill behavior of this task. Various methods were invented which, given some assumptions on the initial probability distribution, try to regularize the unfolding problem. Most of these methods definitely introduce bias into the estimate of the initial probability distribution. We propose a linear iterative method (motivated by the Neumann series / Landweber iteration known in functional analysis), which has the advantage that no assumptions on the initial probability distribution is needed, and the only regularization parameter is the stopping order of the iteration, which can be used to choose the best compromise between the introduced bias and the propagated statistical and systematic errors. The method is consistent: "binwise" convergence to the initial probability distribution is proved in absence of measurement errors under a quite general condition on the response function. This condition holds for practical applications such as convolutions, calorimeter response functions, momentum reconstruction response functions based on tracking in magnetic field etc. In presence of measurement errors, explicit formulae for the propagation of the three important error terms is provided: bias error (distance from the unknown to-be-reconstructed initial distribution at a finite iteration order), statistical error, and systematic error. A trade-off between these three error terms can be used to define an optimal iteration stopping criterion, and the errors can be estimated there. We provide a numerical C library for the implementation of the method, which incorporates automatic statistical error propagation as well. The proposed method is also discussed in the context of other known approaches.

1. Introduction

In data analysis one commonly faces the problem that the probability density function (pdf) of a given physical quantity of interest is to be measured, but some random physical process, such as the intrinsic behavior of the measurement apparatus, smears it. The reconstruction of the pertinent pdf based on the measured smeared pdf and on the response function of the measurement procedure is called unfolding. To be specific, let us have the original unknown pdf $x \mapsto f(x)$ of the undistorted physical quantity which we need to reconstruct, and assume that the actual measured pdf can be expressed of the form $y \mapsto g(y) = \int \rho(y|x)\, f(x)\, \mathrm{d}x$, where $(y, x) \mapsto \rho(y|x)$ describes the smearing effect in a probabilistic manner.[1] Then, it is said that

[1] All pdfs are understood to be real valued non-negative Lebesgue integrable functions over some finite dimensional real vector space X.

the pdf g is the pdf f folded with the response function ρ.[2] Our mathematical task is to solve the above linear integral equation in order to obtain f, given g and ρ. This problem is known not to be a simple numerical task (ill-posed problem), and several methods are used by the data analysis communities in order to regularize the problem (for an overview on the most popular approaches, we refer to [1, 2]).

Let us denote by A_ρ the pertinent folding operator, which acts like $(A_\rho f)(y) = \int \rho(y|x)\, f(x)\, \mathrm{d}x$ on a function f at a point y.[3] Given the measured pdf $g = A_\rho f$, the problem of unfolding can then be formalized as follows: the pdf $f = A_\rho^{-1}(g)$ is to be determined or approximated. The mathematical cause of the numerical ill-posedness of this unfolding problem can then be put forward as: the inverse A_ρ^{-1} of a generic folding operator can be shown not to be continuous despite the forward folding operator A_ρ always being continuous[4] (this phenomenon is discussed in detail e.g. in [9]). The non-continuity of the inverse folding operator A_ρ^{-1} may be also reformulated in a less abstract manner: initially distant functions can be mapped close by the folding operator A_ρ, as illustrated in Figure 1. I.e. one can lose discriminating power between pdfs upon a folding.

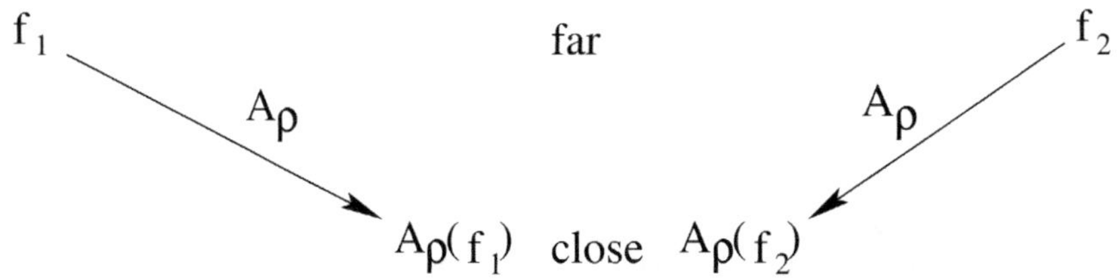

Figure 1. Illustration of the non-continuity of the inverse of a folding operator A_ρ: two distant functions f_1 and f_2 may be mapped close by the folding – distance of functions are here understood as probabilistic distance, i.e. in the $L^1(X)$ function norm.

A further aspect of the numerical ill-posedness of the unfolding problem is that in practice the folded pdf g is often obtained via statistical measurements (e.g. histograming), and therefore is contaminated by statistical errors. I.e. in reality $g = A_\rho f + e$ holds instead of the idealized equation $g = A_\rho f$, where $e(x)$ is a random variable for each point x (or for each histogram bin – in the language of histograms). Thus, when estimating the unfolded pdf as $A_\rho^{-1}(g) = f + A_\rho^{-1}(e)$, the contribution of the second term is not guaranteed to remain small due to the non-continuity of the inverse folding operator A_ρ^{-1} even when e is initially known to be small. On top of this, the statistical error term e may contain modes not within the image of the folding operator A_ρ, on which the evaluation of the inverse operator A_ρ^{-1} is not meaningful if the problem is not initially discrete. These effects are demonstrated in Figure 2, which shows that simple inversion of the discretized folding operator on the measured pdf gives unphysical numerical result: a result very different from the initial pdf, having large negative and positive alternating amplitudes.

In order to regularize the numerical ill-posedness of the unfolding problem, various methods are used. These methods can be divided into three large classes.

[2] Whenever the response function ρ is translation invariant in the sense that for all $x, y, z \in X$ one has $\rho(y+z|x) = \rho(y|x-z)$, the folding is specially called convolution, and in that case ρ may be expressed by a single pdf: $\rho(y|x) = \rho(y-x|0)$.

[3] To be precise, A_ρ is a $L^1(X) \to L^1(X)$ continuous linear operator, where $L^1(X)$ denotes the normed space of complex valued integrable functions over the vector space X. The response function ρ is assumed to be $\rho(\cdot|x) \in L^1(X)$ for all $x \in X$.

[4] Continuous in the $L^1(X) \to L^1(X)$ sense.

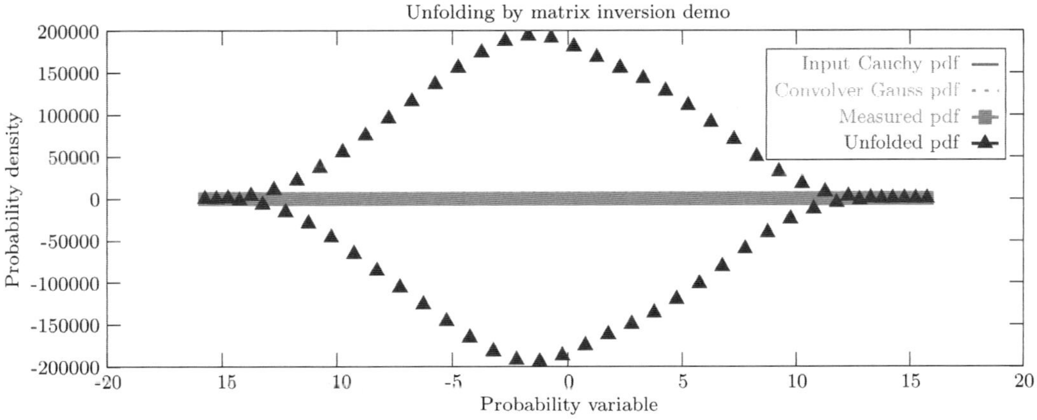

Figure 2. (Color online) Demonstration of the numerical ill-posedness of the unfolding problem: a Cauchy distribution is convolved with a Gauss distribution with Monte Carlo method to generate the measured distribution contaminated with statistical errors. Clearly, the unfolded pdf, obtained by simple numerical inversion of the discretized folding operator on the measured pdf gives physically unreasonable numerical result: large alternating positive / negative amplitude pdf values.

(i) Using a parametric Ansatz for f, and fit parameters, so that $A_\rho f$ gets close to g. This method can be slightly insensitive to the details of the true f (as illustrated in Figure 1), and of course can introduce strong systematic bias on the result if the parametric Ansatz does not hold in an exact manner of the form that was assumed. Such methods are used in general for inclusive particle identification by specific ionization (see e.g. [10]).

(ii) Bin-by-bin fitting of the bin values of the histogramed f, so that $A_\rho f$ gets close to g. This is basically equivalent to the naive inversion of the discretized folding operator, and therefore produces similar oscillatory results, except when an artificial penalty function is added to the χ^2 in order to suppress large local gradients. In that case, the method can provide meaningful answers, but the introduced systematic bias is difficult to quantify. Similarly to the parametric Ansatz method, the fit can be slightly insensitive to the details of the true f. Most popular methods, such as SVD method [3], are based on this idea.

(iii) The iterative method of convergent weights (also known as iterative Bayesian unfolding) of Kondor-Mülthei-Schorr-d'Agostini [4, 5, 6, 7, 8]. This method is, as opposed to the previously mentioned methods, is non-linear. On the other hand, by construction it preserves positivity and integral of the initial pdf, and therefore maps a pdf exactly into a pdf, which does not hold for linear methods, thus, this approach is quite favorable for statistical applications. Regularization is achieved solely by stopping the iteration at a finite order. However, there is no known proof yet if the iterated pdfs converge[5] to the initial to-be-reconstructed pdf in a non-discrete scenario, even in the absence of measurement errors [6]. Also propagation of statistical and systematic errors of the measured pdf to the unfolded pdf has not been investigated, and consequently no generally applicable iteration stopping condition is known.

In a previous paper [9] we proposed a linear iterative unfolding method, for which under certain conditions convergence to the initial pdf was proved analytically for some unfolding problems

[5] In case of an iterative unfolding method it is an absolute must to show that the sequence of iterated unfolded pdfs converge to the initial one, in the absence of measurement errors (consistency of the method).

in probability theory (such as convolutions), and due to the linearity of the method, exact propagation of statistical errors of the measured (folded) pdf to the unfolded pdf was possible. In this paper we propose an improved version of that algorithm, which could be proved to be convergent in quite general cases for unfolding problems in a probability theory setting.[6] The key equality of the convergence proof leads to explicit error propagation formulae for the three important error terms: for the bias error (distance from the true unfolded pdf), for the propagated statistical error, and most notably for the propagated systematic error, which is of great importance in reporting experimental results. An implementation of the algorithm is written as C library, along with application examples [13]. The implementation also incorporates automatic statistical error propagation.

The paper is organized as follows: in Section 2 the algorithm and its convergence theorem shall be formulated, Section 3 is devoted to the corresponding error propagation formulae which help to formulate an optimal stopping criterion and error estimates therein, while in Section 4 we demonstrate our method on examples.

2. A linear iterative unfolding algorithm

We provide now a linear iterative solution for a probability theory unfolding problem of the form $g = A_\rho f$, where f is the initial (unknown) pdf, g is the folded (measured) pdf, and ρ is the response function. Given the response function ρ, one can also define along with the folding operator A_ρ the transpose folding operator A_ρ^T by swapping the variables of the response function.[7] Then, one can attempt to approximate the true unfolded pdf f in the following way: define the function sequence by setting the normalization factor

$$K_\rho = \max_x \int\int \rho(y|z)\,\rho(y|x)\,\mathrm{d}y\,\mathrm{d}z \tag{1}$$

and then taking the

$$\begin{aligned} f_0 &= K_\rho^{-1} A_\rho^T g, \\ f_{N+1} &= f_N + \Big(f_0 - K_\rho^{-1} A_\rho^T A_\rho f_N\Big) \end{aligned} \tag{2}$$

iteration formula. We provide a convergence result on this iterative approximation below in absence of measurement errors on g (which is necessary for the consistency of the method).

Theorem 1. *(Convergence) The function sequence $N \mapsto f_N$ resulting from the above iteration scheme converges to the closest possible function to the true unfolded pdf f in the average over any compact region, whenever the normalization factor K_ρ is finite. I.e. for all compact sets $S \subset X$ one has*

$$\lim_{N\to\infty} \frac{1}{\mathrm{Volume}(S)} \int_S \Big(f - P_{\mathrm{Ker}(A_\rho)} f - f_N\Big)(x)\,\mathrm{d}x = 0. \tag{3}$$

Here, $P_{\mathrm{Ker}(A_\rho)}$ denotes the orthogonal projection operator to the kernel set of A_ρ, and thus $P_{\mathrm{Ker}(A_\rho)} = 0$ holds automatically whenever A_ρ is invertible. In addition, the convergence shall also hold in the space of square-integrable functions, i.e. one has also

$$\lim_{N\to\infty} \int \Big|f - P_{\mathrm{Ker}(A_\rho)} f - f_N\Big|^2 (x)\,\mathrm{d}x = 0. \tag{4}$$

[6] The detailed mathematical proof of convergence shall be published elsewhere: [11]. The proposed iteration scheme was motivated by the so called Neumann series and Landweber iteration [12] known in functional analysis, but the convergence of neither iterations hold, unfortunately, in a probability theory setting in their original form, as one can prove. Our improved iterative algorithm, however, is specially developed to be convergent for unfolding problems in probability theory.

[7] The transpose folding operator is defined by $(A_\rho^T f)(y) = \int \rho(x|y)\,f(x)\,\mathrm{d}x$ for all functions f and points y. Note, that this simply translates to matrix transposition whenever the folding is discretized.

Proof The proof is based on Riesz-Thorin theorem and on the spectral representation of positive operators in the space of complex square-integrable functions over X (to be published in a more mathematically specialized journal: [11]).

The following observations help to shed some further light on properties of the proposed unfolding algorithm.

(i) In case the pdfs are modeled with histograming, the setwise convergence of pdfs means binwise convergence of histograms, i.e. the probability of each histogram bin is restored in the limit of infinite iterations.

(ii) When the inverse of A_ρ exists, the original pdf f is completely restored. Whenever the pertinent inverse does not exist, still the maximum possible information about f is restored, namely the function $f - P_{\mathrm{Ker}(A_\rho)} f$.

(iii) Whenever A_ρ is a convolution, then $K_\rho = 1$ holds automatically, i.e. $K_\rho < \infty$ is satisfied.

(iv) The convergence condition $K_\rho < \infty$ holds provably for a wide class of practically relevant response functions, such as energy response function of calorimeters, momentum response function of track reconstruction in magnetic field etc.

(v) The iteration scheme of the theorem is motivated by the Neumann series known in functional analysis. A similar iterative solution, also referred to as Landweber iteration [12], is known in the theory of Fredholm operators. In probability theory unfolding problems, however, the necessary convergence criteria for Neumann series or for Landweber iteration do not hold in their original form.

(vi) The proposed iterative unfolding algorithm does not necessarily need an initial binning of pdfs. It may be implemented as well by different density estimators than histograms. However, when the pdfs are modeled by histograms, one may recognize that the binning and truncation of histograming domain can also be considered as folding operator. Therefore, the histogram binning and truncation effect may be included in the response function ρ, and then the effect of histograming can be unfolded (to the maximum possible extent) as well. If one wants to numerically implement this, the initial pdf f must be assumed to better approach the continuum pdf, i.e. must be assumed to be an unknown histogram over a larger domain with finer granulation than the folded one. The schematic of such possible rebinning trick is illustrated in Figure 3.

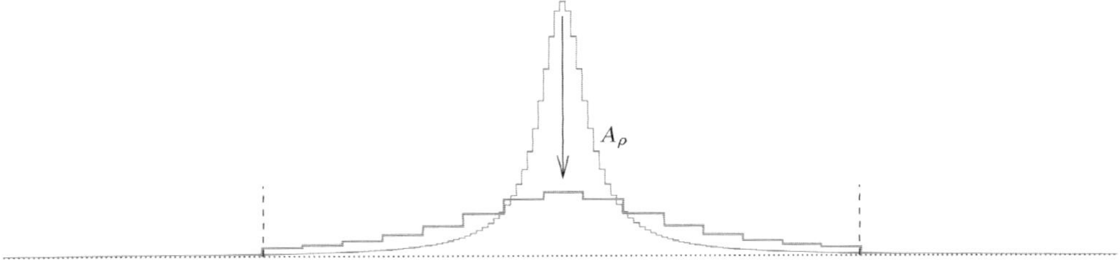

Figure 3. (Color online) Illustration the rebinning trick for unfolding the histogram binning and domain truncation as well (to the maximum possible extent) along with the smearing effect of the response function ρ. For this, the implementation of the folding operator A_ρ must map histograms over a larger domain and with finer graining to histograms with the binning scheme of the measured (folded) pdf.

3. Bias, statistical and systematic errors of the unfolded distribution

In real measurements, the folded pdf g also admits statistical and systematic errors, and the propagation of these terms into the unfolded pdf is necessary to quantify at each finite iteration step. The key equality of the proof of Theorem 1 leads to explicit error propagation formulae for bias error (distance from the true unfolded pdf), statistical error, and systematic error. First we present our result about bias error.

Theorem 2. *(Bias error) Take the iterative solution for the unfolding problem as in Section 2. Then, if the normalization factor K_ρ is finite, the distance of an N-th iterate f_N from the closest possible function to the true unfolded pdf f in the average over a compact region has the following upper bound: for any compact set $S \subset X$ one has*

$$\left|\frac{1}{\text{Volume}(S)}\int_S \left(f - P_{\text{Ker}(A_\rho)}f - f_N\right)(x)\,\mathrm{d}x\right| \leq \frac{1}{\sqrt{\text{Volume}(S)}}\,\frac{1}{N+2}\sqrt{\int |f|^2(x)\,\mathrm{d}x}, \qquad (5)$$

i.e. the residual deviation of f_N from the limit averaged to the set S decreases as $\frac{1}{\sqrt{\text{Volume}(S)}}$ and as $\frac{1}{N+2}$.

The above result, translated to the language of histograms means that the bin-by-bin average deviation from the true unfolded pdf is bound by the right hand side of the inequality in Theorem 2., where Volume(S) is the histogram bin volume, N is the iteration order, and $\sqrt{\int |f|^2(x)\,\mathrm{d}x}$ is an unknown coefficient depending on the true unfolded pdf f. This unknown coefficient, however, may be substituted by the calculable expression $\sqrt{\int |f_N|^2(x)\,\mathrm{d}x}$ for large N. It is seen that the bias error tends to zero with increasing iteration order N.

In practical applications, the pdfs are often measured by statistical methods (e.g. histograming). In that case, the value of the folded pdf g in each histogram bin admits a statistical error. The below theorem states an exact formula for the propagation of this error into the unfolded pdf.

Theorem 3. *(Statistical error) Take the iterative solution for the unfolding problem as in Section 2. Let C be the covariance matrix of the measured pdf g, where g is assumed to be of the form of a histogram. Since a covariance matrix C is positive definite, it is always possible to decompose it – not uniquely – in the form $C = E\,E^T$ for some matrix E, $(\cdot)^T$ being the matrix transpose. (Whenever C is diagonal, construction of such an E is just trivial.) Then, the following iteration calculates the statistical error propagation:*

$$\begin{aligned} E_0 &= K_\rho^{-1}A_\rho^T E, \\ E_{N+1} &= E_N + \left(E_0 - K_\rho^{-1}A_\rho^T A_\rho E_N\right), \end{aligned} \qquad (6)$$

where in each step the covariance matrix of f_N shall be $C_N = E_N\,E_N^T$.

Due to the linearity of the method, the contribution of the propagated statistical error term is exactly calculable by means of the above formulae, if it is known for the measured pdf g. This error term increases with increasing iteration order N. The statistical error of a given histogram bin of the N-th iterate f_N is nothing but the square-root of the corresponding diagonal element of C_N.

Whenever the folded pdf g is a result of an experiment, it may admit a systematic error δg. Also the systematic error $\delta\rho$ of the response function ρ may give a non-zero contribution to it: $A_{\delta\rho}f$. The effect of this initial systematic error on the unfolded pdf is quantified by the following theorem.

Theorem 4. *(Systematic error) Take the iterative solution for the unfolding problem as in Section 2. Assume that δg is the systematic error of g (possibly including contribution from systematic error of the response function). Then the systematic error for the N-th iterate f_N averaged over a compact region has the following upper bound: for any compact set $S \subset X$*

$$\left|\frac{1}{\mathrm{Volume}(S)} \int_S \delta f_N(x)\,\mathrm{d}x\right| \leq \frac{1}{\sqrt{\mathrm{Volume}(S)}} \left(\Psi(N+2)+\gamma\right) \sqrt{\int \left|K_\rho^{-1} A_\rho^T \delta g\right|^2 (x)\,\mathrm{d}x}, \quad (7)$$

i.e. the systematic error of f_N averaged to the set S decreases as $\frac{1}{\sqrt{\mathrm{Volume}(S)}}$ and increases as $\Psi(N+2)+\gamma \approx 1+\ln(N+1)$. ($\Psi$ being the digamma function, γ being Euler's constant.)

The above result, translated to the language of histograms means that the bin-by-bin average systematic error of the N-th iterate f_N is bound by the formula in the right hand side of the inequality in Theorem 4., where Volume(S) is the histogram bin volume, N is the iteration order, and the coefficient $\sqrt{\int \left|K_\rho^{-1} A_\rho^T \delta g\right|^2 (x)\,\mathrm{d}x}$ is calculable knowing the bin-by-bin systematic errors δg of the measured pdf g. Clearly, this contribution increases logarithmically with increasing iteration order N.

As the bias error decreases, while the statistical and systematic error of the N-th iterate f_N increases with the iteration order N, a trade-off between these error terms provides an optimal cutoff criterion[8] in the iteration order N, and error estimates therein. Consequently the true unfolded pdf f can be approximated optimally and the error of this approximation can be put under full control. Thus, the regularization of the numerically ill-posed unfolding problem is achieved, in case of the proposed approach, solely by using an iterative approximation and choosing an optimal iteration stop order taking into account the convergent terms (bias error) and the divergent terms (statistical and systematic errors).

4. Examples

In this section we give two examples to demonstrate our method. In the first example, we take a Cauchy distribution, and convolve it with a Gaussian distribution with Monte Carlo method. The folded pdf is determined by histograming the sum of the Cauchy and Gaussian distribution random numbers, i.e. the measured pdf shall admit Poissonian statistical errors. In this example, a relatively modest statistics of 5000 entries was taken to be able to judge the method in the low statistics limit. The results is shown in Figure 4. It is seen that the original Cauchy pdf is restored, modulo the fluctuations arising from the propagated statistical errors – these are seen as "shoulders" of the unfolded pdf, the amplitude of which decrease with increased statistics. The iteration was stopped when the integral of the statistical error term reached about 5% level.

In the second Monte Carlo example, we generate the energy distribution of transversely emitted hadrons in 7 GeV p+p collisions [14], and we assume that this particle spectrum was measured by the CMS-HCAL calorimeter [15]. The unfolded spectrum, along with the true and measured distribution is shown in Figure 5.

5. Concluding remarks

We proposed a linear iterative spectrum unfolding method for application in data analysis. Convergence to the true unfolded pdf is proved under a quite general condition [11] in absence of measurement errors, and error propagation formulae are derived for bias error, statistical error, and systematic error in the presence of measurement errors. The method is demonstrated on physical examples. A numerical library in C is provided with the implementation of the method [13]. The algorithm could be included in the ROOUnfold package [16] in the future.

[8] E.g. one can take the sum of the three error terms, and stop the iteration when it reaches a minimum.

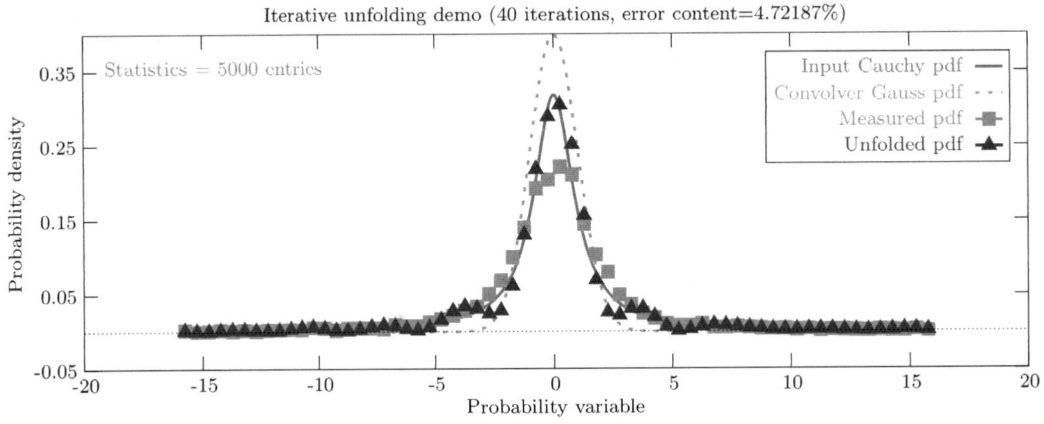

Figure 4. (Color online) Test example with unfolding a Cauchy distribution convolved with a Gaussian distribution. Iteration was stopped when the integral of the statistical error term reached about 5%.

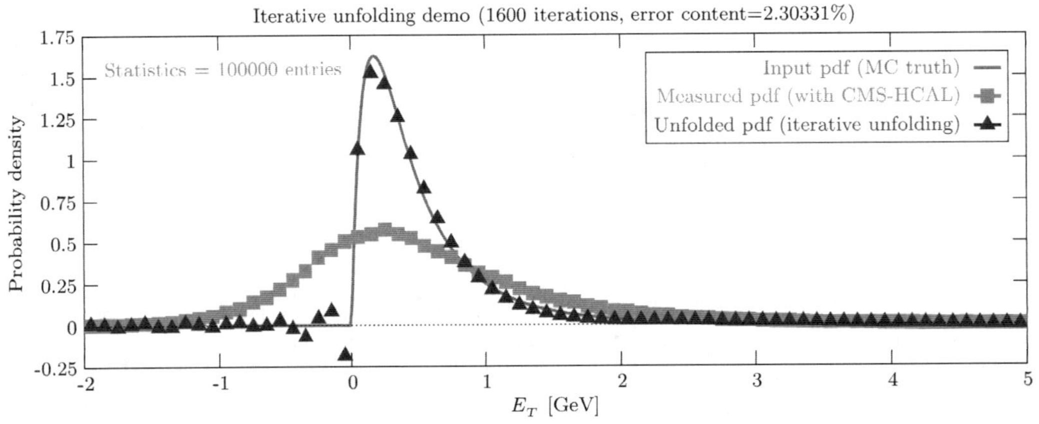

Figure 5. (Color online) A physical example with unfolding energy distribution of charged hadrons measured with hadronic calorimeter. Iteration was stopped when the integral of the statistical error term reached about 2.3%.

Acknowledgments
I would like to thank prof. Günter Zech for valuable discussions.

References
[1] Cowan G 2002 *Proceedings of Conference on Advanced Statistical Techniques in Particle Physics (18-22 March 2002, Durham, United Kingdom)* (IPPP/02/39, Durham) p 248
[2] Blobel V 2008 Unfolding for HEP Experiments (*Talk at DESY Computing Seminar*) `http://www.desy.de/~blobel/DESYcompsem08.pdf`
[3] Hoecker A and Kartvelishvili V 1996 *Nucl. Instr. Meth.* A **372** 469
[4] Shepp L A and Vardi Y 1982 *IEEE Trans. Med. Imag.* **1** 113
[5] Kondor A 1983 *Nucl. Instr. Meth.* **216** 177
[6] Mülthei H N and Schorr B 1987 *Mat. Meth. Appl. Sci.* **9** 137
[7] Mülthei H N and Schorr B 1987 *Nucl. Instr. Meth.* A **257** 371

[8] D'Agostini G 1995 *Nucl. Instr. Meth.* A **362** 487
[9] László A 2006 *J. Phys.* A **39** 13621
[10] László A 2008 *Phys. Rev.* C **77** 034906
[11] László A Convergence results on a linear iterative spectrum unfolding method (*in preparation*)
[12] Landweber L 1951 *Am. J. Math.* **73** 615
[13] László A 2011 The libunfold package (*Source code*)
`http://www.rmki.kfki.hu/~laszloa/downloads/libunfold.tar.gz`
[14] Khachatryan V *et al* 2010 *Phys. Rev. Lett.* **105** 022002
[15] Yazgan E *et al* 2009 *J. Phys. Conf. Ser.* **160** 012056
[16] Adye T *et al* The ROOUnfold package
`http://hepunx.rl.ac.uk/~adye/software/unfold/RooUnfold.html`

Journal of Physics: Conference Series **368** (2012) 012044 doi:10.1088/1742-6596/368/1/012044

An adaptive Monte-Carlo Markov chain algorithm for inference from mixture signals

Rémi Bardenet and Balázs Kégl

Linear Accelerator Laboratory (LAL) & Computer Science Laboratory (LRI)
IN2P3/CNRS & University of Paris-Sud, 91405 Orsay (France)

E-mail: bardenet@lri.fr, balazs.kegl@gmail.com

Abstract. Adaptive Metropolis (AM) is a powerful recent algorithmic tool in numerical Bayesian data analysis. AM builds on a well-known Markov Chain Monte Carlo algorithm but optimizes the rate of convergence to the target distribution by automatically tuning the design parameters of the algorithm on the fly. Label switching is a major problem in inference on mixture models because of the invariance to symmetries. The simplest (non-adaptive) solution is to modify the prior in order to make it select a single permutation of the variables, introducing an identifiability constraint. This solution is known to cause artificial biases by not respecting the topology of the posterior. In this paper we describe an online relabeling procedure which can be incorporated into the AM algorithm. We give elements of convergence of the algorithm and identify the link between its modified target measure and the original posterior distribution of interest. We illustrate the algorithm on a synthetic mixture model inspired by the muonic water Cherenkov signal of the surface detectors in the Pierre Auger Experiment.

1. Introduction

Inferring properties of individual particles in a mixture signal is a challenging problem. The goal of this paper is to describe an adaptive Monte-Carlo Markov chain (MCMC) algorithm that estimates the parameters of individual components from the mixture signal. Our objective is twofold: we aim at "popularizing" an existing technique [1] that eliminates the parameter-tuning step from MCMC, and we propose an extension of the algorithm to the case when the signal exhibits certain symmetries (such as in the case of mixtures). The main idea of the algorithm is to combine adaptive Metropolis [1] with online relabeling [2] with two goals in mind: 1) carrying out inference on individual components and 2) accelerating convergence. More details and further analysis of the algorithm can be found in [3].[1]

The paper is organized as follows. In section 2 we describe a synthetic model inspired by the muonic signal in water Cherenkov tanks of the Pierre Auger Experiment [4]. Section 3 contains the formal description of the Adaptive Metropolis with Online Relabeling (AMOR)

[1] The precise connection of the two papers is the following. The present paper concentrates on a concrete application (it describes the model and provides quantitative results), whereas [3] uses the physics example as an illustration and concentrates on the theoretical analysis of the algorithm (with a proof sketch available as supplementary material). The two papers are written to two almost mutually exclusive communities. The theoretical analysis of the algorithm is of little interest in the physics community whereas the concrete physics model we use in analyzing the Auger signal would draw equally little interest in the statistics community.

algorithm. We validate the algorithm on experiments conducted on synthetic data in section 4, and conclude the paper in section 5.

2. A motivating example

When a muon crosses a water tank, it generates Cherenkov photons and photons coming from other processes (e.g., delta rays) along its track at a rate depending on its energy. Some of these photons are captured by photomultipliers. The resulting photoelectrons (PEs) then generate analog signals that are discretized by an analog-to-digital converter.

In this section, we model the integer photoelectron (PE) count vector $\mathbf{n} = (n_1, \ldots, n_M) \in \mathbb{N}^M$ in the M bins of the signal, which means that we omit the model of the PMT electronics. Formally, n_i is the number of PEs in the i-th bin

$$[t_{i-1}, t_i) \triangleq [t_0 + (i-1)t_\Delta, t_0 + it_\Delta), \tag{1}$$

where t_0 is the absolute starting time of the signal, and $t_\Delta = 25\,\text{ns}$ is the signal resolution (size of one bin). The goal is to parametrize the likelihood $\mathcal{P}(\mathbf{n} \mid t_\mu, A_\mu)$ and the priors $\mathcal{P}(A_\mu)$ and $\mathcal{P}(t_\mu)$, where t_μ is the arrival time of the muon and A_μ is the integrated signal amplitude. A_μ is itself defined by $A_\mu = L_\mu \nu \phi_\mu$, where L_μ is the tracklength of the muon, ν is the mean number of PEs generated by a muon with kinetic energy of 1 GeV on a tracklength of 1 m, and ϕ_μ is the ratio between the number of photons produced by the muon and the expected number of photons generated by a muon with kinetic energy of 1 GeV. In this paper we use the dimensions of the Auger tank ($R = 1.8\,\text{m}$, $h = 1.2\,\text{m}$), and we choose to model signals of muons that arrive at $\theta = 45°$. Furthermore, we use an energy spectrum measured for atmospheric muons [5] and assume that a muon with kinetic energy of 1 GeV generates, on average, $\nu = 228$ PEs per 1 m of tracklength. Putting it all together numerically, the distribution $\mathcal{P}(A_\mu)$ of the total number of PEs deposited by a muon is depicted by figure 1(a).

To model the time response of a muon, we assume that the photon generation is uniform in a window of width $t_d = 10\,\text{ns}$. The decay phase is modeled by an exponential distribution with parameter $\tau = 60\,\text{ns}$. The distribution of PEs in time can be modeled by a convolution of the generative process and the decay process. In our case the convolution can be solved analytically to obtain the time response distribution

$$\mathcal{P}_{\tau,t_d}(t) = \frac{1}{t_d} \cdot \begin{cases} 0 & \text{if } t < 0, \\ 1 - \exp(-t/\tau) & \text{if } 0 \le t < t_d, \\ \exp\big(-(t-t_d)/\tau\big) - \exp(-t/\tau) & \text{if } t_d \le t. \end{cases} \tag{2}$$

depicted by figure 1(b). The time of arrival distribution $\mathcal{P}(t_\mu)$ of the muons in a tank depends on the energy and the geometry of the shower, on the distance of the tank from the shower core, and on other external parameters. In this paper we will use an inverse Gamma distribution, with parameters dependent on several factors (distance of the tank from the shower axis, zeinth angle and energy of the shower). An example is shown in figure 1(c) with parameters 2.5 and 350. All distributions and parameter values were picked to realistically model the Auger water detector signal in certain regimes.

Given the arrival time t_μ of a muon and the associated total number of PEs A_μ, the PE count in the ith bin is a Poisson variate with parameter

$$\bar{n}_i(A_\mu, t_\mu) = A_\mu \int_{t_{i-1}}^{t_i} p_{\tau,t_d}(t - t_\mu)\mathrm{d}t. \tag{3}$$

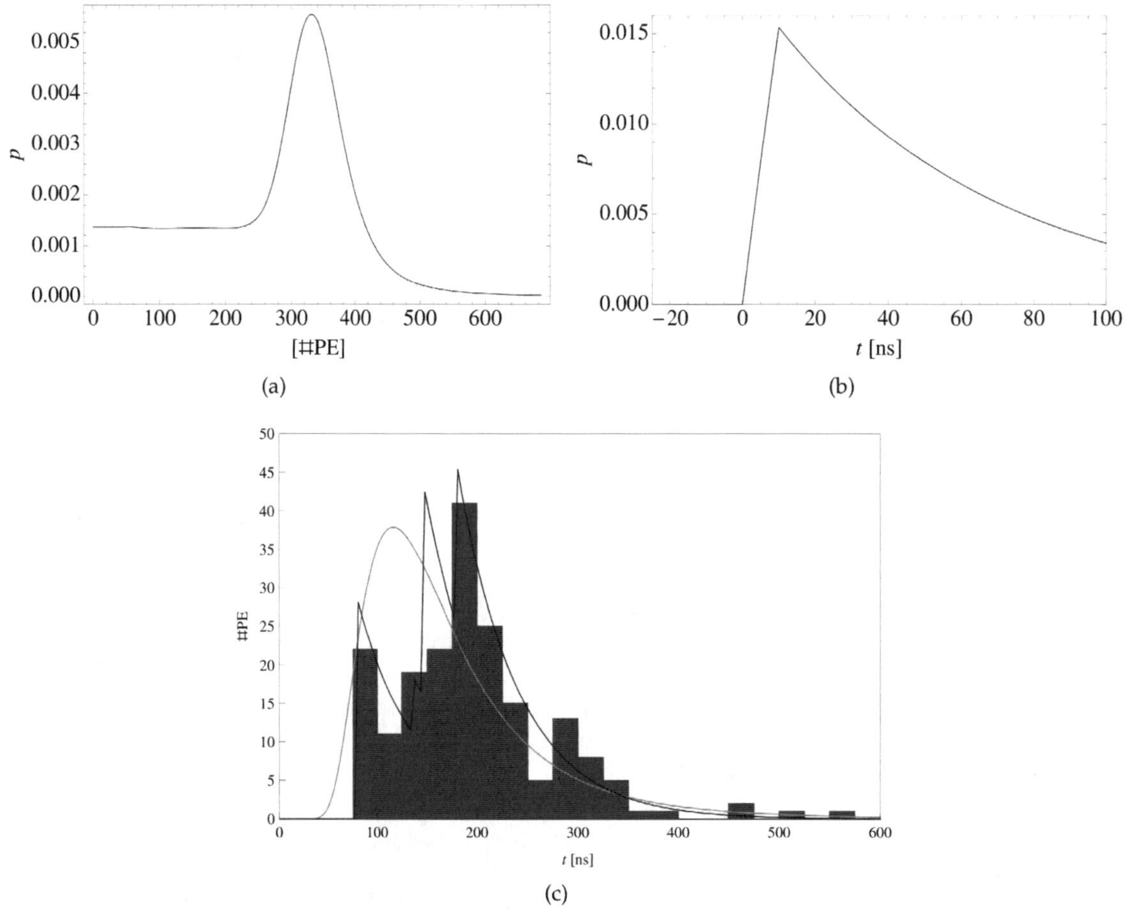

Figure 1. The generative model of the muonic signal. (a) The distribution $\mathcal{P}(A_\mu)$ of the muonic signal amplitude, used as a prior in the inference. (b) The muonic time response model $p_{\tau,t_d}(t)$. (c) An example signal. The green curve is the time-of-arrival distribution $\mathcal{P}(t_\mu)$, used as a prior in the inference. The blue curve is the "ideal" response $\sum_{j=1}^{N} A_{\mu_j} p_{\tau,t_d}(t - t_{\mu_j})$ of $N = 4$ muons drawn randomly from $t_\mu \sim p(t_\mu)$ and $A_\mu \sim \mathcal{P}(A_\mu)$, and the red histogram is the signal (PE count vector) **n**.

Given N muons with signal amplitudes $\mathbf{A}_\mu = (A_{\mu_1}, \ldots, A_{\mu_N})$ and arrival times $\mathbf{t}_\mu = (t_{\mu_1}, \ldots, t_{\mu_N})$, the binwise signal expectation is

$$\bar{n}_i(\mathbf{A}_\mu, \mathbf{t}_\mu) = \sum_{j=1}^{N} \bar{n}_i(A_{\mu_j}, t_{\mu_j}), \tag{4}$$

and so our model is fully specified by the likelihood

$$\mathcal{P}(\bar{\mathbf{n}} \mid \mathbf{A}_\mu, \mathbf{t}_\mu) = \prod_{i=1}^{M} \mathrm{Poi}_{\bar{n}_i(\mathbf{A}_\mu, \mathbf{t}_\mu)}(n_i) \tag{5}$$

and the prior

$$\mathcal{P}(\mathbf{A}_\mu, \mathbf{t}_\mu) = \prod_{j=1}^{N} \mathcal{P}(A_{\mu,j}) \mathcal{P}(t_{\mu,j}), \tag{6}$$

where all bins and muons are assumed independent. Figure 1(c) depicts an example drawn from the model.

3. An adaptive MCMC algorithm with online relabeling

3.1. *The symmetric random walk Metropolis algorithm*

The goal of MCMC algorithms in the Bayesian context is to sample from the posterior distribution

$$\pi(X) \triangleq \pi(\mathbf{A}_\mu, \mathbf{t}_\mu) \propto \mathcal{p}(\bar{\mathbf{n}} \mid \mathbf{A}_\mu, \mathbf{t}_\mu)\mathcal{p}(\mathbf{A}_\mu, \mathbf{t}_\mu). \tag{7}$$

The posterior $\pi(X) = \pi(\mathbf{A}_\mu, \mathbf{t}_\mu)$, whose normalization constant is unknown, can be explored by running a Markov chain $X_t = (\mathbf{A}_\mu, \mathbf{t}_\mu)_t$ with stationary distribution π. In this context, π is also said to be the *target distribution* of the MCMC chain. The symmetric random walk Metropolis algorithm (SRWM [6], corresponding to the blue steps in figure 3) is one of the most popular techniques for simulating such a chain (X_t). In SRWM the user has to provide a symmetric proposal kernel that will be used to propose a new sample $\widetilde{X}$ given the previous sample X_{t-1}. This new sample is then accepted with probability $\min(1, \pi(\widetilde{X})/\pi(X_{t-1}))$. When the posterior is a distribution over a continuous space $\mathcal{X} = \mathbb{R}^d$ (with $d = 2N$ in our case), the most common proposal kernel is a multivariate Gaussian $\mathcal{N}(\,\cdot\,|X_{t-1}, \Sigma)$.

The goal of *adaptive Metropolis* (AM) (corresponding to the blue and green steps in figure 3) is to automatically calibrate the design parameter Σ of SRWM. When the target $\pi(x)$ is multivariate Gaussian with covariance Σ_π, the optimal choice of Σ is of the order of $(2.38)^2\Sigma_\pi/d$ [7, 8]. In practice, Σ_π is unknown thus motivating the use of an estimate of the covariance of the posterior based on samples $(X_1, \ldots, X_{t-1})$ generated so far. From a theoretical point of view, the conditional distribution of X_t given the past then depends on the whole past, rendering the analysis of AM more challenging. The convergence of AM has been recently addressed under quite general conditions (see, e.g., [9, 10] and references therein).

3.2. *The challenge of permutation invariance*

The optimal choice for Σ is appropriate only when the target distribution is strongly unimodal [8]. In our case both the signal likelihood $p(\bar{\mathbf{n}} \mid \mathbf{A}_\mu, \mathbf{t}_\mu)$ and the prior $\mathcal{p}(\mathbf{A}_\mu, \mathbf{t}_\mu)$ are invariant under any permutation of the muons which means that the posterior (7) has, in the general case, $N!$ modes. More precisely, let us assume that the number of muons is fixed to $N = 3$ and drop the μ index for the sake of simplicity, then any parameter vector $X = (A_1, A_2, A_3, t_1, t_2, t_3)^\intercal$ has the same posterior value as the relabeled vector $PX = (A_2, A_3, A_1, t_2, t_3, t_1)^\intercal$, where

$$P = \begin{pmatrix} 0 & 1 & 0 & 0 & 0 & 0 \\ 0 & 0 & 1 & 0 & 0 & 0 \\ 1 & 0 & 0 & 0 & 0 & 0 \\ 0 & 0 & 0 & 0 & 1 & 0 \\ 0 & 0 & 0 & 0 & 0 & 1 \\ 0 & 0 & 0 & 1 & 0 & 0 \end{pmatrix}. \tag{8}$$

Actually, any other permutation $P \in \mathfrak{P}$ of the 3 muons would give the same posterior value. Throughout this paper, $\mathfrak{P}$ will denote the set of all block permutation matrices such as the one shown in (8). The described permutation invariance implies a risk for the MCMC chain to exhibit *label switching*, making componentwise averages meaningless. Figures 2(a)-2(b) illustrate the challenges when running vanilla AM on the example presented in figure 2(d). The red variable gets stuck in one of the mixture components, whereas the blue, green, and brown variables visit all the three remaining components. Marginal estimates computed for the blue, green, and brown variables are then mostly identical as seen on figure 2(b). In addition, the shaded ellipses, depicting the marginal posterior covariances of the two parameters A_μ and t_μ of each muon, indicate that the resulting empirical covariance estimate is very broad, resulting in poor efficiency of the adaptive algorithm.

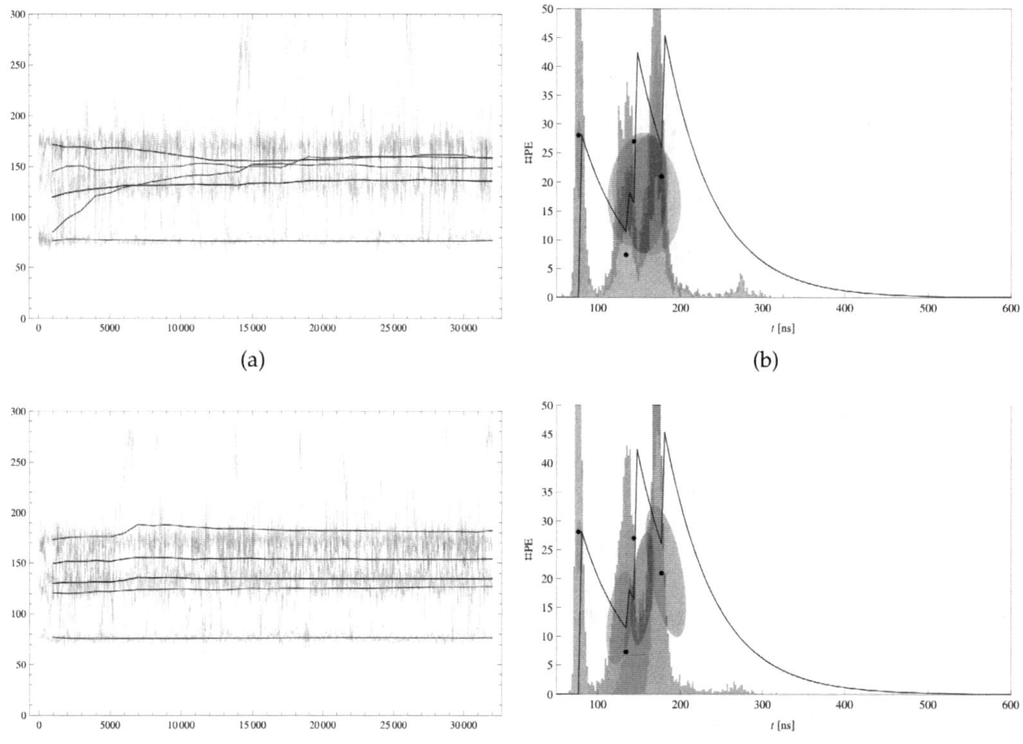

(a) (b) (c) (d)

Figure 2. The results of AM (top row) and AMOR (bottom row) algorithms on the signal example of figure 1(c). The right panels show the parameters of the four muons. Black dots: the x-coordinates are the time-of-arrival parameters t_μ, and the y-coordinates are proportional to the amplitudes A_μ (more precisely, $A'_\mu = A_\mu t_\Delta(1 - e^{t_d/\tau})/t_d$ is the size of the "muon peak"). Colored ellipses are $\exp(1/2)$-level sets of Gaussian distributions: the means are the Bayesian estimates of (t_μ, A'_μ) for each muon, and the covariance is the marginal posterior covariance of each (t_μ, A'_μ) couple. The left panels show the four chains of the time-of-arrival parameters t_μ (light colors), the running means (dark colors), and the mean of the running means (black curve). The AM algorithm shows heavy label switching among the three rightmost muons whereas the AMOR algorithm separates the muons nicely.

Several approaches have been proposed to deal with the label switching problem. The first solution consists in modifying the prior in order to make it select a single permutation of the variables, introducing an *identifiability constraint* [11]. This solution is known to cause artificial biases by not respecting the topology of the posterior [12]. An effort has then been made to adapt to the estimated posterior surface through the design of relabeling algorithms [13, 12] that process the MCMC sample *after* the completion of the simulation run. These techniques look for a permutation P_t of each individual sample point X_t so as to minimize a posterior-based criterion depending on the *whole* chain history. [2] proposed an *online* version of the relabeling procedure in which the simulation of each X_t is followed by a permutation P_t of its components. The permutation P_t is chosen to minimize a user-defined criterion that depends only on the *past* history of the chain up to time t. The major advantage of this online approach is that it is compatible with our objective of solving label switching "on the fly" in order to optimize AM for permutation-invariant models.

3.3. The AMOR algorithm

To prevent AM from label-switching, we go one step further and introduce online adaptive relabeling, the red steps in figure 3, resulting in the *adaptive Metropolis with online relabeling* algorithm (AMOR). The key step is the selection, after each proposal $\widetilde{X} \sim \mathcal{N}(.|X_{t-1}, \Sigma_{t-1})$), of a permutation (relabeling) of the proposed vector by minimizing a quadratic cost function $L_{\mu_{t-1},\Sigma_{t-1}}(x)$ over all permutations of the proposed vector, where L is defined by

$$L_{\mu,\Sigma}(x) = (x - \mu)^{\intercal}\Sigma^{-1}(x - \mu). \tag{9}$$

This step forces the posterior sample to look as unimodal as possible. A side effect is the modification of the acceptance ratio (line 7 of the algorithm in figure 3), which now includes sums over $\mathfrak{P}$. Computing this correction factor for all the permutations in $\mathfrak{P}$ can be prohibitively slow if $|\mathfrak{P}|$ is large. In practice, in this case we can use an approximate term which sums over all single inversions, i.e. permutations of only two components/muons, which decreases the computational cost from $\mathcal{O}(|\mathfrak{P}|!)$ to $\mathcal{O}(|\mathfrak{P}|^2)$. Although this modification does not change the practical performance of the algorithm, the price we pay is the lack of theoretical convergence guarantee.

To illustrate the algorithm, we ran AMOR on the same example we used for AM. The mode selected by AMOR (figure 2(c)) corresponds roughly to ordering the arrival times (i.e., multiplying the target π by the indicator of the set $\{t_{\mu_1} < t_{\mu_2} < t_{\mu_3} < t_{\mu_4}\}$), although AMOR does allow the eventual mixing of neighboring components which is important to avoid the well-known bias of the ordering-based relabeling.

AMOR, similarly to identifiability constraints, aims at approximately selecting one of the many repeated modes of π. However, AMOR has the advantage of 1) selecting its mode automatically, avoiding the need for human intervention and possible bad choices of the constraint, and 2) providing good conditions to the AM algorithm since the sample looks as Gaussian as possible among its relabelings. The double adaptivity of AMOR, both in its proposal and in its selection mechanisms (and hence in its target π), makes its analysis particularly challenging. The convergence proof of AMOR is out of the scope of this paper and has been recently addressed in [3].

4. Experimental validation of AMOR

To validate AMOR and to illustrate the pitfalls of using AM without removing the permutation invariance of π, we ran both algorithms on 1200 tank signals with 20 bins and $N = 4$ muons each, simulated from the model presented in section 2. To create difficult and realistic situations, we set the arrival time distribution to be inverse Gamma with parameters 2 and 100. This had the effect of reducing the variance of the arrival time distribution, thus making simulations exhibit a reasonable number of overlapping muons. Examples of such simulated signals are depicted in figures 4(a), 4(b) and 4(c).

To quantify the performance after T iterations, we first selected the permutation of the running posterior mean components $(\hat{t}^{(T)}_{\mu_i})_{i=1,2,3,4}$ which minimized the sum of the ℓ_2 errors on the four estimates of the times of arrival $t_{\mu_i}, i = 1, 2, 3, 4$, and we considered the sum of squared errors taken at this best permutation τ of the posterior mean to compute an error per

AMOR$(\pi(x), X_0, T, \mu_0, \Sigma_0, c)$

1 $\mathcal{S} \leftarrow \emptyset$
2 **for** $t \leftarrow 1$ **to** T
3 $\Sigma \leftarrow c\Sigma_{t-1}$ ▷ *scaled adaptive covariance*
4 $\widetilde{X} \sim \mathcal{N}(\cdot \mid X_{t-1}, \Sigma)$ ▷ *proposal*
5 $\widetilde{P} \sim \arg\min_{P \in \mathfrak{P}} L_{(\mu_{t-1}, \Sigma_{t-1})}(P\widetilde{X})$ ▷ *pick an optimal permutation*
6 $\widetilde{X} \leftarrow \widetilde{P}\widetilde{X}$ ▷ *permute*
7 **if** $\dfrac{\pi(X)\sum_{P \in \mathfrak{P}} \mathcal{N}(PX_{t-1} \mid X, \Sigma)}{\pi(X_{t-1})\sum_{P \in \mathfrak{P}} \mathcal{N}(PX \mid X_{t-1}, \Sigma)} > \mathcal{U}[0,1]$ **then**
8 $X_t \leftarrow X$ ▷ *accept*
9 **else**
10 $X_t \leftarrow X_{t-1}$ ▷ *reject*
11 $\mathcal{S} \leftarrow \mathcal{S} \cup \{X_t\}$ ▷ *update posterior sample*
12 $\mu_t \leftarrow \mu_{t-1} + \frac{1}{t}(X_t - \mu_{t-1})$ ▷ *update running mean and covariance*
13 $\Sigma_t \leftarrow \Sigma_{t-1} + \frac{1}{t}\left((X_t - \mu_t)(X_t - \mu_t)^\top - \Sigma_{t-1}\right)$
14 **return** $\mathcal{S}$

Figure 3. The pseudocode of the AMOR algorithm. The steps of the classical SRWM algorithm are in blue, the AM algorithm adds the green steps, and the new online relabeling steps are in red. Notice the adaptation of both the proposal (line 4) and the selection mechanism through the dependence of $L_{(\mu,\Sigma)}$ on (μ, Σ). Note that for practical reasons, a small εI_d is often added to the covariance matrix in line 3, but [14] recently confirmed that core AM does not lead to degenerate covariances. Note also that line 5 is usually a simple assignment of the optimal permutation. In case of ties, we draw uniformly from the finite set of optimal permutations.

muon:

$$S_T = \frac{1}{4}\left(\arg\min_{\tau \in \mathfrak{S}_4} \sum_{i=1}^{4} \left(\hat{t}^{(T)}_{\mu_{\tau(i)}} - t_{\mu_i}\right)^2\right)^{1/2}. \tag{10}$$

Figure 4(d) shows a scatterplot of S_T after $T = 3 \times 10^6$ iterations. On each signal, both AM and AMOR started with the same initial point, that is all points in figure 4(d) were lying on the diagonal. AMOR clearly outperformed AM on cases where label switching appears, leading to an estimate of the average error per muon of 17.0 ± 0.1ns versus 18.3 ± 0.1ns on these difficult cases.

The computational complexity of AMOR depends on several factors. The most costly step is the evaluation of the posterior ratio (line 7 in figure 3). In mixture models the evaluation of the posterior ratio is proportional to the number of components. Since it has to be done in every iteration, the number of iterations is also a linear factor. The execution time of our current unoptimized Mathematica implementation with a fixed number of iterations $T = 32000$ was in the order of minutes to tens of minutes, depending on these factors. An optimized C++ implementation would reduce this time by one or two orders of magnitudes.

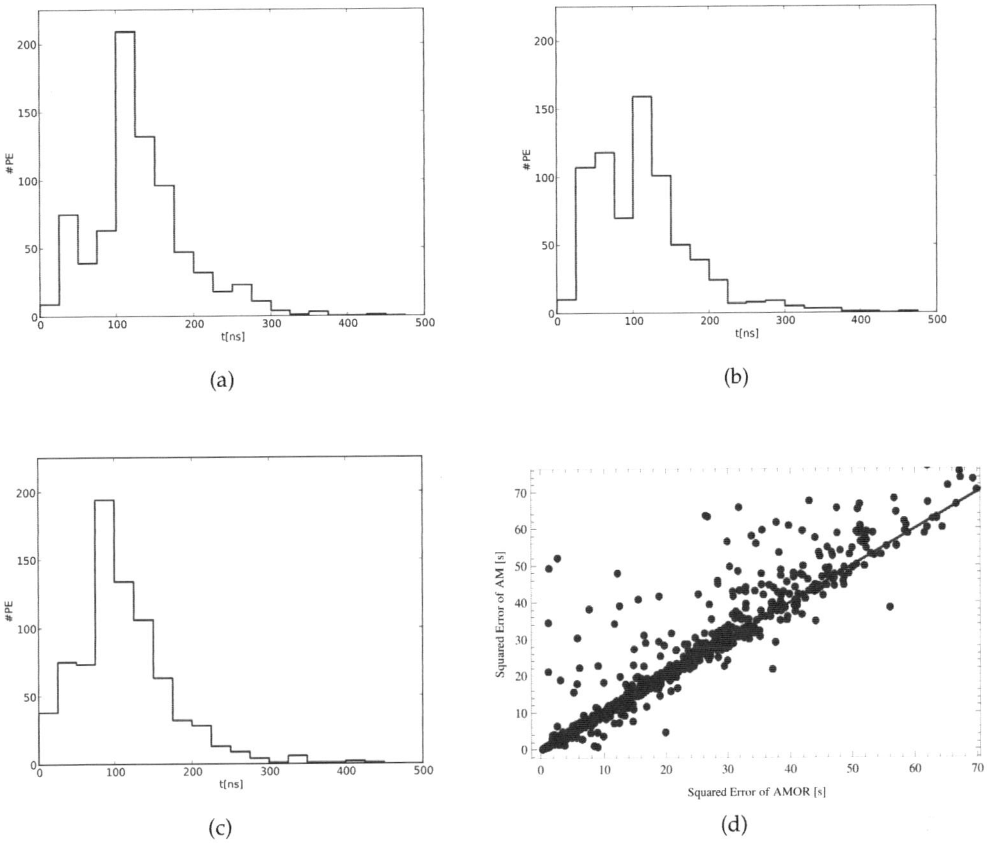

Figure 4. (a),(b),(c) Examples of generated signals, with 20 bins and 4 muons each. Arrival time distribution is $\mathcal{IG}(2, 100)$. (d) S_{3M}^{AM} vs. S_{3M}^{AMOR} results (see text).

5. Conclusion

We presented AMOR, a generic adaptive MCMC algorithm designed to tackle problems with permutation invariance, such as inferring the parameters of the muons of an Auger tank signal, and we demonstrated its capacities on simulated signals. The main advantages of AMOR are that 1) it selects a constraint on the posterior in a fully automatic fashion, avoiding the need for external expert knowledge, and 2) it allows for adaptive proposals which are known to be efficient when the target is unimodal [1]. The first advantage is a key point when it comes to models as complex as the generative model of a cosmic ray shower in Auger, since inferring the parameters of the muons is only meant to be a basic building block of a much larger scale inference algorithm.

Acknowledgments

This work was supported by the ANR-2010-COSI-002 grant of the French National Research Agency.

References

[1] Haario H, Saksman E and Tamminen J 2001 *Bernoulli* **7** 223–242
[2] Celeux G 1998 *COMPSTAT 98* ed Payne R and Green P (Physica-Verlag)

[3] Bardenet R, Cappé O, Fort G and Kégl B 2012 *Proceedings of the International Conference on Artificial Intelligence and Statistics (AISTATS)* vol 22 pp 91–99
[4] Pierre Auger Collaboration 1997 Pierre Auger project design report Tech. rep. Pierre Auger Observatory
[5] Boezio et al M 2003 *Phys. Rev. D* **67**
[6] Metropolis N, Rosenbluth A, Rosenbluth M, Teller A and Teller E 1953 *Journal of Chemical Physics* **21** 1087–1092
[7] Roberts G, Gelman A and Gilks W 1997 *The Annals of Applied Probability* **7** 110–120
[8] Roberts G and Rosenthal J 2001 *Statistical Science* **16** 351–367
[9] Saksman E and Vihola M 2010 *Annals of Applied Probability* **20** 2178–2203
[10] Fort G, Moulines E and Priouret P 2012 *Annals of Statistics* **39** 3262–3289
[11] Richardson S and Green P J 1997 *Journal of the Royal Statistical Society, Series B* **59** 731–792
[12] Marin J, Mengersen K and Robert C 2004 *Handbook of Statisics* **25**
[13] Stephens M 2000 *Journal of the Royal Statistical Society, Series B* **62** 795–809
[14] Vihola M 2011 *Electronic Journal of Probability* **16** 45–75

Two dimensional correlated sampling using alias technique

S Mohanty[1,2], S Banerjee[1,3], J Jose[1,3], D Goyal[1,4], A K Mohanty[1,5] and F Carminati[1]

[1] CERN, CH-1211, Geneve 23, Switzerland
[2] Department of Computer Science, VNIT, Nagpur, India 440010
[3] Department of Computer Science, LNMIIT, Jaipur, India
[4] Department of Electronics and Communication Engineering, LNMIIT, Jaipur, India
[5] Nuclear Physics Division, Bhabha Atomic Research Centre, Mumbai, India 400085

E-mail: siddhant9892@gmail.com, sbanerje@cern.ch, Johny.Jose@cern.ch, Dushyant.Goyal@cern.ch, ajit@cern.ch, Federico.Carminati@cern.ch

Abstract. Monte-Carlo sampling of two dimensional correlated variables (with non zero covariance) has been carried out using an extended alias technique which was originally proposed by A. J. Walker to sample from an one dimensional distribution. Although, the method has been applied to a correlated two dimensional Gaussian data sample, it is quite general and can easily be extended for sampling from a multidimensional correlated data sample of any arbitrary distribution.

1. Introduction

Monte-Carlo technique enables one to generate random samples from distributions with known characteristics and helps to make probability based inferences of the underlying physical processes. Simulation codes based on Monte-Carlo techniques have become an important aspect in science, technology and business. Fast and efficient Monte-Carlo particle transport code particularly for high energy nuclear and particle physics experiments has become an important tool starting from the design and fabrication of detectors to the modeling of the physics outcome as close as the reality. Quite often Monte-Carlo simulations require multivariate random numbers to be generated from correlated data both from normal and non-normal distributions. Although several techniques exist for multivariate correlated samplings of varying degrees of success, the most practical method is the technique that uses the principal component analysis (PCA) of the given correlation matrix for generating multivariate random numbers with specified inter-correlations. In case of multivariate correlated Gaussian, the distribution can be transformed to a different orthogonal basis using PCA technique[1]. After PCA transformation, the new distribution function can be expressed as the product of n independent Gaussian distributions with variance λ_i where λ_is are distinct eigenvalues of the original covariance matrix C. Monte-Carlo sampling is then carried out independently to build the normal random vector $Y = (y_1, y_2 ... y_n)^T$ with variance λ_i which can finally be transformed to the original vector $X = (x_1, x_2 ... x_n)^T$ with mean $\mu_x = E\{X\}$ through the relation $X = A^T Y + \mu_x$. Here A is the matrix consists of eigenvalue vectors of the covariance matrix C. While the above component analysis is suitable for multivariate normal distribution, it fails when the distribution is non-Gaussian. Alternate methods have also been proposed for non-Gaussian multivariate analysis which requires knowledge of higher order moments like skewness and kurtosis in addition to mean and variance [2] thus adding further computational complexity. In this work, we propose an extended alias sampling which was originally proposed by A. J. Walker in 1977 [3] to sample from an one dimensional distribution. This method is quite simple to implement and reproduces the original distribution well (verified through chi-square, co-variance and Kolmogorov-Smirnov goodness fit tests). This method is also quite robust and is applicable to all type of multivariate distribution irrespective of whether the distribution is Gaussian or Non-Gaussian.

Although this method is quite general and can be applied to any dimensions, in this work we have restricted only to two dimensions. The motivation behind this study has been to develop a ROOT based Monte-Carlo application package for low energy neutron transport (down to a few keV) using data from ENDF (Evaluated Nuclear Data File) which is available in ROOT format (for detail on ROOT, refer CERN web page). Work is in progress to apply this new method of alias technique to data set where the angle and energy distributions of neutron emissions are correlated.

2. The alias sampling

Theoretically any random variable $x \epsilon (a, b)$ with probability density $p(x)$ satisfying

$$\int_a^b p(x)dx = 1 \tag{1}$$

and cumulative probability density,

$$Q(x) = \int_a^x p(x')dx' \tag{2}$$

can be sampled using the inverse function method which gives,

$$x = Q^{-1}(u) \tag{3}$$

where u is a random number uniformly distributed in the interval $[0, 1]$. Numerically, $Q^{-1}(u)$ can be divided into N equal probable intervals. Sampling is carried out by selecting a bin randomly between 1 and N with equal probability $1/N$ and with bin number, $j = Nu_1 + 1$ and then randomly selecting a value within the bin with an uniform probability $x = (1 - u_2)x_{j+1} + u_2x_j$ where u_1 and u_2 are random numbers uniformly distributed between 0 and 1. This method is known as Equal Probable Bin (EPB) method which is quite fast and has been adopted in most of the Monte-Carlo programs.

However, it is inaccurate as much of the original distribution is lost by the necessary assumption of uniform likelihood within each bin. Another popular method is alias sampling which is as fast as the equal probable bin and can be made as accurate as table look up. Initially, it was proposed by A. J. Walker in 1977 to sample from a discrete data set [3] which was subsequently modified to generate continuous distribution through the technique known as linear alias sampling [4].

2.1. Alias sampling for discrete distributions

The alias method described by Walker, generates random variables from any discrete distribution with a definite number of outcomes. The alias method very closely resembles the rejection method of generating discrete random numbers, however instead of rejecting a number, a number is either accepted or it is replaced with its alias value (defined later) . The alias technique works by constructing an alias table for a given distribution. Consider we are given a distribution describing the probability of occurrence of N discrete events. The alias technique converts this distribution to a distribution of N separate events which occur with the equal probability of $\frac{1}{N}$. Each of these events E_i consists of two possible outcomes, the standard outcome x_i with probability p_i and the alias outcome Λ_i with the probability of it's occurrence Π_i.

As suggested by Walker, we construct this table by considering any two events with probabilities such that one is more and other is less than the average likelihood $\frac{1}{N}$. In the following, we give an example which has been adopted from [4]. Consider a data set having $N(=6)$ discrete tabulated probability values p_i such that $\sum p_i = 1$ (see table 1 below). The first and second column of the table 1 show the original value of i (in case i represents a bin number, x_i represents the bin content) and it's normalized probability p_i. In Walker's notation, i and p_i are called non-alias outcome and probability. In this example, the average likelihood value is $1/6 = 0.1667$. Consider two non-alias events $i = 1$ and $i = 2$. The probability of event 2 is less from the average by an amount $0.1667 - 0.08 = 0.0867$ which is subtracted from the probability of event 1 to get a new probability $0.24 - 0.0867 = 0.1533$. The alias probability Π is then calculated by multiplying N with 0.08 i.e to the probability of event 2 that gives $\Pi = 0.48$. The corresponding alias outcome Λ is taken as 1 which is the donor event (reduced by donating a value 0.0867). Thus, in the first step, we get alias outcome $\Lambda = 1$ corresponding to the alias probability $\Pi = 0.48$. The corresponding non-alias outcome i and probability p_i are 2 and 0.08 respectively. Next, event 2 is removed from the table and the second iteration begins with remaining events (except 2) with event 1 replaced by the reduced probability of 0.1533. The process continues and gets completed after six iterations (see appendix for remaining iterations). The third and fourth column of table 1 show the alias outcome Λ_i and alias probability Π_i obtained after six iterations.

The outcome is sampled by first randomly selecting an equal probable event i,

$$i = \lceil N u_1 + 1 \rceil, \tag{4}$$

where u_1 is an uniform random number between $[0, 1]$. The method then compares a second uniform random number u_2 against the alias probability to select either the alias or non-alias event. If $u_2 \leq \Pi_i$, then non-alias event i is chosen. Otherwise an alias event Λ_i is selected. Note that in case of i (or Λ_i) represents a bin number, the bin content x is selected by,

$$x = \begin{cases} x[i] & \text{if } u_2 \leq \Pi_i \\ x[\Lambda_i] & \text{otherwise} \end{cases} \tag{5}$$

Table 1. The alias table, as an example.

i	p_i	Λ_i	Π_i
1	0.24	3	0.92
2	0.08	1	0.48
3	0.28	3	1.0
4	0.12	3	0.72
5	0.12	3	0.72
6	0.16	3	0.96

2.2. Alias sampling for continuous distributions

Though the above method is meant to generate random variables which has a discrete distribution, it can be extended to sample continuous probability densities as well [4]. Given a probability distribution function $y(x)$, we estimate the discrete tables through the integral,

$$F_i = \int_{x_i}^{x_{i+1}} y(x)\, dx \tag{6}$$

We now construct the alias table for the discrete distribution (i, F_i) and carry out alias sampling using the set $[F_i]$ as discussed above which results in an interval $[x_i, x_{i+1}]$. Assuming that the distribution is piece-wise linear, a third uniform random number u_3 (between $[0, 1]$) is used to estimate x_1 and x_2 given by,

$$x_1 = (1 - u_3)x_i + u_3 x_{i+1} \tag{7}$$

$$x_2 = u_3 x_i + (1 - u_3)x_{i+1} \tag{8}$$

Using a fourth uniform random number u_4, we generate the required random variable x,

$$x = \begin{cases} x_1 & \text{if } u_4(y_i + y_{i+1}) \leq (1 - u_3)y_i + u_3 y_{i+1} \\ x_2 & \text{otherwise} \end{cases} \tag{9}$$

Thus, it can be noticed that the EPB method requires two random numbers where as the alias sampling requires four random numbers. Another difference is that the alias method stores two tables Π and Λ. Although extra storage requirement is a disadvantage as compared to EPB method, the linear alias method is easy for numerical implementation. This alias technique can also be used for interpolation between two tabulated discrete data sets [5].

In the following, we carry out a comparative study using both EPB and linear alias samplings. As an example, we have considered RENDF/B-VI, file 5 for LF=1 which gives the secondary neutron energy probability distribution as an arbitrary tabulated probability. We select the reaction MT=91 for neutron interacting with ^{238}U via inelastic continuum reaction. Figure 1 shows a typical sampling at $E_n = 8$ MeV and figure 2 shows the similar plot at $E_n = 15$ MeV using both EPB and alias samplings. The filled circles (blue color) are the ENDF tabulated data points. The red curve is the result of EPB samplings which divides the entire range into 25 equal probable bins of width 0.04 each. The blue curve is the result of linear alias sampling which passes exactly though the data points which is by construction. However, in between two data points, a linear alias sampling is carried out as discussed before. If the curve is linear interpolable between two data points, the linear alias sampling will give accurate result. On the other hand, the EPB sampling will not be able to reproduce the distributions if the rise is very fast (see figure 1(b) on log scale) or the distribution has long tail.

2.3. A 2D correlated alias sampling

We have extended the above alias technique to sample from a multi dimensional correlated data set. To demonstrate how it works, we consider a correlated two dimensional Gaussian given by,

$$f(x, y) = e^{-[(x-x_0)^2+(y-y_0)^2+\alpha(x-x_0)(y-y_0)]/\sigma^2} \tag{10}$$

The above function is uncorrelated Gaussian when $\alpha = 0.0$ and can be written as the product of two independent Gaussian. However for non-zero value of α, the above function can not be factorized. This corresponds to a situation when co-variance is non zero. Therefore, we

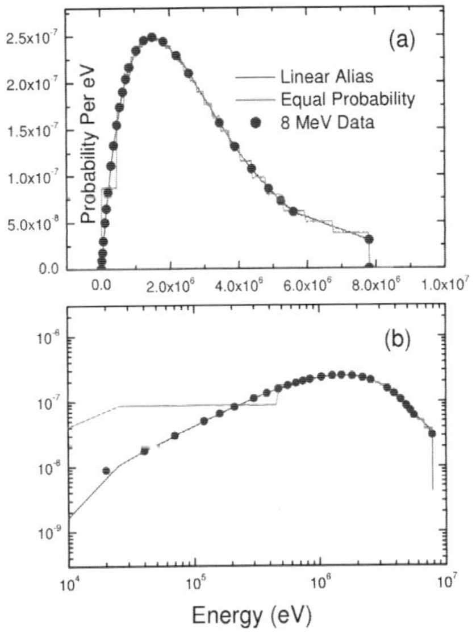

Figure 1. (a) The probability distribution of the out going neutron of the reaction $n +^{238} U \rightarrow n' +^{238}$ at 8 MeV. The data points (blue filled circles) are taken from ENDF data file. The red curve is obtained using EPB samplings where as the blue curve is due to linear alias sampling. (b) Same as (a), i.e. probability per eV versus energy, but shown on log scale. The sampling has been carried out using $10E7$ events.

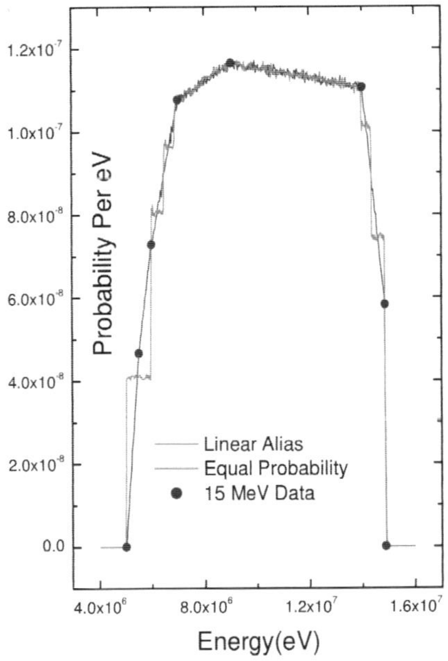

Figure 2. The probability distribution of the out going neutron of the reaction $n +^{238} U \rightarrow n' +^{238}$ at 15 MeV. The data points (blue filled circles) are taken from ENDF data file. The red curve is obtained using EPB samplings where as the blue curve is due to linear alias sampling. The sampling has been carried out using $10E7$ events.

have proposed a new sampling scheme based on the alias technique as described below. In this example, we generate 100 discrete data points $p[i][j]$ corresponding to $i = 0, 1, 2, 3, 4, 5, 6, 7, 8, 9$ and $j = 0, 1, 2, 3, 4, 5, 6, 7, 8, 9$ for discrete sampling. First, we choose a row in random based on the distribution $px[i]$ (using alias technique) where $px[i] = \sum_j p[i][j]$. Once the row I is chosen the corresponding j values are chosen depending on the distribution $p[I][j]$. In the present work, we have considered alias sampling as this method helps to get back the original distribution at each input points. So we can carry out χ^2 , covariance and Kolmogorov-Smirnov (K-S) test to study the quality of the sampling. It can be mentioned here that other methods like equal probability bin can also be implemented with this algorithm.

Figure 3 shows the original and generated distributions after sampling $n = 10^6$ events for $\alpha = 1.0$.

We define χ^2 as

$$\chi^2 = \frac{1}{N} \sum \left[\frac{p[i][j] - g[i][j]}{p[i][j]} \right]^2, \tag{11}$$

where g is the sampled distribution over the N data points. In addition to the χ^2 test, we carry

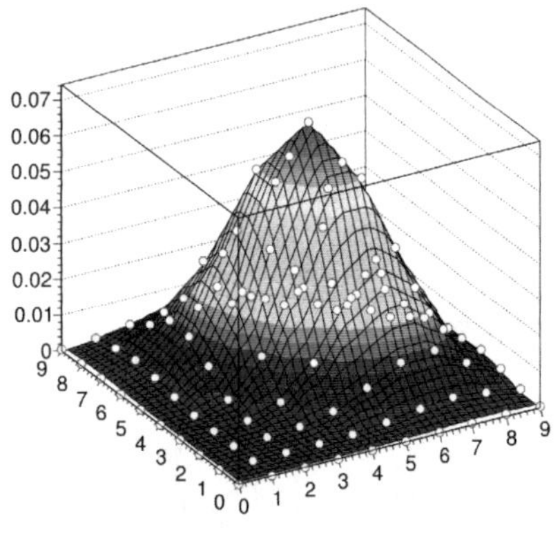

Figure 3. The comparison between actual inputs (circles) and generated data (shown as continuous line) using alias sampling for $\alpha = 1.0$. The X and Y axes are bin numbers (0-9) whereas the height represents the normalized probability distribution of the correlated Gaussian as described in the text.

out co-variance test which is defined as,

$$\sigma_{xy} = \frac{1}{n}\sum(x-\bar{x})(y-\bar{y}) \tag{12}$$

We have also carried out goodness of fit test using similar procedure as that of K-S (one sample test) where we compute the maximum absolute difference D_n of the observed and expected cumulative distribution functions (cdf) $F_o(i)$ and $F_e(i)$,

$$D_n = max|F_o^n(i) - F_e(i)|. \tag{13}$$

where cdfs $F_o^n(i)$ and $F_e(i)$ are given by,

$$F_o^n(i) = \frac{1}{n}\sum_{j=1}^{n} I_{j\leq i} \qquad F_e(i) = \sum_{j=0}^{i} p[i]. \tag{14}$$

In the above, $I_{j\leq i}$ is the indicator function equal to 1 if $j \leq i$ and equal to zero otherwise [6]. Note that since our data set is discrete, we estimate the cdfs at each bin i. As a goodness of fit, we expect $D_n \to 0$ when $n \to \infty$. Thus, χ^2 test makes the comparison using the two pdfs where as D_n test is done using cdfs.

Table 2. The table for χ^2 test using TRandom3 of ROOT package, $N = 100$ is the number of data points and n is the sample size. The parameters of the Gaussian are $x_0 = y_0 = 5$ and $\sigma^2 = 4$.

α	$\chi^2/N,\ n = 10^2K$	$\chi^2/N,\ n = 10^3K$	$\chi^2/N,\ n = 10^4K$
0.0	0.1449	0.0328	0.0032
1.0	0.512	0.112	0.034
2.0	0.38	0.418	0.108
3.0	1.453	0.266	0.491

Table 3. The table for D_n test using TRandom3 of ROOT package, $N = 100$ is the number of data points and n is the sample size. The parameters of the Gaussian are $x_0 = y_0 = 5$ and $\sigma^2 = 4$.

α	$D_n,\ n = 10^2K$	$D_n,\ n = 10^3K$	$D_n,\ n = 10^4K$
0.0	1.8E-3	6.5E-4	4.0E-4
1.0	1.6E-3	1.2E-3	3.0E-4
2.0	2.0E-3	7.0E-4	2.5E-4
3.0	1.0E-3	7.4E-4	2.9E-4

The tables 2 and 3 show the χ^2 and D_n values for several α with different sample size n. The decreasing values of χ^2 and D_n with increasing sample size indicate that the method is working well. Please note that for a given sample size n, the dependence of both χ^2 and D_n on α is bit erratic. This should not be interpreted as the failure of the method as with increasing α, the tail of the distribution decreases sharply and the sample size needs to be increased accordingly for meaningfull comparison. For example, the ratio of the lowest to highest probability becomes as low as 10^{-5} when $\alpha = 0$ and 10^{-12} when $\alpha = 3$. Therefore, for statistical comparisons as a function of α, the sample size should be at least 10^{12} and more. Table 4 shows the co-variance test of the actual inputs versus the sampled outputs. The second and third columns show the input variance and co-variance whereas the fourth and fifth columns show the corresponding sampled variances which are in good agreement.

Table 4. The table for co-variance test using TRandom3 of ROOT package, $N = 100$ is the number of data points and n is the sample size which is fixed at $100K$. The parameters of the Gaussian are $x_0 = y_0 = 5$ and $\sigma^2 = 4$.

α	σ^2	σ^2_{xy}	σ^2	σ^2_{xy}
0.0	1.985	-3.32E-17	1.977	-0.0008
1.0	2.579	-1.264	2.569	-1.255
2.0	6.856	-5.968	6.845	-5.978
3.0	16.30	-15.94	16.31	-15.94

3. Conclusion

In conclusion, we have carried out alias sampling from a discrete data set generated using a two dimensional correlated Gaussian distribution. In case of multivariate correlated Gaussian distribution, principal component analysis is a well known technique quite often used for Monte-Carlo samplings, although the method fails when distribution is strongly non-Gaussian. The present alias sampling is definitely not a superior method as compared to PCA, but quite robust and convenient for numerical implementation. It may be mentioned here that unlike PCA method, this alias technique is quite general and can be applied to any data set irrespective of whether the distribution is Gaussian or non-Gaussian. At present, our approach is confined only to two dimensional discrete sampling. Work is in progress to extend this method to higher dimensions and also to generate continuous distributions using linear alias principle.

Appendix

For the computer implementation, consider the following probability distribution (for N=6):

$$\{p/i\} = \{0.24/1, .08/2, 0.28/3, 0.12/4, 0.12/5, 0.16/6\}$$

Consider any two events having probability greater and smaller than $1/N = 0.1667$ (say events 1 and 2). Event 2 is less by an amount .0867 from the average which can be subtracted from event 1 ($.24 - .0867 = 0.1533$). Thus, in the step 1, the original distribution is updated to:

$$\{p/i\} = \{0.1533/1, 0/2, 0.28/3, 0.12/4, 0.12/5, 0.16/6\}$$

with alias representation,

$$\{\Pi/\Lambda\} = \{-, 0.48/1, -, -, -, -\}$$

In the above, the alias probability Π has been estimated by multiplying N to the original non alias probability 0.08 and the alias outcome Λ is the donor event 1 whose probability is reduced to 0.1533. Next, we can consider events 1 and 3. Since event 1 is less by an amount of 0.0134 from the average which can be subtracted from event 3. Thus, in step 2 the updated distributions become

$$\{p/i\} = \{0/1, 0/2, 0.2667/3, 0.12/4, 0.12/5, 0.16/6\}$$

with alias representation,

$$\{\Pi/\Lambda\} = \{0.92/3, 0.48/1, -, -, -, -\}$$

Like before, the alias probability is N times 0.1533 with alias outcome 3. w

Next consider events 3 and 4. The updated distributions in step 3 now becomes

$$\{p/i\} = \{0/1, 0/2, 0.22/3, 0/4, 0.12/5, 0.16/6\}$$

with alias representation,

$$\{\Pi/\Lambda\} = \{0.92/3, 0.48/1, -, 0.72/3, -, -\}$$

Next consider events 3 and 5. The updated distributions in step 4 now becomes

$$\{p/i\} = \{0/1, 0/2, 0.1733/3, 0/4, 0/5, 0.16/6\}$$

with alias representation,

$$\{\Pi/\Lambda\} = \{0.92/3, 0.48/1, -, 0.72/3, 0.72/3, -\}$$

Next consider events 3 and 6. The updated distributions in step 5 now becomes

$$\{p/i\} = \{0/1, 0/2, 0.1667/3, 0/4, 0/5, 0/6\}$$

with alias representation,

$$\{\Pi/\Lambda\} = \{0.92/3, 0.48/1, -, 0.72/3, 0.72/3, 0.96/3\}$$

Finally, in the 6th step, the updated distribution becomes

$$\{p/i\} = \{0/1, 0/2, 0/3, 0/4, 0/5, 0/6\}$$

with alias representation,

$$\{\Pi/\Lambda\} = \{0.92/3, 0.48/1, 1.0/3, 0.72/3, 0.72/3, 0.96/3\}$$

Note that the above representation is not unique, there can be different non-alias/alias representation Π/Λ depending on how the iteration is carried out.

References

[1] Jolliffe I T 2002 *Principal Component Analysis* (Berlin, Springer-Verlag)
[2] Nagahara Yuichi 2003 *Computational Statistics and Data Analysis* vol 47 p 1-29
[3] Walker A J 1977 *ACM Transactions on Mathematical Software* vol 3 p 253
[4] Edwards A L, Rathkopf J A, and Smidt R K 1991 *Report UCRL-JC-104791*
[5] Popescu L M 2000 *Journal of Computational Physics* vol 160 p 612
[6] Eadie W T, Drijard D, James F E, Roos M and Sadoulet B 1971 *Statistical Methods in Experimental Physics* (Amsterdam, North-Holland)

ACAT 2011 IOP Publishing
Journal of Physics: Conference Series **368** (2012) 012046 doi:10.1088/1742-6596/368/1/012046

Non-parametric comparison of histogrammed two-dimensional data distributions using the Energy Test

Ivan D Reid, Raul H C Lopes and Peter R Hobson

School of Engineering and Design, Brunel University, Uxbridge UB8 3PH, UK

E-mail: ivan.reid@brunel.ac.uk

Abstract. When monitoring complex experiments, comparison is often made between regularly acquired histograms of data and reference histograms which represent the ideal state of the equipment. With the larger HEP experiments now ramping up, there is a need for automation of this task since the volume of comparisons could overwhelm human operators. However, the two-dimensional histogram comparison tools available in ROOT have been noted in the past to exhibit shortcomings. We discuss a newer comparison test for two-dimensional histograms, based on the Energy Test of Aslan and Zech, which provides more conclusive discrimination between histograms of data coming from different distributions than methods provided in a recent ROOT release.

1. Introduction

A traditional task when monitoring HEP experiments has been the comparison between regularly acquired histograms of data, representing some aspect of the machine performance, and reference histograms which reflect the baseline or ideal state of the equipment. Histograms are typically used rather than the raw data points because of the compactness they provide, both in data storage space and in visual presentation. When a discrepancy is seen, it is flagged and the problem passed to an appropriate expert to decide what action is to be taken. With the advent of larger experiments, such as at the LHC, there is the potential for automated comparisons to ease the load on control-room shift-workers faced with a wide array of such histograms.

These types of comparison are called goodness-of-fit (GoF) tests, and can be subdivided into two broad types: the one-sample GoF test considers whether a given data sample is consistent with being generated from some specified distribution, while the two-sample GoF test examines the hypothesis that two data samples are derived from the same parent distribution. In general, similar methods can be applied to both types of tests. However, the problems are ill-posed – only the null hypothesis (that the distributions are the same) is well defined, the alternative hypothesis (that the distributions do not match) is not fully specified. It is important, therefore, to determine the most appropriate GoF method for any given problem.

Methods for comparing one-dimensional data distributions are well known, one of the more widespread being the Kolmogorov-Smirnov (KS) test [1] which compares cumulative distribution functions (CDF) for the two sets of data and takes as a statistic the maximum difference between them. Although this test is intended to be applied to discrete data, it is feasible to apply it to histogrammed data as well, provided that the effects of the binning on the test are taken into account. Applying this test in more than one dimension is problematic since it relies on an ordering of the data to obtain the CDFs, but there are 2^d-1 distinct ways of defining an ordering in a d-dimensional space [2]. Multidimensional GoF tests are also ill-posed in that they lack metric invariance. That is, the choice of scale factor or, in the case of histogrammed data, the number of bins can greatly affect the comparison result.

1.1. Histogram comparisons in ROOT

The widely-used data-handling and analysis package ROOT [3] provides two methods for comparing histograms, `Chi2Test` (χ^2) and `KolmogorovTest` (a KS test). Details of these tests may be found elsewhere [3, 4, 5] but in brief the χ^2 test compares histograms on a bin-by-bin basis while the KS test compares neighbourhoods, using the CDFs as described above. Extension of the χ^2 test to two dimensions (2D) is relatively straightforward, but the 2D-KS test is complicated by the ordering problem. ROOT addresses this by computing two CDFs for each histogram, accumulating the binned data rasterwise, in column- and row-major patterns, so that the comparison yields two maximum differences and the Kolmogorov function is applied to their average to return the probability P of the null hypothesis (i.e., that the two histograms represent selections from the same distribution). However, as 2D histograms are more finely binned, the order in which the binned data are accumulated approaches the order of the discrete data in the most-slowly varying dimension [5]. Consequently the CDFs generated by the ROOT 2D-KS test approach those of the discrete data ordered in one dimension along each coordinate separately. In extreme cases this can lead to false positives as histograms with similar projections onto the axes are compared (figure 1).

1.2. An alternative 2D test

Another method for comparing distributions in more than one dimension is the Energy Test presented in recent years by Aslan and Zech [6, 7, 8]. While this is again originally designed for discrete data, the authors postulated that speed gains may be obtained by applying it to histogrammed or clustered data sets [6]. We have previously shown a version of the Energy Test for histogrammed data within the ROOT framework, and provided some evaluations of

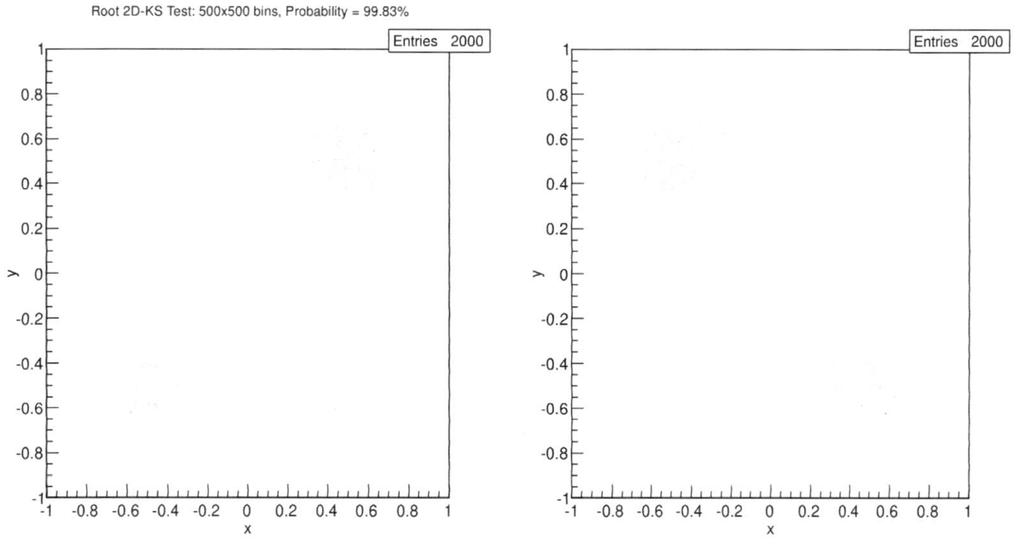

Figure 1. A ROOT 2D-KS comparison of two 2 000-point histograms binned at 500×500. The test returns a high probability (P=99.83%) that the both sets of data come from the same distribution. This is because they each have the same projections onto the axes.

its performance [5]. Some shortcomings in the ROOT implementation of the 2D-χ^2 Test were found, which have since been improved, and we revisit here some comparisons with the latest version of ROOT, in order to introduce the method to a wider audience.

2. The Energy Test

Consider two samples of data points in a d-dimensional domain, A of n points $\mathbf{x}_1,\mathbf{x}_2,\mathbf{x}_3,\ldots,\mathbf{x}_n$ and B of m points $\mathbf{y}_1,\mathbf{y}_2,\mathbf{y}_3,\ldots,\mathbf{y}_m$, whose compatibility with the hypothesis that they arise from the same distribution is to be tested. If A is considered a system of positive charges, each $1/n$, and B a system of negative charges, each -$1/m$ (i.e., each system is normalised to unity charge), then from electrostatics in the limit of $n \to \infty$, $m \to \infty$ the total potential energy of the combined samples, computed for a $1/r$ potential, will be a minimum when both charge samples have the same distribution. This concept is generalised in the Energy Test.

2.1. The test statistic

The test statistic Φ_{nm} for the Energy Test consists of three terms, corresponding to the self-energies (from the repulsive forces between like charges) of the samples A and B (Φ_A and Φ_B, respectively) and the interaction energy (from the attraction between opposite charges) between the samples (Φ_{AB}) – a simple example is shown in figure 2:

$$\Phi_{nm} = \Phi_A + \Phi_B + \Phi_{AB} \tag{1}$$

$$\Phi_A = \frac{1}{n^2}\sum_{i=2}^{n}\sum_{j=1}^{i-1} R(|\mathbf{x}_i - \mathbf{x}_j|),\, \Phi_B = \frac{1}{m^2}\sum_{i=2}^{m}\sum_{j=1}^{i-1} R(|\mathbf{y}_i - \mathbf{y}_j|),\, \Phi_{AB} = -\frac{1}{nm}\sum_{i=1}^{n}\sum_{j=1}^{m} R(|\mathbf{x}_i - \mathbf{y}_j|)$$

where R is a continuous, monotonically-decreasing function of the Euclidean distance r between the charges. Usually $R(r) = -\ln(r+\epsilon)$ is chosen rather than the electrostatic potential $1/r$ because it renders the test scale-invarant (although this is strictly true only if the same scale is applied in all dimensions) and offers a good rejection power against many alternatives to the null hypothesis. The value of the cutoff parameter ϵ, used to avoid singularities, is not critical so long as it is of the order of the mean distance between points at the densest region of the sample distributions.

It has been shown [7] that the test statistic is positive and has a minimum when the two samples are from the same distribution, in the limit of $n \to \infty$, $m \to \infty$, while another argument [6, 8] shows that when the samples have the same number of points, Φ_{nm} has a minimum when the points are pairwise coincident.

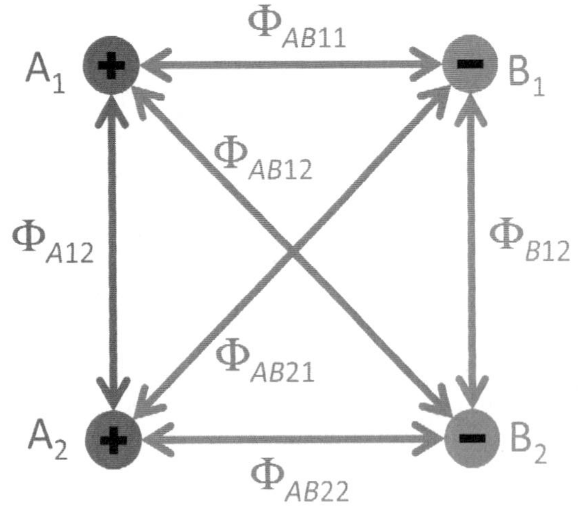

Figure 2. A simple example of the terms involved in the Energy Test. In this two-dimensional case the distributions A and B each have two points. The red arrow illustrates the (repulsive) self-energy between the positive particles in A, and the blue arrow shows that of the negative particles in B. The green arrows illustrate the (attractive) interaction energy between the oppositely-charged particles.

2.2. Implementing the Energy Test for 2D histograms

The test was implemented as a compiled ROOT macro, for equally-binned ($N \times N$) histograms in the first instance. Aslan and Zech [7] suggest that the ranges of the data can be normalised, to equalise the relative scales of the x- and y-coordinates. A similar normalisation is realised here by taking the histogram limits to be zero and unity (i.e., the distance between adjacent rows or columns is set to $1/N$). Underflow and overflow bins (with indices 0 and $N{+}1$, respectively, in ROOT notation) are included with nominal widths of $1/N$ below or above the histogram limits; arguments to the routine determine whether or not they are included in the comparison.

There is a slight complication from the fact that histograms do not preserve positional information about the points within a given bin so they must all be assigned a single position, for example the bin centre. This means that the case where $r=0$ must be treated specially – i.e., when bin (i,j) is being compared to bin (i,j), either when computing Φ_{AB} (different histograms) or when calculating Φ_A and Φ_B (same histogram; unlike the discrete case, the self-energy between points in the same bin must be taken into account). We assume the original points are randomly distributed within the (square) bin limits and take the average distance between pairs of random points in a unit square to calculate an effective cutoff ϵ.

This value is $<r>=\frac{1}{15}(2+\sqrt{2}+5\sinh^{-1}1) = 0.521\,405\,433\ldots$ [9] so we use $\epsilon=<r>/N$ as the distance. For an $N \times M$ histogram one would use instead the expression for the average distance between uniform random points in a $1/N \times 1/M$ rectangle [10],

$$\epsilon = \frac{1}{3}\rho + \frac{a^2}{6b}\ln\frac{b+\rho}{a} + \frac{b^2}{6a}\ln\frac{a+\rho}{b} + \frac{a^5+b^5-\rho^5}{15a^2b^2} \qquad (2)$$

where $a=1/N$, $b=1/M$ and $\rho=\sqrt{a^2+b^2}$. Distances for other bin combinations are calculated simply as the Euclidean distance between bin centres, justified by the proximity of this value to the average distance between random points in the two bins from Monte Carlo simulations.

A minor modification to the calculation of the self-energy of the k points within a given bin is to weight by $k^2/2$ rather than the rigorous $k(k-1)/2$, as this ensures that comparisons between identical histograms return exactly zero analytically. An added benefit is that any scaling factors applied across individual histograms will be cancelled out rather than producing an offset that is dependent on the total histogram content [5]. If desired the rigorous behaviour can be obtained by the use of a parameter flag.

To summarise, the implementation of the three terms in the energy sum when comparing two $N \times N$ ROOT histograms A and B with total contents n and m, respectively, is given by

$$\Phi_A = \frac{1}{n^2}\sum_{i=0}^{N+1}\sum_{j=0}^{N+1} A(i,j)\left(\sum_{k=0}^{i-1}\sum_{l=0}^{N+1} A(k,l)R(i,j,k,l) + \sum_{l=0}^{j-1} A(i,l)D(j,l) + 0.5A(i,j)D_0\right) \qquad (3)$$

$$\Phi_B = \frac{1}{m^2}\sum_{i=0}^{N+1}\sum_{j=0}^{N+1} B(i,j)\left(\sum_{k=0}^{i-1}\sum_{l=0}^{N+1} B(k,l)R(i,j,k,l) + \sum_{l=0}^{j-1} B(i,l)D(j,l) + 0.5B(i,j)D_0\right)$$

$$\Phi_{AB} = -\frac{1}{nm}\sum_{i=0}^{N+1}\sum_{j=0}^{N+1} A(i,j)\sum_{k=0}^{N+1}\sum_{l=0}^{N+1} B(k,l)R(i,j,k,l)$$

where $D_0 = -\ln(<r>/N)$, $R(i,j,k,l) = D_0$ when $(i{=}k,\ j{=}l)$ or $-\frac{1}{2}\ln(((i-k)^2+(j-l)^2)/N^2)$ otherwise, $D(j,l) = R(i,j,i,l) = -\ln(|j-l|/N)$, and $A(i,j)$, $B(i,j)$ are the contents of individual bins within the histograms.

2.3. Speed

The calculation, and thus the time complexity, of the test statistic Φ_{nm} is, by inspection, $\mathbf{O}(n^2)$ so that in terms of histogram dimensions the cost is $\mathbf{O}(N^4)$. Table 1 shows the times taken for

Table 1. Times taken to compare 10^6-point histograms of various binnings with the ROOT 2D-KS and 2D-χ^2 tests and the Energy Test.

Size	ROOT 2D-KS	ROOT 2D-χ^2	Energy Test
25 × 25	<10 ms	<10 ms	<10 ms
50 × 50	<10 ms	<10 ms	10 ms
100 × 100	<10 ms	<10 ms	160 ms
250 × 250	<10 ms	10 ms	6.1 s
500 × 500	30 ms	30 ms	96.3 s

comparison of several pairs of histograms, filled with 10^6 points in a random uniform distribution and as a constant bin content, respectively. Despite all efforts to reduce calculations as much as possible, the increase in CPU time for the Energy Test beyond a binning of about 100×100 is evident. As with all the results reported in this work, the calculations were performed on a 2.67 GHz Xeon X5550 computer using ROOT Version 5.30/00. Outlier bins were always excluded from the comparisons.

3. Performance

3.1. Testing the power

The *power* of a comparison test is its ability to discriminate against non-conforming data, i.e., the fraction of non-compatible data which is rejected based on a selection criterion. In order to determine the power, the confidence level for accepting a test result must first be established. A common criterion is the 95th percentile CL_{95} – the value of a test beyond which only 5% of valid comparisons will lie.

Two references were developed for several tests; a constant distribution (i.e., no statistical fluctuation) of 10 points in each bin of a 100×100 histogram across the unit square, and a continually re-generated sample of 100 000 points randomly and uniformly distributed across the square. 50 000 tests were performed against these references using further samples of 100 000 random points. These are thus, respectively, references for one-sample GoF tests, to determine

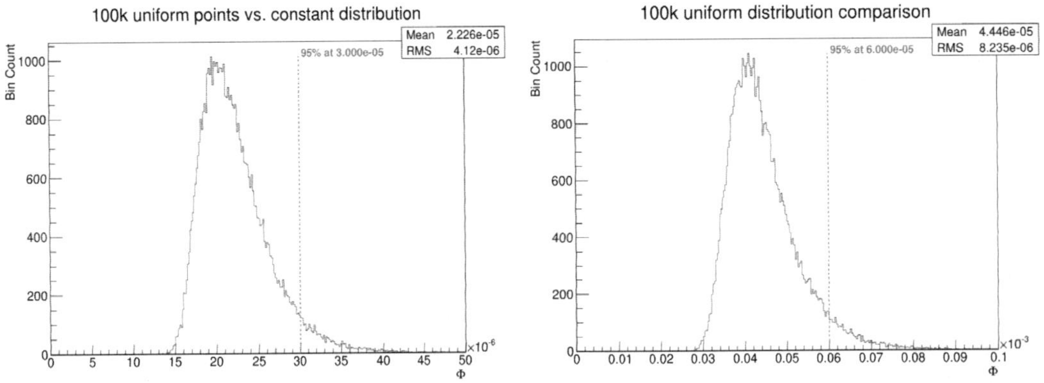

Figure 3. The distribution of results of the histogrammed energy test, comparing 50 000 sets of 100 000 randomly distributed points on the unit square to a constant distribution and to a second uniform distribution at 100×100 binning.

if samples are consistent with arising from a constant distribution, and two-sample GoF tests to test if samples come from the same parent distribution as the random distribution.

The resulting test statistic distributions are shown in figure 3. Aslan and Zech [8, 11] found that their test distribution is well described by a generalised extreme value (GEV) distribution [12] but were unable to calculate the parameters of the distribution from first principles so they recommended determining the distribution by Monte Carlo methods. We also have found that GEV distributions fit well to data such as in figure 3 [5] but have used the experimental distributions rather than fits to determine percentile values. These give values for CL_{95} of 3×10^{-5} for a constant parent and 6×10^{-5} for comparisons against a uniform random distribution, as shown in the figure.

3.1.1. The Cook-Johnson distribution The power of the histogrammed energy test to determine deviations from the constant and random distributions was tested using various levels of the Cook-Johnson distribution, one of the tests used by Aslan and Zech for their discrete energy test [7]. The Cook-Johnson distribution is the multivariate uniform distribution given by

$$(X_1, \ldots, X_d) = ((1 + \frac{E_1}{S})^{-a}, \ldots, (1 + \frac{E_d}{S})^{-a}) \tag{4}$$

where $E_1, \ldots, E_d$ are independent and identically distributed exponential random variables, S is an independent gamma(a) random variable and $a > 0$ is a parameter [13]. For $a \to \infty$ this approaches a uniform distribution within the d-dimensional hypercube; as $a \to 0$ the distribution becomes correlated, $X_1 = \ldots = X_d$ (see figure 4 for examples in two dimensions).

The power of the energy test and the ROOT 2D-KS and 2D-χ^2 tests for comparing the various Cook-Johnson distributions on the unit square against the constant reference and uniform distributions are given in table 2. The selection criteria are the 95% confidence levels established in section 3.1 for the energy test, and a 5% acceptance level for the probability P returned by the ROOT 2D-KS and 2D-χ^2 tests. From the table it is evident that the histogrammed energy test has a much higher power than the ROOT 2D tests, rejecting Cook-Johnson distributions up to $a = 50$, whereas the ROOT tests only reject distributions with much lower values of a. Both of the ROOT tests perform rather better at rejecting the hypothesis that the Cook-Johnson

Table 2. The discrimination power of the histogrammed energy test and the ROOT 2D tests comparing 2D Cook-Johnson distributions to a constant reference distribution and a uniform distribution.

	Constant reference			Uniform reference		
Cook-Johnson parameter a	Energy Test power	2D-KS power	2D-χ^2 power	Energy Test power	2D-KS power	2D-χ^2 power
200	0.092	0.0	0.0	0.076	0.012	0.035
100	0.19	0.0	0.0	0.114	0.012	0.051
50	0.806	0.0	0.0	0.412	0.013	0.05
20	1.0	0.0	0.0	1.0	0.023	0.172
10	1.0	0.0	0.0	1.0	0.018	0.87
5	1.0	0.0	0.0	1.0	0.01	1.0
2	1.0	0.379	1.0	1.0	0.491	1.0
1	1.0	1.0	1.0	1.0	1.0	1.0
0.6	1.0	1.0	1.0	1.0	1.0	1.0

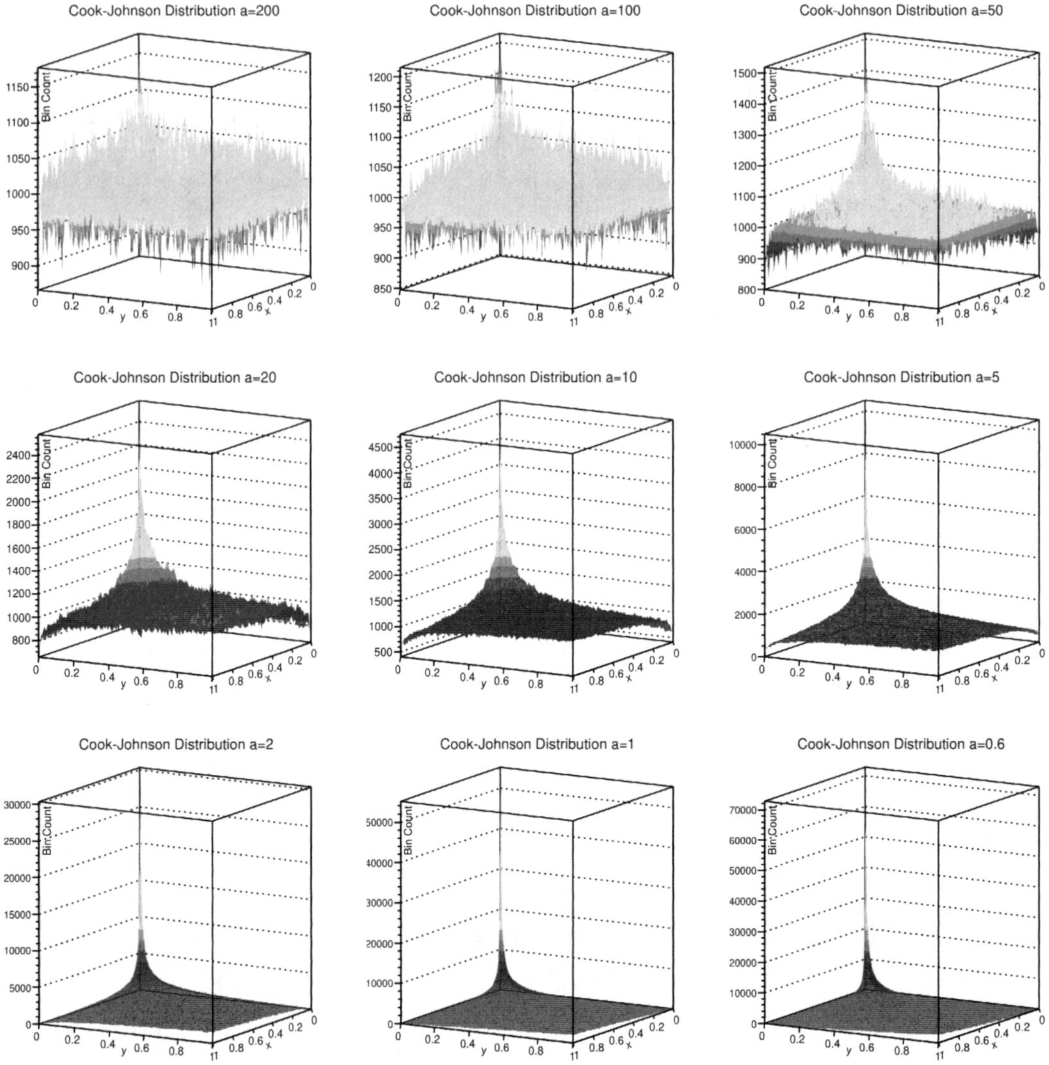

Figure 4. The two-dimensional Cook-Johnson distribution, for parameter a=200, 100, 50, 20, 10, 5, 2, 1 and 0.6, plotted as 100×100 histograms on the unit square. Each histogram contains 10^7 points (i.e., an average of 1 000 points per bin). Note the change in the z range of the distribution as a becomes smaller.

distributions are from a randomly distributed population than in rejecting them as from a constant distribution.

3.1.2. Gaussian contamination As a test of sensitivity to contamination, similar comparisons were made between the constant and uniform reference distributions and 1 000 samples of a uniform distribution where $n\%$ ($n = 0, 1, \ldots, 8$) of the 100 000 points in each sample were replaced by points from a bivariate N(0,1) (Gaussian) distribution (see figure 5). The 100×100 histograms' limits were [-3,3] in each dimension with the bivariate distribution truncated at these limits; because of the normalisation in the energy test, the same CL_{95} values as in section 3.1 are expected for 100 000-point uniform distributions. Acceptance criteria for the ROOT tests were the same as in section 3.1.1.

Table 3 gives the discrimination power of the three tests. As expected, the observed power of the energy test for 0% contamination is consistent with the choices of CL_{95}, which strongly reject distributions with 2% contamination or higher. However, the ROOT 2D-KS test only shows high discrimination power at 3% contamination and above for both hypotheses, while the ROOT 2D-χ^2 test does not reject any distributions in the case of a constant reference but performs slightly better at considering the contaminated samples not to match a uniform distribution.

Table 3. The discrimination power of the histogrammed energy test and the ROOT 2D-KS and 2D-χ^2 tests for comparisons of increasing levels of a bivariate N(0,1) distribution contamination in a uniform distribution in -3<x, y<3 against a constant and a uniformly distributed reference.

	Constant reference			Uniform reference		
Gaussian contamination	Energy Test power	2D-KS power	2D-χ^2 power	Energy Test power	2D-KS power	2D-χ^2 power
0%	0.044	0.0	0.0	0.045	0.005	0.048
1%	0.731	0.003	0.0	0.351	0.083	0.054
2%	1.0	0.194	0.0	0.982	0.560	0.075
3%	1.0	0.964	0.0	1.0	0.983	0.114
4%	1.0	1.0	0.0	1.0	1.0	0.238
5%	1.0	1.0	0.0	1.0	1.0	0.454
6%	1.0	1.0	0.0	1.0	1.0	0.690
7%	1.0	1.0	0.0	1.0	1.0	0.917
8%	1.0	1.0	0.0	1.0	1.0	0.983

3.1.3. Displacement sensitivity The sensitivity of the tests to a shift in the position of a histogrammed sample was investigated by comparing 1 000 pairs of 100 000-point bivariate N(0,1) distributions, in 100 × 100 histograms with a range of [-3,3] in each dimension, while the second distribution was moved away from (0,0) in x-increments of 0.01 (1/6th of a bin width). All

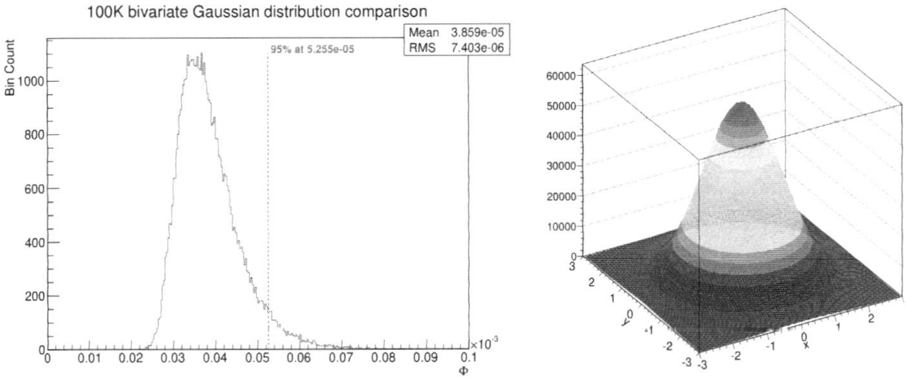

Figure 5. The distribution of results of the histogrammed energy test, comparing 50 000 pairs of histograms, each consisting of 100 000 points drawn from a bivariate N(0,1) distribution (shown on the right) in the region -3<x, y<3.

distributions were truncated at the histogram limits.

For the histogrammed energy test CL_{95} was taken from the test metric distribution obtained from 50 000 pair-wise comparisons at δx=0 which yielded a value of 5.26×10^{-5} (figure 5); acceptance criteria for the ROOT 2D-KS and 2D-χ^2 tests were again $P > 5\%$.

The calculated powers for the tests are given in table 4. The histogrammed energy test provides slightly better rejection than the ROOT 2D-KS test, approaching full rejection at δx=0.02 (1/3rd of a bin width) compared to 0.03 for the 2D-KS test. The ROOT 2D-χ^2 test, however, does not provide high rejection until the separation approaches two bin-widths (0.12).

Table 4. The discrimination power of the histogrammed energy test and the ROOT 2D-KS and 2D-χ^2 tests for comparisons of increasing displacement δx between bivariate N(0,1) distributions in 100 × 100 histograms covering $-3<x, y<3$.

δx	Energy Test power	2D-KS power	2D-χ^2 power	δx	Energy Test power	2D-KS power	2D-χ^2 power
0	0.045	0.009	0.026	0.07	1.0	1.0	0.490
0.01	0.380	0.135	0.021	0.08	1.0	1.0	0.713
0.02	0.965	0.723	0.036	0.09	1.0	1.0	0.884
0.03	1.0	0.990	0.057	0.10	1.0	1.0	0.958
0.04	1.0	1.0	0.100	0.11	1.0	1.0	0.992
0.05	1.0	1.0	0.169	0.12	1.0	1.0	1.0
0.06	1.0	1.0	0.305				

4. Conclusions

We have presented our investigations into a new test for performing goodness-of-fit comparisons between two-dimensional histograms, based upon the Energy Test of Aslan and Zech.

Compared with the two existing ROOT tests for 2D histograms, the histogrammed energy test proves far superior to the ROOT `Chi2Test` and outperformed the ROOT `KolmogorovTest` in our comparisons of synthetic data sets.

The main reason for this ranking in performance seems to be that the histogrammed energy test is a global test, with comparisons between every pair of bins in the histograms entering into the result, while the ROOT 2D-KS is a regional test more influenced by neighborhood variations as the CDFs are built up. The ROOT χ^2 test for its part is strictly a localised test with each bin in the histogram only being compared to its counterpart.

The disadvantage of the histogrammed energy test is that it takes longer to perform, especially at the highest binnings, but for moderately-sized histograms the penalty is slight, particularly when the time taken to construct the histograms is also considered.

Acknowledgments

This work has been funded by the Science and Technology Facilities Council, UK.

References

[1] Chakravati I M, Laha R G and Roy J 1967 *Handbook of Methods of Applied Statistics, Volume I* (New York: John Wiley and Sons), 392–4

[2] Peacock J A 1983 Two-dimensional goodness-of-fit testing in astronomy *Mon. Not. R. Astron. Soc.* **202** 615–27

[3] http://root.cern.ch
[4] Lopes R H C, Hobson P R and Reid I D 2008 Computationally efficient algorithms for the two-dimensional Kolmogorov-Smirnov test, *Int. Conf. on Computing in High Energy and Nuclear Physics* (2–7 September 2007, Victoria BC) *J. Phys.: Conf. Ser.* **119** 042019 http://iopscience.iop.org/1742-6596/119/4/042019/pdf/1742-6596_119_4_042019.pdf
[5] Reid I D, Lopes R H C and Hobson P R 2008 Comparison of two-dimensional binned data distributions using the energy test *CMS NOTE-2008/023* (Geneva: CERN) http://cdsweb.cern.ch/record/1127567/files/NOTE2008_023.pdf
[6] Aslan B and Zech G 2003 A new class of binning-free, multivariate goodness-of-fit tests: the energy tests *Preprint* arXiv:hep-exp/0203010v5
[7] Zech G and Aslan B 2003 A new test for the multivariate two-sample problem based on the concept of minimum energy *Preprint* arXiv:math/0309164v1
[8] Aslan B 2004 *The concept of energy in nonparametric statistics: Goodness-of-Fit problems and deconvolution* Thesis, Universität Siegen http://deposit.ddb.de/cgi-bin/dokserv?idn=972172122
[9] Weisstein E W *Square Line Picking*, From MathWorld – A Wolfram Web Resource. http://mathworld.wolfram.com/SquareLinePicking.html
[10] Gaboune B, Laporte G and Soumis F 1993 Expected distances between two uniformly distributed random points in rectangles and rectangular parallelepipeds *J. Opl Res. Soc.* **44** 513
[11] Aslan B and Zech G 2005 Statistical energy as a tool for binning-free, multivariate goodness-of-fit tests, two-sample comparison and unfolding *Nucl. Instr. and Meth.* A **537** 626–36
[12] Kotz S and Nadarajah S 2000 *Extreme Value Distributions: Theory and Applications* (Singapore: World Scientific)
[13] Devroye L 1986 *Non-Uniform Random Variate Generation* (New York: Springer-Verlag) http://cg.scs.carleton.ca/~luc/rnbookindex.html

Off-line data processing and analysis for the GERDA experiment

M Agostini[1], L Pandola[2] and P Zavarise[2,3]

[1] Physik Department E15, Technische Universität München, Garching, Germany
[2] INFN, Laboratori Nazionali del Gran Sasso, Assergi (AQ), Italy
[3] Dipartimento di Fisica, Università dell'Aquila, L'Aquila, Italy

E-mail: matteo.agostini@ph.tum.de, pandola@lngs.infn.it, zavarise@lngs.infn.it

Abstract. GERDA is an experiment designed to look for the neutrinoless double beta decay of ^{76}Ge. The experiment uses an array of high-purity germanium detectors (enriched in ^{76}Ge) directly immersed in liquid argon. GERDA is presently operating eight enriched coaxial detectors (approximately 15 kg of ^{76}Ge) and about 25 new custom-made enriched BEGe detectors will be deployed in the next phase (additional 20 kg of ^{76}Ge). The paper describes the GERDA off-line analysis of the high-purity germanium detector data. Firstly we present the signal processing flow, focusing on the digital filters and on the algorithms used. Secondly we discuss the rejection of non-physical events and the data quality monitoring. The analysis is performed completely with the GERDA software framework (GELATIO), designed to support a multi-channel processing and to perform a modular analysis of digital signals.

1. Introduction

GERDA [1, 2] is a low-background experiment designed to search for the neutrinoless double beta decay of ^{76}Ge using an array of bare high-purity germanium (HPGe) detectors isotopically enriched in ^{76}Ge. The detector array is operated directly in ultra radiochemically-pure liquid argon, allowing a substantial background reduction at the $Q_{\beta\beta}$-value of ^{76}Ge (2039 keV) with respect to the previous experiments [3–5]. The experiment pursues a staged implementation. In the present phase (Phase I) eight enriched coaxial detectors are being used (approximately 15 kg of ^{76}Ge). In Phase II, about 25 new custom-made enriched BEGe detectors [6] will be deployed providing additional ~20 kg of ^{76}Ge. The experiment is located in the underground Laboratori Nazionali del Gran Sasso of the INFN (Italy).

If no background events are observed in the region of interest, the sensitivity of GERDA will scale with the product of the ^{76}Ge mass and the exposure time, $M \cdot t$. Otherwise, the sensitivity will scale with the square root of this product divided by the background index (BI) and the energy resolution (ΔE) in the region of interest, $\sqrt{(M \cdot t)/(BI \cdot \Delta E)}$. To enhance the experimental sensitivity, an advanced off-line analysis of the HPGe detector signals is performed. Digital filters are applied to reconstruct the event energy with high resolution. In addition the analysis of the signal time-structure is a powerful tool to identify background events [7–10].

To ease this kind of analysis a new software framework (GELATIO) has been recently developed [11]. The framework is presently used to handle the full data analysis flow of GERDA and the R&D activities related to the experiment. It is implemented in C++ and is based on the MGDO library [12]. The framework implements a modular analysis. Each module handles a precise and self-consistent task of the signal processing and is implemented as a dedicated C++ class. The output of the modules, which is either a scalar parameter (e.g. the amplitude of the signals) or a shaped trace, can be used as input for other modules and/or stored to disk. The users can easily create chains of modules and customize the parameters of each algorithm. The software supports also the processing of multi-channel data. While the analysis is performed event-per-event in order to keep the output synchronized, each channel can be processed along its own dedicated and customized chain of modules.

The framework was used for the reference analysis of the data acquired in the GERDA commissioning phase (from June 2010 to October 2011), when up to seven HPGe detectors have been operated simultaneously. The commissioning data were used as a benchmark to validate GELATIO against other independent analysis codes and to prove its suitability for the use in GERDA Phase I [11].

This paper describes the basic off-line analysis of the GERDA data performed with the GELATIO framework. In section 2 we present the flow of the signal processing and analysis along the module chains and the shaping algorithms. Then, in section 3, we discuss the identification of non-physical events or of signals not properly processed along the analysis pipeline. Also, the data quality monitoring will be described. Finally, summary and conclusions are presented in section 4.

2. Signal processing flow

The charge signals of the HPGe detectors deployed in GERDA are composed by pulses with a $\sim 1\,\mu$s long rising part followed by an exponential decay tail ($\sim 100\,\mu$s decay time) folded by the front-end charge-sensitive preamplifier. The rising part (leading edge) is induced by the drift of electron-hole pairs – generated by γ-ray or charge particle interactions – toward the detector electrodes and it contains information concerning the event topology.

The signals are digitized at 100 MHz sampling frequency by 14-bit flash-ADCs (FADC) equipped with integrated anti-aliasing bandwidth filters [13–15]. For each event, the FADC computes in run-time and writes to disk two traces. The first trace is sampled at 100 MHz and is $4\,\mu$s long (high-frequency-short trace). It is used for identifying background events through the analysis of the signal time-structure. The second trace is sampled at 25 MHz and is $160\,\mu$s long (low-frequency-long trace). It is used for those operations, as energy reconstruction, which involve the integration of the pulse.

The two traces are processed along different chains of GELATIO modules, as shown in figure 1. The first module, GEMDTop, extracts from the input file the traces requested by the following modules. Before publishing the traces the module checks, and possibly changes, the pulse polarity in order to always have positive-polarity pulses. The read-out traces are the starting point for two chains: the low-frequency-long trace is processed along Chain1 while the high-frequency-short trace along Chain2.

Chain1 starts with GEMDBaseline. This module analyzes the baseline of the signal by computing the average value, the root-mean-square deviation (RMS) and the linear slope before the leading edge. In addition, the module performs a baseline restoration — a subtraction of the average baseline value to the trace — and provides the new signal to the other modules:

- GEMDTrigger. The module is used to identify the beginning of the pulse leading edge (trigger position). It implements a leading-edge discriminator with threshold defined dynamically as three times the RMS of the signal baseline. After the trigger, the signal has to remain above threshold for at least $40\,\mu$s, otherwise the trigger is rejected.
- GEMDFTTrigger. While the previous module is tuned to determine the trigger position with high precision and stability, this module is used to identify events with multiple physical pulses occurring within the same trace. The module applies to the input signal a $1.5\,\mu$s

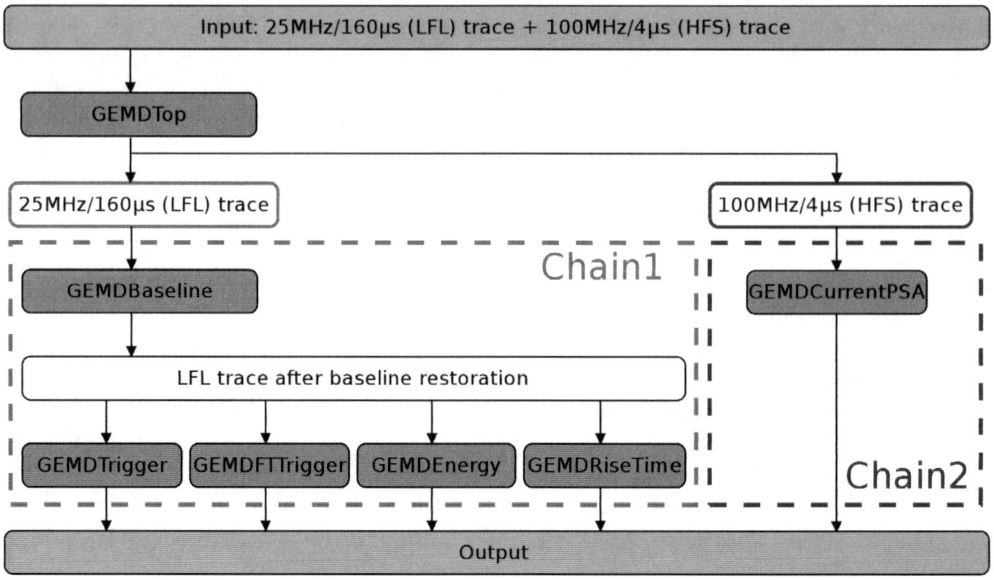

Figure 1. Flow chart of the signal processing. The two traces saved by the digitizer are processed along two different chains of GELATIO analysis modules. The low-frequency-long (LFL) trace is used for reconstructing the energy, the trigger and the rise time. The high-frequency-short (HFS) trace is used to analyse the time-structure of the signals.

moving differentiation filter and a $1\,\mu$s moving average filter for noise reduction (see figure 2). The resulting trace has a peak for each sharp variation of the signal (such as the leading edge of a pulse). The peak width is similar to the size of the moving differentiation and was chosen to maximize the pile-up identification efficiency and to avoid the misidentification of highly-multiple-site events. The number and the position of the peaks are estimated by applying a leading-edge discriminator, whose threshold is four times the RMS of the baseline. After this condition is met, the signal has to remain above the threshold for at least $1\,\mu$s.

- GEMDEnergyGauss. The module reconstructs the event energy using an approximate Gaussian filter [16, 17]. The pulse is differentiated by a moving differentiation filter and then integrated 15 times by a moving average filter to achieve an approximated Gaussian shape[4]. The energy information is eventually stored in the maximum amplitude of the quasi-Gaussian pulse. The width of the moving filters has been set to $10\,\mu$s in order to minimize losses due to ballistic effects. The intermediate steps of the shaping are shown in figure 3.
- GEMDRiseTime. The module computes the rise time between 10% and 90% of the maximum amplitude of the pulse. The maximum amplitude is computed as the difference between the maximum of the pulse and the average baseline value. Then, the first samples below the 10% and 90% of the maximum amplitude are found by moving backwards from the position of the maximum.

The second chain is used to evaluate parameters relevant for the identification of background events and it will be better defined during the future data taking. The chain presently includes

[4] Historically, the energy reconstruction filters for γ-ray spectroscopy also perform a deconvolution of the exponential function which is folded in the signal by the charge sensitive preamplifier. However, the approach suggested in this paper was found to provide better results on the GERDA data with respect to the usual filters.

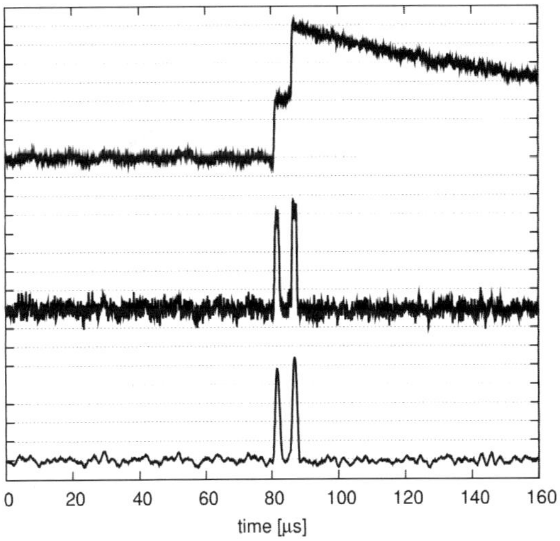

Figure 2. Digital signal processing performed by GEMDFTTrigger. The incoming trace (top trace) is differentiated (middle trace) and integrated (bottom). The output trace has a peak for each pulse in the incoming trace. The illustrative input trace contains two physical pulses with leading edge shifted by $\sim 5\,\mu$s.

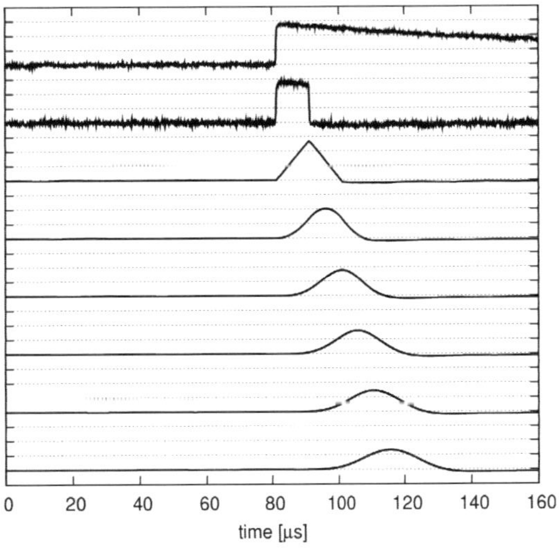

Figure 3. Digital signal processing performed by GEMDEnergyGauss. The incoming signal (top trace) is differentiated (second trace) and then integrated several times (following traces) by a moving average filter. The output signal has a Gaussian shape and its maximum is proportional to the event energy.

only one module, GEMDCurrentPSA, which computes the current signal as the derivative of the charge signal and then extracts the basic features of the current pulse, like rise time, width and area.

3. Data selection and monitoring

In the GERDA data sets there are two main classes of signals that have to be identified and tagged: 1) signals corrupted or produced by non-physical events, i.e. discharges, cross-talk, pick-up noise; 2) signals which are not properly processed along the analysis pipeline, as pile-ups and accidental coincidences.

The first class includes signals with anomalous shape, wrong polarity, extremely short/long rise time or exceeding the dynamic range of the FADC (see figure 4). To identify these events a sequence of cuts based on four parameters is applied. The first parameters are the trigger position computed by GEMDTrigger and the time position of the maximum amplitude of the Gaussian pulse (maxAmpTime) computed by GEMDEnergyGauss. If the signal has a leading edge at the proper position, followed by an exponential decay tail, then the trigger has to be reconstructed roughly in the center of the trace and maxAmpTime has to be in a well-defined range. The third parameter is the 10-90% rise time which can be used to identify signals that are extremely fast or slow, and hence inconsistent with well-behaved physical events. Finally, signals that saturate the dynamic range of the FADC are identified by scanning the individual samples of the traces.

The second class includes signals generated by the superposition of multiple physical pulses, or having the leading edge not aligned with the center of the trace (see figure 5). These signals can be identified using the baseline slope computed by GEMDBaseline, the number of pulses provided by GEMDFTTrigger and the trigger position yielded by GEMDTrigger. Their amount is proportional to the event rate and can reach up to 15% in the calibration data sets, while it

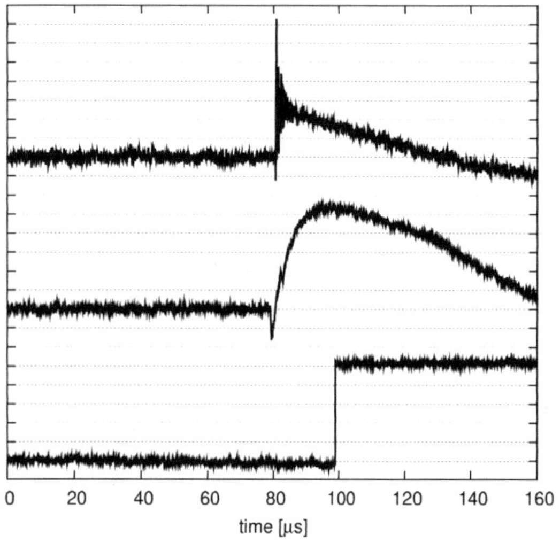

Figure 4. Illustrative traces generated by non-physical events. Note that these pulses do not have the typical exponential decay tail after the leading edge.

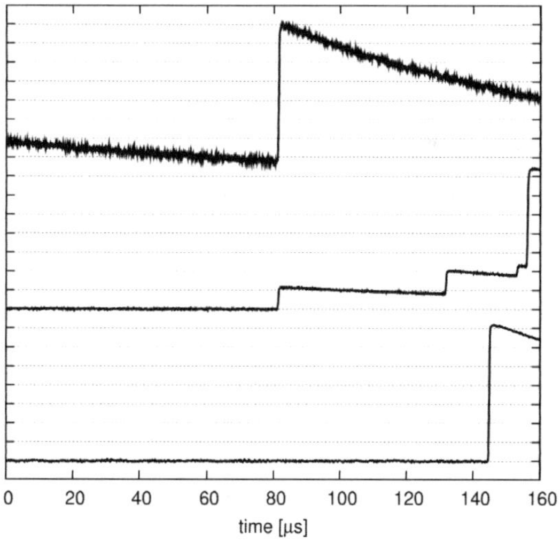

Figure 5. Example of traces generated by pile-up events (top-middle trace) and accidental coincidences (bottom).

is usually negligible in the physics data sets.

The identification of this second class of events is of primary importance for GERDA since the background rejection techniques based on the analysis of the signal time-structure are developed and tuned by using calibration data. Therefore the calibration data have to be filtered to extract samples of events as similar as possible to the physics run data and a small inefficiency could bias the following analysis. The result of the data selection applied to a calibration run with a ^{228}Th source is shown in figure 6. The cuts remove efficiently bad signals and pile-up events, improving the shape of the γ-ray peaks and the agreement with the standard analytical functions used to model γ-line peaks.

Besides cuts for removing undesirable classes of signals, there are also parameters which can be used to monitor the quality of the data taking and the stability of the set-up, the most important being the average value and the RMS of the baseline. These parameters are sensitive to noise changes and to gain variations in the read-out chain. Figure 7 shows these parameters as a function of time for a 10-day commissioning run. The parameters are stable over the whole data taking, except for a few hours during day 7. These instabilities can be correlated with hardware operations in the set-up and the corresponding data can be removed by applying a cut on the two parameters.

4. Conclusions

The GERDA experiment is currently starting the data taking of Phase I. The off-line analysis of the HPGe detector signals will be performed with the GELATIO software framework, a tool specifically developed for GERDA which supports a multi-channel analysis and implements signal processing chains based on a modular approach. A reference analysis pipeline has been defined and optimized. The signal processing is performed along chains of modules and includes the estimate of the trigger position, the amplitude and several basic pulse shape analysis parameters. The digital filters have been improved and optimized during the GERDA commissioning phase and the shaping parameters have been tuned. Also, a set of cuts has been defined to identify

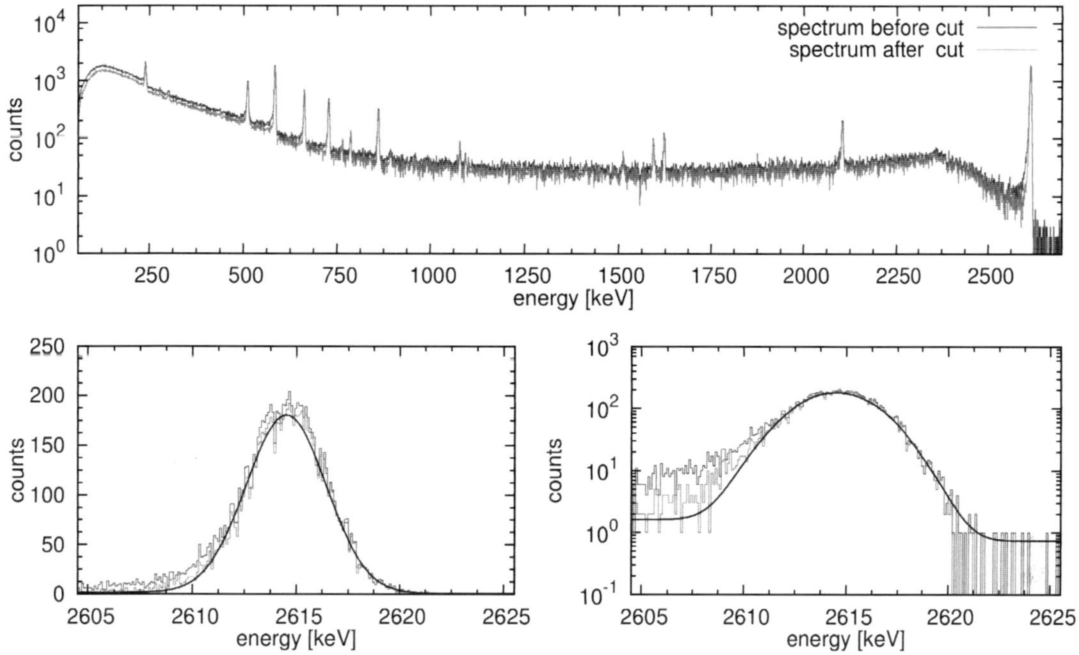

Figure 6. Energy spectrum reconstructed by an enriched detector deployed in the GERDA set-up during a calibration run with a ^{228}Th source. The cuts remove approximately 15% of the total events (~10% in the γ lines). The bottom panels show the peak of the 2.614 MeV γ-ray line in linear and logarithmic scale. The reduction of the tail due to pile-up events improves significantly the fit with the analytical model (the χ^2 per degree of freedom is significantly reduced, from 1.77 to 0.99).

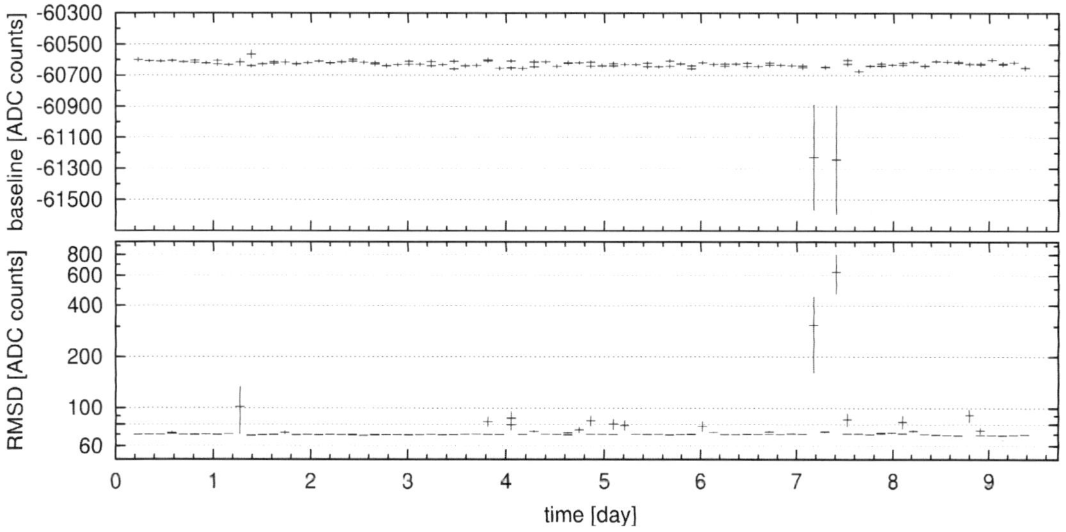

Figure 7. Average value (top panel) and RMS (bottom panel) of the signal baseline vs. time for a detector operated in GERDA during a 10-day run. The bin content represents the mean value of the parameter and the error bars are related to the width of the distribution. The bins are 2.4 hours wide.

signals induced by non-physical events or signals which are not properly processed. In addition, a set of parameters was identified to monitor the data quality and possibly to discriminate corrupted data.

The software and the digital filters have been validated during the GERDA commissioning and in several R&D activities related to the experiment. The new pipeline has been tested and it proved to be stable and ready to be used for the reference analysis of GERDA Phase I data.

Acknowledgments
We would like to acknowledge our colleagues of the GERDA Collaboration, especially D. Budjáš and B. Schwingenheuer, for many invaluable advices concerning signal analysis and data quality control. We want also to thank D. Bazzacco, C. A. Ur and T. Kihm for all the stimulating discussions and suggestions concerning digital signal processing and the algorithms used in γ-ray spectroscopy.

This work was supported in part by the Transregio Sonderforschungsbereich SFB/TR27 "Neutrinos and Beyond" by the Deutsche Forschungsgemeinschaft and by the Munich Cluster of Excellence "Origin and Structure of the Universe".

References

[1] Abt I *et al.* (GERDA Collaboration) 2004 GERDA: The GERmanium Detector Array for the Search of Neutrinoless $\beta\beta$ Decay of ^{76}Ge at LNGS *Laboratori Nazionali del Gran Sasso - Proposal*
[2] Schönert S *et al.* (GERDA Collaboration) 2005 The GERmanium Detector Array (GERDA) for the search of neutrinoless beta beta decays of Ge-76 at LNGS *Nucl. Phys. Proc. Suppl.* B **145** 242-245
[3] Gunther M *et al.* 1997 Heidelberg - Moscow beta-beta experiment with Ge-76: Full setup with five detectors *Phys. Rev.* D **55** 54-67
[4] Klapdor-Kleingrothaus H V, Krivosheina I V, Dietz A and Chkvorets O 2004 Search for neutrinoless double beta decay with enriched Ge-76 in Gran Sasso 1990-2003 *Phys. Lett.* B **586** 198-212
[5] Aalseth C E *et al.* (IGEX Collaboration) 2002 The IGEX Ge-76 neutrinoless double beta decay experiment: Prospects for next generation experiments *Phys. Rev.* D **65** 092007
[6] CANBERRA Broad Energy Ge (BEGe) Detector, URL http://www.canberra.com/products/485.asp
[7] Budjáš D, Barnabé Heider M, Chkvorets O, Khanbekov N and Schönert S 2009 Pulse shape discrimination studies with a Broad-Energy Germanium detector for signal identification and background suppression in the GERDA double beta decay experiment *JINST* **4** P10007 (*Preprint* arXiv:0909.4044)
[8] Agostini M *et al.* 2011 Signal modeling of HPGe detectors with a small read-out electrode and application to neutrinoless double beta decay search in Ge-76 *JINST* **6** P03005 (*Preprint* arXiv:1012.4300)
[9] Gonzalez D *et al.* 2003 Pulse shape discrimination in the IGEX experiment *Nucl. Instrum. Meth.* A **515** 634-643
[10] Hellmig J and Klapdor-Kleingrothaus H V 2000 Identification of single-site events in germanium detectors by digital pulse shape analysis *Nucl. Instrum. Meth.* A **455** 638-644
[11] Agostini M, Pandola L, Zavarise P and Volynets O 2011 GELATIO: a general framework for modular digital analysis of high-purity Ge detector signals *JINST* **6** P08013 (*Preprint* arXiv:1106.1780)
[12] Agostini M *et al.* 2011 The MGDO software libraries for Germanium neutrinoless double beta decay experiments to appear in *J. Phys.: Conf. Series* XII International Conference on Topics in Astroparticle and Underground Physics (TAUP 2011), September 5-9 2011, Munich, Germany
[13] MIZZI Computer Software GmbH, URL www.mizzi-computer.de
[14] Isocrate R *et al.* 2004 MD2S - Digital sampling electronics for the MARS detector, A data acquistion system for the MD2S digital sampling electronics *Laboratori Nazionali di Legnaro - Annual Report* 226-227
[15] Ur C A *et al.* 2004 A data acquistion system for the MD2S digital sampling electronics *Laboratori Nazionali di Legnaro - Annual Report* 228-229
[16] Smith S W 1999 *The Scientist and Engineer's Guide to Digital Signal Processing* (California: California Technical Publishing San Diego)
[17] Van Trees H L 1968 *Detection, Estimation, and Modulation Theory* (New York: John Wiley and Sons)

Journal of Physics: Conference Series **368** (2012) 012048 doi:10.1088/1742-6596/368/1/012048

Ten years of Object-Oriented analysis on H1

Paul Laycock

Department of High Energy Physics, Oliver Lodge Laboratory, University of Liverpool, Liverpool, L69 7ZE, UK

E-mail: laycock@hep.ph.liv.ac.uk

Abstract. Over a decade ago, the H1 Collaboration decided to embrace the object-oriented paradigm and completely redesign its data analysis model and data storage format. The event data model, based on the ROOT framework, consists of three layers - tracks and calorimeter clusters, identified particles and finally event summary data - with a singleton class providing unified access. This original solution was then augmented with a fourth layer containing user-defined objects. This contribution will summarise the history of the solutions used, from modifications to the original design, to the evolution of the high-level end-user analysis object framework which is used by H1 today. Several important issues are addressed - the portability of expert knowledge to increase the efficiency of data analysis, the flexibility of the framework to incorporate new analyses, the performance and ease of use, and lessons learned for future projects.

1. Introduction

The H1 experiment analysed electron-proton data from the HERA machine at DESY, Hamburg. During the luminosity upgrade at the turn of the millennium, H1 decided to change their analysis framework to an object-oriented paradigm, concentrating on the data storage model [1, 2, 3, 4].

2. The Analysis Models

2.1. The old analysis model

Figure 1 shows the old data analysis model of H1. All of the triggered RAW data from the detector was reconstructed using the H1 reconstruction software package written in Fortran into a BOS [5] bank format. The entire reconstruction output was stored as Physics Output Tape (POT), a summary of which was written to (a more easily accessible) Data Summary Tape (DST). The actual physical storage media of course changed with time, while the names did not.

The DST was processed privately by private ntuple production code to produce the analysis data format. This was usually the HBOOK format, and physics analysis was performed using PAW [6]. The scope for sharing analysis information was fairly limited, as the variable definitions depended on the privately maintained ntuple production code which varied from group of data analysts to another. Meanwhile, the private production of ntuples was usually quite inefficient, resulting in many copies of essentially the same data, wasting resources.

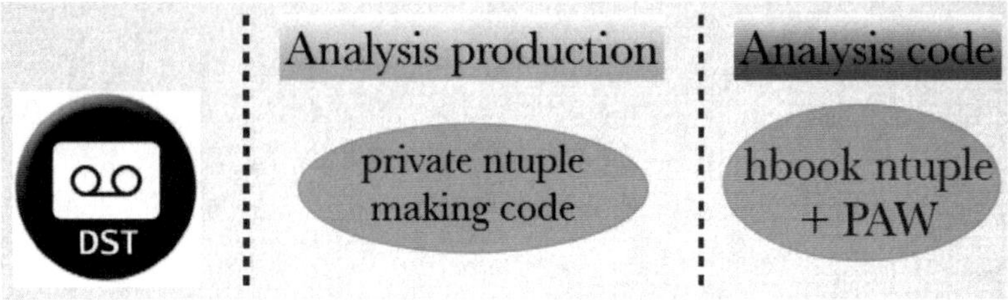

Figure 1. The old data analysis model.

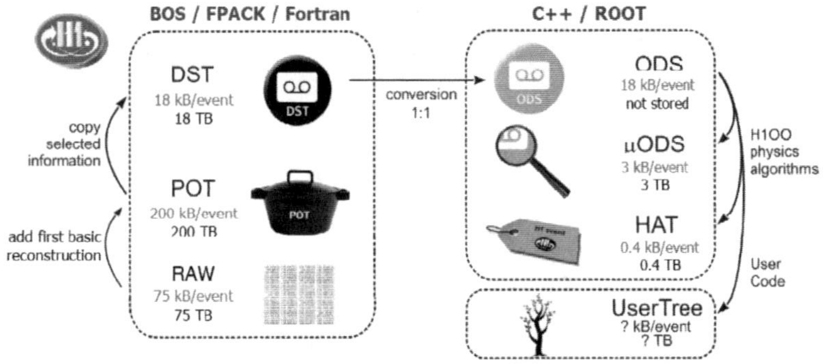

Figure 2. The new data storage model of H1.

2.2. The H1OO data analysis model

Figure 2 shows the object-oriented data analysis model. The aim was to centrally produce the analysis data storage format, replacing the custom ntuples used previously. Private ntuple production was not prohibited and was used by some individuals, but the vast majority of the collaboration enjoyed and still enjoys central production of their analysis data format.

2.3. The H1OO data storage model

The DST remains the source of all derived analysis formats. The object-oriented data format consists of several layers, and all data storage layers are encapsulated in the H1Tree interface. This object provides smart access to the data, such that the user doesn't need to know which layer is accessed.

The Object Data Storage (ODS) layer is entirely equivalent to the DST, but now Track and Cluster classes are used for the data representation. Object-oriented bank classes also exist, providing DST, POT and RAW data access if and as necessary. The ODS was generally not used for analysis, as the content of the two derived layers described next was usually found to be sufficient. However, it can be produced and stored if necessary (if the ODS content will be repeatedly analysed), or created on-the-fly from DST (more performant for rare access).

Particle finders, shown in figure 3, are used to produce the content stored in the next object-oriented layer, the micro-ODS or MODS. The particle-finding algorithms used represent the best knowledge of H1 by definition, and thus all analyses benefit from this best knowledge. Event summary information, e.g. the primary vertex coordinates and the number of each type of particle stored on the MODS, is stored in the third layer, the H1 Analysis Tag or HAT. The HAT also contains experimental conditions such as beam energies and other information critical for analysis calculations. Together, the MODS and HAT layers are the analysis data format,

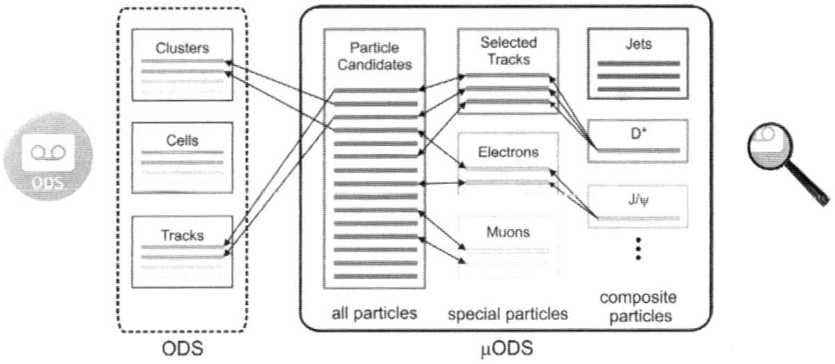

Figure 3. The Particle Finder model.

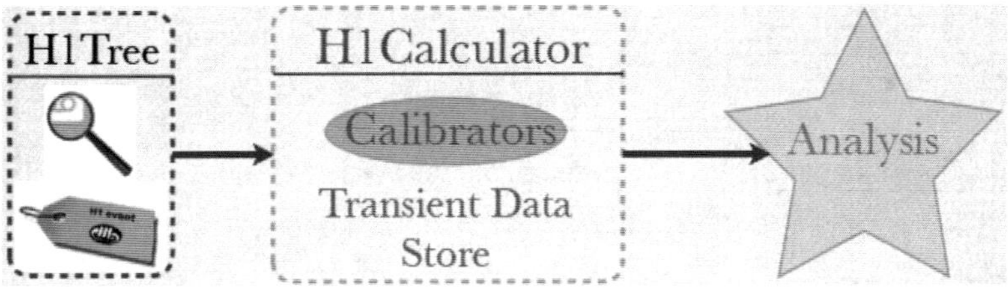

Figure 4. The transient data interface, H1Calculator.

centrally produced, for the vast majority of H1 analyses.

A fourth and final optional layer, the UserTree, allows ultimate flexibility by allowing a user-defined storage layer. If this layer was found to be useful for several groups, it entered the central production framework and was centrally maintained. It's worth noting that there are currently three UserTree packages in the core H1OO framework, suggesting that this flexibility was both necessary and useful.

3. Transient Data

The original H1OO concept concentrated on unifying the data analysis format and optimising resource usage for its production, and this was very successful. However, there were one or two limitations to this model. Chief among these was the lack of treatment of transient data, i.e. quantities calculated or recalculated at run time. This was found to be particularly relevant for the evaluation of systematic uncertainties, where derived quantities need to be recalculated based on a systematically shifted base quantity. It was also relevant when analysts wanted to apply the latest calibrations on-the-fly, rather than wait for a full-scale production of data and Monte Carlo.

3.1. Transient data interface

To address these problems, a transient data interface called the H1Calculator was designed and implemented [7, 8], as shown in figure 4. This reads the persistent data provided by the H1Tree, and stores transient derived quantities. The latest calibration can then be applied easily, and systematic uncertainty evaluation simply requires recalculating the derived quantities. All data access in user analysis code goes through this interface, guaranteeing a consistent treatment of the data.

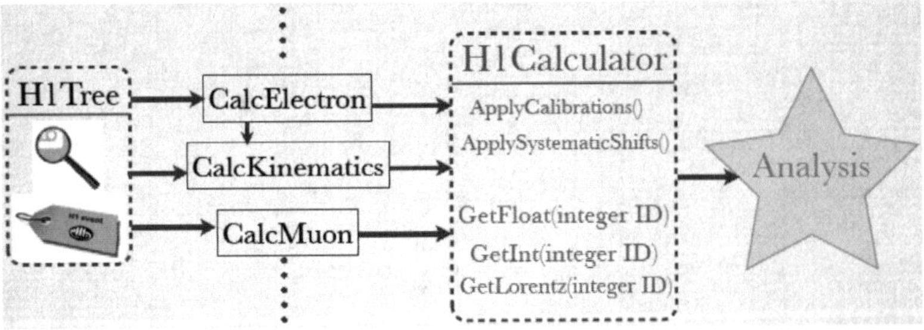

Figure 5. A more detailed view of the transient data interface.

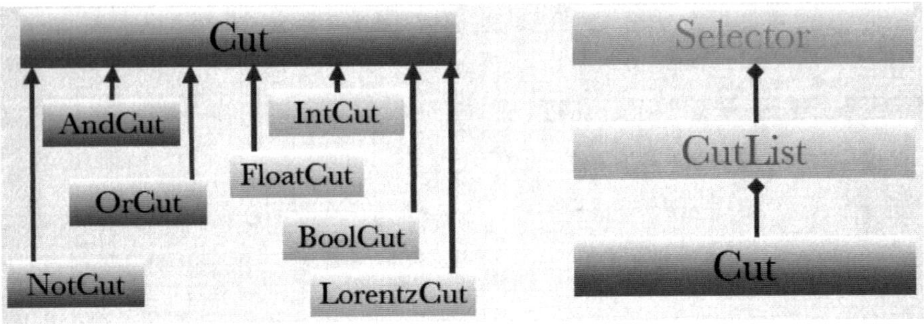

Figure 6. The event selector classes.

In practice, the scope for derived quantities for all H1 analyses is very large, and the H1Calculator is composed of several smaller, themed calculator classes which deal with specific quantities, e.g. one for electron quantities, another for event kinematics. A generic interface to the data provides access to variables by type (integer, TLorentzVector, etc.), which then allows user classes to be decoupled from the details of this structure. The main H1Calculator class itself then provides a simple interface to variables by integer ID (the generic interface) as well as switches to apply calibrations and systematic shifts, as shown in figure 5.

4. Higher Level Analysis Objects

Following the generic interface to the transient data, more end-user classes could be defined. One particular highlight are the event selector classes [7], composed of lists of cut objects. A cut object returns a boolean answer based either on a variable read from the transient data interface or on a logical combination of other cut objects. This simple but very useful set of classes is shown in figure 6. They also provide detailed debugging information and cutflow statistics. These classes could be passed from analyst to analyst to apply particular event selections, together with other simple classes responsible for histogram management [7], i.e. classes which book and fill sets of (related) histograms. These simple organisational aides proved to be very useful, not least in the context of data quality and software validation, as well as in physics analysis.

4.1. Analysis objects

Figure 7 shows a flow diagram of a simplified physics analysis. The selector and histogram manager objects can be matched to "Event Selection" and "Fill Histogram", respectively. The decisions made by the selector are passed on to an object or process which "Bins the event",

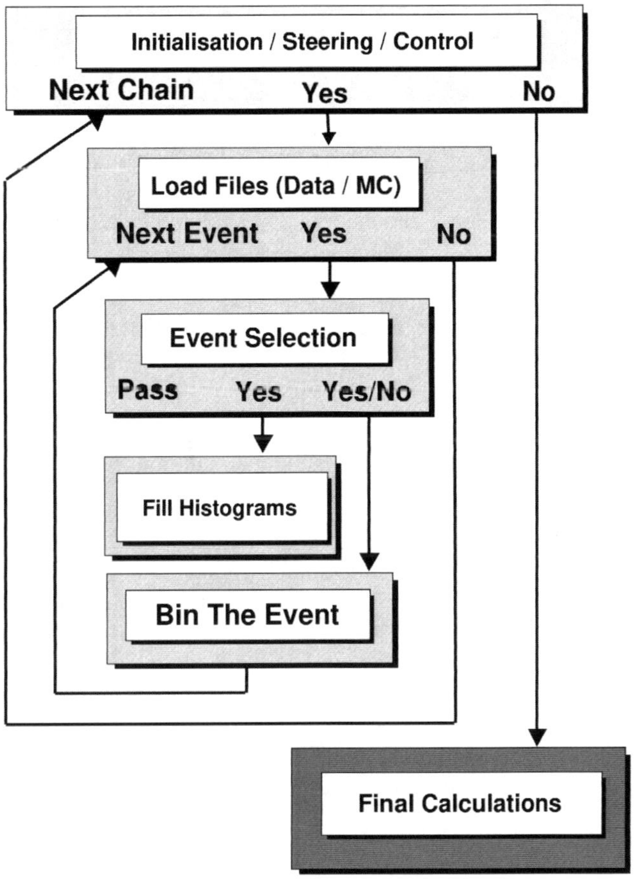

Figure 7. Flow diagram of physics analysis.

determining the reconstructed and truth-level bins in order to correct the data to a final physics measurement (taken care of in "Final Calculations"). The event information itself is made available to these various objects via the H1Calculator.

Two other organisational structures can also be seen, namely at the level of looping over one set of data or Monte Carlo, where we used the term "chain" from PAW, and finally the analysis level itself.

Correspondingly, Analysis and AnalysisChain objects also proved to be useful organisational classes [7]. An Analysis object is composed of several AnalysisChains, which each contain one or more Histogram Managers. The Event Selector is an object defined at the Analysis level. Data access goes through the H1Calculator. This simple model is shown in figure 8 and allows a common framework for nearly all stages of a physics analysis. This in turn allowed better collaboration between physics groups, and the easy exchange of high-level analysis code.

5. Conclusions

During the luminosity upgrade at the turn of the millennium, H1 decided to change their analysis framework to an object-oriented paradigm, concentrating on the data storage model. This move was very successful in unifying the data storage model and analysis formats of H1. The flexibility to have a user defined data storage layer proved crucial in several analyses.

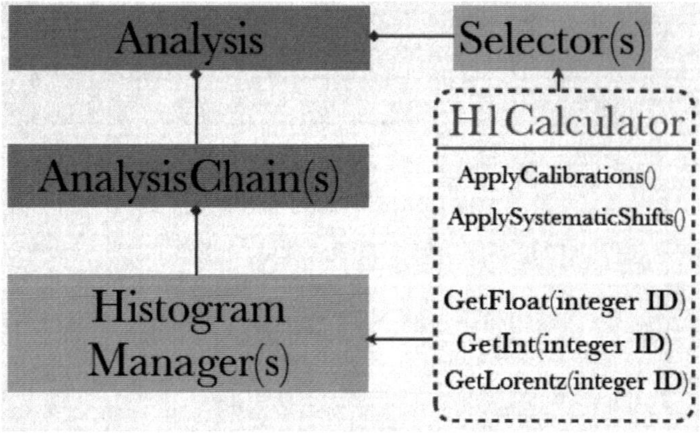

Figure 8. Analysis object model.

The development of a transient data interface improved physics analysis in many cases, especially those where access to the latest calibrations was critical and/or complicated systematic effects had to be evaluated. It paved the way for further developments which allowed for a more efficient exchange of higher-level physics analysis tools, up to and including entire analyses.

Acknowledgments
The material presented here is the culmination of many years of work by members of the H1 Collaboration. I'd like to thank all of those who worked on the H1OO project and in particular my friend and collaborator Dave South.

References

[1] Berthon U, Benisch T, Gerhards R, Grab C, Hadig T and Van Mechelen P, New data analysis environment in H1, *International Conference on Computing in High-Energy Physics and Nuclear Physics (CHEP 2000), Padova, Italy, 7-11 Feb 2000*

[2] Benisch T, Berthon U, Gerhards R, Grab C and Hadig T, New Data Storage Model For H1, *International Conference on Computing in High-Energy Physics and Nuclear Physics (CHEP 2000), Padova, Italy, 7-11 Feb 2000*

[3] Peez M [H1 Collaboration], The New object oriented analysis framework for H1, eConfC **0303241** (2003) THLT007 (*Preprint* physics/0306124).

[4] Katzy J, H1OO - an analysis framework for H1, *Computing in High-Energy Physics (CHEP '04), Interlaken, Switzerland, 27 Sep - 1 Oct 2004*

[5] Blobel V, BOS and related packages, *Erice 1990, Proceedings, Data structures for particle physics experiments, 1-6*

[6] Brun R, Couet O, Vandoni C E and Zanarini P, Paw: A General Purpose Portable Software Tool For Data Analysis And Presentation, *Comput. Phys. Commun.* **57** (1989) 432.

[7] Laycock P, A Measurement of the Diffractive Reduced Cross-Section $\sigma_r^{D(3)}$ at High Q^2 with the H1 Detector at HERA, PhD thesis (2003), *https://www-h1.desy.de/psfiles/theses/h1th-321.pdf*

[8] South D, Events with Isolated Leptons and Missing Transverse Momentum in e+p Collisions at HERA, PhD thesis (2003), *https://www-h1.desy.de/psfiles/theses/h1th-314.pdf*

The toolbox of modern multi-loop calculations: novel analytic and semi-analytic techniques

Alexey Pak
Institut für Theoretische Teilchenphysik, KIT, 76128 Karlsruhe, Germany
E-mail: apak@particle.uni-karlsruhe.de

Abstract. We describe three algorithms for computer-aided symbolic multi-loop calculations that facilitated some recent novel results. First, we discuss an algorithm to derive the canonical form of an arbitrary Feynman integral in order to facilitate their identification. Second, we present a practical solution to the problem of multi-loop analytical tensor reduction. Finally, we discuss the partial fractioning of polynomials with external linear relations between the variables. All algorithms have been tested and used in real calculations.

1. Introduction

A higher-order calculation is a multi-stage process, with details strongly dependent on the physics problem. There is no generic one-size-fits-all approach, but usually one first generates the diagrams, then performs Dirac algebra and/or projections on scalar integrals, and then manipulates large expressions that depend on scalar products of loop and external momenta until they can be reduced to known integrals and computed. Each calculation has its own "key" stage which requires the most efforts. In many cases, those are the algebraic reduction of integrals (with integration-by-parts identities [1, 2, 3]) to a small subset of "master" integrals, and the evaluation of the latter. Those problems have been the focus of active research for many years and gave rise to a plethora of various methods and algorithms [4]. However, as the three-, four- and higher-loop problems become the norm in the LHC era, even the traditionally "simple" steps become impossible to perform manually or with the ad-hoc methods.

In this contribution we present the three algorithms that automate the tasks usually considered "routine" and not worth mentioning. First, we discuss the classification of integrals as instances of different "topologies", or families of integrals that differ by the exponents of the (fixed) denominator factors. When the number of such families reaches a few hundred, automation becomes a necessity.

Second, we consider the tensor reduction, that may e.g. help re-arrange numerators of factorizable topologies in order to perform integration independently. In certain cases, involving complicated asymptotic expansions, this stage may dominate the total computation time of a diagram. In addition, we provide a practical solution for the situations where general formulas are not known. Finally, we present a (relatively straightforward) re-formulation of the partial fractioning problem and solve it using Gröbner bases.

The primary use of these methods is to generate the code for the computer algebra system that processes the actual diagrams (e.g. FORM [5]), and thus the runtime efficiency is not the

Figure 1. Graphs corresponding to integrals I_1 and I_2.

highest priority. Nevertheless, we find that our solutions are rather efficient and do not require enormous times for real-life problems.

2. Identification of Feynman integrals

In multi-loop calculations, one often has to solve a problem of identifying individual Feynman integrals or deciding whether an integral belongs to a given family ("topology") which helps reduce the number of integrals to compute. In general, in order to say whether two integrals are equal one has to compute both. However, for some integrals it is possible to elaborate a simple transformation of loop momenta and establish the identity of integrands. In simple cases, such transformations can be derived manually just by looking at graphs, but in multi-loop multi-scale problems with non-trivial kinematic constraints this is a daunting task.

Let us consider the two forward-scattering integrals shown in figure 1:

$$I_1 = \int \frac{d^D k_1\, d^D k_2}{D_1 D_2 D_3 D_4 D_5}, \quad \begin{array}{l} D_1 = (k_1+p_1)^2,\ D_2 = k_1^2,\ D_3 = k_2^2, \\ D_4 = (k_2+p_2)^2,\ D_5 = (k_1+k_2)^2 + m^2, \end{array} \tag{1}$$

$$I_2 = \int \frac{d^D q_1\, d^D q_2}{E_1 E_2 E_3 E_4 E_5}, \quad \begin{array}{l} E_1 = (q_1+p_1)^2,\ E_2 = q_1^2,\ E_3 = q_2^2, \\ E_4 = (q_2+p_2)^2,\ E_5 = (q_1+q_2+p_1+p_2)^2 + m^2 \end{array} \tag{2}$$

with the common additional constraints $p_1^2 = p_2^2 = 0$, $(p_1+p_2)^2 = -1$.

The integrands of I_1 and I_2 obviously cannot be related by only renaming symbols. However, the value of e.g. I_1 is invariant with respect to a shift of loop momenta, e.g. $k_1 \to k_1' = k_1 + p_1$, or a permutation of denominator factors, e.g. $D_1 \leftrightarrow D_2$, etc. There may exist a transformation $D_i(K) = E_{p(i)}(Q)$ where $K = MQ$ is loop re-parametrization and $p(i)$ is some permutation of indices $1, 2, ..., 5$, such that

$$K = \begin{pmatrix} k_1 \\ k_2 \\ p_1 \\ p_2 \end{pmatrix}, \quad Q = \begin{pmatrix} q_1 \\ q_2 \\ p_1 \\ p_2 \end{pmatrix}, \quad M = \left(\begin{array}{c|c} A & B \\ \hline 0 & I \end{array} \right), \tag{3}$$

with A and B being 2×2 matrices, $|A| = 1$[1] , and I a 2×2 identity matrix.

One practical way to compare two topologies is to compare graphs. In figure 1 we present the graphs corresponding to I_1 and I_2, and one can see that the graphs do not match. However, some integrals may have no corresponding graphs (for example, integrals originating from some effective theories may have linear denominators, e.g. $2p_1k_1 + i0$, which cannot be interpreted as graph lines), and some integrals may have multiple corresponding graphs (as in the given case). In addition, (sub)graph isomorphism is itself a computationally hard problem[2].

[1] or in general $|A| = d \neq 0$, in which case integrals are equal modulo corresponding Jacobian.

[2] Subgraph isomorphism problem is NP-complete, while graph isomorphism belongs to the NP complexity class.

Instead, in order to compare integrals, one could use alpha-representation [6], which is explicitly covariant, very closely related to the definition of dimensionally regularized Feynman integrals and can be derived for any set of quadratic denominators. The structure of any Feynman integral is encoded by the two homogeneous polynomials, U and F, which do not depend on the exponents of denominator factors and the dimensionality of space D. In particular, we have:

$$I_i = C \int dx_1 \; ... \; dx_5 \; \delta\left(x_1 + ... + x_5 - 1\right) \; U_i^a F_i^b, \tag{4}$$

$$U_1 = U_2 = x_5(x_3 + x_4) + (x_1 + x_2)(x_3 + x_4 + x_5),$$

$$F_1 = m^2 x_5 U_1 - x_1 x_4 x_5, \quad F_2 = m^2 x_5 U_1 - x_2 x_3 x_5.$$

with constants C, a and b (irrelevant here) depending only on D.

The only freedom that is left when comparing $\{U_1, F_1\}$ with $\{U_2, F_2\}$ is a permutation of lines, or equivalently, variables x_i. One then has to find a permutation $p(i)$ such that $U_1(\vec{x}_i) = U_2(\vec{x}_{p(i)})$ and $F_1(\vec{x}_i) = F_2(\vec{x}_{p(i)})$. Naturally, before looking at permutations, one may compare more mundane properties of polynomials common to all permutations, such as the total number of terms or the number of terms proportional to m^2.

An obvious way to find $p(i)$ when comparing two integrals would be to try all $5! = 120$ permutations of x_i. Since usually one has more than two candidates to compare, it is beneficial to introduce a "canonical" ordering, maximizing some metric over all permutations. The canonical ordering of lines should then be derived only once per integral, and the identity of canonical alpha-representations would provide the definitive answer.

In practice, we found that it is sufficient to build a metric in the space of not pairs $\{U, F\}$ but products UF. One example of a suitable metric is given by the following rules (we assume that a unique ordering of coefficients is available, as it is in any computer algebra system):

(i) Turn the polynomial of n variables with m terms into a matrix with rows corresponding to monomials. The first column contains coefficients, and the subsequent columns contain the (non-negative integer) exponents of variables $x_1, ..., x_n$.

(ii) Make n copies of the table. In the i-th copy, exchange the second column (corresponding to x_1) with the i-th column (originally corresponding to x_i).

(iii) In all copies, sort rows lexicographically by the first two columns (i.e. compare only the first two entries in each row).

(iv) Extract the second column from each copy (as vectors of length m), and determine the lexicographically largest vector, comparing all m elements.

(v) In the table copies with the maximized second column, continue recursively: produce $n - 1$ copies-of-copies, in each select a different third column, sort by the three first entries, find the maximum third column, discard non-maximal entries, etc.

(vi) The permutations of columns in the copies maximizing all columns (there can be a few due to symmetries) can be taken as the "canonical permutations" of x_i.

While there is a theoretical possibility of combinatorial growth in this strategy, we have found this approach quite fast and practical to at least five-loop integrals. In our example, after the canonical re-ordering, one easily establishes that $U_1 F_1(x_5, x_2, x_3, x_1, x_4) = U_2 F_2(x_5, x_1, x_4, x_2, x_3)$, i.e. I_1 and I_2 are indeed identical. After this identification it is easy to find the relation between the loop momenta. The graphs, however, may differ as shown in figure 2.

Figure 2. Graphs corresponding to integrals I_1 and I_2 after reparametrization.

3. Analytic tensor reduction

A traditional approach to computing multi-loop integrals with tensor structures in numerators is to decompose the result into all tensors allowed by the symmetries and then determine the coefficients of the decomposition. Let us consider a simple one-loop example:

$$\int k^{\mu_1}k^{\mu_2}k^{\mu_3}k^{\mu_4}\cdot\frac{d^Dk}{k^2+m^2}=\langle k^{\mu_1}...k^{\mu_4}\rangle=\sum_i C_iT_i^{\mu_1...\mu_4}. \tag{5}$$

The tensors $T_i^{\mu_1...\mu_4}$ cannot depend on any external momenta, thus they may only be composed of metric tensors. There are only three unique products: $T_1=g^{\mu_1\mu_2}g^{\mu_3\mu_4}$, $T_2=g^{\mu_1\mu_3}g^{\mu_2\mu_4}$, and $T_3=g^{\mu_1\mu_4}g^{\mu_2\mu_3}$.

The coefficients C_i can be found by multiplication of the original equation 5 with T_1, T_2, and T_3 and solving the system of equations:

$$\begin{aligned}
T_{1\mu_1...\mu_4}\langle k^{\mu_1}...k^{\mu_4}\rangle &= \langle (k^2)^2\rangle = C_1D^2+C_2D+C_3D, \\
T_{2\mu_1...\mu_4}\langle k^{\mu_1}...k^{\mu_4}\rangle &= \langle (k^2)^2\rangle = C_1D+C_2D^2+C_3D, \\
T_{3\mu_1...\mu_4}\langle k^{\mu_1}...k^{\mu_4}\rangle &= \langle (k^2)^2\rangle = C_1D+C_2D+C_3D^2
\end{aligned} \tag{6}$$

The equality of all products on the left hand side implies that $C_1=C_2=C_3=C$, thus reducing the number of independent equations to one, and we find $C=[D(D+2)]^{-1}\langle (k^2)^2\rangle$.

In this case, it is not difficult to generalize the formula to a general number of indices: the result will be represented in terms of symmeterized product of metric tensors, and the coefficients will contain a number of gamma-functions. In a higher number of loops, however, there exist only a few general formulas. The two-loop generalization of equation 5, first given in [7] involves Gegenbauer polynomials, traceless tensors, and is much more computationally expensive due to multiple nested summations. Tensor reduction may also be derived in terms of dimension shifts and differential operators [8], but that is not very convenient in real calculations.

Instead, we suggest to directly use the method as outlined above for the fixed number of indices. For illustration, let us consider a two-loop propagator-type integral with loop momenta k_1 and k_2, and the external momentum p, where the formulas of Davydychev-Tausk still apply (although with some effort), and limit ourselves with six open indices in the numerator.

In the left-hand side we can have seven distinct distributions of indices over the loop momenta (e.g., $\langle k_1^{\mu_1}k_1^{\mu_2}k_1^{\mu_3}k_1^{\mu_4}k_2^{\mu_5}k_2^{\mu_6}\rangle$). On the right-hand side we may build 76 possible tensors out of metric tensors and the components of the external momentum p (e.g., $g^{\mu_1\mu_2}p^{\mu_3}g^{\mu_4\mu_5}p^{\mu_6}$).

For each of the seven LHS combinations, the coefficients in front of the RHS tensors have to be determined separately. Naively, in every case one would need to solve 76 equations with 76 variables. However, the example above gives us a hint: one may reduce the number of independent variables by exploiting the symmetries. If the products of an LHS tensor with the two RHS tensors coincide, so will the corresponding coefficients. In this case, one has at most 10 independent coefficients to determine (each time from 76 equations).

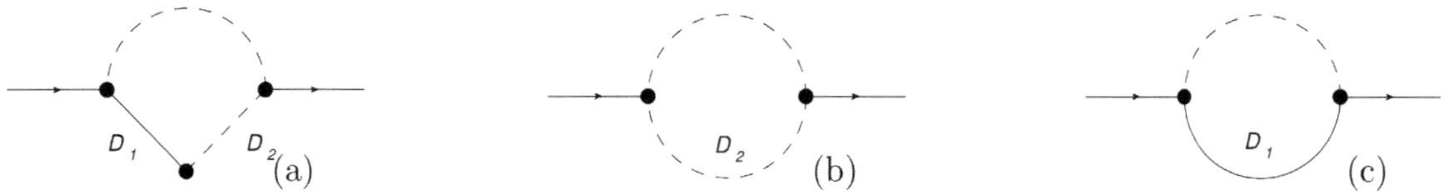

Figure 3. Original (linearly dependent, (a)) and fraction-decomposed ((b) and (c)) topologies.

This procedure does not depend on the actual denominators of the integral: we only derive a decomposition of the numerators. Thus, it is possible to pre-compute the tables with the reductions for a sufficiently large number of indices and re-use them in different computations. One only needs a practical way to solve the systems of linear equations, where coefficients depend on dimensionality D and possibly kinematic invariants. In our setup, we reuse the components from the Laporta algorithm that perform Gauss reduction.

We have checked that this approach is indeed very practical: for example, our implementation builds a table for all 4-loop propagator-type integrals with up to 6 indices in the numerator in less than one hour with a regular PC.

4. Partial fractioning

Normally after the reduction to scalar integrals one has a long polynomial with terms that differ in exponents of denominator factors, e.g.

$$\begin{array}{l} \text{diagram} = D_1^2 D_2 + 33 D_1^{-1} D_2^{-2} - 45/16 D_1 D_2^{-1} + (\epsilon+2) D_2 + ..., \\ D_1 = k^2 + m^2, \quad D_2 = k^2, \quad \text{i.e. linear dependence: } D_1 - D_2 - m^2 = 0. \end{array} \tag{7}$$

(the corresponding graph is in figure 3 (a)). In what follows we will discuss general transformations of monomials $D_1^a D_2^b$, where a and b are positive or negative integer numbers.

The condition that D_1 and D_2 are dependent allows us to simplify such monomials. For example, we may use the "classical" fraction decomposition relation

$$\frac{1}{D_1 D_2} \to \frac{1}{m^2 D_2} - \frac{1}{m^2 D_1} \tag{8}$$

repeatedly to reduce any monomial with negative exponents a and b to a linear combination of terms where at least one of a or b is non-negative.

The form where one of D_1 and D_2 is absent is preferable for the further computations. In this case, the goal of the fraction decomposition is to identically re-arrange equation 7 so that each term in the final expression could be identified with one of the "simpler" topologies in figure 3(b) and (c).

However, the rule of equation 8 alone does not achieve this goal. In particular, it does not apply when $b > 0$, $a < 0$ and one has to apply the direct decomposition rule $D_2 \to D_1 - m^2$. In the symmetric case $b < 0$, $a > 0$ it could be convenient to use $D_1 \to D_2 + m^2$, but applying both those rules to monomials with $a > 0$ and $b > 0$ will not terminate. After the inspection of all relevant cases one can find that only a system of the three rules

$$(D_1 D_2)^{-1} \to \left(m^2\right)^{-1} \left(D_2^{-1} - D_1^{-1}\right), \quad D_2 \to D_1 - m^2, \quad D_1/D_2 \to m^2/D_2 + 1 \tag{9}$$

provides the complete decomposition.

The example above is relatively straightforward and can be easily done manually. However, in practice one may have a much more complicated case for linearly dependent factors. In figure 4

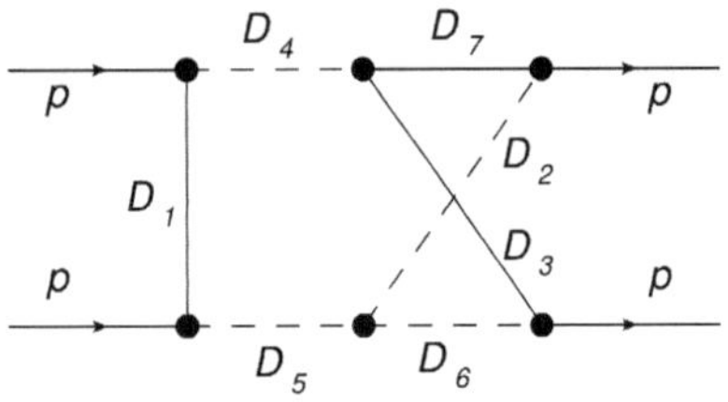

Figure 4. Linearly dependent topology with two relations between denominator factors.

is a two-loop example of a topology in special kinematics, where $p^2 = -m^2$ and each massive line has mass m. Here one has two relations between the seven factors:

$$2D_1 - D_4 - D_5 - 4m^2 = 0, \qquad (10)$$
$$2D_2 - 2D_3 + D_4 - D_5 + 2D_6 - 2D_7 + 4m^2 = 0.$$

Let us formulate the problem in a slightly different way. First, let us get rid of the negative powers of D_i by introducing variables $Y_i = D_i^{-1}$. Second, for all monomials $D_1^{a_1}...D_2^{a_7}Y_1^{a_8}...Y_7^{a_{14}}$, $a_j \geq 0$, let us introduce an ordering that incorporates our understanding of what is "simpler". In particular, it is beneficial to use linear weighting of exponent vectors $(a_1, ..., a_{14})$. Given two such vectors $\vec{a}$ and $\vec{b}$, we decide which of them is the "largest" by lexicographically comparing the products $M\vec{a}$ and $M\vec{b}$, where M is some square 14×14 matrix. One reasonable choice is $M_{ij} = \begin{cases} 1, & \text{if } i \geq j \\ 0, & \text{otherwise} \end{cases}$. The sum of all exponents is here the primary criterion, which agrees with the intuitive definition that the monomials with fewer non-zero exponent are "simpler".

Given this ordering, the problem is to re-arrange polynomial conditions of equation 10 and the additional relations $D_iY_i - 1 = 0$ into a set of equivalent relations that would unambiguously reduce any given monomial to the "simplest" form (and would not lead to loops during the repeated application).

As formulated above, exactly this problem is solved by the so-called Buchberger algorithm, that generates a set of polynomials known as the Gröbner basis. We then need to interpret each element of this basis (a polynomial $p = 0$) as a rule to substitute its most "complex" monomial (according to ordering M) with the remaining terms. Quite conveniently, Mathematica has a function `GroebnerBasis[...]` that is rather efficient and has an option `MonomialOrder` to select the proper weight matrix M. In this particular case, this function produces a Gröbner basis consisting of 14 polynomials that would be very difficult to derive manually. This output can be directly translated to FORM code and used to decompose expressions of any complexity.

Acknowledgements
The author is indebted to A. Smirnov, K. Chetyrkin and M. Steinhauser for useful discussions.

References

[1] 't Hooft G and Veltman M J G 1972 *Nucl. Phys.* **B44** 189–213
[2] Tkachov F V 1981 *Phys. Lett.* **B100** 65–68
[3] Chetyrkin K G and Tkachov F V 1981 *Nucl. Phys.* **B192** 159–204
[4] Smirnov V 2006 *Feynman integral calculus* (Berlin, Germany: Springer)
[5] Vermaseren J A M 2000 *Preprint* `math-ph/0010025`
[6] Bogolyubov N N and Shirkov D V 1959 *Introduction to the theory of quantized fields* (*Intersci. Monogr. Phys. Astron.* vol 3) (New York, USA: Interscience)
[7] Davydychev A I and Tausk J 1996 *Nucl.Phys.* **B465** 507–520 (*Preprint* `hep-ph/9511261`)
[8] Tarasov O 1998 *Acta Phys.Polon.* **B29** 2655 (*Preprint* `hep-ph/9812250`)

DRA method: Powerful tool for the calculation of the loop integrals

R N Lee
The Budker Institute of Nuclear Physics
e-mail: r.n.lee@inp.nsk.su

Abstract. We review the method of the calculation of multiloop integrals suggested in [1].

1. Introduction

The calculation of the multiloop integrals is important for the applications of the perturbative quantum field theory. A number of the calculational methods has been developed for this purpose. They roughly fall in to two categories: direct and indirect methods. The former include Feynman parameterization technique, Mellin-Barnes representation method, Gegenbauer polynomials expansion in coordinate space. The latter methods make use of the equations which are satisfied by the integrals. The construction of these equations relies on the possibility to make the IBP reduction of the integrals under consideration to some finite set of the integrals, called "master integrals". Using this technique one can derive the differential equation for the integrals with several scales, or difference equation for the integrals with one scale. These equations determine the integrals up to the solution of their homogeneous parts.

In [1] a method of multiloop integrals evaluation based on $\mathcal{D}$ recurrence relations [2] and $\mathcal{D}$-analyticity was suggested (DRA method). Since then it was applied to the calculation of various complicated integrals [3, 4, 5, 6, 7, 8]. In this contribution we review this method.

2. DRA method

2.1. Dimensional recurrence relation

Assume that we are interested in the calculation of the L-loop integral depending on E linearly independent external momenta $p_1, \ldots, p_E$. There are $N = L(L+1)/2 + LE$ scalar products depending on the loop momenta l_i:

$$s_{ij} = s_{ji} = l_i \cdot q_j\,; \quad i = 1, \ldots, L; \quad j = 1, \ldots, K, \tag{1}$$

where $q_{1,\ldots,L} = l_{1,\ldots,L}$, $q_{L+1,\ldots,K} = p_{1,\ldots,E}$, and $K = L + E$.

The loop integral has the form

$$J^{(\mathcal{D})}(n_1, \ldots, n_N) = \int \frac{d^{\mathcal{D}} l_L \ldots d^{\mathcal{D}} l_1}{\pi^{L\mathcal{D}/2} D_1^{n_1} D_2^{n_2} \ldots D_N^{n_N}} \tag{2}$$

where the scalar functions D_α are linear polynomials with respect to s_{ij}. The functions D_α are assumed to be linearly independent and to form a complete basis in the sense that any non-zero linear combination of them depends on the loop momenta, and any s_{ij} can be expressed in terms of D_α.

The integral $J^{(\mathcal{D})}(n_1,\ldots,n_N) = J^{(\mathcal{D})}(\mathbf{n})$ can be considered as a value of function of integer N-dimensional vector $\mathbf{n} = (n_1,\ldots,n_N)$. The set of points in $\mathbb{Z}^N$ (and the corresponding set of integrals), having the same set of positive coordinates, form a *sector*. Thus the whole space is split into 2^N different sectors which can be conveniently labeled by their simplest elements. E.g., $(1,1,0,0,\ldots,0)$ denotes the sector in which all points have exactly two first coordinates positive. The points (and the corresponding integrals) in $\mathbb{Z}^N$ and the sectors form a partially ordered set with respect to the following relation. We say that the point $\mathbf{n}$ is *simpler* than the point $\mathbf{m}$ iff all coordinates, positive in $\mathbf{n}$, are also positive in $\mathbf{m}$, but not vice versa.

It is convenient [1] to introduce the operators A_i, B_i which act on such functions as

$$\begin{aligned}(A_i f)(\ldots,n_i,\ldots) &= n_i f(\ldots,n_i+1,\ldots)\,,\\ (B_i f)(\ldots,n_i,\ldots) &= f(\ldots,n_i-1,\ldots)\,.\end{aligned} \tag{3}$$

These operators respect the notion of sectors in a sense that the values of functions $A_i f$ and $B_i f$ in the point of some sector are expressed via the values of function f in the same or simpler sectors. Using these operators, we can express the integral in $\mathcal{D}\pm 2$ dimensions via the integrals in $\mathcal{D}$ dimensions [9]:

$$J^{(\mathcal{D}+2)}(\mathbf{n}) = \frac{(2\mu)^L \left[\det\{p_i\cdot p_j\}_{i,j=1\ldots E}\right]^{-1}}{(\mathcal{D}-E-L+1)_L} P(B_1,\ldots,B_N)\,J^{(\mathcal{D})}(\mathbf{n})\,, \tag{4}$$

$$J^{(\mathcal{D}-2)}(\mathbf{n}) = (\mu/2)^L Q(A_1,\ldots,A_N)\,J^{(\mathcal{D})}(\mathbf{n})\,, \tag{5}$$

where $\mu = \pm 1$ for the Euclidean/pseudoEuclidean case, and polynomials P and Q are determined as

$$\begin{aligned}P(D_1,\ldots,D_N) &= \det\{q_i\cdot q_j\}_{i,j=1\ldots K}\,,\\ Q\left(\frac{\partial}{\partial D_1},\ldots,\frac{\partial}{\partial D_N}\right) &= \det\left\{2^{\delta_{ij}}\frac{\partial}{\partial s_{ij}}\right\}_{i,j=1\ldots L}\,.\end{aligned} \tag{6}$$

These relations are the most useful when the integral in the left-hand side is master. Then, making the IBP reduction of the right-hand side of equations (4),(5), one obtains difference equations for the master integrals. After this, we obtain a linear combination of the integrals of the same, or simpler, sectors as the integral in the left-hand side of equations (4),(5). We assume that the integrals of the the simpler sectors are already known at this stage, either by the same, or by some other method.

The IBP reduction makes equations (4),(5) equivalent, so in what follows we consider only the lowering dimensional recurrence relation, obtained from equation (4). We also change notations, writing

$$J^{(\mathcal{D})} = J(\nu), \tag{7}$$

where $\nu = \mathcal{D}/2$. If a sector of the integral in the left-hand side contains only one master integral, the general form of the equation is the following:

$$J(\nu+1) = C(\nu)J(\nu) + R(\nu), \tag{8}$$

where $C(\nu)$ is some rational function of ν and $R(\nu)$ is a non-homogeneous part constructed of the simpler master integrals in $\mathcal{D} = 2\nu$ dimensions. If a sector contains several master integrals, it is convenient to consider them as a column-vector integral $\mathbf{J}(\nu)$ and to write the equations for them in a matrix form:

$$\mathbf{J}(\nu+1) = \mathbb{C}(\nu)\mathbf{J}(\nu) + \mathbf{R}(\nu), \tag{9}$$

where $\mathbb{C}(\nu)$ is some rational matrix (i.e., the matrix with coefficients being rational functions of ν).

2.2. General solution

In order to write the general solution of this equation, we introduce the notion of indefinite sum and indefinite product. Given the function $F(\nu)$, the *indefinite sum* $\Sigma\, F(\nu)$ is any function, satisfying the relation

$$\Sigma\, F(\nu+1) - \Sigma\, F(\nu) = F(\nu) \tag{10}$$

The indefinite sum $\Sigma\, F(\nu)$ is defined up to the addition of arbitrary periodic function of ν. If the function $F(\nu)$ decreases faster than $1/\nu$ when $\nu \to +\infty$ and/or $\nu \to -\infty$, one can write

$$\Sigma\, F(\nu) = \underset{+\infty}{\Sigma}\, F(\nu) + \omega(z) \overset{\text{def}}{=} -\sum_{n=0}^{+\infty} F(\nu+n) + \omega(z) \tag{11}$$

and/or

$$\Sigma\, F(\nu) = \underset{-\infty}{\Sigma}\, F(\nu) + \omega(z) \overset{\text{def}}{=} \sum_{n=-\infty}^{-1} F(\nu+n) + \omega(z), \tag{12}$$

where $\omega(z) = \omega(\exp(2\pi i\nu))$ is arbitrary periodic function.

The *indefinite product* $\Pi F(\nu)$ is any function, satisfying the relation

$$\Pi F(\nu+1)/\Pi F(\nu) = F(\nu) \tag{13}$$

$\Pi F(\nu)$ is defined up to the multiplication by arbitrary periodic function of ν. Note that for any rational function $F(\nu)$ one can express $\Pi F(\nu)$ as a product of Γ-functions multiplied by arbitrary periodic function, see [1]. We will determine the indefinite product $\Pi\mathbb{F}(\nu)$ also for the matrix functions $\mathbb{F}(\nu)$ as a matrix function, satisfying the relation

$$\Pi\mathbb{F}(\nu+1)(\Pi\mathbb{F}(\nu))^{-1} = \mathbb{F}(\nu) \tag{14}$$

Obviously, this defines $\Pi\mathbb{F}(\nu)$ up to the multiplication from the right by arbitrary periodic matrix.

In these notations one can write down the general solution of equation (9) in the form

$$\mathbf{J}(\nu) = \mathbb{S}(\nu)^{-1}\, \Sigma\, \mathbb{S}(\nu)\mathbf{R}(\nu), \tag{15}$$

where $\mathbb{S}(\nu) = [\Pi\mathbb{C}(\nu)]^{-1}$ is called a *summing factor*. If the summand in equation (16) decreases faster than $1/\nu$ when $\nu \to +\infty$ and/or $\nu \to -\infty$, one can use equations (11),(12) to write

$$\mathbf{J}(\nu) = \mathbb{S}(\nu)^{-1}\boldsymbol{\omega}(z) + \mathbb{S}(\nu)^{-1} \underset{\pm\infty}{\Sigma}\, \mathbb{S}(\nu)\mathbf{R}(\nu)\,. \tag{16}$$

Note that the second term in the right-hand side is invariant with respect to the choice of the summing factor. Thus, in order to find the integral, we need to fix the function $\boldsymbol{\omega}(z)$ for any choice of the summing factor. Here we note the essential difference between the general solutions of *differential* and *difference* equations. The homogeneous part of the general solution of the differential equation is parameterized by finite number of constants, while the homogeneous part of the general solution of the difference equation is parameterized by the periodic function(s).

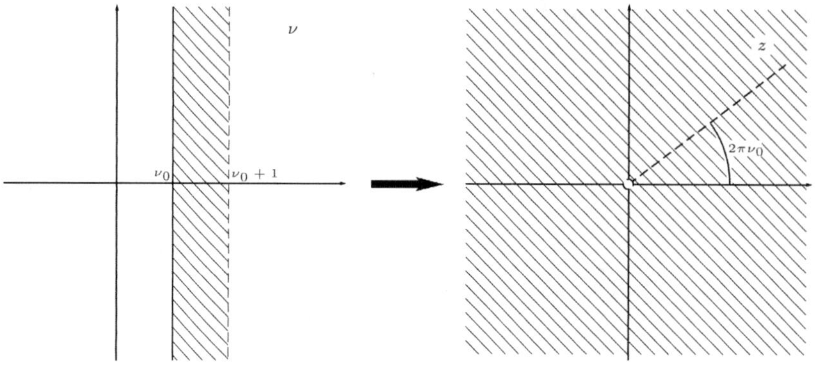

Figure 1. Mapping $\nu \to z$

While fixing the finite number of constants is a relatively simple task, fixing the periodic function may be quite difficult. The key point of the DRA approach is the usage of the analytical properties of the integral in order to fix this function. Let us rewrite equation (16) as

$$\boldsymbol{\omega}(z) = \mathbb{S}(\nu)\mathbf{J}(\nu) - \sum_{\pm\infty} \mathbb{S}(\nu)\mathbf{R}(\nu) . \tag{17}$$

Suppose that we can show that the right-hand side of this equation is a meromorphic function of z in the Riemann sphere. Then $\boldsymbol{\omega}(z)$ is determined, up to a constant, by the principal parts of its Laurent series around singular points. The analysis of the analytical properties of the right-hand side of equation (17) can be performed in any vertical stripe of unit length in the complex plane of ν, as demonstrated in figure 1. There is no cut on the ray $\arg(z) = 2\pi\nu_0$ because the right-hand side is necessarily a periodic function of ν. The proper choice of this stripe (*basic stripe*) can simplify the evaluation of the integral.

2.3. Analytical properties of integrals

Usually, one is interested in the analytical properties of the integral as a function of some external invariant. Here we are interested instead in the analytical properties of the integral as a function of $\nu = \mathcal{D}/2$. It appears that the latter properties are much simpler than the former.

Let us consider the parametric representation of some L-loop integral in Euclidean space with I internal lines (see, e.g., [10]):

$$J(\nu) = \Gamma(I - L\nu) \int dx_1 \dots dx_I \delta\left(1 - \sum x_i\right) \frac{[q(x)]^{\nu L - I}}{[p(x)]^{\nu(L+1) - I}} . \tag{18}$$

The polynomials $q(x)$ and $p(x)$ are determined by the graph. For our consideration it is important only that both these functions are nonnegative in the whole integration region,

$$q(x) \geqslant 0, \qquad p(x) \geqslant 0.$$

Suppose that the parametric representation converges for all ν in some interval (ν_1, ν_2). Then it is easy to see that it converges on the whole stripe $S = \{\nu, \quad \mathrm{Re}\,\nu \in (\nu_1, \nu_2)\}$. Indeed, we can estimate

$$|J(\nu)| \leqslant \frac{|\Gamma(I - L\nu)|}{\Gamma(I - L\,\mathrm{Re}\,\nu)} J(\mathrm{Re}\,\nu) . \tag{19}$$

Therefore, equation (18) determines $J(\nu)$ as a holomorphic function on the whole stripe S. In fact, in [11] a more general fact was proved: the representation (18) determines the meromorphic

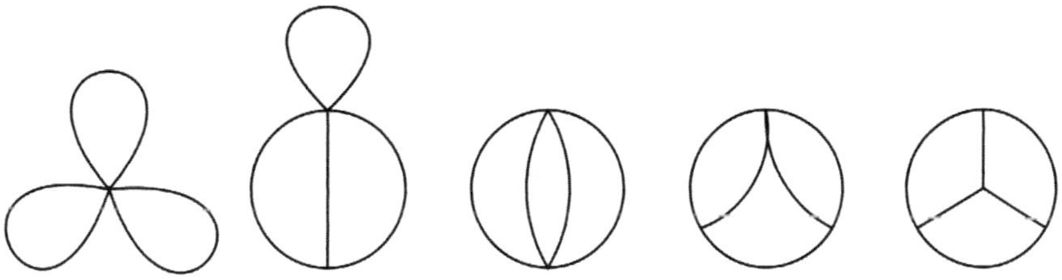

$T_{3,1}(c.l = 0)$ $T_{4,1}(c.l. = 1)$ $T_{4,2}(c.l. = 1)$ $T_{5,1}(c.l. = 2)$ $T_{6,1}(c.l. = 3)$

Figure 2. Three-loop all-massive tadpole master ntegrals.

function on the whole complex ν-plane. The determination of poles position can be automatized thanks to `FIESTA` [12].

As $\operatorname{Im}\nu$ tends to $\pm\infty$ while $\operatorname{Re}\nu$ is fixed, equation (19) shows that

$$|J(\nu)| \lesssim \text{const} \times e^{-\frac{\pi L|\operatorname{Im}\nu|}{2}} |\operatorname{Im}\nu|^{I-1/2-L\operatorname{Re}\nu} < \text{const} \times |z|^{\mp(L+0)/4} . \tag{20}$$

Thus, the parametric representation can be used for the determination of the analytical properties of the first term in the right-hand side of equation (17). Since the second term in the right-hand side of equation (17) is constructed of the integrals already known at this stage, we, indeed, can determine the position of the poles of function $\omega(z)$.

2.4. Numerical issues

In order to demonstrate how the representations obtained within the DRA method can be treated numerically, let us consider the three-loop master integrals depicted in figure 2. Direct application of the DRA method gives the following results:

$$T_{3,1} = \Gamma(1-\nu)^3 \tag{21}$$

$$T_{4,1} = \frac{1}{S_{4,1}}\left\{\frac{4\pi}{3\sqrt{3}} - \underset{-\infty}{\Sigma} S_{4,1}T_{3,1}\right\}, \quad S_{4,1} = \frac{3^{-\nu}}{\Gamma(2-2\nu)\Gamma(1-\nu)} \tag{22}$$

$$T_{4,2} = \frac{1}{S_{4,2}}\left\{\frac{3\pi^2}{16} - \underset{-\infty}{\Sigma} \frac{(11\nu-8)S_{4,2}T_{3,1}}{8(\nu-1)}\right\}, \quad S_{4,2} = \frac{4^{1-3\nu}\Gamma(2-\nu)^2}{\Gamma(3-3\nu)\Gamma(3-2\nu)} \tag{23}$$

$$T_{5,1} = \frac{1}{S_{5,1}}\left\{\frac{16\pi^2}{27} + \underset{-\infty}{\Sigma} S_{5,1}\frac{4(7\nu-4)T_{3,1} - 36(3\nu-2)T_{4,1} - (2+13\nu)T_{4,2}}{27(3\nu-2)}\right\},$$
$$S_{5,1}(\nu) = \frac{9^{-\nu}\Gamma(1-\nu)}{\Gamma(2-2\nu)^2} \tag{24}$$

$$T_{6,1} = \frac{1}{S_{6,1}}\left\{2\pi^2 + \underset{-\infty}{\Sigma} \frac{S_{6,1}}{2}(T_{4,2} - 3T_{5,1})\right\}, \quad S_{6,1} = \frac{2^{-\nu}}{\Gamma(2-2\nu)\Gamma(2-\nu)} \tag{25}$$

The above results for the integrals $T_{5,1}$ and $T_{6,1}$ contain repeated sums of the form:

$$s = \sum_{n_1=0}^{\infty}\sum_{n_2=n_1}^{\infty}\ldots\sum_{n_k=n_{k-1}}^{\infty} f_1(n_1)f_2(n_2)\ldots f_k(n_k) \tag{26}$$

From the viewpoint of numerical calculation, it is important that the dependence on the summation variables is factorized in the summand. We note that it is not the case for the

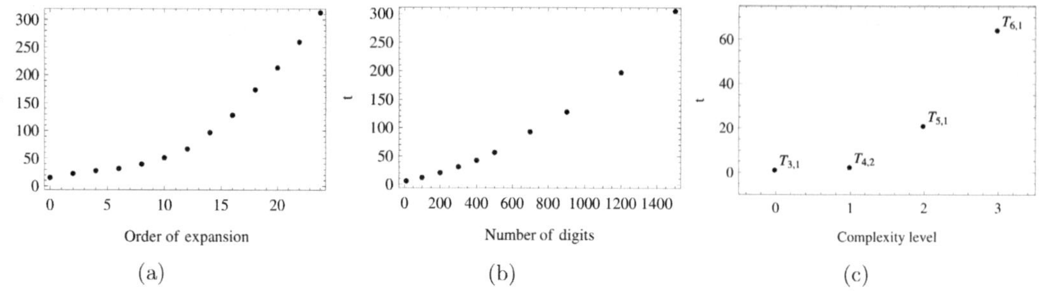

Figure 3. Time of calculation (sec) as a function of (a) requested ϵ-order ($T_{6,1}(4-2\epsilon)$ with 200-digit precision); (b) requested precision ($T_{6,1}(4-2\epsilon)$ up to $O(\epsilon^3)$); (c) complexity level (expansion up to $O(\epsilon^4)$ with 500-digits precision)

Mellin-Barnes multiple sums. Let us show that the calculation of such factorized sums can be performed without nested loops. Indeed, let us define k sequences $s_1(n), \ldots, s_k(n)$ as follows:

$$s_i(-1) = 0\,, \quad s_1(n \geqslant 0) = s_1(n-1) + f_1(n)\,, \quad s_{i>1}(n \geqslant 0) = s_i(n-1) + f_i(n)s_{i-1}(n) \quad (27)$$

It is easy to check that the last sequence, $s_k(n)$, converges to s. Obviously, the iterative calculation of these quantities can be organized in one loop. Therefore, the complexity of the numerical calculation grows moderately with the complexity level of the integrals. The dependence of the calculation time on the order of the expansion, required precision, and complexity level of the integrals is shown in figure 3. We see that for all three cases the method shows quite moderate growth of the evaluation time.

3. Conclusion

We have briefly reviewed the method of calculation of multiloop integrals based on the $\mathcal{D}$-recurrence and $\mathcal{D}$-analyticity. The method appears to be powerful enough to deal with the most complicated cases.

Acknowledgments

This work was supported by RFBR (grants Nos. 10-02-01238, 11-02-01196, 11-02-00220, 11-02-08379) and by Federal special-purpose program "Scientific and scientific-pedagogical personnel of innovative Russia". I appreciate the organizers support for the participation in the ACAT conference. I am grateful to Y. Schröder for helpful remarks.

References

[1] Lee R N 2010 *Nucl. Phys. B* **830** 474 (*Preprint* arXiv:0911.0252)
[2] Tarasov O V 1996 *Phys. Rev. D* **54** 6479 (*Preprint* hep-th/9606018)
[3] Lee R N, Smirnov A V and Smirnov V A 2010 *JHEP* **1004** 020 (*Preprint* arXiv:1001.2887 [hep-ph])
[4] Lee R N, Smirnov A V and Smirnov V A 2010 *Nucl. Phys. Proc. Suppl.* **205-206** 308 (*Preprint* arXiv:1005.0362 [hep-ph])
[5] Lee R N and Smirnov V A 2011 *JHEP* **1102** 102 (*Preprint* arXiv:1010.1334 [hep-ph])
[6] Lee R N and Terekhov I S 2011 *JHEP* **1101** 068 (*Preprint* arXiv:1010.6117 [hep-ph])
[7] Lee R N, Smirnov A V and Smirnov V A 2011 *Eur. Phys. J. C* **71** 1708 (*Preprint* arXiv:1103.3409 [hep-th])
[8] Lee R N, Smirnov A V and Smirnov V A 2012 *Nucl. Phys.B* **856** 95 (*Preprint* arXiv:1108.0732 [hep-th])
[9] Lee R N 2010 *Nucl. Phys. Proc. Suppl.* **205-206**, 135. [arXiv:1007.2256 [hep-ph]].
[10] Itzykson C and Zuber J B *Quantum field theory* McGraw Hill, New York, 1980.
[11] Bernshtein I N and Gelfand S I 1969 *Funkcional. Anal. i Priložen.* **3** 84 (Russian)
[12] Smirnov A V, Smirnov V A and Tentyukov M 2011 *Comput. Phys. Commun.* **182** 790

SecDec: A tool for numerical multi-loop/leg calculations

S Borowka[1], J Carter[2] and G Heinrich[1]
[1]Max-Planck Institute for Physics, Munich, Germany
[2]Institute for Particle Physics Phenomenology, University of Durham, UK
E-mail: j.p.carter@durham.ac.uk, sborowka@mpp.mpg.de, gudrun@mpp.mpg.de

Abstract. The new version of the program SecDec is described, which can be used for the extraction of poles within dimensional regularisation from multi-loop integrals as well as phase space integrals. The numerical evaluation of the resulting finite functions is also done by the program in an automated way, with no restriction on the kinematics in the case of loop integrals.

1. Introduction

Nowadays we are in the long awaited situation of being confronted with a wealth of high energy collider physics data, enabling us to explore physics at the TeV scale. However, for the analysis and interpretation of these data, precise theory predictions are mandatory. In most cases, this means that calculations beyond the leading order in perturbation theory are necessary. Such calculations either involve integrations over loop momenta for the virtual corrections, or integrations over phase spaces for the real radiation corrections. In both cases, multi-dimensional parameter integrals need to be evaluated, which can contain ultraviolet singularities and, in the presence of massless particles, contain soft and/or collinear singularities. These singularities, if regulated by dimensional regularisation, appear as poles in $1/\epsilon$, but factorising the poles from complicated multi-loop or multi-parameter integrals is a highly non-trivial task. The program SecDec, presented in [1], performs this task in an automated way, based on the algorithm of sector decomposition [2, 3, 4]. Other implementations of sector decomposition in public programs can be found in [5, 6, 7]. The method already has been applied in various calculations, for a review see e.g. [8], and combined with other techniques like non-linear transformations [9] or the use of the Bernstein-Tkachov theorem [10, 11, 12]. In this talk the emphasis is on the presentation of the new features of the program SecDec.

2. The Algorithm

2.1. General framework

The decomposition algorithm is described in detail in [8], here we only describe the basic concepts. Consider an N-dimensional parameter integral, for example

$$I_N = \int_0^1 dx_1 \ldots \int_0^1 dx_N \, x_1^{-1-\epsilon} \frac{g_3(\vec{x})}{[x_1\, g_1(\vec{x}) + x_2\, g_2(\vec{x})]} \,, \tag{1}$$

where g_1, g_2 are polynomial functions which do not vanish for $x_1, x_2 \to 0$. However, we would like to factorize the integrand such that the term in square brackets is non-vanishing in the limit

$x_1 \to 0, x_2 \to 0$. This can be achieved by a decomposition into sectors where x_1 and x_2 are ordered: we multiply equation (1) with $1 = \Theta(x_1 - x_2) + \Theta(x_2 - x_1)$ and substitute $x_2 \to x_1 x_2$ in the first sector and $x_1 \to x_2 x_1$ in the second sector, to arrive at

$$I_N = \int_0^1 dx_1 \ldots \int_0^1 dx_N \, x_1^{-1-\epsilon} \left\{ \frac{\tilde{g}_3}{[\tilde{g}_1(\vec{x}) + x_2\, \tilde{g}_2(\vec{x})]} + x_2^{-1-\epsilon} \frac{\bar{g}_3}{[x_1\, \bar{g}_1(\vec{x}) + \bar{g}_2(\vec{x})]} \right\} , \qquad (2)$$

where $\tilde{g}_i, \bar{g}_i$ are functions of the transformed variables. As g_1, g_2 are non-vanishing at the origin, the terms in square brackets cannot lead to singularities anymore; the singularities are factored out into the monomials of type $x_i^{-1-\epsilon}$ instead, and subtraction of the poles and expansion in ϵ are straightforward in this form. In more complicated cases, the decomposition procedure may have to be iterated to achieve a full factorisation. A detailed description can be found e.g. in [2, 8].

2.2. Loop integrals

A scalar Feynman integral G in D dimensions at L loops with N propagators, where the propagators can have arbitrary, not necessarily integer powers ν_j, has the following representation in momentum space:

$$G = \int \prod_{l=1}^{L} \mathrm{d}^D\kappa_l \, \frac{1}{\prod\limits_{j=1}^{N} P_j^{\nu_j}(\{k\},\{p\},m_j^2)}$$

$$\mathrm{d}^D\kappa_l = \frac{\mu^{4-D}}{i\pi^{\frac{D}{2}}}\, \mathrm{d}^D k_l \; , \; P_j(\{k\},\{p\},m_j^2) = q_j^2 - m_j^2 + i\delta \; , \qquad (3)$$

where the q_j are linear combinations of external momenta p_i and loop momenta k_l. Introducing Feynman parameters leads to

$$G = \frac{(-1)^{N_\nu}}{\prod_{j=1}^{N}\Gamma(\nu_j)} \Gamma(N_\nu - LD/2) \int\limits_0^\infty \prod_{j=1}^{N} dx_j \; x_j^{\nu_j - 1}\, \delta(1 - \sum_{l=1}^{N} x_l) \frac{\mathcal{U}^{N_\nu - (L+1)D/2}}{\mathcal{F}^{N_\nu - LD/2}} \; , \; N_\nu = \sum_{j=1}^{N} \nu_j \; , \quad (4)$$

where $\mathcal{U}$ is a polynomial of degree L in the Feynman parameters, while $\mathcal{F}$ is of degree $L+1$ in the Feynman parameters, and also contains the Lorentz invariants which can be formed from the external momenta of the diagram, as well as propagator masses. As a simple example, consider the function $\mathcal{F}$ for a massless one-loop box diagram:

$$\mathcal{F} = -s_{12}\, x_2 x_4 - s_{23}\, x_1 x_3 - i\,\delta \; . \qquad (5)$$

The functions $\mathcal{U}$ and $\mathcal{F}$ can also be constructed directly from the topology of the corresponding Feynman graph [13, 14], and the implementation of this construction in SecDec version 2.0 is one of the new features of the program.

For a diagram with massless propagators, none of the Feynman parameters occurs quadratically in the function $\mathcal{F} = \mathcal{F}_0$. If massive internal lines are present, $\mathcal{F}$ gets an additional term $\mathcal{F}(\vec{x}) = \mathcal{F}_0(\vec{x}) + \mathcal{U}(\vec{x}) \sum\limits_{j=1}^{N} x_j m_j^2$. $\mathcal{U}$ is a positive semi-definite function. A vanishing $\mathcal{U}$ function is related to the UV sub-divergences of the graph. In the region where all invariants formed from external momenta are negative, which we will call the *Euclidean region* in the following, $\mathcal{F}$ is also a positive semi-definite function. Its vanishing does not necessarily lead to an IR singularity. Only if some of the invariants are zero, for example if some of the external momenta are light-like, the vanishing of $\mathcal{F}$ may induce an IR divergence. Thus it depends on the *kinematics* and not only on the topology (like in the UV case) whether a zero of $\mathcal{F}$ leads to a divergence or not. The necessary (but not sufficient) conditions for a divergence are given by the Landau equations [15, 16, 13]:

$$x_j\,(q_j^2 - m_j^2) = 0 \quad \forall\, j \tag{6}$$

$$\frac{\partial}{\partial k_l^\mu} \sum_j x_j\,\left(q_j^2(k,p) - m_j^2\right) = 0\,. \tag{7}$$

If all kinematic invariants formed by external momenta are of the same sign, the necessary condition $\mathcal{F} = 0$ for an IR divergence can only be fulfilled if some of the parameters x_i go to zero. These singularities can be regulated by dimensional regularisation and factored out of the function $\mathcal{F}$ using sector decomposition. The same holds for dimensionally regulated UV singularities contained in $\mathcal{U}$. However, after these singularities, appearing as poles in $1/\epsilon$, have been extracted using sector decomposition, for non-Euclidean kinematics we are still faced with integrable singularities related to kinematic thresholds. How we deal with these singularities will be described in Section 4.4.

2.3. General parameter integrals

The program can also deal with parameter integrals which are more general than the ones related to multi-loop integrals, for example phase space integrals involving massless particles, where the regions in phase space corresponding to soft and/or collinear configurations lead to singularities which can be extracted as poles in $1/\epsilon$. The only restrictions are that the integration domain should be the unit hypercube, and the singularities should be only endpoint singularities, i.e. should be located at zero or one. We assume that the singularities are regulated by non-integer powers of the integration parameters, where the non-integer part is the ϵ of dimensional regularisation or some other regulator. The general form of the integrals is

$$I = \int_0^1 dx_1 \ldots \int_0^1 dx_N \prod_{i=1}^{m} P_i(\vec{x}, \{\alpha\})^{\nu_i}\,, \tag{8}$$

where $P_i(\vec{x}, \{\alpha\})$ are polynomial functions of the parameters x_j, which can also contain some symbolic constants $\{\alpha\}$. The user can leave the parameters $\{\alpha\}$ symbolic during the decomposition, specifying numerical values only for the numerical integration step. This way the decomposition and subtraction steps do not have to be redone if the values for the constants are changed. The ν_i are powers of the form $\nu_i = a_i + b_i\epsilon$ (with a_i such that the integral is convergent). Note that half integer powers are also possible.

3. The SecDec program

The program consists of two parts, an algebraic part and a numerical part. The algebraic part uses code written in Mathematica [17] and does the decomposition into sectors, the subtraction of the singularities, the expansion in ϵ and the generation of the files necessary for the numerical integration. In the numerical part, Fortran or C++ functions forming the coefficient of each term in the Laurent series in ϵ are integrated using the Monte Carlo integration programs contained in the CUBA library [18, 19], or BASES [20]. The different subtasks are handled by perl scripts. The flowchart of the program is shown in Fig. 1 for the basic flow of input/output streams.

Journal of Physics: Conference Series **368** (2012) 012051 doi:10.1088/1742-6596/368/1/012051

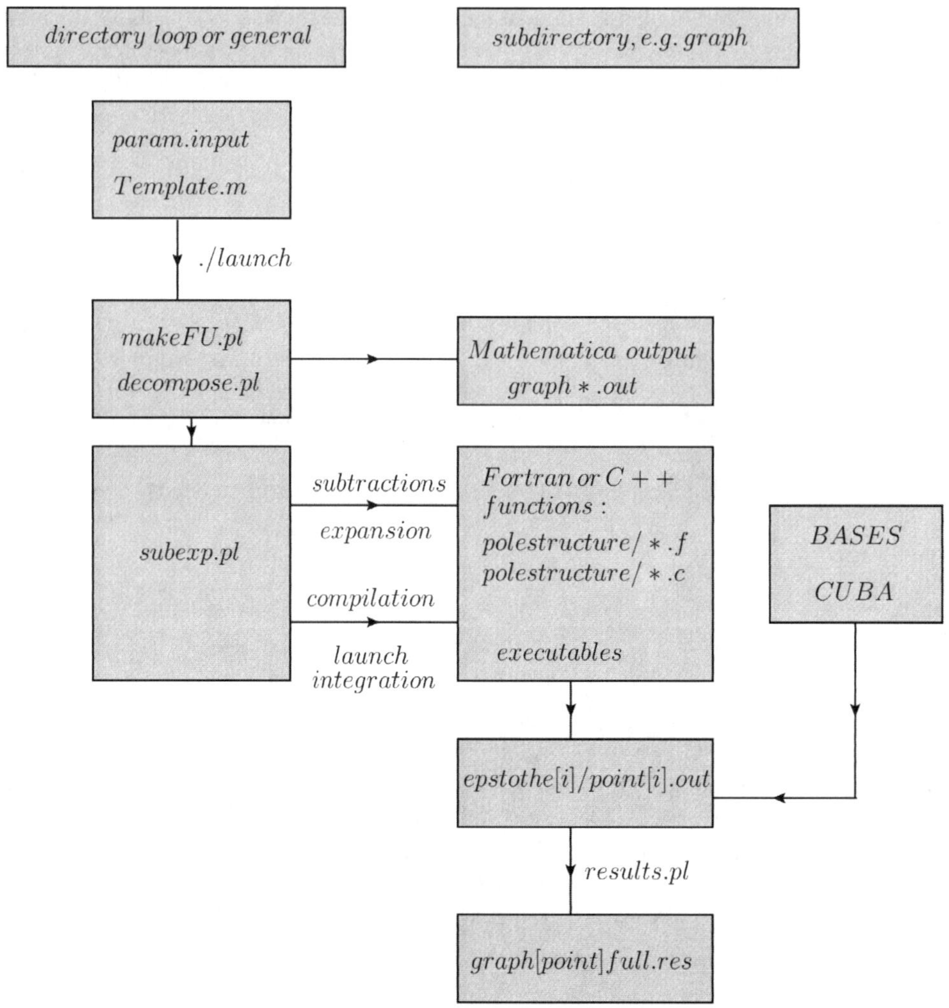

Figure 1. Flowchart showing the main steps the program performs to produce the result files. In each of the subdirectories `loop` or `general`, the file `Template.m` can be used to define the integrand. The produced files are written to a subdirectory created according to the settings given in `param.input`. By default, a subdirectory with the name of the graph or integrand is created to store the produced functions. This directory will contain subdirectories according to the pole structure of the integrand.

The directories `loop` and `general` have the same global structure, only some of the individual files are specific to loops or to more general parametric functions. The directories contain a number of perl scripts steering the decomposition and the numerical integration. The scripts use perl modules contained in the subdirectory `perlsrc`.

The Mathematica source files are located in the subdirectories `src/deco` (files used for the decomposition), `src/subexp` (files used for the pole subtraction and expansion in ϵ) and `src/util` (miscellaneous useful functions). The documentation, created by *robodoc* [21] is contained in the subdirectory `doc`. It contains an index to look up documentation of the source code in html format by loading `masterindex.html` into a browser.

In order to use the program, the user only has to edit the following two files:

- `param.input`: (text file)
  specification of paths, type of integrand, order in ϵ, output format, parameters for numerical integration, further options
- `Template.m`: (Mathematica syntax)
  - for loop integrals: specification of loop momenta and propagators, resp. of the topology; optionally numerator, non-standard propagator powers, space-time dimensions
  - for general functions: specification of integration variables, integrand, variables to be split

The program comes with example input and template files in the subdiretories `loop/demos` respectively `general/demos`, described in detail in [1].

4. Installation and usage

4.1. Installation

The program can be downloaded from
`http://projects.hepforge.org/secdec`.

Unpacking the tar archive via *tar xzvf SecDec.tar.gz* will create a directory called `SecDec` with the subdirectories as described above. Then change to the `SecDec` directory and run *./install*.

Prerequisites are Mathematica, version 6 or above, perl (installed by default on most Unix/Linux systems), a Fortran compiler (e.g. gfortran, ifort), and a C++ compiler if the C++ option is used.

4.2. Usage

(i) Change to the subdirectory `loop` or `general`, depending on whether you would like to calculate a loop integral or a more general parameter integral.

(ii) Copy the files `param.input` and `Template.m` to create your own parameter and template files `myparamfile`, `mytemplatefile`.

(iii) Set the desired parameters in `myparamfile` and specify the integrand in `mytemplatefile`.

(iv) Execute the command *./launch -p myparamfile -t mytemplatefile* in the shell.
If you omit the option *-p myparamfile*, the file `param.input` will be taken as default. Likewise, if you omit the option *-t mytemplatefile*, the file `Template.m` will be taken as default. If your files *myparamfile, mytemplatefile* are in a different directory, say, *myworkingdir*, use the option **-d myworkingdir**, i.e. the full command then looks like *./launch -d myworkingdir -p myparamfile -t mytemplatefile*, executed from the directory `SecDec/loop` or `SecDec/general`.

(v) Collect the results. Depending on whether you have used a single machine or submitted the jobs to a cluster, the following actions will be performed:

- If the calculations are done sequentially on a single machine, the results will be collected automatically (via `results.pl` called by `launch`). The output file will be displayed with your specified text editor.
- If the jobs have been submitted to a cluster, when all jobs have finished, execute the command *./results.pl [-d myworkingdir -p myparamfile]*. This will create the files containing the final results in the `graph` subdirectory specified in the input file.

(vi) After the calculation and the collection of the results is completed, you can use the shell command *./launchclean[graph]* to remove obsolete files.

It should be mentioned that the code starts working first on the most complicated pole structure, which takes longest. This is because in case the jobs are sent to a cluster, it is advantageous to first submit the jobs which are expected to take longest.

4.3. New Features

Version 2.0 of SecDec contains the following new features, some of which will be illustrated by examples in Section 5. More details are given in [22].

- Multi-scale loop integrals can be evaluated without restricting the kinematics to the Euclidean region.
- The possibility to loop over ranges of parameter values is automated, with the option of outputting results in a format suitable for plotting.
- The user can define functions at a symbolic level and specify them only later after the integrand has been transformed into a set of finite functions for each order in ϵ.
- The regulator of the parameter integrals can be different from the dimensional regulator ϵ. This is particularly useful to define e.g. measurement functions at a later stage of the calculation.
- For scalar multi-loop integrals, the integrand can be constructed directly from the topology of the diagram. The user only has to provide the labels of the vertices connected by the propagators and the propagator masses, but does not have to provide the momentum flow.
- The files for the numerical integration of multi-scale loop integrals are written in C++ rather than Fortran. For integrations in Euclidean space, both the Fortran and the C++ versions are supported.

4.4. Implementation of the contour deformation

Unless the function $\mathcal{F}$ in equation (4) is of definite sign for all possible values of invariants and Feynman parameters, the integrand of a multi-loop integral will vanish within the integration region on a hypersurface given by the solutions of the Landau equations (6),(7). However, we can avoid the poles on the real axis by a deformation of the integration contour into the complex plane. As long as the deformation is in accordance with the causal $i\delta$ prescription of the Feynman propagators, and no poles are crossed while changing the integration path, we can make use of Cauchy's theorem to choose an integration contour such that the integration is convergent. The $i\delta$ prescription for the Feynman propagators tells us that the contour deformation into the complex plane should be such that the imaginary part of $\mathcal{F}$ should always be negative. For real masses and Mandelstam invariants s_{ij}, the following Ansatz [23, 24, 25] is convenient:

$$
\begin{aligned}
\vec{z}(\vec{x}) &= \vec{x} - i\,\vec{\tau}(\vec{x}) \\
\tau_k &= \lambda\, x_k(1-x_k) \sum_{j=1}^{N-1} \frac{\partial \mathcal{F}}{\partial x_j}\,.
\end{aligned}
\qquad (9)
$$

In terms of the new variables, we thus obtain

$$\mathcal{F}(\vec{z}(\vec{x})) = \mathcal{F}(\vec{x}) - i\,\lambda \sum_{j=1}^{N-1} \left(\frac{\partial \mathcal{F}}{\partial x_j}\right)^2 + \mathcal{O}(\lambda^2)\,, \tag{10}$$

such that $\mathcal{F}$ acquires a negative imaginary part of order λ.

The convergence of the numerical integration can be improved significantly by choosing an "optimal" value for λ. Values of λ which are too small lead to contours which are too close to the poles on the real axis and therefore lead to bad convergence. Too large values of λ can modify the real part of the function to an unacceptable extent and could even change the sign of the imaginary part if the terms of order λ^3 get larger than the terms linear in λ. This would lead to a wrong result. Therefore we implemented a four-step procedure to optimize the value of λ, consisting of

- ratio check: To make sure that the terms of order λ^3 in equation (10) do not spoil the sign of the imaginary part, we evaluate the ratio of the terms linear and cubic in λ for a quasi-randomly chosen set of sample points to determine a maximal allowed $\lambda = \lambda_{max}$.
- modulus check: The imaginary part is vital at the points where the real part of $\mathcal{F}$ is vanishing. In these regions, the deformation should be large enough to avoid large numerical fluctuations due to a highly peaked integrand. Therefore we check the modulus of each subsector function $\mathcal{F}_i$ at a number of sample points, and pick the fraction of the value of λ_{max} maximising the function with the minimal modulus, i.e. the value of lambda which keeps $\mathcal{F}$ furthest from zero.
- individual $\lambda(i,j)$ adjustments: If the values of $\frac{\partial \mathcal{F}_i}{\partial x_j}$ are very different in magnitude, it can be convenient to have an individual parameter $\lambda(i,j)$ for each subsector function $\mathcal{F}_i$ and each Feynman parameter x_j.
- sign check: After the above adjustments to λ have been made, the sign of $\mathrm{Im}(\mathcal{F})$ is again checked for a number of sample points. If the sign is ever positive, this value of λ is disallowed.

5. Examples and Results

In this example, we will demonstrate three of the new features of the SecDec 2.0 program: the construction of $\mathcal{F}$,$\mathcal{U}$ directly from the topology of the graph, the evaluation of the graph in the physical region, and how results for a whole set of different numerical values for the invariants can be produced and plotted in an automated way. We will use the two-loop diagram shown in Fig. 2 as an example. Numerical results for this diagram have been produced in [26, 27], and analytically it has been calculated in [28, 29].

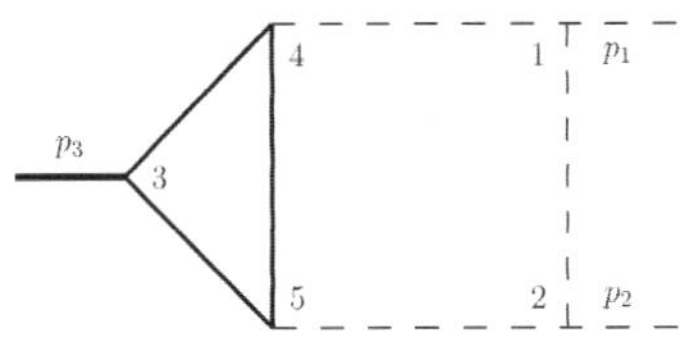

Figure 2. Two-loop vertex graph called P_{126}, containing a massive triangle loop. Bold lines are massive, dashed lines are massless. The vertices are labeled to match the construction of the integrand from the topology as explained in the text.

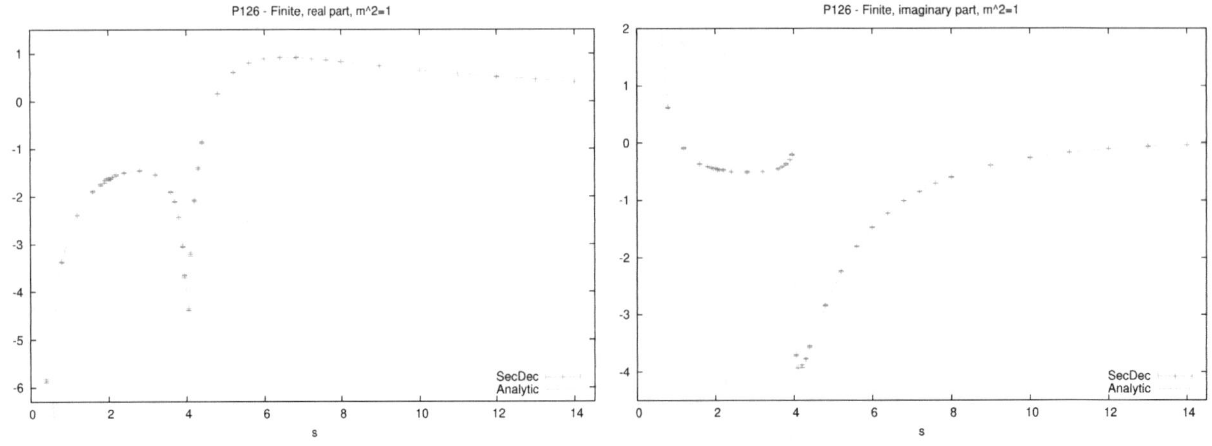

Figure 3. Comparison of analytic and numerical results for the diagram P_{126}.

5.1. Topology-based construction of the integrand

The template file `templateP126.m` in the `demos` subdirectory contains the following lines:
$proplist = \{\{ms[1], \{3,4\}\}, \{ms[1], \{4,5\}\}, \{ms[1], \{5,3\}\}, \{0, \{1,2\}\}, \{0, \{1,4\}\}, \{0, \{2,5\}\}\};$
$onshell = \{ssp[1] \to 0, ssp[2] \to 0, ssp[3] \to sp[1,2]\};$
where each entry in *proplist* corresponds to a propagator of the diagram; the first entry is the mass of the propagator, and the second entry contains the labels of the two vertices which the propagator connects. Note that if an external momentum p_k is flowing into the vertex, the vertex also must have the label k. For vertices containing only internal propagators the labeling is arbitrary. The on-shell conditions in the above example state that $p_1^2 = p_2^2 = 0$, $p_3^2 = s$.

5.2. Results in the physical region

To run this example, from the `loop` directory, issue the command `./launch -d demos -p paramP126.input -t templateP126.m`. The timings for the finite part and a relative accuracy of 1%, using CUBA-3.0 [19], are about 12 secs for a typical point far from the $s = 4m^2$ threshold, and 13.6 secs for a point close to threshold ($s/m^2 = 3.9$) on an Intel(R) Core i7 CPU at 2.67GHz.

5.3. Producing data files for sets of numerical values

To loop over a set of numerical values for the invariants s and m^2 once the C++ files are created, issue the command
`./multinumerics.pl -d demos -p multiparamP126.input`. This will run the numerical integrations for the values of s and m^2 specified in the file `demos/multiparamP126.input`. The files containing the results will be found in `demos/2loop/P126`, and the files `p-2.gpdat`, `p-1.gpdat` and `p0.gpdat` will contain the data files for each point, corresponding to the coefficients of $\epsilon^{-2}, \epsilon^{-1}$ and ϵ^0 respectively. These files can be used to plot the results against the analytic results using gnuplot. This will produce the files `P126R0.ps`, `P126I0.ps` which will look like Fig. 3.

6. Conclusions

We have presented the program SECDEC 2.0, which can be used to factorise dimensionally regulated singularities and numerically calculate multi-loop integrals in an automated way. As a new feature of the program, it now can deal with fully physical kinematics, i.e. is not restricted to the Euclidean region anymore. A new construction of the integrand, based entirely on topological rules, is also included, and the new features are demonstrated for the examples of a massive two-loop three-point function. In addition, the program can produce numerical results for more general parameter integrals, as they occur for example in phase space integrals for multi-particle production with several unresolved massless particles, and offers the possibility to include symbolic functions which can be used to define measurement functions like e.g. jet algorithms at a later stage. We are looking forward to applications of the program for the calculation of higher order corrections to various observables.

Acknowledgements

J.C. was supported by the British Science and Technology Facilities Council (STFC). We also acknowledge the support of the Research Executive Agency (REA) of the European Union under the Grant Agreement number PITN-GA-2010-264564 (LHCPhenoNet).

References

[1] Carter J and Heinrich G 2011 *Comput.Phys.Commun.* **182** 1566–1581 (*Preprint* 1011.5493)
[2] Binoth T and Heinrich G 2000 *Nucl. Phys.* **B585** 741–759 (*Preprint* hep-ph/0004013)
[3] Roth M and Denner A 1996 *Nucl. Phys.* **B479** 495–514 (*Preprint* hep-ph/9605420)
[4] Hepp K 1966 *Commun. Math. Phys.* **2** 301–326
[5] Smirnov A V, Smirnov V A and Tentyukov M 2009 (*Preprint* 0912.0158)
[6] Bogner C and Weinzierl S 2008 *Comput. Phys. Commun.* **178** 596–610 (*Preprint* 0709.4092)
[7] Gluza J, Kajda K, Riemann T and Yundin V 2011 *Eur.Phys.J.* **C71** 1516 (*Preprint* 1010.1667)
[8] Heinrich G 2008 *Int. J. Mod. Phys.* **A23** 1457–1486 (*Preprint* 0803.4177)
[9] Anastasiou C, Herzog F and Lazopoulos A 2010 (*Preprint* 1011.4867)
[10] Ferroglia A, Passera M, Passarino G and Uccirati S 2003 *Nucl.Phys.* **B650** 162–228 (*Preprint* hep-ph/0209219)
[11] Bernstein L 1972 66
[12] Tkachov F V 1997 *Nucl.Instrum.Meth.* **A389** 309–313 (*Preprint* hep-ph/9609429)
[13] Nakanishi N 1971 *Graph Theory and Feynman Integrals* (Gordon and Breach, New York)
[14] Tarasov O V 1996 *Phys. Rev.* **D54** 6479–6490 (*Preprint* hep-th/9606018)
[15] Landau L D 1959 *Nucl. Phys.* **13** 181–192
[16] Eden R J, Landshoff P V, Olive D I and Polkinghorne J C 1966 *The Analytic S-Matrix* (Cambridge University Press)
[17] Wolfram S , Mathematica, Copyright by Wolfram Research
[18] Hahn T 2005 *Comput. Phys. Commun.* **168** 78–95 (*Preprint* hep-ph/0404043)
[19] Agrawal S, Hahn T and Mirabella E 2011 (*Preprint* 1112.0124)
[20] Kawabata S 1995 *Comp. Phys. Commun.* **88** 309–326
[21] Slothouber F and et al *http://www.xs4all.nl/ rfsber/Robo/robodoc.html*
[22] Borowka S, Carter J and Heinrich G 2012 (*Preprint* 1204.4152)
[23] Soper D E 2000 *Phys. Rev.* **D62** 014009 (*Preprint* hep-ph/9910292)
[24] Nagy Z and Soper D E 2006 *Phys. Rev.* **D74** 093006 (*Preprint* hep-ph/0610028)
[25] Binoth T, Guillet J P, Heinrich G, Pilon E and Schubert C 2005 *JHEP* **10** 015 (*Preprint* hep-ph/0504267)
[26] Bonciani R, Mastrolia P and Remiddi E 2004 *Nucl.Phys.* **B690** 138–176 (*Preprint* hep-ph/0311145)
[27] Ferroglia A, Passera M, Passarino G and Uccirati S 2004 *Nucl.Phys.* **B680** 199–270 (*Preprint* hep-ph/0311186)
[28] Fleischer J, Kotikov A and Veretin O 1998 *Phys.Lett.* **B417** 163–172 (*Preprint* hep-ph/9707492)
[29] Davydychev A I and Kalmykov M 2004 *Nucl.Phys.* **B699** 3–64 (*Preprint* hep-th/0303162)

Multiloop calculations in supersymmetric theories with the higher covariant derivative regularization

K V Stepanyantz

Department of Theoretical Physics, Physical Faculty, Moscow State University, Moscow, Russia

E-mail: stepanyantz@mail.ru

Abstract. Most calculations of quantum corrections in supersymmetric theories are made with the dimensional reduction, which is a modification of the dimensional regularization. However, it is well known that the dimensional reduction is not self-consistent. A consistent regularization, which does not break the supersymmetry, is the higher covariant derivative regularization. However, the integrals obtained with this regularization can not be usually calculated analytically. We discuss application of this regularization to the calculations in supersymmetric theories. In particular, it is demonstrated that integrals defining the β-function are possibly integrals of total derivatives. This feature allows to explain the origin of the exact NSVZ β-function, relating the β-function with the anomalous dimensions of the matter superfields. However, integrals for the anomalous dimension should be calculated numerically.

1. Introduction

In order to deal with divergent expressions in the quantum field theory, it is necessary to regularize a theory. A proper choice of a regularization can simplify the calculations or reveal some features of quantum corrections. Most calculations in the quantum field theory were made with the dimensional regularization [1] in $\overline{MS}$-scheme [2]. However, the dimensional regularization breaks the supersymmetry and is not convenient for calculations in supersymmetric theories. That is why most calculations in supersymmetric theories were made with the dimensional reduction [3]. For example, the β-function in supersymmetric theories was calculated up to the four-loop approximation [4, 5, 6, 7, 8]. After a special redefinition of the coupling constant [9, 10] the result coincides with the exact NSVZ β-function, proposed in [11, 12, 13, 14].

However, it is well known that the dimensional reduction is not self-consistent [15]. Removing the inconsistencies one breaks the explicit supersymmetry [16, 17]. Then the supersymmetry can be broken by quantum corrections in higher loops [18, 19, 20]. In the $N = 2$ SYM theory this already occurs in the three-loop approximation [18, 20], while in the $N = 4$ SYM theory the supersymmetry is not broken even in the four-loop approximation [21]. Thus, a problem of regularization in supersymmetric theories is rather nontrivial [22].

For supersymmetric theories one can use the higher covariant derivative regularization, proposed by A.A.Slavnov [23, 24]. Different versions of this regularization for supersymmetric theories, which do not break the supersymmetry, were proposed in [25, 26]. Unlike the dimensional reduction, the higher covariant derivative regularization is consistent. However, it was not often applied to explicit calculations of quantum corrections, because the corresponding

Published under licence by IOP Publishing Ltd

integrals have very complicated structure, and it is not easy to calculate them analytically, especially in higher loops. Moreover, some theoretical subtleties can raise nontrivial questions even in the simplest calculations [27, 28, 29].

However, we argue that for supersymmetric theories this regularization has some very attractive features and can be used for the calculations. Namely, the integrals defining the β-function are integrals of double total derivatives [30, 31, 32, 33], and one of them can be calculated analytically. As a result, the β-function is related with the anomalous dimension, producing the exact NSVZ β-function without redefinition of the coupling constant. In this paper we demonstrate how this can proved in $N=1$ SQED in all loops and for the general renormalizable $N=1$ SYM in the two-loop approximation.

2. Quantum corrections in N=1 SQED, regularized by higher derivatives

2.1. Higher derivative regularization

The action of the massless $N=1$ SQED in terms of superfields [34, 35] is written as

$$S = \frac{1}{4e^2}\mathrm{Re}\int d^4x\, d^2\theta\, W_a C^{ab} W_b + \frac{1}{4}\int d^4x\, d^4\theta\,\Big(\phi^* e^{2V}\phi + \tilde\phi^* e^{-2V}\tilde\phi\Big). \tag{1}$$

The theory is regularized by adding the term with the higher derivatives

$$S_\Lambda = \frac{1}{4e^2}\mathrm{Re}\int d^4x\, d^2\theta\, W_a C^{ab}\Big(R\Big(\frac{\partial^2}{\Lambda^2}\Big)-1\Big)W_b, \tag{2}$$

where $R(0)=1$ and $R(\infty)=\infty$. For example, it is possible to choose $R=1+\partial^{2n}/\Lambda^{2n}$. The gauge is fixed by adding

$$S_{\mathrm{gf}} = -\frac{1}{64e^2}\int d^4x\, d^4\theta\,\Big(VR\Big(\frac{\partial^2}{\Lambda^2}\Big)D^2\bar D^2 V + VR\Big(\frac{\partial^2}{\Lambda^2}\Big)\bar D^2 D^2 V\Big). \tag{3}$$

Then the propagator will contain large degrees of the momentum in the denominator, and all loop diagrams beyond the one-loop approximation become convergent. The remaining one loop diagrams are regularized by inserting the Pauli–Villars determinants [36]

$$\prod_I\Big(\int D\phi_I^* D\phi_I e^{iS_I}\Big)^{-c_I} \tag{4}$$

into the generating functional, where

$$S_I = \frac{1}{4}\int d^4x\, d^4\theta\,\Big(\phi_I^* e^{2V}\phi_I + \tilde\phi_I^* e^{-2V}\tilde\phi_I\Big) + \Big(\frac{1}{2}\int d^4x\, d^4\theta\, M_I\phi_I\tilde\phi_I + \mathrm{h.c.}\Big) \tag{5}$$

and $\sum c_I = 1$, $\sum c_I M_I^2 = 0$. It is important that the masses M_I are proportional to the parameter Λ.

2.2. Three-loop β-function

In order to find the β-function we consider

$$\Gamma^{(2)}_V = -\frac{1}{16\pi}\int \frac{d^4p}{(2\pi)^4}\, d^4\theta\, V(-p)\,\partial^2\Pi_{1/2}V(p)\, d^{-1}(\alpha,\mu/p) \tag{6}$$

and calculate

$$\frac{d}{d\ln\Lambda}\Big(d^{-1}(\alpha_0,\Lambda/p)-\alpha_0^{-1}\Big)\Big|_{p=0} = -\frac{d\alpha_0^{-1}}{d\ln\Lambda} = \frac{\beta(\alpha_0)}{\alpha_0^2}. \tag{7}$$

In the three-loop approximation the result can be written as ($R_k \equiv R(k^2/\Lambda^2)$)

$$\frac{\beta(\alpha_0)}{\alpha_0^2} = 2\pi \frac{d}{d\ln\Lambda}\sum_I c_I \int \frac{d^4q}{(2\pi)^4}\frac{\partial}{\partial q^\mu}\frac{\partial}{\partial q_\mu}\frac{\ln(q^2+M^2)}{q^2} + 4\pi\frac{d}{d\ln\Lambda}\int\frac{d^4q}{(2\pi)^4}\frac{d^4k}{(2\pi)^4}\frac{e^2}{k^2R_k^2}$$
$$\times\frac{\partial}{\partial q^\mu}\frac{\partial}{\partial q_\mu}\Bigg(\frac{1}{q^2(k+q)^2} - \sum_I c_I \frac{1}{(q^2+M_I^2)((k+q)^2+M_I^2)}\Bigg)\Bigg[R_k\Big(1+\frac{e^2}{4\pi^2}\ln\frac{\Lambda}{\mu}\Big)$$
$$-2e^2\Bigg(\int\frac{d^4t}{(2\pi)^4}\frac{1}{t^2(k+t)^2} - \sum_J c_J\int\frac{d^4t}{(2\pi)^4}\frac{1}{(t^2+M_J^2)((k+t)^2+M_J^2)}\Bigg)\Bigg]$$
$$+4\pi\frac{d}{d\ln\Lambda}\int\frac{d^4q}{(2\pi)^4}\frac{d^4k}{(2\pi)^4}\frac{d^4l}{(2\pi)^4}\frac{e^4}{k^2R_kl^2R_l}\frac{\partial}{\partial q^\mu}\frac{\partial}{\partial q_\mu}\Bigg\{\Bigg(-\frac{2k^2}{q^2(q+k)^2(q+l)^2(q+k+l)^2}$$
$$+\frac{2}{q^2(q+k)^2(q+l)^2}\Bigg) - \sum_I c_I\Bigg(-\frac{2(k^2+M_I^2)}{(q^2+M_I^2)((q+k)^2+M_I^2)((q+l)^2+M_I^2)}$$
$$\frac{1}{((q+k+l)^2+M_I^2)} + \frac{2}{(q^2+M_I^2)((q+k)^2+M_I^2)((q+l)^2+M_I^2)} - \frac{1}{(q^2+M_I^2)^2}$$
$$\times\frac{4M_I^2}{((q+k)^2+M_I^2)((q+l)^2+M_I^2)}\Bigg)\Bigg\}. \qquad (8)$$

From this expression we see that the β-function is defined by the integrals of (double) total derivatives [30, 31, 32]. Such a structure allows to calculate one of the loop integrals analytically and, thus, to reduce a number of integrations. This can be done using the identity

$$\int\frac{d^4q}{(2\pi)^4}\frac{\partial}{\partial q^\mu}\frac{\partial}{\partial q_\mu}\Big(\frac{f(q^2)}{q^2}\Big) = \lim_{\varepsilon\to 0}\int\limits_{S_\varepsilon}\frac{dS_\mu}{(2\pi)^4}\frac{(-2)q^\mu f(q^2)}{q^4} = \frac{1}{4\pi^2}f(0), \qquad (9)$$

where f is a nonsingular function which rapidly tends to 0 at the infinity. It is equivalent to the identity

$$\int\frac{d^4q}{(2\pi)^4}\frac{1}{q^2}\frac{d}{dq^2}f(q^2) = \frac{1}{16\pi^2}\Big(f(\infty)-f(0)\Big) = -\frac{1}{16\pi^2}f(0). \qquad (10)$$

(This is a total derivative in the four-dimensional spherical coordinates.) Taking the integrals in equation (8) and comparing the result with the expression for the two-loop anomalous dimension of the matter superfield

$$\gamma(\alpha_0) = -2e^2\int\frac{d^4k}{(2\pi)^4}\frac{d}{d\ln\Lambda}\frac{1}{k^4R_k^2}\Bigg[R_k\Big(1+\frac{e^2}{4\pi^2}\ln\frac{\Lambda}{\mu}\Big) - \int\frac{d^4t}{(2\pi)^4}\frac{2e^2}{t^2(k+t)^2} \qquad (11)$$
$$+\sum_I c_I\int\frac{d^4t}{(2\pi)^4}\frac{2e^2}{(t^2+M_I^2)((k+t)^2+M_I^2)}\Bigg] - \int\frac{d^4k}{(2\pi)^4}\frac{d^4l}{(2\pi)^4}\frac{d}{d\ln\Lambda}\frac{4e^4k_\mu l_\mu}{k^4R_k\,l^4R_l(k+l)^2},$$

in the considered approximation we obtain the exact NSVZ β-function for the $N=1$ SQED [14]:

$$\beta(\alpha_0) = \frac{\alpha_0^2}{\pi}\Big(1-\gamma(\alpha_0)\Big) + O(\alpha_0^5). \qquad (12)$$

This equality of the well-defined integrals is obtained without any redefinitions of the coupling constant. Thus, with the higher derivative regularization the NSVZ scheme can be naturally defined. However, it is much more complicated problem to calculate analytically the integrals for the anomalous dimension. Possibly, in the lowest orders this can be done analytically, but it seems that in higher loops for this purpose one should use numerical methods.

2.3. The exact result

For the $N=1$ SQED it is possible to demonstrate that the features discussed in the previous section take place in all loops. In particular, the integrals defining the β-function are integrals of double total derivatives, and the β-function coincides with the NSVZ expression without a redefinition of the coupling constant [37].

For this purpose, first, we make the substitution

$$V \to \bar\theta^a \bar\theta_a \theta^b \theta_b \equiv \theta^4, \tag{13}$$

which allows to extract the function d^{-1} in equation (6). Then we will try to present the sum of Feynman diagrams as integrals of total derivatives, which in the coordinate representation are given by

$$\mathrm{Tr}\Big([x^\mu, \mathrm{Something}]\Big) = 0. \tag{14}$$

We start with the expression for the part of the effective action corresponding to the two-point Green function of the gauge superfield [37]

$$\Delta\Gamma^{(2)}_{\mathbf{V}} = \Big\langle -2i\Big(\mathrm{Tr}(\mathbf{V}J_0*)\Big)^2 - 2i\mathrm{Tr}(\mathbf{V}J_0*\mathbf{V}J_0*) - 2i\mathrm{Tr}(\mathbf{V}^2J_0*)\Big\rangle + \text{terms with } \widetilde{*} + (PV), \tag{15}$$

where

$$* \equiv \frac{1}{1-(e^{2V}-1)\bar D^2 D^2/16\partial^2}, \qquad \widetilde{*} = \frac{1}{1-(e^{-2V}-1)\bar D^2 D^2/16\partial^2} \tag{16}$$

encode sequences of vertexes and propagators on the matter line, and $J_0 = e^{2V}\bar D^2 D^2/16\partial^2$ is the effective vertex. (PV) denotes contributions of the Pauli–Villars fields. The first term in equation (15) is a sum of diagrams in which external lines are attached to different loops of the matter superfields. The second term is a sum of diagrams in which external lines are attached to a single line of the matter superfields. The last term is not transversal. The sum of such terms vanishes due to the Ward identities.

After substitution (13) and some algebraic transformations [37] the first term in equation (15) gives the contribution

$$-2i\frac{d}{d\ln\Lambda}\Big\langle\Big(\mathrm{Tr}\Big(-2\theta^c\theta_c\bar\theta^d[\bar\theta_d, \ln(*)-\ln(\widetilde{*})] + i\bar\theta^c(\gamma^\nu)_c{}^d\theta_d[y^*_\nu, \ln(*)-\ln(\widetilde{*})]\Big) + (PV)\Big)^2\Big\rangle. \tag{17}$$

Similarly, the second term in equation (15) gives

$$i\frac{d}{d\ln\Lambda}\mathrm{Tr}\Big\langle\theta^4\Big[y^*_\mu, \Big[(y^\mu)^*, \ln(*)+\ln(\widetilde{*})\Big]\Big]\Big\rangle + (PV) - \text{terms with a } \delta\text{-function}. \tag{18}$$

The third term in equation (15) vanishes after substitution (13). From expressions (17) and (18) we see that in all orders the β-function is given by integrals of double total derivatives. A different method to see this [31] is based on the covariant Feynman rules in the background field method [38, 39].

Expressions (17) and (18) can be calculated explicitly in the three-loop approximation. The result coincides with equation (8).

Terms with the δ-function in equation (18) appear due to the identity

$$[x^\mu, \frac{\partial_\mu}{\partial^4}] = [-i\frac{\partial}{\partial p_\mu}, -\frac{ip^\mu}{p^4}] = -2\pi^2\delta^4(p_E) = -2\pi^2 i\delta^4(p). \tag{19}$$

Due to this δ-function one of loop integrals can be calculated and a number of integrations is reduced. Qualitatively, we consider all diagrams in which two external gauge lines are attached to the same graph. Integration of the δ-function corresponds to cutting a matter line in this graph [31]. This gives diagrams with two external matter lines, defining the anomalous dimension. Thus, the β-function in a certain loop is reduced to the anomalous dimension in the previous loop. These arguments can formulated rigorously [37] in all orders and allow to obtain the exact NSVZ β-function

$$\beta(\alpha) = \frac{\alpha^2}{\pi}\Big(1 - \gamma(\alpha)\Big). \tag{20}$$

Note that deriving equation (20) one does not redefine the coupling constant, as in the case of the dimensional reduction [9, 10]. Therefore, with the higher derivative regularization we can naturally define the NSVZ scheme for the $N = 1$ SQED.

3. Two-loop β-function with the higher covariant derivative regularization in the non-Abelian case

Let us consider a general renormalizable $N = 1$ supersymmetric Yang–Mills theory with a gauge group G and matter superfields ϕ_i in a representation R, in the massless limit:

$$S = \frac{1}{2e^2}\mbox{Re}\,\mbox{tr}\int d^4x\, d^2\theta\, W_a C^{ab} W_b + \frac{1}{4}\int d^4x\, d^4\theta\, (\phi^*)^i (e^{2V})_i{}^j \phi_j + \\ +\left(\frac{1}{6}\int d^4x\, d^2\theta\, \lambda^{ijk}\phi_i\phi_j\phi_k + \mbox{h.c.}\right) \tag{21}$$

This theory is invariant under the gauge transformation if

$$(T^A)_m{}^i\lambda^{mjk} + (T^A)_m{}^j\lambda^{imk} + (T^A)_m{}^k\lambda^{ijm} = 0. \tag{22}$$

Below we assume that this condition is satisfied. In order to introduce the higher covariant derivative regularization and calculate the β-function in this case we will use the background field method. In the supersymmetric case [34, 35] we split the gauge superfield (which is below denoted by V') into the quantum part V and the background field $\mathbf{\Omega}$ according to the prescription $e^{2V'} \equiv e^{\mathbf{\Omega}^+} e^{2V} e^{\mathbf{\Omega}}$. Then the gauge can be fixed without breaking the background gauge invariance:

$$S_{\rm gf} = -\frac{1}{32e^2}\,\mbox{tr}\int d^4x\, d^4\theta\,\Big(V\boldsymbol{D}^2\bar{\boldsymbol{D}}^2 V + V\bar{\boldsymbol{D}}^2\boldsymbol{D}^2 V\Big). \tag{23}$$

(In our notation $\boldsymbol{D}$, $\bar{\boldsymbol{D}}$, and $\boldsymbol{D}_\alpha$ are the background covariant derivatives.) Certainly, the gauge fixing procedure also requires introducing the Faddeev–Popov and Nielsen–Kallosh ghosts. The higher covariant derivative regularization can be also introduced without breaking the background gauge invariance. This can be done by different ways. For example, it is possible to add

$$S_\Lambda = \frac{1}{2e^2}\mathrm{tr\,Re}\int d^4x\,d^4\theta\,V\frac{(\boldsymbol{D}_\mu^2)^{n+1}}{\Lambda^{2n}}V + \frac{1}{8}\int d^4x\,d^4\theta\left((\phi^*)^i\left[e^{\boldsymbol{\Omega}^+}e^{2V}\frac{(\boldsymbol{D}_\alpha^2)^m}{\Lambda^{2m}}e^{\boldsymbol{\Omega}}\right]_i{}^j\phi_j + \right.$$
$$\left. +(\phi^*)^i\left[e^{\boldsymbol{\Omega}^+}\frac{(\boldsymbol{D}_\alpha^2)^m}{\Lambda^{2m}}e^{2V}e^{\boldsymbol{\Omega}}\right]_i{}^j\phi_j\right) \qquad (24)$$

to the action, assuming that $n > m$. (It is important that for a theory with a nontrivial cubic superpotential a term with the higher covariant derivatives should be also introduced for the matter superfields.)

As in the case of $N = 1$ SQED, the higher covariant derivative term does not remove divergences in the one-loop approximation, and in order to regularize them the Pauli–Villars determinants should be inserted into the generating functional. The Pauli–Villars fields should be introduced for the matter superfields and all ghosts. (A contribution of the gauge superfields in the one-loop approximation vanishes.) Masses of the Pauli–Villars superfields ϕ_I are proportional to the parameter Λ: $M_I^{ij} = a_I^{ij}\Lambda$ and satisfy the relation $M_I^{ij}(M_I^*)_{jk} = M_I^2\delta_k^i$.

Using the background gauge invariance it is possible to choose $\boldsymbol{\Omega} = \boldsymbol{\Omega}^+ = \mathbf{V}$. Then the function d^{-1} is defined by

$$\Gamma_V^{(2)} = -\frac{1}{8\pi}\mathrm{tr}\int\frac{d^4p}{(2\pi)^4}\,d^4\theta\,\mathbf{V}(-p)\,\partial^2\Pi_{1/2}\mathbf{V}(p)\,d^{-1}(\alpha,\lambda,\mu/p). \qquad (25)$$

In order to find the β-function we again use prescription (7) and find the derivative of d^{-1} with respect to $\ln\Lambda$ in the limit of the vanishing external momentum. In the non-Abelian case the result is given by

$$\beta(\alpha,\lambda) = -\frac{3\alpha^2}{2\pi}C_2 + \alpha^2 T(R)I_0 + \alpha^3 C_2^2 I_1 + \frac{\alpha^3}{r}C(R)_i{}^j C(R)_j{}^i I_2$$
$$+\alpha^3 T(R)C_2 I_3 + \alpha^2 C(R)_i{}^j\frac{\lambda^*_{jkl}\lambda^{ikl}}{4\pi r}I_4 + \ldots, \qquad (26)$$

where

$$\mathrm{tr}\,(T^A T^B) \equiv T(R)\,\delta^{AB}; \qquad (T^A)_i{}^k(T^A)_k{}^j \equiv C(R)_i{}^j;$$
$$f^{ACD}f^{BCD} \equiv C_2\delta^{AB}; \qquad r \equiv \delta_{AA}. \qquad (27)$$

The integrals defining the β-function are given by

$$I_i = I_i(0) - \sum_I c_I I_i(M_I), \quad i = 0,2,3, \qquad (28)$$

where for simplicity we do not write the ghost constributions (they are also given by integrals of double total derivatives) and

$$I_0(M) = -\pi\int\frac{d^4q}{(2\pi)^4}\frac{d}{d\ln\Lambda}\frac{\partial}{\partial q^\mu}\frac{\partial}{\partial q_\mu}\left\{\frac{1}{q^2}\ln\left(q^2(1+q^{2m}/\Lambda^{2m})^2 + M^2\right)\right\}; \qquad (29)$$

$$I_1 = -12\pi^2\int\frac{d^4q}{(2\pi)^4}\frac{d^4k}{(2\pi)^4}\frac{d}{d\ln\Lambda}\frac{\partial}{\partial k^\mu}\frac{\partial}{\partial k_\mu}\left\{\frac{1}{k^2(1+k^{2n}/\Lambda^{2n})q^2(1+q^{2n}/\Lambda^{2n})(q+k)^2}\right.$$

$$\times\frac{1}{(1+(q+k)^{2n}/\Lambda^{2n})}\Big\}; \qquad (30)$$

$$I_2(M)=2\pi^2\int\frac{d^4q}{(2\pi)^4}\frac{d^4k}{(2\pi)^4}\frac{d}{d\ln\Lambda}\frac{\partial}{\partial q^\mu}\frac{\partial}{\partial q_\mu}\Big\{\frac{(2+(q+k)^{2m}/\Lambda^{2m}+q^{2m}/\Lambda^{2m})^2}{k^2(1+k^{2n}/\Lambda^{2n})}$$
$$\times\frac{(1+q^{2m}/\Lambda^{2m})(1+(q+k)^{2m}/\Lambda^{2m})}{\Big(q^2(1+q^{2m}/\Lambda^{2m})^2+M^2\Big)\Big((q+k)^2(1+(q+k)^{2m}/\Lambda^{2m})^2+M^2\Big)}\Big\}; \qquad (31)$$

$$I_3(M)=2\pi^2\int\frac{d^4q}{(2\pi)^4}\frac{d^4k}{(2\pi)^4}\frac{d}{d\ln\Lambda}\frac{\partial}{\partial q^\mu}\frac{\partial}{\partial k_\mu}\Big\{\frac{(2+k^{2m}/\Lambda^{2m}+q^{2m}/\Lambda^{2m})^2}{(k+q)^2(1+(q+k)^{2n}/\Lambda^{2n})}$$
$$\times\frac{(1+k^{2m}/\Lambda^{2m})(1+q^{2m}/\Lambda^{2m})}{\Big(k^2(1+k^{2m}/\Lambda^{2m})^2+M^2\Big)\Big(q^2(1+q^{2m}/\Lambda^{2m})^2+M^2\Big)}\Big\}; \qquad (32)$$

$$I_4=-8\pi^2\int\frac{d^4q}{(2\pi)^4}\frac{d^4k}{(2\pi)^4}\frac{d}{d\ln\Lambda}\frac{\partial}{\partial q^\mu}\frac{\partial}{\partial q_\mu}\Big\{\frac{1}{k^2(1+k^{2m}/\Lambda^{2m})q^2(1+q^{2m}/\Lambda^{2m})(q+k)^2}$$
$$\times\frac{1}{(1+(q+k)^{2m}/\Lambda^{2m})}\Big\}. \qquad (33)$$

Thus, we see that the β-function is given by integrals of double total derivatives. In the considered (two-loop) approximation they can be easily calculated analytically using equation (9):

$$\beta=-\frac{\alpha^2}{2\pi}\Big(3C_2-T(R)\Big)+\frac{\alpha^3}{(2\pi)^2}\Big(-3C_2^2+T(R)C_2+\frac{2}{r}C(R)_i{}^jC(R)_j{}^i\Big)-\frac{\alpha^2C(R)_i{}^j\lambda^*_{jkl}\lambda^{ikl}}{8\pi^3r}+\ldots \qquad (34)$$

This expression should be compared with the one-loop anomalous dimension

$$\gamma_i{}^j(\alpha)=-\frac{\alpha C(R)_i{}^j}{\pi}+\frac{\lambda^*_{ikl}\lambda^{jkl}}{4\pi^2}+\ldots \qquad (35)$$

Then we see that in the considered approximation the β-function agrees with the exact NSVZ β-function

$$\beta(\alpha)=-\frac{\alpha^2\Big[3C_2-T(R)+C(R)_i{}^j\gamma_j{}^i(\alpha)/r\Big]}{2\pi(1-C_2\alpha/2\pi)}. \qquad (36)$$

4. Conclusion

Although it is generally believed that the integrals appearing with the higher covariant derivative regularization have too complicated structure, we see that in the supersymmetric case some of them can be calculated analytically. This makes possible analytical multiloop calculations in supersymmetric theories with this regularization. In principle, in the lowest loops it is not very difficult to construct integrals corresponding to various Green functions.

A very attractive feature of the higher covariant derivative regularization is that all integrals defining the β-function in the supersymmetric case seem to be integrals of double total derivatives. As a consequence, one of them can be calculated analytically. (For $N=1$ SQED this was proved exactly in all loops. For the general renormalizable $N=1$ SYM theory this was so far verified only in the two-loop approximation.) In both considered cases the factorization of integrands into total derivatives allows to obtain the exact NSVZ β-function, and for $N = 1$ SQED for this purpose it is not necessary to redefine the coupling constant. (In the non-Abelian case we cannot so far make this conclusion, because the calculation was made only in the two-loop approximation, where the β-function is scheme-independent.)

Acknowledgments

This work was supported by Russian Foundation for Basic Research grants No 11-01-00296-a and 11-02-08451-z. I would like to thank the organizers of the conference ACAT 2011 for supporting my participation. I am also very grateful to Prof. A.L.Kataev for valuable discussions.

References

[1] t'Hooft G and Veltman M 1972 *Nucl.Phys.* B **44** 189
[2] Bardeen W A, Buras A J, Duke D W and Muta T 1978, *Phys.Rev.* **D18** 3998
[3] Siegel W 1979 *Phys.Lett.* B **84** 193
[4] Ferrara S and Zumino B 1974 *Nucl.Phys.* B **79** 413
[5] Jones D R T 1975 *Nucl.Phys.* B **87** 127
[6] Avdeev L V and Tarasov O V 1982 *Phys.Lett.* B **112** 356
[7] Jack I, Jones D R T and North C G 1996 *Phys.Lett* B **386** 138
[8] Harlander R V, Jones D R T, Kant P, Mihaila L and Steinhauser M 2006 *JHEP* **0612** 024
[9] Jack I, Jones D R T and North C G 1997 *Nucl.Phys.* **B 486** 479
[10] Jack I, Jones D R T and Pickering A 1998 *Phys.Lett.* B **435** 61
[11] Novikov V A, Shifman M A, Vainshtein A I and Zakharov V I 1983 *Nucl.Phys.* B **229** 381
[12] Novikov V A, Shifman M A, Vainshtein A I and Zakharov V I 1985 *Phys.Lett.*B **166** 329
[13] Shifman M A and Vainshtein A I 1986 *Nucl.Phys.* B **277** 456
[14] Vainshtein A I, Zakharov V I and Shifman M A 1985 *JETP Lett.* **42** 224
[15] Siegel W 1980 *Phys.Lett.* B **94** 37
[16] Avdeev L V, Chochia G A and Vladimirov A A 1981 *Phys.Lett.* B **105** 272
[17] Stöckinger D 2005 *JHEP* **0503** 076
[18] Avdeev L V 1982 *Phys.Lett.* B **117** 317
[19] Avdeev L V and Vladimirov A A 1983 *Nucl.Phys.* B **219** 262
[20] Velizhanin V N 2009 *Nucl.Phys.* B **818** 95
[21] Velizhanin V N 2011 *Phys.Lett.* B **696** 560
[22] Jack I and Jones D R T 1997 Regularisation of supersymmetric theories *Preprint* hep-ph/9707278
[23] Slavnov A A 1971 *Nucl.Phys.* B **31** 301
[24] Slavnov A A 1972 *Theor.Math.Phys.* **13** 1064
[25] Krivoshchekov V K 1978 *Theor.Math.Phys.* **36** 745
[26] West P 1986 *Nucl.Phys.* B **268** 113
[27] Martin C and Ruiz Ruiz F 1995 *Nucl.Phys.* B **436** 645
[28] Asorey M and Falceto F 1996 *Phys.Rev.* D **54** 5290
[29] Bakeyev T and Slavnov A A 1996 *Mod.Phys.Lett.* A **11** 1539
[30] Soloshenko A A and Stepanyantz K V 2004 *Theor.Math.Phys.* **140** 1264
[31] Smilga A and Vainshtein A 2005 *Nucl.Phys.* B **704** 445
[32] Stepanyantz K V 2005 *Theor.Math.Phys.* **142** 29
[33] Pimenov A B, Shevtsova E S and Stepanyantz K V 2010 *Phys.Lett.* B **686** 293
[34] West P 1986 *Introduction to supersymmetry and supergravity* (World Scientific)
[35] Buchbinder I L and Kuzenko S M 1998 *Ideas and methods of supersymmetry and supergravity* (Bristol and Philadelphia, Institute of Physics Publishing)
[36] Faddeev L D and Slavnov A A 1990 *Gauge fields, introduction to quantum theory* (Benjamin Reading)
[37] Stepanyantz K V 2011 *Nucl.Phys.* B **852** 71
[38] Grisaru M T and Zanon D 1985 *Nucl.Phys.* B **252** 578
[39] Grisaru M T, Milewski B and Zanon D 1986 *Nucl.Phys.* B **266** 589

ACAT 2011 IOP Publishing
Journal of Physics: Conference Series **368** (2012) 012053 doi:10.1088/1742-6596/368/1/012053

Three-loop calculation of the higgs boson mass in supersymmetry

Philipp Kant
Humboldt-Universität zu Berlin
E-mail: philipp.kant@physik.hu-berlin.de

Abstract. A Key feature of the minimal supersymmetric extension of the Standard Model (MSSM) is the existence of a light Higgs boson, the mass of which is not a free parameter but an observable that can be predicted from the theory. Given that the LHC is able to measure the mass of a light Higgs with very good accuracy, a lot of effort has been put into a precise theoretical prediction. We present a calculation of the SUSY-QCD corrections to this observable to three-loop order. We perform multiple asymptotic expansions in order to deal with the multi-scale three-loop diagrams, making heavy use of computer algebra and keeping a keen eye on the numerical error introduced. We provide a computer code in the form of a Mathematica package that combines our three-loop SUSY-QCD calculation with the literature of one- and two-loop corrections to the Higgs mass, providing a state-of-the-art prediction for this important observable.

1. Introduction

The minimal supersymmetric extension of the Standard Model (MSSM) is a promising candidate for physics beyond the Standard Model. Its Higgs sector is a two-Higgs doublet model with the additional constraint that supersymmetry relates the quartic Higgs couplings to the gauge couplings of the theory. This increases the predictiveness of the model and allows the Higgs sector to be parametrised by just two new parameters, the mass M_A of the pseudoscalar Higgs and the ratio $\tan\beta$ of the vacuum expectation values of the Higgs doublets.

In particular, the mass M_h of the light Higgs boson is not a free parameter, but can be predicted. At tree level, M_h is bounded above by M_Z, but radiative corrections shift the value significantly. Since M_h will be a precision observable once the Higgs is found at the Large Hadron Collider (LHC), it is imperative to have a precise theoretical prediction. Consequently, a lot of effort has been put into the calculation of radiative corrections to M_h at the one- and two-loop level (see, for example [1, 2, 3, 4, 5, 6, 7, 8]). The remaining theoretical uncertainty has been estimated to be about $3 - 5\,\mathrm{GeV}$ [9, 10], which is confirmed by a study of the leading and next-to-leading terms in $\ln(M_{SUSY}/M_t)$, where M_{SUSY} is the typical scale of SUSY particle masses, at three-loop order [11].

This uncertainty of the theoretical prediction justifies a calculation of the next term in the perturbative expansion. A study of the corrections to M_h shows that the contributions from loops of top quarks and their superpartners, the stops, are dominant at the one- and two-loop level. In [12, 13], we have calculated three-loop SUSY-QCD corrections to these diagrams. The calculation of these terms is the subject of this talk.

2. Calculation of the Three-Loop Corrections

Motivated by the discussion above, we restrict the calculation at the three-loop level to the SUSY-QCD corrections where the Higgs couples to a top quark or its superpartners. In the perturbative expansion of M_h, these are the terms of $\mathcal{O}\left(\alpha_t\alpha_s^2\right)$, where α_s is the strong coupling and α_t denotes the coupling of the Higgs to the top quark.

We thus have to evaluate three-loop propagator diagrams in SUSY-QCD, which faces us with two complications. First, we need a convenient regulator that respects supersymmetry, and second, a lot of masses appear in our diagrams.

2.1. Regularisation by Dimensional Reduction

The regularisation workhorse of multi-loop calculations is dimensional regularisation (DREG) [14], where the number of spacetime dimensions is altered from 4 to $d = 4 - 2\varepsilon$, and the divergences of the loop integrals are manifest as poles in ε.

Unfortunately, DREG does not respect supersymmetry. An easy way to see this is that changing the number of spacetime dimensions also changes the number of degrees of freedoms of the vector fields, and supersymmetry requires an equal number of bosonic and fermionic degrees of freedom. While it is possible (but tedious) to manually construct finite counterterms that restore the supersymmetric ward identities, a more convenient approach was suggested by Siegel under the name of dimensional reduction (DRED) [15]. The main idea is that the change of dimensions need only affect momenta in order to regularise the loop integrals. In DRED, one splits the four dimensional space into orthogonal spaces of dimension 2ε and $d = 4 - 2\varepsilon$, and restricts the loop momenta to the d dimensional subspace while all gauge fields are kept four dimensional. The 2ε components of the gauge fields transform as a 2ε tuple of scalar fields and are called ε-scalars.

The consistent formulation of DRED is more involved than sketched here [16], and the question whether the inclusion of the ε-scalars truly restores the supersymmetric ward identities in all cases is not yet resolved. However, DRED has successfully been applied in many multi-loop calculations[17, 18, 19, 20, 21, 22, 23, 24, 25, 26], so far without observing a violation of the ward identities.

2.2. Asymptotic Expansion in the Masses

The difficulty of calculating a Feynman diagram rises with the number of loops, the number of external momenta and the number of masses in the diagram. To obtain the $\mathcal{O}\left(\alpha_t\alpha_s^2\right)$ terms of M_h, we have to evaluate diagrams with three loops, two legs and a lot of masses: treating the light quarks as massless and their superpartners as mass degenerate with mass $M_{\tilde{q}}$, we are left with five masses: the top quark mass M_t, the masses of its superpartners $M_{\tilde{t}_1}$, $M_{\tilde{t}_2}$, the mass of the gluino $M_{\tilde{g}}$ and $M_{\tilde{q}}$. Calculating these diagrams without any approximations is not feasible with contemporary methods.

Given the mild dependence of the one-loop corrections to M_h on the external momentum, it is reasonable to expand the diagrams in small external momentum, reducing the problem to the evaluation of vacuum diagrams. A further simplification is possible by performing nested asymptotic expansions [27, 28, 29, 30], which expresses the multi-mass diagrams as a series in ratios of the masses and logarithms of these ratios. The coefficients in the series involve only one-scale integrals. Assuming that the ratios of the masses are small, the series should give good approximations to the original diagrams.

Of course, before carrying out the asymptotic expansions, one has to identify which mass ratios are small. Given that so far none of the superpartner masses are known, this is an undecidable problem. So, instead of calculating with a fixed, known, hierarchy among the masses, one has to systematically consider various possible mass hierarchies and carry out the calculation for each of these. To get a prediction of M_h for a specific point in the parameter

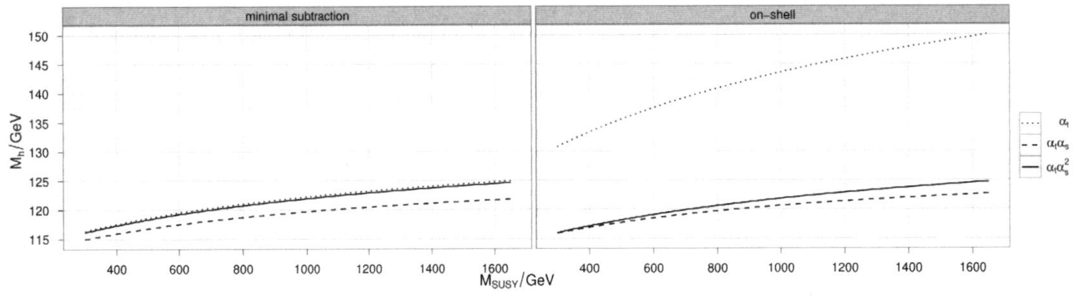

Figure 1. Behaviour of the perturbative series for M_h using the on-shell and the minimal subtraction scheme (for degenerate superpartner masses). The short-dashed, long-dashed and solid lines are the one-, two-, and three-loop approximation to M_h, respectively. In the on-shell scheme, the distance between one- and two-loop approximation is much larger than in the $\overline{\text{DR}}$ scheme.

space of the MSSM, where the masses of the superpartners have specific values, one has to choose whichever hierarchy fits these values best.

We carry out the calculation using the following setup: In a first step, all diagrams contributing to M_h at $\mathcal{O}(\alpha_t\alpha_s^2)$ are found using the program QGRAF [31]. There are 30.717 diagrams. For each hierarchy, these diagrams are expanded asymptotically using Q2E/EXP [32]. The one-scale integrals are then calculated using the FORM [33] program MATAD [34].

2.3. Renormalisation

For the renormalisation of the parameters entering our calculation, we adopt a variation of the $\overline{\text{DR}}$-scheme, i.e. minimal subtraction using dimensional reduction. This leads to a much better behaviour of the perturbative series compared to using on-shell renormalisation (see figure 1).

3. Computer Code

In order to provide a precise prediction for the value of M_h within the MSSM, our $\mathcal{O}(\alpha_t\alpha_s^2)$ terms have to be combined with the wealth of corrections from other sectors of the MSSM at one- and two-loop level that are available in the literature [1, 2, 3, 4, 5, 6, 7, 8]. Also, the choice of hierarchy mentioned in 2.2 should be automatised. In [13], we have presented the MATHEMATICA package H3M, which is publicly available, to address these points. For convenience, it implements the Susy Les Houches Accord SLHA [35] and has an interface to the spectrum generators SOFTSUSY [36], SuSPECT [37] and SPHENO [38].

The program uses FEYNHIGGS [39, 9] to get a two-loop approximation of M_h. An obstacle for adding our terms of $\mathcal{O}(\alpha_t\alpha_s^2)$ to the result delivered by FEYNHIGGS is the usage of on-shell renormalisation in FeynHiggs. In order to be consistent, we have to convert the $\mathcal{O}(\alpha_t)$ and $\mathcal{O}(\alpha_t\alpha_s)$ terms that contain on-shell parameters to the $\overline{\text{DR}}$ scheme. Thanks to [7], which gives a compact expression for these terms both in the $\overline{\text{DR}}$ and in the on-shell scheme, this obstacle is easily overcome.

The corrections to M_h depend very strongly on the mass M_t of the top quark and the strong coupling α_s. These parameters must be known, within SUSY-QCD renormalised in the $\overline{\text{DR}}$ scheme, as precisely as possible. In [40], the relation between the top mass in the $\overline{\text{DR}}$- and the on-shell scheme in SUSY-QCD has been calculated to $\mathcal{O}(\alpha_s^2)$. Using the library TSIL [41], this relation can be used to obtain $M_t^{\overline{\text{DR}}}$ from the known value of the top quark pole mass.

We determine α_s in SUSY-QCD from the value of α_s in five-flavour QCD at the Z mass following [42] by running, within five-flavour QCD, to the decoupling scale where we perform the transition to the full theory. We then run, within SUSY-QCD, to the desired value of the renormalisation scale.

In order to choose the most appropriate hierarchy and get an estimate for the error introduced by the asymptotic expansion, we compute an approximation to the two-loop corrections of $\mathcal{O}(\alpha_t\alpha_s)$ using asymptotic expansions in the mass ratios and compare this approximation to the result of [7]. The hierarchy that matches the exact two-loop result best is chosen for the calculation of the three-loop term, and the error at two-loop level is recorded to get a handle on the error in the three-loop result (see figure 4).

4. Numerical Results

In this chapter, we present a state-of-the-art numerical prediction for M_h in the MSUGRA model that has been obtained with H3M. figure 2 shows the dependence of M_h on the parameters $m_{1/2}$, m_0, $\tan\beta$, and A_0 for $\mu > 0$. The shaded bands around the individual curve show the variation of M_h when M_t is varied between 171.4 GeV and 174.4 GeV. We restrict the plot to values of $m_{1/2} > 300$ GeV in light of the latest exclusions from the LHC experiments [43, 44, 45, 46].

We observe that of all MSUGRA parameters, varying $m_{1/2}$ has by far the largest impact on the Higgs mass. We also observe that the present uncertainty of the top mass directly translates to a parametric uncertainty of about one GeV of the Higgs mass.

To estimate the effect of unknown higher orders, figure 3 shows the two- and three-loop corrections to M_h. We observe a partial cancellation between the two- and three-loop correction. We also observe that while the higher order corrections do decrease in magnitude, they do not do so by a large factor. Thus, we should be careful when estimating the magnitude of the missing higher order contributions. We choose to assign 50 % of the three-loop contribution as a theoretical error. This amounts to a hundred MeV for low values of $m_{1/2}$ and to about one GeV for $m_{1/2}$ above one GeV. Using the scale variation as an error estimate would lead to a smaller error [13].

By performing asymptotic expansions, we have introduced an additional source of uncertainty to the prediction of M_h. With figure 4, we can analyse how large this uncertainty is. It shows the error that we would have introduced had we made the same approximation at the two-loop level. We see that this error is typically at or below 100 MeV, or 200 MeV for low values of $m_{1/2}$ and $\tan\beta$. Since the three-loop corrections are smaller than the two-loop corrections, we expect the actual error due to asymptotic expansions to be below 100 MeV.

5. Conclusions

We have presented a calculation of the $\mathcal{O}(\alpha_t\alpha_s^2)$ corrections to the mass M_h of the light Higgs boson in the MSSM. These contributions shift the value of M_h by $1 - 3$ GeV, depending on the mass spectrum of the superpartners.

The results have been implemented in the program H3M, which is freely available [13].

Using our results, we improve the theoretical error significantly. The theoretical error is now on the order of about 100 MeV for light and 1 GeV for heavy superpartner masses. This is comparable to the parametric uncertainty with the top mass and α_s.

Acknowledgements

This work was supported by the DFG through SFB/TR-9 and by the Helmholtz Alliance "Physics at the Terascale". We thank Robert Harlander, Luminita Mihaila and Matthias Steinhauser, with whom this work was done, Pietro Slavich and Steve Martin for providing us with private code that implements results from [7] and [40], respectively, and Thomas Hahn for useful discussions about some technical details of FeynHiggs.

Journal of Physics: Conference Series **368** (2012) 012053 doi:10.1088/1742-6596/368/1/012053

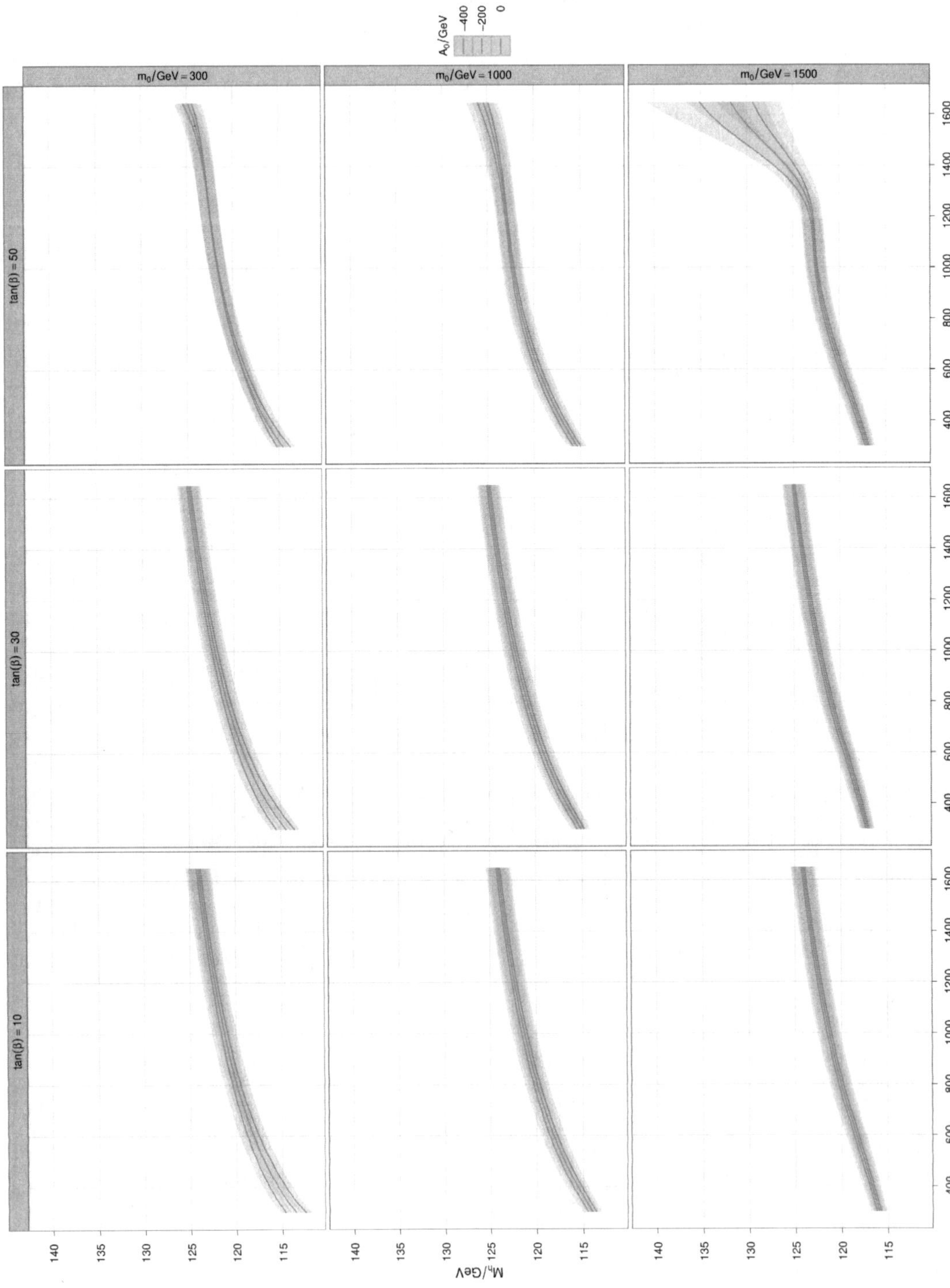

Figure 2. Theoretical prediction for the light Higgs mass in the MSUGRA model. We plot M_h over the fermion mass parameter $m_{1/2}$. The panels show, from left to right, an increasing value of $\tan\beta$, and, from top to bottom, of the scalar mass parameter m_0. We show three curves according to different values of the trilinear coupling A_0.

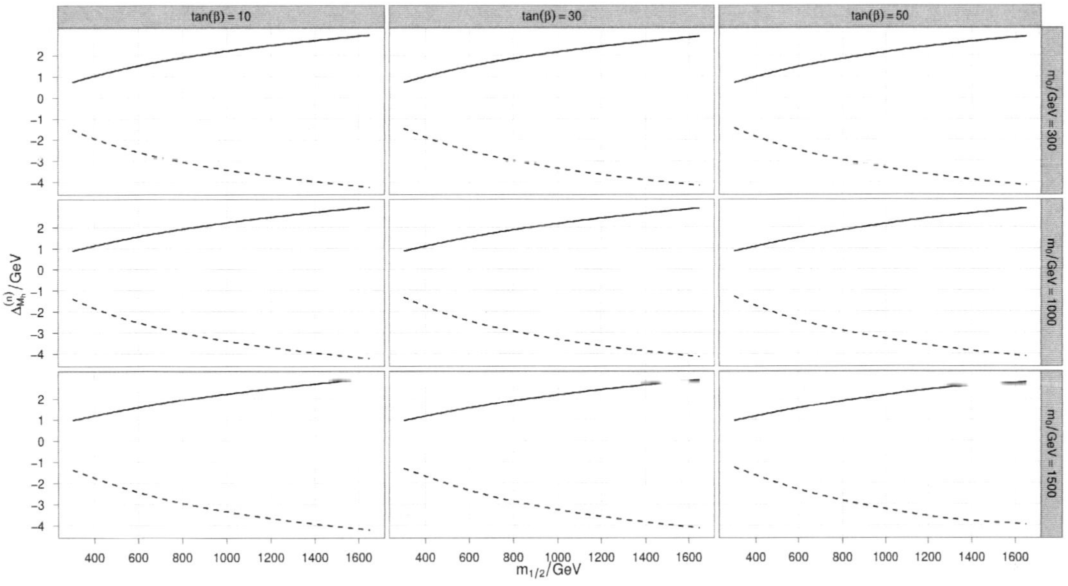

Figure 3. Two-loop (dashed) and three-loop (solid) contributions to M_h. The same conventions as in figure 2 are adopted. For lucidity, only the curves for $A_0 = 0$ and $M_t = 172.9\,\text{GeV}$ are shown. We observe that the three-loop terms are smaller than the two-loop terms, but not by a large factor. This suggests a conservative approach when estimating higher order effects.

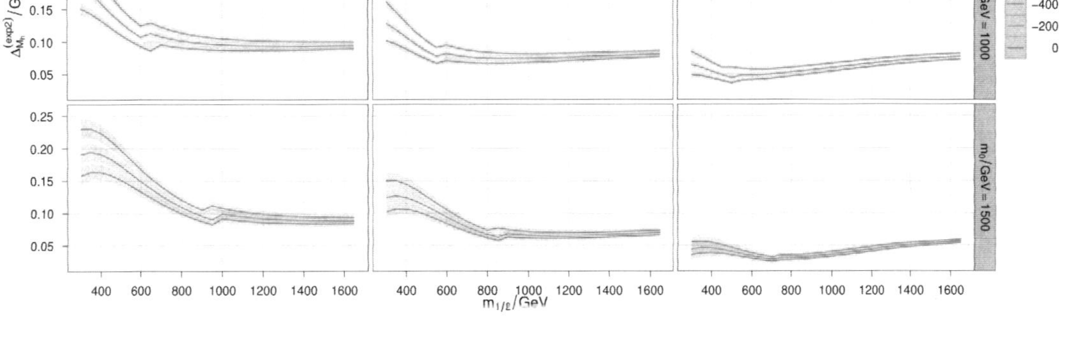

Figure 4. Deviation of the expansion in masses from the exact result at two loops. The conventions are the same as in figure 2. The error introduced by the asymptotic expansions is well under control. The little bumps in the curves are due to a change from one hierarchy to another.

References

[1] Hempfling R and Hoang A H 1994 *Phys. Lett.* **B331** 99–106 (*Preprint* hep-ph/9401219)
Haber H E, Hempfling R and Hoang A H 1997 *Z. Phys.* **C75** 539–554 (*Preprint* hep-ph/9609331)
[2] Zhang R 1999 *Phys. Lett.* **B447** 89–97 (*Preprint* hep-ph/9808299)
[3] Heinemeyer S, Hollik W and Weiglein G 1998 *Phys. Rev.* **D58** 091701 (*Preprint* hep-ph/9803277)
[4] Pilaftsis A and Wagner C E M 1999 *Nucl. Phys.* **B553** 3–42 (*Preprint* hep-ph/9902371)
[5] Carena M S *et al.* 2000 *Nucl. Phys.* **B580** 29–57 (*Preprint* hep-ph/0001002)
[6] Espinosa J R and Navarro I 2001 *Nucl. Phys.* **B615** 82–116 (*Preprint* hep-ph/0104047)
[7] Degrassi G, Slavich P and Zwirner F 2001 *Nucl. Phys.* **B611** 403–422 (*Preprint* hep-ph/0105096)
[8] Martin S P 2003 *Phys. Rev.* **D67** 095012 (*Preprint* hep-ph/0211366)
[9] Degrassi G, Heinemeyer S, Hollik W, Slavich P and Weiglein G 2003 *Eur. Phys. J.* **C28** 133–143 (*Preprint* hep-ph/0212020)
[10] Allanach B, Djouadi A, Kneur J, Porod W and Slavich P 2004 *JHEP* **0409** 044 (*Preprint* hep-ph/0406166)
[11] Martin S P 2007 *Phys. Rev.* **D75** 055005 (*Preprint* hep-ph/0701051)
[12] Harlander R V, Kant P, Mihaila L and Steinhauser M 2008 *Phys. Rev. Lett.* **100** 191602 (*Preprint* 0803.0672)
[13] Kant P, Harlander R V, Mihaila L and Steinhauser M 2010 *JHEP* **08** 104 (*Preprint* 1005.5709)
[14] Wilson K G 1973 *Phys. Rev.* **D7** 2911–2926
't Hooft G and Veltman M J G 1972 *Nucl. Phys.* **B44** 189–213
Breitenlohner P and Maison D 1977 *Commun. Math. Phys.* **52** 11–38
[15] Siegel W 1979 *Phys. Lett.* **B84** 193
[16] Stöckinger D 2005 *JHEP* **03** 076 (*Preprint* hep-ph/0503129)
[17] Harlander R V, Jones D R T, Kant P, Mihaila L and Steinhauser M 2006 *JHEP* **12** 024 (*Preprint* hep-ph/0610206)
[18] Marquard P, Mihaila L, Piclum J and Steinhauser M 2007 *Nucl.Phys.* **B773** 1–18 (*Preprint* hep-ph/0702185)
[19] Bednyakov A 2007 *Int.J.Mod.Phys.* **A22** 5245–5277 (*Preprint* 0707.0650)
[20] Jack I, Jones D R T, Kant P and Mihaila L 2007 *JHEP* **09** 058 (*Preprint* 0707.3055)
[21] Signer A and Stockinger D 2009 *Nucl.Phys.* **B808** 88–120 (*Preprint* 0807.4424)
[22] Mihaila L 2009 *Phys.Lett.* **B681** 52–59 (*Preprint* 0908.3403)
[23] Pak A, Steinhauser M and Zerf N 2011 *Eur.Phys.J.* **C71** 1602 (*Preprint* 1012.0639)
[24] Harlander R V, Hofmann F and Mantler H 2011 *JHEP* **1102** 055 (*Preprint* 1012.3361)
[25] Kilgore W B 2011 *Phys.Rev.* **D83** 114005 (*Preprint* 1102.5353)
[26] Hermann T, Mihaila L and Steinhauser M 2011 *Phys.Lett.* **B703** 51–59 (*Preprint* 1106.1060)
[27] Tkachov F V 1993 *Int. J. Mod. Phys.* **A8** 2047–2117 (*Preprint* hep-ph/9612284)
Pivovarov G B and Tkachov F V 1993 *Int. J. Mod. Phys.* **A8** 2241–2286 (*Preprint* hep-ph/9612287)
[28] Gorishnii S G and Larin S A 1987 *Nucl. Phys.* **B283** 452
[29] Smirnov V A and Chetyrkin K G 1987 *Proceedings, Problems of quantum field theory (Alushta)* 48-53; Chetyrkin K G MPI-PH-91-13
[30] Smirnov V A 1990 *Commun. Math. Phys.* **134** 109–137
Smirnov V A Basel, Switzerland: Birkhaeuser (1991) 380 p. (Progress in physics, 14)
[31] Nogueira P 1993 *J. Comput. Phys.* **105** 279–289
[32] Seidensticker T 1999 (*Preprint* hep-ph/9905298)
Harlander R, Seidensticker T and Steinhauser M 1998 *Phys. Lett.* **B426** 125–132 (*Preprint* hep-ph/9712228)
[33] Vermaseren J A M 2000 (*Preprint* math-ph/0010025)
[34] Steinhauser M 2001 *Comput. Phys. Commun.* **134** 335–364 (*Preprint* hep-ph/0009029)
[35] Skands P Z *et al.* 2004 *JHEP* **07** 036 (*Preprint* hep-ph/0311123)
[36] Allanach B 2002 *Comput.Phys.Commun.* **143** 305–331 (*Preprint* hep-ph/0104145)
[37] Djouadi A, Kneur J L and Moultaka G 2007 *Comput.Phys.Commun.* **176** 426–455 (*Preprint* hep-ph/0211331)
[38] Porod W and Staub F 2011 *Preprint* 1104.1573
Porod W 2003 *Comput.Phys.Commun.* **153** 275–315 (*Preprint* hep-ph/0301101)
[39] Heinemeyer S, Hollik W, Weiglein G 2000 *Comput. Phys. Commun.* **124** 76–89 (*Preprint* hep-ph/9812320)
Heinemeyer S, Hollik W and Weiglein G 1999 *Eur. Phys. J.* **C9** 343–366 (*Preprint* hep-ph/9812472)
Frank M *et al.* 2007 *JHEP* **02** 047 (*Preprint* hep-ph/0611326)
[40] Martin S P 2005 *Phys. Rev.* **D72** 096008 (*Preprint* hep-ph/0509115)
[41] Martin S P and Robertson D G 2006 *Comput. Phys. Commun.* **174** 133–151 (*Preprint* hep-ph/0501132)
[42] Harlander R V, Mihaila L and Steinhauser M 2007 *Phys. Rev.* **D76** 055002 (*Preprint* 0706.2953)
[43] ATLAS Collaboration 2011 *Preprint* 1110.2299
[44] ATLAS Collaboration 2011 *Phys.Rev.D* **D85** 012006 (*Preprint* 1109.6606)
[45] Padhi S (for the CMS Collaboration) 2011 *Preprint* 1111.2733
[46] Schettler H (for the CMS Collaboration) 2011 *Preprint* 1111.2256

FormCalc 7

S. Agrawal[1], **T. Hahn**[2], **E. Mirabella**[2]

[1] LNM IIT, Rupa Ki Nagal, Post-Sumel, Via Jamdoli, Jaipur–302031, India,
[2] MPI für Physik, Föhringer Ring 6, D 80805 Munich, Germany

E-mail: shivam@mpp.mpg.de, hahn@mpp.mpg.de, mirabell@mpp.mpg.de

Abstract. We present additions and improvements in Version 7 of FormCalc, most notably analytic tensor reduction, choice of OPP methods, and MSSM initialization via FeynHiggs, as well as a parallelized Cuba library for numerical integration.

1. Introduction

The Mathematica package FormCalc [1] simplifies Feynman diagrams generated with FeynArts [2] up to one-loop order. It provides both the analytical results and can generate Fortran code for the numerical evaluation of the squared matrix element.

This note presents the following improvements and additions in FormCalc 7 and the numerical integration package Cuba [3], which is included in FormCalc:

- Analytic tensor reduction,
- Unitarity methods (OPP),
- Improved code generation,
- Command-line parameters for model initialization,
- MSSM initialization via FeynHiggs,
- Built-in parallelization in Cuba.

2. Analytic tensor reduction

Despite the hype that surrounds unitarity methods today, the Passarino–Veltman decomposition of tensor one-loop integrals [5] remains a valuable technique, also because it admits a fully analytic reduction. The complete tensor reduction consists of two steps:

- The Lorentz-covariant decomposition of the tensors of the loop momentum appearing in the numerator into linear combinations of tensors constructed from $g_{\mu\nu}$ and the external momenta with coefficient functions, e.g.

$$\int \mathrm{d}^4 q \frac{q_\mu q_\nu}{D_0 D_1} \sim B_{\mu\nu} = g_{\mu\nu} B_{00} + p_\mu p_\nu B_{11} \,.$$

  This part has always been performed in FormCalc, as the actual tensors are rather unwieldy objects for further evaluation.
- Solving the linear system that determines the coefficient functions, i.e. expressing the coefficient functions through scalar integrals.

FormCalc has for long included the add-on `FormCalc'btensor'` package which analytically reduces one- and two-point functions when loaded, but higher-point functions could be reduced only indirectly through FeynCalc [4], i.e. the user had to save the amplitudes with `FeynCalcPut`, run FeynCalc in a different Mathematica session, and load the reduced expressions into FormCalc again with `FeynCalcGet`. This procedure was not only suboptimal in terms of user-friendliness but also did not take advantage of the field levels of FeynArts, i.e. FeynCalc always operated on the fully inserted amplitudes rather than the (typically much fewer) Generic amplitudes.

The analytic tensor reduction is meanwhile properly available in FormCalc and can be turned on through the option

```
CalcFeynAmp[..., PaVeReduce → True]
```

Our code implements the reduction formulas of Denner and Dittmaier [6]. While these are fully worked out, it nevertheless took considerable effort to program them in FORM due to at first sight trivial issues, e.g. that there is no straightforward way to obtain the N-th argument of a function. Adding the reduction code to the Mathematica part of FormCalc instead was not an option, however, as we wanted to operate on the Generic amplitudes, before the substitution of the insertions, and this happens in FORM.

Inverse Gram determinants, which appear as a by-product of inverting the coefficient-function system, may lead to instabilities in the numerical evaluation later on and therefore FormCalc tries to cancel them as much as possible. The ones that cannot be cancelled immediately are returned as `IGram[`x`]` $(= 1/x)$ and so can easily be found and processed further in Mathematica.

3. Unitarity methods (OPP)

We employ the OPP (Ossola, Papadopoulos, Pittau [7]) unitarity method as implemented in the two libraries CutTools [8] and Samurai [9]. Instead of introducing tensor coefficients as in the last section, in this case the whole numerator is placed in a subroutine, as in:

$$\varepsilon_1^\mu \varepsilon_2^\nu B_{\mu\nu}(p, m_1^2, m_2^2) = B_{\text{cut}}(2, N, p, m_1^2, m_2^2)\,, \qquad \text{where} \qquad N(q_\mu) = (\varepsilon_1 \cdot q)\,(\varepsilon_2 \cdot q)\,.$$

The numerator subroutine N will be sampled by the OPP function (B_{cut} in this example). The first argument of B_{cut}, 2, refers to the maximum power of q in N.

Subexpressions of the numerator function (coefficients, summands, etc.) which do not depend on q are pulled out and computed once, ahead of invoking the OPP function, using FormCalc's abbreviationing machinery [10]. In particular in BSM theories, these coefficients can be lengthy such that pulling them out significantly increases performance.

Our implementation admits mixing Passarino–Veltman decomposition with OPP in the sense that one chooses an integer n starting from which an n-point function is treated with OPP methods. For example, `OPP → 4` means that A, B, C functions are treated with Passarino–Veltman and D and up with OPP. A negative n indicates that the rational terms for the OPP integrals shall be added analytically whereas else their computation is left to the OPP package.

The CutTools and Samurai libraries have minor differences in calling conventions but are otherwise similar enough to let the preprocessor handle the switching. That is, one does not need to re-generate the Fortran code in order to link with the other library. Specifically, the following steps must be taken in order to use the OPP method in FormCalc:

- The amplitudes must be prepared with `CalcFeynAmp[..., OPP →` n`]` $(n < \infty)$.
- In the generated code, the OPP library (CutTools or Samurai) must be chosen and the declarations in `opp.h` be included. This is most conveniently done in `user.h`, in the following structure:

```
#ifndef USER_H
#define USER_H
* declarations for the whole file (e.g. preprocessor defs)
#define SAMURAI                    (or CUTTOOLS)
#else
* declarations for every subroutine
#include "opp.h"                   (necessary for OPP)
#include "model_sm.h"
#endif
```

As for testing, we have currently compared a handful of $2 \to 2$ and $2 \to 3$ scattering reactions, both QCD and electroweak, and found agreement to about 10 digits between Passarino–Veltman and OPP, with CutTools and Samurai delivering similar results. This shows that the method is working.

Performance is somewhat wanting as of now, however. For the tested processes, quite frankly, Passarino–Veltman beats OPP hands down. To be fair, OPP was in the first place designed to increase the reach of one-loop calculations to higher-leg multiplicities and not so much to speed up the ones with not so many legs such as those we have looked at so far. Nevertheless, there seems to be room for improvement on a technical level and we are certainly continuing work on optimizing the performance.

4. Improved code generation

The main subroutine `SquaredME.F` is now sectioned by comments, to aid automated substitution e.g. with `sed`. For example, the variable declarations are enclosed in

```
* BEGIN VARDECL
...
* END VARDECL
```

Furthermore, the generated code and the driver files are consistently formulated in terms of the newly introduced `RealType` and `ComplexType` data types, by default equivalent to `double precision` and `double complex`, respectively. Note that capitalization matters as these words are substituted by the preprocessor. This introduces a level of abstraction which makes it easier to e.g. work with a different precision.

More importantly, work is ongoing to produce output in C (C99 to be precise, for support of complex numbers), in addition to the Fortran mode. The advantages are threefold:

- It makes integration of generated code into existing C/C++ packages still easier (no linking hassles) while not impeding use in Fortran, i.e. even though it is not planned to remove the native Fortran output in a future version of FormCalc, a Fortran user will be able to link with the C object files directly without any changes to his code.
- It makes GPU programming feasible without reverting to commercial Fortran compilers.
- One can take advantage of C's `long double` data type which, at least on Intel x86 hardware, gives an additional 2–3 digits of precision at essentially no extra cost. Extended real data types in Fortran, if available at all, are usually IEEE-754-compliant `REAL*16` emulated in software.

5. Command-line parameters for model initialization

FormCalc includes a suite of so-called driver programs to manage the automatically generated code for computing the squared matrix element. They parse the command line, initialize model constants, set up phase space, etc.

Journal of Physics: Conference Series **368** (2012) 012054 doi:10.1088/1742-6596/368/1/012054

In particular the driver modules for the initialization of the model parameters and luminosity calculation (which includes e.g. the setup of PDFs used in hadronic reactions) had no access to the command-line arguments so far and could use only variables supplied by the user in the main control program `run.F`. In other words, the model inputs and PDF selections were 'compiled in' and the executable had to be re-built every time those values changed.

The present command-line parser accepts so-called colon arguments (arguments starting with a ':') before the usual ones on the command line, as in:

run :arg_1 :arg_2 ... uuuuu 0,1000

The colon arguments are read into an array (sans colon) and handed to the model-initialization and luminosity-calculation subroutines:

```
subroutine ModelDefaults(argc, argv)

subroutine LumiDefaults(argc, argv)
integer argc
character*128 argv(*)
```

Note that, unlike in C (`char **argv`), fixed-length strings are passed in `argv` since there are no pointers in Fortran 77. It is up to the `ModelDefaults` and `LumiDefaults` subroutines to handle the arguments. In Fortran it is furthermore no fatal error to have no formal arguments in the `ModelDefaults` and `LumiDefaults` subroutines (as in previous FormCalc versions), so old code will compile and run without change.

6. MSSM initialization via FeynHiggs

The colon arguments of the previous section are immediately put to use for the initialization of the MSSM through FeynHiggs [11]. The default MSSM initialization is a stand-alone routine (i.e. requires no external library to be linked), but is not quite as thorough as FeynHiggs when it comes to the corrections included e.g. in the computation of the Higgs masses.

From FeynHiggs version 2.8.1 on not only the computational engine but the entire Frontend functionality is available through library routines so that the colon arguments can simply be passed to a FeynHiggs subroutine to make FeynHiggs initialize itself as if invoked from its own command-line Frontend. The FormCalc-generated code inherits thus the ability to read parameter files in either native FeynHiggs or SLHA format, and of course obtains all MSSM parameters and Higgs observables from FeynHiggs. There is no duplication of initialization code this way, and moreover the parameters are consistent between the Higgs-mass and the cross-section calculations.

To use the FeynHiggs initialization, one chooses the model-initialization module `model_fh.F` instead of `model_mssm.F`. The compiled code is invoked as

run :*parafile* :*flags* uuuuu 0,1000

The colon arguments are just the ones of the FeynHiggs Frontend: *parafile* is the name of the parameter file and the optional *flags* allows to override the default flags of FeynHiggs.

7. Built-in parallelization in Cuba

Cuba is a library for multidimensional numerical integration which is integrated in FormCalc but of course can be used independently, too. Only the Mathematica interface was able to compute in parallel so far, by redefining the function `MapSample` with e.g. `ParallelMap`. In the latest release we added parallelization also to the C/C++ and Fortran interfaces.

We attempt no parallelization across the network, say via MPI. That is, we restrict ourselves to parallelization on one computer, using operating-system functions only, hence no extra software is required. A common setup these days, even on laptops, is a single CPU with a number of cores, typically 4 or 8. Utilizing many more compute nodes, as one could potentially do with MPI, is more of a theoretical option anyway since the speed-ups cannot be expected to grow linearly.

We use `fork/wait` rather than the `pthread*` functions. The latter are slightly more efficient at communicating data between parent and child because they share the same memory space (the `fork/wait` version uses pipes instead), but for the same reason they also require a reentrant integrand function, and apart from the extra work this takes, a programmer may not even have control over reentrancy in his language, e.g. Fortran's I/O is typically non-reentrant. `fork` on the other hand creates a completely independent copy of the running process and thus works for any integrand function.

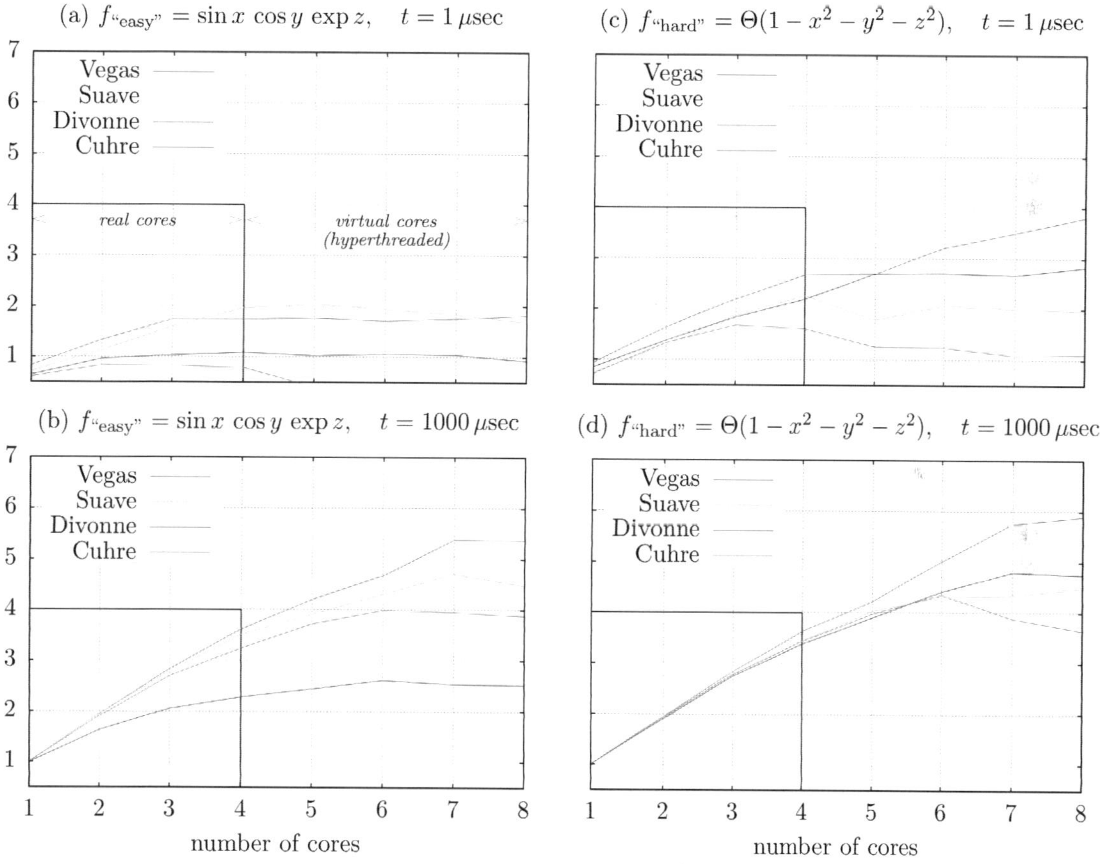

Figure 1. Cuba speed-ups for a three-dimensional integral on an i7-2600 Linux system (2.6.37) with 4 real/8 virtual (hyperthreaded) cores. The vertical line at 4 cores marks the cross-over.
(a), (b): 'easy' integrand, (c), (d): 'hard' integrand.
(a), (c): 'fast' integrand ($1\,\mu$sec), (b), (d): 'slow' integrand ($1000\,\mu$sec per evaluation).
Note that also in the one-core case a parallel version is used (one master, one worker), which explains why the timings normalized to the serial version are below 1, in the top row visibly so. The measured speed-ups are, if anything, on the conservative side as the i7 CPUs boast a feature named Turbo Boost which allows the CPU to scale up the frequency if not all cores are loaded, i.e. the serial version will likely have run at a somewhat higher CPU frequency.

Changing the number of cores to use does not require a re-compile, which is particularly useful as the program image should be able to run on several computers (with possibly different numbers of cores) simultaneously. Cuba determines the number of cores from the environment variable `CUBACORES`, or if this is unset, takes the idle cores on the present system (total cores minus load average). That is, unless the user explicitly sets `CUBACORES`, a program calling a Cuba routine will automatically parallelize on the available cores. A master process orchestrates the parallelization but does not count towards the number of cores, e.g. `CUBACORES = 4` means four workers and one master. Very importantly, the samples are generated by the master process only and distributed to the workers, such that random numbers are never used more than once.

Parallelization entails a certain overhead as usual, so the efficiency will depend on the 'cost' of an integrand evaluation, i.e. the more 'expensive' (time-consuming) it is to sample the integrand, the better the speed-up will be. To give an idea of the values that can be attained, Fig. 1 shows the speed-ups for an 'easy' and a 'hard' one of the 11 integrands of the demo program included in the Cuba package for two different integrand delays. To tune the 'cost' of the integrands, we introduced a calibrated delay loop into the integrand functions (which are simple one-liners and for our purposes 'infinitely' fast). The calibration and the timing measurements are rather delicate and shall not be discussed here.

The first, expected, observation is that parallelization is worthwhile only for not-too-fast integrands. This is not a major showstopper, however, as many integrands in particle physics (one-loop cross-sections, for example) easily fall into the 1000-μsec-and-beyond category.

The second observation is that parallelization works best for 'simple-minded' integrators, e.g. Vegas. The 'intelligent' algorithms are generally much harder to parallelize because they don't just do mechanical sampling but take into account intermediate results, make extra checks on the integrand (e.g. try to find extrema), etc. This is particularly true for Divonne, where we spent significant time to lift the speed-up curve even this far above 1 and which is still clearly lagging in parallelization efficiency (see e.g. the bottom-left plot of Fig. 1). Then again, the 'intelligent' algorithms are usually faster to start with (i.e. converge with fewer points sampled), which compensates for the lack of parallelizability.

8. Summary

FormCalc 7 (`http://feynarts.de/formcalc`) has many new and improved features, most notably the analytic tensor reduction, choice of OPP methods, and the link with FeynHiggs. The Cuba library (`http://feynarts.de/cuba`), also included in FormCalc, parallelizes integrations automatically and achieves decent speed-ups for typical cross-section integrands.

[1] Hahn T and Pérez-Victoria M 1999 *Comput. Phys. Commun.* **118** 153
[2] Hahn T 2001 *Comput. Phys. Commun.* **140** 418
[3] Hahn T 2005 *Comput. Phys. Commun.* **168** 78
[4] Mertig R, Böhm M and Denner A 1991 *Comput. Phys. Commun.* **64** 345
[5] Passarino G, Veltman M 1979 *Nucl. Phys. B* **160** 151
[6] Denner A and Dittmaier S 2006 *Nucl. Phys. B* **734** 62
[7] Ossola G, Papadopoulos C, and Pittau R 2007 *Nucl. Phys. B* **763** 147
[8] Ossola G, Papadopoulos C, and Pittau R 2008 *JHEP* **0803** 042
[9] Mastrolia P, Ossola G, Reiter T and Tramontano F 2010 *JHEP* **1008** 080
[10] Hahn T 2010 *PoS ACAT 2010* 078
[11] Frank M, Hahn T, Heinemeyer S, Hollik W, Rzehak H and Weiglein G 2007 *JHEP* **0702** 047

ACAT 2011 IOP Publishing
Journal of Physics: Conference Series **368** (2012) 012055 doi:10.1088/1742-6596/368/1/012055

Numerical evaluation of one-loop QCD amplitudes

Simon Badger [1], **Benedikt Biedermann** [2,3] **and Peter Uwer** [2]

[1] The Niels Bohr International Academy and Discovery Center, The Niels Bohr Institute, Blegdamsvej 17, DK-2100 Copenhagen, Denmark
[2] Humboldt-Universität zu Berlin, Institut für Physik, Newtonstraße 15, D-12489 Berlin, Germany

E-mail: Simon.Badger@nbi.dk, Benedikt.Biedermann@physik.hu-berlin.de, Peter.Uwer@physik.hu-berlin.de

Abstract. We present the publicly available program NGluon allowing the numerical evaluation of primitive amplitudes at one-loop order in massless QCD. The program allows the computation of one-loop amplitudes for an arbitrary number of gluons. The focus of the present article is the extension to one-loop amplitudes including an arbitrary number of massless quark pairs. We discuss in detail the algorithmic differences to the pure gluonic case and present cross checks to validate our implementation. The numerical accuracy is investigated in detail.

1. Introduction

The automation of next-to-leading (NLO) corrections to multi-particle processes in the Standard Model is an important step in making precision predictions for signal and background reactions studied at the Large Hadron Collider (LHC) at CERN. Recent years have seen considerable progress in simplifying this complex task into a definite algorithm for arbitrary processes [1, 2, 3, 4]. Full NLO distributions for a growing number of $2 \to 4$ [5, 6, 7, 8, 9, 10, 11, 12] and even $2 \to 5$ [13, 14, 15] processes have now been achieved. The degree of automation to which the virtual corrections to NLO observables can be computed has steadily improved [16, 17, 18, 19, 20, 21, 22, 23].

In this article we review the computations of multi-parton amplitudes in massless QCD using the NGluon c++ library [24]. The algorithm employs the generalised unitarity cutting procedure to construct one-loop amplitudes from on-shell tree-level building blocks which we review in section 2. We then describe the inclusion of amplitudes with multiple fermion pairs in section 3 and present a detailed analysis of the performance and validity of our approach in section 4. We finish by presenting our conclusions in section 5.

2. Review of Generalised Unitarity

The method of generalised unitarity has been studied extensively over the last few years. Based on a purely algebraic approach, the method can be used to formulate a numerical algorithm for the computation of one-loop amplitudes [26, 25, 4]. Detailed and helpful reviews on the subject can be found in [27, 28, 29].

[3] Speaker

The one-loop amplitudes we consider in this article are colour ordered QCD primitive amplitudes where both colour generators and internal colour flow structure have been stripped off leaving the simplest gauge invariant building blocks of the full amplitude. Each of these terms has a well defined ordering of external legs and internal propagators.

A general one-loop primitive amplitude with no external massive particles, regulated in $D = 4 - 2\epsilon$ dimensions, can be expressed in terms of a basis of scalar two-, three- and four-point integral functions together with rational terms [25],

$$A^{(1)}(p_1, \cdots, p_n) = \sum_{i<j<k<l} c_{4;i,j,k,l} I_4(s_{i,k-1}, s_{j,l-1}; s_{i,j-1}, s_{j,k-1}, s_{k,l-1}, s_{l,i-1}) \tag{1}$$

$$+ \sum_{i<j<k} c_{3;i|k|k} I_3(s_{i,j-1}, s_{j,k-1}, s_{k,i-1}) + \sum_{i<j-1} c_{2;i|j} I_2(s_{i,j-1}) \tag{2}$$

$$- \frac{1}{6} \sum_{i<j<k<l} c^{[4]}_{4;i,j,k,l} - \frac{1}{2} \sum_{i<j<k} c^{[2]}_{3;i,j,k} - \frac{1}{6} \sum_{i<j-1} s_{i,j-1} c^{[2]}_{2;i,j} + \mathcal{O}(\epsilon). \tag{3}$$

Here $s_{i,j} = (p_i + \cdots + p_j)^2$ are the Lorentz invariants of the amplitude. I_4, I_3 and I_2 are the well known scalar integral functions that can be written in terms of logarithms, dilogarithms and the dimensional regulator, ϵ. Thanks to public programs for the numerical evaluation of the scalar integrals for arbitrary kinematics, e.g. FF [30], OneLOop [31] or QCDLoop [32] [4], the only process dependent quantities are the coefficients $c_{k;X}$.

The box coefficients, $c_{4;X}$, can be extracted by applying maximal cuts to the one-loop primitive amplitude, factorises into a product of four tree-level amplitudes. The lower point integral coefficients are then computed systematically by considering fewer cuts and subtracting the contribution from the higher point functions previously evaluated. In each case the integrand can be parametrised by a polynomial of the scalar products between the loop momentum and spurious vectors that can be neatly described by the van Neerven-Vermaseren basis. The construction in `NGluon` follows the description of [33]. The coefficients of the aforementioned polynomial can be efficiently computed using a discrete Fourier projection.

For the rational terms we need to consider cuts in five, or more, dimensions. In the Four Dimensional Helicity (FDH) scheme this can be achieved using a mass-shifted representation of the amplitude, where the coefficients $c_4^{[4]}, c_3^{[2]}$ and $c_2^{[2]}$ can be extracted from a discrete Fourier projection over the additional mass parameter [34]. The mass shifted integrands factorise into tree-level amplitudes with massive scalar particles and massive fermions.

3. Extension to Multiple Fermion

The extension of the pure gluonic case to multiple fermions has basically two fundamental new features:

(i) more complicated tree-level amplitudes

(ii) restrictions when sewing tree-level amplitudes together to reconstruct the integrand

In the pure gluonic case, all tree-level amplitudes that can occur in a unitarity cut can be assigned a unique colour ordering. This means that they can always be computed with the well known Berends-Giele recursion relations [36]. The simple reason for this is the fact that restoring the colour to the primitive amplitude both external legs and loop lines live in the adjoint representation.

With the inclusion of quarks in the fundamental representation, this one-to-one correspondence is in general lost. It happens that through unitarity cuts basic building blocks

[4] QCDLoop is the default choice in `NGluon`

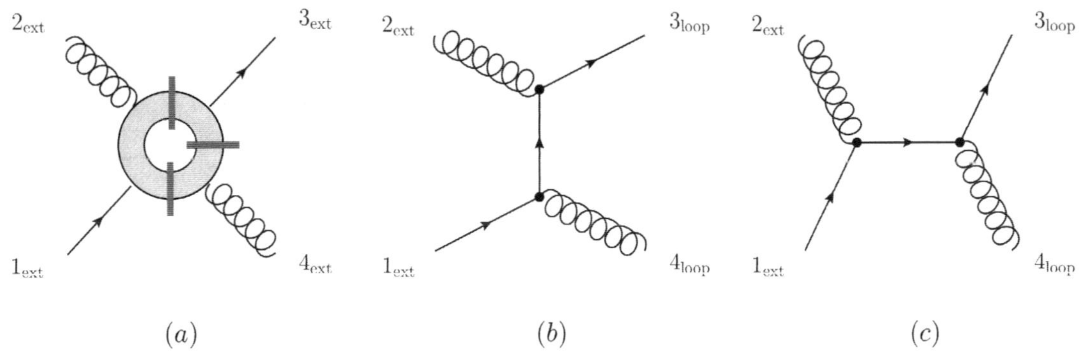

Figure 1. A triple cut of the primitive amplitude $A^{\text{prim}}(\overline{Q}_1; g_2; Q_3; g_4)$ as shown in (a) with two three-point tree amplitudes and one four-point tree amplitude $A^{\text{tree}}(\overline{Q}_{1\,\text{ext}}; g_{2\,\text{ext}}; Q_{3\,\text{loop}}; g_{4\,\text{loop}})$. The latter consists of the above two ordered tree diagrams (b) and (c). "ext" resp. "loop" denote external legs, resp. on-shell loop legs. Note the absence of any gluon self interactions.

appear that do not belong to any colour structure which would occur in the computation of a Born amplitude. This can be seen, for instance, at those quark-gluon primitive amplitudes contributing solely to the subleading colour part of the full amplitude. As an example, consider the primitive amplitude $A^{\text{prim}}(\overline{Q}_1; g_2; Q_3; g_4)$ where the fixed ordering of external particles is such that a quark anti-quark pair is separated by one gluon. In certain bubble and triangle cuts, for example, among others, the following building block arises: $A^{\text{tree}}(\overline{Q}_1; g_2; Q_{\text{loop}}; g_{\text{loop}})$ as illustrated in figure 1. In $A^{\text{tree}}(\overline{Q}_1; g_2; Q_{\text{loop}}; g_{\text{loop}})$, the non-abelian gluon interactions are missing and in addition, the structure is not colour-ordered. It is possible to compute these building blocks as a linear combination of colour-ordered tree amplitudes. However, in `NGluon` all tree amplitudes are computed directly with help of colour stripped Feynman rules [35]. One has to specify only the external leg order and its particle content. Tree amplitudes are then evaluated independently on whether they can be assigned a definite colour structure or not.

Technically, we follow for the trees an off-shell bottom-up approach. This means that after the order of n external legs with appropriate particle content has been fixed, one starts computing all one-point currents, i.e. polarisation vectors for gluons and spinors for quarks. Subsequently, higher point currents are computed systematically from lower point currents. The n-point amplitude is computed by contracting the $(n-1)$-point current with the nth one-point current. This procedure has the advantage that many currents can be reused during the computation. The most complicated vertex, the four gluon vertex, dictates the asymptotic scaling behaviour, which can be shown to be polynomial of order $O(n^4)$.

In the context of generalised unitarity, an additional reduction in complexity can be achieved: For different topologies, those currents that involve exclusively external legs never change and can be computed once in the beginning for all possible cuts. An amplitude evaluation is then equivalent to joining new currents that involve loop legs with external, already known currents. This procedure reduces the polynomial scaling even further to order $O(n^3)$ [24].

In the highly symmetric pure gluonic case, every cut is possible. In other words, there is always a loop propagator which connects directly two external legs. This is not necessarily the case if more then one fermion line is involved in the amplitude. Take for example a closed quark loop and one external $\bar{q}q$-pair: The external quark line will never enter the loop and can't be put on shell.

In order to describe this problem algorithmically, we use the concept of a parent diagram. In our framework, a parent diagram of an n-point amplitude is an abstract one-loop diagram with n fictitious three-point vertices and n propagators. The particle content of its propagators labels the possible on-shell settings. In addition, it knows also which propagators do not exist. The propagator content of the parent diagram is therefore the backbone to compute systematically all integral coefficients. The procedure is exactly like in the pure gluonic case with the exception that if a topology involves a non-existing propagator, the topology is simply skipped.

In order to determine the parent diagram, one draws an abstract loop without specifying any particle content yet and attaches all ordered external legs to it. One starts with the specification of the particle content of an arbitrary initial propagator in the loop and determines with the external leg attached to this propagator the content of the next propagator in the loop. This procedure is repeated for the whole diagram until one hits the initial propagator.

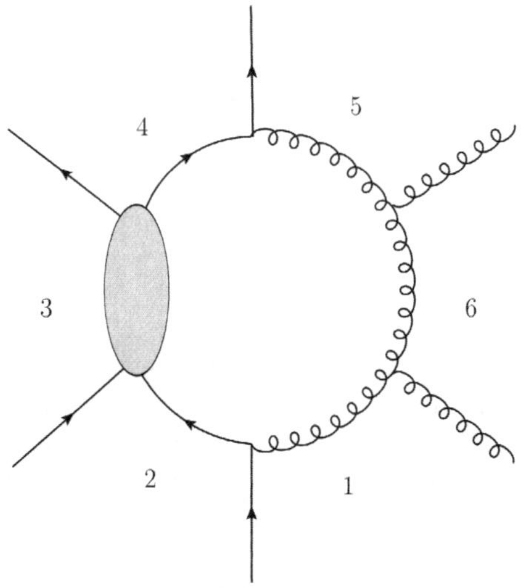

Figure 2. A six-point parent diagram for a primitive with two external fermion lines. The first propagator is initialised with a gluon. The third one is then unphysical and represented by a blob.

One sees immediately that certain propagators of the parent diagram can't be assigned properly in case we are dealing with either 1) a closed fermion loop plus at least one external fermion line or 2) two (or more) fermion lines where a quark-antiquark pair is separated by another quark-antiquark pair. An example is given in figure 2. Literally speaking, this means the following: under the grey blob in figure 2 live two fermion lines which can't be represented by a valid QCD propagator. As soon as the enclosed quark line "leaves" the blob, the parent diagram follows the same pattern as before. With the knowledge of the parent diagram and the external legs, the primitive is uniquely determined.

Another subtlety occurs for certain subleading colour contributions where a distinct cut requires a tree consisting of two fermion lines which belong, however, inside the one-loop amplitude to the same fermion line. Formally, such trees must be treated like two different flavour lines since otherwise one would include contributions that belong to amplitudes with a closed fermion loop, a different class of amplitudes. A simple four-point example is given in figure 3.

4. Performance and Cross Checks

The first detailed verification of the implementation comes from comparison with the well known universal Infra-Red and Ultra-Violet poles in the dimensional regulator, ϵ. This analytic cross check tests the four-dimensional (or cut-constructible) parts of the amplitude and can also give some hints as to a loss of numerical precision.

In order to get an overall estimate for the numerical accuracy, we use the following observation: A re-scaling of the external momenta is equivalent to a simple change of units and should therefore be physically equivalent. Since the floating point arithmetic at the hardware level changes, the difference between the two evaluations — rescaled and un-rescaled — can be used to estimate the numerical uncertainty. The validity of this approach is investigated in detail in [24]. This test is very convenient to detect numerical instabilities and has in addition the advantage that it is independent from any analytic input. Of course, it is not at all suited to check the correctness of the implementation itself and has the drawback of a factor of two in the

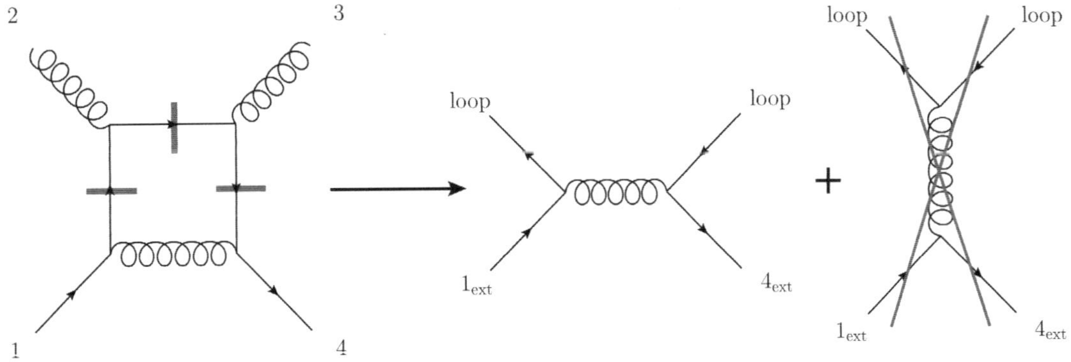

Figure 3. Triple cut with a two fermion tree amplitude. Although at one-loop level, there is only one fermion line, they must be treated at tree-level as two distinct flavours in order to avoid contributions from a closed quark loop (right diagram crossed out).

runtime. In case, the internal accuracy checks do fail, the phase space point can be reprocessed using higher precision with the qd package described in [39].

Besides the poles, we made cross checks against known analytical and numerical results. For the trivial extension of the pure gluonic case to a closed quark loop with exclusively external gluons, we find full agreement with the reference phase space points in [37]. The correctness of the implementation for primitives with one external quark line has been tested intensively against all formulae of the $\bar{q}qggg$ results given in [35]. The agreement of these 5-point cases holds both for different helicity configurations and quark-antiquark separations. For higher n-point functions, we checked the IR-finite amplitudes $A(\bar{q}_1, g_2, ..., q_i, ...g_n)$ with helicity configuration $-++...+$ given analytically in [38] for all possible quark-antiquark separations and find numerical agreement. An example for an 8-point function is shown in figure 4. Only a very small fraction of events show an accuracy of less then 3 digits which is enough for most practical calculations. While the accuracy for different quark-antiquark separations does not differ much in the mixed quark-gluon loop, it does for a closed quark loop. This is simply a consequence of the fermion loop primitive amplitudes having fewer allowed propagators as the $\bar{q}q$-separation increases. This is also the reason why the computation time differs by roughly one order of magnitude depending on the separation of the quark and the antiquark. An example for the runtime difference is shown in table 1. Due to the vanishing of massless tadpoles, the largest two separations in the closed quark loop do not contribute.

For the multiple fermion case, we find agreement in the poles for all primitives that we have tested (up to 5 external quark-antiquark pairs).

When computing the full amplitude out of primitives, many tree-level amplitudes can be reused. The necessary conditions are: 1) the loop momentum for a topology agrees, 2) both loop and external flavours agree and 3) both loop and external helicities agree. It is important to stress that for one primitive with fixed helicity, no information can be reused. The cache system starts first its work when one is dealing with either different permutations of primitive amplitudes or with different helicity amplitudes (if, for example, an average over all helicities is carried out). Take for instance a bubble cut with two trees: flipping one helicity in one tree does not affect the other tree which can be recycled. Similar considerations apply to permutations that affect only trees inside one cut: At any time, the other tree is left untouched and can be reused. Note that we reuse only full tree amplitudes and no off-shell currents. Especially when dealing with permutations, a cache that works on current level could be constructed. The book-keeping for such an approach is however much more involved. As a consequence this possibility

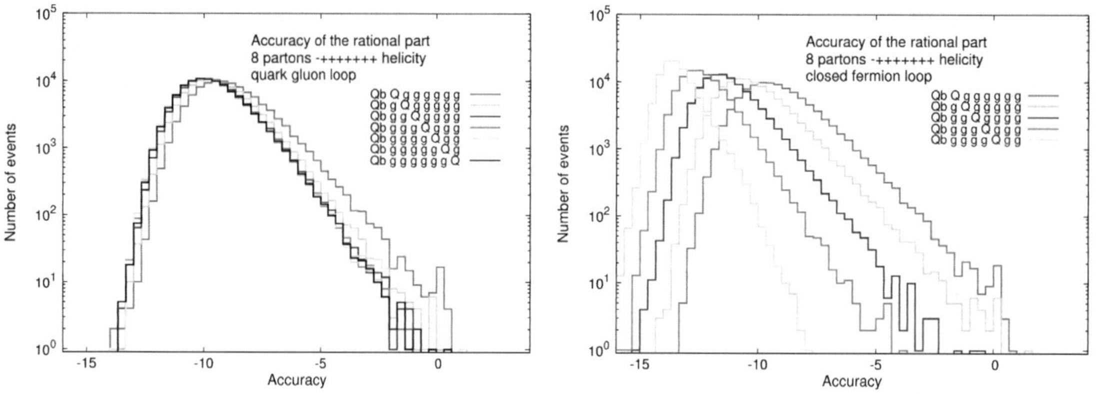

(a) (b)

Figure 4. Accuracy distribution for the rational part of $(\overline{q}q + 6\,g)$-primitives for the mixed quark-gluon loop (a) and the closed quark loop (b). The histograms plot the number of events against the decadic logarithm of the relative accuracy (# of valid digits) between the result from `NGluon` and the analytic formulae in [38] for all possible separations of the two external quark legs.

Table 1. Runtime in milli seconds for primitives with one quark antiquark pair on an Intel(R) Core(TM)2 Duo CPU E8400 @ 3.00GHz processor. While in the mixed quark-gluon loop the runtime increases with larger $\overline{q}q$-separation, the contrary is the case for the closed quark loop.

6-point example:

$\overline{q}q$-separation	0	1	2	3	4
mixed quark-gluon loop	4.85	5.95	7.27	8.77	10.30
closed quark loop	6.13	2.61	0.72	0.01	0.01

8-point example:

$\overline{q}q$-separation	0	1	2	3	4	5	6
mixed quark-gluon loop	19.84	22.08	23.81	27.35	30.54	34.13	38.07
closed quark loop	26.31	14.82	7.57	3.14	0.91	0.03	0.03

is not yet used in `NGluon` .

We have implemented the cache system via a binary tree which is part of the c++ standard template library. Due to the internal index system, the cache is restricted to 13 external legs[5] with six (different flavour) external quark pairs. The asymptotic behaviour is estimated via the computation time of $(n-1)!/2$ permutations of primitives. For low multiplicity (in our case four point amplitudes), there is an overhead using such a system since the stored amplitudes are usually only three-point or at most four point functions and very quickly evaluated. From five external legs on until eight one gains speed up factors between 2 and 3. For higher multiplicities, however, the amount of storage information is so large that the cache starts to slow down the

[5] This is restricted to 6 legs on 32 bit machines

computation compared to the direct evaluation without cache. It is expected that for the leading colour approximation where far less primitives are needed, the performance of the cache is more efficient.

5. Conclusions

`NGluon` is a publicly available program to compute primitive amplitudes with a fixed order of external legs in pure gauge theory. We have shown the extension of the program to an arbitrary number of external massless quarks: For a fixed order of helicity and flavour of the external particles, the primitive amplitude is evaluated in a fully automatic way. As internal accuracy check, we find that the "scaling test" gives reliable estimates. We have made several cross checks with known results from the literature and find numerical agreement with our implementation. An additional cache system which recycles tree amplitudes for the unitarity cut can lead to speed up factors between two and three. First phenomenological applications are in preparation.

References

[1] Bern Z, Dixon L, Dunbar D and Kosower D 1994 *Nucl. Phys. B* **425** 217 (*Preprint* hep-ph/9403226)
[2] Britto R, Cachazo F and Feng B 2005 *Nucl. Phys. B* **725** 275 (*Preprint* hep-th/0412103)
[3] Ossola G, Papadopoulos C and Pittau R 2007 *Nucl. Phys. B* **763** 147 (*Preprint* hep-ph/0609007)
[4] Berger C, Bern Z, Dixon L, Febres Cordero F, Forde D, Ita H, Kosower D and Maitre D, 2008 *Phys. Rev. D* **78** 036003 (*Preprint* 0803.4180)
[5] Berger C, Bern Z, Dixon L, Febres Cordero F, Forde D, Gleisberg T, Ita H, Kosower D and Maitre D 2010 *Phys. Rev. D* **82** 074002 (*Preprint* 1004.1659)
[6] Berger C, Bern Z, Dixon L, Febres Cordero F, Forde D, Gleisberg T, Ita H, Kosower D and Maitre D 2009 *Phys. Rev. D* **80** 074036 (*Preprint* 0907.1984)
[7] Berger C, Bern Z, Dixon L, Febres Cordero F, Forde D, Gleisberg T, Ita H, Kosower D and Maitre D 2009 *Phys. Rev. Lett.* **102** 222001 (*Preprint* 0902.2760)
[8] Greiner N, Guffanti A, Reiter T and Reuter J 2011 *Phys. Rev. Lett.* **107** 102002 (*Preprint* 1105.3624)
[9] Bevilacqua G, Czakon M, Papadopoulos C and Worek M 2010 *Phys. Rev. Lett.* **104** 162002 (*Preprint* 1002.4009)
[10] Denner A, Dittmaier S, Kallweit S and Pozzorini S 2011 *Phys. Rev. Lett.* **106** 052001 (*Preprint* 1012.3975)
[11] Bredenstein A, Denner A, Dittmaier S and Pozzorini S 2009 *Phys. Rev. Lett.* **103** 012002 (*Preprint* 0905.0110)
[12] Melia T, Melnikov K, Rontsch R and Zanderighi G 2011 *Phys. Rev. D* **83** 114043 (*Preprint* 1104.2327)
[13] Frederix R, Frixione S, Melnikov K and Zanderighi G 2010 *J. High Energy Phys.* **1011** 050 (*Preprint* 1008.5313)
[14] Berger C, Bern Z, Dixon L, Febres Cordero F, Forde D, Gleisberg T, Ita H, Kosower D and Maitre D 2011 *Phys. Rev. Lett.* **106** 092001 (*Preprint* 1009.2338)
[15] Ita H, Bern Z, Dixon L, Cordero F, Kosower D and Maitre D 2011 *Phys. Rev. D* **85** 031501 (*Preprint* 1108.2229)
[16] Giele W and Zanderighi G 2008 *J. High Energy Phys.* **0806** 038 (*Preprint* 0805.2152)
[17] Lazopoulos A 2008 *Preprint* 0812.2998
[18] Giele W, Kunszt Z and Winter J 2010 *Nucl. Phys. B* **840** 214 (*Preprint* 0911.1962)
[19] Mastrolia P, Ossola G, Reiter T and Tramontano F 2010 *J. High Energy Phys.* **1008** 080 (*Preprint* 1006.0710)
[20] Hirschi V, Frederix R, Frixione S, Garzelli M, Maltoni F and Pittau R 2011 *J. High Energy Phys.* **1105** 044 (*Preprint* 1103.0621)
[21] Bevilacqua G, Czakon M, Garzelli M, van Hameren A, Kardos A, Papadopoulos C, Pittau R and Worek M 2011 *Preprint* 1110.1499
[22] Cullen G, Greiner N, Heinrich G, Luisoni G, Mastrolia P, Ossola G, Reiter T and Tramontano F 2012 *Eur. Phys. J. C* **72** 1889 (*Preprint* 1111.2034)
[23] Hirschi V 2011 *Preprint* 1111.2708
[24] Badger S, Biedermann B and Uwer P 2011 *Comput. Phys. Commun.* **182** 1674 (*Preprint* 1011.2900)
[25] Giele W, Kunszt Z and Melnikov K 2008 *J. High Energy Phys.* **0804** 049 (*Preprint* 0801.2237)
[26] Ellis R, Giele W and Kunszt Z 2008 *J. High Energy Phys.* **0803** 003 (*Preprint* 0708.2398)
[27] Britto R 2011 *J. Phys. A* **A44** 454006 (*Preprint* 1012.4493)
[28] Ellis R, Kunszt Z, Melnikov K and Zanderighi G 2011 *Preprint* 1105.4319
[29] Ita H 2011 *J. Phys. A* **44** 454005 (*Preprint* 1109.6527)
[30] van Oldenborgh G 1991 *Comput. Phys. Commun.* **66** 1

[31] van Hameren A 2011 *Comput. Phys. Commun.* **182** 2427 (*Preprint* 1007.4716)
[32] Ellis R and Zanderighi G 2008 *J. High Energy Phys.* **0802** 002 (*Preprint* 0712.1851)
[33] Ellis R, Giele W and Kunszt Z 2008 *J. High Energy Phys.* **0803** 003 (*Preprint* 0708.2398)
[34] Badger S 2009 *J. High Energy Phys.* **0901** 049 (*Preprint* 0806.4600)
[35] Bern Z, Dixon L and Kosower D 1995 *Nucl. Phys. B* **437** 259 (*Preprint* hep-ph/9409393)
[36] Berends F and Giele W 1988 *Nucl. Phys. B* **306** 759
[37] van Hameren A 2009 *J. High Energy Phys.* **0907** 088 (*Preprint* 0905.1005)
[38] Bern Z, Dixon L and Kosower D 2005 *Phys. Rev. D* **72** 125003 (*Preprint* hep-ph/0505055)
[39] Hida Y, Li X and Bailey D 2008 *Lawrence Berkeley National Laboratory Technical Report* LBNL-46996 http://crd-legacy.lbl.gov/~dhbailey/mpdist/

GoSam: A program for automated one-loop calculations

G Cullen[1], N Greiner[2], G Heinrich[2], G Luisoni[3], P Mastrolia[2,4], G Ossola[5], T Reiter[2], F Tramontano[6]

[1]Deutsches Elektronen-Synchrotron DESY, Zeuthen, Germany
[2]Max-Planck-Institute for Physics, Munich, Germany
[3]Institute for Particle Physics Phenomenology, University of Durham, UK
[4]Dipartimento di Fisica, Universita di Padova, Italy
[5]New York City College of Technology, City University of New York
[6]CERN, Geneva, Switzerland

Abstract. The program package GoSam is presented which aims at the automated calculation of one-loop amplitudes for multi-particle processes. The amplitudes are generated in terms of Feynman diagrams and can be reduced using either D-dimensional integrand-level decomposition or tensor reduction, or a combination of both. GoSam can be used to calculate one-loop corrections to both QCD and electroweak theory, and model files for theories Beyond the Standard Model can be linked as well. A standard interface to programs calculating real radiation is also included. The flexibility of the program is demonstrated by various examples.

1. Introduction

Precise theory predictions play an important role in the analysis and the interpretation of collider physics data, in the search for new particles as well as to constrain model parameters at a later stage. Therefore it is desirable to have predictions at next-to-leading order (NLO) in perturbation theory as a standard at the LHC. However, this can only be achieved if a level of automation for NLO predictions is reached which comes close to the one we have for leading order tools.

The need for an automation of NLO calculations has been noticed some time ago and lead to public programs like FeynArts [1] and QGraf [2] for diagram generation and FormCalc/LoopTools [3] and GRACE [4] for the automated calculation of NLO corrections. However, calculations of one-loop amplitudes with more than four external legs were still tedious case-by-case calculations. Only very recently, conceptual and technical advances in multi-leg one-loop calculations opened the door to the possibility of an *automated* generation and evaluation of multi-leg one-loop amplitudes. As a consequence, public NLO tools containing a collection of hard-coded individual processes, like e.g. MCFM [5, 6], VBFNLO [7, 8, 9], MC@NLO [10, 11], POWHEG-Box [12, 13, 14], POWHEL [15, 16, 17], are now being complemented by flexible automated tools [18, 19, 20, 21, 22] which have as their aim that basically any process which may turn out to be important for the comparison of LHC findings to theory can be evaluated at NLO accuracy.

In this talk, we present the program package GoSam[18] which allows the automated calculation of one-loop amplitudes for multi-particle processes. Amplitudes are expressed in terms of Feynman diagrams, where the integrand is generated analytically using QGRAF [2], FORM [23], spinney [24] and haggies [25]. The individual program tasks are steered via python scripts, while the user only needs to edit an "input card" to specify the details of the process to be calculated, and launch the generation of the source code and its compilation, without having to worry about internal details of the code.

The program offers the option to use different reduction techniques: either the unitarity-based integrand reduction [26, 27] as implemented in SAMURAI [28] or traditional tensor reduction as implemented in golem95C [29, 30] interfaced through tensorial reconstruction at the integrand level [31], or a combination of both. It can be used to calculate one-loop corrections within both QCD and electroweak theory. Beyond the Standard Model theories can be interfaced using FeynRules [32] or LanHEP [33]. The Binoth Les Houches interface [34] to programs providing the real radiation contributions is also included.

2. Theoretical framework

2.1. Generation of the diagrams

For the diagram generation we use the program QGRAF [2] and supplement it by adding another level of analyzing and filtering the diagrams, written in Python. This allows for example to drop diagrams whose colour factor turns out to be zero, or to determine the signs for diagrams with Majorana fermions.

Within our framework, QGRAF generates three sets of output files: an expression for each diagram to be processed with FORM [23], Python code to draw all diagrams, and Python code to compute the properties of each diagram. The information about the model is either read from the built-in Standard Model file or is generated from a user defined LanHEP [33] or Universal FeynRules Output (UFO) [32] file.

The program also produces a LaTeX file which contains graphical representations of all diagrams as well as a summary of the helicity and colour basis used.

2.2. Code generation

The amplitude is generated in terms of algebraic expressions based on Feynman diagrams and then processed with a FORM program, using spinney [24] for the spinor algebra. In GoSam we have implemented the 't Hooft-Veltman scheme (HV) and dimensional reduction (DRED). In both schemes all external vectors (momenta and polarisation vectors) are kept in four dimensions, while internal vectors are kept in the n-dimensional vector space ($n = 4 - 2\epsilon$). We adopt the conventions used in [24], where $\hat{k}$ denotes the four-dimensional projection of an n-dimensional vector k. The $(n-4)$-dimensional orthogonal projection is denoted by $\tilde{k}$. For the loop momentum q we introduce the symbol $\mu^2 = -\tilde{q}^2$, such that

$$q^2 = \hat{q}^2 + \tilde{q}^2 = \hat{q}^2 - \mu^2. \tag{1}$$

To prepare the numerator functions of the one-loop diagrams for their numerical evaluation, we separate the symbol μ^2 and dot products involving the momentum $\hat{q}$ from all other factors. All subexpressions which do not depend on either $\hat{q}$ or μ^2 are substituted by abbreviations, which are evaluated only once per phase space point. Each of the two parts is then processed by haggies [25], which generates optimised Fortran95 code for the numerical evaluation. For each diagram we generate an interface to SAMURAI [28], golem95C [30] and/or PJFRY [35, 36].

Our standard choice for the reduction is to use SAMURAI [28], which provides a very fast and stable reduction in a large part of the phase space. Furthermore, SAMURAI reports to the client code if the quality of the reconstruction of the numerator suffices the numerical requirements. In GoSam we use this information to trigger an alternative reduction with either golem95C [30] or PJFRY [35] whenever these reconstruction tests fail. This combination of on-shell techniques and traditional tensor reduction is achieved using tensorial reconstruction at the integrand level [31]. The tensorial reconstruction not only can cure numerical instabilities, but in some cases also can reduce the computational cost of the reduction. Since the reconstructed numerator is typically of a form where kinematics and loop momentum dependence are already separated, the use of a reconstructed numerator tends to be faster than the original procedure, in particular in cases with a large number of legs and low rank.

2.3. Rational terms

Terms containing the symbols μ^2 or ε in the numerator of the integrands can lead to a so-called R_2 term [37], which contributes to the rational part of the amplitude. We generate the R_2 part along with all other contributions without the need to separate the different parts. In addition, we provide an *implicit* and an *explicit* construction of the R_2 terms, using the fact that there are two ways of splitting the numerator function:

$$\mathcal{N}(\hat{q},\mu^2,\varepsilon) = \mathcal{N}_0(\hat{q},\mu^2) + \varepsilon\mathcal{N}_1(\hat{q},\mu^2) + \varepsilon^2\mathcal{N}_2(\hat{q},\mu^2) \tag{2}$$

or, alternatively,

$$\mathcal{N}(\hat{q},\mu^2,\varepsilon) = \hat{\mathcal{N}}(\hat{q}) + \tilde{\mathcal{N}}(\hat{q},\mu^2,\varepsilon). \tag{3}$$

The implicit construction uses the splitting of equation (2) and treats all three numerator functions $\mathcal{N}_i$ on equal grounds. Each of the three terms is reduced separately in a numerical reduction and the Laurent series of the three results are added up taking into account the powers of ε.

The explicit construction of R_2 is based on the assumption that each term in $\tilde{\mathcal{N}}$ in equation (3) contains at least one power of μ^2 or ε. The expressions for those integrals are relatively simple and known explicitly. Hence, the part of the amplitude which originates from $\tilde{\mathcal{N}}$ is computed analytically whereas the purely four-dimensional part $\hat{\mathcal{N}}$ is passed to the numerical reduction.

In the program, possible options for R_2 are `r2=implicit,explicit`, `off` and `only`. Using `r2=only` discards everything but the R_2 term and puts GoSAM in the position of providing R_2 terms to supplement other codes which work entirely in four dimensions.

2.4. Conventions

To be specific, we consider the case where the user wants to compute QCD corrections. In the case of electroweak corrections, the analogous conventions apply except that the strong coupling g_s is replaced by e. In the QCD case, the tree-level matrix element squared can be written as

$$|\mathcal{M}|^2_{\text{tree}} = \mathcal{A}_0^\dagger\mathcal{A}_0 = (g_s)^{2b}\cdot a_0\,, \tag{4}$$

where $b=0$ is also possible. The matrix element at one-loop level, i.e. the interference term between tree-level and one-loop, can be written as

$$|\mathcal{M}|^2_{\text{1-loop}} = \mathcal{A}_1^\dagger\mathcal{A}_0 + \mathcal{A}_0^\dagger\mathcal{A}_1 = 2\cdot\Re(\mathcal{A}_0^\dagger\mathcal{A}_1) = \frac{\alpha_s(\mu)}{2\pi}\frac{(4\pi)^\varepsilon}{\Gamma(1-\varepsilon)}(g_s)^{2b}\left[c_0 + \frac{c_{-1}}{\varepsilon} + \frac{c_{-2}}{\varepsilon^2} + \mathcal{O}(\varepsilon)\right]\,. \tag{5}$$

A call to the subroutine `samplitude` returns an array consisting of the four numbers $(a_0, c_0, c_{-1}, c_{-2})$ in this order. The average over initial state colours and helicities is included in the default setup. Renormalisation is included depending on the options chosen by the user, for a more detailed description we refer to [18]. In cases where the process is loop induced, i.e. the tree level amplitude is absent, the program returns the values for $\mathcal{A}_1^\dagger\mathcal{A}_1$, where a factor

$$\left(\frac{\alpha_s(\mu)}{2\pi}\frac{(4\pi)^\varepsilon}{\Gamma(1-\varepsilon)}\right)^2$$

has been extracted.

Table 1. Part of the process card defining the initial and final state and the type and order of the coupling at LO and NLO.

process_path=	qqgz
in=	u,u~
out=	g, e-, e+
helicities=	$+-+-+, +---+, -++-+, -+--+$
order=	QCD, 1, 3

3. Installation and Usage

3.1. Installation

The user can download the code GoSam either as a tar-ball or from the subversion repository at

`http://projects.hepforge.org/gosam/` .

The build process and installation of GoSam is controlled by `Python Distutils`, while the build process for the libraries Samurai and `golem95C` is controlled by `Autotools`. To install GoSam, the user needs to run

```
python setup.py install --prefix MYPATH
```

If `MYPATH` is different from the system default, the environment variables `PATH`, `LD_LIBRARY_PATH` and `PYTHONPATH` might have to be set accordingly. For more details we direct the user to the GoSam reference manual coming with the code.

The program is designed to run in any modern Linux/Unix environment; we expect that `Python` (≥ 2.6), Java (≥ 1.5) and Make are installed on the system. Furthermore, a Fortran 95 compiler is required in order to compile the generated code.

On top of a standard Linux environment, the programs `FORM` [23], version $\geq$ 3.3, and `QGRAF` [2] need to be installed on the system. Further, at least one of the libraries Samurai [28] and `golem95C` [30] needs to be present at compile time of the generated code. For the user's convenience we have prepared a package containing Samurai and `golem95C` together with the integral libraries `OneLOop` [38], `QCDLoop` [39] and `FF` [40]. The package `gosam-contrib-1.0.tar.gz` containing all these libraries is available for download from `http://projects.hepforge.org/gosam/`.

3.2. Usage

In order to generate the code for a process, the user needs to prepare an input file, called *process card* in the following, which contains

- process specific information, such as a list of initial and final state particles, their helicities (optional) and the order of the coupling constants;
- scheme specific information and approximations, such as the regularisation and renormalisation schemes, the underlying model, masses and widths which are set to zero, the selection of subsets of diagrams, etc;
- system specific information, such as paths to programs and libraries or compiler options;
- optional information for optimisations which control the code generation.

For example, to compute the QCD corrections for the process $u\bar{u} \to gZ^0 \to g\, e^- e^+$, the first part of the process card looks as shown in table 1: The first line defines the (relative) path to the directory where the process files will be generated. GoSam expects that this directory has already been created. Lines 2 and 3 define the initial and final state of the process in terms of field names, which are defined in the model file. Besides the field names one can also use PDG codes [41, 42] instead. For more details we refer to [18] and the reference manual.

3.3. Code generation

If the process card is called `gosam.in`, it can be invoked by `gosam.py`:

```
mkdir qqgz
gosam.py gosam.in
cd qqgz
```

All further steps are controlled by the generated make files; in order to generate and compile all files relevant for the matrix element one needs to invoke

```
make compile
```

The generated code can be tested with the program `matrix/test.f90`. The following sequence of commands will compile and run the test program:

```
cd matrix
make test.exe
./test.exe
```

The last lines of the program output should look as follows.

```
#              LO:   0.3450350717601E-06
# NLO, finite part    -10.77604823456547
# NLO, single pole    -19.98478948141949
# NLO, double pole    -5.666666665861926
# IR,  single pole    -19.98478948439310
# IR,  double pole    -5.666666666666666
```

The printed numbers are, in this order, a_0, c_0/a_0, c_{-1}/a_0, c_{-2}/a_0 and the pole parts calculated from the infrared insertion operator [43, 44]. One can also generate a pictorial representation of all generated diagrams using the command

```
make doc
```

which generates the file `doc/process.ps`

3.4. Interfacing the code

3.4.1. Using the BLHA Interface The so-called *Binoth Les Houches Accord* (BLHA) [34] defines an interface for a standardized communication between one-loop programs (OLP) and Monte Carlo (MC) tools. GoSam can act as an OLP in the framework of the BLHA, such that the calculation of complete cross sections is straightforward.

In general, the MC writes an order file, called for example `olp_order.lh`, and invokes the script `gosam.py` as follows:

```
gosam.py --olp olp_order.lh
```

The invocation of `gosam.py` generates a set of files which can be compiled with a generated make file. The BLHA routines are defined in the `Fortran` module `olp_module` but can also be accessed from `C` programs.

3.4.2. Using External Model Files GoSAM can also make use of model files generated by either `FeynRules` [45] in the `UFO` format [32] or by LanHEP [33]. The particles can be specified by their PDG code. Details about how to import the model files are described in the GoSAM reference manual. Precompiled `MSSM_UFO` and `MSSM_LHEP` files can also be found in the subdirectory `examples/model`. The `examples` directory also contains examples where `UFO` or LanHEP model files are imported.

4. Examples

The code generated by GoSAM has been compared to a considerable number of processes available in the literature, as listed in table 2. Many of these processes are also included as examples in the code, including reference values.

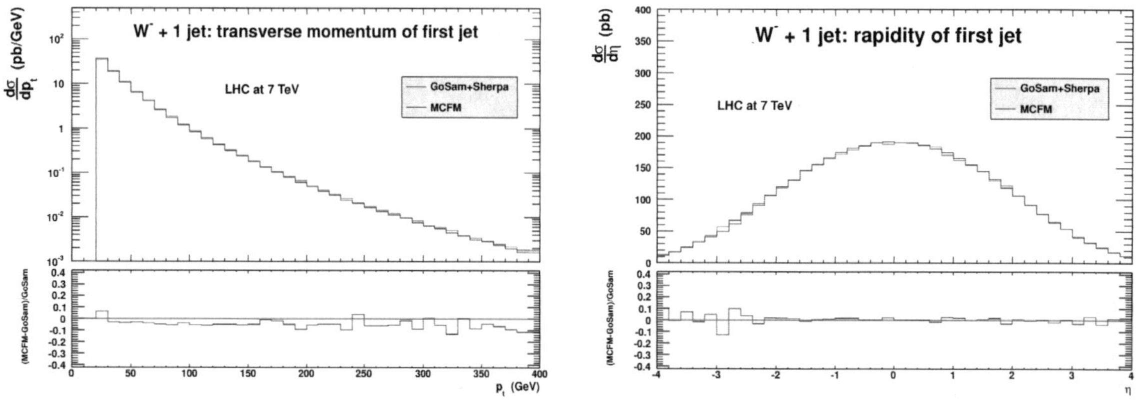

Figure 1. Transverse momentum and pseudo-rapidity distributions of the leading jet for W^-+jet production at LHC with $\sqrt{s} = 7$ TeV, based on an NLO calculation using GoSAM interfaced with `SHERPA` via the Binoth Les Houches interface and compared to MCFM.

As an example for the *Binoth Les Houches Accord* interface of GoSAM we present results for the QCD corrections to $W^- + 1$ jet production, obtained by linking GoSAM with `SHERPA` [61]. Furthermore, `SHERPA` offers the possibility to match NLO calculations with a parton shower [62, 63]. Results for the transverse momentum and rapidity distribution of the leading jet are shown in figure 1. The comparison with MCFM [5, 6] shows perfect agreement.

5. Conclusions

We have presented the program GoSAM which can produce code for the evaluation of one-loop matrix elements for multi-particle processes in an automated way. The program is publicly available at `http://projects.hepforge.org/gosam/` and can be used to calculate one-loop amplitudes within QCD, electroweak theory, or other models which can be imported via an interface to LanHEP or FeynRules. Monte Carlo programs for the real radiation can be easily linked via the interface defined by the *Binoth Les Houches Accord*.

The amplitudes are generated in terms of Feynman diagrams and can be reduced by unitarity based reduction at integrand level or traditional tensor reduction, or a combination of the two approaches. This feature makes the program extremely robust against numerical instabilities. The user can choose among different libraries for the master integrals, and the setup is such that other libraries can be linked easily.

The calculation of the rational terms can proceed either together with the same numerical reduction as the rest of the amplitude, or before any reduction, using analytic information on

Table 2. Processes for which GoSam has been compared to the literature.

process	**checked with Ref.**
$e^+e^- \to u\bar{u}$	[46]
$e^+e^- \to t\bar{t}$	[47, 48], own analytic calculation
$u\bar{u} \to d\bar{d}$	[49, 20]
$gg \to gg$	[50]
$gg \to gZ$	[51]
$d\bar{d} \to t\bar{t}$	[20], MCFM [5, 52]
$gg \to t\bar{t}$	[20], MCFM [5, 52]
$bg \to H\,b$	[53, 20]
$\gamma\gamma \to \gamma\gamma$	[54]
$u\bar{d} \to e^-\bar{\nu}_e$	[20]
$u\bar{d} \to e^-\bar{\nu}_e\,g$	[20]
$e^+e^- \to e^+e^-\gamma$ (QED)	[55]
$pp \to H\,t\bar{t}$	[20]
$pp \to W^+W^+jj$	[56, v3]
$pp \to W^\pm j$ (QCD corr.)	MCFM [5, 52]
$pp \to W^\pm j$ (EW corr.)	for IR poles: equations (67),(70) of [57], [58]
$pp \to b\bar{b}b\bar{b}$	[59, 60]
$pp \to W^+W^-b\bar{b}$	[19, 20]
$u\bar{u} \to t\bar{t}b\bar{b}$	[19, 20]
$gg \to t\bar{t}b\bar{b}$	[19, 20]
$u\bar{d} \to W^+ggg$	[19]

the integrals which can potentially give rise to a rational part. Moreover, the GoSam generator can produce code for processes which include unstable particles, i.e. intermediate states with complex masses.

GoSam is very well suited for the automated matching of Monte Carlo programs to NLO virtual amplitudes, and therefore can be used as a module to produce differential cross sections for multi-particle processes which can be compared directly to experiment. Thus GoSam can contribute to the goal of automating NLO corrections to an extent where NLO tools become a standard for data analysis at the LHC.

Acknowledgements

G.C. and G.L. were supported by the British Science and Technology Facilities Council (STFC). The work of G.C. was supported by DFG Sonderforschungsbereich Transregio 9, Computergestützte Theoretische Teilchenphysik. N.G. was supported in part by the U.S. Department of Energy under contract No. DE-FG02-91ER40677. P.M. and T.R. were supported by the Alexander von Humboldt Foundation, in the framework of the Sofja Kovaleskaja Award Project "Advanced Mathematical Methods for Particle Physics", endowed by the German Federal Ministry of Education and Research. The work of G.O. was supported in part by the National Science Foundation under Grant PHY-0855489 and PHY-1068550. The research of F.T. is supported by Marie-Curie-IEF, project: "SAMURAI-Apps". We also acknowledge the support of the Research Executive Agency (REA) of the European Union under the Grant Agreement number PITN-GA-2010-264564 (LHCPhenoNet).

References

[1] Hahn T 2001 *Comput.Phys.Commun.* **140** 418–431 (*Preprint* hep-ph/0012260)
[2] Nogueira P 1993 *J.Comput.Phys.* **105** 279–289
[3] Hahn T and Perez-Victoria M 1999 *Comput.Phys.Commun.* **118** 153–165 (*Preprint* hep-ph/9807565)
[4] Belanger G, Boudjema F, Fujimoto J, Ishikawa T, Kaneko T *et al.* 2006 *Phys.Rept.* **430** 117–209 (*Preprint* hep-ph/0308080)
[5] Campbell J M and Ellis R 1999 *Phys.Rev.* **D60** 113006 (*Preprint* hep-ph/9905386)
[6] Campbell J M, Ellis R and Williams C 2011 *JHEP* **1107** 018 (*Preprint* 1105.0020)
[7] Arnold K, Bahr M, Bozzi G, Campanario F, Englert C *et al.* 2009 *Comput.Phys.Commun.* **180** 1661–1670 (*Preprint* 0811.4559)
[8] Arnold K, Bellm J, Bozzi G, Brieg M, Campanario F *et al.* 2011 (*Preprint* 1107.4038)
[9] Campanario F 2011 *JHEP* **1110** 070 (*Preprint* 1105.0920)
[10] Frixione S and Webber B R 2002 *JHEP* **0206** 029 (*Preprint* hep-ph/0204244)
[11] Frixione S, Stoeckli F, Torrielli P, Webber B R and White C D 2010 (*Preprint* 1010.0819)
[12] Frixione S, Nason P and Oleari C 2007 *JHEP* **0711** 070 (*Preprint* 0709.2092)
[13] Alioli S, Nason P, Oleari C and Re E 2010 *JHEP* **1006** 043 (*Preprint* 1002.2581)
[14] Alioli S, Moch S O and Uwer P 2011, (*Preprint* 1110.5251)
[15] Kardos A, Papadopoulos C and Trocsanyi Z 2011 *Phys.Lett.* **B705** 76–81 (*Preprint* 1101.2672)
[16] Garzelli M, Kardos A, Papadopoulos C and Trocsanyi Z 2011 *Europhys.Lett.* **96** 11001 (*Preprint* 1108.0387)
[17] Kardos A, Papadopoulos C and Trocsanyi Z 2011 (*Preprint* 1111.0610)
[18] Cullen G, Greiner N, Heinrich G, Luisoni G, Mastrolia P *et al.* 2011 (*Preprint* 1111.2034)
[19] van Hameren A, Papadopoulos C and Pittau R 2009 *JHEP* **0909** 106 (*Preprint* 0903.4665)
[20] Hirschi V, Frederix R, Frixione S, Garzelli M V, Maltoni F *et al.* 2011 *JHEP* **1105** 044 (*Preprint* 1103.0621)
[21] Bevilacqua G, Czakon M, Garzelli M, van Hameren A, Kardos A *et al.* 2011 (*Preprint* 1110.1499)
[22] Hahn T, *these proceedings.*
[23] Vermaseren J 2000 (*Preprint* math-ph/0010025)
[24] Cullen G, Koch-Janusz M and Reiter T 2011 *Comput.Phys.Commun.* **182** 2368–2387 (*Preprint* 1008.0803)
[25] Reiter T 2010 *Comput.Phys.Commun.* **181** 1301–1331 (*Preprint* 0907.3714)
[26] Ossola G, Papadopoulos C G and Pittau R 2007 *Nucl.Phys.* **B763** 147–169 (*Preprint* hep-ph/0609007)
[27] Ellis R, Giele W T, Kunszt Z and Melnikov K 2009 *Nucl.Phys.* **B822** 270–282 (*Preprint* 0806.3467)
[28] Mastrolia P, Ossola G, Reiter T and Tramontano F 2010 *JHEP* **1008** 080 (*Preprint* 1006.0710)
[29] Binoth T, Guillet J P, Heinrich G, Pilon E and Reiter T 2009 *Comput.Phys.Commun.* **180** 2317–2330 (*Preprint* 0810.0992)
[30] Cullen G, Guillet J, Heinrich G, Kleinschmidt T, Pilon E *et al.* 2011 *Comput.Phys.Commun.* **182** 2276–2284 (*Preprint* 1101.5595)
[31] Heinrich G, Ossola G, Reiter T and Tramontano F 2010 *JHEP* **1010** 105 (*Preprint* 1008.2441)
[32] Degrande C, Duhr C, Fuks B, Grellscheid D, Mattelaer O *et al.* 2011 (*Preprint* 1108.2040)
[33] Semenov A 2010 (*Preprint* 1005.1909)
[34] Binoth T, Boudjema F, Dissertori G, Lazopoulos A, Denner A *et al.* 2010 *Comput.Phys.Commun.* **181** 1612–1622 (*Preprint* 1001.1307)
[35] Fleischer J and Riemann T 2011 *Phys.Rev.* **D83** 073004 (*Preprint* 1009.4436)
[36] Yundin V, PJFRY, *program available from* https://github.com/Vayu/PJFry/
[37] Ossola G, Papadopoulos C G and Pittau R 2008 *JHEP* **0805** 004 (*Preprint* 0802.1876)
[38] Van Hameren A 2011 *Comput.Phys.Commun.* **182** 2427–2438 (*Preprint* 1007.4716)
[39] Ellis R K and Zanderighi G 2008 *JHEP* **02** 002 (*Preprint* 0712.1851)
[40] Van Oldenborgh G and Vermaseren J 1990 *Z.Phys.* **C46** 425–438
[41] Yost G *et al.* (Particle Data Group) 1988 *Phys.Lett.* **B204** 1–486
[42] Caso C *et al.* (Particle Data Group) 1998 *Eur.Phys.J.* **C3** 1–794
[43] Catani S and Seymour M 1997 *Nucl.Phys.* **B485** 291–419 (*Preprint* hep-ph/9605323)
[44] Catani S, Dittmaier S and Trocsanyi Z 2001 *Phys.Lett.* **B500** 149–160 (*Preprint* hep-ph/0011222)
[45] Christensen N D and Duhr C 2009 *Comput. Phys. Commun.* **180** 1614–1641 (*Preprint* 0806.4194)
[46] Ellis R, Stirling W and Webber B 1996 *Camb.Monogr.Part.Phys.Nucl.Phys.Cosmol.* **8** 1–435
[47] Jersak J, Laermann E and Zerwas P 1982 *Phys.Rev.* **D25** 1218
[48] Catani S, Dittmaier S, Seymour M H and Trocsanyi Z 2002 *Nucl.Phys.* **B627** 189–265 (*Preprint* hep-ph/0201036)
[49] Ellis R and Sexton J 1986 *Nucl.Phys.* **B269** 445
[50] Binoth T, Guillet J and Heinrich G 2007 *JHEP* **0702** 013 (*Preprint* hep-ph/0609054)
[51] van der Bij J and Glover E 1989 *Nucl.Phys.* **B313** 237
[52] Campbell J M and Ellis R K 2000 *Phys. Rev.* **D62** 114012 (*Preprint* hep-ph/0006304)

[53] Campbell J M, Ellis R, Maltoni F and Willenbrock S 2003 *Phys.Rev.* **D67** 095002 (*Preprint* hep-ph/0204093)
[54] Gounaris G, Porfyriadis P and Renard F 1999 *Eur.Phys.J.* **C9** 673–686 (*Preprint* hep-ph/9902230)
[55] Actis S, Mastrolia P and Ossola G 2010 *Phys.Lett.* **B682** 419–427 (*Preprint* 0909.1750)
[56] Melia T, Melnikov K, Rontsch R and Zanderighi G 2010 *JHEP* **1012** 053 (*Preprint* 1007.5313)
[57] Kuhn J H, Kulesza A, Pozzorini S and Schulze M 2008 *Nucl.Phys.* **B797** 27–77 (*Preprint* 0708.0476)
[58] Gehrmann T and Greiner N 2010 *JHEP* **12** 050 (*Preprint* 1011.0321)
[59] Binoth T, Greiner N, Guffanti A, Reuter J, Guillet J P *et al.* 2010 *Phys.Lett.* **B685** 293–296 (*Preprint* 0910.4379)
[60] Greiner N, Guffanti A, Reiter T and Reuter J 2011 *Phys.Rev.Lett.* **107** 102002 (*Preprint* 1105.3624)
[61] Gleisberg T, Höche S, Krauss F, Schönherr M, Schumann S *et al.* 2009 *JHEP* **0902** 007 (*Preprint* 0811.4622)
[62] Höche S, Krauss F, Schönherr M and Siegert F 2011 *JHEP* **1104** 024 (*Preprint* 1008.5399)
[63] Höche S, Krauss F, Schönherr M and Siegert F 2011 (*Preprint* 1111.1220)

One-loop tensor Feynman integral reduction with signed minors

Jochem Fleischer[1], **Tord Riemann**[2] **and Valery Yundin**[3]

[1] Fakultät für Physik, Universität Bielefeld, Universitätsstr. 25, 33615 Bielefeld, Germany
[2] Deutsches Elektronen-Synchrotron, DESY, Platanenallee 6, 15738 Zeuthen, Germany
[3] Niels Bohr International Academy and Discovery Center, Niels Bohr Institute, University of Copenhagen, Blegdamsvej 17, DK-2100, Copenhagen, Denmark

E-mail: `fleischer@physik.uni-bielefeld.de`, `Tord.Riemann@desy.de`, `yundin@nbi.dk`

Abstract. We present an algebraic approach to one-loop tensor integral reduction. The integrals are presented in terms of scalar one- to four-point functions. The reduction is worked out explicitly until five-point functions of rank five. The numerical C++ package PJFry evaluates tensor coefficients in terms of a basis of scalar integrals, which is provided by an external library, e.g. QCDLoop. We shortly describe installation and use of PJFry. Examples for numerical results are shown, including a special treatment for small or vanishing inverse four-point Gram determinants. An extremely efficient application of the formalism is the immediate evaluation of complete contractions of the tensor integrals with external momenta. This leads to the problem of evaluating sums over products of signed minors with scalar products of chords. Chords are differences of external momenta. These sums may be evaluated analytically in a systematic way. The final expressions for the numerical evaluation are then compact combinations of the contributing basic scalar functions.

1. Tensor reductions

The reduction of tensorial Feynman integrals to scalar Feynman integrals is an old technique. A systematic approach has been developed for one- to four-point functions in [1]. The simple case of a self-energy vector

$$\begin{aligned} I_2^\mu &= \int \frac{d^dk}{i\pi^{d/2}} \frac{k^\mu}{[k^2-M_1^2]\,[(k+p)^2-M_2^2]} \\ &= p^\mu B_1 \end{aligned} \tag{1}$$

is evaluated in terms of scalar functions A_0, B_0 as follows:

$$\begin{aligned} p_\mu I_2^\mu &= p^2 B_1(p,M_1,M_2) \\ &= \int \frac{d^dk}{i\pi^{d/2}} \frac{pk}{[k^2-M_1^2]\,[(k+p)^2-M_2^2]} \equiv \int \frac{d^dk}{i\pi^{d/2}} \frac{pk}{D_1\,D_2} \\ &= \frac{1}{2}\int \frac{d^dk}{i\pi^{d/2}} \left[\frac{D_2-(p^2-M_2^2-M_1^2)-D_1}{D_1\,D_2}\right], \end{aligned} \tag{2}$$

$$B_1(p,M_1,M_2) = \frac{1}{2p^2}\left[A_0(M_1)-A_0(M_2)-\frac{1}{2}(p^2-M_2^2-M_1^2)B_0(p,M_1,M_2)\right]. \tag{3}$$

This works fine for many situations, but for higher-point functions and for specific kinematical situations other approaches are more useful. We advocate here Davydychev's approach [2] and represent n-point tensors by n-point scalars in higher dimensions and with higher indices (powers of the propagators $D_i = 1/[(k-q_i)^2 - m_i^2]$). An example is:

$$\begin{aligned} I_5^{\mu\nu\lambda} &= \int \frac{d^dk}{i\pi^{d/2}} \frac{k^\mu k^\nu k^\lambda}{D_1 \cdots D_5} \\ &= -\sum_{i,j,k=1}^{4} q_i^\mu q_j^\nu q_k^\lambda n_{ijk} I_{5,ijk}^{[d+]^3} + \frac{1}{2}\sum_{i=1}^{n-1} (g^{\mu\nu} q_i^\lambda + g^{\mu\lambda} q_i^\nu + g^{\nu\lambda} q_i^\mu) I_{5,i}^{[d+]^2}, \end{aligned} \tag{4}$$

with $n_{ijk} = (1+\delta_{ij})(1+\delta_{ik}+\delta_{jk})$. The $I_{5,ijk}^{[d+]^L}$ is defined here in $D = d - 2\epsilon + 2L$ dimensions and propagators of lines i, j, k have a power raised by one unit. In a next step Tarasov's dimensional recurrences [3, 4] may be applied in order to diminish the dimensions of the scalar integrals:[4]

$$\nu_j (\mathbf{j}^+ I_5^{[d+]}) = \frac{1}{()_5}\left[-\binom{j}{0}_5 + \sum_{k=1}^{5} \binom{j}{k}_5 \mathbf{k}^- \right] I_5, \tag{5}$$

and

$$(d - \sum_{i=1}^{5} \nu_i + 1) I_5^{[d+]} = \frac{1}{()_5}\left[\binom{0}{0}_5 - \sum_{k=1}^{5} \binom{0}{k}_5 \mathbf{k}^- \right] I_5, \tag{6}$$

where the operators $\mathbf{i}^\pm, \mathbf{j}^\pm, \mathbf{k}^\pm$ act by shifting the indices ν_i, ν_j, ν_k by ± 1.

After dedicated simplifications [5, 6], the tensor of rank $R = 3$ may be written as follows:

$$I_5^{\mu\,\nu\,\lambda} = \sum_{i,j,k=1}^{4} q_i^\mu\, q_j^\nu\, q_k^\lambda E_{ijk} + \sum_{k-1}^{4} g^{[\mu\nu} q_k^{\lambda]} E_{00k}, \tag{7}$$

with the tensor coefficients

$$E_{00j} = \sum_{s=1}^{5} \frac{1}{\binom{0}{0}_5} \left[\frac{1}{2} \binom{0s}{0j}_5 I_4^{[d+],s} - \frac{d-1}{3} \binom{s}{j}_5 I_4^{[d+]^2,s} \right], \tag{8}$$

$$E_{ijk} = -\sum_{s=1}^{5} \frac{1}{\binom{0}{0}_5} \left\{ \left[\binom{0j}{sk}_5 I_{4,i}^{[d+]^2,s} + (i \leftrightarrow j) \right] + \binom{0s}{0k}_5 \nu_{ij} I_{4,ij}^{[d+]^2,s} \right\}. \tag{9}$$

At this stage, we have reached:

- All scalar 5-point integrals are eliminated.
- No inverse Gram determinants $()_5$ have appeared.
- Scalar 4-point integrals in higher dimensions and/or with higher indices are present yet, e.g. $I_{4,ij}^{[d+]^2,s}$.
- Inverse Cayley determinants $\binom{0}{0}_5$ have appeared.

The integrals $I_{4,ij}^{[d+]^2,s}$ arise from 5-point integrals by shrinking line s.

[4] The explanation of notations may be found in any of the articles [5, 3, 4, 6].For Gram determinants and signed minors see Appendix A.

By further, nontrivial manipulations [6] we have avoided shifted indices and isolated the inverse Gram determinants $()_4$ in the higher-dimensional scalar 4-point functions; see e.g. in (9):

$$I_{4,i}^{[d+],s} = \frac{1}{\binom{0s}{0s}_5}\left[-\binom{0s}{is}_5 (d-3) I_4^{[d+],s} + \sum_{t=1}^{5} \binom{0st}{0si}_5 I_3^{st}\right], \tag{10}$$

$$\begin{aligned} \nu_{ij} I_{4,ij}^{[d+]^2} &= \frac{\binom{0}{i}_4 \binom{0}{j}_4}{\binom{0}{0}_4 \binom{0}{0}_4}(d-2)(d-1) I_4^{[d+]^2} + \frac{\binom{0i}{0j}_4}{\binom{0}{0}_4} I_4^{[d+]} \\ &- \frac{\binom{0}{j}_4}{\binom{0}{0}_4} \frac{d-2}{\binom{0}{0}_4} \sum_{t=1}^{4} \binom{0t}{0i}_4 I_3^{[d+],t} + \frac{1}{\binom{0}{0}_4} \sum_{t=1}^{4} \binom{0t}{0j}_4 I_{3,i}^{[d+],t}. \end{aligned} \tag{11}$$

These equations contain yet the generic 4-point and (partly indexed) 3-point functions in higher dimensions, $I_4^{[d+],s}$, $I_{3,i}^{[d+],t}$, etc. Several strategies are now possible:

- Evaluate them analytically in $d+2l-2\epsilon$ dimensions – if you may do that.
- Evaluate them numerically in $d+2l-2\epsilon$ dimensions.
- Reduce them further by recurrences $\rightarrow$ apply (6) – but buy some towers of $1/()_4$ at this stage.
- Make a small Gram determinant expansion $\rightarrow$ apply (6) the other way round

The last two items are done here.

For non-small 4-point Gram determinants, the direct, iterative use of (6) yields

$$I_4^{[d+]^l} = \left[\frac{\binom{0}{0}_4}{()_4} I_4^{[d+]^{l-1}} - \sum_{t=1}^{4} \frac{\binom{t}{0}_4}{()_4} I_3^{[d+]^{l-1},t}\right] \frac{1}{d+2l-5}, \tag{12}$$

$$I_3^{[d+]^l,t} = \left[\frac{\binom{0t}{0t}_4}{\binom{t}{t}_4} I_3^{[d+]^{l-1},t} - \sum_{u=1,u\neq t}^{4} \frac{\binom{ut}{0t}_4}{\binom{t}{t}_4} I_2^{[d+]^{l-1},tu}\right] \frac{1}{d+2l-4}, \tag{13}$$

and we are done. This works fine if $()_4$ is not small [and also the $\binom{t}{t}_4$].

For small $()_4$, one may apply the recurrence relation the other way round and get:

$$\begin{aligned} I_n^D &= \sum_k^n \frac{\binom{0}{k}_n}{\binom{0}{0}_n} I_{n-1}^{D,k} \\ &- \frac{()_n}{\binom{0}{0}_n}[(D+1) - \sum_i^n \nu_i] \\ &\times \left[\sum_k^n \frac{\binom{0}{k}_n}{\binom{0}{0}_n} I_{n-1}^{D+2,k} - \frac{()_n}{\binom{0}{0}_n}[(D+3) - \sum_i^n \nu_i] I_n^{D+4}\right] + \cdots \end{aligned} \tag{14}$$

This works in fact fine, especially when improved by Pade approximations. For explicit examples see [6] and section 2 on the PJFry package here. The approach has been independently implemented and used by Reina et al. [7].

2. PJFry

The goal of the C++ package PJFry is a stable and fast open-source implementation of tensor reduction, suitable for any physically relevant kinematics.[5] It performs the reduction of 5-point 1-loop tensor integrals up to rank 5. The 4- and 3-point tensor integrals are obtained as a by-product. Main features are:

- Any combination of (real) internal or external masses
- Automatic selection of optimal formula for each coefficient
- Leading $()_5$ are eliminated in the reduction
- Small $()_4$ are avoided using asymptotic expansions where appropriate
- Cache system for tensor coefficients and signed minors
- Interfaces for C, C++, FORTRAN and Mathematica
- Uses QCDLoop [8, 9] or OneLOop [10] for 4-dim scalar integrals
- Available from the project webpage `https://github.com/Vayu/PJFry/` [11, 6]

The installation of PJFry may be performed following the instructions given at the project webpage.
The project subdirectories are
./src - the library source code
./mlink - the MathLink interface
./examples - the FORTRAN examples of library use, built with make check

A build on Unix/Linux and similar systems is done in a standard way by sequential performing ./configure, make, make install. See the INSTALL file for a detailed description of the ./configure options.

The PJFry is used as one option of the GoSam package [12].

The functions for tensor coeffcients for up to rank-5 pentagon integrals are declared in the Mathematica interface:

```
In:= Names["PJFry‘*"]Names[PJFry]

{A0v0, B0v0, B0v1, B0v2, C0v0, C0v1, C0v2, C0v3, \
D0v0, D0v1, D0v2, D0v3, D0v4, E0v0, E0v1, E0v2, \
E0v3, E0v4, E0v5, GetMu2, SetMu2}

Out= {A0v0, B0v0, B0v1, B0v2, C0v0, C0v1, C0v2, C0v3, \
D0v0, D0v1, D0v2, D0v3, D0v4, E0v0, E0v1, E0v2, \
E0v3, E0v4, E0v5, GetMu2, SetMu2}
```

The syntax is very close to that of e.g. LoopTools/FF:

```
E0v3[i,j,k,p1s,p2s,p3s,p4s,p5s,s12,s23,s34,s45,s15,m1s,m2s,m3s,m4s,m5s,ep=0]
```

where:
i, j, k are indices of the tensor coefficient $(0 < i \leq j \leq k < n)$,
$p1s, p2s, \ldots$ are squared external masses p_i^2,
$s12, s23, \ldots$ are Mandelstam invariants $(p_i + p_j)^2$,
$m1s, m2s, \ldots$ are squared internal masses m_i^2,
$ep = 0, -1, -2$ selects the coefficient of the ϵ-expansion.

[5] The presentation relies partly on: V. Yundin, One loop tensor reduction program PJFRY, talk at meeting of SFB/TR9, 15 Nov. 2011, Aachen, Germany.

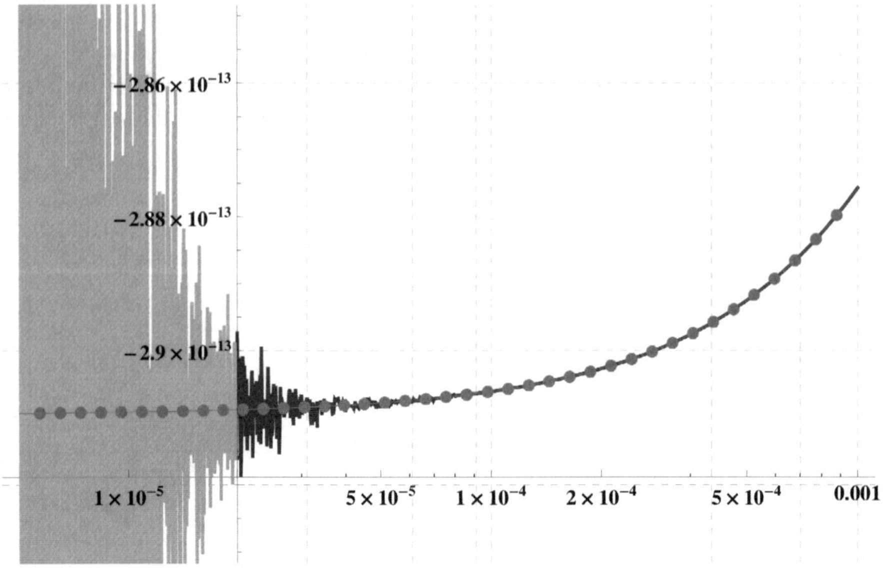

Figure 1. Absolute accuracy of E_{3333} near the region of small Gram determinant. Blue curve: conventional Passarino-Veltman reduction, red curve: PJFry.

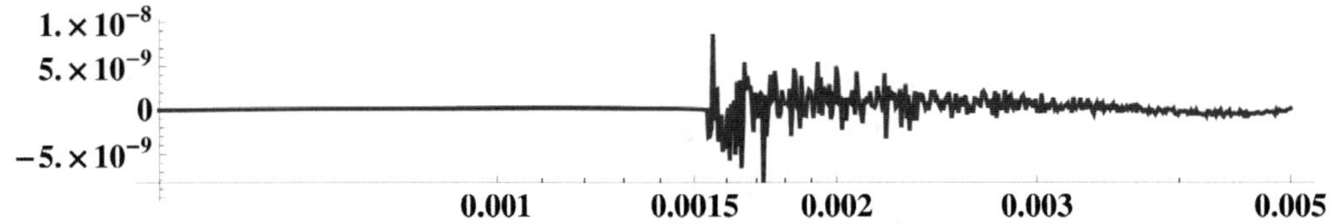

Figure 2. Relative accuracy of E_{3333} near the region of small Gram determinant. At $x \sim 0.0015$ PJFry switched to the asymtotic expansion.

The average evaluation time per phase-space point on a 2 GHz Core 2 laptop for the evaluation of all 81 rank 5 tensor form-factors amounts to 2 ms.

Numerical examples are shown in figures 1 and 2 for a rank $R = 4$ tensor coefficient in a region, where one of the 4-point sub-Gram determinants becomes small

$$E_{3333}(0, 0, -6\times10^4(x+1), 0, 0, 10^4, -3.5\times10^4, 2\times10^4, -4\times10^4, 1.5\times10^4, 0, 6550, 0, 0, 8315)$$

When $x = 0$, the 4-point Gram determinant vanishes; see Appendix A for details.

3. Contractions of tensor integrals with external momenta

The contents of this section is worked out in detail in [13]. A starting point are representations of tensors in terms of basis functions I_n^D, which are independent of the tensor indices $\{i, j, k, ...\}$. Then, one may use contractions with external momenta in order to perform all the sums over $\{i, j, k, ...\}$. This will evidently lead to a significant simplification and shortening of calculations.

To demonstrate the idea, let us look at the most involved 5-point tensor studied:

$$I_5^{\mu\nu\lambda\rho\sigma} = \sum_{s=1}^{5}\left[\sum_{i,j,k,l,m=1}^{5} q_i^\mu q_j^\nu q_k^\lambda q_l^\rho q_m^\sigma E_{ijklm}^s + \sum_{i,j,k=1}^{5} g^{[\mu\nu} q_i^\lambda q_j^\rho q_k^{\sigma]} E_{00ijk}^s + \sum_{i=1}^{5} g^{[\mu\nu} g^{\lambda\rho} q_i^{\sigma]} E_{0000i}^s\right]. \tag{15}$$

In the approach avoiding inverse Gram determinants $1/()_5$, we have

$$E^s_{ijklm} = -\frac{1}{\binom{0}{0}_5}\left\{\left[\binom{0l}{sm}_5 n_{ijk} I^{[d+]^4,s}_{4,ijk} + (i\leftrightarrow l) + (j\leftrightarrow l) + (k\leftrightarrow l)\right] + \binom{0s}{0m}_5 n_{ijkl} I^{[d+]^4,s}_{4,ijkl}\right\} \tag{16}$$

and simpler expressions. Now, in a next step, one may avoid the appearance of inverse sub-Gram determinants $()_4$ and eliminate the higher indices of propagators $\{i,j,k,\ldots\}$. Then, the complete dependence on the indices i of the tensor coefficients is contained in the pre-factors with signed minors: *the tensor indices decouple from the integrals.*

As an example, we reproduce the 4-point part of $I^{[d+]^4}_{4,ijkl}$:

$$\begin{aligned} n_{ijkl} I^{[d+]^4}_{4,ijkl} &= \frac{\binom{0}{i}\binom{0}{j}\binom{0}{k}\binom{0}{l}}{\binom{0}{0}\binom{0}{0}\binom{0}{0}\binom{0}{0}} d(d+1)(d+2)(d+3) I^{[d+]^4}_4 \\ &+ \frac{\binom{0i}{0j}\binom{0}{k}\binom{0}{l} + \binom{0i}{0k}\binom{0}{j}\binom{0}{l} + \binom{0j}{0k}\binom{0}{i}\binom{0}{l} + \binom{0i}{0l}\binom{0}{j}\binom{0}{k} + \binom{0j}{0l}\binom{0}{i}\binom{0}{k} + \binom{0k}{0l}\binom{0}{i}\binom{0}{j}}{\binom{0}{0}^3} d(d+1) I^{[d+]^3}_4 \\ &+ \frac{\binom{0i}{0l}\binom{0j}{0k} + \binom{0j}{0l}\binom{0i}{0k} + \binom{0k}{0l}\binom{0i}{0j}}{\binom{0}{0}^2} I^{[d+]^2}_4 + \cdots \end{aligned} \tag{17}$$

In (17), the 3-point terms are not reproduced, and one has to understand the 4-point integrals to carry the corresponding index s and the signed minors are $\binom{0}{k} \to \binom{0s}{ks}_5$ etc.

A chord is the momentum shift of an internal line due to external momenta, $D_i = (k-q_i)^2 - m_i^2 + i\epsilon$, and $q_i = (p_1 + p_2 + \; + p_i)$, with $q_n = 0$. The tensor 5-point integral of rank $R=1$ yields, when contracted with a chord,

$$q_{a\mu} I^\mu_5 = -\frac{1}{\binom{0}{0}_5}\sum_{s=1}^5\left[\sum_{i=1}^4 (q_a\cdot q_i)\binom{0i}{0s}_5\right] I^s_4. \tag{18}$$

In fact, the sum over i may be performed explicitly:

$$\Sigma^{1,s}_a \equiv \sum_{i=1}^4 (q_a\cdot q_i)\binom{0s}{0i}_5 = +\frac{1}{2}\left\{\binom{s}{0}_5 (Y_{a5} - Y_{55}) + \binom{0}{0}_5 (\delta_{as} - \delta_{5s})\right\}, \tag{19}$$

and we get immediately the desired compact result for the contraction of chords (or external momenta) and tensor integrals:

$$q_{a\mu} I^\mu_5 = -\frac{1}{\binom{0}{0}_5}\sum_{s=1}^5 \Sigma^{1,s}_a \; I^s_4. \tag{20}$$

Similarly, the tensor 5-point integral of rank $R=2$ may be treated:

$$I^{\mu\,\nu}_5 = \sum_{i,j=1}^4 q_i^\mu\, q_j^\nu E_{ij} + g^{\mu\nu} E_{00}, \tag{21}$$

It has the following tensor coefficients free of $1/()_5$:

$$E_{00} = -\sum_{s=1}^5 \frac{1}{2}\frac{1}{\binom{0}{0}_5}\binom{s}{0}_5 I^{[d+],s}_4, \tag{22}$$

$$E_{ij} = \sum_{s=1}^5 \frac{1}{\binom{0}{0}_5}\left[\binom{0i}{sj}_5 I^{[d+],s}_4 + \binom{0s}{0j}_5 I^{[d+],s}_{4,i}\right]. \tag{23}$$

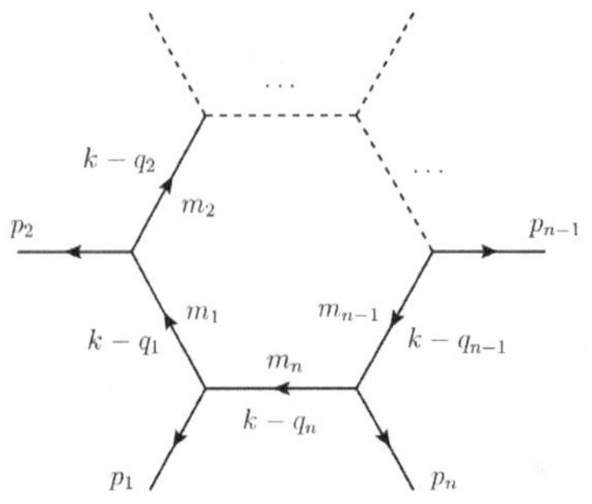

Figure 3. Momenta definitions.

Equation (21) yields for the contractions with chords:

$$q_{a\mu}q_{b\nu}I_5^{\mu\,\nu} = \sum_{i,j=1}^{4}(q_a\cdot q_i)(q_b\cdot q_j)E_{ij} + (q_a\cdot q_b)E_{00}, \tag{24}$$

and finally (24) simply reads

$$\begin{aligned} q_{a\mu}q_{b\nu}I_5^{\mu\,\nu} &= \frac{1}{4}\sum_{s=1}^{5}\left\{\frac{\binom{s}{0}_5}{\binom{0s}{0s}_5}(\delta_{ab}\delta_{as}+\delta_{5s}) + \frac{\binom{s}{s}_5}{\binom{0s}{0s}_5}\Big[(\delta_{as}-\delta_{5s})(Y_{b5}-Y_{55})\right. \\ &\left. + \ (\delta_{bs}-\delta_{5s})(Y_{a5}-Y_{55}) + \frac{\binom{s}{0}_5}{\binom{0}{0}_5}(Y_{a5}-Y_{55})(Y_{b5}-Y_{55})\Big]\right\} I_4^{[d+],s} \\ &+ \frac{1}{\binom{0}{0}_5}\sum_{s=1}^{5}\frac{\Sigma_b^{1,s}}{\binom{0s}{0s}_5}\sum_{t=1}^{5}\Sigma_a^{2,st}I_3^{st}, \end{aligned} \tag{25}$$

with $\Sigma_b^{1,s}$ from (19) and

$$\begin{aligned} \Sigma_a^{2,st} &\equiv \sum_{i=1}^{4}(q_a\cdot q_i)\binom{0st}{0si}_5 \\ &= \frac{1}{2}(1-\delta_{st})\left\{\binom{ts}{0s}_5(Y_{a5}-Y_{55}) + \binom{0s}{0s}_5(\delta_{at}-\delta_{5t}) - \binom{0s}{0t}_5(\delta_{as}-\delta_{5s})\right\}, \end{aligned} \tag{26}$$

This can be extended also to higher ranks. We need at most double sums, e.g.:

$$\begin{aligned} \Sigma_{ab}^{2,s} &\equiv \sum_{i,j=1}^{4}(q_a\cdot q_i)(q_b\cdot q_j)\binom{si}{sj}_5 \\ &= \frac{1}{2}(q_a\cdot q_b)\binom{s}{s}_5 - \frac{1}{4}()_5\,(\delta_{ab}\delta_{as}+\delta_{5s}). \end{aligned} \tag{27}$$

Many of the sums over signed minors, weighted with scalar products of chords are given in [13], and an almost complete list may be obtained on request from the authors.

Journal of Physics: Conference Series **368** (2012) 012057 doi:10.1088/1742-6596/368/1/012057

4. Summary

We gave an introductory to the one-loop tensor reduction package PJFry and the underlying algorithms. PJFry may be used for Feynman integrals with up to five external legs. The treatment of 6-point functions is straightforward, while the expressions for contracted tensor integrals, exemplified in section 3, have not yet been completely worked out. Both 6-point functions and contractions are foreseen to be included into PJFry in the near future. The n-point functions with $n > 6$ can be conveniently treated following the approach advocated in [14, 15]. This was worked out after ACAT 2011 in [16].

We gave a presentation of tensor reduction with some focus on recent develoments by our group. Of course, several other approaches had been worked out in recent years which also deserve notice, usually treating reductions for arbitrary n. Let us mention few of them, just as an introduction to the literature. The authors of [17] construct combinations of n-point and $(n-1)$-point scalar integrals that are finite in the limit of vanishing Gram determinant. The representations are obtained by differentiation with respect to external parameters or by expanding the scalar integrals in $D = 6 - 2\epsilon$ or higher dimensions. An algorithm proposed in [18], based on the so-called generalized Bernstein functional relation, is applied and worked out in [19] for one-loop (tensor) Feynman integrals with n legs. In [20], a set of algorithms is worked out which are a basis for a fast and reliable numerical calculation of one-loop multi-leg Feynman diagrams with up to 6 external legs; also with a special attention to potentially singular points in phase space. In the approach in [21], six 4-point functions in $D = 6$, one 3-point function in $D = 4$, and one 2-point function in D dimensions are used for a general one-loop tensor reduction. All the soft and collinear singularities of general one-loop n-point integrals are worked out explicitly in [22]. The singular parts are expressed in terms of 3-point functions. Finally, we like to mention the approach in [23] for n-point functions. The infrared and ultraviolet divergences are separated analytically from the finite one-loop contributions, and the latter can then be evaluated numerically using recursion relations.

To summarize, there are a lot of competing algorithms available in principle and worked out in detail. One may only hope that most of them will be realized in publicly available open-source software packages for the use by others.

Appendix A. Determinants

The Gram determinant G_N of the diagram shown in 3 is

$$G_N \;=\; |2q_iq_j|, i,j = 1,\dots N. \tag{A.1}$$

We are also using the modified Cayley determinant $()_N$ of a diagram with N internal lines. For the choice $q_N = 0$, both determinants are related:

$$()_N \;\equiv\; \begin{vmatrix} 0 & 1 & 1 & \dots & 1 \\ 1 & Y_{11} & Y_{12} & \dots & Y_{1N} \\ 1 & Y_{12} & Y_{22} & \dots & Y_{2N} \\ \vdots & \vdots & \vdots & \ddots & \vdots \\ 1 & Y_{1N} & Y_{2N} & \dots & Y_{NN} \end{vmatrix} = -G_{N-1}, \tag{A.2}$$

where the matrix elements are defined by

$$Y_{ij} = -(q_i - q_j)^2 + m_i^2 + m_j^2, \quad (i,j = 1\dots N). \tag{A.3}$$

Evidently, the Gram determinant G_N does not depend on the masses, and so doesn't $()_N$.

Signed minors of $()_N$ are constructed by deleting m rows and m columns from $()_N$, and multiplying it with a sign factor:

$$\begin{pmatrix} j_1 & j_2 & \cdots & j_m \\ k_1 & k_2 & \cdots & k_m \end{pmatrix}_N \equiv (-1)^{\sum_l (j_l+k_l)} \mathrm{sgn}_{\{j\}}\, \mathrm{sgn}_{\{k\}} \begin{vmatrix} \text{rows } j_1 \cdots j_m \text{ deleted} \\ \text{columns } k_1 \cdots k_m \text{ deleted} \end{vmatrix}, \tag{A.4}$$

where $\mathrm{sgn}_{\{j\}}$ and $\mathrm{sgn}_{\{k\}}$ are the signs of permutations that sort the deleted rows $j_1 \cdots j_m$ and columns $k_1 \cdots k_m$ into ascending order.

Example:

$$\begin{pmatrix} 0 \\ 0 \end{pmatrix}_N \equiv \begin{vmatrix} Y_{11} & Y_{12} & \dots & Y_{1N} \\ Y_{12} & Y_{22} & \dots & Y_{2N} \\ \vdots & \vdots & \ddots & \vdots \\ Y_{1N} & Y_{2N} & \dots & Y_{NN} \end{vmatrix}, \tag{A.5}$$

To be definite, take fig. A1 as a starting point. The example is taken from [24]. The corresponding 4-point tensor integrals are, in LoopTools/FF [25, 26, 9] notation:

$$\mathrm{D0i}(\mathrm{id}, 0, 0, s_{\bar{\nu}u}, t_{ed}, t_{\bar{e}\mu}, s_{\mu\bar{\nu}u}, 0, M_Z^2, 0, 0). \tag{A.6}$$

The Gram determinant is

$$()_4 \;=\; -\, 2t_{\bar{e}\mu}[s_{\mu\bar{\nu}u}^2 + s_{\bar{\nu}u}t_{ed} - s_{\mu\bar{\nu}u}(s_{\bar{\nu}u} + t_{ed} - t_{\bar{e}\mu})], \tag{A.7}$$

and it vanishes if:

$$t_{ed} \to t_{ed,\mathrm{crit}} = \frac{s_{\mu\bar{\nu}u}(s_{\mu\bar{\nu}u} - s_{\bar{\nu}u} + t_{\bar{e}\mu})}{s_{\mu\bar{\nu}u} - s_{\bar{\nu}u}}. \tag{A.8}$$

In terms of a dimensionless scaling parameter x,

$$t_{ed} \;=\; (1 + x)t_{ed,\mathrm{crit}}, \tag{A.9}$$

the Gram determinant becomes:

$$()_4 = 2\; x\; s_{\mu\bar{\nu}u}t_{\bar{e}\mu}(s_{\mu\bar{\nu}u} - s_{\bar{\nu}u} + t_{\bar{e}\mu}). \tag{A.10}$$

The modified Cayley determinant

$$\begin{aligned} \begin{pmatrix} 0 \\ 0 \end{pmatrix}_4 &= \begin{pmatrix} 2M_Z^2 & M_Z^2 & M_Z^2 - s_{\mu\bar{\nu}u} & M_Z^2 \\ M_Z^2 & 0 & -s_{\bar{\nu}u} & M_Z^2 \\ M_Z^2 - s_{\mu\bar{\nu}u} & -s_{\bar{\nu}u} & 0 & -t_{ed} \\ M_Z^2 & -t_{\bar{e}\mu} & -t_{ed} & 0 \end{pmatrix} \\ &= s_{\mu\bar{\nu}u}^2 t_{\bar{e}\mu}^2 + 2\; M_Z^2 t_{\bar{e}\mu}[-2s_{\bar{\nu}u}t_{ed} + s_{\mu\bar{\nu}u}(s_{\bar{\nu}u} + t_{ed} - t_{\bar{e}\mu})] \\ &\quad + M_Z^4(s_{\bar{\nu}u}^2 + (t_{ed} - t_{\bar{e}\mu})^2 - 2s_{\bar{\nu}u}(t_{ed} + t_{\bar{e}\mu})) \end{aligned} \tag{A.11}$$

has another dimension. From (14) we see that a small Gram determinant expansion will be useful when the following dimensionless parameter becomes small:

$$R \;=\; \frac{()_4}{\binom{0}{0}_4} \times S, \tag{A.12}$$

where S is a typical scale of the process, e.g. we will choose $S = s_{\mu\bar{\nu}u}$.

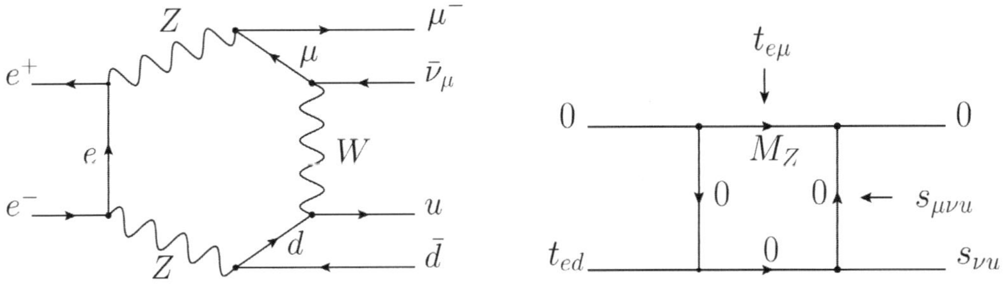

Figure A1. A six-point topology (a) leading to four-point functions (b) with realistically vanishing Gram determinants.

Acknowledgements
J.F. thanks DESY for kind hospitality. Work is supported in part by Sonderforschungsbereich/Transregio SFB/TRR 9 of DFG "Computergestützte Theoretische Teilchenphysik" and European Initial Training Network LHCPHENOnet PITN-GA-2010-264564.

References

[1] Passarino G and Veltman M 1979 *Nucl. Phys.* **B160** 151 doi:10.1016/0550-3213(79)90234-7
[2] Davydychev A I 1991 *Phys. Lett.* **B263** 107–111 doi:10.1016/0370-2693(91)91715-8
[3] Tarasov O 1996 *Phys.Rev.* **D54** 6479–6490 (*Preprint* hep-th/9606018)
[4] Fleischer J, Jegerlehner F and Tarasov O 2000 *Nucl. Phys.* **B566** 423–440 (*Preprint* hep-ph/9907327)
[5] Diakonidis T, Fleischer J, Gluza J, Kajda K, Riemann T and Tausk J 2009 *Phys. Rev.* **D80** 036003 (*Preprint* arXiv:0812.2134)
[6] Fleischer J and Riemann T 2011 *Phys. Rev.* **D83** 073004 (*Preprint* arXiv:1009.4436)
[7] Reina L and Schutzmeier T 2011 (*Preprint* 1110.4438)
[8] Ellis R K and Zanderighi G 2008 *JHEP* **02** 002 (*Preprint* arXiv:0712.1851)
[9] van Oldenborgh G J 1991 *Comput. Phys. Commun.* **66** 1–15 doi:10.1016/0010-4655(91)90002-3, scanned version at http://ccdb3fs.kek.jp/cgi-bin/img_index?9004168
[10] van Hameren A 2011 *Comput. Phys. Commun.* **182** 2427–2438 (*Preprint* arXiv:1007.4716)
[11] Yundin V c++ package PJFry, available at the webpage https://github.com/Vayu/PJFry/
[12] Cullen G, Greiner N, Heinrich G, Luisoni G, Mastrolia P, Ossola G, Reiter T and Tramontano F 2011 (*Preprint* 1111.2034)
[13] Fleischer J and Riemann T 2011 *Phys.Lett.* **B701** 646–653 (*Preprint* arXiv:1104.4067)
[14] Bern Z, Dixon L J and Kosower D A 1994 *Nucl. Phys.* **B412** 751–816 (*Preprint* hep-ph/9306240)
[15] Binoth T, Guillet J, Heinrich G, Pilon E and Schubert C 2005 *JHEP* **10** 015 (*Preprint* hep-ph/0504267)
[16] Fleischer J and Riemann T 2011 (*Preprint* arXiv:1111.5821)
[17] Campbell J, Glover E W N and Miller D 1997 *Nucl. Phys.* **B498** 397–442 (*Preprint* hep-ph/9612413)
[18] Tkachov F V 1997 *Nucl.Instrum.Meth.* **A389** 309–313 (*Preprint* hep-ph/9609429)
[19] Passarino G 2001 *Nucl.Phys.* **B619** 257–312 (*Preprint* hep-ph/0108252)
[20] Ferroglia A, Passera M, Passarino G and Uccirati S 2003 *Nucl.Phys.* **B650** 162–228 (*Preprint* hep-ph/0209219)
[21] Duplancic G and Nizic B 2004 *Eur. Phys. J.* **C35** 105–118 (*Preprint* hep-ph/0303184)
[22] Dittmaier S 2003 (*Preprint* hep-ph/0308246)
[23] Giele W T and Glover E W N 2004 *JHEP* **04** 029 (*Preprint* hep-ph/0402152)
[24] Denner A Introductory
Lecture at DESY Theory Workshop on Collider Phenomenology, Hamburg, 29 Sep - 2 Oct 2009
[25] Hahn T and Perez-Victoria M 1999 *Comput.Phys.Commun.* **118** 153–165 (*Preprint* hep-ph/9807565)
[26] Hahn T LoopTools 2.5 User's Guide, available from http://www.feynarts.de/looptools/LT25Guide.pdf

Progress in automated Next-to-Leading-Order calculations

Francesco Tramontano
CERN Theory Group, Geneve 23, CH-1211, Switzerland
E-mail: Francesco.Tramontano@cern.ch

Abstract. We review the recent progress towards automation in the computation of the next-to-leading corrections to scattering amplitudes. Such progress allows for the construction of quite general, flexible and fully automated packages that would be of major importance for the Higgs boson and beyond the Standard Model physics searches at high energy particle colliders.

1. Introduction

The search for the Higgs boson and signals of physics beyond the Standard Model of elementary particles (SM) is underway at the CERN Large Hadron Collider (LHC) with considerable efficiency. The Higgs boson is an essential ingredient of the SM, but its existence has not been proved yet. Radiative corrections are however sensitive to the Higgs boson mass and the combination of the precision electroweak tests and the LEP limit set the upper bound $m_H < 185\ GeV$ at the 95% of confidence level. On the other hand, more and more severe exclusion limits are delivered by the Tevatron at the Fermilab and the Large Hadron Collider at CERN and the Higgs boson discovery could be around the corner.

The high luminosity and high center of mass energy of the collisions at the LHC make of course discoveries possible, but through very challenging searches. This is because of the many background processes that have an experimental signature which overlaps with the one of the signal processes and that are hard to disentangle, but also for the possible underlying events in proton-proton collisions and the pile up of collisions happening during the same bunch crossing. From the phenomenological perspective the need of precise predictions obtainable through flexible tools able to produce distributions or events for any desired process and kinematic cuts is mandatory.

Many predictions are currently performed with tools merging the exact leading-order (LO) calculation of the partonic matrix elements (ME) with the evolution, provided by so-called shower Monte Carlo (MC) codes, of the partonic shower (PS) and the subsequent hadronization of the partons into physical hadrons [1, 2, 3, 4]. This is because they can readily be tuned with data and because there are many very flexible LO tools on the market. Further, the parton shower performs a resummation of the collinear emission giving a first reliable approximation in certain regions of phase space for important measured kinematical distributions where fixed order calculations fails. On the other hand the Next-to-Leading-Order (NLO) computations give more precise predictions on total rates, a better control of the jet structure of events and more reliable estimation of the theoretical error related to the renormalization and factorization scale dependences. Also, the uncertainty related to the Parton Distributions Functions can be estimated more reliably within NLO computations. Therefore NLO matrix element generators at parton level paired with a parton shower represent the ideal instruments to address the phenomenology of the colliders. The Next-to-Next-to-Leading-Order computations are quite far from automation because the structure of the divergences has still not allowed for an easy and fully general implementation of a subtraction scheme and, as a result, only very few differential computations are available for hadron colliders processes. There exist a number of excellent NLO public tools containing many hard-coded processes and a full setup to produce phenomenological results, among them MCFM [5, 6], VBF@NLO [7, 8], MC@NLO [9, 10] and POWHEG [11, 12, 13]. Nevertheless, nowadays it is possible to flank these packages with fully automated tools able to generate any process up to NLO accuracy within one and the same framework, as it is already possible at LO with MADGRAPH/MADEVENT [14, 15] for example. Such automated codes are under construction by several groups partly interfacing well tested tools for subtasks of the full package. In the following, I will describe corrections within Quantum Chromodynamics (QCD), as these are the most relevant at present hadron colliders, although the most recent developments are quite general and apply to other one-loop computations as well.

2. Ingredients of NLO computations

The actual description of the hadron collisions characterized by a hard scale or a large momentum transfer among the hadron constituents, is based on the factorization formula:

$$\sigma_{AB} = \int dx_a dx_b f_{a/A}(x_a, Q^2) f_{b/B}(x_b, Q^2) \hat{\sigma}_{ab \to X}. \tag{1}$$

This equation expresses a dynamical property of the QCD that, when there is a hard collision, meaning a collision at large momentum transfer, the cross section σ for the scattering of initial hadrons A and B, can be computed as the convolution of two factors: the product of the parton distribution probability functions PDF's inside the hadrons $f_{a/A} f_{b/B}$, that represents the result of the low momentum scale dynamic, and the hard scale cross section among the partons $\hat{\sigma}$, summed over all the possible parton pairs that could scatter in the specific hadron-hadron collision experiment. The convolution is over the fractional momentum carried by the partons inside the hadrons. The actual knowledge of QCD does not allow to derive the PDF's, they have to be evaluated through dedicated experiments, but the formula is general so that the PDF's measured in an experiment can be used to assess prediction for another experiment. On the other hand the partonic cross section is calculable in perturbation theory as a series expansion of the QCD coupling constant.

The computation of the parton level matrix element up to NLO consists of several pieces: first there is of course the LO matrix element made of tree level Feynman diagrams, to get the NLO correction to this, diagrams with one extra parton in the final state have to be included, as well as virtual one-loop diagrams with the same number of external particles as the Born process. These two processes are of the same order in the coupling constants and in general separately divergent. The virtual part is both ultraviolet and infrared divergent, the UV divergences being removed through renormalization. The infrared divergences of the virtual part are cancelled by the ones of the real part leaving only the initial state mass singularities to be subtracted "renormalising" the parton distribution functions of the partons in the incoming hadrons. A complication here is related to the fact that the cancellation among the infrared singularities involves integrals over different phase space. One way to overcome this problem is the implementation of the so called subtraction method [16, 17, 18, 19], in which one designs counter matrix elements that map the singularities of the real part point-wise and that are analytically integrable over the extra radiation degrees of freedom, to be added to the virtual part as poles times born-like matrix elements. So that numerically the NLO contribution amounts to construct two distinguished integrals well defined and numerically integrable in four dimensions.

Tools for the computation of tree level matrix elements contributing to the Born and the real correction processes have existed for many years and to a very high level of automation. The full formulation of the subtraction method has been known since many years and recently several packages have been constructed which, for a given process, perform the full generation of the counter terms and their integrated version [20, 21, 22, 23, 24, 25].

The need for an automation of NLO virtual calculations has been noticed some time ago and lead to public programs like FeynArts [26] and QGraf [27] for diagram generation and FormCalc/LoopTools [28, 29] and GRACE [30] for the automated calculation of NLO corrections, primarily in the electroweak sector. Nevertheless, until recently the computation with more than two particles in the final state were still an involved case by case computation. The number of particles in the final state sets the difficulty of NLO computations for basically two reasons. First the virtual computation even if conceptually straightforward becomes very extensive and difficult to keep under proper control. Second, the integration over the real matrix element phase space that has the extra parton becomes more and more time consuming. The problems related to the virtual part is however still considered the bottleneck for the automated next to leading order computation. The recent development in this last bit of automation will be the subject of the next section.

3. Recent developments

Improved reduction techniques for the tensor integrals have been proposed in [31] and [32], the first group also released a public library for the numerical computation of loop tensor integrals recently updated in [33]. It was then possible to construct quite general algorithms [34, 35] to reconstruct the tensor structure of a loop integrand in terms of the external particle kinematic tensors convoluted with the loop tensor integrals and then express the latter numerically thanks to the tensor loop integral libraries. Further, we have recently experienced major advances in the activity of constructing packages for fully automated one-loop calculations, see e.g. [36, 37, 38, 39, 40, 41]. The concepts that lead to these advances have been recently reviewed in [42]. Among the most important developments are the integrand-reduction technique [43, 44] and the generalized unitarity in dimensional regularization [45]. Their main outcome is a numerical reconstruction of a representation of the tensor structure of any one-loop integrand where the multi-particle pole configuration is manifest. As a consequence, decomposing one-loop amplitudes in terms of basic integrals becomes equivalent to reconstructing the polynomial forms of the residues to all multi-particle cuts. Within this algorithm, the integrand of a given scattering amplitude, carrying complete and explicit information on the chosen dimensional-regularisation scheme, is the only input required. In fact, the integration is substituted by a much simpler operation, namely by polynomial fitting, which requires the sampling of the integrand on the solutions of generalised on-shell conditions.

This kind of algorithm is very flexible, allowing for the construction of the integrand both by sewing of tree level on-shell amplitudes or through Feynman diagrams. The first approach works with gauge invariant building blocks that in principle allows for the construction of more stable and faster codes. Nevertheless it is still not clear how to automate the construction of the rational terms in general. The integrands constructed from Feynman diagrams contains of course all the information on the rational terms and so can easily be used for general purposes matrix element generators, but single diagrams are of course not gauge invariant and in principle there could be large cancellations among different diagrams.

The Ngluon [46] code is publically available and represents an example of the efficiency of the methods discussed above for the computation of pure QCD one loop virtual amplitudes, see the contribution by Biederman at this conference. The Blackhat [47] and Rocket [48] codes are

Table 1. Sample of processes for which GoSAM has been compared to the literature.

process	**checked with Ref.**
$e^+e^- \to u\bar{u}$	[50]
$e^+e^- \to t\bar{t}$	[51, 52], own analytic calculation
$u\bar{u} \to d\bar{d}$	[53, 37]
$gg \to gg$	[54]
$gg \to gZ$	[55], own analytic calculation
$pp \to t\bar{t}$	[37], [5, 6]
$bg \to H\,b$	[56, 37]
$\gamma\gamma \to \gamma\gamma$ (W loop)	[57]
$pp \to W^\pm j$ (QCD corr.)	[5, 6]
$pp \to W^\pm j$ (EW corr.)	for IR poles: Eqs. (67),(70) of [58], [59]
$pp \to W^\pm t$	[5, 6]
$pp \to W^\pm jj$	[5, 6]
$pp \to W^\pm b\bar{b}$ (massive b's)	[5, 6]
$e^+e^- \to e^+e^-\gamma$ (QED)	[60]
$pp \to t\bar{t}H$	[37]
$pp \to t\bar{t}Z$	[40]
$\gamma\gamma \to \gamma\gamma\gamma\gamma$ (fermion loop)	[62]
$pp \to W^+W^+jj$	[61, v3]
$pp \to b\bar{b}b\bar{b}$	[63, 64]
$pp \to W^+W^-b\bar{b}$	[36, 37]
$pp \to t\bar{t}b\bar{b}$	[36, 37]
$u\bar{d} \to W^+ggg$	[36]

other examples. Further, other three codes for the automated generation of the virtual matrix elements have been recently made available: the HELAC-NLO [40] and the GoSam [49] codes have been released, while comparison with the MadLoop [37] code can be done through a web interface. In table 1 there is a list of SM processes tested with the GoSam code generator, see the contribution by G. Heinrich at this conference for the GoSam collaboration for more details.

4. Selection of recent full physical results obtained exploiting the new developments

The big progress in the technical treatment of one loop correction allowed to perform computations of incredible complexity. Among the most relevant NLO results there has been the production of $t\bar{t}b\bar{b}$ by two collaborations [65, 66], of W^+W^- plus two jets [67] and W^+W^- plus two b-jets [68, 69] at the hadron colliders. All these processes include six-point one loop diagrams. Further, the Blackhat collaboration pioneered the computation of the NLO QCD corrections to processes with seven external particles, producing precise predictions for the associate production of an electroweak gauge boson $W/Z/\gamma^*$ and four jets [70, 71].

Regarding the results obtained through fully automated setup I will mention three recent achievements. The HELAC-NLO group produced a full study for the top-antitop plus two jets final state [72]. A sample of the long list of distributions presented in [72] is given in figure 1. These results have been obtained with a fully automated setup constructed out a suite of programs that perform the generation of the matrix elements, the needed subtractions and integrate everything to produce physical predictions. The second example is the study of the

Journal of Physics: Conference Series **368** (2012) 012058 doi:10.1088/1742-6596/368/1/012058

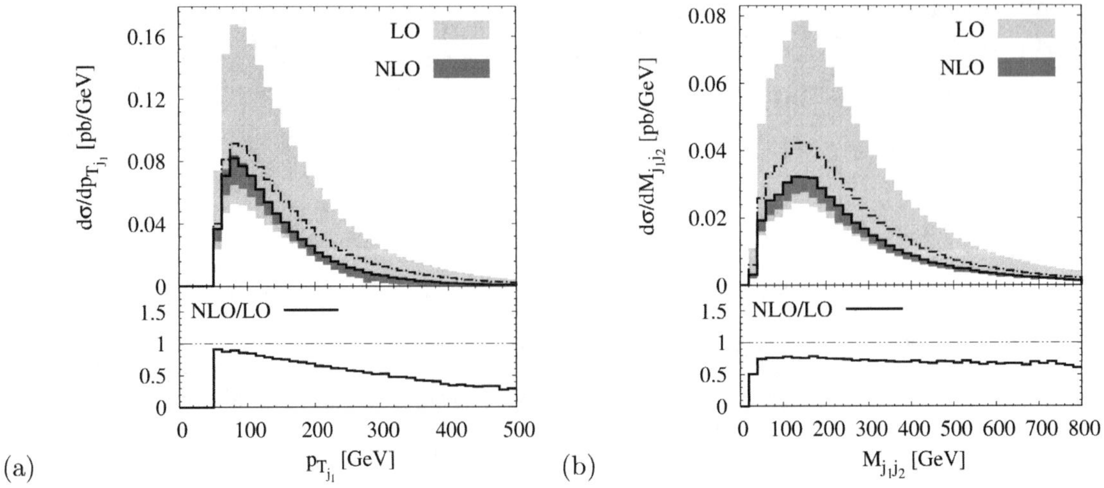

Figure 1. (a) p_T of the first jet (top) and (b) dijet mass (bottom) distributions for the process $pp \to t\bar{t}jj$ at the LHC with $\sqrt{s} = 7\ TeV$, see [72] for the cut selection and the other details of the analysis.

product of Wjj final state at the Tevatron by the aMC@NLO collaboration [73]. In this case the fully automated NLO parton level code produced merging MadLoop and MadFKS has been automatically interfaced with MC@NLO to produce event samples. The same strategy linking automatically Helac-NLO [40] with the Powheg-box [13] has been followed in [74] producing top-antitop final states in association with other particles, the most complicated example is the one including a Z boson [74]. As the last example I mention the four b quark jet production at

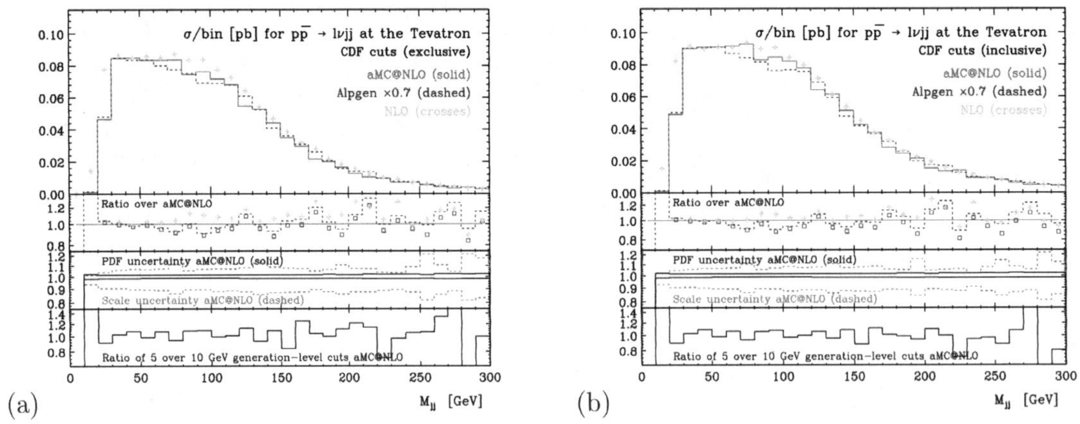

Figure 2. Dijet invariant mass distribution for the process $p\bar{p} \to Wjj$ at the Tevatron energy studied by the aMC@NLO collaboration, see [73] for the detail of the analysis, in the bottom panel (b) a veto on extra jet radiation present in the top (a) is removed.

the LHC energies reported in [64]. In this case the generation of the virtual matrix elements has been done through the GoSam package, while the tree level matrix elements, the subtractions and the integration has been worked out with MadGraph/MadEvent and MadDipole. In this case the simplicity of the Binoth Les Houchs interface [75] allowed for the interface of the GoSam virtual code generator with the MadGraph/MadEvent/MadDipole suite of programs.

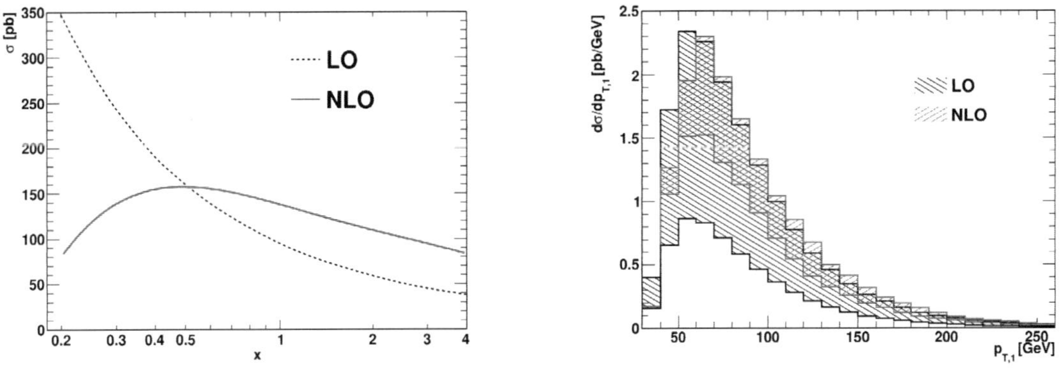

Figure 3. (a) Scale variation (top) and (b) p_T distribution of the hardest jet (bottom) for the process $pp \to b\bar{b}b\bar{b}$, see [64] for the details of the analysis.

5. Conclusions

Thanks to the recent progress in the computation of the virtual corrections for scattering amplitudes, the full automation for packages performing the NLO computation of distributions related to multiparticle final states in hadronic collisions is now feasible. Several groups are working towards this goal. Once such generators are publically available they allow to perform every desired analysis at the NLO accuracy for both the signal and the main background processes to the experimental signature under study. The bottleneck for this achievement has been the automated computation of the virtual matrix elements. Nowadays, thanks to the implementation of modern techniques, examples of full automation in the production of NLO computations have been provided by the aMC@NLO, the HELAC-NLO and the GoSam+MadEvent+MadDipole combination of tools respectively. With perfect timing, extraordinarily powerful tools are being developed that hopefully will be of help to disentangle New Physics signals from the large source of backgrounds at the LHC.

Acknowledgments

The author wishes to acknowledge N. Glover and G. Heinrich for the invitation to give this presentation.

References

[1] Mangano M L, Moretti M, Piccinini F, Pittau R and Polosa A D 2003 *JHEP* **0307** 001 (*Preprint* hep-ph/0206293)
[2] Sjostrand T, Mrenna S and Skands P Z 2008 *Comput.Phys.Commun.* **178** 852 (*Preprint* 0710.3820)
[3] Corcella G, Knowles I G, Marchesini G, Moretti S, Odagiri K, Richardson P, Seymour M H and Webber B R 2001 *JHEP* **0101** 010 (*Preprint* hep-ph/0011363)
[4] Catani S, Krauss F, Kuhn R and Webber B R 2001 *JHEP* **0111** 063 (*Preprint* hep-ph/0109231)
[5] Campbell J M and Ellis R K 1999 *Phys.Rev.* **D60** 113006 (*Preprint* hep-ph/9905386)
[6] Campbell J M, Ellis R K, Williams C 2011 *JHEP* **1107** 018 (*Preprint* 1105.0020)
[7] Arnold K, Bahr M, Bozzi G, Campanario F, Englert C, *et al.* 2009 *Comput.Phys.Commun.* **180** 1661-1670 (*Preprint* 0811.4559)
[8] Arnold K, Bellm J, Bozzi G, Brieg M, Campanario F, *et al.* 2011 (*Preprint* 1107.4038)
[9] Frixione S and Webber B R 2002 *JHEP* **0206** 029 (*Preprint* hep-ph/0204244)
[10] Frixione S, Stoeckli F, Torrielli P, Webber B R, and White C D 2010 (*Preprint* 1010.0819)
[11] Nason P 2004 *JHEP* **0411** 040 (*Preprint* hep-ph/0409146)
[12] Frixione S, Nason P and Oleari C 2007 *JHEP* **0711** 070 (*Preprint* 0709.2092)
[13] Alioli S, Nason P, Oleari C and Re E 2010 *JHEP* **1006** 043 (*Preprint* 1002.2581)

[14] Maltoni F and Stelzer T 2003 *JHEP* **0302** 027 (*Preprint* hep-ph/0208156)
[15] Alwall J, Herquet M, Maltoni F, Mattelaer O and Stelzer T 2011 *JHEP* **1106** 128 (*Preprint* 1106.0522)
[16] Ellis R K, Ross D A and Terrano A E 1981 *Nucl.Phys.* **B178** 421
[17] Catani S and Seymour M H 1997 *Nucl.Phys.* **B485** 421, 1998 **B510** 503, (*Preprint* hep-ph/9605323)
[18] Catani S, Dittmaier S, Seymour M H and Trocsanyi Z 2002 *Nucl.Phys.* **B627** 189 (*Preprint* hep-ph/0201036)
[19] Frixione S, Kunszt Z and Signer S 1996 *Nucl.Phys.* **B467** 399 (*Preprint* hep-ph/9512328)
[20] Frederix R, Gehrmann T and Greiner N 2008 *JHEP* **0809** 122 (*Preprint* 0808.2128)
[21] Frederix R, Gehrmann T and Greiner N 2010 *JHEP* **1006** 086 (*Preprint* 1004.2905)
[22] Czakon M, Papadopoulos C G and Worek M 2009 *JHEP* **0908** 085 (*Preprint* 0905.0883)
[23] Hasegawa K, Moch S and Uwer P 2010 *Comput.Phys.Commun.* **181** 1802 (*Preprint* 0911.4371)
[24] Gleisberg T and Krauss F 2008 *Eur.Phys.J.* **C53** 501 (*Preprint* 0709.2881)
[25] Frederix R, Frixione S, Maltoni F and Stelzer T 2009 *JHEP* **0910** 003 (*Preprint* 0908.4272)
[26] Hahn T 2001 *Comput.Phys.Commun.* **140** 418–431 (*Preprint* hep-ph/0012260)
[27] Nogueira P 1993 *J.Comput.Phys.* **105** 279–289.
[28] Hahn T and Perez-Victoria M 1999 *Comput.Phys.Commun.* **118** 153–165 (*Preprint* hep-ph/9807565)
[29] Hahn T, *Talk given at the International Workshop on Advanced Computing and Analysis Techniques in Physics Research (ACAT)*, Uxbridge, London, September 2011
[30] Belanger G, Boudjema F, Fujimoto J, Ishikawa T, Kaneko T, *et al.* 2006 *Phys.Rept.* **430** 117-209 (*Preprint* hep-ph/0308080)
[31] Binoth T, Guillet J P, Heinrich G, Pilon E and Schubert C 2005 *JHEP* **0510** 015 (*Preprint* hep-ph/0504267)
[32] Denner A and Dittmaier S 2006 *Nucl.Phys.* **B734** 62 (*Preprint* hep-ph/0509141)
[33] Cullen G, Guillet J P, Heinrich G, Kleinschmidt T, Pilon E, Reiter T and Rodgers M 2011 *Comput.Phys.Commun.* **182** 2276 (*Preprint* 1101.5595)
[34] Heinrich G, Ossola G, Reiter T and Tramontano F 2010 *JHEP* **1010** 105 (*Preprint* 1008.2441)
[35] Cascioli F, Maierhofer P and Pozzorini S 2011 (*Preprint* 1111.5206)
[36] van Hameren A, Papadopoulos C and Pittau R 2009 *JHEP* **0909** 106 (*Preprint* 0903.4665)
[37] Hirschi V, Frederix R, Frixione S, Garzelli M V, Maltoni F *et al.* 2011 *JHEP* **1105** 044 (*Preprint* 1103.0621)
[38] Mastrolia P, Ossola G, Reiter T and Tramontano F 2010 *JHEP* **1008** 080 (*Preprint* 1006.0710)
[39] Cullen G, Greiner N, Guffanti A, Guillet J P, Heinrich G, *et al.* 2010 *Nucl.Phys.Proc.Suppl.* **205-206** 67–73 (*Preprint* 1007.3580)
[40] Bevilacqua G, Czakon M, Garzelli M, van Hameren A, Kardos A, *et al.* 2011 (*Preprint* 1110.1499)
[41] Reina L and Schutzmeier T 2011 (*Preprint* 1110.4438)
[42] Ellis R K, Kunszt Z, Melnikov K and Zanderighi G 2011 (*Preprint* 1105.4319)
[43] Ossola G, Papadopoulos C G and Pittau R 2007 *Nucl.Phys.* **B763** 147–169 (*Preprint* hep-ph/0609007)
[44] Ossola G, Papadopoulos C G and Pittau R 2007 *JHEP* **07** 085 (*Preprint* 0704.1271)
[45] Ellis R K, Giele W T, Kunszt Z and Melnikov K 2009 *Nucl.Phys.* **B822** 270–282 (*Preprint* 0806.3467)
[46] Badger S, Biedermann B and Uwer P 2011 *Comput.Phys.Commun.* **182** 1674 (*Preprint* 1011.2900)
[47] Berger C F, Bern Z, Dixon L J, Febres Cordero F, Forde D, Ita H, Kosower D A and Maitre D 2008 *Phys.Rev.* **D78** 036003 (*Preprint* 0803.4180)
[48] Ellis R K, Giele W T, Kunszt Z, Melnikov K and Zanderighi G 2009 *JHEP* **0901** 012 (*Preprint* 0810.2762)
[49] Cullen G, Greiner N, Heinrich G, Luisoni G, Mastrolia P, Ossola G, Reiter T and Tramontano F 2011 (*Preprint* 1111.2034)
[50] Ellis R, Stirling W and Webber B 1996 *Camb.Monogr.Part.Phys.Nucl.Phys.Cosmol.* **8** 1–435
[51] Jersak J, Laermann E and Zerwas P 1982 *Phys.Rev.* **D25** 1218
[52] Catani S, Dittmaier S, Seymour M H and Trocsanyi Z 2002 *Nucl.Phys.* **B627** 189–265 (*Preprint* hep-ph/0201036)
[53] Ellis R and Sexton J 1986 *Nucl.Phys.* **B269** 445
[54] Binoth T, Guillet J and Heinrich G 2007 *JHEP* **0702** 013 (*Preprint* hep-ph/0609054)
[55] van der Bij J and Glover E 1989 *Nucl.Phys.* **B313** 237
[56] Campbell J M, Ellis R, Maltoni F and Willenbrock S 2003 *Phys.Rev.* **D67** 095002 (*Preprint* hep-ph/0204093)
[57] Gounaris G, Porfyriadis P and Renard F 1999 *Eur.Phys.J.* **C9** 673–686 (*Preprint* hep-ph/9902230)
[58] Kuhn J H, Kulesza A, Pozzorini S and Schulze M 2008 *Nucl.Phys.* **B797** 27–77 (*Preprint* 0708.0476)
[59] Gehrmann T and Greiner N 2010 *JHEP* **12** 050 (*Preprint* 1011.0321)
[60] Actis S, Mastrolia P and Ossola G 2010 *Phys.Lett.* **B682** 419–427 (*Preprint* 0909.1750)
[61] Melia T, Melnikov K, Rontsch R and Zanderighi G 2010 *JHEP* **1012** 053 (*Preprint* 1007.5313)
[62] Bernicot C 2008 (*Preprint* 0804.1315)
[63] Binoth T, Greiner N, Guffanti A, Reuter J, Guillet J P *et al.* 2010 *Phys.Lett.* **B685** 293–296 (*Preprint* 0910.4379)
[64] Greiner N, Guffanti A, Reiter T and Reuter J 2011 *Phys.Rev.Lett.* **107** 102002 (*Preprint* 1105.3624)

[65] Bredenstein A, Denner A, Dittmaier S and Pozzorini S 2009 *Phys.Rev.Lett.* **103** 012002 (*Preprint* 0905.0110)
[66] Bevilacqua G, Czakon M, Papadopoulos C G, Pittau R and Worek M 2009 *JHEP* **0909** 109 (*Preprint* 0907.4723)
[67] Melia T, Melnikov K, Rontsch R and Zanderighi G 2011 *Phys.Rev.* **D83** 114043 (*Preprint* 1104.2327)
[68] Denner A, Dittmaier S, Kallweit S and Pozzorini S 2011 *Phys.Rev.Lett.* **106** 102002 (*Preprint* 1012.3975)
[69] Bevilacqua G, Czakon M, van Hameren A, Papadopoulos C G and Worek M 2011 *JHEP* **1102** 083 (*Preprint* 1012.4230)
[70] Berger C F, Bern Z, Dixon L J, Febres Cordero F, Forde D, Gleisberg T, Ita H, Kosower D A *et al.* 2011 *Phys.Rev.Lett.* **106** 092001 (*Preprint* 1009.2328)
[71] Ita H, Bern Z, Dixon L J, Febres Cordero F, Kosower D A and Maitre D 2011 (*Preprint* 1108.2229)
[72] Bevilacqua G, Czakon M, Papadopoulos C G and Worek M 2011 (*Preprint* 1108.2851)
[73] Frederix R, Frixione S, Hirschi V, Maltoni F, Pittau R and Torrielli P 2011 (*Preprint* 1110.5502)
[74] Kardos A, Papadopoulos C and Trocsanyi Z 2011 (*Preprint* 1111.0610)
[75] Binoth T, Boudjema F, Dissertori G, Lazopoulos A, Denner A *et al.* 2010 *Comput.Phys.Commun.* **181** 1612–1622 dedicated to the memory of, and in tribute to, Thomas Binoth, who led the effort to develop this proposal for Les Houches 2009 (*Preprint* 1001.1307)

Regularization schemes and higher order corrections

William B Kilgore

Physics Department, Brookhaven National Laboratory, Upton, New York 11973, U.S.A.

E-mail: kilgore@bnl.gov

Abstract. I apply commonly used regularization schemes to a multiloop calculation to examine the properties of the schemes at higher orders. I find complete consistency between the conventional dimensional regularization scheme and dimensional reduction, but I find that the four-dimensional helicity scheme produces incorrect results at next-to-next-to-leading order and singular results at next-to-next-to-next-to-leading order. It is not, therefore, a unitary regularization scheme.

1. Regularization Schemes

In recent years, there has been a great deal of progress in the calculation of higher-order corrections. At one loop, especially, there are many new techniques being developed. It is important to understand whether these new techniques are reliable tools of quantum field theory that can be applied to multi-loop calculations or if they are just short-cuts that are only valid at one loop.

One of the workhorses of the effort to compute one-loop helicity amplitudes in QCD is the Four Dimensional Helicity (FDH) regularization scheme [1, 2]. In a recent paper I have shown that the FDH is not a unitary regularization scheme (for non-supersymmetric theories) and that it generates incorrect results beyond one loop.

1.1. Dimensional Regularization

Dimensional Regularization [3] is the basis for most regularization schemes in use today. Among its many favorable properties are that it respects gauge invariance, respects Lorentz invariance and handles both UV and IR divergences.

The application of Dimensional Regularization to different kinds of problems has led to the development of a variety of regularization schemes which share the dimensional regularization of momentum integrals but differ in their handling of observed states and spin degrees of freedom.

I will be discussing four different regularization schemes which commonly appear in the literature, the 't Hooft-Veltman (HV) Scheme [3], the Conventional Dimensional Regularization (CDR) Scheme [4], the Dimensional Reduction (DRED) Scheme [5] and the FDH Scheme [1, 2].

The first two are closely related and yield identical results in the calculation that I will be describing. Superficially at least, the second two are also closely related in much the same way, but yield very different results.

1.2. The 't Hooft-Veltman Scheme

The original formulation of dimensional regularization (the HV scheme) specifies that external (observed) states are treated as four-dimensional, while internal states are to be treated as $D_m = 4 - 2\varepsilon$ dimensional. The D_m-dimensional vector space is *larger* than 4-dimensional space-time:

Journal of Physics: Conference Series **368** (2012) 012059 doi:10.1088/1742-6596/368/1/012059

$$g^{\mu\nu}\, g^{\alpha}_{\nu} = g^{\mu\alpha}\,, \qquad g^{\mu\nu}\, \eta^{\alpha}_{\nu} = \eta^{\mu\alpha}\,, \qquad \eta^{\mu\nu}\, \eta^{\alpha}_{\nu} = \eta^{\mu\alpha}\,,$$
$$g^{\mu\nu}\, g_{\mu\nu} = D_m\,, \qquad \eta^{\mu\nu}\, \eta_{\mu\nu} = 4\,. \tag{1}$$

In HV, internal gluons have $D_m - 2 = 2 - 2\varepsilon$ spin degrees of freedom. Internal fermions, however, still have exactly 2 spin degrees of freedom.

1.3. The Conventional Dimensional Regularization Scheme

In the CDR scheme, all states (observed or internal) are continued to $D_m = 4 - 2\varepsilon$ dimensions. This is in many ways simpler than the HV scheme, especially when dealing with infrared sensitive theories like QCD. In HV, if external states have an infrared overlap, they must be treated as internal (D_m-dimensional). In CDR, all states are already D_m-dimensional, so the overlap is automatically treated properly.

The HV and CDR schemes are closely related. Their behaviors under the renormalization group (β-functions, anomalous dimensions) is identical and in the calculations I will present they give identical results.

1.4. The Dimensional Reduction Scheme

In the DRED scheme, one starts from 4-dimensional space-time and compactifies to a *smaller* vector space of dimension $D_m = 4 - 2\varepsilon$ in which momenta take values.

$$g^{\mu\nu}\, g^{\alpha}_{\nu} = g^{\mu\alpha}\,, \qquad g^{\mu\nu}\, \eta^{\alpha}_{\nu} = g^{\mu\alpha}\,, \qquad \eta^{\mu\nu}\, \eta^{\alpha}_{\nu} = \eta^{\mu\alpha}\,. \tag{2}$$

Particles in the spectrum retain their 2 spin degrees of freedom from 4 dimensions. This preserves supersymmetry.

BUT: The Ward Identity only applies to the vector subspace in which momenta are defined!

In non-SUSY theories, the "evanescent" (2ε-dimensional) gluons are independent degrees of freedom from the D_m-dimensional gluons. The fields and their couplings renormalize independently!

1.5. The Four Dimensional Helicity Scheme

The FDH takes the D_m-dimensional space where momenta take values to be *larger* than 4-dimensional space-time, but also defines a *still larger* D_s-dimensional vector space where spin degrees of freedom take values. D_s is taken to be equal to 4 so that particles have the same number of spin degrees of freedom as they have in 4 dimensions.

$$g^{\mu\nu}\, g_{\mu\nu} = D_s\,, \qquad \hat{g}^{\mu\nu}\, \hat{g}_{\mu\nu} = D_m\,, \qquad \eta^{\mu\nu}\, \eta_{\mu\nu} = 4\,,$$
$$g^{\mu\nu}\hat{g}^{\rho}_{\nu} = \hat{g}^{\mu\rho}\,, \qquad g^{\mu\nu}\eta^{\rho}_{\nu} = \eta^{\mu\rho}\,, \qquad \hat{g}^{\mu\nu}\eta^{\rho}_{\nu} = \eta^{\mu\rho}\,, \tag{3}$$
$$g^{\mu\nu}\delta^{\rho}_{\nu} = \delta^{\mu\rho}\,, \qquad \hat{g}^{\mu\nu}\delta^{\rho}_{\nu} = 0\,, \qquad \eta^{\mu\nu}\delta^{\rho}_{\nu} = 0\,.$$

One might expect that my remarks about the Ward Identity and evanescent states for DRED would apply to FDH, but that is not the way the scheme has been used.

Instead, FDH calculations are performed using the following rules.

(i) All momentum integrals are D_m dimensional.

(ii) All "observed" external states are taken to be four-dimensional.

(iii) All "unobserved" or internal states are treated as D_s dimensional, and the D_s dimensional vector space is taken to be larger than the D_m dimensional vector space.

(iv) Both the D_s and D_m dimensional vector spaces are larger than the standard four-dimensional space-time.

All degrees of freedom that originate from the gauge symmetry are treated as parts of the gauge bosons, NOT as independent degrees of freedom with independent couplings.

The claim is that the crucial difference between FDH and DRED that allows this treatment of the evanescent components is that $D_s > 4$.

2. Test Calculation

I will test the reliability of computing high-order corrections in these schemes by recalculating a physical quantity that is known to very high order: the inclusive cross section for an electron and positron to annihilate and produce hadrons.

I will perform these calculations by means of the optical theorem, taking the imaginary part of the forward scattering amplitudes. This means taking the imaginary part of the photon vacuum polarization tensor sandwiched between external states.

Since the optical theorem is a direct consequence of the unitarity of the S-matrix, any unitary regularization scheme must give the same result, once one expands in terms of a standard coupling.

The basic Lagrangian (4-dimensional) is

$$\begin{aligned}\mathscr{L} = & -\frac{1}{2}A^a_\mu\left(\partial^\mu\partial^\nu(1-\xi^{-1}) - g^{\mu\nu}\Box\right)A^a_\nu \\ & - g f^{abc}(\partial^\mu A^{a\nu})A^b_\mu A^c_\nu - \frac{g^2}{4} f^{abc} f^{ade} A^{b\mu} A^{c\nu} A^d_\mu A^e_\nu \\ & + i\sum_f \overline{\Psi}^i_f\left(\delta_{ij}\partial\!\!\!/ - i g t^a_{ij} A\!\!\!/^a - i g_V Q_f V\!\!\!/\right)\psi^j_f - \overline{c}^a\Box c^a + g f^{abc}\left(\partial_\mu \overline{c}^a\right) A^{b\mu} c^c .\end{aligned} \tag{4}$$

Some sample diagrams at 1, 2, and 3 loops are I will also compute the N_f^2 terms at 4 loops, The full

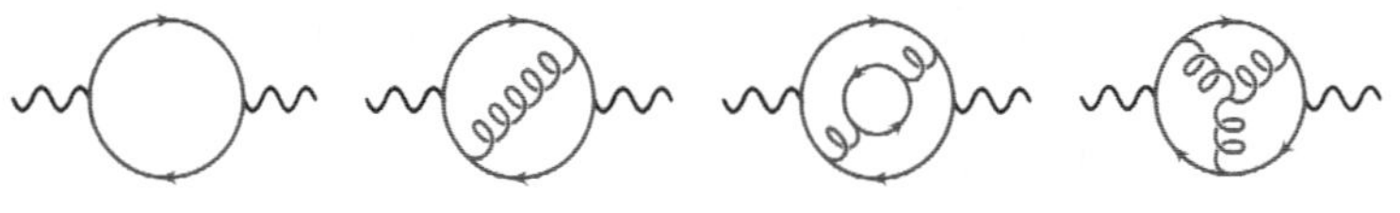

Figure 1. Sample diagrams at 1, 2 and 3 loops.

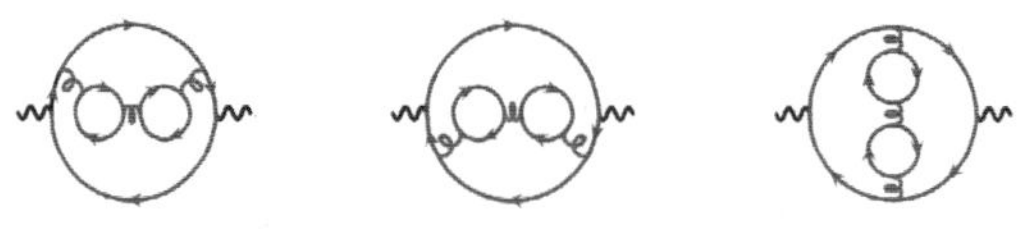

Figure 2. Sample diagrams at 4 loops.

result through order α_s^3 is well known [6, 7, 8, 9, 10]

$$
\begin{aligned}
\sigma^{e^+e^-\to\,\mathrm{had}}(Q^2) &= \frac{4\pi\alpha^2}{3Q^2} N_c \sum_f Q_f^2 \left\{ 1 + \left(\frac{\alpha_s^{\overline{\mathrm{MS}}}}{\pi}\right) C_F \frac{3}{4} \right. \\
&+ \left(\frac{\alpha_s^{\overline{\mathrm{MS}}}}{\pi}\right)^2 \left[\left(-C_F^2 \frac{3}{32} + C_F C_A \left(\frac{123}{32} - \frac{11}{4}\zeta_3 \right) + C_F N_f \left(-\frac{11}{16} + \frac{1}{2}\zeta_3 \right) \right) \right. \\
&+ \left(\frac{\alpha_s^{\overline{\mathrm{MS}}}}{\pi}\right)^3 \left[-C_F^3 \frac{69}{128} + C_F^2 C_A \left(-\frac{127}{64} - \frac{143}{16}\zeta_3 + \frac{55}{4}\zeta_5 \right) \right. \\
&\quad + C_F C_A^2 \left(\frac{90445}{3456} - \frac{2737}{144}\zeta_3 - \frac{55}{24}\zeta_5 \right) \\
&\quad + C_F^2 N_f \left(-\frac{29}{128} + \frac{19}{8}\zeta_3 - \frac{5}{2}\zeta_5 \right) + C_F C_A N_f \left(-\frac{485}{54} + \frac{56}{9}\zeta_3 + \frac{5}{12}\zeta_5 \right) \\
&\quad \left. \left. + C_F N_f^2 \left(\frac{151}{216} - \frac{19}{36}\zeta_3 \right) - \frac{1}{4}\pi^2 C_F \beta_0^{\overline{\mathrm{MS}}^2} \right] \right\} .
\end{aligned} \tag{5}
$$

3. Renormalization

In order to obtain the correct result, it is essential that we properly renormalize the theory. In CDR, this just means carrying out the standard $\overline{\mathrm{MS}}$ renormalization.

In DRED, we must follow a more elaborate program. Naïve application of minimal subtraction to the scattering amplitudes does not properly renormalize the evanescent terms [11]. Instead we must renormalize so that the evanescent Green functions are finite before we sum over the spin degrees of freedom [12, 13].

3.1. CDR Renormalization

In the CDR scheme, the Lagrangian has the same form as in 4 dimensions and the needed renormalizations are

$$
\begin{aligned}
&\Gamma^{(B)}_{AAA} = Z_1 \Gamma_{AAA}\,, && \psi_f^{(B)\,i} = Z_2^{\frac{1}{2}} \psi_f^i\,, && A_\mu^{(B)\,a} = Z_3^{\frac{1}{2}} A_\mu^a \\
&\Gamma^{(B)}_{c\bar{c}A} = \widetilde{Z}_1 \Gamma_{q\bar{q}A}\,, && c^{(B)\,a} = \widetilde{Z}_3^{\frac{1}{2}} c^a\,, && \bar{c}^{(B)\,a} = \widetilde{Z}_3^{\frac{1}{2}} \bar{c}^a\,, \\
&\Gamma^{(B)}_{q\bar{q}A} = Z_{1F} \Gamma_{q\bar{q}A}\,, && \xi^{(B)} = Z_3\, \xi\,. &&
\end{aligned} \tag{6}
$$

To remove sub-divergences in the calculation of the photon vacuum polarization, the QCD coupling needs to be renormalized, which requires the self-energy and vertex renormalization constants.

$$
\alpha_s^B = \left(\frac{\mu^2 e^{\gamma_E}}{4\pi} \right)^{\varepsilon} Z_{\alpha_s^{\overline{\mathrm{MS}}}}\, \alpha_s^{\overline{\mathrm{MS}}}\,, \qquad Z_{\alpha_s^{\overline{\mathrm{MS}}}} = \frac{Z_1^2}{Z_3^3} = \frac{Z_{1F}^2}{Z_2^2 Z_3} = \frac{\widetilde{Z}_1^2}{\widetilde{Z}_3^2 Z_3}\,. \tag{7}
$$

3.2. DRED Renormalization

Because the evanescent gauge bosons and their couplings are independent, the DRED Lagrangian and the resulting renormalization is far more complicated.

$$
\begin{aligned}
&\Gamma^{(B)}_{AAA} = Z_1 \Gamma_{AAA}\,, && \psi^{(B)\,i}_f = Z_2^{\frac{1}{2}}\,\psi^i_f\,, && A^{(B)\,a}_\mu = Z_3^{\frac{1}{2}}\,A^a_\mu\,,\\
&\Gamma^{(B)}_{c\bar{c}A} = \widetilde{Z}_1 \Gamma_{q\bar{q}A}\,, && c^{(B)\,a} = \widetilde{Z}_3^{\frac{1}{2}}\,c^a\,, && \bar{c}^{(B)\,a} = \widetilde{Z}_3^{\frac{1}{2}}\,\bar{c}^a\,,\\
&\Gamma^{(B)}_{q\bar{q}A} = Z_{1F} \Gamma_{q\bar{q}A}\,, && \xi^{(B)} = Z_3\,\xi\,, && \\
&\Gamma^{(B)}_{q\bar{q}e} = Z_{1e} \Gamma_{q\bar{q}e}\,, && A^{(B)\,a}_{e\,\mu} = Z_{3e}^{\frac{1}{2}} A^a_{e\,\mu}\,, && \Gamma^{(B)\,i}_{eeee} = Z^i_{1\,eeee}\,\Gamma^i_{eeee}\,,\\
&\Gamma^{(B)}_{q\bar{q}V_e} = Z_{1\,Ve} \Gamma_{q\bar{q}V_e}\,, && V^{(B)}_{e\,\mu} = Z_{3\,Ve}^{\frac{1}{2}}\,V_{e\,\mu}\,. &&
\end{aligned}
\tag{8}
$$

Note that we also need to compute the wavefunction and vertex corrections of the evanescent photon!

As in CDR, subdivergences are removed through coupling constant renormalizations, but in DRED, there are many couplings to renormalize.

$$
\begin{aligned}
\alpha_s^B &= \left(\frac{\mu^2 e^{\gamma_E}}{4\pi}\right)^{\varepsilon} Z_{\alpha_s^{\overline{\rm DR}}}\,\alpha_s^{\overline{\rm DR}}\,, & Z_{\alpha_s^{\overline{\rm DR}}} &= \frac{Z_1^2}{Z_3^3} = \frac{Z_{1F}^2}{Z_2^2 Z_3} = \frac{\widetilde{Z}_1^2}{\widetilde{Z}_3^2 Z_3}\,,\\
\alpha_e^B &= \left(\frac{\mu^2 e^{\gamma_E}}{4\pi}\right)^{\varepsilon} Z_{\alpha_e^{\overline{\rm DR}}}\,\alpha_e^{\overline{\rm DR}}\,, & Z_{\alpha_e^{\overline{\rm DR}}} &= \frac{Z_{1e}^2}{Z_2^2 Z_{3e}}\,,\\
\eta_i^B &= \left(\frac{\mu^2 e^{\gamma_E}}{4\pi}\right)^{\varepsilon} Z_{\eta_i^{\overline{\rm DR}}}\,\eta_i^{\overline{\rm DR}}\,, & Z_{\eta_i^{\overline{\rm DR}}} &= \frac{\left(Z^i_{1\,eeee}\right)^2}{Z_{3e}^4}\,,\\
\alpha_{Ve}^B &= \left(\frac{\mu^2 e^{\gamma_E}}{4\pi}\right)^{\varepsilon} Z_{\alpha_{Ve}^{\overline{\rm DR}}}\,\alpha_{Ve}^{\overline{\rm DR}}\,, & Z_{\alpha_{Ve}^{\overline{\rm DR}}} &= \frac{Z_{1Ve}^2}{Z_2^2 Z_{3Ve}}\,.
\end{aligned}
\tag{9}
$$

3.3. FDH Renormalization

As in the CDR scheme, the Lagrangian in FDH has the same form as in 4 dimensions and the needed renormalizations are

$$
\begin{aligned}
&\Gamma^{(B)}_{AAA} = Z_1 \Gamma_{AAA}\,, && \psi^{(B)\,i}_f = Z_2^{\frac{1}{2}}\,\psi^i_f\,, && A^{(B)\,a}_\mu = Z_3^{\frac{1}{2}}\,A^a_\mu\,,\\
&\Gamma^{(B)}_{c\bar{c}A} = \widetilde{Z}_1 \Gamma_{q\bar{q}A}\,, && c^{(B)\,a} = \widetilde{Z}_3^{\frac{1}{2}}\,c^a\,, && \bar{c}^{(B)\,a} = \widetilde{Z}_3^{\frac{1}{2}}\,\bar{c}^a\,,\\
&\Gamma^{(B)}_{q\bar{q}A} = Z_{1F} \Gamma_{q\bar{q}A}\,, && \xi^{(B)} = Z_3\,\xi\,, &&
\end{aligned}
\tag{10}
$$

Again as in CDR, it would seem that only the QCD coupling needs to be renormalized.

$$
\alpha_s^B = \left(\frac{\mu^2 e^{\gamma_E}}{4\pi}\right)^{\varepsilon} Z_{\alpha_s^{\overline{\rm FDH}}}\,\alpha_s^{\overline{\rm FDH}}\,, \qquad Z_{\alpha_s^{\overline{\rm FDH}}} = \frac{Z_1^2}{Z_3^3} = \frac{Z_{1F}^2}{Z_2^2 Z_3} = \frac{\widetilde{Z}_1^2}{\widetilde{Z}_3^2 Z_3}\,. \tag{11}
$$

4. Results

4.1. Methods

For each of the regularization schemes, I generate the necessary Feynman diagrams using QGRAF [14]. I then use the symbolic algebra program FORM [15] to identify the integral topology and implement the Feynman rules. The resulting Feynman integrals are reduced to master integrals using the program REDUZE [16].

The master integrals that occur in this calculation are: Most of the master integrals are trivial iterated-bubble diagrams and the others were evaluated long ago [17, 18].

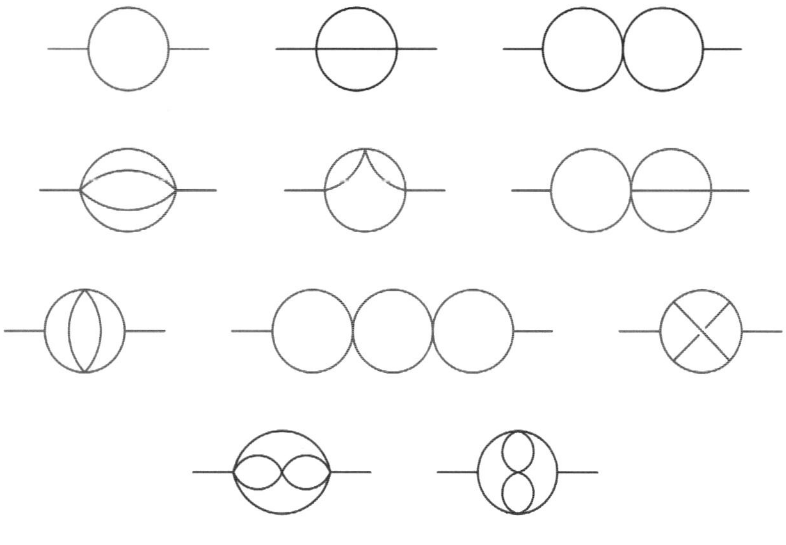

Figure 3. Master Integrals that appear in this calculation.

4.2. Conventional Dimensional Regularization

The imaginary part of the unrenormalized vacuum polarization tensor in the CDR scheme is

$$
\begin{aligned}
\Im\left[\Pi^{(B)}_{\mu\nu}(Q)\Big|_{CDR}\right] = \; & \frac{-Q^2 g_{\mu\nu} + Q_\mu Q_\nu}{3}\,\alpha_V^B N_c \sum_f Q_f^2 \left(\frac{4\pi}{Q^2 e^{\gamma_E}}\right)^{\varepsilon} \Bigg\{ \\
& 1 + \left(\frac{\alpha_s^B}{\pi}\right)\left(\frac{4\pi}{Q^2 e^{\gamma_E}}\right)^{\varepsilon} C_F \left[\frac{3}{4} + \varepsilon\left(\frac{55}{8} - 6\zeta_3\right) + \varepsilon^2\left(\frac{1711}{48} - \frac{15}{4}\zeta_2 - 19\zeta_3 - 9\zeta_4\right) + \mathcal{O}(\varepsilon^3)\right] \\
& + \left(\frac{\alpha_s^B}{\pi}\right)^2 \left(\frac{4\pi}{Q^2 e^{\gamma_E}}\right)^{2\varepsilon} \Bigg[\frac{1}{\varepsilon}\left(\frac{11}{16} C_F C_A - \frac{1}{8} C_F N_f\right) \\
& - \frac{3}{32} C_F^2 + C_F C_A \left(\frac{487}{48} - \frac{33}{4}\zeta_3\right) + C_F N_f \left(-\frac{11}{6} + \frac{3}{2}\zeta_3\right) \\
& + \varepsilon \left(C_F^2 \left(-\frac{143}{32} - \frac{111}{8}\zeta_3 + \frac{45}{2}\zeta_5\right) + C_F C_A \left(\frac{50339}{576} - \frac{231}{32}\zeta_2 - \frac{109}{2}\zeta_3 - \frac{99}{8}\zeta_4 - \frac{15}{4}\zeta_5\right) \right. \\
& \left. + C_F N_f \left(-\frac{4417}{288} + \frac{21}{16}\zeta_2 + \frac{19}{2}\zeta_3 + \frac{9}{4}\zeta_4\right)\right) + \mathcal{O}(\varepsilon^2)\Bigg] \\
& + \left(\frac{\alpha_s^B}{\pi}\right)^3 \left(\frac{4\pi}{Q^2 e^{\gamma_E}}\right)^{3\varepsilon} C_F N_f^2 \left[\frac{1}{48\varepsilon^2} + \frac{1}{\varepsilon}\left(\frac{121}{288} - \frac{1}{3}\zeta_3\right) + \frac{2777}{576} - \frac{3}{8}\zeta_2 - \frac{19}{6}\zeta_3 - \frac{1}{2}\zeta_4\right] + \ldots \Bigg\} .
\end{aligned} \tag{12}
$$

Upon renormalizing the couplings, I find

$$\Im\left[\Pi_{\mu\nu}(Q)\big|_{CDR}\right] = \frac{-Q^2 g_{\mu\nu} + Q_\mu Q_\nu}{3}\alpha_V N_c \sum_f Q_f^2 \left\{ 1 + \left(\frac{\alpha_s^{\overline{\rm MS}}}{\pi}\right) C_F \frac{3}{4} \right.$$
$$+ \left(\frac{\alpha_s^{\overline{\rm MS}}}{\pi}\right)^2 \left[-C_F^2 \frac{3}{32} + C_F C_A \left(\frac{123}{32} - \frac{11}{4}\zeta_3\right) + C_F N_f \left(-\frac{11}{16} + \frac{1}{2}\zeta_3\right)\right] \quad (13)$$
$$\left. + \left(\frac{\alpha_s^{\overline{\rm MS}}}{\pi}\right)^3 C_F N_f^2 \left[\frac{151}{216} - \frac{1}{24}\zeta_2 - \frac{19}{36}\zeta_3\right] + \ldots \right\}.$$

Sandwiching the vacuum polarization between external states, I obtain the expected result that I showed in equation (5).

4.3. Dimensional Reduction

In DRED, there are two independent vacuum polarization tensors to compute, corresponding to the photon and the evanescent photon.

$$\Im\left[\Pi_{\mu\nu}^{(B)}(Q)\Big|_{DRED}\right] = \frac{-Q^2 \hat{g}_{\mu\nu} + Q_\mu Q_\nu}{3}\Im\left[\Pi_A^{(B)}(Q)\Big|_{DRED}\right] - Q^2 \frac{\delta_{\mu\nu}}{2\varepsilon}\Im\left[\Pi_B^{(B)}(Q)\Big|_{DRED}\right]. \quad (14)$$

Before renormalization, both components are singular and depend on the QCD coupling and the various evanescent couplings.

$$\Im\left[\Pi_A^{(B)}(Q)\Big|_{DRED}\right] = \alpha_V^B N_c \sum_f Q_f^2 \left(\frac{4\pi}{Q^2 e^{\gamma_E}}\right)^\varepsilon \left\{ \right.$$
$$1 + \left(\frac{\alpha_s^B}{\pi}\right)\left(\frac{4\pi}{Q^2 e^{\gamma_E}}\right)^\varepsilon C_F \left[\frac{3}{4} + \varepsilon\left(\frac{51}{8} - 6\zeta_3\right) + \varepsilon^2\left(\frac{497}{16} - \frac{15}{4}\zeta_2 - 15\zeta_3 - 9\zeta_4\right) + \mathcal{O}(\varepsilon^3)\right]$$
$$+ \left(\frac{\alpha_e^B}{\pi}\right)\left(\frac{4\pi}{Q^2 e^{\gamma_E}}\right)^\varepsilon C_F \left[-\varepsilon\frac{3}{4} - \varepsilon^2\frac{29}{8} + \mathcal{O}(\varepsilon^3)\right]$$
$$+ \left(\frac{\alpha_s^B}{\pi}\right)^2 \left(\frac{4\pi}{Q^2 e^{\gamma_E}}\right)^{2\varepsilon} \left[\frac{1}{\varepsilon}\left(\frac{11}{16}C_F C_A - \frac{1}{8}C_F N_f\right) - \frac{3}{32}C_F^2 + \left(\frac{77}{8} - \frac{33}{4}\zeta_3\right) C_F C_A - \left(\frac{7}{4} - \frac{3}{2}\zeta_3\right) C_F N_f \right.$$
$$+ \varepsilon\left(C_F^2\left(-\frac{141}{32} - \frac{111}{8}\zeta_3 + \frac{45}{2}\zeta_5\right) + C_F C_A\left(\frac{15301}{192} - \frac{231}{32}\zeta_2 - \frac{193}{4}\zeta_3 - \frac{99}{8}\zeta_4 - \frac{15}{4}\zeta_5\right)\right.$$
$$\left.\left. + C_F N_f\left(-\frac{1355}{96} + \frac{21}{16}\zeta_2 + \frac{17}{2}\zeta_3 + \frac{9}{4}\zeta_4\right)\right) + \mathcal{O}(\varepsilon^2)\right]$$
$$+ \left(\frac{\alpha_e^B}{\pi}\right)^2 \left(\frac{4\pi}{Q^2 e^{\gamma_E}}\right)^{2\varepsilon} \left[\frac{3}{4}C_F^2 - \frac{3}{8}C_F C_A + \frac{3}{16}C_F N_f - \varepsilon\left(\frac{47}{8}C_F^2 - \frac{11}{4}C_F C_A + \frac{7}{4}C_F N_f\right) + \mathcal{O}(\varepsilon^2)\right]$$
$$+ \left(\frac{\alpha_s^B}{\pi}\right)\left(\frac{\alpha_e^B}{\pi}\right) \left(\frac{4\pi}{Q^2 e^{\gamma_E}}\right)^{2\varepsilon} \left[-\frac{9}{8}C_F^2 - \varepsilon\left(\frac{141}{16}C_F^2 + \frac{21}{16}C_F C_A\right) + \mathcal{O}(\varepsilon^2)\right]$$
$$+ \left(\frac{\alpha_s^B}{\pi}\right)^3 \left(\frac{4\pi}{Q^2 e^{\gamma_E}}\right)^{3\varepsilon} C_F N_f^2 \left[\frac{1}{48\varepsilon^2} + \frac{1}{\varepsilon}\left(\frac{13}{32} - \frac{1}{3}\zeta_3\right) + \frac{7847}{1728} - \frac{3}{8}\zeta_2 - \frac{53}{18}\zeta_3 - \frac{1}{2}\zeta_4\right]$$
$$\left. + \left(\frac{\alpha_e^B}{\pi}\right)^3 \left(\frac{4\pi}{Q^2 e^{\gamma_E}}\right)^{3\varepsilon} C_F N_f^2 \left[-\frac{1}{\varepsilon}\frac{3}{64} - \frac{83}{128}\right] + \ldots \right\},$$

(15)

$$
\begin{aligned}
\Im\left[\Pi_B^{(B)}(Q)\Big|_{DRED}\right] &= \alpha_{Ve}^B N_c \sum_f Q_f^2 \left(\frac{4\pi}{Q^2 e^{\gamma_E}}\right)^{\varepsilon} \Bigg\{ \varepsilon + 2\varepsilon^2 + \left(4 - \frac{3}{2}\zeta_2\right)\varepsilon^3 + \mathcal{O}(\varepsilon^4) \\
&+ \left(\frac{\alpha_s^B}{\pi}\right)\left(\frac{4\pi}{Q^2 e^{\gamma_E}}\right)^{\varepsilon} C_F \left[\frac{3}{2} + \varepsilon\frac{29}{4} + \varepsilon^2\left(\frac{227}{8} - \frac{15}{2}\zeta_2 - 6\zeta_3\right) + \mathcal{O}(\varepsilon^3)\right] \\
&+ \left(\frac{\alpha_e^B}{\pi}\right)\left(\frac{4\pi}{Q^2 e^{\gamma_E}}\right)^{\varepsilon} C_F \left[-1 - 4\varepsilon - \varepsilon^2\left(\frac{27}{2} - 5\zeta_2\right) + \mathcal{O}(\varepsilon^3)\right] \\
&+ \left(\frac{\alpha_s^B}{\pi}\right)^2 \left(\frac{4\pi}{Q^2 e^{\gamma_E}}\right)^{2\varepsilon} \Bigg[\frac{1}{\varepsilon}\left(\frac{9}{8}C_F^2 + \frac{11}{16}C_F C_A - \frac{1}{8}C_F N_f\right) + \frac{279}{32}C_F^2 + \frac{199}{32}C_F C_A - \frac{17}{16}C_F N_f \\
&\quad + \varepsilon\left(C_F^2\left(\frac{3139}{64} - \frac{189}{16}\zeta_2 - \frac{45}{4}\zeta_3\right) + C_F C_A\left(\frac{2473}{64} - \frac{231}{32}\zeta_2 - \frac{75}{8}\zeta_3\right)\right. \\
&\quad \left. + C_F N_f\left(-\frac{207}{32} + \frac{21}{16}\zeta_2 + \frac{3}{2}\zeta_3\right)\right) + \mathcal{O}(\varepsilon^2)\Bigg] \\
&+ \left(\frac{\alpha_s^B}{\pi}\right)\left(\frac{\alpha_e^B}{\pi}\right)\left(\frac{4\pi}{Q^2 e^{\gamma_E}}\right)^{2\varepsilon} \Bigg[-\frac{1}{\varepsilon}\frac{9}{4}C_F^2 - \frac{129}{8}C_F^2 - \frac{3}{8}C_F C_A \\
&\quad - \varepsilon\left(\left(\frac{671}{8} - \frac{189}{8}\zeta_2 - 9\zeta_3\right)C_F^2 + \frac{53}{16}C_F C_A\right) + \mathcal{O}(\varepsilon^2)\Bigg] \\
&+ \left(\frac{\alpha_e^B}{\pi}\right)^2 \left(\frac{4\pi}{Q^2 e^{\gamma_E}}\right)^{2\varepsilon} \Bigg[\frac{1}{\varepsilon}\left(C_F^2 - \frac{1}{4}C_F C_A + \frac{1}{8}C_F N_f\right) + \frac{13}{2}C_F^2 - \frac{3}{2}C_F C_A + \frac{15}{16}C_F N_f \\
&\quad + \varepsilon\left(\left(31 - \frac{21}{2}\zeta_2 - \frac{3}{4}\zeta_3\right)C_F^2 - \left(\frac{53}{8} - \frac{21}{8}\zeta_2 - \frac{3}{8}\zeta_3\right)C_F C_A + \left(\frac{157}{32} - \frac{21}{16}\zeta_2\right)C_F N_f\right) + \mathcal{O}(\varepsilon^2)\Bigg] \\
&+ \left(\frac{\alpha_s^B}{\pi}\right)^3 \left(\frac{4\pi}{Q^2 e^{\gamma_E}}\right)^{3\varepsilon} C_F N_f^2 \left[\frac{1}{72\varepsilon^2} + \frac{1}{\varepsilon}\frac{73}{432} + \frac{3595}{2592} - \frac{1}{4}\zeta_2 - \frac{1}{3}\zeta_3\right] \\
&+ \left(\frac{\alpha_e^B}{\pi}\right)^3 \left(\frac{4\pi}{Q^2 e^{\gamma_E}}\right)^{3\varepsilon} C_F N_f^2 \left[-\frac{1}{48\varepsilon^2} - \frac{1}{\varepsilon}\frac{11}{48} - \frac{155}{96} + \frac{3}{8}\zeta_2\right] \qquad + \ldots \qquad \Bigg\}.
\end{aligned} \tag{16}
$$

Renormalizing the many couplings, including that of the evanescent photon, including the finite renormalization [19, 20] which shifts $\alpha_s^{\overline{\mathrm{DR}}}$ to $\alpha_s^{\overline{\mathrm{MS}}}$, I obtain

$$
\begin{aligned}
\Im\left[\Pi_A(Q)|_{DRED}\right] &= \alpha_V N_c \sum_f Q_f^2 \Bigg\{1 + \left(\frac{\alpha_s^{\overline{\mathrm{MS}}}}{\pi}\right) C_F \frac{3}{4} \\
&+ \left(\frac{\alpha_s^{\overline{\mathrm{MS}}}}{\pi}\right)^2 \left[-C_F^2\frac{3}{32} + C_F C_A\left(\frac{123}{32} - \frac{11}{4}\zeta_3\right) + C_F N_f\left(-\frac{11}{16} + \frac{1}{2}\zeta_3\right)\right] \\
&+ \left(\frac{\alpha_s^{\overline{\mathrm{MS}}}}{\pi}\right)^3 C_F N_f^2 \left[\frac{151}{216} - \frac{1}{24}\zeta_2 - \frac{19}{36}\zeta_3\right]\Bigg\}, \\
\Im\left[\Pi_B(Q)|_{DRED}\right] &= \mathcal{O}(\varepsilon).
\end{aligned} \tag{17}
$$

The evanescent vacuum polarization does not contribute to the cross section, while the gluon vacuum polarization produces exactly the expected result.

4.4. Four Dimensional Helicity

In FDH, however, the calculation is in trouble from the very beginning. The calculation is term-by-term identical to the DRED calculation except that evanescent terms are identified with gauge terms. So, as in DRED, the vacuum polarization tensor splits into two independent components; a D_m-dimensional component and a D_x-dimensional ($D_x = D_s - D_m$) component. For the photon vacuum polarization, the demand that external states be 4-dimensional means that we only need the D_m-dimensional component.

The gluon vacuum polarization, however, is a problem, since we need to extract the renormalization constant to determine the β-function. At one loop, averaging over degrees of freedom means that only the D_m-dimensional piece contributes, and we get the usual QCD β-function. At two loops, the D_x-dimensional piece is still singular after spin averaging. Only by dropping the D_x term do I get the usual two-loop contribution to the QCD β-function,

$$
\begin{aligned}
\Im\left[\Pi_A^{(B)}(Q)\Big|_{FDH}\right] &= \alpha_V^B N_c \sum_f Q_f^2 \left(\frac{4\pi}{Q^2 e^{\gamma_E}}\right)^{\varepsilon} \Bigg\{ \\
& 1+\left(\frac{\alpha_s^B}{\pi}\right)\left(\frac{4\pi}{Q^2 e^{\gamma_E}}\right)^{\varepsilon} C_F \left[\frac{3}{4}+\varepsilon\left(\frac{45}{8}-6\zeta_3\right)+\varepsilon^2\left(\frac{439}{16}-\frac{15}{4}\zeta_2-15\zeta_3-9\zeta_4\right)+\mathcal{O}(\varepsilon^3)\right] \\
& +\left(\frac{\alpha_s^B}{\pi}\right)^2\left(\frac{4\pi}{Q^2 e^{\gamma_E}}\right)^{2\varepsilon}\left[\frac{1}{\varepsilon}\left(\frac{11}{16}C_F C_A-\frac{1}{8}C_F N_f\right)-\frac{15}{32}C_F^2+\left(\frac{37}{4}-\frac{33}{4}\zeta_3\right)C_F C_A\right. \\
& -\left(\frac{25}{16}-\frac{3}{2}\zeta_3\right)C_F N_f+\varepsilon\left(C_F^2\left(-\frac{235}{32}-\frac{111}{8}\zeta_3+\frac{45}{2}\zeta_5\right)\right. \\
& +C_F C_A\left(\frac{14521}{192}-\frac{231}{32}\zeta_2-\frac{193}{4}\zeta_3-\frac{99}{8}\zeta_4-\frac{15}{4}\zeta_5\right) \\
& \left.\left.+C_F N_f\left(-\frac{1187}{96}+\frac{21}{16}\zeta_2+\frac{17}{2}\zeta_3+\frac{9}{4}\zeta_4\right)\right)+\mathcal{O}(\varepsilon^2)\right] \\
& +\left(\frac{\alpha_s^B}{\pi}\right)^3\left(\frac{4\pi}{Q^2 e^{\gamma_E}}\right)^{3\varepsilon} C_F N_f^2\left[\frac{1}{48\varepsilon^2}+\frac{1}{\varepsilon}\left(\frac{23}{64}-\frac{1}{3}\zeta_3\right)+\frac{13453}{3456}-\frac{3}{8}\zeta_2-\frac{53}{18}\zeta_3-\frac{1}{2}\zeta_4\right]\Bigg\}.
\end{aligned} \tag{18}
$$

I only need the leading term in the β-function to renormalize these terms.

$$
\begin{aligned}
\Im\left[\Pi_A(Q)|_{FDH}\right] &= \alpha_V N_c \sum_f Q_f^2 \Bigg\{ 1+\left(\frac{\alpha_s^{\mathrm{FDH}}}{\pi}\right)\frac{3}{4}C_F \\
& +\left(\frac{\alpha_s^{\mathrm{FDH}}}{\pi}\right)^2\left[-C_F^2\frac{15}{32}+C_F C_A\left(\frac{131}{32}-\frac{11}{4}\zeta_3\right)+C_F N_f\left(-\frac{5}{8}+\frac{1}{2}\zeta_3\right)\right] \\
& +\left(\frac{\alpha_s^{\mathrm{FDH}}}{\pi}\right)^3 C_F N_f^2\left[-\frac{1}{192\varepsilon}+\frac{1843}{3456}-\frac{1}{24}\zeta_2-\frac{19}{36}\zeta_3\right]\Bigg\}.
\end{aligned} \tag{19}
$$

Even after the finite transformation of $\alpha_s^{\overline{\mathrm{FDH}}} \to \alpha_s^{\overline{\mathrm{MS}}}$, the NNLO term is incorrect and no finite transformation can repair the fact that the $\mathrm{N^3LO}$ term is singular.

One must conclude that the renormalization program of the FDH scheme has failed, resulting in the violation of unitarity.

Is this the end of the FDH?

No! The FDH greatly simplifies loop calculations and is therefore a useful tool. The fact that it is non-unitary and generates spurious terms beyond one-loop simply means that we must find ways to correct for the errors [21].

Because the breakdown is a failure of the renormalization program, the spurious terms are proportional to the lower-loop contributions and can be determined from them. Boughezal, Melnikov and Petriello have proposed a method that they call "dimensional reconstruction" which involves calculating the lower-loop terms in various integer dimensions, allowing one to solve for the correct renormalizations.

5. Conclusions

I have studied the behavior of several regularization schemes in high-order radiative corrections. I find that the CDR and DRED schemes are correct and equivalent ways of performing QCD calculations through N^3LO. The FDH scheme, however, has been shown to be incorrect and to violate unitarity beyond NLO when applied to nonsupersymmetric theories.

The FDH scheme is not a unitary regularization scheme because its renormalization program fails to remove all of the ultraviolet singularities. This failure can be overcome by solving for the terms which must multiply the lower-loop results to remove the uncanceled ultraviolet singularities.

References

[1] Bern Z and Kosower D A 1992 *Nucl. Phys.* **B379** 451–561
[2] Bern Z, De Freitas A, Dixon L J and Wong H L 2002 *Phys. Rev.* **D66** 085002 (*Preprint* hep-ph/0202271)
[3] 't Hooft G and Veltman M J G 1972 *Nucl. Phys.* **B44** 189–213
[4] Collins J 1984 *Renormalization* (Cambridge, England: Cambridge University Press)
[5] Siegel W 1979 *Phys. Lett.* **B84** 193
[6] Chetyrkin K G, Kataev A L and Tkachov F V 1979 *Phys. Lett.* **B85** 277
[7] Dine M and Sapirstein J R 1979 *Phys. Rev. Lett.* **43** 668
[8] Celmaster W and Gonsalves R J 1980 *Phys. Rev.* **D21** 3112
[9] Gorishnii S G, Kataev A L and Larin S A 1988 *Phys. Lett.* **B212** 238–244
[10] Gorishnii S G, Kataev A L and Larin S A 1991 *Phys. Lett.* **B259** 144–150
[11] van Damme R and 't Hooft G 1985 *Phys. Lett.* **B150** 133
[12] Jack I, Jones D R T and Roberts K L 1994 *Z. Phys.* **C62** 161–166 (*Preprint* hep-ph/9310301)
[13] Jack I, Jones D R T and Roberts K L 1994 *Z. Phys.* **C63** 151–160 (*Preprint* hep-ph/9401349)
[14] Nogueira P 1993 *J. Comput. Phys.* **105** 279–289
[15] Vermaseren J A M 2000 New features of FORM Report No. NIKHEF-00-0032 (*Preprint* math-ph/0010025)
[16] Studerus C 2010 *Comput. Phys. Commun.* **181** 1293–1300 (*Preprint* 0912.2546)
[17] Chetyrkin K G, Kataev A L and Tkachov F V 1980 *Nucl. Phys.* **B174** 345–377
[18] Kazakov D I 1984 *Theor. Math. Phys.* **58** 223–230
[19] Kunszt Z, Signer A and Trocsanyi Z 1994 *Nucl.Phys.* **B411** 397–442 (*Preprint* hep-ph/9305239)
[20] Harlander R, Kant P, Mihaila L and Steinhauser M 2006 *JHEP* **09** 053 (*Preprint* hep-ph/0607240)
[21] Boughezal R, Melnikov K and Petriello F 2011 *Phys.Rev.* **D84** 034044 (*Preprint* 1106.5520)

Journal of Physics: Conference Series **368** (2012) 012060 doi:10.1088/1742-6596/368/1/012060

Regularization of IR divergent loop integrals

E de Doncker[1], F Yuasa[2] and Y Kurihara[2]

[1] Department of Computer Science, Western Michigan University, Kalamazoo MI 49008, U. S.
[2] High Energy Accelerator Research Organization (KEK), Oho 1-1, Tsukuba, Ibaraki, 305-0801, Japan

E-mail: elise.dedoncker@wmich.edu, fukuko.yuasa@kek.jp, yoshimasa.kurihara@kek.jp

Abstract. We report results of a new numerical regularization technique for *infrared (IR) divergent* loop integrals using *dimensional regularization*, where a positive regularization parameter ε, satisfying that the dimension $d = 4 + 2\varepsilon$, is introduced in the integrand to keep the integral from diverging as long as $\varepsilon > 0$. A sequence of integrals is computed for decreasing values of ε, in order to carry out a *linear extrapolation as* $\varepsilon \to 0$. Each integral in the sequence is calculated according to the *Direct Computation Method (DCM)* to handle (threshold) integrand singularities in the interior of the domain. The technique of this paper is applied to one-loop N-point functions. In order to simplify the computation of the integrals for small ε, particularly in the case of a threshold singularity, a *reduction* of the N-point function is performed *numerically* to a set of 3-point and 4-point integrals, and *DCM* is applied to the resulting vertex and box integrals.

1. Introduction

The integral for an n-dimensional N-point Feynman diagram with L loops can be represented as

$$\mathcal{I} = (-1)^N \Gamma(N - Ln/2) \int_0^\infty d^N x \, \delta(1 - \sum_{j=1}^N x_j) \, \frac{\mathcal{U}^{N-(L+1)n/2}}{\mathcal{F}^{N-Ln/2}}. \tag{1}$$

Here the functions $\mathcal{F}$ and $\mathcal{U}$ are polynomials determined by the topology of the corresponding Feynman diagram [1]. A mechanism for the evaluation of *infrared (IR) divergent* multi-loop integrals of the form (1) by sector decomposition and dimensional regularization is given in [2].

We consider the one-loop N-point function, denoted in [3] by

$$\mathcal{I} = I_N^n = (-1)^N \Gamma(N - n/2) \int_0^\infty d^N x \, \frac{\delta(1 - \sum_{j=1}^N x_j)}{(x \cdot S^{(N)} \cdot x/2)^{N-n/2}} \tag{2}$$

with

$$S_{k\ell}^{(N)} = -(r_\ell - r_k)^2 + m_\ell^2 + m_k^2, \;\; 1 \le k, \ell \le N \tag{3}$$

where $r_\ell = \sum_{j=1}^\ell p_j$; the p_j and m_j are the external loop momenta and the corresponding particle masses, respectively. In order to account for IR divergence, the formalism of dimensional regularization can be applied by setting the dimension $n = 4 + 2\varepsilon$. The integral in equation (2) then generally *diverges* as $\varepsilon \to 0$ *(*$n \to 4$*)*. It is expanded with respect to ε, via sector decompositions of the integration domain and expansion of the integrand around $\varepsilon = 0$ in the resulting sector functions.

The expansion coefficients are calculated in [4] using FORM [5], as sums of multivariate integrals assembled from contributions over the sectors. The coefficients are derived in [6] by expanding the expressions for the integrals directly, based on properties of hypergeometric and related functions. We make use of the asymptotic behavior to compute the coefficients numerically by a linear extrapolation.

We construct a *linear system* of equations, which incorporates values of the integral for decreasing ε in the right hand side of the system. The linear system is extended by one equation at a time and solved for the leading coefficients of the Laurent expansion of the integral. This gives rise to an extrapolation as ε tends to zero. The solutions can be obtained by solving the systems directly or by a *recursive method* [7]. We will make use of *Bulirsch* [8] type sequences of the form $b_\ell = 2, 3, 4, 6, 8, 12, 16, 24, \ldots$, in order to set $\varepsilon = \varepsilon_\ell = 1/b_\ell$.

In previous work [9, 10] we applied the *dimensional extrapolation* to infrared cases where the integrand denominator does not vanish in the interior of the integration domain. In [11] we dealt with vanishing denominators for vertex (triangle) integrals. At present we target one-loop N-point functions after their reduction to vertex and box integrals according to [3, 12], and sector decompositions to disentangle overlapping singularities along multiple coordinate axes. These will allow for an efficient numerical computation of integrals which suffer from both threshold singularities inside the domain, and IR divergence causing singular behavior at the boundaries of the integration region.

All computations are numerical and performed with automatic integration code after reduction and sector decompositions, as implemented using QUADPACK in [13] for non-IR divergent N-point functions through the hexagon. The basic vertex and box functions are computed with techniques of the *Direct computation method (DCM)* [10, 14, 15]:

- *iterated multivariate integration*, which uses one-dimensional or low-dimensional integration techniques repeatedly [14, 16, 17];
- *nonlinear extrapolation* by the ϵ-algorithm [18, 19], to handle singularities inside the domain.

DCM relies on the one-dimensional QUADPACK [20] numerical methods DQAG and DQAGS, which are deterministic and adaptive. DQAGS also uses a nonlinear extrapolation internally, for treating algebraic and algebraic-logarithmic integrand singularities. *DCM* introduces a term $i\delta$ with finite $\delta > 0$ in the integrand denominator, and extrapolates a sequence of integral values for decreasing values of δ.

It is our goal in this paper to extend *DCM* to IR divergent integrals which may have threshold singularities, in particular for the basic triangle and box elements in the decomposition, by resorting to a *double extrapolation* or regularization (with respect to δ and ε) if needed.

In section 2 we review the *reduction* and *sector decomposition* formalism of one-loop N-point functions into triangles and boxes. Section 3 explains the dimensional extrapolation by solving a linear system, or by using a recursive method developed for Vandermonde type systems in [7]. The use of the linear extrapolation is validated by the underlying asymptotic expansions for the vertex and box sub-problems, which leads to the double extrapolation for IR divergent problems where the denominator also vanishes in the interior of the integration domain. Numerical results of the procedures are given for sample problems in section 4.

2. Reduction overview

Through the reduction formalism applied in [3], the n-dimensional hexagon, pentagon and box functions are expressed in terms of n-dimensional triangle and $(n+2)$-dimensional box functions.

Assuming $\det(S) \neq 0$,

$$I_N^n = \sum_{k=1}^{N} B_k I_{N-1,k} + (N-n-1)\,\frac{\det(G)}{\det(S)}\, I_N^{n+2}, \tag{4}$$

where S is defined by equation (3), G is the Gram matrix, $G_{kl} = 2r_l r_k,\ k, \ell = 1, \cdots, N$; and the reduction coefficients $B_k = -\sum_{l=1}^{N} S_{kl}^{-1}$ can be obtained by solving the system of linear equations

$\sum_{\ell=1}^{N} S_{k\ell} B_\ell = -1,\ k = 1, \dots, N$. Infrared singularities emerge through poles in the integral expansions as a function of ε where $n = 4+2\varepsilon$. Thus the n-dimensional N-point function is decomposed as a linear combination of n-dimensional $(N-1)$-point functions, and a remainder term which involves a $(n+2)$-dimensional N-point function. It is noted in [3] that the latter is IR finite, even in the massless case. In particular, denoting the Mandelstam variables $s_j = p_j^2$ and $s_{ij\dots} = (p_i + p_j + \dots)^2$, the IR behavior in

$$\begin{aligned} I_4^n(s_{12}, s_{23}, s_1, s_2, s_3, s_4, m_1^2, m_2^2, m_3^2, m_4^2) = \\ + B_1 I_3^n(s_{12}, s_3, s_4, m_2, m_3, m_4) + B_2 I_3^n(s_{23}, s_4, s_1, m_3, m_4, m_1) \\ + B_3 I_3^n(s_{12}, s_1, s_2, m_4, m_1, m_2) + B_4 I_3^n(s_{23}, s_2, s_3, m_1, m_2, m_3) \\ + (n-3)(B_1 + B_2 + B_3 + B_4)\, I_4^{n+2}(s_{12}, s_{23}, s_1, s_2, s_3, s_4, m_1^2, m_2^2, m_3^2, m_4^2) \end{aligned} \tag{5}$$

is confined to the triangle integrals.

The integrals I_3^n and I_4^{n+2} are split into sector functions of the form

$$I_3^n(s_1, s_2, s_3, m_1^2, m_2^2, m_3^2) = -\Gamma(3 - n/2) \sum_{\mathcal{P}_{(1,2,3)}} S_{Tri}^n(s_1, s_2, s_3, m_1^2, m_2^2, m_3^2) \tag{6}$$

where $\mathcal{P}_{(1,2,3)}$ is the set of the cyclic permutations of $(1, 2, 3)$, and

$$\begin{aligned} I_4^{n+2}(s_{12}, s_{23}, s_1, s_2, s_3, s_4, m_1^2, m_2^2, m_3^2, m_4^2) = -\Gamma(4 - n/2)\times \\ (S_{Box}^{n+2}(s_{23}, s_{12}, s_2, s_3, s_4, s_1, m_2^2, m_3^2, m_4^2, m_1^2) + S_{Box}^{n+2}(s_{12}, s_{23}, s_3, s_4, s_1, s_2, m_3^2, m_4^2, m_1^2, m_2^2) \\ + S_{Box}^{n+2}(s_{23}, s_{12}, s_4, s_1, s_2, s_3, m_4^2, m_1^2, m_2^2, m_3^2) + S_{Box}^{n+2}(s_{12}, s_{23}, s_1, s_2, s_3, s_4, m_1^2, m_2^2, m_3^2, m_4^2)\,). \end{aligned} \tag{7}$$

The basic triangle and box functions satisfy the integral representations

$$S_{Tri}^n(s_1, s_2, s_3, m_1^2, m_2^2, m_3^2) = \int_0^1 dt_1 dt_2 \frac{(1 + t_1 + t_2)^{3-n}}{(\mathcal{D}_{Tri})^{3-n/2}} \tag{8}$$

$$S_{Box}^{n+2}(s_{12}, s_{23}, s_1, s_2, s_3, s_4, m_1^2, m_2^2, m_3^2, m_4^2) = \int_0^1 dt_1 dt_2 dt_3 \frac{(1 + t_1 + t_2 + t_3)^{2-n}}{(\mathcal{D}_{Box})^{3-n/2}} \tag{9}$$

where the functions $\mathcal{D}_{Tri}$ and $\mathcal{D}_{Box}$ are quadratics,

$$\begin{aligned} \mathcal{D}_{Tri} &= (-s_1)t_1 + (-s_2)t_1t_2 + (-s_3)t_2 + (1 + t_1 + t_2)(t_1 m_1^2 + t_2 m_2^2 + m_3^2) - i0 \\ \mathcal{D}_{Box} &= (-s_{12})t_2 + (-s_{23})t_1t_3 + (-s_1)t_1 + (-s_2)t_1t_2 + (-s3)t_2t_3 + (-s_4)t_3 \\ &\quad + (1 + t_1 + t_2 + t_3)(t_1 m_1^2 + t_2 m_2^2 + t_3 m_3^2 + m_4^2) - i0 \end{aligned} \tag{10}$$

The functions (8) and (9) will be computed with the integration technique of *DCM*, and a *nonlinear extrapolation* will be performed when denominators vanish within the integration region. This calculation will be carried out for each value of ε needed in the *dimensional extrapolation as* $\varepsilon \to 0$.

3. Linear extrapolation

Consider a function $S(\varepsilon)$ which satisfies an *asymptotic expansion*

$$S(\varepsilon) \sim \mathcal{S} + a_1\, \varphi_1(\varepsilon) + a_2\, \varphi_2(\varepsilon) + \dots, \quad \text{as } \varepsilon \to 0,$$

where the functions $\varphi_j(\varepsilon)$ are known and in decreasing order, in the sense that $\lim_{\varepsilon\to 0} \frac{\varphi_{j+1}(\varepsilon)}{\varphi_j(\varepsilon)} = 0$.

Then for a linear extrapolation we construct a sequence $S(\varepsilon_\ell),\ \ell = 0, 1, \cdots$, and solve linear systems of the form

$$a_0 + \varphi_1(\varepsilon_\ell)\, a_1 + \ldots \varphi_\nu(\varepsilon_\ell)\, a_\nu = S(\varepsilon_\ell), \quad \ell = 0, \ldots, \nu, \tag{11}$$

in the unknowns $a_j,\ j = 0, \cdots, \nu,$ of order $(\nu + 1) \times (\nu + 1)$, for increasing values of ν and for decreasing ε_ℓ.

This process can be related to the Neville algorithm as follows (see, e.g., [7]). The interpolating polynomial $P_{k,d}(x)$ of degree d for a function $f(x)$ is defined as the polynomial of degree d which interpolates at the points x_ℓ, so that

$$P_{k,d}(x_\ell) = f(x_\ell), \quad \text{for } \ell = k - d,\ k - d + 1, \cdots, k. \tag{12}$$

Denoting the polynomial coefficients by $c_{k,d}^{(j)}$, the value of the polynomial at a particular point x,

$$P_{k,d}(x) = \sum_{j=0}^{d} c_{k,d}^{(j)} x^j, \tag{13}$$

can be calculated recursively with the Neville algorithm, using

$$\begin{aligned} P_{k,0}(x) &= f(x_k), \\ P_{k,d}(x) &= \frac{(x_k - x)P_{k-1,d-1}(x) - (x_{k-d} - x)P_{k,d-1}(x)}{x_k - x_{k-d}}, \quad d = 1, \cdots, k. \end{aligned} \tag{14}$$

Furthermore, by substituting the formula $P_{k,d}^{(j)}(0) = c_{k,d}^{(j)}\, j!$ into equation (14), formulas are obtained for the coefficients,

$$\begin{aligned} c_{k,0}^{(0)} &= f(x_k), \\ c_{k,d}^{(j)} &= \frac{x_k\, c_{k-1,d-1}^{(j)} - x_{k-d}\, c_{k,d-1}^{(j)} + c_{k,d-1}^{(j-1)} - c_{k-1,d-1}^{(j-1)}}{x_k - x_{k-d}}, \\ &\qquad \text{for } d = 1, \cdots, k \ \text{ and } j = 1, \cdots, d-1. \end{aligned} \tag{15}$$

This gives rise to an extrapolation table, where the values for $j = 0$ and $j = d$ can be included by setting $c_{k,d}^{(-1)} = c_{k,d}^{(d+1)} = 0$. The table allows calculating the coefficients of the interpolating polynomial $P_{k,k}(x)$ of degree k, coinciding with $f(x)$ at the points $x = x_0, \cdots, x_k$. Note that $f,\ x$ and k can play the role of $S,\ \varepsilon$ and ν in equation (11) above, for $\varphi_j(\varepsilon) = \varepsilon^j$. In this case, the system of equation (12) combined with equation (13), or equation (11) is a Vandermonde type system, and (15) yields an algorithm for its solution.

4. Asymptotics

This section establishes the underlying asymptotic expansions which validate the linear (dimensional) extrapolation for various classes of vertex and box diagrams, and gives results for sample problems.

4.1. Vertex diagrams

Kurihara [6] derives expansions with respect to ε for the tensor integral of the massless one-loop 3-point integral of rank $M \le 3$,

$$T^{(3)}_{\mu\cdots\nu} = \sum_k C^k_{\mu\cdots\nu} J^k_3(p_1^2, p_2^2, p_3^2; n_x^{(k)} n_y^{(k)}),$$

with

$$
\begin{aligned}
J_3^k(p_1^2,p_2^2,p_3^2;n_x^{(k)},n_y^{(k)}) &= \frac{1}{(4\pi)^2}\frac{\varepsilon\Gamma(-\varepsilon)}{(4\pi\mu_R^2)^\varepsilon}\int_0^1 dx\int_0^{1-x} dy\frac{x^{n_x^{(k)}}y^{n_y^{(k)}}}{\mathcal{D}^{1-\varepsilon}}, \\
\mathcal{D} &= (p_1x-p_2y)^2-\rho xy-p_1^2x-p_2^2y-i0, \\
\rho &= p_3^2-(p_1+p_2)^2.
\end{aligned}
\tag{16}
$$

In the case $p_1^2 = p_2^2 = 0$ and $p_3^2 \neq 0$ with $n_x = n_y = 0$, (16) gives

$$
J_3(0,0,p_3^2;0,0;\varepsilon) = \frac{1}{(4\pi)^2}\frac{\varepsilon\Gamma(-\varepsilon)}{(4\pi\mu_R^2)^\varepsilon}\int_0^1 dx\int_0^{1-x} dy\frac{1}{(-p_3^2xy-i0)^{1-\varepsilon}} \tag{17}
$$

where μ_R is the renormalization energy scale. As $\varepsilon \to 0$, J_3 of equation (17) satisfies an expansion of the form

$$
J_3(0,0,p_3^2;0,0;\varepsilon) \sim \frac{C_{-2}}{\varepsilon^2}+\frac{C_{-1}}{\varepsilon}+C_0+\mathcal{O}(\varepsilon). \tag{18}
$$

The linear extrapolation method of section 3 can be applied to the system of equation (11) with $S(\varepsilon) = \varepsilon^2 J_3(0,0,p_3^2;0,0;\varepsilon)$. We gave numerical results for the coefficients C_{-2}, C_{-1} and C_0 of equation (18) in [10].

If, in (16) with $p_1^2 = p_2^2 = 0$, $p_3^2 \neq 0$, one of n_x and n_y is zero and the other is not, e.g., $n_x = \eta > 0$ and $n_y = 0$, then the asymptotic behavior is given by

$$
J_3(0,0,p_3^2;\eta,0;\varepsilon) \sim \frac{1}{(4\pi)^2 p_3^2}\left(\frac{C_{-1}}{\varepsilon}+C_0+\mathcal{O}(\varepsilon)\right), \tag{19}
$$

with coefficients

$$
C_{-1} = \frac{1}{\eta}, \quad C_0 = -\frac{2}{\eta^2}+\frac{1}{\eta}\left(\log(-p_3^2)-\sum_{j=1}^{\eta-1}\frac{1}{j}\right).
$$

Values for the coefficients obtained in double precision for (generating and solving) the system of equation (11) with $S(\varepsilon) = \varepsilon J_3(0,0,p_3^2;\eta,0;\varepsilon)$ were reported in [10] for $n_x = \eta = 2$, $n_y = 0$ and $p_3^2 = 100$, a tolerated relative error of 10^{-13} for (repeated) integration with the QUADPACK program DQAGS, using a sequence of $\varepsilon = \varepsilon_\ell = \frac{1}{b_\ell}$, where $\{b_\ell\}$ is a Bulirsch type sequence. The reported integration times ranged between 0.03s and 0.09s, for $b_\ell = 12, 16, 24, \cdots, 384$. The contribution of the times for the actual extrapolation procedure were found to be negligible. Table 1 lists results for the coefficients of equation (19) obtained in a quadruple precision run for this problem.

For $p_1^2 = 0$, $p_2^2 \neq 0$ and $p_3^2 \neq 0$ in equation (16), IR divergence occurs when $n_x = n_y = 0$ or $n_x \neq 0, n_y = 0$, leading to

$$
J_3(0,p_2^2,p_3^2;\eta,0;\varepsilon) \sim \frac{1}{(4\pi)^2 p_3^2}\left(\frac{C_{-1}}{\varepsilon}+C_0+\mathcal{O}(\varepsilon)\right). \tag{20}
$$

with coefficients given in [6]. Apart from the IR divergence as indicated by the ε-dependence in (20), a singularity may appear inside the integration region through a zero denominator function $\mathcal{D}$ in (16). For each value of ε in the sequence, the interior singularity can be handled in *DCM*, by replacing $i0$ in $\mathcal{D}$ by $i\delta$ and performing a nonlinear extrapolation as $\delta \to 0$. Results of the double extrapolation procedure are given in [11] for a diagram with $n_x = n_y = 0$ and $p_2^2 = 40$, $p_3^2 = -100$.

Table 1. IR vertex of equation (19) coefficients, $n_x = \eta = 2,\ n_y = 0$ and $p_3^2 = 100$. Integration with $(\text{DQAGS})^2$, for target relative integration error tolerance 5×10^{-24}. Extrapolated real values (quad. precision) are given; $\varepsilon_\ell = 1/b_\ell$.

b_ℓ	C_{-1}	C_0
12	0.4976817315764365660976698223580	1.3769024472521615918683765304
16	0.5001453140694736901417736134445	1.3030216329875385877494858112
24	0.5000001921131322347567104032762	1.3025651116288535549087530061
32	0.4999999959467322713652940448364	1.3025850791321435704903401609
48	0.4999999999993260163866288358759	1.3025850932043285642085160900
64	0.5000000000000053069685193174512	1.3025850929940632251166716065
96	0.4999999999999999722450728860463	1.3025850929940632251166716065
128	0.5000000000000000001089023336541	1.3025850929940455873126312708
192	0.4999999999999999999997158886319	1.3025850929940456843793808956
256	0.5000000000000000000000005564493	1.3025850929940456840169987640
384	0.4999999999999999999999999992988	1.3025850929940456840179932776
512	0.4999999999999999999999999999980	1.3025850929940456840179914927
Exact :	0.5	1.3025850929940456840179914547

4.2. Box diagrams

The IR divergent integral $J_4\,(p_1^2, p_2^2, p_3^2, p_4^2; n_x, n_y, n_z)$ occurs in the tensor integral of a massless one-loop box of rank $M \leq 4$,

$$T^{(4)}_{\mu\ldots\nu} = \sum_k C^k_{\mu\ldots\nu} J^k_4(p_1^2, p_2^2, p_3^2; n_x^{(k)}, n_y^{(k)}, n_z^{(k)})$$

and is expressed in [6] using dimensional regularization as

$$J_4^k\,(s, t, p_1^2, p_2^2, p_3^2, p_4^2; n_x^{(k)}, n_y^{(k)}, n_z^{(k)}) = \frac{\Gamma(2-\epsilon)}{(4\pi)^2\,(4\pi\mu_R^2)^\epsilon} \int_0^1 dx \int_0^{1-x} dy \int_0^{1-x-y} dz\, \frac{x^{n_x^{(k)}}\, y^{n_y^{(k)}}\, z^{n_z^{(k)}}}{\mathcal{D}^{2-\epsilon}} \tag{21}$$

where

$$\mathcal{D} = -s\,x\,z - t\,y\,(1-x-y-z) - p_1^2\,x\,y - p_2^2\,y\,z - p_3^2\,z\,(1-x-y-z) - p_4^2\,x\,(1-y-z) - i0.$$

We consider the case where $p_1^2 = p_2^2 = p_3^2 = p_4^2 = 0$ (all particles on-shell) and $n_x = n_y = n_z = 0$. The integral of the form I_4^n in equation (2), is IR divergent, and its asymptotic behavior is captured in

$$\begin{aligned}
&\frac{(4\pi)^2\,(4\pi\mu_R^2)^\varepsilon}{\Gamma(2-\varepsilon)} J_4\,(s, t; 0, 0, 0, 0; 0, 0, 0; \varepsilon) \\
&= \int_0^1 dx \int_0^{1-x} dy \int_0^{1-x-y} dz\, \frac{1}{(-s\,x\,z - t\,y\,(1-x-y-z)\; - i0)^{2-\varepsilon}}, \qquad (22) \\
&\sim \frac{c_{-2}}{\varepsilon^2} + \frac{c_{-1}}{\varepsilon} + c_0 + \mathcal{O}(\varepsilon) \qquad (23)
\end{aligned}$$

For $s = t = -1$, the integrand denominator of of equation (22) does not vanish in the interior of the integration region, so that only the dimensional regularization is performed ($\delta = 0$). Note that, for example, only one of the Mandelstam variables occurring in the parameter list of the triangle functions I_3^n of equation (5) is -1 and the other two are 0. The same holds for the Mandelstam parameters of the triangle denominator S^n_{Tri} in equation (8).

Table 2 gives coefficient values resulting from the numerical reduction according to (4), applied for $N = 4$. The linear (Bulirsch type) extrapolation sequence is started at $\varepsilon = 1/8$. Thus the first (3×3) system is solved at $\varepsilon = 1/16$. The requested accuracy was set to about 10 digits for the triple integration of the $(n+2)$-dimensional box function in equation (5), whereas the triangle integrations were performed to a tolerated error of about 12 digits. Note that the $(n+2)$-dimensional box function is IR finite; thus

Table 2. IR box of equation (23) coefficients, $n_x = n_y = n_z = 0$; $s = t = -1$. Triple iterated integration with $(\text{DQAGS})^3$, for target relative integration error tolerance of 10^{-10}. Extrapolation real values (double precision) are given; $\varepsilon_\ell = 1/b_\ell$.

b_ℓ	c_{-2}	c_{-1}	c_0
16	3.97866911577	4.741597564	-20.3973583
24	4.00113602366	3.932788880	-11.0511246
32	3.99996917536	4.002799778	-12.5446904
48	4.00000076575	3.999893462	-12.4436012
64	3.99999998946	4.000002142	-12.4495134
128	3.99999999980	4.000000034	-12.4493421
	Exact: 4.0	4.0	-12.4493407

Table 3. IR box equation (23) coefficients, $n_x = n_y = n_z = 0$; $s = 123, t = -200$. Triple iterated integration with $(\text{DQAGS})^3$, for target relative integration error tolerance 10^{-9}. Extrapolation real values (double precision) are given; $\varepsilon_\ell = 1/b_\ell$.

b_ℓ	c_{-2}	c_{-1}	c_0
16	-0.1649247432e-03	-0.902352425e-03	-0.33055285e-02
24	-0.1625429760e-03	-0.988096045e-03	-0.23147133e-02
32	-0.1626021617e-03	-0.984544906e-03	-0.23904710e-02
48	-0.1626016138e-03	-0.984595309e-03	-0.23887178e-02
64	-0.1626016266e-03	-0.984593520e-03	-0.23888151e-02
96	-0.1626016260e-03	-0.984593638e-03	-0.23888055e-02
128	-0.1626016260e-03	-0.984593640e-03	-0.23888053e-02
Ex:	-0.1626016260e-03	-0.984593636e-03	-0.23888058e-02

it is integrated with $\varepsilon = 0$. The time incurred in the triple integrations dominates the total computation time. The computation to produce table 2 takes about 13s (user execution time) on a MacBook Pro laptop (Intel Core Duo processor, 3.06 GHz).

The integrand denominator of the box function in equation (22) vanishes in the interior of the integration region for $s = 123$, $t = -200$; thus a double extrapolation is required. However, in this case, the triangle function denominator in equation (8) does not vanish in the interior of the integration domain. Results for the real parts of the coefficients in (23) are given in table 3. The computation for table 3 takes about 406s.

5. Concluding remarks

In this contribution we reported on a new numerical regularization technique for infrared (IR) divergent loop integrals using dimensional regularization. This method relies on a numerical extrapolation to derive the behavior of the integral around four-dimensions.

In order to achieve this, a *double extrapolation* is required, to deal with singular integrand behavior inside the integration domain as well as IR divergence. We resort to a reduction and sector decompositions, applied numerically, where the Direct Computation Method (*DCM*) is applied on the basic box and triangle levels. Since the method is fully numerical, it is viable without change for various problem types.

Many integrals result, rendering the procedure a good candidate for parallel/distributed implementations. For difficult box calculations we are developing a multi-core parallelization on the (outer) function evaluation levels in the iterated integration procedure.

Acknowledgments

This work is partially supported by Grant-in-Aid for Scientific Research (20340063) of JSPS and by the CPIS program of Sokendai (Japan). We also acknowledge the support from Western Michigan University and the College of Engineering and Applied Sciences (CEAS), and from the Numerical Algorithms Group Inc.

References

[1] Nakanishi N 1971 *Graph Theory and Feynman Integrals* (New York: Gordon and Breach)
[2] Binoth T and Heinrich G 2004 *Nucl. Phys. B* **680** 375–88 (*Preprint* hep-ph/0305234v1)
[3] Binoth T, Heinrich G and Kauer N 2003 *Nuclear Physics B* **654** 277–300 (*Preprint* hep-ph/0210023)
[4] Fujimoto J and Ueda T 2008 New implementation of the sector decomposition on FORM *Proc. XII Adv. Comp. and Anal. Tech. in Phys. Res. (Erice, Sicily)* PoS (ACAT08) 120 (*Preprint* ArXiv:0902.2656v1 [hep-ph])
[5] Vermaseren J A M 2000 New features of FORM *Preprint* math-ph/0010025, http://www.nikhef.nl/ form
[6] Kurihara Y 2006 *Eur. Phys. J. C* **45** 427–44 (*Preprint* hep-ph/0504251 v3)
[7] Lyness J N and Moler C B 1966 *Numer. Math.* **8** 458–64
[8] Bulirsch R 1964 *Numerische Mathematik* **6** 6–16
[9] de Doncker E, Fujimoto J, Hamaguchi N, Ishikawa T, Kurihara Y, Shimizu Y and Yuasa F 2010 *Springer Lecture Notes in Computer Science (LNCS)* **6017** 139–54
[10] de Doncker E, Fujimoto J, Hamaguchi N, Ishikawa T, Kurihara Y, Shimizu Y and Yuasa F 2011 *Journal of Computational Science (JoCS)* doi:10.1016/j.jocs.2011.06.003
[11] de Doncker E, Fujimoto J, Hamaguchi N, Ishikawa T, Kurihara Y, Ljucovic M, Shimizu Y and Yuasa F 2010 Extrapolation algorithms for infrared divergent integrals PoS (CPP2010) ©Copyright owned by the author(s) under the terms of the Creative Commons Attribution-NonCommercial-ShareAlikeLicence (*Preprint* arXiv:hep-ph/1110.3587)
[12] Binoth T and Heinrich G 2000 *Nuclear Physics B* **585** 741–59 (*Preprint* hep-ph/0004013v2)
[13] de Doncker E, Fujimoto J, Kurihara Y, Hamaguchi N, Ishikawa T, Shimizu Y and Yuasa F 2010 Recursive box and vertex integrations for the one-loop hexagon reduction in the physical region *Proc. XIV Adv. Comp. and Anal. Tech. in Phys. Res. (Jaipur, India)* PoS (ACAT10) 073
[14] de Doncker E, Shimizu Y, Fujimoto J and Yuasa F 2004 *Computer Physics Communications* **159** 145–56
[15] Yuasa F, Ishikawa T, Fujimoto J, Hamaguchi N, de Doncker E and Shimizu Y 2008 Numerical evaluation of Feynman integrals by a direct computation method *Proc. XII Adv. Comp. and Anal. Tech. in Phys. Res. (Erice, Sicily)* PoS (ACAT08) 122 (*Preprint* arXiv:0904.2823)
[16] Li S, de Doncker E and Kaugars K 2004 *Springer Lecture Notes in Computer Science (LNCS)* **3514** 123–30
[17] de Doncker E and Kaugars K 2010 *Procedia Computer Science* **1**
[18] Shanks D 1955 *J. Math. and Phys.* **34** 1–42
[19] Wynn P 1956 *Mathematical Tables and Aids to Computing* **10** 91–6
[20] Piessens R, de Doncker E, Überhuber C W and Kahaner D K 1983 *QUADPACK, A Subroutine Package for Automatic Integration* Springer Series in Computational Mathematics (Berlin: Springer-Verlag)

Polynomial Algebra in Form 4

J Kuipers[1]

[1] Nikhef, Science Park 105, 1098 XG Amsterdam, The Netherlands

E-mail: jkuipers@nikhef.nl

Abstract. New features of the symbolic algebra package Form 4 are discussed. Most importantly, these features include polynomial factorization and polynomial gcd computation. Examples of their use are shown. One of them is an exact version of Mincer which gives answers in terms of rational polynomials and 5 master integrals.

1. Introduction

Form is a symbolic algebra package that has been developed by J. Vermaseren et al. for the past decades [1]. Its fast manipulation of large expressions makes it an ideal tool to attack the big problems that occur in particle phenomenology. Recently, version 4β has been released, that includes many new features including polynomial algebra. This new feature can be divided in three different operations, namely greatest common divisors (gcds), factorization and rational polynomials. This paper discusses these operations and provides examples of Form code to show how to use them.

2. Greatest common divisors

The greatest common divisor of two polynomials is a polynomial that divides both of them and that is divisable by every other polynomial that divides both. In Form 4, this polynomial can be determined by function gcd_ in the following way.

```
Symbols x,y;
Local E = gcd_(x^2+x*y, y^2+x*y);
Print;
.end

E=
   y + x;
```

In previous version of Form, the gcd_ function would return E=1 for this particular example.

To determine the gcd, Form uses the following algorithms. For small polynomials, a heuristic that substitutes integers and performs integer gcd calculations is used [2]. For large polynomials, Zippel's modular algorithm is used [3]. This latter algorithm is fast for both sparse and dense polynomials. The speed of Form's gcd algorithms is comparable to Mathematica's.

3. Factorization

In Form, there are a few statements that can be used to factorize a polynomial over the rational numbers. First, there exists FactArg_, which is used to factorize a function argument. Its usage is as follows.

```
Symbols k,i,f,e,h;
CFunction N;
Local E = N(k*f+k*e+k*h+i*f+i*e+i*h);
FactArg N;
Print;
.end

E-
  N(i + k,h + e + f);
```

Note that this changes the previous behaviour of FactArg_. To get the old behaviour, a switch On OldFactArg has been implemented.

The second variety of polynomial factorization is for dollar variables and can be invoked as follows.

```
Symbols x,y;
#$a = x^2-y^2;
#Factorize $a;
#do i=1, `$a[0]'
  #write "%$", $a[`i'];
#enddo

-y+x
y+x
```

Note that $a[0] results in the number of factors, while $a[i] yield the actual factors.

The third way of using the factorization routines works for expressions. The usage is as follows:

```
Symbols x;
Local E = x^2-1;
Factorize E;
.end

E =
     ( -1 + x )
   * ( 1 + x );
```

The number of factors of a factorized expression is given by numfactors(E) and the individual factors can be addressed to by E[factor_^i].

To factorize polynomials, Form uses the following algorithms. For univariate polynomials, Berlekamp's algorithm is used [4]. Multivariate polynomials are reduced to univariate ones by substuting integers for all but one variable and factorized as such. Afterwards, the multivariate result is obtained by Hensel lifting [5]. For big random polynomials that factorize into two factors, speedups of upto a factor of 1 000 have been measured compared to Mathematica.

4. Rational polynomials

In Form, the statement PolyRatFun turns a function into a rational polynomial with the first argument as numerator and the second argument as denominator. The function also serves as coefficient of the terms, just like the PolyFun in previous versions of Form. The following code shows how it can be used.

```
Symbols x,y,z;
CFunction f;
PolyRatFun f;
Local E = x * f(y,z) + x * f(y,1-z)
        + x^2 * f(y^2-1,y-1);
Print;
.end
```

```
E =
    x*f( - y,z^2 - z) + x^2*f(y + 1,1);
```

5. Application: MincerExact

As a first application of Form's polynomial algebra, a new version of Mincer [6] has been written. Mincer is a program for three-loop massless propagator diagrams and it works in expansions in $\epsilon = (4 - D)/2$. The original version keeps track of expansions typically up to the sixth power. Big tables with expansions of Pochhammer symbols, inverse gamma functions and alike are needed for the calculations.

The new version, MincerExact, does not expand in ϵ, but stores the result as a rational polynomial in it. In this way, no big tables are needed, which results in much shorter and cleaner source code. The resulting program is not much slower than the original version.

References

[1] Vermaseren J A M 2000 New features of FORM *Preprint* math-ph/0010025v2
[2] Char B W, Geddes K O and Gonnet G H 1989 *J. Symbolic Comp.* **9** 31–48
[3] Zippel R E 1979 *Probabilistic algorithms for sparse polynomials* (M.I.T. PhD thesis)
[4] Berlekamp E R 1967 *Bell System Technical Journal* **46** 1853–1959
[5] Geddes K O, Czapor S R, and Labahn G 1992 *Algorithms for Computer Algebra* (Boston: Kluwer)
[6] Larin S A, Tkachov F V and Vermaseren J A M 1991 The FORM version of Mincer *Preprint* NIKHEF-H/91-18

GPU Linear algebra extensions for GNU/Octave

L B Bosi[1], M Mariotti[2] and A Santocchia[3]
[1] Dipartimento di Fisica, Camerino, IT
[2] Dipartimento di Fisica, Perugia, IT
[3] INFN and Dipartimento di Fisica, Perugia, IT

E-mail: leonebosi@gmail.com, mirko.mariotti@fisica.unipg.it, attilio.santocchia@pg.infn.it

Abstract. Octave is one of the most widely used open source tools for numerical analysis and liner algebra. Our project aims to improve Octave by introducing support for GPU computing in order to speed up some linear algebra operations. The core of our work is a C library that executes some BLAS operations concerning vector- vector, vector matrix and matrix-matrix functions on the GPU. OpenCL functions are used to program GPU kernels, which are bound within the GNU/octave framework. We report the project implementation design and some preliminary results about performance.

1. Introduction

GPU devices are impacting enormously the science computing, mainly due to their very interesting performances in terms of GFlops per Watt per Euro. Just to have an idea, looking at the top 500 list [1] we discover that the first and the third ranked supercomputers in the word are GPU powered. The strong invasion of GPU hardware in the field of science computing has been stimulated and supported by a lot of important GPU software projects born all around the world.

The main purpose of our work is to make the computational power of modern GPUs accessible within octave, the popular free software scientific tool.

2. Octave

Octave [2] provides functions to solve numerical solution of linear and nonlinear problems, performing other numerical experiments, plotting facilities. Octave can be user via command line, using a Matlab [3] like language and also via program scripts for a non–interactive execution.

GNU Octave was written by John W. Eaton and many others and it is also freely redistributable software and can be redistribute and/or modified under the terms of the GNU General Public License (GPL) as published by the Free Software Foundation.

Octave is a free software and for that the Octave community encourages to help making the software more useful. Contributions can be provided by writing additional functions for it, and by reporting any problem occurred.

3. OpenCL

OpenCL [4] means Open Computing Language. This language has been developed to permit writing programs across multi-platform realized with CPU, GPU and other processors.

Journal of Physics: Conference Series **368** (2012) 012062 doi:10.1088/1742-6596/368/1/012062

OpenCL was initially developed by Apple, which holds trademark rights, and refined into an initial proposal in collaboration with technical teams at AMD, IBM, Intel, ARM and Nvidia. Apple submitted this initial proposal to the Khronos Group. On June 16, 2008 the Khronos Compute Working Group was formed with representatives from CPU, GPU, embedded-processor, and software companies. This group worked for five months to finish the technical details of the specification for OpenCL 1.0 by November 18, 2008. This technical specification was reviewed by the Khronos members and approved for public release on December 8, 2008.

OpenCL is analogous to the open industry standards OpenGL and OpenAL, for 3D graphics and computer audio, respectively. OpenCL is managed by the non-profit technology consortium Khronos Group. The OpenCL include a language based on C99 for writing the functions that are executed on OpenCL devices. These functions are called kernels. Moreover the framework provides an API used to define and control the various platforms. OpenCL gives to the applications access to the graphics-processing unit for non-graphical computing, and a task-based or data-based parallel computing models.

We use OpenCL to develop the GPU linear algebra library extension for octave.

4. Extending octave

The aim of our work is to introduce the GPU computing inside the octave framework, bringing the OpenCL power within GNU Octave. This approach promises much better performance on big matrices/vectors size context, where GPU devices can express better their characteristic.

In particular we start analysing the feasibility of this integration, optimizing the GPU context handling and producing the first prototype version of the octave-gpu library. By design, this version provides simple vector-vector, vector-matrix, matrix-matrix operation such as fundamental operators and determinants. The work has been complicated due to the octave and OpenCL peculiarity, but also due to the poor octave documentation on object extensions, forcing as on analysing octave sources and pre-processed code. This work permitted to reconstruct enough advanced information on octave design, needed to introduce our optimized OpenCL support.

The standard way to extend Octave is using the so-called OCT files. An OCT file is essentially a loadable-shared library that the interpreter may load at runtime and use. Octave provides a utility to create OCT files from sources: mkoctfile.

Let's take a look at a trivial OCT file:

```
#include <octave/oct.h>
#include <stdio.h>
#include <stdlib.h>
DEFUN_DLD(testme, args, , "Test OCT file")
{
  int nargin = args.length ();
  return octave_value(nargin);
}
```

This example adds an octave function which purpose is counting and returns the number of its own arguments. The DEFUN_DLD macro is needed to create the interpreter entry point of a new octave function (testme is this example). The files has to be named the same as the function it declares, in this case testme.cc will be the source and testme.oct the OCT file. The octave interpreter expects to find a called external function in an OCT file named in the same way. If we would like to put two or more functions within the same source file we have to take care that every function has its own file (eventually creating symbolic links or copying a single OCT file to various files). Creating the OCT file is straightforward: "mkoctfile testme.cc", and to use it from an octave interpreter (opened in the same directory where the OCT file is):

```
octave:1> testme(3,4,5)
ans = 3
```

4.1. The first extension test "vector_add"

Our first test to put OpenCL code within Octave is to create a vector_add.cc that take two octave arrays, copies them to a GPU device, perform the sum and copy the result back to octave. All the OpenCL Context, device and kernel initialization is inside such function. This was a feasibility level zero test about the OpenCL-octave that demonstrated the possibility of the integration and provided preciouses information for the next step of the library that will be reported on the next chapters.

4.1.1. "vector_add" benchmarking. We wrote a simple benchmark under Octave to the vector_add level zero code. The code loops several times over different vector length the "vector add" command, where the length is changed in order to test the code increasing computational complexity. The GPU is expected to best work under high computing density. Moreover we loop each length in order to have some statistics. We develop more implementation of the benchmark, in order to have number to compare together:

- Octave native vector sum: this use the native octave "+"
- Vector_add: this is the first GPU implementation
- Vector_add_precompiled_kernel: this implements a precompiled kernel loading
- Vector_add_static: this is not called from the octave interpreter but is called as a function (we use it as gain upper limit)

The result are reported below, where we report execution time versus vector size. The hardware used for the test is a CPU Intel Xeon E5520 2.27GHz and a GPU Nvidia Tesla C1060.

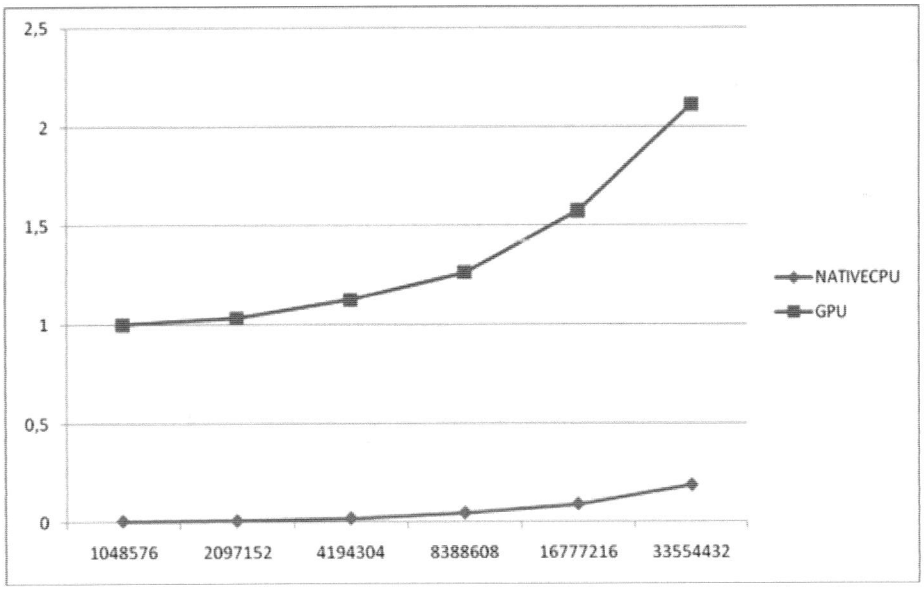

Figure 1. Comparison between GPU and CPU performance, reporting the "vector_add" average processing time (s) vs vector size.

With the zero level implementation we show the octave-GPU link feasibility and the implementation perspectives and first evident criticality, because the GPU performance are much worsted than the CPU. Some overheads were clear at design time and others were shown analysing the octave data/object handling. At the first order:

- The OpenCL initialization occurs every time an operation is performed.
- The OpenCL kernel compilation occurs every time an operation is performed.
- The Vector data are copied every time from host/host (double buffering) and from the GPU/host memory.

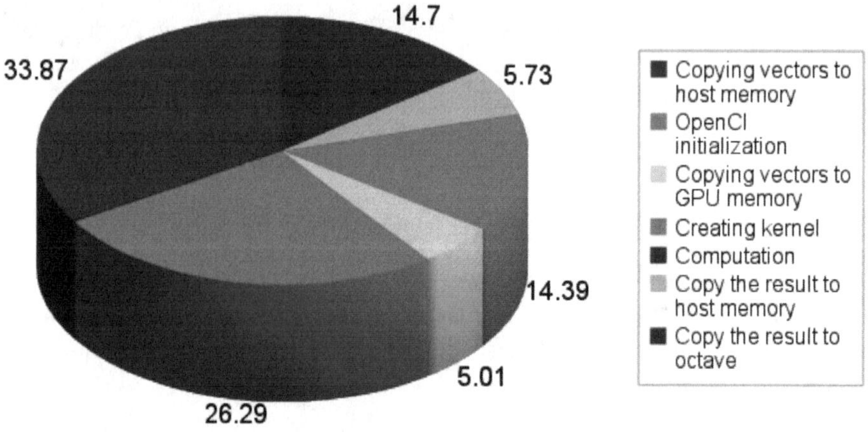

Figure 2. "vector_add" profiling, reporting the percentage of time spent on the specific step.

The result analysis shows at first sight the main latency causes, mainly due to OpenCL initialization, and OpenCL/octave IO overheads.

We discovered that several overheads are added from OpenCL to the octave code. Especially the copy operations in and from the octave native elements bring many delays in. The analysis on such results bring us to the version one implementation, introducing important optimization actions:

- Definition of a permanent GPU object to host OpenCL status variables and status object.
- Kernels pre-compilation and bufferization
- OpenCL device and memory with once initialization approach

The our GPU-octave implementation consist of two main object, GPUdevice and GPUobject, where a GPUdevice is the GPU representation inside the octave space, while a GPUobject is a GPU data representation inside the octave space.

Moreover we already introduced at the epoch of this article the overloading of the most common operators: +,-,*

4.1.2. The GPUdevice octave object. The purpose of the GPUdevice is to store all the OpenCL initialization data for a given GPU device id. This code implementation supports the multi device support, where the device id is given as initialization parameter. A GPUdevice object, such as oGPU, can be declared using the following sintax:

```
oGPU=GPUdevice(0)
```

Where "0" mean the GPU device 0, in our case the C1060. When a GPU device object is declared the code execute the following operations:

- Compile and load into memory all the defined OpenCL kernels
- Initialize the GPU device
- Initialize the GPU context

4.1.3. The GPUobject octave object. Once a GPUdevice has been declared, the user can allocate several GPUobjects starting by that. The GPUobject are the objects used by the octave user to perform the vector/matrix manipulation. The syntax introduced for this purpose is:

```
A=GPUobject(oGPU,[1:100000])
```

The command creates a GPUobject, embedding the OpenCL context and pre-initialized data through the GPUdevice - GPUobject initial linkage. Moreover with this command the GPU memory is allocated and a vector of 100000 elements is stored, containing number from 1 to 100000.

The user can declare and initialize several GPUobjects starting from the same GPUdevice and the purpose of the library is to provide the common octave syntax for the vectors operations. For example the user can declare two objects:

```
A=GPUobject(oGPU,[1:100000])
B=GPUobject(oGPU,[10:100010])
```

And perform a sum, difference or element-by-element multiplication:

```
C=A+B
D=A-B
E=A*B
```

4.1.4. Benchmarking. We benchmark the sum and product functions of the new version of the library following the similar approach of the zero level version. In detail we estimate the gain factor between CPU and GPU timing on vector size increasing about:

A+B (SUM)
A*B (PRODUCT)
A+B+C (SUM3)
A+B*C (TRIAD)

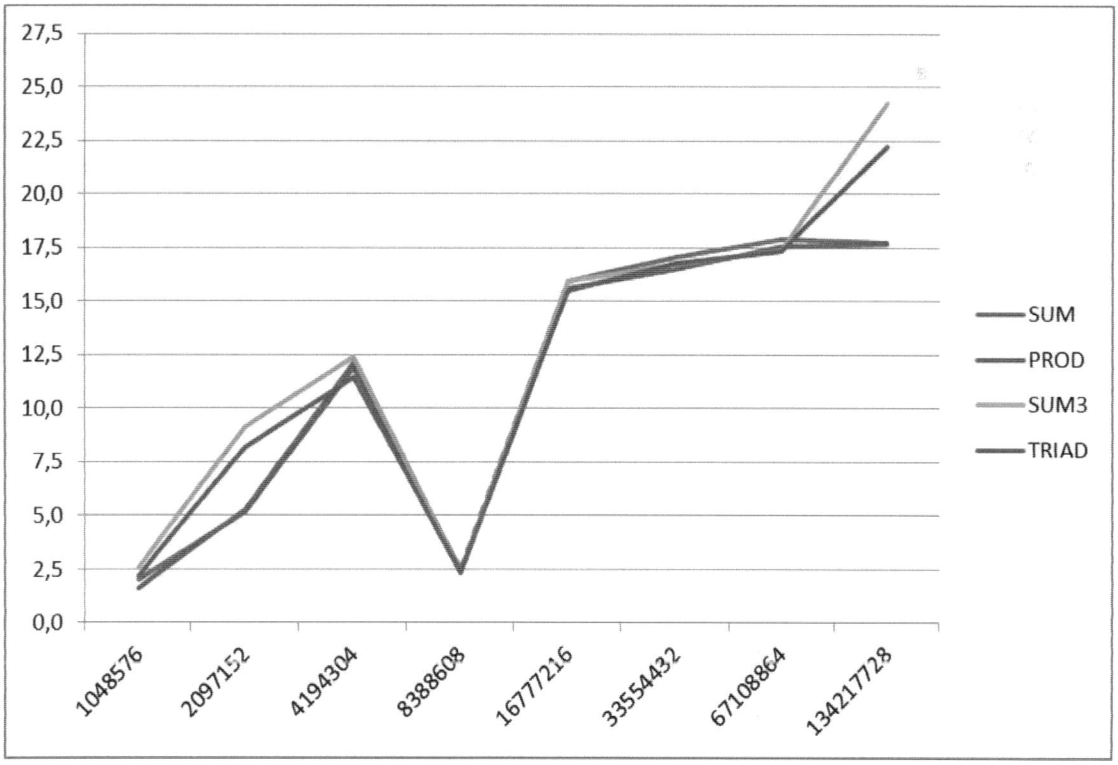

Figure 3. Processing gain factors of sum (a+b), mul (a*b), sum 3 terms (a+b+c), triad (a+b*c) operations between CPU and GPU on vector length change.

The first main result of this library version is the gain on data processing using GPU respect to the normal octave CPU implementation. This consideration is true in the range of the tests. We observe, as expected, a gain factor that roughly increase proportionally with processed vectors length. We observe a drop at 2^23 that needs investigation, probably due to the GPU computing blocks size that it hasn't been optimized at this step.

Sum and product report very similar performances with a gain between 2 and 17. The SUM3 and triad report a higher gain between 2.5 and 24. So it means that increasing the GPU computing volume the gain factor increases. This consideration lead us to deduce that the measured timing has still a systematic dominant delay component due to some overhead that can be again object of optimization.

5. Conclusions

Our work demonstrates the feasibility of GPU (OpenCL)-octave integration. Our library design allows getting a significant gain from the GPU implementation. In this version of the library we report gains between 2 and 24 respect to the normal CPU-octave computation. Now that the design responds positively, the work will continue finalizing the version 1 of the library implementing more operators and type to matrix.

6. References

[1] http://www.top500.org
[2] http://www.gnu.org/software/octave/index.html
[3] http://www.mathworks.com/products/matlab/
[4] http://www.khronos.org/opencl/

Journal of Physics: Conference Series **368** (2012) 012063 doi:10.1088/1742-6596/368/1/012063

The NNPDF2.2 parton set

F Cerutti [1] **and N P Hartland** [2,]

[1] Universitat de Barcelona, Departament d'Estructura i Constituents de la Matèria, Av. Diagonal 647, Barcelona 08028, España
[2] School of Physics and Astronomy, University of Edinburgh, JCMB, KB, Mayfield Rd, Edinburgh EH9 3JZ, Scotland

E-mail: francesco.cerutti@gmail.com, N.P.Hartland@sms.ed.ac.uk

Abstract. We present a method developed by the NNPDF Collaboration that allows the inclusion of new experimental data into an existing set of parton distribution functions without the need for a complete refit. A Monte Carlo ensemble of PDFs may be updated by assigning each member of the ensemble a unique weight determined by Bayesian inference. The reweighted ensemble therefore represents the probability density of PDFs conditional on both the old and new data. This method is applied to the inclusion of W-lepton asymmetry data into the NNPDF2.1 fit producing a new PDF set, NNPDF2.2.

1. NNPDF Approach

The NNPDF methodology differs from other parton fitting approaches in two main aspects, the choice of parameterization and the treatment of uncertainties. Neural networks are used to parametrize the PDFs at the initial scale. These are chosen because they provide a robust, unbiased interpolation and permit a very flexible parameterization. An NNPDF fit has almost three hundred free parameters and is therefore free from the bias associated with the choice of an inflexible functional form. The flexibility of the neural network fit has been demonstrated in several ways. For example, the fit has been shown to be stable under changes in network architecture [1] and under the addition of new PDFs [2].

In order to obtain an accurate representation of PDF uncertainty, NNPDF fits are performed on pseudo data samples generated from the original experimental data by Monte Carlo sampling. Each sample, or 'replica' forms the basis for a separate fit, resulting in an ensemble of PDF replicas that faithfully represent the uncertainty in the experimental data, without the need for a tolerance criterion.

These features of the NNPDF approach allow for an interesting exploitation of Bayesian inference to determine the impact of new datasets.

2. Reweighting Method

We shall discuss how an existing probability distribution in the space of PDFs may be updated with information from new data. To include the new data, one can of course perform a fit with the new, enlarged dataset. However this is a time consuming task, particularly for observables where no fast code is available. It is therefore desirable to have a faster method of including new data in order to assess it's impact rapidly without the need for a full refit. NNPDF parton sets are supplied as an ensemble of N parton distribution replicas $\mathcal{E}$, representing the probability density in PDFs $\mathcal{P}(f)$ based upon the data in the existing fit. We can therefore include new data by weighting each replica f_k in the ensemble by an associated weight w_k. If the replica weights are computed correctly, then reweighting is completely equivalent to a refit.

In order to illustrate the reweighting method, consider the computation of the expected value of a PDF-dependant observable $\mathcal{O}[f]$. We note that as the NNPDF Monte Carlo ensemble is a good representation of the probability density $\mathcal{P}(f)$, the expectation value $\langle \mathcal{O}[f] \rangle$ can be calculated as a simple average,

$$\langle \mathcal{O} \rangle = \int \mathcal{O}[f]\, \mathcal{P}(f)\, Df = \tfrac{1}{N} \sum_{k=1}^{N} \mathcal{O}[f_k]\,.$$

New data can be included into the existing ensemble by assigning each replica a unique weight w. This weight assesses the agreement between the replica and the new data. The reweighted ensemble now forms a representation of the probability distribution of PDFs $\mathcal{P}_{\text{new}}(f)$ conditional on both the existing and new data. The mean value of the observable $\mathcal{O}$ taking account of the new data is then given by the weighted average

$$\langle \mathcal{O} \rangle_{new} = \int \mathcal{O}[f]\, \mathcal{P}_{new}(f)\, Df = \tfrac{1}{N} \sum_{k=1}^{N} w_k \mathcal{O}[f_k],$$

where the weights are given in terms of the individual replica χ^2 to new data by

$$w_k = \frac{(\chi_k^2)^{(n-1)/2} e^{-\frac{1}{2}\chi_k^2}}{\frac{1}{N}\sum_{k=1}^{N} (\chi_k^2)^{(n-1)/2} e^{-\frac{1}{2}\chi_k^2}} = \mathcal{N}_\chi \mathcal{P}(\chi^2|f_k)\,.$$

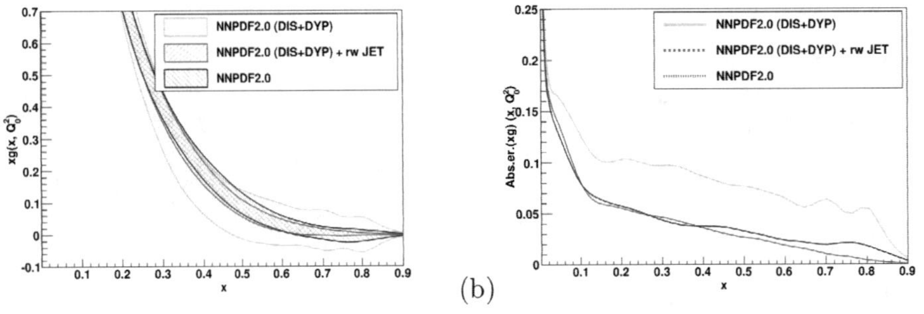

(a) (b)

Figure 1. (a) The gluon distribution and (b) its uncertainty for the NNPDF2.0(DIS+DY) fit before and after reweighting with the inclusive jet data compared to the refitted gluon from NNPDF2.0

Note that after reweighting a given ensemble of N PDF replicas we no longer have the same efficiency in describing the distribution of PDFs. The reweighting procedure will often assign replicas very small weights, therefore these replicas no longer contribute to the ensemble. The efficiency of the representation of the underlying distribution $\mathcal{P}_{new}(f)$ will therefore be less than it would be in a new fit. The loss of information due to reweighting can be quantified using the Shannon entropy to determine the effective number of replicas in the reweighted set:

$$N_{\text{eff}} \equiv \exp\{\tfrac{1}{N} \sum_{k=1}^{N} w_k \ln(N/w_k)\}.$$

3. Unweighting

Once we have a reweighted PDF set, we would like to be able to produce a new PDF ensemble with the same probability distribution as a reweighted set, but without the need to include the weight information. A method of unweighting has therefore been developed, whereby the new set is constructed by deterministically sampling with replacement the weighted probability distribution. This means that replicas with a very small weight will no longer appear in the final unweighted set while replicas with large weight will occur repeatedly.

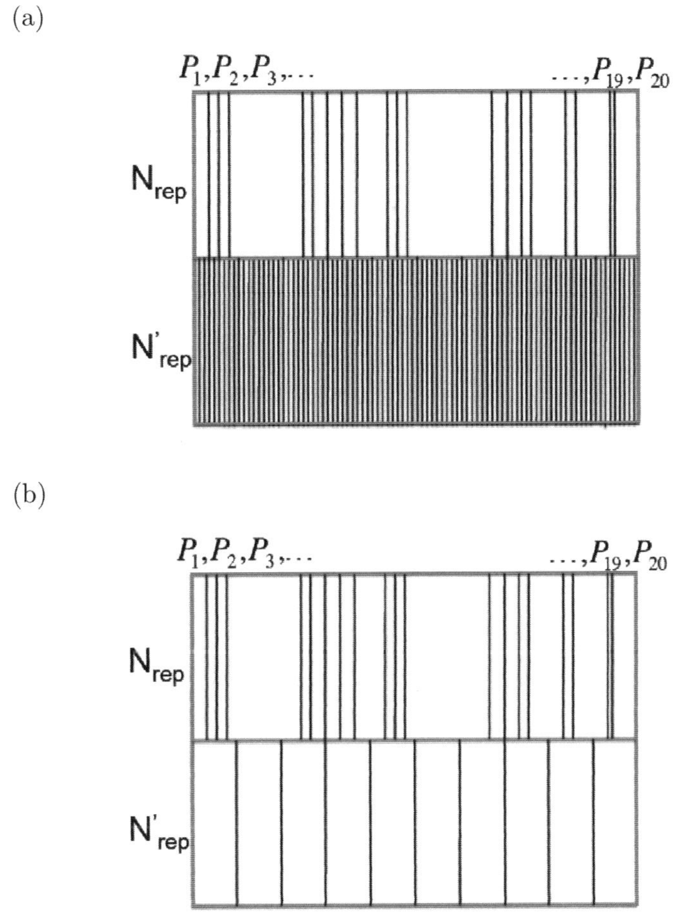

Figure 2. Graphical representation of the construction of a set of N'_{rep} unweighted replicas from a set of $N_{rep} = 20$ weighted ones. Each segment is in one-to-one correspondence to a replica, and its length is proportional to the weight of the replica. The cases of $N'_{rep} \gg N_{rep}$ (a) and $N'_{rep} = 10$ (b) are shown.

If we define the probability for each replica, and the probability cumulants as

$$p_k = \frac{w_k}{N_{rep}} \qquad P_k \equiv P_{k-1} + p_k = \sum_{j=0}^{k} p_j \, .$$

we can quantitatively describe the unweighting procedure. Starting with N_{rep} replicas with weights w_k, we determine N_{rep} new weights w'_k:

$$w'_k = \sum_{j=1}^{N'_{rep}} \theta(\frac{j}{N'_{rep}} - P_{k-1})\theta(P_k - \frac{j}{N'_{rep}}) .$$

These weights are therefore either zero or a positive integer. By construction they satisfy:

$$N'_{\rm rep} \equiv \sum_{k=1}^{N_{\rm rep}} w'_k .$$

i.e the new unweighted set consists of $N'_{\rm rep}$ replicas, simply constructed by taking w'_k copies of the k-th replica, for all $k = 1, ..., N_{rep}$. This procedure is illustrated graphically in figure 2.

4. Testing Reweighting and Unweighting

To verify the effectiveness of the reweighting procedure, we will show that including datasets by reweighting produces an ensemble of PDF replicas statistically equivalent to a full refit. We begin by producing a new NNPDF2.0 fit, including only DIS and Drell-Yan data. The data left out of the fit (Tevatron Run II inclusive jet data) is then reintroduced by reweighting. The resulting reweighted ensemble is then compared to the full NNPDF2.0 fit.

In figure 1 we see the gluon PDF for the three sets; the prior fit NNPDF2.0(DIS+DY), the reweighted set NNPDF2.0(DIS+DY) with jet data included, and the refitted full set NNPDF2.0. The figure shows excellent agreement between the reweighted set and the full fit. Differences stay well below statistical fluctuations.

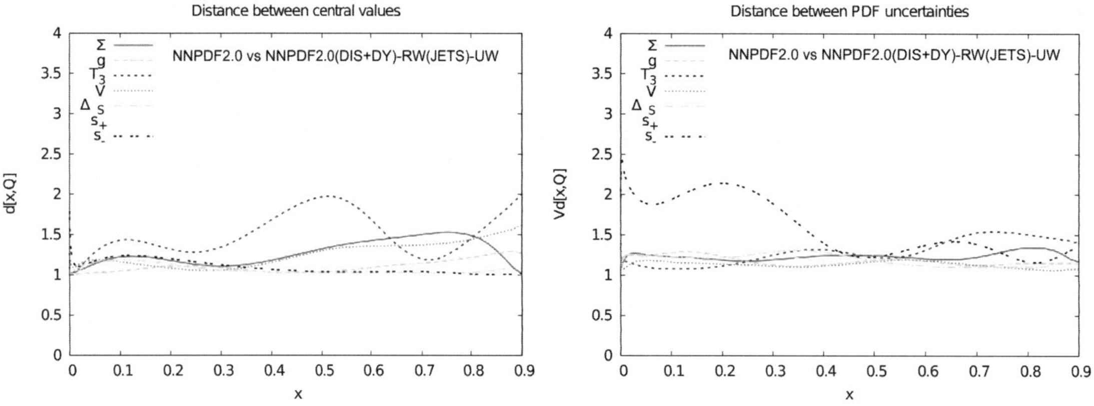

(a) (b)

Figure 3. Distance between central values (a) and uncertainties (b) of the NNPDF2.0 PDFs and the NNPDF2.0 DIS+DY PDFs reweighted with Tevatron jet data and then unweighted

To obtain a more precise estimation of the statistical equivalence of the refitted and reweighted parton sets, and also to test the unweighting procedure, we may examine the statistical distances between the new unweighted distributions and the refitted set. The distance formulae are defined in Appendix A of [3]. If two sets give a description of the same underlying probability distribution and so are statistically equivalent, the distance between them will fluctuate around a value of one. At $d \sim 7$ the discrepancy between the two sets is at the one sigma level. In the case of the Tevatron jets reweighting exercise, we see in figure 3 that these distances oscillate around one. The reweighted set is therefore equivalent to the refit and there is no significant loss of accuracy in the unweighting procedure.

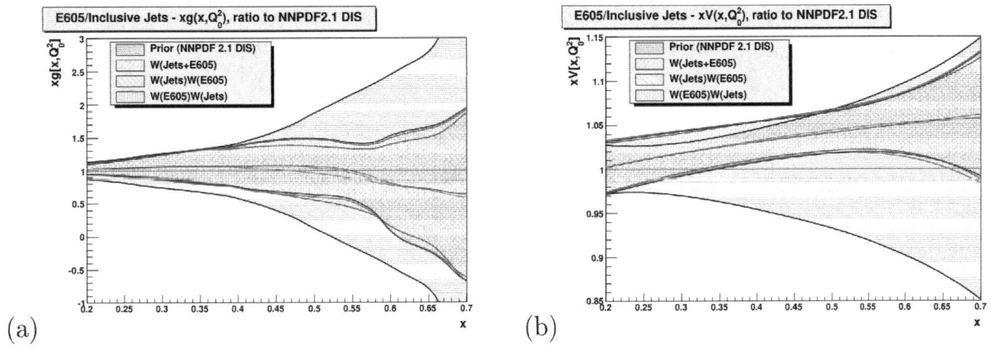

Figure 4. Multiple reweighting demonstration. Plots of gluon PDF (a) and valence PDF (b)

Having developed the unweighting procedure, we can perform another check on the consistency of the reweighting method. When adding more than one set of data by reweighting, our method must satisfy combination and commutation properties. Reweighting with both sets must be equivalent to reweighting with one, unweighting then reweighting with the other. Of course switching the order in which we reweight must produce an equivalent distribution.

To check that our procedure satisfies these properties, we perform a test using the Tevatron jet data as the first dataset and E605 fixed target Drell-Yan data as the second. In figure 4 we compare the inclusion of the combined set with the inclusion of one set after the other. The result is clearly independent from the order in which the inclusion of single datasets is performed. A distance analysis performed on the three produced sets confirms that the reweighting method satisfies the combination and commutation requirements.

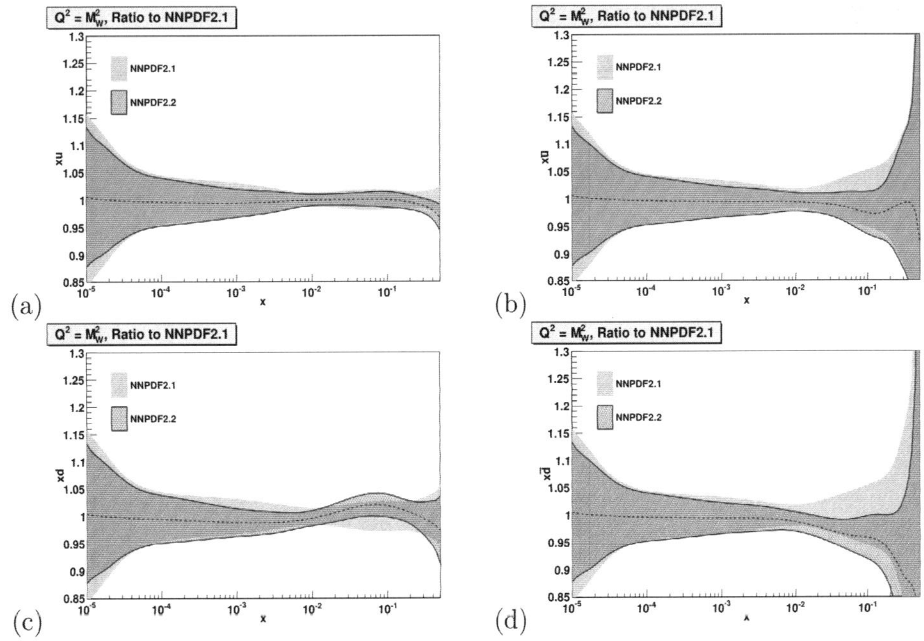

Figure 5. Comparison of (a) up (b) anti-up (c) down and (d) anti-down distributions at the scale $Q^2 = M_W^2$ from the global NNPDF2.1 and NNPDF2.2 global fits. Parton densities are plotted normalized to the NNPDF2.1 central value

5. NNPDF2.2

The Bayesian reweighting method has been used to construct a new NNPDF parton set: NNPDF2.2[5]. In this set we take as a prior ensemble the NNPDF2.1 fit and include by reweighting the W-lepton charge asymmetry measurements of the ATLAS, CMS and D0 collaborations.

The NNPDF2.1 set provides a reasonable description of the new measurements, with $\chi^2_{tot}/N_{dat} = 2.22$. After reweighting with the new data this improves to an excellent level of agreement with $\chi^2_{tot}/N_{dat} = 0.81$. Having reweighted a prior set with $N_{\rm rep} = 1000$ initial replicas, 181 remain, indicating that the data provides a substantial constraint. Using the unweighting procedure outlined above, we have produced the new PDF set with $N_{\rm rep} = 100$.

Figure 5 demonstrates the impact of the new data on the light quark and antiquark PDFs. The uncertainties are significantly constrained by the data in two main regions, there is a reduction of around 20% at $x \sim 10^{-3}$ and 30% in the region $x \sim 10^{-2}$ to $x \sim 10^{-1}$. The overall fit quality improves slightly, from a total χ^2_{tot}/N_{dat} of 1.165 with NNPDF2.1 to 1.157. The constraints demonstrated here are the first such constraints upon parton distributions from LHC data.

Such constraints are particularly important given the discrepancies between global parton distribution fits in flavour separation at medium to large x. The W-lepton charge asymmetry data included here may prove useful in resolving some of these discrepancies. Future LHC data will no doubt provide further constraints upon PDFs in this region.

References

[1] Ball R D *et al.* [The NNPDF Collaboration], *Preprint* 1110.1863

[2] Ball R D *et al.* [The NNPDF Collaboration], 2009 *Nucl. Phys. B* **823** 195 (*Preprint* 0906.1958)

[3] Ball R D *et al.* [The NNPDF Collaboration], 2010 *Nucl. Phys. B* **838** 136 (*Preprint* 1002.4407)

[4] Ball R D *et al.* [The NNPDF Collaboration], 2011 *Nucl. Phys. B* **849** 112, 2012 Erratum *Nucl. Phys. B* **854** 926, 2012 Erratum *Nucl. Phys. B* **855** 927 (*Preprint* 1012.0836)

[5] Ball R D *et al.* [The NNPDF Collaboration], 2012 *Nucl. Phys. B* **855** 608 (*Preprint* 1108.1758)

Journal of Physics: Conference Series **368** (2012) 012064 doi:10.1088/1742-6596/368/1/012064

Self-organizing maps algorithm for parton distribution functions extraction

Simonetta Liuti[1], Katherine A Holcomb[2], and Evan Askanazi[1]
[1]University of Virginia - Physics Department, 382, McCormick Rd., Charlottesville, VA 22904, U.S.A.
[2]University of Virginia - UVACSE, 112, Albert H. Small Building, Charlottesville, Virginia 22904, U.S.A.

E-mail: sl4y@virginia.edu, kah3f@virginia.edu, ema9u@virginia.edu

Abstract. We describe a new method to extract parton distribution functions from hard scattering processes based on Self-Organizing Maps. The extension to a larger, and more complex class of soft matrix elements, including generalized parton distributions is also discussed.

1. Introduction

In the past twenty years Artificial Neural Networks (ANNs) have remarkably established their role as a computational tool in high energy physics analyses. Important applications have been developed that provide, for instance, classification methods for off-line jet identification and tracking, non-classification-type tools for on-line process control/event trigger and mass reconstruction, and optimization techniques in *e.g.* track finding [1]. More recently, ANNs have been proposed as a theoretical tool to address the problem of the extraction of Parton Distribution Functions (PDFs) from high energy scattering processes. In a series of papers [2, 3, 4] the authors developed the Neural Networks PDFs (NNPDFs). NNPDFs differ from standard global analyses in that they avoid the bias that is introduced when a particular parametric functional form for the PDFs is adopted. Relying on an initial PDF form is instead a fundamental assumption in standard analyses. In NNPDFs each PDF is parameterized with a redundant functional form given by a neural network with 37 free parameters represented by the ANNs weights. The parametric forms are subsequently fitted to the experimental data, and their χ^2 is minimized using a genetic algorithm [3].

What distinguishes this approach from the standard ones is, besides an improved statistical analysis, the automated minimization procedure which has the *advantage* of eliminating the user's bias in extracting PDFs from experimental data. In fact, no physical assumption goes into determining the shape of the parametrization, rather the physical behavior is inferred directly from the data in terms of smooth PDF curves. This feature turns, however, into a *disadvantage* in kinematical regions where there is no experimental information, or in between the data points if the data are sparse: due to the inherent unbiasedness NNPDFs cannot be sensibly extrapolated to kinematical regions with few or no data, starting from their behavior in regions where data do exist. In other words, since for NNPDFs the effect of modifying individual

NN parameters is unknown – the weights are in a non tunable, hidden layer – the result might not be under control in the extrapolation region.

The ability to extrapolate is, however, often a desirable feature in the performance of a fit to high energy physics data. It is even more important for extracting and processing physics information from a new generation of experiments that go beyond the inclusive scattering ones intensively studied so far. The newly accessible experiments include besides the LHC high energy frontier, semi-inclusive and exclusive scattering reactions with both unpolarized and polarized variables in the kinematical regimes accessible at RHIC and at Jefferson Lab@12 GeV.

In the new LHC era it is of the utmost importance to provide methods to carefully extract the soft matrix elements, including PDFs and their uncertainties with methods that are apt to handle and interpret the increasingly complicated, and diverse sets of observations. In this contribution we present an alternative approach based on Self-Organizing Maps (SOMs) [5]. SOMs are also ANNs where, however, the basic learning algorithm is replaced by an *unsupervised learning* one that makes use of the underlying *self-organizing* properties of the network's input vectors – in our case the PDFs. Unsupervised learning works without knowledge about what features characterize the output: data are organized according to certain attributes without any "teaching". A most important aspect of the SOM algorithm is in its ability of projecting high dimensional *input* data onto lower dimensions representations while preserving the topological features present in the *training* data. Results using this technique can be represented as 2D geometrical configurations where one can visualize a neighborhood of nodes being simultaneously updated while reproducing the clustering of the data's features. The 2D visualization features are characteristic of a map.

In this contribution we present results obtained by restructuring the original SOMPDF code [6] in such a way that we obtain directly smooth, continuous types of solutions for which a fully quantitative error analysis can be implemented. Also, our new code is now sufficiently flexible to allow for analyses of different observables including both the unpolarized and polarized inclusive type structure functions, and the matrix elements for deeply virtual exclusive processes (Generalized Parton Distributions, GPDs), and semi-inclusive processes (Transverse Momentum Distributions, TMDs). Our first quantitative results for the unpolarized case using Next-to-Leading-Order (NLO) perturbative QCD were shown in [7, 8]. Here we present some of the intermediate results defining the working of SOMPDF.

2. SOM algorithm

The SOM algorithm consists of three stages: *i) Initialization*; *ii) Training*; *iii) Mapping*. Each cell (neuron) is sensitized to a different domain of vectors.

i) Initialization

During the initialization procedure weight vectors of dimension n are associated to each cell, i,

$$V_i = [v_i^{(1)}, ..., v_i^{(n)}]$$

V_i are given spatial coordinates, *i.e.* one defines the geometry/topology of a 2D map that gets populated randomly with V_i.

i) Training

For the training, a set of input data,

$$\xi = [\xi^{(1)}, ..., \xi^{(n)}],$$

(isomorphic to V_i) is then presented to V_i, or compared by calculating the distance, D_i, between each pair of vectors via a "similarity metric" that we choose to be L_2 (Euclidean distance)

$$D_i = \sqrt{\sum_{j=1,n} (v_i^{(j)} - \xi^{(j)})^2}$$

The most similar weight vector is the Best Matching Unit (BMU). SOMs are based on unsupervised and "competitive" learning. This means that the cells that are closest to the BMU activate each other in order to "learn" from ξ. Practically, the BMU, and the neighboring map vectors within a radius defined adjust their values according to a learning principle defined by,

$$V_i(n+1) \quad = \quad V_i(n) + C(n)\alpha(n)[\xi(n) - V_i(n)] \tag{1}$$

where n is the iteration number; $\alpha(n)$ is the *learning function*; $C(n)$ is the *neighborhood function* for the BMU defined as a circle of decreasing radius. Both $C(n)$, and $\alpha(n)$, decrease monotonically with n, so that an initial global ordering is established at the first iteration, while the subsequent iterations yield more specific adjustments. In our case we use square maps of size L_{MAP}, and

$$\alpha(n) \quad = \quad L_{MAP}\left(\frac{n_{train} - n}{n_{train}}\right) \tag{2}$$

$$R \quad = \quad R_{MAP}\left(\frac{n_{train} - n}{n_{train}}\right) \tag{3}$$

$$C(n) \quad = \quad \exp\left(-\frac{D_i^2}{2R^2}\right) \tag{4}$$

where in our case, $L_{MAP} = 1$, n_{train} is the (variable) total number of iterations, R_{MAP} defines the monotonically decreasing radius, and $K_R = 1.5$.

iii) Mapping
At the end of a properly trained SOM, cells that are topologically close to each other will contain data which are similar to each other. In the final phase the actual data gets distributed on the map and clusters emerge. Since each map vector now represent a class of similar objects, the SOM is an ideal tool to visualize high-dimensional data, by projecting it onto a low-dimensional map clustered according to some desired similar feature. In what follows we apply these ideas to PDF fitting.

3. SOMPDF parametrization

As an illustrative example we fit a subset of all available DIS data. Our fit uses SOMs to generate and classify possible "candidate" distribution functions. A Genetic Algorithm (GA) is subsequently applied.

In the initialization stage, a set of database/input PDFs is formed by randomly selecting them from a range of functional forms tabulated as functions of (x, Q^2). The input vectors could be in principle chosen completely at random. However, in all practical SOM applications, some initial ordering limiting the type and number of possible map vectors is given [5]. In our case, the goal is to obtain a stochastic ensemble that still preserves the smoothness of the curves. The initial PDFs are constructed to be linear combinations of functions of a similar form to the ones provided by existing global analyses parametrization sets [9, 10, 11, 12]. For each parameterization set we perform random variations on the *parameters*, instead than on the functions' values themselves. Baryon number and momentum sum rules are imposed at every

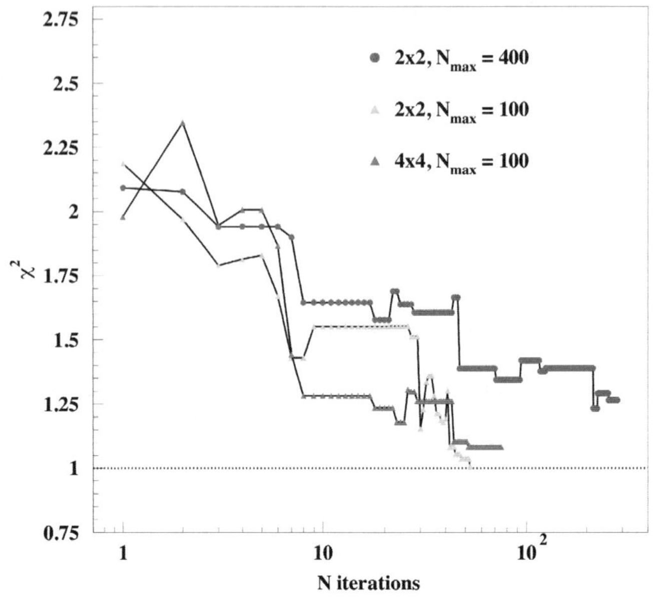

Figure 1. χ^2 vs. number of iterations in the GA, obtained by fixing the values of the algorithm's parameters such as the maximum number of iterations, the maximum value of χ^2 allowed, and the size of the map.

step. From our initial analysis we obtained that it is sufficient to combine three different PDF types in order to achive both the required smoothness and variety/randomness. This constitutes a considerable improvement over the various initialization procedures tested in [6] in that no smoothing techniques are now required. The input PDFs, are connected through evolution, and convoluted with the pQCD coefficient functions at NLO. A subset of the input PDFs is finally placed on the map.

For the training stage, a different subset of the input PDFs is used. The similarity between our map cell vectors and the code vectors is computed by comparing the PDFs, according to equations (1,2,3,4). at given x and Q^2 values. The new map PDFs are obtained by averaging the neighboring PDFs with the BMU. Once the map is trained, the GA is implemented. The χ^2 per input PDFs is calculated with respect to the experimental data. We then take a subset of these functions with the best χ^2, and use them as seeds to form a new set of input PDFs. We train the map with the new set of input PDFs and repeat the process. The χ^2 was found to decrease towards $\chi^2 = 1$ with every GA iteration. Our stopping criterion is established when the χ^2 stops varying (its curve flattens). In figure 1 we show results on the dependence of the χ^2 on the number of iterations in the GA, obtained using small size maps for illustrative purposes. The role of the different values of the algorithm's parameters can be seen.

In figure 2 we illustrate the shape and size of the envelopes constituting the initial set of map PDFs. A "test" set of data from BCDMS + is used.

Finally, in figure 3 we show results of runs using more complete data sets, consistent with the sets used in [2, 9, 10]. The data sets chosen were from BCDMS, H1, NMC, SLAC and ZEUS (see *e.g.* references in [3]).

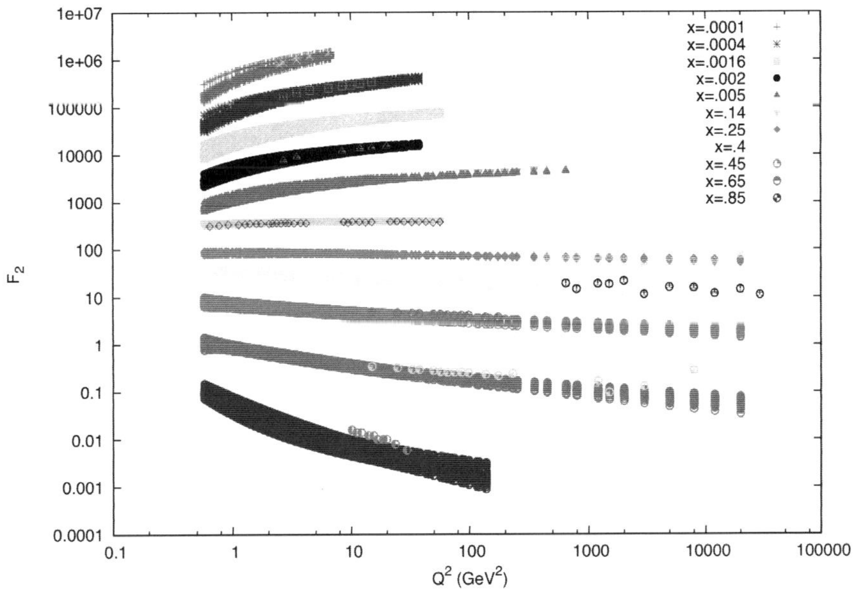

Figure 2. A set of map vectors from the PDF envelope characterizing the SOMPDF initialization stage. Data from SLAC, BCDMS, H1 and ZEUS are shown.

4. Conclusions and future developments

In [6] a new approach, SOMPDF, was presented as an alternative to the purely automated fitting procedure of NNPDF. Initial results were aimed at proving the viability of the method as a fitting procedure, by successfully generating χ^2 values that decreased with the number of iterations. These studies did not focus on the specific clustering properties of the map. The clustering features are, however, important, especially when extending the fitting procedure to a wider set of semi-inclusive and exclusive scattering experimental data which purport to extract the Transverse Momentum Distributions (TMDs) and Generalized Parton Distributions (GPDs), respectively. We believe that SOMPDFs are an ideal tool for analyzing these rather elusive observables because of the following reasons:

i) the extrapolation to regions where data are scarce is feasible, since the SOM method does not rely on outputs for the learning process;

ii) visualization advantages are a key feature of SOMs [5, 13];

iii) one obtains an improved physical insight on multivariable dependent phenomena. This is accomplished by first singling out the main features of the data through the clustering property of the SOM, [3], and by subsequently inspecting the content of the map cells that are sensitive to those features (see *e.g.* [13]).

We therefore refined the SOMPDF approach of [6] and restructured the original code in such a way that on one side, smooth fitting curves are obtained and a fully quantitative error analysis can be now implemented, and on the other we have sufficient flexibility to allow for analyses of different observables, including the matrix elements for deeply virtual exclusive and

[3] SOMs are in fact also defined as a non linear extension of Principal Component Analysis (PCA).

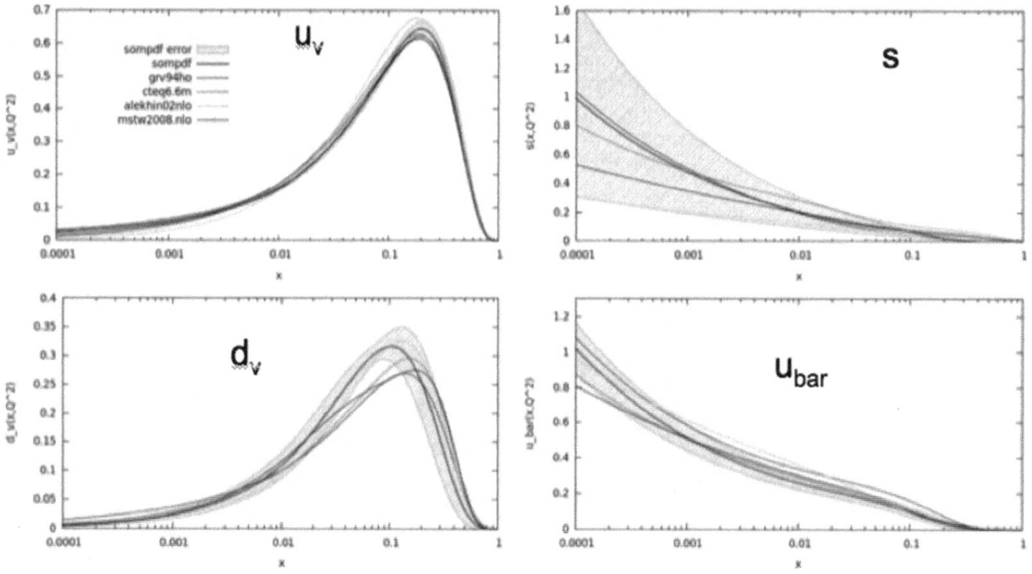

Figure 3. Test results using a 5×5 map for a set of 43 runs. The panels represent (clockwise from upper left)the u_v, s, d_v, $\bar{u}$ distributions, respectively, at $Q^2 = 7.5$ GeV^2. The shaded areas are our results including the error analys outlined in the text. For comparison we show also results from several other parametrizations.

semi-inclusive processes. Our first quantitative results for the unpolarized case using Next-to-Leading-Order (NLO) perturbative QCD in the $\overline{MS}$ scheme were presented in [7, 8] and at this workshop.

Based on our results we are now confident that the SOMPDF method stands as a viable tool for PDF analyses, which is complementary to NNPDF. It is also an essential tool to be used in more complex multivariable analyses.

More work is in progress that will refine our initial results, and eventually provide PDFs with an improved error analysis. As, in fact, recently articulated in [11], the evaluation of the PDFs' uncertainty is complicated because it originates from different experimental and theoretical sources. In particular, the treatment of the theoretical errors cannot be uniquely defined, since their correlations are not well known. Our approach is currently defining statistical errors on an ensemble of SOMPDF runs. An evaluation using the Lagrange multiplier method is in progress.

References

[1] Peterson C, Rognvaldsson T, Lonnblad L (1994) Comput. Phys. Commun. **81**, 185-220; Peterson C (1992) in Proceedings of Workshop on *"New Computing Techniques in Physics Research II"*, Edited by D. Perret-Gallix. River Edge, N.J., World Scientific.

[2] Ball R D, Del Debbio L, Forte S, Guffanti A, Latorre J, Rojo J and Ubiali M, (2010) Nucl. Phys. B**838** 136; (2010) JHEP **1005** 075; (2009) Nucl. Phys. B**823** 195 ; (2009) Nucl. Phys. B**809**, 1; Erratum Nucl. Phys. B**816** 293

[3] Del Debbio L, Forte S, Latorre J, Piccione A and Rojo J [NNPDF Collaboration] (2007) JHEP **0703** 039; (2005) JHEP **0503** 080

[4] Forte S, Garrido L, Latorre J and Piccione A (2002) JHEP **0205** 062

[5] Kohonen T (2001) Self-organizing Maps (Springer, New York) 3rd. ed.

[6] Honkanen H, Liuti S, Carnahan J, Loitiere Y, Reynolds P R (2009) Phys. Rev. D **79** 034022

[7] Perry D Z, Holcomb K and Liuti S (2010) *Preprint* arXiv:1008.2137 [hep-ph]

[8] Holcomb K, Liuti S and Perry D Z (2010) *Preprint* arXiv:1008.4197 [hep-ph]

[9] Pumplin J, Stump D, Huston J, Lai H, Nadolsky P, Tung W K (2001) Phys. Rev. D **65**, 014013
[10] Alekhin S, Melnikov K, Petriello F (2006) Phys. Rev. D **74** 054033
[11] Lai H, Huston J, Li Z, Nadolsky P, Pumplin J, Stump D and Yuan C P, *Preprint* arXiv:1004.4624 [hep-ph]
[12] Gluck M, Reya E and Vogt A (1998) Eur. Phys. J. C **5** 461
[13] Lonnblad L, Peterson C, Pi H and Rognvaldsson T (1991) Comput. Phys. Commun. **67** 193-209

AUTHOR INDEX

AUTHOR INDEX

Institute of Physics
Dirac House, Temple Back
Bristol BS1 6BE UK

CURRAN ASSOCIATES INC
proceedings.com